AR

ANNUAL REVIEW OF ECOLOGY AND SYSTEMATICS

AR

ANNUAL REVIEW OF ECOLOGY AND SYSTEMATICS

VOLUME 23, 1992

DAPHNE GAIL FAUTIN, *Editor*
University of Kansas

DOUGLAS J. FUTUYMA, *Associate Editor*
State University of New York at Stony Brook

FRANCES C. JAMES, *Associate Editor*
Florida State University

ANNUAL REVIEWS INC. 4139 EL CAMINO WAY P.O. BOX 10139 PALO ALTO, CALIFORNIA 94303-0897

ANNUAL REVIEWS INC.
Palo Alto, California, USA

International Standard Serial Number: 0066–4162
International Standard Book Number: 0–8243–1423-9
Library of Congress Catalog Card Number: 71-135616

∞ The paper used in this publication meets the minimum requirements of American National Standard for Information Sciences—Permanence of Paper for Printed Library Materials, ANSI Z39.48-1984.

Typesetting by Kachina Typesetting Inc., Tempe, Arizona; John Olson, President; Jeannie Kaarle, Typesetting Coordinator; and by the Annual Reviews Inc. Editorial Staff

PRINTED AND BOUND IN THE UNITED STATES OF AMERICA

DEDICATION

This twenty-third volume of the *Annual Review of Ecology and Systematics*—the last planned under the guidance of Richard F. Johnston, editor, and Peter Frank and Charles D. Michener, associate editors—is dedicated to these three men who conceived of ARES and shaped it and fostered it for nearly a quarter of a century. Modest men, they might attribute the success of the volume to the quality and range of people who wrote for it. In truth it was their vision at the start, and their commitment to excellence and to breadth of coverage, that has established ARES as one of the most successful books in the Annual Reviews series. Indeed ARES taps a wide array of disciplines and has welcomed an unusual number and variety of volunteer authors. The result is a journal, one of the most frequently cited in the field, that is read and respected across all the scientific disciplines it covers. For this solid foundation, we thank Richard F. Johnston, Peter Frank, and Charles D. Michener.

DAPHNE GAIL FAUTIN
EDITOR

PREFACE

Volume 23 of the *Annual Review of Ecology and Systematics* (ARES) represents a departure from previous volumes in several respects. One is in arrangement of its contents. Its first eight chapters center on a common theme—global change—and comprise a unit, introduced by Peter Vitousek. The issue of global change itself represents something of a departure for ARES, with human beings squarely at the heart of the matter. Indeed, our species has intruded itself increasingly into all aspects of our science as numbers and technology have allowed us to expand our access to and influence upon every environment of the world. This rapid ascent to center stage is reflected in the portrayal of humans in textbooks of ecology: from virtual invisibility early on, to a factor perturbing the "natural order" some decades ago, to a major force with the potential to alter Earth's very habitability. The role of humans in precipitating and/or accelerating global change is still debated in some circles. That we will be living with its consequences is beyond question. And that has profound implications for the role of human beings in our global ecosystem, how we do our science, and what science we choose to do. Hence the prominence we accord global change in this volume.

Volume 23 contains, in addition, 11 chapters on a variety of other topical issues in ecology and systematics. This edition of ARES is the last planned under the triumvirate of Richard F. Johnston, Peter W. Frank, and Charles D. Michener, who comprised, until last year, the only editorial board ARES had known. Johnston dedicated Volume 22 to Michener, providing in his introduction a brief history of ARES. With Volume 23 the succession from the pioneers to the next seral stage has been completed. This seems an appropriate opportunity to renew the invitation to "members of the ecology and systematics communities" last issued in Volume 14 (1983) for "suggestions for reviews that we might not think of soliciting ourselves." The names on the spine and title have changed, but the objectives have not.

DAPHNE GAIL FAUTIN
EDITOR

Annual Review of Ecology and Systematics
Volume 23, 1992

CONTENTS

 (*continued*)

RELATED ARTICLES FROM OTHER ANNUAL REVIEWS

From the *Annual Review of Energy and the Environment,* Volume 17 (1992)

Interconnections between Energy and the Environment, R. W. Fri and J. Darmstadter

From the *Annual Review of Entomology,* Volume 37 (1992)

The Analysis of Parasite Transmission by Bloodsucking Insects, C. Dye
Small Ermine Moths (Yponomeuta): *Their Host Relations and Evolution,* S. B. J. Menken, W. M. Herrebout, J. T. Wiebes
The Chemical Ecology of Aphids, J. A. Pickett, L. J. Wadhams, C. M. Woodcock, and J. Hardie
Ecology of Infochemical Use by Natural Enemies in a Tritrophic Context, L. E. M. Vet and M. Dicke
The Evolution of Aphid Life Cycles, N. A. Moran
Frugivory, Seed Predation, and Insect-Vertebrate Interactions, R. Sallabanks and S. P. Courtney
Odor Plumes and How Insects Use Them, J. Murlis, J. S. Elkinton, R. T. Cardé

From the *Annual Review of Genetics,* Volume 26 (1992)

Translational Accuracy and the Fitness of Bacteria, C. G. Kurland

From the *Annual Review of Phytopathology,* Volume 30 (1992)

Changing Concepts in the Taxonomy of Plant Pathogenic Bacteria, J. M. Young, Y. Takikawa, L. Gardan and D. E. Stead
Evolutionary Biology of Phytophthora, C. M. Brasier
Evolution of Cyst and Noncyst-Forming Heteroderinae, J. G. Baldwin
Role of Abiotic Stress in the Decline of Red Spruce in High Elevation Forests of the Eastern United States, A. H. Johnson
Phenolic Compounds and Their Role in Disease Resistance, R. L. Nicholson, R. E. Hammerschmidt

From the *Annual Review of Plant Physiology and Plant Molecular Biology,* Volume 43 (1992)

Chronicles from the Agrobacterium-*Plant Cell DNA Transfer Story,* P. Zambryski

From the *Annual Review of Sociology,* Volume 18 (1992)

Changing Fertility Patterns and Policies in the Third World, G. McNicoll

ANNUAL REVIEWS INC. is a nonprofit scientific publisher established to promote the advancement of the sciences. Beginning in 1932 with the *Annual Review of Biochemistry,* the Company has pursued as its principal function the publication of high quality, reasonably priced *Annual Review* volumes. The volumes are organized by Editors and Editorial Committees who invite qualified authors to contribute critical articles reviewing significant developments within each major discipline. The Editor-in-Chief invites those interested in serving as future Editorial Committee members to communicate directly with him. Annual Reviews Inc. is administered by a Board of Directors, whose members serve without compensation.

ANNUAL REVIEWS OF
Anthropology
Astronomy and Astrophysics
Biochemistry
Biophysics and Biomolecular Structure
Cell Biology
Computer Science
Earth and Planetary Sciences
Ecology and Systematics
Energy and the Environment
Entomology
Fluid Mechanics
Genetics
Immunology
Materials Science
Medicine
Microbiology
Neuroscience
Nuclear and Particle Science
Nutrition
Pharmacology and Toxicology
Physical Chemistry
Physiology
Phytopathology
Plant Physiology and Plant Molecular Biology
Psychology
Public Health
Sociology

SPECIAL PUBLICATIONS

Excitement and Fascination of Science, Vols. 1, 2, and 3

Intelligence and Affectivity, by Jean Piaget

For the convenience of readers, a detachable order form/envelope is bound into the back of this volume.

Annu. Rev. Ecol. Syst. 1992. 23:1–14

GLOBAL ENVIRONMENTAL CHANGE: An Introduction

Peter M. Vitousek
Department of Biological Sciences, Stanford University, Stanford, California 94305

KEYWORDS: land use, human population, greenhouse gas, climate change, biological diversity

INTRODUCTION

This issue of the *Annual Review of Ecology and Systematics* includes a set of papers on the ecological causes and consequences of global environmental change, and on the methods by which these can be studied. Global studies are likely to be among the most important concerns of ecological research and teaching for some decades. While the *Annual Review of Ecology and Systematics* has published a number of papers in this area (cf 4, 7, 22, 41, 44), the Editorial Committee believed that it would be useful to bring together a set of reviews selected to illustrate the breadth of ecologists' contributions to research on global environmental change. This collection is designed to demonstrate that a very wide array of ecological research is crucial to the analysis of global change—and that both our basic work as ecologists and our contribution to the understanding of Earth as a system will be enhanced if we keep that relevance in mind.

Earth is a dynamic system; global environmental change has always been a part of its functioning. A recent and prominent example is the glacial/interglacial cycles of the past 2 million years (12, 13, 16). The current interest in global change arises from the fact that some components of human-caused global change have reached a magnitude at least equal to that of natural changes—and the human-caused changes are often more rapid than and beyond the bounds of natural change, at least for the past millions of years.

COMPONENTS OF GLOBAL ENVIRONMENTAL CHANGE

Global changes are defined as those that alter the well-mixed fluid envelopes of the Earth system (the atmosphere and oceans) and hence are experienced

0066-4162/92/1120-0001$02.00

globally, and those that occur in discrete sites but are so widespread as to constitute a global change. Examples of the first category include: change in the composition of the atmosphere, climate change, decreased stratospheric ozone concentrations and increased ultraviolet input. The second type of global change is exemplified by land use change, loss of biological diversity, biological invasions, and changes in atmospheric chemistry. Each of these is discussed below.

Change in the Composition of the Atmosphere

Increases in global concentrations of carbon dioxide have been documented very clearly. Careful atmospheric measurements have been carried out since 1957; carbon dioxide concentrations increased from 315 ppm then to over 350 ppm in 1988 (29), and the rate of increase itself increased over that interval. Could this relatively short record have happened to coincide with a natural fluctuation in carbon dioxide concentrations? No—a record of earlier concentrations is preserved in polar ice, and analyses of cores taken from this ice demonstrate that concentrations were quite stable near 280 ppm for at least 1000 years, then began increasing around the year 1750 (63). A longer-term record covering the past 160,000 years (3) shows that there are natural fluctuations in carbon dioxide concentrations through glacial-interglacial cycles—but the recent increase began at a time when concentrations already were high and took them outside the bounds of known Pleistocene variation. The recent increase is now similar in magnitude to, and much more rapid than, the fluctuations occurring over glacial-interglacial time (Figure 1).

There is no doubt that current increases in carbon dioxide are caused primarily by fossil fuel combustion and secondarily by changes in land use.

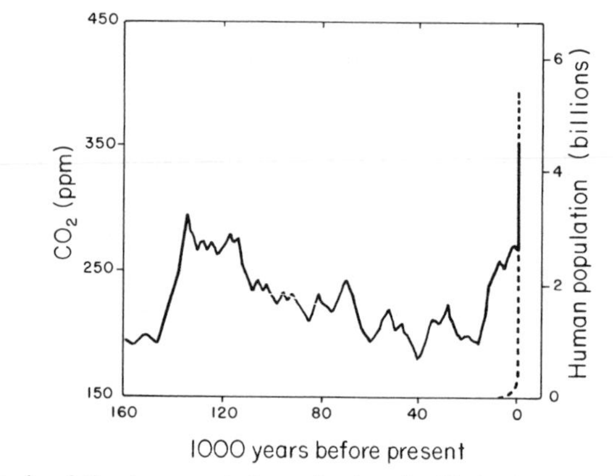

Figure 1 Natural variation in concentrations of carbon dioxide for the past 160,000 years from the Vostok ice core (3), together with recent anthropogenic increases in carbon dioxide concentrations and the growth of the human population.

The timing and magnitude of fossil fuel combustion alone are more than sufficient to account for the global increase (51). Moreover, the atmospheric abundances of the carbon isotopes ^{13}C and ^{14}C relative to ^{12}C have decreased over time in a pattern that reflects dilution of the atmospheric pool by carbon released from fossil fuel combustion (which is ^{14}C-free and ^{13}C-depleted) and loss of terrestrial biomass (^{13}C-depleted) (56, 57). The increased concentrations of carbon dioxide are expected to have substantial direct effects on plants, herbivores, and ecosystems (4, 38).

A number of other gases are also increasing globally (40, 63). The wholly anthropogenic chlorofluorocarbons (CFCs) have been increasing most rapidly (66). Concentrations of methane have more than doubled since 1750; this increase is believed to be caused by a combination of agricultural and industrial activities (11). Nitrous oxide is increasing more slowly; the reasons for its increase are less certain, but they are probably primarily related to tropical land use change and agricultural activity (34). The history of increase in tropospheric concentrations of these four relatively stable gases is summarized in Figure 2; an excellent review of their sources and dynamics can be found in a recent Intergovernmental Panel on Climate Change (IPCC) report (63).

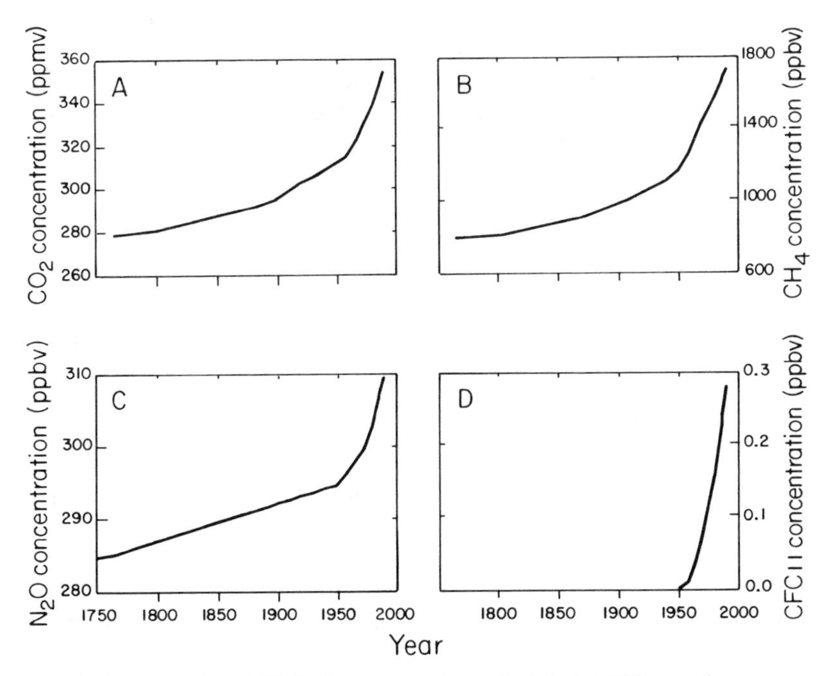

Figure 2 Increases since 1750 in the concentrations of relatively stable greenhouse gases: carbon dioxide (A), methane (B), nitrous oxide (C), and chlorofluorocarbon-11 (D). From (27)

Climate Change

The four gases whose status is summarized in Figure 2 are all greenhouse gases, in that they are transparent to incoming solar radiation but absorb outgoing infrared radiation (IR) emitted by Earth's surface and atmosphere. The continuing increase in their concentrations therefore will enhance the naturally occurring greenhouse effect, which has maintained the average temperature of Earth's surface more than 30°C warmer than it would have been in the absence of atmospheric absorption of infrared radiation (24).

That the greenhouse effect will be enhanced by anthropogenic increases in these gases is wholly uncontroversial, and that this enhancement will cause global warming is nearly so. A recent "best estimate" for the temperature increase likely to result from a doubling of carbon dioxide (and equivalent forcing from the other gases) is 2.5°C globally, with greater changes towards the poles (27). This corresponds to about half of the average temperature change from full-glacial to interglacial conditions; its effect on the distribution and dynamics of natural and agricultural ecosystems would be substantial (6). However, more detailed predictions of the rate, magnitude, and distributions of changes in temperature and (particularly) precipitation contain substantial uncertainties. These reflect: (i) an imperfect understanding of some fundamental processes in the climate system; (ii) the inherent complexity of the climate system and the difficulty of representing it in general circulation models, no matter how detailed and demanding these are; (iii) the existence of both positive and negative feedbacks in the response of the climate system to altered conditions—even where all of the feedbacks can be identified, it is difficult to determine their net effects; and (iv) the influence of offsetting human activities such as the generation of aerosols and cloud condensation nuclei from anthropogenic sulfur dioxide emissions (10).

A number of current summaries of the certainties and uncertainties of climate change and the likely effects of climate change on ecosystems are available; these range from moderately technical articles (53) and books (52) to the IPCC Scientific Assessment discussed above (15, 35, 36).

Decreased Stratospheric Ozone Concentrations and Increased Ultraviolet Input

Ozone in the stratosphere absorbs incoming ultraviolet radiation (UV), thereby reducing the amount reaching Earth's surface. Since 1974, it has been known that chlorofluorocarbons break down in the stratosphere, liberating chlorine and thereby catalyzing the breakdown of stratospheric ozone (37). These findings, and models based on them, sparked substantial discussion and debate—but the discovery in the mid-1980s of a growing springtime Antarctic ozone hole (20) came as a nasty surprise to all parties. Subsequent proof that CFCs cause Antarctic ozone destruction through a previously unsuspected series of processes (cf 45, 47) illustrates that global environmental change is not always a gradual, predictable process.

More recently, significant stratospheric ozone depletion has been documented over a much wider area of Earth. The consequent increase in UV radiation is likely to affect plants, ecosystems, and human health (9, 58), although to date experimental studies of these effects at realistic levels of increased UV are relatively sparse. This component of global environmental change is unique in that widespread concern over the effects of increased UV has led to an international agreement to eliminate CFC emissions globally by early in the next century (5).

Another type of global environmental change occurs in enough discrete sites or regions to constitute a global change. Examples of this type are land use change, loss of biological diversity, biological invasions, and changes in atmospheric chemistry.

Land Use Change

Changes in land use are reviewed in this volume by Meyer & Turner; they are discussed only briefly here. Unlike the atmosphere or oceans, terrestrial ecosystems do not mix on any time scale relevant to global change—and hence any global effect of land use change must be a consequence of many local changes. In fact, human-caused land use change alters enough local ecosystems substantially enough to affect global budgets of greenhouse gases (as discussed above; also see 26), and to alter regional climate (17, 55), regional atmospheric chemistry (2, 30), and the dynamics of major river systems (60). Nevertheless, its most important global effect may be simply the alteration of so many local ecosystems (59); some major ecosystem-types (such as tall grass prairie and tropical deciduous forests) have virtually disappeared (cf 28), and many more have been significantly altered and/or fragmented.

Loss of Biological Diversity

Human activity is now causing the extinction of species and genetically distinct populations at a rate far above background (65). Most species are ephemeral features of the Earth system; under normal conditions (between rare episodes of mass extinction), an average species lasts perhaps 10 million years (19). Human activity is now making many species much more ephemeral; observations of well-studied groups such as birds, together with calculations based on species/area curves, suggest that current rates of extinction are several orders of magnitude above background rates. So far the increase is due primarily to land use change (secondarily to biological invasions, below), but other components of global environmental change are likely to contribute increasingly in the future.

If these increased rates of extinction continue, the next few centuries will witness a mass extinction of a magnitude not seen since the Cretaceous-Tertiary boundary—and this time, there will be no need to argue about the cause of the episode. The causes, rates, and consequences of the loss of biological diversity at the species and population levels are discussed briefly in Ehrlich & Wilson (19), and at greater length in Wilson & Peter (65).

Biological Invasions

One consequence of humanity's extraordinary mobility is an increase in the mobility of numerous species that we transport either deliberately or inadvertently. The resultant biological invasions reduce the biological distinctiveness of the various continents and islands, leading toward a homogenization of the biota of Earth (18, 39). This component of global environmental change, and its consequences for the composition, structure, and functioning of terrestrial ecosystems, is reviewed in this issue by D'Antonio & Vitousek.

Changes in Atmospheric Chemistry

A syndrome of elevated tropospheric ozone concentrations, acidic precipitation, and elevated nitrogen deposition occurs over most economically developed regions of Earth. Similar changes are now being observed seasonally in developing tropical regions (21). Increases in tropospheric ozone are driven by emissions of reactive nitrogen oxide gases from internal combustion engines, agriculture, and other human activities (43); biomass burning is the most important source in developing tropical regions (2, 30). These nitrogen oxides interact with anthropogenic or natural hydrocarbons and carbon monoxide to lead to the photochemical production of tropospheric ozone (14, 32). Acid precipitation (and dry deposition) results from the oxidation of nitrogen and sulfur oxides to nitric and sulfuric acid, and their subsequent rain-out or deposition downwind. This deposition, and that of ammonia emitted from agricultural systems, can increase atmospheric inputs of nitrogen several-fold for hundreds to thousands of kilometers downwind of highly developed regions.

These alterations to regional atmospheric chemistry occur more-or-less independently in a number of large regions; they can therefore be considered a global environmental change. The effects of altered atmospheric chemistry include threats to human health (23), observed reductions in agricultural productivity (46), and changes in the dynamics of terrestrial and aquatic ecosystems (1, 41, 50 54).

IMPORTANCE AND INTERACTIONS OF THE COMPONENTS

The relative importance of the various components of global environmental change is a matter for debate. A number of factors must go into any such weighting, including the relative impact of plausible changes in each component, the relative certainty of that change, and the reversibility of change. Of these, the relative certainty of change can be addressed straightforwardly. Changes in the atmosphere, especially increased carbon dioxide, methane, and nitrous oxide concentrations in the troposphere and decreased ozone concentrations in the stratosphere, already are well documented; globally significant land use change clearly has occurred; and an ongoing loss of biological diversity can be inferred with confidence. On the other hand, human-caused global climate change has not been detected unequivocally as yet (64). Our understanding of the climate system is sufficient to say that it will occur, but the complexity of that system is such that the magnitude, timing, and geographic distribution of change all remain uncertain.

The reversibility of components of global environmental change—the rate at which they would return to previous levels if human forcing were reduced or eliminated—differs widely. The greenhouse gas methane has an atmospheric lifetime of about a decade, and so its concentrations could return to background levels relatively rapidly. The other major greenhouse gases have effective atmospheric lifetimes in excess of a century (31). Reversing the effects of land use change is a successional process (decades to centuries)—although where soils have been degraded substantially, some millennia may be required for recovery (assuming, of course, that it is feasible to take land out of food production). Extinction, however, is much less reversible than the other components. While overall levels of diversity might be reestablished in a very few million years following an episode of extinction, the loss of particular species and the information they contain is wholly irreversible. In any case, except for CFCs and ozone depletion, reversibility is a moot point. Most components of global environmental change are becoming more important rather than less so, and some will continue to do so for the foreseeable future.

The probable impact of the various components of global change is more difficult to assess, in part because it is dependent on the time scale of concern. Ultimately, climate change probably has the greatest potential to alter the functioning of the Earth system; its direct effects on natural and managed systems ultimately could become overwhelming, and it interacts strongly with most of the other components of global change, as discussed below. It is particularly worrisome that climate change may not be a gradual phenomenon; rather, the Earth system may switch rapidly from one climate mode to another (8). Nevertheless, the major effects of climate change are mostly in the future, while most of the others are already with us. There is a reasonably strong consensus among terrestrial ecologists that for the next several decades, land use change is likely to be the most certain and the most significant component of global change, followed by changes in the composition of the atmosphere.

Analyses of the causes and significance of global environmental change are complicated greatly by the fact that many of the components of change interact with each other directly and are modified further by feedbacks from their effects on ecosystems. For example, the increasing concentrations of carbon dioxide will drive climatic change by enhancing the greenhouse effect, which in turn is likely to alter the balance between rates of net primary production and decomposition in terrestrial ecosystems (42, 48, 49). This change can feed back to alter carbon dioxide concentrations in the atmosphere—if this process is not offset by direct effects of elevated carbon dioxide on carbon storage (4), or the indirect effects of climatic warming on the biological availability of soil nitrogen (42, 49). Similarly, land use change is the major process responsible for the extinction of species and genetically distinct populations (so far), but the loss of genetic diversity is itself a global change that could have significant effects on land use (19).

The complexity of causal interactions among components of global change and their consequences, together with our uncertainty concerning some of these effects, makes simple, universal statements concerning chains of causation difficult. Nevertheless, there is no uncertainty concerning the ultimate cause of most current global change—it occurs because the scale of human activity has become large relative not just to that of other species, but to the flow of energy and materials on a global scale. For example, humanity consumes less than 1% of the terrestrial net primary productivity of Earth but in doing so dominates or destroys nearly 40% of the total (61). All of the other tens of millions of species must either adapt to us or subsist on the remainder. Our activity more than doubles rates of nitrogen fixation globally (62) and alters aspects of the global sulfur cycle to a greater extent (10). Certain components of the global influence of humanity are amenable to technological fixes—for example, we can control chlorofluorocarbon emissions, and hence ultimately stratospheric ozone depletion, without altering our way of life fundamentally. Nevertheless, the human population will continue to cause—but not control—global environmental change as long as our population and level of activity loom so large on the planet.

ECOLOGY AND GLOBAL CHANGE

The reviews in this volume cover diverse aspects of global environmental change. Webb & Bartlein examine the rich record of past global change—a record that helps place the ongoing human-driven changes in context and also provides a basis for testing the validity of modern climate models. Much of this record is a biological history, one that is read through an understanding of the ecology of marine foraminifera or the pollen of terrestrial plants.

D'Antonio & Vitousek consider an often-neglected component of global environmental change—the loss of the regional distinctiveness of Earth's biota brought about by human-caused breakdown of biogeographic barriers to species dispersal. They use the widespread introduction of exotic grasses, often from Africa, as an illustration. Invasions by these grasses are driven in part by physiological and demographic characteristics of the grasses themselves and in part by biotic interactions in the invaded communities. However, grass invasions have substantial, interactive effects on fire regimes and other aspects of ecosystem function; they thereby represent a significant threat to the maintenance of biological diversity and to regional biogeochemistry.

The next three reviews, by Field et al, Carpenter et al, and Smith & Buddemeier, examine the interactive effects of global change on terrestrial ecosystems, freshwater ecosystems, and coral reefs respectively. All three demonstrate that ecosystems are far from passive participants in global environmental change. Rather, changes to ecosystems caused by one process can alter other components of global change. For example, land use change in both upland and wetland systems alters microbial processes that cause emissions of radiatively active (greenhouse) and chemically reactive trace gases. These in turn can drive changes in the composition of the atmosphere and ultimately in global climate.

In addition, Smith & Buddemeier suggest that while substantial attention has been paid to the effects of climate change on coral reefs, the effects of land use change (sedimentation and nutrient loading) and elevated carbon dioxide may be more significant for the next several decades. This is not to downplay climate change, which is a real concern, but to suggest that the other components of global environmental change may be even more important. The similarity of their ranking to that of terrestrial ecologists (above) is striking.

Wessman discusses a problem that is common to all scientific efforts to evaluate global phenomena—the disparity between the spatial and temporal scale of most ecological measurements and that of the Earth system that we seek to analyze. How can we "scale up" from local measurements to global dynamics? Wessman illustrates how this disparity can be addressed in ecological theory and through a variety of technological developments that include satellite- and aircraft-based remote sensing, coarse-scale measurements of gas fluxes, and global tracer models. These new technologies have made a true science of the biosphere into an attainable goal.

Shugart et al discuss a set of models of plant community dynamics that is based on the replacement dynamics of individual trees. They consider how physiological and ecosystem processes have been and are being incorporated into these models. The resultant syntheses can be used to evaluate the effects of interacting components of global change on the distribution and dynamics of terrestrial ecosystems. The models are also useful in the analysis of regional and global feedbacks resulting from alterations to ecosystems. Because these models are based on the mechanisms that control species replacement and consequent changes in ecosystem function, they are more useful for dealing with wholly novel conditions of land use, climate, and atmospheric composition than are correlative models.

Finally, Meyer & Turner evaluate global changes in land use and land cover, and demonstrate the importance of understanding the human dimension of global environmental change. They point to the distinction between land conversion (often the focus of ecologists) and agricultural intensification; the latter may be more significant in its global consequences. They also discuss the mixed evidence for the role that human population growth per se plays in driving land use change. Most importantly, they show that an understanding of the overall causes of land use change, and hence any rational attempt to affect this most important component of global environmental change, must be based on an understanding of the reasons for the human actions that drive it. Social science, and social scientists, therefore must be integral parts of any global change research program.

A few other reviews were commissioned but for various reasons not received; as with any collection, the ones that got away are particularly missed. These included reviews about the use of globally distributed measurements and models to determine the global sources and sinks of carbon dioxide, the continental-scale analysis of ecosystem function, the current rate of loss of biological diversity, and pelagic marine ecosystems.

Even if these reviews had been included, the papers in this volume would be far from a complete introduction to global environmental change. Further information can be obtained from the many references cited herein. Readers interested in the directions taken by global research programs concerned with ecological aspects of global change should consult the planning and operational documents of the International Geosphere-Biosphere Program (IGBP). IGBP is the coordinating body for a number of global projects that include Global Change in Terrestrial Ecosystems (GCTE), International Global Atmospheric Chemistry project (IGAC), Joint Global Ocean Flux Study (JGOFS), Global Changes of the Past (PAGES), Biospheric Aspects of the Hydrologic Cycle (BAHC), and Land-Ocean Interactions in the Coastal Zones (LOICZ). All of these include greater or lesser amounts of ecological research in their core program. The addresses of the project offices are appended to this introduction.

In addition, the Ecological Society of America has suggested a set of research priorities that includes a focus on ecological causes and consequences of global environmental change, on biological diversity, and on the development of sustainable human uses of the biosphere. Their document (33) includes a rich array of suggestions for ecological research that will further the understanding, and perhaps ultimately the management, of global environmental change.

ACKNOWLEDGMENTS

I thank J. M. Melillo for his contribution to putting together the set of reviews, and C. B. Field, H. A. Mooney, and A. R. Townsend for comments on an earlier draft. C. Nakashima prepared the manuscript for publication. Research was supported by a Pew Foundation Fellowship.

Appendix: Addresses Of International Geosphere-Biosphere Program Core Project Offices

Global Change in Terrestrial Ecosystems
Dr. Brian Walker
CSIRO Division of Wildlife and Ecology
P. O. Box 84
Lyneham, Canberra, ACT 2602 Australia

Joint Global Ocean Flux Study
Dr. Geoffrey Evans
Institut für Meereskunde
Düsternbrooker Weg 20
D-2300 Kiel, Germany

International Global Atmospheric Chemistry Project
Professor Ronald Prinn
Department of Earth, Atmosphere, and Planetary Sciences
Room 54-1312
Massachusetts Institute of Technology
Cambridge, MA 02139 USA

Global Changes of the Past
Dr. Hans Oeschger
Institute of Physics
University of Bern
Silderstrasse 5
CH-3012 Bern, Switzerland

Land-Ocean Interactions in the Coastal Zone
Chair, Dr. Patrick Holligan
Natural Environmental Research Council
Plymouth Marine Laboratory
Prospect Place
West-Hoe, Plymouth PL1 3DH, United Kingdom

Biospheric Aspects of the Hydrologic Cycle
Institut für Meteorologie
Freie Universität Berlin
Dietrich-Schäfer-Weg 6-10
D-1000 Berlin 41, Germany

Literature Cited

1. Aber, J. D., Nadelhoffer, K. J., Steudler, P., Melillo, J. M. 1989. Nitrogen saturation in northern forest ecosystems. *Bioscience* 39:378–86
2. Andreae, M. O., Browell, E. V., Garstang, M., Gregory, G. L., Harriss, R. C., et al. 1988. Biomass burning and associated haze layers over Amazonia. *J. Geophys. Res.* 93:1509–27
3. Barnola, J. M., Raynaud, D., Korotkevich, Y. S., Lorius, C. 1987. Vostok ice core: A 160,000 year record of atmospheric CO_2. *Nature* 329:408–14
4. Bazzaz, F. A. 1990. The response of natural ecosystems to rising global CO_2 levels. *Annu. Rev. Ecol. Syst.* 21:167–96
5. Benedick, R. E. 1991. *Ozone Diplomacy: New Directions in Safeguarding the Planet*. Cambridge, Mass: Harvard Univ. Press
6. Bolin, B., Döös, B. R., Jager, J., Warrick, R. A., eds. 1986. *The Greenhouse Effect, Climate Change and Ecosystems*. Chichester: Wiley
7. Bonan, G. B., Shugart, H. H. 1989. Environmental factors and ecological processes in boreal forests. *Annu. Rev. Ecol. Syst.* 20:1–28
8. Broecker, W. S. 1987. Unpleasant surprises in the greenhouse. *Nature* 328: 123–26
9. Caldwell, M. M., Teramura, A. H., Tevini, M. 1989. The changing solar ultraviolet climate and the ecological consequences for higher plants. *Trends Ecol. Evol.* 4:363–67
10. Charlson, R. J., Schwartz, S. E., Hales, J. M., Cess, R. D., Coakley, J. A. Jr. 1992. Climate forcing by anthropogenic aerosols. *Science* 255:423–31
11. Cicerone, R. J., Oremland, R. 1988. Biogeochemical aspects of atmospheric methane. *Global Biogeochemical Cycles* 2:299–327
12. CLIMAP Project. 1976. The surface of the ice age earth. Science 191:1131–36
13. COHMAP Project. 1988. Climatic changes of the last 18,000 years: Observations and model simulations. *Science* 241:1043–52
14. Crutzen, P. J., Zimmermann, P. H. 1991. The changing photochemistry of the atmosphere. *Tellus* 43:136–51
15. Cubasch. U., Cess, R. D. 1990. Processes and modelling. See Ref 25, p. 69–91
16. Davis, M. B. 1990. Biology and paleobiology of global climate change: Introduction. *Trends Ecol. Evol.* 5:269–70
17. Dickinson, R. E. 1991. Global change and terrestrial hydrology: a review. *Tellus* 43AB:176–81
18. Drake, J., DiCastri, F., Groves, R., Kruger, F., Mooney, H. A., et al, eds. 1989. *Biological Invasions: A Global Perspective*. Chichester: Wiley
19. Ehrlich, P. R., Wilson. E. O. 1991. Biodiversity studies: science and policy. *Science* 253:758–62
20. Farman, J. C., Gardiner, B. G., Shanklin, J. D. 1985. Large losses of total ozone in Antarctica reveal seasonal $C10_x/NO_x$ interaction. *Nature* 315:207–10
21. Fishman, J., Fakhruzzaman, K., Cros, B., Nganga, D. 1991. Identification of widespread pollution in the southern hemisphere deduced from satellite analyses. *Science* 252:1693–96
22. Glynn, P. W. 1988. El Niño-Southern Oscillation 1982–1983: Nearshore population, community, and ecosystem responses. *Annu. Rev. Ecol. Syst.* 19: 309–45
23. Hall, J. V., Winer, A. M., Kleinman, M. T., Lurmann, F. W., Brajer, V., et al. 1992. Valuing the health benefits of clean air. *Science* 255:812–17
24. Henderson-Sellers, A. 1987. *A Climate Modelling Primer*. Chichester: Wiley
25. Houghton, J. T., Jenkins, G. J., Ephraums, J. J. eds. 1990. *Climate Change: The IPCC Scientific Assessment*. Cambridge: Cambridge Univ. Press
26. Houghton, R. A., Boone, R. D., Fruci, J. R., Hobbie, J. E., Melillo, J. M. 1987. The flux of carbon from terres-

trial ecosystems to the atmosphere in 1980 due to changes in land use: geographic distribution of the global flux. *Tellus* 39B:122–39
27. IPCC. 1990. Policymakers summary. See Ref. 25, p. xi-xxxix
28. Janzen, D. H. 1988. Tropical dry forests: the most endangered major tropical ecosystem. See Ref. 65, p. 130–37
29. Keeling, C. D., Bacastow, R. B., Carter, A. F., Piper, S. C., Whorf, T. P. 1989. A three dimensional model of atmospheric CO2 transport based on observed winds: 1. Analysis of observational data. *Geophys. Monogr.* 55: 165–236
30. Keller, M., Jacob, D. J., Wofsy, S. C., Harris, R. C. 1991. Effects of tropical deforestation on global and regional atmospheric chemistry. *Climatic Change* 19:145–58
31. Lashof, D. A., Ahuja, D. R. 1990. Relative contributions of greenhouse gas emissions to global warming. *Nature* 344:529–31
32. Logan, J. A. 1985. Tropospheric ozone: Seasonal behavior, trends, and anthropogenic influence. *J. Geophys. Res.* 90:10,463–82
33. Lubchenco, J., Olson, A. M., Brubaker, L. B., Carpenter, S. R., Holland, M.M., et al. 1991. The substainable biosphere initiative: an ecological research agenda. *Ecology* 72:371–412
34. Matson, P. A., Vitousek, P. M. 1990. Ecosystem approach to a global nitrous oxide budget. *Bioscience* 40:667–72
35. Melillo, J. M., Callaghan, T. V., Woodward, F. I., Salati, E., Sinha, S. K. 1990. Effects on ecosystems. See Ref. 25, p. 283–310
36. Mitchell, J. F. B., Manabe, S., Tokioka, T., Meleshko, V. 1990. Equilibrium climate change. See Ref. 25, p. 131–72
37. Molina, M. J., Rowland, F. S. 1974. Stratospheric sink for chlorofluoromethanes: chlorine atomic catalysed destruction of ozone. *Nature* 249:810–12
38. Mooney, H. A., Drake, B. C., Luxmoore, R. J., Oechel, W. C., Pitelka, L. F. 1991. Predicting ecosystem responses to elevated CO_2 concentrations. *Bioscience* 41:96–104
39. Mooney, H. A., Drake, J. eds. 1986. *Biological Invasions of North America and Hawaii*. New York: Springer-Verlag
40. Mooney, H. A., Vitousek, P. M., Matson, P. A. 1987. Exchange of materials between terrestrial ecosystems and the atmosphere. *Science* 238:926–32
41. Morris, J. T. 1991. Effects of nitrogen loading on terrestrial ecosystems with particular reference to atmospheric deposition. *Annu. Rev. Ecol. Syst.* 22: 257–79
42. Pastor, J., Post, W. M. 1988. Response of northern forests to CO_2-induced climate change. *Nature* 334:55–57
43. Penner, J. E., Atherton, C. S., Dignon, J., Chan, S. J., Walton, J. J. 1991. Tropospheric nitrogen: A three-dimensional study of sources, distributions, and deposition. J. Geophys. Res. 96:959– 90
44. Peterson, B. J., Fry, B. 1987. Stable isotopes in ecosystem studies. *Annu. Rev. Ecol. Syst.* 18:293–320
45. Prather, M. J., Watson, R. T. 1990. Stratospheric ozone depletion and future levels of atmospheric chlorine and bromine. *Nature* 344:729–34
46. Reich, P. B., Amundson, R. B. 1985. Ambient levels of ozone reduce net photosynthesis in tree and crop species. *Science* 230:566–70
47. Rowland, F. S. 1989. Chlorofluorocarbons and the depletion of stratospheric ozone. *Am. Sci.* 77:42–44
48. Schimel, D. S., Kittel, T. G. F., Parton, W. J. 1991. Terrestrial biogeochemical cycles: Global interactions with the atmosphere and hydrology. *Tellus* 43AB:188–203
49. Schimel, D. S., Parton, W. J., Cole, C. V., Ojima, D. S., Kittel, T. G. F. 1990. Grassland biogeochemistry: Links to atmospheric processes. *Climatic Change* 17:13–25
50. Schindler, D. W. 1988. Effects of acid rain on freshwater ecosystems. *Science* 239:149–57
51. Schlesinger, W. H. 1991. *Biogeochemistry: An Analysis of Global Change*. San Diego, Calif: Academic
52. Schneider, S. H. 1989. *Global Warming: Are We Entering the Greenhouse Century?* New York: Sierra Club
53. Schneider, S. H. 1990. The changing climate. In *Managing Planet Earth: Readings from Scientific American*, ed. pp. 24–36. New York: Freeman
54. Schulze, E.-D. 1989. Air pollution and forest decline in a spruce (*Picea abies*) forest. *Science* 244:776–83
55. Shukla, J., Nobre, C., Sellers, P. 1990. Amazonian deforestation and climate change. *Science* 247:1322–24
56. Siegenthaler, U., Oeschger, H. 1987. Biospheric CO_2 emissions during the past 200 years reconstructed by deconvolution of ice core data. *Tellus* 39B:140–54

57. Stuiver, M. 1978. Atmospheric carbon dioxide and carbon resevoir changes. *Science* 199:253–58
58. Titus, J. G., ed. 1986. *Effects of Changes in Stratospheric Ozone and Global Climate*. Washington, DC: Environmental Protection Agency
59. Turner, B. L. II, Clark, W. C., Kates, R. W., Richards, J. F., Matthews, J. T. 1990. *The Earth as Transformed by Human Action*. Cambridge: Cambridge Univ. Press
60. Turner, R. E., Rabalais, N. N. 1991. Changes in Mississippi River water quality in this century. *BioScience* 41: 140–47
61. Vitousek, P. M., Ehrlich, P. R., Ehrlich, A. H., Matson, P. A. 1986. Human appropriation of the products of photosynthesis. *Bioscience* 36:368–73
62. Vitousek, P. M., Matson, P. A. 1992. Agriculture, the global nitrogen cycle, and trace gas flux. In *Biogeochemistry of Global Change: Radiative Trace Gases*, ed. R. Oremland, In press. New York: Chapman & Hall. In press
63. Watson, R. T., Rodhe, H., Oeschger, H., Siegenthaler, U. 1990. Greenhouse gases and aerosols. See Ref. 25, p. 1–40
64. Wigley, T. M. L., Barnett, T. P. 1990. Detection of the greenhouse effect in the observations. See Ref. 25, p. 239–55
65. Wilson, E. O., Peter, F. M., eds. 1988. *Biodiversity*. Washington, DC: Natl Acad. Sci.
66. WMO. 1989. Scientific assessment of stratospheric ozone: 1989. Global ozone research and monitoring project. *Report 20*, Geneva

Annu. Rev. Ecol. Syst. 1992. 23:15–38

THE POTENTIAL FOR APPLICATION OF INDIVIDUAL-BASED SIMULATION MODELS FOR ASSESSING THE EFFECTS OF GLOBAL CHANGE*

H. H. Shugart and T. M. Smith
Department of Environmental Sciences, The University of Virginia, Charlottesville, Virginia 22901

W. M. Post
Environmental Sciences Division, Oak Ridge National Laboratory, Oak Ridge, Tennessee 37830

KEYWORDS: succession, forest dynamics, climate, vegetation

INTRODUCTION

The environmental changes that could result from human activities are sufficiently large to be characterized as an "uncontrolled global experiment" (7). The obvious lack of experimental control when complex changes are global in scale, along with a myriad of other logistic difficulties, confound the evaluation of global consequences. This has created a need for reasoned extrapolations from experimental and observational data and the application of computer models to predict the ecological response of the terrestrial surface to changes in the environment.

The first model-based evaluations of global changes for terrestrial vegetation focused on identifying which phenomena need to be considered and at what temporal and spatial scales (88, 33). Then the emphasis on plant physiology and biophysics (22, 27, 41, 71, 72, 118) versus the emphasis on individual-plant natural history and demography (34, 43, 44, 55, 56, 76, 121) formed an important dichotomy for the models simulating the vegetation dynamics in response to environmental change. This dichotomy has blurred as a new generation of models developed that incorporate the dominant features of each approach (55). The interpretation of robust consequences of a spectrum of models is replacing an earlier interest in discerning the "best" model for interpreting potential global responses.

The computer models with potential to provide some insight into global change issues are legion as are the potential schemes to classify them. We focus this review on an important class of models that take the simulation of the birth, growth, and dynamics of individual plants on a small area as their basis (55). These models are later related to others with respect to their role in assessing global change.

Model-based global assessments are most typically made at two scales, site and global scale; there is also considerable recent interest in assessments at an intermediate, regional scale. Global models, in the sense we are using the term, are relatively aggregated models of material fluxes from major reservoirs of carbon, nitrogen, and other important elements.

At the site scale (0.1 to ca 10 ha), a diverse array of models is usually based on assumptions of local homogeneity of many or all of the process controlling environmental factors. Often these models are interpreted at the continental or global scale by considering a large number of independent cases with a representative sampling of important controlling environmental variables. There are analogous experimental and observational approaches typically involving rates of production (17, 18, 60, 90) or decomposition (66). Most individual-based models function at this spatial scale (93).

The most problematic spatial scale for modeling vegetation change involves the classes of problems that require a consideration of temporal and spatial dynamics over linear dimensions greater than 100 m to about 1 km. At these scales we must consider explicit spatial interactions among vegetation units, environmental heterogeneity, and processes that propagate in space [over relatively long periods of time in the case of plant migration (31, 87) and over short intervals in the case of the propagation of wildfires or insect outbreaks]. Such considerations create classes of problems that are generally tractable only for particular cases. There have been some applications of individual-based simulation models at regional scales. This is particularly the case for the more recent spatially explicit versions of the models discussed below.

INDIVIDUAL-BASED MODELS OF VEGETATION CHANGE

The importance (or at least the potential importance) of individual plants for understanding vegetation/environment feedbacks implies a related importance of vegetation dynamics in global change issues. The principal individual-based models under consideration for global applications originate with models of forest dynamics based on the changes of individual trees. Such models have become increasingly prominent over the past two decades.

Trees can grow to sufficient size to locally alter their own microenvironment and that of subordinate trees (17, 18, 90, 120). The magnitude of this effect can depend on the species, shapes, and sizes of trees. The environment, in turn, has a profound influence on the success of different species, shapes, and sizes of trees. In particular, there can be a feedback from the canopy tree to the local microenvironment and subsequently to the seedling and sapling regeneration that may become the next canopy (93, 94). The environmental alteration from a canopy tree is most easily observed for the forest light environment (47). The nature of the leaf area profile and canopy geometry are dominant factors in the amount and pattern of the light at the forest floor (5, 26, 28). Other tree/environment interactions are also potentially important, and plants can alter their local environment with respect to other variables, for example, throughfall (49, 124), soil moisture (21, 92, 109), soil nutrient availability (21, 124), and soil thermal regime (9, 12, 22, 37).

Origins of Individual-Based Forest Models

The earliest model to simulate the dynamics of a forest by following the fates of each individual tree was developed by Newnham (73) in 1964. Foresters at several institutions (6, 8, 48, 65, 69) proceeded to develop additional models of forest change that involved using a digital computer to dynamically change a map of the sizes and positions of each tree in a forest. These models have become increasingly used as the computer power available to ecologists has increased (72, 100).

Huston et al (55) note that an advantage of such models is that two implicit assumptions normally required in applying traditional population models are not necessary. These are the assumptions that: (i) The unique features of individuals are sufficiently unimportant to the degree that individuals are assumed to be identical, and (ii) the population is "perfectly mixed" so that there are no local spatial interactions of any important magnitude. These assumptions seem particularly inappropriate for trees that are sessile and that vary greatly in size over their life A class of individual-organism–based tree models that have been widely used in ecological (as opposed to traditional forestry) applications are the so-called "gap" models (100). The models are constructed by assuming horizontal homogeneity inside the simulated plot, and the size of the simulated plot is critical. The spatial scale at which the models operate is an area corresponding to the zone of influence of a single individual of maximum size. This means that an individual growing on the plot can achieve maximum size while allowing for the death of a large individual to significantly influence the light environment on the plot.

Individual-based forest gap models simulate the establishment, diameter growth, and mortality of trees usually at an annual time-step and on a plot of defined size (ca 0.1 ha). This general approach has been applied to a wide variety of forested systems around the world (Figure 1) implying a generality that is useful in global or regionally extensive studies. Most of the models indicated in Figure 1 are variants of the FORET model (99).

The FORET model was originally derived from the JABOWA model (14) which, in turn, is an ecological version of the earlier forestry models based on individual tree dynamics (6, 8, 48, 63, 65, 69, 73). The JABOWA model is not spatially explicit (as were the earlier forestry models). Its principal differences from these models are the relatively simple protocols for estimating the model parameters.

There is a considerable body of information on the performance of individual trees (growth rates, establishment requirements, height/diameter relations) that can be used directly in estimating the parameters of such models. Most gap models are parsimonious in that they are based on simple rules for interactions among individuals (e.g. shading, competition for limiting resources, etc) in conjunction with simple rules for birth, death, and growth of individuals.

Structure of Gap Models

Many of the individual-based simulation models for plants are similar in their structure, and a brief description of the functioning of these models is appropriate. Individual-based forest gap models (100) simulate the establishment, diameter growth, and mortality of trees usually at an annual time-step and on a plot of defined size (ca 0.1 ha).

There is a considerable body of information on the dynamics of individual trees (growth rates, establishment requirements, height/diameter relations) that are used in estimating the parameters of such models. Most gap models are parsimonious in that they are based on simple rules for interactions among individuals (e.g. shading, competition for limiting resources, etc) in conjunction with simple rules for birth, death, and growth of individuals. The simplicity of the models is an advantage in the sense that it has allowed development of a large number of such models in different parts of the world (Figure 1); it is a disadvantage in that some complex physiological and ecological mechanisms are approximated with relatively simple functions.

The models differ in their inclusion of processes that may be important in the dynamics of particular sites being simulated (e.g. hurricane disturbance, flooding) but share a common set of characteristics:

MODEL STRUCTURE Each individual plant is modeled as a unique entity with respect to the processes of establishment, growth, and mortality. Thus, the models have appropriate information to allow computation of species- and size-specific demographic effects.

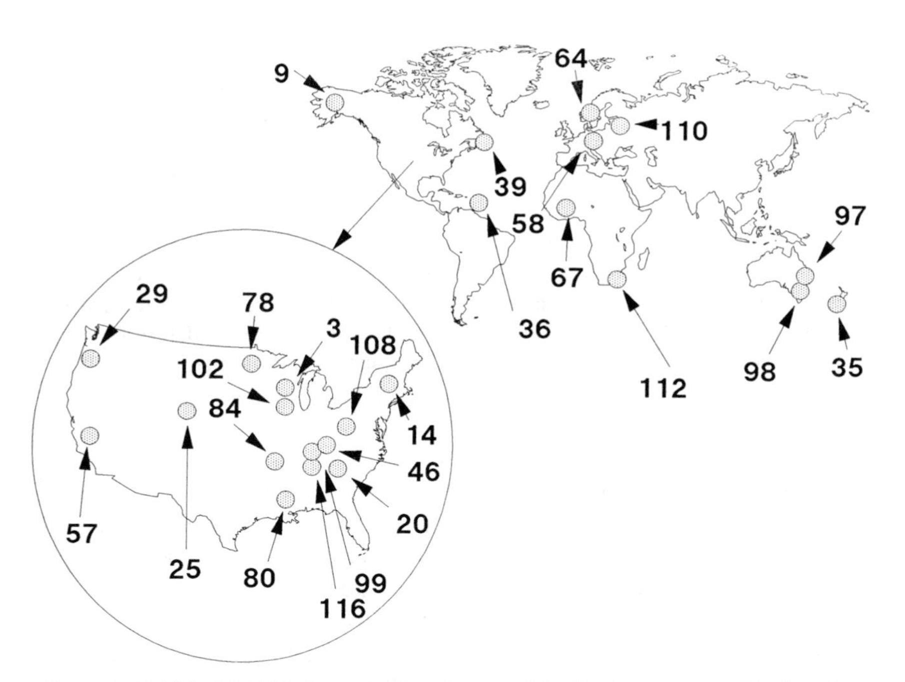

Figure 1 Published individual-organism-based gap models. Numbers correspond both to the numbers on the figure and to the citation. Mnemonic for each model and a broad characterization of the system that is simulated is also provided.

	Mnemonic	Ecological System Simulated
3.	FORTNITE	Wisc. Mixed Wood Forest.
9	LOKI	North American Spruce-Fir Forest.
14.	JABOWA	Northern Hardwood Forest.
20.	FORANAK	Appalachian Spruce/Fir Forest.
25.	STEPPE	North American Short-grass Prairie.
29.	CLIMACS	Pacific Northwest Coniferous Forest.
35.	FORENZ	New Zealand Forest.
36.	FORICO	Puerto Rican Montane Rain Forest.
39.	SMAFS	Eastern Canadian Mixed Wood Forest.
46.	OVALIS	Appalachian Oak-Hickory Forest.
57.	SILVA	Mixed Conifer Forest.
58.	FORECE	Central European Forest.
64.	FORSKA	Scandinavian Forest.
67.		West African Humid Savanna
78.	LINKAGES	Temperate/Boreal Forest Transition.
80.	FORFLO	Southern USA Floodplain Forest
84.	SWAMP	Arkansas Floodplain Forest.
97.	KIAMBRAM	Australian Subtropical Rain Forest.
98.	BRIND	Australian Eucalyptus Forest.
99.	FORET	Southern Appalachian Deciduous Forest.
102.	ZELIG	Temperate Deciduous Forest.
108.	FORENA	Forests of Eastern North America.
110.	SJABO	Estonian Conifer Forest.
112.	OUTENIQUA	South African Temperate Rain Forest.
116.	FORCAT	Southern Oak/Hickory Forest.

SPACIAL SCALING The vertical position of each individual on the plot is influenced by (and influences) the growth of all other individuals on the plot. Although horizontal homogeneity on the plot is assumed, the vertical structure of the canopy is modeled explicitly. The height and leaf area of each individual are computed from allometric functions of the tree diameter and are used to construct a vertical leaf-area profile. Using a light extinction equation, the vertical profile of available light is then calculated so that the light environment for each individual can be defined. The models differ in the degree to which the canopy geometry and canopy shapes are computed: in the earlier models (e.g. JABOWA and FORET), the height of the tree determines the insolation and shading; the newer models (FORSKA, BOFORS, and ZELIG) compute the canopy depth of each tree as well as complex interactions such as side lighting.

ENVIRONMENTAL CONSTRAINTS AND RESOURCE COMPETITION All gap models simulate individual tree response to light availability at height intervals on the plot. Other environmental features incorporated in different models include soil moisture, fertility, nitrogen, and temperature as well as disturbances such as fires, hurricanes, floods, and windthrow. Environmental responses of each individual plant are modeled by making the assumption that each tree has a maximum potential performance under optimal conditions (i.e. maximum diameter increment, survivorship, or establishment rate). This optimum is reduced according to the environmental context of the plot (e.g. shading, drought), to yield the realized behavior under ambient conditions. Typically, the curves describing species response to environmental resources are scaled between 0.0 and 1.0, and species are often categorized into a small number of functional types (Figures 2a-d).

Competition in the model depends on the relative performance of different trees under the environmental conditions on the model plot. These environmental conditions may be influenced by the trees themselves (e.g. a tree's leaf area influences light available beneath it), or may be modeled as extrinsic and not influenced by the trees (e.g. temperature). Competition operates in two modes: One-way competition such as competition for light—a tree at a given height absorbs light and reduces the resource available to trees at lower positions in the canopy, and two-way competition such as for below-ground resources (water and nutrients—23, 38)—each tree experiences a resource level common to the plot, and a resource shortage reduces the growth of all the trees on the plot. Competitive ability depends strongly on the context of the modeled gap; the tree that has the best performance relative to other trees on the plot is the most successful. Competitive success depends on the environmental conditions on the plot, on which species are present, and on the relative sizes of the trees; each of these varies through time in the model.

GROWTH The growth of an individual tree is calculated using a species-specific function that predicts the expected diameter increment under optimal conditions and given the tree's diameter (Figure 3). This increment is modified by the environmental response functions (Figure 2) to determine the realized increment added to the tree.

MORTALITY The death of individual trees is simulated as a stochastic process with two components, an intrinsic probability of mortality and an extrinsic probability of mortality. The intrinsic probability is usually a function of the expected longevity for the species. For typical species, annual survivorship is around 1–2%. For computation of extrinsic mortality, if the diameter increment is below a minimal diameter increment (often 10% of optimum), the individuals are subjected to an elevated mortality rate.

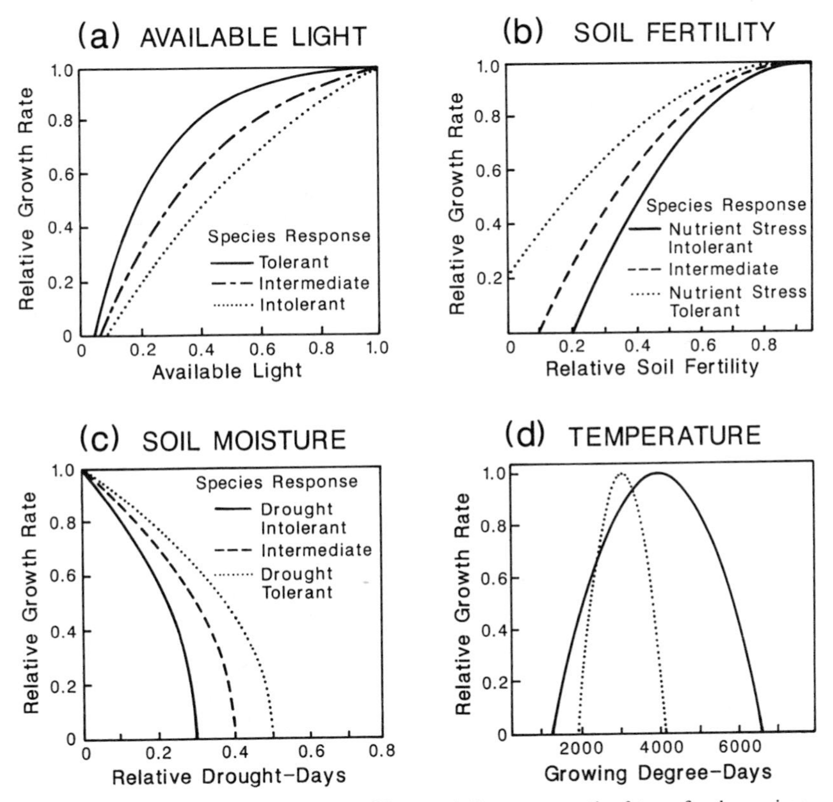

Figure 2 Response functions used to modify annual diameter growth of trees for the environmental constraints of *(a)* light, *(b)* soil fertility, *(c)* soil moisture, and *(d)* temperature. Most species can be assigned to response categories for light, nutrients, and moisture; temperature responses are based on geographic ranges of species.

ESTABLISHMENT Most gap models are designed to treat regeneration in trees from a pragmatic view that the factors influencing the establishment of seedlings can be usefully grouped in broad classes (34, 45, 61, 62, 114). Tree establishment and regeneration are largely stochastic, with maximum potential establishment rates constrained by the same environmental factors that modify tree growth. Each simulation year, a pool of potential recruits is filtered through the environmental context of the plot, and a few new individuals are established.

The model structure outlined above includes two features important to a dynamic description of vegetation pattern: (i) the response of the individual plant to the prevailing environmental conditions, and (ii) how the individual modifies those environmental conditions. The models are hierarchical in that the higher-level patterns observed (i.e. population, community, and ecosystem) are the integration of plant responses to the environmental constraints defined at the level of the individual.

Tests of Gap Model Performance

Interest in individual-based models for assessments of the effects of global environmental change stems in part from the degree of past model testing and the apparent ability of the models to predict ecosystem patterns and dynamics under novel circumstances (tabulated up until 1983 in Ref. 93) as discussed

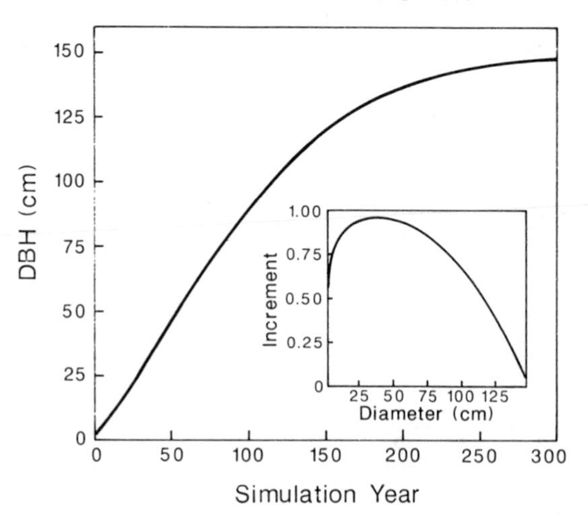

Figure 3 Annual diameter growth as simulated in gap models: (*a*) optimal growth through time, and (*b*, inset) transformed to predict potential annual diameter increment based on current diameter. Potential increment is multiplied by environmental response functions associated with environmental limitation to growth.

below. Simulation models, in general, and individual-based gap models, in particular, have several features that incline scientists to expect a high standard of model testing. The models seem to lack the parsimony that is expected in models; mathematically, the models are not open to formal analysis as are many simpler population models; the high level of detail in model output implies a concomitant degree of detail in testing of models. Further, since the models are often used in application domains outside the range of conditions used in model development, good scientific practice requires continual model testing and assessment. Successful model tests generally involve nontrivial comparisons of model response with reality (93).

The parameters in the gap models are constrained to a degree by the requirement of realistic parameter values (93). For example, a growth rate parameter must produce a maximum diameter increment of an appropriate size; the maximum heights and diameters of the trees used in allometric functions must be reasonable; the light extinction coefficient in the models must be within measured ranges. Nonetheless, there is a degree of flexibility in the ranges of the model parameters. The typical protocol for developing these models is to estimate virtually all model parameters prior to any model/data comparisons. The models have been tested in several different forests with a diverse range of model/data intercomparisons.

Usually a simulator is initially judged through inspection as to whether the model, when parametrized a priori with reasonable species parameters for individual plant performance, can reproduce forest-level features such as total biomass, leaf area, stem density or average tree diameter for forests of differing ages. Such tests have been used in a variety of cases [e.g. Estonian conifer forest (110), New Zealand southern beech forest (35), floodplain forests in the southern United States (80, 84), Pacific coast Douglas-fir forests (29), etc (1, 39, 46, 57, 97, 116)]. Models are sometimes calibrated by producing an appropriate structure for stands of a given age and then tested as to whether they can independently predict the structure for forests of another age (without recalibration). Shugart & West (99) predicted composition and biomass for current forest in Tennessee and then projected the expected composition of mature forests. These comparisons were given increased rigor in that the historically documented mature forest had viable American chestnut (*Castanea dentata*) codominants that were not present in today's forests (due to the chestnut blight, *Endotheca parasitica*). Busing & Clebsch (20) tested their spruce-fir forest simulator by inspecting model performance for old-growth stands and then predicting the forest structure for stands of an intermediate age that had been clear-cut 40 to 70 years earlier. Doyle (36) used structural comparisons with stands of known age as one of several tests on a tropical rain forest simulator, and Shugart et al (97) compared model predictions with forest composition in stands of differing ages for Australian rain forest. The comparison of model performance after long periods of simulated time to existing mature forests has been used to test gap models in several cases (9, 29, 35, 39, 58, 64, 110, 116). The scientific value of mature forests as a test of the long-term responses of models (and the synthesis of biological understanding that they imply) is considerable.

Independent predictions of tree diameter increment and hence forest growth were used to test a model of temperate African rain forest (112). Some of the most detailed data compilations on forest structural dynamics can be found in forest yield tables. These tables compile the volume, number, mean diameters, heights, and other structural quantifiers of forests for stands of different ages growing on sites of varying quality for well over 100 years in some cases. Comparison with yield tables was used as an independent test of the BRIND model of montane eucalyptus communities (98). Leemans & Prentice made use of Swedish yield tables to inspect the performance of the FORSKA model of Scandinavian forests (64), and Kienast & Kuhn (58) made similar application of Swiss yield tables.

One important class of model examinations involves the comparison of model predictions for different environmental input conditions, followed by an inspection of the responses of forests to natural gradients deemed comparable to the input conditions. The earliest gap model, JABOWA (14), was tested on its ability to predict the approximate altitude of the transition from northern hardwood forest to coniferous forest in New Hampshire. Pearlstine et al (80) looked at the responses of simulated floodplain forest composition at different flood frequencies to test the FORFLO model. The BRIND model was tested in its ability to characterize montane zonations of eucalyptus forests in the Brindabella Mountains in Australia (98) as a function of wildfire frequency and altitude. Kercher & Axelrod (57) inspected model predictions for fire frequencies, altitude, and aspect in the mixed-conifer montane forests of California. Doyle (36) used simulations at different hurricane frequencies to assess the performance of the FORICO model of Puerto Rican *Dacroides-Sloanea* forests.

In more recent gap models, the simulation of model responses to single controlling variables has evolved to more complex cases in which the responses are to multiple gradients and the reconstruction of forest landscapes. Harrison & Shugart (46) compared ordinated forest simulation for different slopes and ages to plot data from an Appalachian watershed. Bonan (9) developed a simulator of forest composition, structure, successional response to wildfire, and the depth to permafrost (which is under a degree of control from feedbacks from forest condition) for boreal forests in the vicinity of Fairbanks, Alaska. The FORECE model simulated the forest patterns on different altitudes, slopes, and soils in the Swiss Alps (58). Pastor & Post (78), building on earlier work by Aber et al (1), modeled forest pattern in response to soil fertility and texture for forest in the northern hardwood–to–conifer transition.

Simulations involving forest response to complex gradients and the applications of the models to simulate prehistoric forests (discussed below) are germane to the use of the models in the assessment of global environmental change. While gap models have been tested to a relatively greater degree than most ecological models, the models are under a continual revision as new phenomena are incorporated, and they should be continually tested. Further, any model is an abstraction and as such may not incorporate all the relevant phenomena for a given assessment. Any model extrapolations of a future vegetation should be regarded with caution and compared with other equivalent evaluations from other bases.

Non-Forest Individual-Based Vegetation Models

Although a considerable number of individual-based simulation models have been developed for forest systems, the relatively smaller number of individual-based simulators developed for other ecosystems show a healthy diversity of approaches. There are several problems that attend the development of individual plant simulators for non-forest environments. The recognition of individuals in plant life forms that are prone to clonal propagation, tillering, and other vegetative modes of regeneration can be problematic. Of course, this problem occurs for trees as well, but the forestry and ecological traditions of collecting data on the state of forests by measuring tree diameters as if the trees were individuals make individual plant models for forests conceptually consistent with most data collection and data analysis protocols. The competitive interactions in non-forest ecosystems are less well characterized (in general) than for forests and seem to have a greater range of potential factors (light competition, root competition, competition for nutrients, etc) in a given situation. Further, the consideration of interactions between monopodal tall perennials (trees) seems considerably less demanding than, for example, the interactions between perennials and annuals in which spatial heterogeneity in the competitive interactions is confounded by differences in the temporal domain.

There has been a historical interest in the factors that control the geographical distributions of plants of different life forms at geographical scales. One can trace this interest from the encyclopedic works of Von Humboldt through to his successors, the classic German plant geographers, from the start of this century (Drude, Graebner, Warming, and Schimper), and hence to present-day ecologists. The relationships are those between climate and plant structure whose causalities are still being explored today at both the plant (42) and vegetation (16, 50, 122) level. These considerations along with the problem of quantifying the competitive interaction among individuals of differing life forms and the attendant differences in zone of influence, use of resources, and temporal dynamics are central themes in the development of individual-based models for a range of vegetation types.

A favored approach in individual-based models of non-forests is the application of Markov models that use the probability that an individual of a given species will be present after some time interval at a particular location (113) to predict change. This probability is based on the species presence (or absence) at the location and in some cases on the environmental condition that is associated with a particular location. In the development of Markov models, one can elaborate biological and functional realism in several dimensions. For example, one can relate the estimation of the transition probabilities (probabilities that a given species of plant will be replaced by another species over a fixed time interval) to biological attributes of the plants involved. This approach is probably best known from applications relating tree species transition probabilities to canopy geometry (51–53).

Other analogous approaches involve altering transition probabilities based on plant physiology (113) or spatial location of seed sources (115). Noble & Slatyer (75, 76) developed a scheme for estimating the Markov transition probabilities for plants of different life forms using a "vital attributes approach." This is a theoretical framework for inferring the transition probabilities for plants based on such life history attributes as longevity, age of reproduction, duration of various vegetative stages, and response to disturbance (74). The vital attributes approach has been incorporated into an interactive model called the FATE model and applied to several Australian forest and woodland communities (70).

The interaction of trees and other life forms (59, 68, 91, 123) is an important topic with respect to global environmental change because of the implications of these models for shifts in transition zones between forests and other systems (for example, tundra—10). The FATE model (70) and another model of savanna in the Côte d'Ivoire (67) have been applied to the evaluation of changes in fire disturbance regimes on mixtures of grasses, shrubs and trees in savanna ecosystems. The DUNE model (86) has been applied to simulate vegetation and coastal landform change for Virginia coastal barrier islands and also features the interaction and placement of plants of different life forms along a dynamically changing environmental gradient. Such evaluations are important in understanding the structural dynamics of terrestrial systems under climatic change. Recent approaches appear to be moving away from the Markovian modeling approaches used to simulate the presences and absence of different species on small landscape elements. Recently, models of non-forested ecosystems are being developed based on approaches that are more traditional in forestry (67; see discussion above). Some models of grasslands (24, 25, 91, 123) have been developed using the FORET model (99) as a starting point.

APPLICATIONS OF INDIVIDUAL-BASED MODELS TO GLOBAL CHANGE ISSUES

The earliest applications of individual-based simulators to climatic change of vegetation were not in the context of future environmental change but were in the testing of the models on ability to reproduce the vegetation reconstructions based on palynological evidence. Such tests on the ability to reconstruct paleoecological pattern under climates of the geological past have been proposed as a need in testing models for their utility in global change assessments (54). The first such application was using the FORET model (93, 99) to reconstruct a 16,000 year (BP to present day) fossil pollen record from eastern Tennessee (103, 104). Subsequent applications have involved the reconstruction of forest communities in eastern Tennessee during the full glacial condition (93, 105). Currently, the application of simulation models to inspect the consistency of model results, inferred climates, and reconstructed paleoecological communities is an accepted methodology (106), and some of the shortcomings in the methods are better understood (107).

The principal difficulties perceived for such applications involve the apparent circularity of the tests. This is because the paleoclimates are often inferred from the palynological evidence. However, most palynological data are resolved only to the genus level, and the models simulate species composition, so that the possibility of further testing the more specific model results against macrofossils can to some degree obviate this problem. In some cases the paleoclimate can be inferred from other sources such as Bonan & Hayden's (11) reconstruction of ice age vegetation of Virginia that is tested against palynological evidence but based on climate reconstructions derived from marine sediments. Also model applications can also provide a considerable degree of insight into theoretical responses of vegetation to climatic change, and in such applications the model testing is less important than the insight provided (32).

Prediction of vegetation response under future climates has developed in the direction of prediction of continental responses of forests to changed conditions over large areas (13, 79, 108). Simulated responses from a spatially explicit version of the FORET model (93, 99), called ZELIG, and the JABOWA model (14) were an integral part of a recent US assessment of potential climate change effects (101). In general this class of models has also been valuable in identifying potential complex interactions between vegetation and soils (102) and vegetation and disturbance regimes (77) in the context of an altered climate.

Until Solomon (108), gap models were developed for particular sites incorporating species and processes that were important for fairly limited geographical regions. The step was taken with FORENA to include all the dominant tree species of eastern North America. The only variables left to be specified in a particular model run were climate and soil parameters. When the climate was specified for 21 locations distributed over eastern North America, the model reproduced the spatial aspects of eastern deciduous, coniferous, and boreal forests remarkably well. The model was physiognomically correct, simulating stunted woodland where scattered trees grow, simulating forests where forest grow, and simulating growth of nothing where trees are absent on the modern landscape. Major forest types were also correctly simulated. The correct species dominants appear in the simulations for the climate at corresponding eastern North American locations.

Solomon (108) used FORENA to simulate the response of forests to CO_2-induced climate changes. The simulated effects of changing temperature and precipitation at a constant rate as CO_2 concentration doubled resulted in a distinctive dieback of extant trees at most locations. This is caused by climate changes, particularly temperature increases, resulting in conditions less favorable for established species and an insufficient time for individuals of other species to become established before forest decline occurs. Transient responses in species composition and carbon storage continued for as long as 300 years after simulated climate changes ceased. This was one of the first demonstrations that forest tree population dynamics can influence carbon dynamics on a large spatial scale.

While FORENA computes the effects of soil moisture on tree growth, the simulations of Solomon (108) used a very deep mesic silt-loam at all sites that minimized the effects of precipitation deficits. Even more drastic forest changes are predicted under CO_2-induced climate change if shallower, coarser textured soils are used in the simulations. Similar model exercises using ZELIG (111) with soils with less water holding capacity predicted increased evaporative demand with increased temperatures. The resulting increase in drought severely restricted tree growth even if precipitation increased. In Urban & Shugart's (111) results, 18 tree species they considered could no longer grow in the southeastern United States, and much of the southern half of the Southeast would not support trees. Simulated biomass accumulations under CO_2-induced climate change in the Great Lakes region decreased to 23–54% of their present value (15). On poor sites in both regions, forests could be converted to grassland or savanna with very low productivity and carbon storage in biomass.

Climate change simulations with similar models also show changes in species composition, biomass, or both. In some cases species composition shifts with little change in stand biomass as is the case for simulations of forests in the Pacific Northwest (30). Simulations of Wisconsin and Quebec forests (77) showed that forest productivity changed with climate but resulted in small changes in species composition. When increased disturbance frequency was added (several climate-model results indicate increased rates of forest disturbance as a result of weather more likely to cause forest fires, convective wind storms, coastal flooding and hurricanes), then significant changes in forest composition as well as changes in biomass was projected by the simulations. Simulation results for the Great Lakes region (15, 79) found significant changes in species composition without additional disturbances. As in simulations for the southeastern United States, dry sites could lose the capability to grow most tree species and be converted to oak savannas or even prairies under a doubled CO_2 climate.

Whether or not climate change results in species composition shifts, changes in biomass, or both depends largely on whether or not the climate changes introduced into the model inputs are sufficient to change the climate to temperature and moisture conditions that result in dominant species either not being able to grow at all under the new conditions or dominant species losing out in the competition for light to other species that grow faster under the new climate conditions.

So far we have concentrated on model simulations that project the effects of climate change on forest species composition, aboveground biomass, and net primary production. In forest ecosystems, on average, there is nearly twice as much carbon (globally, 927 Pg—Ref. 82) stored in soil organic matter and litter as in aboveground biomass (globally, 515 Pg—Ref. 81). Most of many essential nutrients are supplied by decomposition of this material. Thus it is critical that litter decomposition and soil organic matter dynamics be considered as well as biomass dynamics in determining the forest responses to global change. It is possible to model decomposition and soil organic matter dynamics in a way that is consistent with aboveground population dynamics of individual-based models. This is done by literally extending the population dynamics of individuals into the soil (83). Carbon and nitrogen changes of leaves, woody material, and boles shed by individual trees can be followed in annual cohorts. These dead parts of individuals can affect resource availability to living individuals in the same way that living leaves influence light availability in the canopy.

Decomposing litter material is the major source of organic matter and many nutrients for microbial populations in soils. The microbes convert organic forms of nutrients to inorganic forms that trees are able to take up, but the microbes also compete with the trees for these nutrients. The ratio of nutrients to carbon in leaves and other dead tree parts, especially in the form of the difficult-to-decompose material lignin, along with other factors such as temperature and moisture, determine the rate at which organic materials decompose and nutrients become available to trees. The lignin:nitrogen ratio in fresh litter, a convenient index of the more general concept of litter quality (4), has been shown to be strongly correlated with the amount of nitrogen immobilization, rate of organic matter loss, and the length of time before net nitrogen mineralization begins (2, 66) during the decomposition process. The concentration of nutrients and lignin in litter material varies among species and to a lesser extent among individuals of the same species. As a result species composition has a strong impact on decomposition dynamics and nutrient cycling.

Individual-based models that keep track of dead material from different individuals as the material is decomposed have been developed (3, 78, 119). In LINKAGES (78), the nitrogen cycle forms a positive feedback loop. Nitrogen cycling partially controls the rate of organic matter production and species composition (through competition for available nitrogen). In turn the rate of available nitrogen production depends on the amount and quality of organic matter produced. The simulations that Solomon (108) performed were repeated using LINKAGES (79). The effect of the nitrogen positive feedback was to amplify further the effects of climate change. The direct effect of temperature and precipitation on forest tree growth involves both increases and decreases in productivity, largely depending on whether soil moisture becomes more or less limiting. Changes due to direct climate effects are then accentuated by changes in species composition, altering carbon-nitrogen interactions that increase the initial effects through the positive feedback of the nitrogen cycle (23). Indirect species change effects will be most significant near vegetation zone boundaries such as the current boreal-cool temperate border. Forest simulation responses to climate change are as sensitive to the indirect effects of climate and vegetation on soil properties as they are to direct effects of temperature on tree growth (38).

SCALING INDIVIDUAL-BASED MODELS UP TO GLOBAL SCALES

As we have discussed above, vegetation is not only a product of its environment, it also modifies the local environment. Climate, light availability, and soil properties determine the ability of individual plants to become established and grow. These plants in turn modify the environment by creating shade, altering soil properties, cycling essential nutrients, and providing local sources of propagules. Vegetation also removes CO_2 from the atmosphere, stores it in organic forms, and releases CO_2 (and other trace gases such as methane and nitrous oxide) back to the atmosphere.

The collective impact of all terrestrial ecosystems on the annual global cycle of carbon is tremendous. Annual fixation of carbon into organic material by terrestrial ecosystems is around 60 Pg (81). If a global equilibrium is assumed, then decomposition returns on average the same amount of carbon back to the atmosphere each year. If global change, either through climate change or more directly through effects of elevated CO_2 on photosynthesis, results in a shift in the balance between production and decomposition globally, by even a few percent, then terrestrial ecosystems can alter the concentration of CO_2 in the atmosphere by 2 or 3 Pg/yr. This is the same order of magnitude as current fossil fuel emissions (81). Terrestrial ecosystems are therefore a component of the global environment and respond to and alter global conditions just as they have the capability to alter local conditions.

There is a significant computational problem in being able to model the dynamics of individual plants responding to and modifying local conditions and using these models to model the role of terrestrial systems in the global carbon cycle. While it is conceivable that sometime in the future computational hardware and software may be available to allow an individual-by-individual plant simulation of the terrestrial biosphere, current models are very computationally demanding and have not been fully developed for many important types of ecosystems. For the next decade or so different strategies must be employed to extend responses of ecosystems to global change captured in current individual-based models to global scales. There are several approaches that have been outlined, but none of these have been implemented to date.

Stratified Sampling

Rather than dividing the land surface into 1.2 billion one-twelfth hectare plots and running simulations for each, a coarser terrestrial classification can be identified in which large regions contain broadly uniform constraints on plant growth (monthly temperatures, precipitation, and variables derived from these); character and variation in soil texture, water holding capacity and fertility; topographic diversity; solar insolation and latitude effects). Within these coarser regions, individual-based stand models can be run to sample the nature of variations in environmental factors within these regions. This approach is most fully developed in concept by Prentice et al (85). A flow chart of their scheme is shown in Figure 4. Using a ecophysiological model for a uniform age lodgepole pine ecosystem, Running et al (89) simulated evapotranspiration and photosynthesis for a mountainous region in western Montana, employing a similar approach. Traditionally, gap model simulations have been performed with a series of environmentally identical replicate plots, and the results have been averaged to produce a consistent estimate of stand dynamics. In this expanded approach, instead of using identical plots, the plots are assigned fixed values of environmental variables to represent a statistical sample of the environmental conditions and disturbance regimes within a coarser region. The average of these simulations is a landscape scale description of vegetation for large regions that can be summed directly to determine the net exchange of CO_2 with the atmosphere, for example.

But even with supercomputers and parallel machines, it would be a large computational task to model a geographic array of landscapes across an entire biome or continent. It is even more difficult if large-scale spatial processes need to be incorporated, such as wildfire, pest outbreaks, or long-distance migrations of species in response to global changes. Under these cases, geographic location is important, and simulations of different spatial locations cannot be considered independently.

Aggregated Models Analogous to Gap Models

The results of individual-based simulation models can be used to estimate the

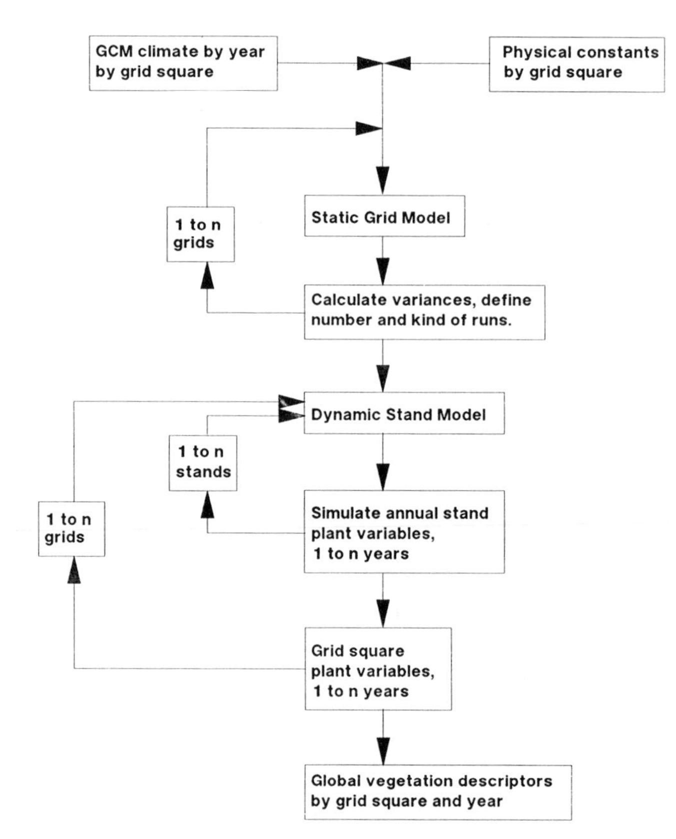

Figure 4 Flow diagram of a global vegetation change model. Static grid model component provides the geographically realistic spatial dynamics while the dynamics stand model characterizes the timing of spatial change within the global vegetation model. (From 85)

parameters of models using more highly aggregated state variables that have much smaller computational demands. As global conditions change, the parameter estimates could be changed in a piecewise fashion according to a calibration scheme that may involve rerunning the individual-based models under new conditions. These aggregated models could be simple box models for different types of biomes or ecosystems. Markov models of plant community dynamics may be more appropriate approximate analogs.

Markov models are constructed by determining the probability that the vegetation on a prescribed area will be in some other vegetation state after a given time interval (53, 95, 96, 117, 115). Results of individual-based simulations could be classified into appropriate states, and the transition probabilities calculated directly. Another approach would be to interpret the individual-based model parameters in the context of a vital attributes modeling approach to obtain a Markov representation of the species mixture (74, 76). The vital attributes concept uses regeneration, response to disturbance, and longevity of plants to determine the parameters of a Markov model. The physiological attributes identified by Huston & Smith (56) and used in an individual-based model could also be used in a vital attributes approach to develop a Markov model that may be compatible with an individual-based simulator.

Analog models would, however, be simplifications of any individual-based model on which they were based. As such, they may not have the ability to reproduce the total range of dynamics of the model on which they are based. However, they may be useful enough for projecting the impact of vegetation change on global carbon dynamics for time scales of several decades.

CONCLUSIONS

We have attempted to demonstrate the appropriateness of using simulation models based on individual plants as a basis for large-scale to global evaluation of terrestrial ecosystems responding to and contributing to global changes. It is clear that important phenomena in the context of global change in terrestrial ecosystems are captured in individual-based models that are not expressed in other modeling approaches. These involve the ability of individual-based models to change composition in response to physical disturbance and environmental changes. These compositional changes may have significant impacts on global carbon dynamics in addition to the direct effects of global change on current vegetation structure.

There are other important phenomena in terrestrial ecosystems that, because of their spatial and temporal scales, are not easily incorporated into such models. Important large issues that must be addressed in future research are: (i) An increased understanding of the transition between different vegetation types (e.g. grassland and forest types) and the associated problem of understanding multiple life-form competition along complex gradients. (ii) Development of ecological studies and a hierarchy of consistent models with nested space- and time-scales to allow interpretation of strongly scale-disconnected phenomena in terrestrial ecosystems. Two important problems involve the coupling of plant physiology at a mechanistic scale with models simulating community dynamics (85, 89), and inclusion of large-spatial phenomena (particularly wildfires, insect outbreaks, and species migration) in ecological models.

ACKNOWLEDGMENTS

Research was supported in part by the US National Aeronautics and Space Administration (Grant NAG-5-1018), the US Environmental Protection Agency (Grant CR-814610-01-0), and the US National Science Foundation (Grant BSR-8702333 and BSR-9020204) and in part by US Department of Energy, Carbon Dioxide Research Division, Atmospheric and Climate Research Division under contract DE-AC05-80OR21400 with Martin Marietta Energy Systems, Inc.

Literature Cited

1. Aber, J. D., Botkin, D. B., Melillo, J. M. 1978. Predicting the effects of differing harvest regimes on forest floor dynamics in northern hardwoods. *Can. J. Forest Res.* 8:306–15
2. Aber, J. D., Melillo, J. M. 1982. Nitrogen immobilization in decaying hardwood leaf litter as a function of initial nitrogen and lignin content. *Can. J. Bot.* 58:416–21
3. Aber, J. D., Melillo, J. M. 1982. FORTNITE: A computer model of organic matter and nitrogen dynamics in forest ecosystems. *Univ. Wisc. Res. Bull. R3130*
4. Agren, G. I., Bosatta, E.. 1991. Dynamics of carbon and nitrogen in the organic matter of the soil: A generic theory. *Am. Nat.* 138:227–45
5. Anderson, M. C. 1964. Light relations of terrestrial plant communities and their measurement. *Biol. Rev.* 39:425–86
6. Arney, J. D. 1974. An individual tree model for stand simulation in Douglas-fir. In *Growth Models for Tree and Stand Simulation,* ed. J. Fries, pp. 38–46. Res. Notes 30, Dep. Forest Yield Res., Royal Coll. Forestry, Stockholm
7. Baes, C. F. Jr., Goeller, H. E., Olson, J. S., Rotty, R. M. 1977. Carbon dioxide and climate: the uncontrolled experiment. *Am. Sci.* 65:310–20
8. Bella, I. E. 1971. A new competition model for individual trees. *For. Sci.* 17:364–72
9. Bonan, G. B. 1989. Environmental factors and ecological processes controlling vegetation patterns in boreal forests. *Landscape Ecol.* 3:111–30
10. Bonan, G. B., Korzukin, M. D. 1989. Simulation of moss and tree dynamics in the boreal forest of interior Alaska. *Vegetatio* 84:31–44
11. Bonan, G. B., Hayden, B. P. 1990. Using a forest stand simulation model to examine the ecological and climatic significance of the late-Quaternary pine/spruce pollen zone in eastern Virginia. *Quaternary Res.* 33:204–18
12. Bonan, G. B., Shugart, H. H. 1989. Environmental factors and ecological processes in boreal forests. *Annu. Rev. Ecol. Syst.* 20:1–28
13. Bonan, G. B., Shugart, H. H., Urban, D. L. 1990. The sensitivity of some high-latitude boreal forests to climatic parameters. *Climatic Change* 16:9–29
14. Botkin, D. B., Janak, J. F., Wallis, J. R. 1972. Some ecological consequences of a computer model of forest growth. *J. Ecol.* 60:849–72
15. Botkin, D. A., Nisbet, R. A., Reynales, T. E. 1989. Effects of climate change on forests of the Great Lake states. In *The Potential Effects of Global Climate*

Change on the United States: Appendix D - Forests, ed. J. B. Smith, D. A. Tirpak. Off. Policy, Planning, Eval. US Environ. Protection Agency, Washington, DC
16. Box, E. O. 1981. *Macroclimate and Plant Forms: An Introduction to Predictive Modeling in Phytogeography.* The Hague:D. R. Junk
17. Brokaw, N. V. L. 1985a. Gap-phase regeneration in a tropical forest. *Ecology* 66:682–87
18. Brokaw, N. V. L. 1985b. Treefalls, regrowth, and community structure in tropical forests. In *The Ecology of Natural Disturbance and Patch Dynamics,* ed. S. T. A. Pickett & P. S. White, pp. 101–8. New York: Academic
19. Busing, R. T. 1991. A spatial model of forest dynamics. *Vegetatio* 92:167–79
20. Busing, R. T., Clebsch, E. E. C. 1987. Application of a spruce-fir forest canopy gap model. *Forest Ecol. Manage.* 20:151–69
21. Challinor, D. 1968. Alteration of soil surface characteristics by four tree species. *Ecology* 49:286–90
22. Chapin, F. S., Bloom, A. J., Field, C. B., Waring, R. H. 1987. Plant responses to multiple environmental factors. *Bioscience* 37:49–57
23. Chapin, F. S., Vitousek, P. M. & Van Cleve, K. 1986. The nature of nutrient limitation in plant communities. *Am. Nat.* 127:48–58
24. Coffin, D. P., Lauenroth, W. K. 1989. A gap dynamics simulation model of succession in a semiarid grassland. *Ecol. Modelling* 49:229–66
25. Coffin, D. P., Lauenroth, W. K. 1989. Disturbances and gap dynamics in a semiarid grassland: a landscape-level approach. *Landscape Ecol.* 3:19–27
26. Cowan, I. R. 1968. The interception and absorption of radiation in plants. *J. Appl. Ecol.* 5:367- 79
27. Cowan, I. R. 1986. Economics of carbon fixation in higher plants. In *On the Economy of Plant Form and Function,* ed. T. J. Givnish, pp. 133–70. Cambridge: Cambridge Univ. Press
28. Cowan, I. R. 1971. Light in plant stands with horizontal foliage. *J. Appl. Ecol.* 8:579–80
29. Dale, V. H., Hemstrom, M. A. 1984. CLIMACS: A computer model of forest stand development for western Oregon and Washington. *USDA/Forest Service Res. Paper. PNW-327*
30. Dale, V. H., Franklin, J. F. 1989. Potential effects of climate change on stand development in the Pacific Northwest. *Can. J. For. Res.* 19:1581–90
31. Davis, M. B. 1989. Lags in vegetation response to greenhouse warming, *Climatic Change* 15:75–82
32. Davis, M. B. and D. B. Botkin. 1985. Sensitivity of cool-temperate forests and their fossil pollen record to rapid temperature change. *Quaternary Res.* 23:327–40
33. Delcourt, H. R., Delcourt, P. A., Webb, T. III. 1983. Dynamic plant ecology: The spectrum of vegetation change in space and time. *Quaternary Sci. Rev.* 1:153–75
34. Denslow, J. S. 1980. Gap partitioning among tropical rainforest trees. *Biotropica* 12:47–55 (Suppl.)
35. Develice, R. L. 1988. Test of a forest dynamics simulator in New Zealand. *NZ J. Bot.* 26:387–92
36. Doyle, T. W. 1981. The role of disturbance in the gap dynamics of a montane rain forest: An application of a tropical forest succession model. In *Forest Succession: Concepts and Application,* ed. D. C. West, H. H. Shugart, D. B. Botkin, pp. 56–73. New York: Springer-Verlag
37. Dyrness, C. T. 1982. Control of depth to permafrost and soil temperature by the forest floor in black spruce/feathermoss communities. *US For. Serv. Res. Note PNW-396.* 19 pp.
38. Dyrness, C. T., Grigal, D. F. 1979. Vegetation-soils relationships along a spruce forest transect in interior Alaska. *Can. J. Bot.* 57:2644–56
39. El Bayoumi, M. A., Shugart, H. H., Wein, R. W. 1984. Modeling succession of the eastern Canadian mixed-wood forest. *Ecol. Modelling* 21:175–98
40. Deleted in proof
41. Givnish, T. J. 1979. On the adaptive significance of leaf form. In *Topics in Plant Population Biology,* ed. O. T. Solbrig, S. Jain, G. B. Johnson, P. H. Raven, pp. 375–407. New York: Columbia Univ. Press
42. Givnish, I. J. 1986. *On the Economy of Plant Form and Function.* Cambridge: Cambridge Univ. Press
43. Grime, J. P. 1977. Evidence for the existence of three primary strategies in plants and its relevance to ecological and evolutionary theory. *Am. Nat.* 111: 1169–94
44. Grime, J. P. 1979. *Plant Strategies and Vegetation Processes.* New York: Wiley
45. Grubb, P. J. 1977. The maintenance of species-richness in plant communi-

ties : The importance of the regeneration niche. *Biol. Rev.* 52:107–45
46. Harrison, E. A., Shugart, H. H. 1990. Evaluating performance on an Appalachian oak forest dynamics model. *Vegetatio* 86:1–13
47. Hartshorn, G. S. 1978. Tree falls and tropical forest dynamics. In *Tropical Trees as Living Systems,* ed. P. B. Tomlinson, M. H. Zimmerman, pp. 617–38. Cambridge: Cambridge Univ. Press
48. Hatch, C. R. 1971. *Simulation of an even-aged red pine stand in northern Minnesota.* PhD thesis. Univ. Minn., Minneapolis
49. Helvey, J. D. & Patric, J. H. 1965. Canopy and litter interception of rainfall by hardwoods of the eastern United States. *Water Resources Res.* 1:193–290
50. Holdridge, L. R. 1967. *Life Zone Ecology.* Tropical Sci. Cent., San Jose, Costa Rica
51. Horn, H. S. 1975a. Forest succession. *Sci. Am.* 232:90–98
52. Horn, H. S. 1975b. Markovian properties of forest succession. In *Ecology and Evolution in Communities,* ed. M. L. Cody, J. M. Diamond, pp. 196–211. Cambridge, Mass:Harvard Univ. Press
53. Horn, H. S. 1976. Succession. In *Theoretical Ecology,* ed. R. M. May, pp. 187–204. Oxford: Blackwell Sci.
54. Huntley, B. 1990. Studying global change—The contribution of Quaternary palynology. *Global Planetary Change* 82:53–61
55. Huston,M., DeAngelis, D., Post, W. M. 1988. New computer models unify ecological theory. *BioScience* 38:682–91
56. Huston, M. A., Smith, T. M. 1987. Plant succession: Life history and competition. *Am. Nat.* 130:168–98
57. Kercher, J. R., Axelrod, M. X. 1984. A process model of fire ecology and succession in a mixed-conifer forest. *Ecology* 65:1725–42
58. Kienast, F., Kuhn, N. 1989. Simulating forest succession along ecological gradients in southern central Europe. *Vegetatio* 79:7–20
59. Knoop, W. J., Walker, B. H. 1985. Interactions of woody and herbaceous vegetation in a southern African savanna. *J. Ecol.* 73:235–53
60. Koop, H., Hilgen, P. 1987. Forest dynamics and regeneration mosaic shifts in unexploited beech (*Fagus sylvatica)* stands at Fontainebleau (France). *For. Ecol. Manage.* 320:135–50
61. Kozlowski, T. T. 1971. *Growth and Development of Trees.* Vol. 1. *Seed Germination, Ontogeny and Shoot Growth.* New York: Academic
62. Kozlowski, T. T. 1971. *Growth and Development of Trees.* Vol II. *Cambial Growth, Root Growth and Reproductive Growth.* New York: Academic
63. Lee, Y. 1967. Stand models for lodgepole pine and limits to their application. *For. Chron.* 43:387–88
64. Leemans, R., Prentice, I. C. 1987. Description and simulation of tree-layer composition and size distributions in a primaeval *Picea-Pinus* forest. *Vegetatio* 69:147–56
65. Lin, J. Y. 1970. *Growing space index and stand simulation of young Western hemlock in Oregon.* PhD thesis. Duke Univ., Durham, NC
66. Meentenmeyer, V. 1978. Macroclimate and lignin control of litter decomposition rates. *Ecology* 59:465–72
67. Menaut, J. C., Gignoux, J., Prado, C., Clobert, J. 1990. Tree community dynamics in a humid savanna of the Côte-d'Ivoire: Modelling the effects of fire and competition with grass and neighbors. *J. Biogeography* 17:471–81
68. McMurtrie, R., Wolf, L. 1983. A model of competition between trees and grass for radiation, water and nutrients. *Ann. Bot.* 52:449–58
69. Mitchell, K. J. 1969. Simulation of growth of even-aged stands of white spruce. *Yale Univ. Sch. For. Bull.* 75:1–48
70. Moore, A. D., Noble, I. R. 1990. An individualistic model of vegetation stand dynamics. *J. Environ. Manage.* 31:61–81
71. Mooney, H. A. & Gulmon, S. L. 1979. Environmental and evolutionary constraints on the photosynthetic characteristics of higher plants. In *Topics in Plant Population Biology,* O. T. Solbrig, S. Jain, G. B. Johnson, P. H. Raven, pp. 316–37. New York: Columbia Univ. Press
72. Munro, D. D. 1974. Forest growth models: A prognosis. In *Growth Models for Tree and Stand Simulation,* ed. J. Fries, pp. 7–21. *Res. Note 30, Royal Coll. For., Stockholm.*
73. Newnham, R. M. 1964. *The development of a stand model for Douglas-fir.* PhD thesis. Univ. Br. Columbia, Vancouver
74. Noble, I. R., Moore, A. D., Strasser, M. J. 1988. Predicting vegetation dynamics based on structural and functional attributes. *Proc. Int. Symp. Vegetation Structure,* Utrecht
75. Noble, I. R., Slatyer, R. O. 1978. The

effect of disturbances on plant succession. *Proc. Ecol. Soc. Austr.* 10:135–45
76. Noble, I. R., Slatyer, R. O. 1980. The use of vital attributes to predict successional changes in plant communities subject to recurrent disturbances. *Vegetatio* 43:5–21
77. Overpeck, J. T., Rind, D., Goldberg, R. 1990. Climate-induced changes in forest disturbance and vegetation. *Nature* 343:51–53
78. Pastor, J., Post, W. M. 1986. Influence of climate, soil moisture, and succession on forest carbon and nitrogen cycles. *Biogeochemistry* 2:3–27
79. Pastor, J., Post, W. M. 1988. Response of northern forests to CO_2 induced climate change. *Nature* 334:55–58
80. Pearlstine, L., McKellar, H., Kitchens, W. 1985. Modeling the impacts of river diversion on bottomland forest communities in the Santee River Floodplain, South Carolina. *Ecol. Modelling* 29:283–302
81. Post, W. M., Peng, T-H., Emanuel, W. R., King, A. W., Dale, V. H., DeAngelis, D. L. 1990. The global carbon cycle. *Am. Sci.* 78:310–26
82. Post, W. M., Pastor, J., Zinke, P. J., Stangenberger, A. G. 1985. Global patterns of soil nitrogen. *Nature* 317: 613–16
83. Post, W. M., Pastor, J. 1990. An individual-based forest ecosystem model for projecting forest response to nutrient cycling and climate changes. In *Forest Simulation Systems: Proceedings of the IUFRO Conference, Berkeley, California, November 2–5, 1988*, ed. L. Wensel, G. Biging, pp. 61–74. *Univ. Calif., Div. Agric. Nat. Resources, Bull. 1927, Berkeley*
84. Phipps, R. L. 1979. Simulation of wetlands forest dynamics. *Ecol. Modelling* 7:257–88
85. Prentice, I. C., Webb, R. S., Ter-Mikhaelian, A. M., Solomon, T. M., Smith, S. E., et al. 1989. *Developing a global vegetation dynamics model: Results of a IIASA summer workshop*. Int. Inst. Appl. Syst. Anal., Laxenburg, Austria
86. Rastetter, E. B. 1991. A spatially explicit model of vegetation-habitat interactions on barrier islands. In *Quantitative Methods in Landscape Ecol*, ed. M. G. Turner, R. H. Gardner, pp. 353–78. New York: Springer-Verlag
87. Roberts, Leslie 1989. How fast can trees migrate? *Science* 243:735–37
88. Rosswall T., Woodmansee, R. G., Risser, P. G., ed. 1988. *Scales and Global Changes: Spatial and Temporal Variability in Biospheric and Geospheric Processes*. London: Wiley
89. Running, S. W., Nemani, R. R., Peterson, D. L., Band, L. E., Potts, D. F., Pierce, L. L., Spanner, M. A. 1989. Mapping regional forest evapotranspiration and photosynthesis by coupling satellite data with ecosystem simulation. *Ecology* 70:1090–1101
90. Silvertown, J., Smith, B. 1988. Gaps in the canopy: the missing dimension in vegetation dynamics, *Vegetatio* 77: 57–60
91. Sharpe, P. J. H., Walker, J., Penridge, L. K., Wu, H. 1985. A physiologically-based continuous-time Markov approach to plant growth modeling in semi-arid woodlands. *Ecol. Modelling* 29:189–213
92. Shear, G. M., Stewart, W. D. 1934. Moisture and pH studies of the soil under forest trees. *Ecology* 15:350–58
93. Shugart, H. H. 1984. *A Theory of Forest Dynamics: The Ecological Implications of Forest Succession Models*. New York: Springer-Verlag. 278 pp
94. Shugart, H. H. 1987. Dynamic ecosystem consequences of tree birth and death patterns. *BioScience* 37:596–602
95. Shugart, H. H., Bonan, G. B., Rastetter, E. B. 1988. Niche theory and community organization. *Can. J. Bot.* 66: 2634–39
96. Shugart, H. H., Crow, T. R., Hett, J. M. 1973. Forest succession models: A rationale and methodology for modeling forest succession over large regions. *For. Sci.* 19:203–12
97. Shugart, H. H., Hopkins, M. S., Burgess, I. P., Mortlock, A. T. 1980. The development of a succession model for subtropical rain forest and its application to assess the effects of timber harvest at Wiangaree State Forest, New South Wales. *J. Environ. Manage.* 11:243–65
98. Shugart, H. H., Noble, I. R. 1981. A computer model of succession and fire response to the high altitude eucalyptus forest of the Brindabella Range, Australian Capital Territory. Australian *J. Ecol.* 6:149-64
99. Shugart, H. H., West, D. C. 1977. Development and application of an Appalachian deciduous forest succession model. *J. Environ. Manage.* 5:161–79
100. Shugart, H. H., West, D. C. 1980. Forest succession models. *BioScience* 30:308–13
101. Smith, J. B., Tirpak, D. A., eds. 1989. *The Potential Effects of Global Climate*

Change on the U. S. : Appendix D - Forests. Off. Policy, Planning Eval. Washington, DC: US Environ. Protection Agency
102. Smith, T. M., Urban, D. L. 1988. Scale and the resolution of forest structural pattern. *Vegetatio* 74:143–50
103. Solomon, A. M., Delcourt, H. R., West, D. C., Blasing, T. J. 1980. Testing a simulation model for reconstruction of prehistoric forest-stand dynamics. *Quaternary Res.* 14:275–93
104. Solomon, A. M., West, D. C., Solomon, J. A. 1981. Simulating the role of climate change and species immigration in forest succession. In *Forest Succession: Concepts and Application,* ed. D. C. West, H. H. Shugart, D. B. Botkin, pp. 154–77. New York: Springer-Verlag
105. Solomon, A. M., Shugart, H. H. 1984. Integrating forest stand simulations with paleoecological records to examine the long-term forest dynamics. pp. 333–56. *In State and Change of Forest Ecosystems—Indicators in Current Research,* ed. G. I. Agren. *Swedish Univ. Agric. Sci., Dept. Ecol. Environ. Res. Rep. 13*. Uppsala, Sweden
106. Solomon, A. M. and T. Webb III. 1985. Computer-aided reconstruction of late Quaternary landscape dynamics. *Annu. Rev. Ecol. Syst.* 16:63–84
107. Solomon, A. M. 1986. Comparison of taxon calibrations, modern analog techniques, and forest-stand simulation models for the quantitative reconstruction of past vegetation: A critique. *Earth Surface Processes Landforms* 11:681–85
108. Solomon, A. M. 1986. Transient response of forests to CO_2-induced climate change: Simulation experiments in eastern North America. *Oecologia* 68: 567–79
109. Swift, L. W., Swank, W. T., Mankin, J. B., Luxmore, R. J., Goldstein, R. A. 1979. Simulation of evapotranspiration and drainage from mature and clearcut deciduous forests and young pine plantation. *Water Resources Res.* 11:667–73
110. Tonu, O. 1983. Metsa suktsessiooni ja tasandilise struktuuri imiteerimisest. *Yearbk. Estonian Naturalist Soc.* 69: 110–17
111. Urban, D. L., Shugart, H. H. 1989. Forest response to climatic change: a simulation study for southeastern forests. In *The Potential Effects of Global Climate Change on the United States: Appendix D - Forests,* ed. J. B. Smith, D. A. Tirpak. Off. Policy, Planning, Evaluation, US Environmental Protection Agency, Washington, DC
112. van Daalen, J. C., Shugart, H. H. 1989. OUTENIQUA—A computer model to simulate succession in the mixed evergreen forests of the southern Cape, South Africa, *Landscape Ecol.* 24:255–67
113. Van Hulst, R. 1979. On the dynamics of vegetation: Markov chains as models of succession. *Vegetatio* 40:3–14
114. van der Pijl, L. 1972. *Principles of Dispersal in Higher Plants*. Berlin: Springer-Verlag
115. van Tongeren, O., Prentice, I. C. 1986. A spatial simulation model for vegetation dynamics. *Vegetatio* 65:163–73
116. Waldrop, T. A., Buckner, E. R., Shugart, H. H., McGee, C. E. 1986. FORCAT: A single tree model of stand development on the Cumberland Plateau. *For. Sci.* 32:297–317
117. Waggoner and Stephens. 1971. AU: Pls complete. Journal 00:000–00
118. Waring, R. H. 1983. Growth efficiency. *Adv. Ecol. Res.* 13:327–54
119. Weinstein, D. A., Shugart, H. H., West, D. C. 1982. *The Long-Term Nutrient Retention Properties of Forest Ecosystems: A Simulation Investigation. ORNL/TM-8472*. Oak Ridge Natl. Lab., Oak Ridge, Tenn.
120. Whitmore, T. C. 1982. On pattern and process in forests. In *The Plant Community as a Working Mechanism. Special Publ. No. 1, Br. Ecol. Soc.* ed. E. I. Newman, pp. 45–59. Oxford: Blackwell Sci.
121. Whittaker, R. H., Goodman, D. 1979. Classifying species according to their demographic strategy. 1. Population fluctuations and environmental heterogeneity. *Am. Nat.* 113:185–200
122. Woodward, F. I. 1987. *Climate and Plant Distribution*. Cambridge: Cambridge Univ. Press
123. Wu, H., Sharpe, P. J., Walker, J., Pendridge, L. K. 1985. Ecological field theory. *Ecol. Modelling* 29:215–43
124. Zinke, P. J. 1962. The pattern of influence of individual trees on soil properties. *Ecology* 43:130–33

Annu. Rev. Ecol. Syst. 1992. 23:39–61

HUMAN POPULATION GROWTH AND GLOBAL LAND-USE/COVER CHANGE

William B. Meyer and B. L. Turner II
Graduate School of Geography and George Perkins Marsh Institute, Clark University, 950 Main Street, Worcester, Massachusetts 01610-1477

KEYWORDS: global environmental change, driving forces, conversion, modification, cultivation

INTRODUCTION

Contemporary interdisciplinary research on human-induced global environmental change recognizes two broad and overlapping fields of study (67). That of industrial metabolism investigates the flow of materials and energy through the chain of extraction, production, consumption, and disposal of modern industrial society. That of land-use/land-cover change, our concern here, deals with the alteration of the land surface and its biotic cover. Environmental changes of either kind become global change in one of two ways (106): by affecting a globally fluid system (the atmosphere, world climate, sea level) or by occurring in a localized or patchwork fashion in enough places to sum up to a globally significant total. Land-use change contributes to both kinds of global change: to such systemic changes as trace-gas accumulation and to such cumulative or patchwork impacts as biodiversity loss, soil degradation, and hydrological change.

Land-use/land-cover change is a hybrid category. Land use denotes the human employment of the land and is studied largely by social scientists. Land cover denotes the physical and biotic character of the land surface and is studied largely by natural scientists. Connecting the two are proximate sources of change: human activities that directly alter the physical environment. These activities reflect human goals that are shaped by underlying social driving forces. Proximate sources change the land cover, with further environmental consequences that may ultimately feed back to affect land use.

Contemporary global environmental change is clearly unique. The human reshaping of the earth has reached a truly global scale, is unprecedented in its magnitude and rate, and increasingly involves significant impacts on the

0066-4162/92/1120-0039$02.00

biogeochemical systems that sustain the biosphere. On the other hand, the land covers being reshaped have long been modified by human action (6, 44, 84, 93, 99, 101, 104). Such terms as "native ecosystem" or "virgin forest" are of questionable usefulness.

The antiquity of land-cover changes is reflected in their prominence in the early classics of environmental science. George Perkins Marsh's *Man and Nature* (58) was a monumental assessment of data and theories, many dating back much earlier, on the effects of land-cover changes, particularly deforestation. Global inventories of arable land date back at least a century (77) and those of forest resources almost as far (119). Specialized studies of land change have proliferated, but broader syntheses remain rare. Surveys of global change such as the World Resources Institute reports and the recent volume *The Earth as Transformed by Human Action* (104) assemble much historical and statistical material and outline the broad global and regional trends. A SCOPE volume on *Land Transformation in Agriculture* (117) covers the principal agricultural impacts on land cover.

Several recent and on-going global change research initiatives have dealt in whole or part with land use/cover issues. An ad hoc working group of the Committee on Global Change (National Research Council) offered an initial outline of study of global land-use/cover change (42). Building on it, the Land-Use Change Working Group of the Social Science Research Council Committee on Global Environmental Change (94) identified the priority forms of change and the contributions that the social sciences might make to understanding them; an appendix (118) assessed the principal global data sources and undertook some simple tests of candidate human driving forces of change. The forthcoming volume of the 1991 Global Change Institute of the Office of Interdisciplinary Earth Studies (61) is devoted to global land-use/cover change; the chapters deal with the major recent trends in change, their environmental consequences, their human causes, and problems in data and modeling. Drawing on these efforts, the International Geosphere-Biosphere Programme and the International Social Science Council are contemplating the development of a land-use/cover change study; a report by their ad hoc working group was planned for spring of 1992.

For many types of change, both the global condition itself and its implications for nature and society remain extremely ill-documented (10). Crosson (18), for instance, writes that "we have very little reliable information about the amount of soil erosion in [the developing] countries and even less about its productivity and water quality effects," and that a similar gap exists for the problems of tropical deforestation, aquifer depletion, and soil degradation. An overview of what we know about major global land-cover changes, their environmental significance, and their human causes is in order. We assess first the broad changes brought about by human activity in five categories of land cover: cultivated land, forest/tree cover, grassland/pasture, wetlands, and settlement. A review follows of the major secondary environmental consequences that have been attributed to these cover changes. Finally, we examine the theoretical and empirical literature that investigates the role of human population growth and other social factors in driving land-use/cover change.

LAND-COVER CHANGE

Land-cover changes take two forms: conversion from one category of land cover to another and modification of condition within a category. Conversion is the better documented and more readily monitored of the two, but too great an emphasis on it obscures important forms of land-cover modification. The problem will vary with the categories of cover used; the broader and fewer the categories, the fewer the instances of conversion from one to another. If one's classes are as coarse as forest/woodland, permanent pasture, cultivation, and "other lands," for example, forest thinning, replacement of old forests with tree plantations, intensification of cultivation, and severe overgrazing will not register as conversion, nor as land-cover change if conversion totals alone are used to measure change.

These four classes—forest/woodland, permanent pasture, cultivation, and other lands—are those of the most widely used global figures, published in the UN Food and Agriculture Organization's *Production Yearbooks*. They purport to show national-level change year by year since the 1950s. The FAO data are so often used, however, less because of their quality than because of their convenience. The FAO does not gather the data independently for the *Yearbooks;* it collates numbers reported by member states. Hence the data quality varies greatly by country, and country size determines the scale at which the (national-level) data are presented. The FAO class of "other lands" combines several important and distinct forms of land use/cover. Definitions of all classes (save "other lands," which is essentially a residual category) have been at once too loose (not clearly specifying the criteria for identifying each class) and too rigid (accounting poorly for sporadic or fluctuating uses of some lands). In the Organization's own words, "it should be borne in mind that definitions used by reporting countries vary considerably and items classified under the same category often relate to greatly differing kinds of land" (109). Global data compilations by the USDA and the CIA draw principally on the FAO numbers and share their problems.

The climatic modeling literature is another source of global land data. Matthews (59) constructed data bases of presumed preagricultural vegetation types and of present-day land cover on $1\frac{1}{2}$ by $1\frac{1}{2}$ cells: her work has received extensive use in modeling efforts. Work on the global carbon flux (35–38) has necessarily involved detailed reconstruction of change in land-cover patterns and the critical assessment of a range of data sources from archival materials to remote sensing. The carbon-modeling literature served as the basis for Richards' (80) continental-scale reconstruction of land-use changes from 1700–1980 (using as categories cultivation, forest/woodland, and grassland/pasture). A valuable feature of this literature is that it deals with ecosystem modification ("degradation" of forests) as well as conversion (e.g. 34). These efforts, however, treat land use and cover as broad, simple categories and at a scale that inhibits connections with social variables. Other digitized data and maps deal with ecoregions (5, 70) and life zones (51), but these data are not transferable to land use or cover. Several Soviet maps of global land use and cover demarcate the world into an intricate array of land use and cover types readily linked to social variables, but problems with validation and data format have thus far prevented their ready use.

Issues of data quality and comparability make global-scale assessments of land-use/cover change difficult. We note in the following sections (see also Table 1) the principal trends in change in major cover categories as they can be assessed from the literature.

Cultivated Land

DEFINITION Cultivated lands are those regularly used to grow domesticated plants, ranging from long-fallow, land-rotational systems to permanent,

Table 1 Global human-induced conversions in selected land covers

Cover	Date	Area ($\times 10^6 km^2$)	Date	Area ($\times 10^6 km^2$)	% Change
Cropland	1700	2.65	1980	15.01	+466
	1700	3.0	1980	14.75	+392
Irrigated cropland	1800	0.08	1989	2.00[a]	+2400
Closed forest	pre-agricultural	46.28	1983	39.27	−15.1
Forest and woodland	pre-agricultural	61.51	1983	52.37	−14.9
Grassland/pasture	1700	68.60	1980	67.88	−1[b]
Lands drained			1985	1.606	
Urban settlement			1985	2.47[c]	
Rural settlement			1990	2.09	

[a] See text for another estimate
[b] The change is small; the data errors are large.
[c] Includes substantial areas not built up; see text for discussion.

multicropping systems. Their full range is difficult to measure because some cultivation occurs in very small units and some in settings not easily distinguished from other cover types. Rotational cultivation and agroforestry (including plantations) are often classified as forest, leading perhaps to an underestimate of cultivated land. Further complications arise in distinguishing cultivated from agricultural land, a broader term that can include land used for livestock production. Perhaps the most common distinction recognizes fodder species grown for livestock as cultivated land and improved pasture as grassland and pasture cover. It is not clear how strictly such distinctions are followed in the data.

CHANGE AND PROXIMATE SOURCES The world total of cultivated land is estimated to have increased by 466% from 1700 to 1980: during this time, a net area of more than 12×10^6 km^2 of land was brought into cultivation (80). This expansion did not occur evenly across the world; the USSR, Southeast Asia, Latin America, and North America all experienced greater expansion of cultivation than the world average. The increase in North America, for example, was 6,666% (80). Two estimates of current cultivated land are 14.75×10^6 km^2 and 15.00×10^6 km^2 (47, 80).

Global expansion of cultivated land (conversion) is accelerating, as is the intensification of use of lands already cultivated (modification). With a few noteworthy exceptions, most prime environments for rainfed cultivation have been consumed. The area suitable for rainfed agriculture is estimated by some to be about 18.74×10^6 km^2, only 3.75 to 4.00×10^6 km^2 above the area currently taken to represent this land cover (47). Land expansion will increasingly occur in environments assumed to be more marginal and fragile for cultivation (102). Tropical forests and grasslands and, to a lesser extent, boreal forests are under increasing pressure from agricultural expansion, as are wetlands.

While the global pattern is one of the expansion of cultivated land, some regions have experienced losses, from either the abandonment of cultivated land or its degradation promoting nonuse. Cultivated land has been decreasing, for example, in Europe (by 3.5%, 1973–1988) (47), to be replaced by settlement and forest. Such benign losses are more than matched by forced abandonment owing to degradation. The FAO estimates that 5.44×10^6 km^2 of rainfed cultivated land have been lost worldwide to degradation; another study estimates that 20.00×10^6 km^2 of former cultivated lands have been irreversibly lost due to degrading uses and to permanent cover changes (e.g. water impoundments, settlements) (85). Both estimates are controversial.

These changes in land cover have been driven by real and perceived needs for expanded agricultural production. Regardless of the underpinning causes (see below), the major proximate sources of conversion have been fire and clear-cut timbering (in forested areas), tillage technologies (in grasslands and heavy soils), drainage (in wetlands; see below), and irrigation (in arid lands or where paddy is used). Irrigated land has expanded, according to one estimate, from 0.08×10^6 km^2 in 1800 to 2.00×10^6 km^2 in 1989 (4). Perhaps no other form of cultivation is so disputed in terms of its current area, however; estimates range from 2.00×10^6 km^2 to 4.58×10^6 km^2 (56). The major sources of the modification of cultivated land cover have been switches in cultigens (among cereals, root crops, agroforestry) and the intensification of cultivation through green revolution hybrid crops, synthetic inputs, and, more recently, biotechnology. Unfortunately, the available data do not allow assessments of the spatial magnitude of these changes.

Forest/Tree Cover

DEFINITION The term "forest" has a variety of meanings that have not yet been standardized in global change studies. Some writers confine it to closed as opposed to open woodland; others would include savanna environments and lands used in fallow agriculture within a forest/woodland category. Some ecological usages would exclude tree plantations from forest. The FAO's tropical forest inventories (49, 50) use a figure of 10% canopy cover to separate forested from deforested areas. Because these problems remain largely unresolved (116), care must be taken in comparing data that may use different definitions.

CHANGE AND PROXIMATE SOURCES Changes in the world's tree cover are of two kinds: clearance and conversion to another land cover (whether cultivation, grassland, or settlement), and change of condition (e.g. forest thinning without outright conversion). The literature focuses overwhelmingly on the former process, although carbon modellers have begun to pay considerable attention to the implications of the latter (34).

The world's current area of closed forest, based on comparison of such sources as the FAO data and a World Resources Institute inventory, is estimated to be around 29×10^6 km^2, or 21% of the world's land area. "Open woodland," an ill-defined category that overlaps significantly with other standard cover types, adds about 18×10^6 km^2 (13% of the land area). Drawing on Matthews's (59) reconstruction of preagricultural land cover, Williams (116) estimates that an original 62×10^6 km^2 of forest and woodland has been reduced by 9×10^6 km^2, of which 7×10^6 represents loss of closed forest (a net global decrease of about 15%). Estimates of contemporary change vary considerably and are hampered by problems of data, definition, and method, though the broad pattern is one of stability or even net gain in the developed countries and rapid, if fluctuating, loss in the developing tropics, adding up to an annual net global loss on the order of 0.1–0.2 × 10^6 km^2 (116).

Goals and proximate sources of forest change differ considerably across the world (116). Clearance for cultivation, often associated with planned and spontaneous frontiers of colonization, is probably the most widespread. Ranching and pasture development have been significant causes of clearance in Central and Latin America. Timber extraction in excess of regrowth is exemplified in Southeast Asian contexts and to some extent in the tropical forests of western Africa; fuelwood extraction exceeds regrowth in Africa and some portions, especially mountainous ones, of Latin America and the Indian subcontinent. Waldsterben or forest decline associated with atmospheric pollutants and other stresses is a relatively new source of damage in the developed world (North America and western Europe). Finally, reforestation or afforestation can result naturally from land abandonment or can be undertaken deliberately by state or private action; many governments have instituted such programs.

Grassland/Pasture

DEFINITION Grassland/pasture represents land having a ground story of vegetation cover in which grasses are the dominant life form (25). The natural distribution and extent of grasslands on the earth's surface is controlled by climate, soil, and fire, only the last of which has been significantly altered by human actions.

CHANGE AND PROXIMATE SOURCES The world area of grassland and pasture in 1700 is estimated at 68.60 × 10^6 km^2, and today at the nearly identical value of 67.88 × 10^6 km^2 (80). Factors increasing and decreasing the global area of grassland have maintained a rough balance over this long period. The FAO figures indicate a decrease of less than 1% from 1971 to 1986 (25). The principal processes of change are loss through conversion to cropland and gain through deforestation. The former process has led to marked net decreases from 1700 to 1980 in grassland/pasture in Europe (−27.4%), North America (−13.7%), and Southeast Asia (−26.4%). The latter has driven increases in tropical Africa by 10.1% and in Latin America by 26.2% (80).

Changes of condition of grasslands are perhaps of greater significance. Desertification has been widely identified as a major human-induced global change associated with excessive pressure on grasslands. The now-vast literature on this subject received a particular stimulus from the 1977 United Nations Conference on Desertification. The UNCOD report identified 6% of the world's area as "man-made deserts" and close to a quarter of the world's surface as threatened by desertification. The 1984 UNEP assessment further estimated the annual degradation of land "to desert-like conditions" as 60,000 km^2 and the area annually "reduced to zero or negative net economic productivity" as more than 200,000 km^2 (57, 76).

Desertification remains an ill-defined concept. In the meaning that has been used for some time by UNEP, desertification is "the diminution or destruction of the biological potential of the land, and can lead ultimately to desertlike conditions." Mortimore (64) prefers to define desertification as the loss of primary productivity of ecosystems in arid or semi-arid regions. The problems in measurement to achieve reliable and current global estimates are enormous; the UNEP literature routinely acknowledges them and has not shown that they can be overcome. Mortimore (64) and others (30a, 111) criticize several other pervasive features of the desertification literature, notably the vast claims made about the extent of land "destroyed" and the assumptions that desertification—however defined—is irreversible and is largely the product of human action. These assumptions, it is argued, lack a solid empirical base, and they ignore what detailed local studies suggest: the resilience of many arid and semi-arid ecosystems and the possible climatic origin of much desertification. Recent studies of the expansion and contraction of the Sahara during the 1980s underline the last point (100). While questioning many of the global figures offered for human-induced desertification, however, Graetz (25) concludes that degradation of grasslands by overgrazing is a globally significant change of threatening proportions.

Wetlands

DEFINITION The term "wetlands," a recent coinage (110), has been associated with more than its share of ambiguities, but recent work has helped clarify the concept. Cowardin et al's (18) definition, "lands transitional between terrestrial and aquatic systems where the water table is usually at or near the surface or the land is covered by shallow water" is widely accepted. The category includes such subtypes as marshes, swamps (including mangrove swamps), peat lands, and riparian floodplains. Wetlands as usually defined are thought to cover about 6% of the world's land area (24, 114), but some suggest as much as 10% (83). It is not, however, exclusive of the other principal land-cover categories; most forested wetlands, for instance, are included in the standard forest/woodland category. By Williams's (114) estimate, three quarters of the world's wetlands represent inland freshwater systems and the rest, coastal ones.

CHANGE AND PROXIMATE SOURCES Conversion through drainage is the key proximate source of wetland change. Overall, "the overwhelming amount of wetland conversion is undertaken for agricultural purposes," perhaps 85–95%

(115). Urban-industrial-port expansion (74) accounts for much of the rest. The global area of artificially drained lands is estimated at 1.606×10^6 km^2 (114). Wetland losses, however, cannot be computed directly from drainage figures because not all areas affected by drainage were wetlands initially. Gosselink & Maltby (24) discuss the global picture of losses and gains. Past conversion is thought to have been most rapid in the developed world, and current conversion in the developing nations; reliable global data, though, are unavailable.

Settlement

DEFINITION The term settlement can denote a form of land cover or a form of land use; the area thus indicated may differ in a number of regards. The category of settlement as a land use includes areas devoted to human habitation, transportation, and industry. As land cover, it incorporates highly altered surfaces such as buildings and pavement, but such cover represents only a portion of the total area that a land-use classification might accord to settlement. Suburban land in the developing world, for example, usually incorporates a large proportion of tree cover. Estimates of the world area of settlement as land use do not claim high precision. Rozanov et al (85) offer the figure of 6% of the world's land surface as lands "radically changed ... for productive human uses, such as settlement, roads, and reservoirs." For the more restricted category of settlement (land "occupied for human living space"), Douglas (20) suggests 2.5%, 60% of it representing rural settlement and the rest urban; rural settlements in the developed world take up 1.03×10^6 km^2 and in the developing world 1.06×10^6 (20). The United Nations (108) estimated the global area of urban land at 2.47×10^6 km^2. As intensively occupied land cover, such settlements would account for a much smaller area. Of the land area of cities, less than half is built up even in large agglomerations; the built-up area has been estimated at only about 10% of urban land worldwide (20, 26).

CHANGE AND PROXIMATE SOURCES The urban share of the world population has increased from 14% at the beginning of the twentieth century to almost half today: it is highest by far in the developed countries. Definitional and data problems even with current or recent settlement areas make estimates of typical rates of change in urban land difficult. The striking growth in absolute urban numbers has, however, during this century, been clearly exceeded by the growth in the land area required by cities, given improvements in transportation and the spread of low-density forms of settlement. The countries of most rapid urban growth today are in the developing tropics, where many large and expanding metropoli—though of relatively high density compared to developed-world cities—are responsible for the most striking urban transformations of the land surface (8, 20).

ENVIRONMENTAL CONSEQUENCES OF LAND-COVER CHANGE

Each category of land-cover change is associated with a number of secondary environmental consequences: wetland drainage, for instance, can affect biodiversity, trace gas emissions, soil, and hydrological balance. In economic terms, these effects often represent the externalities of land-cover changes, the costs or benefits passed on to others by the land user. Secondary impacts of land-cover change can be difficult to distinguish from natural variation. Climatic change and water flows are cases in point.

Land-cover change impacts on biodiversity are not covered in this volume. We examine here five other classes of impacts, focusing on those fairly directly connected to land-cover change: trace-gas emissions, changes in water quality and water flows, soil alteration, and climatic change. Apart from increases in long-lived trace gases in the atmosphere (and their possible global climatic effects), the main secondary changes that can be traced to land-cover change are widely repeated cumulative changes rather than systemic ones.

Trace-Gas Emissions

Much of the human contribution to atmospheric trace species occurs through the processes of industrial metabolism, but land-cover changes significantly contribute to increases in a number of important components. A recent review by Penner (72), on which the following discussion is based, assesses the literature on the major trace components and provides ranges of estimates both for the human inputs as opposed to the natural and for the contribution of land-cover change as opposed to industry.

Atmospheric trace gases affected by human action fall into two broad categories: the relatively reactive and nonreactive, or the short- and long-lived. Several of the greenhouse gases implicated in global climatic change are the key long-lived trace components increased by land-cover change: CO_2 from forest clearance and soil carbon oxidation as well as from fossil fuel burning, methane from rice paddies, landfills, biomass burning, and livestock, and N_2O from soils, fertilizers, and biomass burning. The relative importance of land and industrial sources is a matter of much controversy as regards CO_2 emissions; current estimates of the former's share range from 10% to 50% of present anthropogenic releases. Matters are further complicated by the "missing sink" problem in global models of the carbon budget: the failure of modeled sinks to account for the estimated releases. Methane has doubled in atmospheric concentration since 1800, but the high diversity of sources both natural and anthropogenic leaves the possible contribution of land-cover change to total (natural plus human) emissions anywhere between 25% and 80%. N_2O contributes both to greenhouse forcing in the troposphere and to ozone depletion in the stratosphere; land-cover and land-use changes are known to be the principal (about 70%) human sources of emissions.

Reactive species significantly affected by land transformation include CO, nonmethane hydrocarbons, nitrogen oxides, and tropospheric aerosols. The last are of particular global importance for their role in scattering or absorbing incoming solar radiation; biomass burning is the principal source associated with land-cover change and fossil fuel combustion with industrial metabolism.

Hydrological Change

Hydrological (surface and groundwater) impacts of land-cover and land-use changes include changes in water quality and in water flows. Water pollution due directly to land-cover change stems from cultivation (principally application of fertilizers and pesticides) and settlement (urban sewage). It is largely pollution from non-point or dispersed sources (save for urban sewage plant releases) and as such is more difficult to control than point sources: the techniques exist for managing it, however, where the benefits to be gained justify the expense (83, 92).

Changes in water quantity and flow associated with land alteration result both from deliberate withdrawals and from land-cover changes such as deforestation. Irrigation—largely a consumptive use—is by far the largest element of global withdrawal from the hydrologic cycle; it accounts for about 75% of demand, though the share varies greatly across the globe (highest in Asia and minimal in many countries). Industry and energy generation account for most of the rest of the water withdrawn (56). Secondary hydrologic effects of irrigation include depletion of downstream rivers and water bodies; the desiccation of the Aral Sea in Soviet Central Asia is an extreme case (45, 62, 75). Many groundwater aquifers have been severely depleted, but the worldwide human impact is unknown; "there is no readily available set of data to indicate their magnitude and depletion rates on a continental scale" (56).

Claims have long been made that deforestation, especially in highlands, increases the frequency and severity of flooding downstream. Some of the most notable examples asserted in the literature have been called into question: the Amazon Basin (81, 96) and the Himalaya Mountains-Ganges Basin (40). While most claims of such dramatic and large-scale consequences now seem dubious (40), forest clearance or thinning generally increases annual flow while also making its distribution more uneven. Grassland change has analogous effects, while increases in flood height due to wetland drainage are well documented locally (83).

Soil and Sediment Impacts

Many of the global human impacts on soil erosion and sediment transport are extremely difficult to sum up at the global or even regional scale because of problems of spatial and temporal scale and natural background processes. Worldwide, soil loss and degradation and sediment transport have undoubtedly been increased greatly as a consequence of land-cover change (19a, 85). Most global-scale studies, however, emphasize the uncertainties, local variations, and methodological problems associated with creating reliable estimates of net human impact.

Important processes of soil degradation include erosion, salinization (as a result of irrigation), waterlogging, compaction, acidification, nutrient impoverishment, and dehumification (14). Figures on global change are most readily available for salinization. One study (85) estimates the area of land destroyed for productive use by salinization at one million km^2, with about the same area significantly lowered in productivity by the same processes. Soil changes are assessed in a recent *World Map of the Status of Human-Induced Soil Degradation* (69), which is to be made available in digitized form.

Climatic Change

Various microclimatic changes as a result of land-cover changes are clear and well-documented; claims for regional effects are more controversial, and possible global changes, if they occur, would occur through the effects of trace gas emissions. Suggestions that deforestation may affect global temperature through albedo change, largely discounted (32, 33), have been cautiously reopened by way of regional-level modeling (19, 31).

The urban heat island effect is the best-studied micro-to-meso climatic consequence of settlement expansion. Landsberg's (48) volume on the urban climate remains standard, but recent research on tropical and third world cities has suggested a more varied picture of the relations of city size and growth to heat island magnitude (1, 65). The claim that curves of global temperature rise may in large part be artifacts of the near-urban locations of recording stations has focused attention on urban effects.

Claims of regional impact of land-cover change on climate are in more dispute. The long-discredited (73) claim that forests increase rainfall has been revived in recent years in reference to the tropical rainforests, especially those of Amazonia. Salati & Vose (87) claim that regional forest clearance would have severe climatic consequences because of the high proportion of water recycled by the rainforest; deforestation would significantly lessen rainfall and increase temperatures. Possible regional effects on temperature and precipitation of vegetation loss through overgrazing also remain a lively focus of research.

HUMAN DRIVING FORCES OF CHANGE

The proximate sources of change—the human actions that directly alter land cover—reflect underlying human driving forces. Its importance notwithstanding, research on the human driving forces of global change is in considerable intellectual disarray. "Many studies in the literature are . . . either weakly connected (shopping lists of 'causes' unrelated to one another) or unduly hypothetical (plausible arguments unsupported by case-specific data)" (64). Two extremes in approach are ultra-empiricism and ultra-theoreticism. In the former vein, Newell & Marcus (68) present a correlation of +0.9985 between world population growth and tropospheric CO_2 levels since the 1950s as proof of population's fundamental role. In the latter, Harvey (27) deprecated neo-Malthusian arguments for the primacy of population growth in resource depletion as politically founded. The former is apt to mistake correlation for cause (or absence of correlation for causal unimportance) in a highly complex area; the latter is apt to narrow excessively the scope of investigation without recourse to data. Theory—including theory from the social sciences—and evidence need both to be drawn upon.

Three important points must be kept in mind. First, the driving forces of change may vary with the type of change involved; forces that drive some changes may lessen others. Southgate (95) provides a simple illustration: rising interest rates or agricultural prices will increase deforestation because they provide an incentive for further clearing; at the same time, they will decrease soil erosion on cultivated land because they provide an incentive to adopt soil conservation measures, whose value to the farmer or landowner would be enhanced. Second, the same kind of land-cover change can have different sources in different areas even within particular world regions. Deforestation in Borneo and parts of the Philippines is largely the result of timber extraction for export (13, 46); in peninsular Malaysia, agricultural clearing is the main proximate source (13). Third, in the dynamics of underlying "causes," no agreement exists on the level at which adequate explanation is achieved. To one, deforestation by agricultural expansion may be driven by population growth; to another, that population growth can only be explained in terms of certain sociopolitical and economic conditions that promote it (e.g. an economic context that makes large families valuable to subsistence cultivators); while yet others would emphasize the role of agricultural expansion in causing population growth.

The single comprehensive approach to the question of driving forces is the $I = PAT$ equation used by Ehrlich & Ehrlich (21), and by Commoner (16, 17), where I represents environmental impact, taken to be the product of P (population), A (affluence), and T (technology). Human impact is a product of the number of people, the level at which they consume, and the character of material and energy flows in production and consumption. The *IPAT* formula, however, suffers from the handicap of a mismatch between its categories of driving forces (apart from population) and the categories customarily used in the social sciences. Neither "affluence" nor "technology" as defined is associated with a substantial body of social science theory; any bridges between the *IPAT* and other approaches would have to be built between these categories and those better-conceptualized aspects of behavior and social structure that may drive and limit changes in production and consumption.

To detail the possible human driving forces is a formidable task, not only because of the array that can be found in the literature (66, 107), but because little evidence exists for their pan-global associations with land-cover/use change. We emphasize population because of its prominence in the literature and because it is one of the few candidate driving forces for which there is a simple measure.

Population

Population as a driving force of environmental change is unique in its plausibility and ease of quantification. It incorporates the basic level of resources required per capita for survival and reproduction (biological demand). This role of population is not in dispute: what is controversial is its relative importance among the other forces generating environmental pressures and the conceivably positive character of its role in resource use. We can simplify the various arguments into several broad positions.

In the neo-Malthusian position, global population increases are accorded primary importance in most environmental change because of the resources required to sustain the demands of five billion people. Population growth is seen as having exceeded the capacity of the biosphere, as managed by society, to sustain it (21, 22). Directly opposed is the cornucopian position (89, 90), which holds that population increases stimulate technological and social advances that improve the conditions of life: greater numbers can transform the environment for the better.

Population is relegated to secondary status in other theories in which it is held only to worsen degradation stemming from other factors. The Faustian position holds that runaway or careless use of technology is primary to environmental degradation, though population increase may exacerbate the problems created (16). Neoclassical political economy locates the cause of environmental damage in obstacles to the proper allocation of costs. These obstacles may lie in the distortion of efficient solutions by government policies or property institutions or in imperfect information regarding resources (7). Neo-marxist political economy emphasizes the role of the means of production in the global economy of international capitalism. Profit-seeking and capital accumulation require the unsustainable exploitation of natural resources, and socioeconomic differentiation creates a situation in which the "haves" place heavy demands on the world's resources, driving environmental change, while the subsistence needs of the "have-nots" put marginal environments under stress (9, 54, 79, 98, 102).

The role given to population in any of these positions reflects less conflicting evidence than conflicting interpretations of the same evidence. Case studies of the long-term records indicate major ebbs and flows in regional populations through time, but with varying types of transformation, longevity of impacts, and recoverability of the environment, suggesting caution in assuming a simple population-transformation association (112). This does not undermine the role of population as an important driving force in environmental change, but underlines its significance in the context of the technology and sociocultural organization in question.

It is this context—not the experience of a population outbreak alone —that makes the contemporary global environmental situation novel. The past 300 years, but particularly the latter half of this century, have witnessed an unparalleled magnitude of human-induced environmental changes, including those of land cover (104). These changes have coincided not only with unparalleled numbers of our species, but with major shifts in global affluence, technological capacity, and socioeconomic organization. A pilot study (using the FAO categories and data) reports a strong global-level correlation between modern population growth and annual change in land cover (forest, cropland, and pasture use), but finds equally strong correlations with surrogate variables for technology and affluence (118).

Comparative assessments assume that if population is a key driver of environmental change, then the pressures of population (e.g. density) should closely match the magnitude of various kinds of environmental change across regions and locales. Do such regional or national-level (the unit with which most land-use/cover data are generated and reported) comparisons show that the proportion of change in forest, cropland, and pasture lands correlates well with increases in population densities? The same pilot study did not find significant correlations in these relationships examined at the national level across the world (118).

In contrast, significant correlations between population and land-cover change have been found when investigation is restricted to regions possessing similar socioenvironmental characteristics. Several comparative studies offer statistical evidence supporting the claim that population growth drives or strongly contributes to forest clearance (2, 55, 71, 86). All of them deal only with part or all of the developing tropics, and the results would probably not apply elsewhere. Case studies even in the tropical developing world have contested the primacy of population growth as a driving force of clearance, attacking the "persistent myth that tropical deforestation is caused by overpopulation" (3) and emphasizing such factors as uneven land distribution and a complex array of policies, institutions, and economic forces that promote and reward clearing (29, 30, 97). A recent work analyzes the interactions of population growth and land cover change in several developing countries, and concludes that population growth is an important factor but one significantly modified by natural and institutional context (8a).

Too much must not be made of the comparative studies cited. The accuracy of the data used is highly questionable. Almost all of the statistical studies use the FAO *Production Yearbooks* data, whose drawbacks have been noted. Kummer (46) cites other characteristics of studies of forest-cover change that may undermine the results. In some cases, country-wide population and forest cover change are correlated, masking the fact that population growth and deforestation occurred in different locales; in others, current changes in population are misleadingly compared to changes in forest cover that have transpired over longer periods of human occupation (41, 91).

These cautions aside, population remains one of the few candidate driving forces that is readily measured and for which statistical associations have been found. Surely better data—particularly the spatial congruence of population with particular land covers—and assessments will clarify its role. The evidence to date, however, suggests an interesting insight: that population is an important macro-scale (global) variable in that there is a direct relationship between total world population and total biological demand for resources. Its connections to land-cover change become weaker at increasingly smaller spatial scales because of the importance of other variables that affect demand or spatially deflect its impacts. It is these other variables that must be incorporated in order to improve our understanding of the human causes of land-use/cover change.

In contrast to these studies, which have sought statistical linkages without a well-developed theory, others stress the importance of examining data only within a theoretical framework. Studies of population-agricultural relationships are exemplary. Among the various kinds of environmental change, none is so plausibly linked to population as the land use and cover changes associated with agriculture. Neoclassical economics typically accounts for the role of population change through its influence on demand as manifested through the market. As market signals change, so does land use. But some themes linking population more directly to agriculture have gained considerable attention over the last quarter century: those that identify population pressure as a driving force of agricultural change or that incorporate it as a force in theories of induced innovation or intensification.

The architecture of the modern population-pressure theory is found in the works of Boserup (11, 12; also 15). To simplify, this thesis attributes agricultural development, including its technological base and kind (subsistence to market), to the pressures for production that mount from a growing population and that operate through the process of intensification (increased output per unit area and time). Given sustained growth in population, agricultural land use expands and intensifies, involving conversion and modification of land cover. Sustained population decline has the opposite effect. Many studies among farmers whose production includes a substantial subsistence component have supported the principle that, all else equal, growth in population leads to growth in agriculture [see Turner & Brush (103) for an extensive review: this theme also applies to livestock production (23)]. Intensification is held ultimately to stimulate specialization and market development, shifting the role of agricultural land-use change to one more consistent with neoclassical theories. This portion of the thesis, however, has had less success, in part because of the complexity of factors that influence the development of a full market economy, including the resistance among peasants to entering one completely (39, 53).

The population pressure thesis has been elaborated by a number of "induced intensification" themes (43, 103). Induced intensification explains expansion and intensification by a farming unit as responses to a larger set of driving and mediating forces, the latter including environmental opportunities and constraints. Farmers display subsistence behaviors in response to the food needs of the farming unit and market behaviors in the desire to increase their material standard of living. The results are hybrid farming behaviors that all culminate, in the face of increasing demand (other things equal), in similar trajectories of agricultural land-use/cover change. Taken to a national and international scale, themes of induced innovation also link changes in agricultural land use, particularly in well-developed market systems, to changes in agricultural technology and management resulting from increased levels of research and development (28).

The lessons are that agricultural land uses and covers have much to do with the character of the economy, but that a key element of demand in the economy is the level of population. This population linkage tends to be spatially congruent with land use in economies with a strong subsistence component, and the population-land use connection is most easily demonstrated in local and regional studies where subsistence agriculture remains strong, although some major anomalies have developed in Africa (52, 105). The connection is far harder to trace in market economies, not only because of the spatial incongruencies noted but because the element of demand from population growth alone is difficult to disentangle from other factors.

Other Candidate Forces

Supplementing or contesting the primacy of demographic explanations, numerous candidate driving forces of land-cover (and, more broadly, environmental) change have been advanced in the literature. Turner & Meyer (66, 107) class the principal arguments into four categories: those emphasizing (i) technological change, (ii) socioeconomic organization, (iii) level of economic development, and (iv) culture. Literature reviews and assessments with regard to land-cover change are provided by Grübler (26) on technology, Sanderson (88) and Morrisette (63) on socioeconomic organization (institutions and political structure respectively), and Rockwell (82) on cultural attitudes. All conclude that little can yet be said with confidence. Many of these categories pose methodological problems of defining appropriate and measurable variables and of resolving the problems of interaction and covariance with other candidate forces. We know of no global aggregate or comparative assessments of these driving forces with land-cover change comparable to those undertaken for population.

Most localized studies (regional and historical) that address the complex interlinkages among the causes of change do not employ general and replicable categories. For example, Skole's (91) explanation for the high rates of deforestation observed in the state of Rondônia (Amazonia) in the 1980s invoked the conjuncture of such elements as the 1970s OPEC oil price increases, the spread of commercialized farming in Brazil financed by the consequent oil profits, resulting smallholder dispossession and migration to the Amazonian frontier, and the concurrent efforts of the Brazilian military government to develop the Amazon. Many regional/local case studies of land-cover change offer similar examples of "conjunctural" explanations or of what are termed in the philosophy of science "Cournot processes," where an event is produced by the intersection of two or more independent causal chains. Because of the independence of the causes and the dependence of the outcome on the specific point at which they meet, there is a large element of chance and unpredictability about the result and a consequently low likelihood of successfully generalizing the findings to other situations or of forecasting change in the future.

CONCLUSION

The broad patterns of major land-use/cover changes are known with some confidence, and the literature is rich in contending explanations for them. The area of driving forces requires the most attention. To advance, we need much more precise and spatially congruent data. The quality of the physical data, if still inadequate, is far better than that on the human variables, a disparity that will only be widened by the proposed space platform projects. The reluctance of national and international institutions to fund the collection of social data is an obstacle that must be addressed.

Several other avenues for progress on the issue of driving forces recommend themselves. One is that of a wiring diagram, based on the Bretherton framework of earth system science, to outline the structure of interaction between society and the land (78). A wiring diagram is not a causal model (though it might develop into one), but a structure of cooperation and information flows among disciplines such that the relevant models and theories of different fields can be wired together in integrated studies. A second is to seek a middle scale between the global and the local at which to address driving force-change relationships. The identification of a set of world-regional situations, defined by both socioeconomic and evnironmental variables, may make possible generalizations that cannot be made at the global scale (60).

A number of projects are under way that address directly or indirectly the causes of global land-cover change. They include the Project on Critical Environmental Zones (Roger Kasperson, PI) of the George Perkins Marsh Institute, Clark University, which is comparing nine case studies, and the Working Group on Critical Environmental Zones (International Geographical Union). A proposed joint effort of the IGBP and the ISSC on the study of the human causes of global land-use/cover change offers the opportunity for integrative assessments merging global average and comparative studies addressed to the problems noted above.

Literature Cited

1. Adebayo, Y. R. 1987. A note on the effects of urbanization on temperature in Ibadan. *J. Clim.* 7:185–92
2. Allen, J. C., Barnes, D. F. 1985. The causes of deforestation in developing countries. *Ann. Assoc. Am. Geo.* 75: 163–84
3. Anderson, A. B. 1990. Deforestation in Amazonia: dynamics, causes, and alternatives. In *Alternatives to Deforestation: Steps toward Sustainable Use of the Amazon Rain Forest,* ed. A. B. Anderson, pp. 1–23. New York: Columbia Univ. Press
4. Arnold, R. W., Szabolcs, I., Targulian, V. O. 1990. *Global Soil Change. Int. Inst. Appl. Syst. Anal.* Laxenberg, Austria
5. Bailey, R. G. 1989. Ecoregions of the continents (plus explanatory supplement). *Environ. Conserv.* 16:307–9
6. Balée, W. 1989. The culture of Amazonian forests. *Adv. Econ. Bot.* 7:78–96
7. Baumol, W. J., Oates, W. E. . 1988. *The Theory of Environmental Policy.* New York: Cambridge Univ. Press
8. Berry, B. J. L. 1990. Urbanization. See Ref. 104, pp. 103–19

8a. Bilsborrow, R. E., Okoth-Ogendo, H. W. O. 1992. Population driven changes in land use in developing countries. *Ambio* 21:37–45

9. Blaikie, P. M. 1985. *The Political Economy of Soil Erosion in Developing Countries.* Harlow, Essex: Longman Sci. Tech.
10. Blaikie, P., Brookfield, H. 1987. *Land Degradation and Society.* London: Methuen
11. Boserup. E. 1965. *The Conditions of Agricultural Growth.* Chicago: Aldine.
12. Boserup, E. 1981. *Population and Technological Change: A Study of Long-Term Trends.* Chicago: Univ. Chicago Press.
13. Brookfield, H. C., Lian, F. J., Low, K-S., Potter, L. 1990. Borneo and the Malay Peninsula. See Ref. 104, pp. 495–512
14. Buol, S. W. 1992. Role of soils in land-cover change. See Ref. 61, forthcoming
15. Chayanov, A. V. 1966. Peasant farm organization. In *A. V. Chayanov and the Theory of Peasant Economy,* ed. D. Thorner, B. Kerblay, R. E. F. Smith, pp. 29–269. Homewood, Ill:Erwin
16. Commoner, B. 1972. *The Closing Circle.* New York: Knopf

17. Commoner, B. 1990. *Making Peace with the Planet*. New York: Pantheon
18. Cowardin, L. M., Carter, V., Golet, F. C., LaRoe, E. T. 1979. *Classification of Wetlands and Deepwater Habitats in the United States. FWS\OBS-79/31*. Washington, DC: US Dep. Interior, Fish Wildlife Serv.
18a. Crosson, P. 1990. Arresting renewable resource degradation in the Third World: Discussion. *Am. J. Agric. Econ.* 72:1276–77
19. Dickinson, R. E., Henderson-Sellers, A. 1988. A study of GCM land-surface parametrizations. *Q. J. Roy. Met. Soc.* 114:439–62
19a. Douglas, I. 1990. Sediment transfer and siltation. See Ref. 61, pp. 215–33
20. Douglas, 1. 1992. Human settlements. See Ref. 61, forthcoming
21. Ehrlich, P. R., Ehrlich, A. H. 1990. *The Population Explosion*. New York: Simon & Schuster
22. Ehrlich, P. R., Holdren, J. P., eds. 1988. *The Cassandra Conference: Resources and the Human Predicament*. College Station, Tex: Texas A & M Univ. Press
23. Galaty, J. G., Johnson, D. . L. 1990. *The World of Pastoralism: Herding Systems in Comparative Perspective*. New York: Guilford
24. Gosselink, J. G., Maltby, E. 1990. Wetland losses and gains. See Ref. 113, pp. 296–322
25. Graetz, D. 1992. The grasslands: past, present, and future. See Ref. 61, forthcoming
26. Grübler, A. 1992. Technology and global change: land-use, past and present. See Ref. 61, forthcoming
27. Harvey, D. 1974. Population, resources, and the ideology of science. *Econ. Geogr.* 50:256–77
28. Hayami, Y., Ruttan, V. W. 1971. *Agricultural Development: An International Perspective*. Baltimore, Md: Johns Hopkins Univ. Press
29. Hecht, S. 1985. Environment, development and politics: capital accumulation and the livestock sector in eastern Amazonia. *World Dev.* 13:663–84
30. Hecht, S., Cockburn, A. 1989. *The Fate of the Forest: Developers, Destroyers and Defenders of the Amazon*. London: Verso
30a. Henndén, V. 1991. Desertification—time for an assessment. *Ambio* 20:372–83
31. Henderson-Sellers, A. 1991. A commentary on: Tropical deforestation: albedo and the surface-energy balance. *Clim. Change* 19:135–38
32. Henderson-Sellers, A., Gornitz, V. 1984. Possible climatic impacts of land cover transformations, with particular emphasis on tropical deforestation. *Clim. Change* 6:231–57
33. Henderson-Sellers, A., Wilson, A. 1983. Surface albedo data for climate modelling. *Rev. Geophys. Space Phys.* 21: 1743–48
34. Houghton, R. A. 1991. Releases of carbon to the atmosphere from degradation of forests in tropical Asia. *Can. J. For. Res.* 21:132–42
35. Houghton, R. A., Boone, R. D., Fruci, J. R., Hobbie, J. E., Melillo, J. M., et al. 1987. The flux of carbon from terrestrial ecosystems to the atmosphere in 1980 due to changes in land use; geographic distribution of the global flux. *Tellus* 39I3:122–39
36. Houghton, R. A., Boone, R. D., Melillo, J. M., Palm, C. A., Woodwell, G. M., et al. 1985. Net flux of CO_2 from tropical forests in 1980. *Nature* 316: 617–209
37. Houghton, R. A., Hobble, J. E., Melillo, J. M., Moore, B., Peterson, B. J., Shaver, G. R., Woodwell, G. M. 1983. Changes in the carbon content of terrestrial biota and soils between 1860 and 1980: a net release of CO_2 to the atmosphere. *Ecol. Monogr.* 53:235–62
38. Houghton, R. A., Skole, D. L. 1990. See Ref. 104, pp. 393–408
39. Hyden, G. 1980. *Beyond Ujamaa in Tanzania: Underdevelopment and an Uncaptured Peasantry*. Berkeley: Univ. Calif. Press
40. Ives, J. D., Messerli, B. 1989. *The Himalayan Dilemma: Reconciling Development and Conservation*. London: Routledge
41. Kartawinata, K., Vayda, A. P. 1984. Forest conversion in East Kalimantan, Indonesia: the activities and impact of timber companies, shifting cultivators, migrant pepper-farmers, and others. In *Ecology in Practice*. Vol. 1 *Ecosystem Management*, ed. F. DiCastri, F. W. G. Baker, M. Hadley, pp. 98–121. Dublin and Paris: Tycooly and UNESCO
42. Kates, R. W., Clark, W. C., Norberg-Bohm, V., Turner, B. L. II. 1990. *Human Sources of Global Change: A Report on Priority Research Initiatives for 1990–1995. Discussion Paper G-90–08, July, 1990*. Global Environ. Policy Project, John F. Kennedy Sch. Govt., Harvard Univ.; also *Occasional Paper No. 3*, Inst. Int. Stud., Brown Univ.
43. Kates, R. W., Hyden, G., Turner, B.

L. II. 1992. Theory, evidence, study design. See Ref. 105, forthcoming
44. Kershaw, A. P. 1986. Climatic change and aboriginal burning in north-east Australia during the last two glacial/interglacial cycles. *Nature* 322:47–9
45. Kotlyakov, V. M. 1991. The Aral Sea Basin: a critical environmental zone. *Environment* 33(1):4–9, 36–38
46. Kummer, D. 1992. *Deforestation in the Postwar Philippines. Univ. Chicago Dep. Geogr. Res. Pap. No. 234*
47. Lal, R. 1992. Cultivated land. See Ref. 61, forthcoming
48. Landsberg, H. 1981. *The Urban Climate*. New York: Academic
49. Lanly, J.-P. 1982. *Tropical Forest Resources*. Rome: FAO
50. Lanly, J. -P., Singh, K. D., Janz, K. 1991. FAO's 1990 reassessment of tropical forest cover. *Nat. Resources* 27(2):21–26
51. Leemans, R. 1990. *Global Holdridge Life Zone Classifications*. Digital Data. IIASA. Laxenburg, Austria. .26 MB
52. Lele, U., Stone, S. B. 1989. Population pressure, the environment, and agricultural intensification: variations on the Boserup hypothesis. In *Managing Agricultural Development in Africa*. Washington, DC: World Bank
53. Lipton, M. 1977. *Why Poor People Stay Poor: Urban Bias in World Development*. Cambridge, Mass: Harvard Univ. Press
54. Little, P. D., Horowitz, M. M., Nyerges, A. E., eds. 1987. *Lands at Risk in the Third World: Local-level Perspectives*. Boulder, Colo: Westview
55. Lugo, A., Schmidt, R., Brown, S. 1981. Tropical forests in the Caribbean. *Ambio* 10:318–24
56. L'vovich, M. I., White, G. F. 1990. Use and transformation of terrestrial water systems. See Ref. 104, pp. 235–52
57. Mabbutt, J. A. 1985. Desertification of the world's rangelands. *Desertification Control Bull.* 12:5–11
58. Marsh, G. P. 1864. *Man and Nature: or, the Earth as Modified by Human Action*. New York: Scribners
59. Matthews, E. 1983. Global vegetation and land use: new high resolution data bases for climate studies. *J. Clim. App. Met.* 22:474–87
60. McNeill, J. et al. 1992. Towards a typology and regionalization of land cover and global land use changes. See Ref. 61, forthcoming
61. Meyer, W. B., Turner, B. L. II. eds. 1992. *Global Land-Use/Land-Cover Change*. Boulder: OIES
62. Micklin, P. P. 1988. Desiccation of the Aral Sea: a water management disaster in the Soviet Union. *Science* 241:1170–76
63. Morrisette, P. M. 1992. Developing a political typology of global patterns of land and resource use. See Ref. 61, forthcoming
64. Mortimore, M. 1989. *Adapting to Drought: Farmers, Famines and Desertification in West Africa*. Cambridge: Cambridge Univ. Press
65. Nasrallah, H. A., Brazel, A. J., Balling, R. C. Jr. 1990. Analysis of the Kuwait City heat island. *Int. J. Clim.* 10:401–5
66. National Research Council, Commission on Behavioral and Social Sciences and Education, Committee on the Human Dimensions of Global Change. 1992. *Global Environmental Change: Understanding the Human Dimensions,*. ed. P. Stern, 0. R. Young, D. Druckman. Washington, DC: Natl. Acad. Press
67. National Research Council, Committee on Global Change. 1990. *Research Strategies for the US. Global Change Research Program*. Washington, DC: Natl. Acad. Press
68. Newell, N. D., Marcus, L. 1987. Carbon dioxide and people. *Palaios* 2:101–3
69. Oldeman, L. R., Hakkeling, R. T. A., Sombroek, W. G. 1990. *World Map of the Status of Human-Induced Soil Degradation*. Nairobi: UNEP
70. Olson, J. S. 1989. *World Ecosystems (WE 1.4). Digital Data. NOAA/NGDC- EDC-A*. Boulder, Colo. 2.3 MB
71. Panayotou, T. 1989. *An Econometric Study of the Causes of Tropical Deforestation: The Case of Northeast Thailand. Dev. Discuss. Pap. No. 284*. Cambridge, Mass: Harvard Inst. Int. Dev.
72. Penner, J. 1992. The role of human activity and land use change in atmospheric chemistry and air quality. See Ref. 61, forthcoming
73. Pereira, H. C. 1973. *Land Use and Water Resources in Temperate and Tropical Climates*. Cambridge, Eng: Cambridge Univ. Press.
74. Pinder, D. A., Witherick, M. E. 1990. Port industrialization, urbanization and wetland loss. See Ref. 113, pp. 234–66
75. Precoda, N. 1991. Requiem for the Aral Sea. *Ambio* 20:109–14
76. Rapp, A. 1987. Desertification. In *Human Activity and Environmental Process,* ed. K. J. Gregory, D. E. Walling, pp. 425–43. Chichester: Wiley

77. Ravenstein, E. G. 1890. Lands of the globe still available for European settlement. *Scottish Geogr. Mag.* 6:541–46
78. Rayner, S. et al 1992. A land use/cover change wiring diagram. See Ref. 61, forthcoming
79. Redclift, M. 1987. *Sustainable Development: Exploring the Contradictions*. London: Methuen.
80. Richards, J. F. 1990. Land transformation. See Ref. 104, pp. 163–78
81. Richey, J. E., Nobre, C., Deser, C. 1989. Amazon River discharge and climate variability: 1903 to 1985. *Science* 246:101–3
82. Rockwell, R. 1992. Culture and cultural change as driving forces in global land-use/cover changes. See Ref. 61, forthcoming
83. Rogers, P. 1992. Impacts of land use change on hydrology and water quality. See Ref. 61, forthcoming
84. Roosevelt, A. 1989. Resource management in Amazonia before the conquest: beyond ethnographic projection. *Adv. Econ. Bot.* 7:30–62
85. Rozanov, B. G., Targulian, V., Orlov, D. S. 1990. Soils. See Ref. 104, pp. 203–14
86. Rudel. T. K. 1989. Population, development, and tropical deforestation: A cross-national study. *Rural Sociol.* 54:327–38
87. Salati, E., Vose, P. B. 1984. Amazon Basin: a system in equilibrium. *Science* 225:129–38
88. Sanderson, S. 1992. Institutional dynamics behind land use change. See Ref. 61, forthcoming
89. Simon, J. 1981. *The Ultimate Resource*. Princeton: Princeton Univ. Press
90. Simon, J, Kahn, H., ed. 1984. *The Resourceful Earth*. Oxford: Basil Blackwell
91. Skole, D. L. 1992. Data on global land cover change: acquisition, assessment, and analysis. See Ref. 61, forthcoming
92. Smil, V. 1990. Nitrogen and phosphorus. See Ref. 104, pp. 423–36
93. Smith, N. J. H. 1980. Anthrosols and human carrying capacity in Amazonia. *Ann. Assoc. Am. Geogr.* 70:553–66
94. Social Science Research Council. 1991. *Rep. Work. Group on Land-use Change, Commit. for Res. on Global Environ. Change*
95. Southgate, D. 1990. The causes of land degradation along'spontaneously' expanding agricultural frontiers in the Third World. *Land Econ.* 66:93–101
96. Sternberg, H. 1987. Aggravation of floods in the Amazon River as a consequence of deforestation? *Geografiska Annaler* 69A:201–19
97. Stonich, S. C. 1989. The dynamics of social processes and environmental destruction: a Central American case study. *Popul. Dev. Rev.* 15:269–96
98. Sweezy, P., Magdoff, H. 1989. Capitalism and the environment. *Monthly Rev.* 41(2):1–10
99. Thomas, W. L. Jr., ed. 1956. *Man's Role in Changing the Face of the Earth*. Chicago: Univ. Chicago Press
100. Tucker, C. J., Dregne, H. E., Newcomb, W. W. 1991. Expansion and contraction of the Sahara Desert from 1980 to 1990. *Science* 253:299–301
101. Turner, B. L. II. 1983. *Once Beneath the Forest: Prehistoric Terracing in the Rio Bec Region of the Maya Lowlands*. Boulder, Colo: Westview
102. Turner, B. L. II, Benjamin, P. A. 1991. *Fragile lands: identification and use for agriculture*. Pap. pres. Conf. Institutional Innovations for Sustainable Agric. Dev. into the Twenty-First Century, Oct. 14–18. Bellagio, Italy
103. Turner, B. L. II, Brush, S. B., eds. 1987. *Comparative Farming Systems*. New York: Guilford
104. Turner B. L. II, Clark, W. C., Kates, R. W., Richards, J. F., Mathews, J. T., Meyer, W. B., ed. 1990. *The Earth as Transformed by Human Action*. Cambridge: Cambridge Univ. Press
105. Turner, B. L. II, Hyden, G., Kates, R. W. ed. 1992. *Population Growth and Agricultural Intensification: Studies from Densely Settled Areas in Sub-Saharan Africa*. Gainesville, Fla: Univ. Florida Press, forthcoming
106. Turner, B. L. II, Kasperson, R. E., Meyer, W. B., Dow, K. M., Golding, D., et al. 1990. Two types of global environmental change: definitional and spatial scale issues in their human dimensions. *Global Environ. Change* 1:14–22
107. Turner, B. L. II, Meyer, W. B. 1991. Land use and land cover in global environmental change: considerations for study. *Int. Soc. Sci. J.* 130:669–792
108. UN. 1985. *Compendium of Human Settlement Statistics 1983*. New York: United Nations
109. UN/FAO. 1988. *Production Yearbook*. Rome: FAO
110. Walker, H. J. 1990. The coastal zone. See Ref. 104, pp. 271–94
111. Watts, M. J. 1985. Social theory and environmental degradation. In *Desert*

Development: Man and Technology in Sparselands, ed. Y. Gradus, 14–32. Dordrecht: D. Reidel

112. Whitmore, T. M., Turner, B. L. II, Johnson, D. L., Kates, R. W., Gottschang, T. R. 1990. Long-term population change. See Ref. 104, pp. 25–39
113. Williams, M., ed. 1990. *Wetlands: A Threatened Landscape*. Oxford: Basil Blackwell
114. Williams, M. 1990. Understanding wetlands. See Ref. 113, pp. 1–41
115. Williams, M. 1990. Agricultural impacts in temperate wetlands. See Ref. 113, pp. 181–216
116. Williams, M. 1992. Forests and tree cover. See Ref. 61, forthcoming
117. Wolman, M. G., Fournier, F. G. A., ed. 1987. *Land Transformation in Agriculture. SCOPE 32*. New York: John Wiley
118. Young, S., Benjamin, P., Jokisch, B., Ogneva, Y., Garren, A. 1991. *Global Land Use/Cover: Assessment of Data and Some General Relationships. Report to the Land-Use Working Group*. Committee Res. Global Environ. Change, Social Sci. Res. Council
119. Zon, R., Sparhawk, W. N. 1923. *Forest Resources of the World*. New York: McGraw-Hill

Annu. Rev. Ecol. Syst. 1992. 23:63–87

BIOLOGICAL INVASIONS BY EXOTIC GRASSES, THE GRASS/FIRE CYCLE, AND GLOBAL CHANGE

Carla M. D'Antonio
Department of Integrative Biology, University of California, Berkeley, California 94720

Peter M. Vitousek
Department of Biological Sciences, Stanford University, Stanford, California 94305

KEYWORDS: alien species, land-use change, competitive effects, ecosystem processes, grass-fueled fires

INTRODUCTION

Biological invasions into wholly new regions are a consequence of a far reaching but underappreciated component of global environmental change, the human-caused breakdown of biogeographic barriers to species dispersal. Human activity moves species from place to place both accidentally and deliberately—and it does so at rates that are without precedent in the last tens of millions of years. As a result, taxa that evolved in isolation from each other are being forced into contact in an instant of evolutionary time.

This human-caused breakdown of barriers to dispersal sets in motion changes that may seem less important than the changing composition of the atmosphere, climate change, or tropical deforestation—but they are significant for several reasons. First, to date, biological invasions have caused more species extinctions than have resulted from human-caused climatic change or the changing composition of the atmosphere. Only land use change probably has caused more extinction, and (as we later discuss) land use change interacts strongly with biological invasions. Second, the effects of human-caused biological invasions are long-term: changes in climate, the atmosphere, and land use may be reversible in hundreds to thousands of years, but the breakdown of biogeographic barriers has resulted in self-maintaining and evolving

0066-4162/92/1120-0063$02.00

populations in regions they could not otherwise have reached. Many of these changes must be considered irreversible (32). Finally, some biological invasions alter ecosystem processes in invaded areas, thereby causing functional as well as compositional change.

The fraction of successful invasions that alter ecosystem processes (defined here as whole-system fluxes of energy, the amount and pathway of inputs, outputs, and cycling of materials, and the ways that these vary in time) is not known. Such effects have been evaluated in very few cases, and most of those doubtless were selected for study because they were thought to involve changes in ecosystem processes. Vitousek (167) discussed three ways that biological invasion could alter ecosystems: (i) Invading species could alter system-level rates of resource supply. For example, the actinorrhizal nitrogen-fixer *Myrica faya* invades and dominates nitrogen-limited primary successional sites and increases both nitrogen inputs and the biological availability of nitrogen several-fold (168, 170). Other examples include *Tamarix* spp. in arid lands of the United States and Australia (65, 90, 112), *Mesembryanthemum crystallinum* in California and Australia (83, 171), *Carpobrotus edulis* in California (41), and Australian Acacia species in South Africa (185). (ii) Invading species could alter the trophic structure of the invaded area. Adding (or removing) a top carnivore can have disproportionate effects on ecosystem structure and function. This is the original basis of the keystone species concept (114), and it continues to be important in the development of trophic theory (28). Examples include the establishment of lamprey in the Saint Lawrence Great Lakes (5) and the introduction of the brown tree snake on Guam (135). (iii) Invading species can alter the disturbance regime (type, frequency, and/or intensity) of the invaded area. For example, feral pigs, *Sus scrofa,* alter rates of decomposition, nutrient cycling (166), and even watershed-level nutrient losses by "rooting" through soil (145). Plant invasions can alter fire frequency and intensity (92, 149, 164), including the introduced grasses that are the major topic of this review.

Invasions that alter ecosystem processes are important to ecological theory because such effects are less well characterized than are population or community level effects of invasion, and they represent a clear example of single species control over ecosystem processes. In addition, invasions that alter ecosystems represent a particularly significant threat to native populations and communities: they don't merely compete with or consume native species, they change the rules of the game by altering environmental conditions or resource availability. Finally, invasions that alter ecosystem processes over large areas could feed back to alter other components of global change (e.g. climate, atmospheric composition, and land use).

Grasses are one set of invading species that in the aggregate may be sufficiently widespread and effective to alter regional and even global aspects of ecosystem function. Grass invasions are important for several reasons: (i) As a group, grasses are moved actively by humans so invasions are common (e.g. 115); (ii) exotic grasses compete effectively with native species in a wide range of ecosystems; (iii) where they dominate sites, grasses can alter ecosystem processes from nutrient cycling to regional microclimate; and (iv) many species of grasses tolerate or even enhance fire, and many respond to fire with rapid growth. In turn, fire is a significant agent of both land use and atmospheric change regionally and globally (38, 81).

In this paper we make the case that grass invasions are widespread, that grasses are effective and aggressive competitors with native species, and that grass invasions have substantial ecosystem-level effects. We then discuss the regional and global significance of invasions in the context of the land use changes that are driving much of the earth's surface toward dominance by fire and grasses.

GEOGRAPHIC PATTERNS OF GRASS INVASION

General

Examples of alien grass invasion can be found on all continents, although examples from Eurasia and Africa are rare. These invasions can be divided into three categories: (i) spread of alien grasses into largely undisturbed native vegetation, (ii) spread of grasses into disturbed vegetation, and (iii) the longterm persistence of grasses in areas where they were originally seeded. Invasions into undisturbed sites are the most interesting from the point of view of developing a basic understanding of effects of individual species on ecosystems. However, invasions into human-disturbed areas and even the persistence of once-seeded grasses can be equally effective agents of local, regional, and global change.

North America

Alien grass invasions are most severe in the arid and semi-arid west and include invasions by European annual grasses and by perennial bunchgrasses of African, Eurasian, and South American origin.

Annual grass invasions began with the arrival of Europeans and were largely unplanned. Mack (94) details the expansion of *Bromus tectorum* (cheatgrass) throughout the Great Basin in conjunction with the introduction of sheep and cattle. Other European annual grasses whose invasion appears to be tied to grazing include *Taeniatherum asperum* (medusahead) (186) and *Bromus rubens* (red brome). Also, *Bromus mollis, B. diandrus,* and *Avena* spp. now dominate valley grasslands in California where they have replaced grazing-intolerant native bunchgrasses (10, 68). Other European annual species including *Poa pratensis* and *Bromus inermis* are common invaders of disturbed prairie throughout the Great Plains including Canada (156, 183, 184). In more mesic portions of North America the Asian annual grass *Microstegium vimineum* often forms dense monospecific stands on floodplains and adjacent disturbed and undisturbed mesic slopes throughout 14 eastern states (9).

Unlike the European annual grasses, perennial grass invaders were purposefully introduced as livestock forage or to prevent soil erosion. The Eurasian *Agropyron desertorum* (crested wheat grass) was seeded throughout the sagebrush steppe region because it was more tolerant of grazing than its native congener *A. spicatum* (24, 131). *A. desertorum* has maintained itself and spread into nearby shrublands (75, 124). The South African grasses *Cenchrus ciliarus* (bufflegrass) and *Eragrostis lehmanniana* (Lehmann lovegrass) were also seeded onto arid lands, maintained themselves in seeded sites even after livestock removal (16), and spread into nearby areas even in the absence of grazing (23, 37, 98). *Pennisetum setaceum,* also from Africa, is well established in areas of the Sonoran Desert (90).

Cortaderia jubata (Pampas grass) and *Ehrharta calycina* (Veldt grass) are perennial grasses that are common invaders of coastal habitats in California (138). *C. jubata* (from Argentina) is common in logged coastal forests and maritime chaparral (97, 138) and can invade without obvious human disruption of the habitat, particularly in areas with natural canopy gaps or after fire. *Ehrharta calycina* (from South Africa) invades disturbed and undisturbed coastal habitats in California (138). *Ammophila arenaria* (European beachgrass), also introduced for erosion control, has spread extensively so that it now dominates most beaches and foredunes in northern California, Oregon, and Washington (7, 177).

Central and South America

Alien grasses have invaded both natural and derived savannas (12, 15, 54, 115) in South America. Derived savannas result from forest clearing followed by burning; African grasses were introduced into them because of their tolerance of grazing (115, 144). Invasion of native grasslands by these grasses has been documented in Brazil, Colombia, and Venezuela where *Hyparrhenia rufa* (Jaragua) and *Melinis minutiflora* (Molasses grass) have displaced native pasture grasses such as *Trachypogon plumosus* (11, 36, 54, 113, 115, 144). Ecophysiological studies demonstrate that these grasses tolerate frequent defoliation better than native grasses (144). *Hyparrhenia* has also invaded Central American woodlands and pastures (77, 157). European annual grasses often dominate native grassland in Argentina after range degradation by cattle (39, 53).

Oceania

Alien perennial grasses of African origin are common throughout Oceania. Presently large portions of the Hawaiian islands are dominated by introduced grasses, including *Paspalum, Pennisetum, Melinus minutiflora, Hyparrhenia rufa, Cenchrus ciliarus,* and *Digitaria decumbens* (148). Two North American bunchgrass species, *Andropogon virginicus* and *Schizachyrium condensatum* are also common. Most of these grasses were introduced to support livestock; their spread may have been facilitated by the activity of feral goats which were also brought by early European colonists. Today these grasses persist and often spread in the absence of ungulates. Many of these same grasses are present on other islands in the Indo-Pacific region. For example, *Pennisetum polystachyon* dominates some lowland areas in Fiji (109). *Hyparrhenia rufa* and *Pennisetum clandestinum* are common on La Reunion (93), and *Pennisetum purpureum* covers 5000 ha in the Galapagos (89, 140).

In New Zealand, *Cortaderia jubata* along with its congener, *C. selloana*, has invaded thousands of hectares of forest plantations, but invasion appears limited to cut-over forests (57). *Ammophila arenaria* has also spread away from areas where it was planted and now dominates large sections of coastal dunes (79).

Australia

European annual grasses and African bunchgrasses are common alien species in Australia. Annual grasses invade both disturbed and undisturbed habitats and are largely successional (70, 71, 123). Perennial grasses of African origin were introduced as livestock forage. The most common of these, *Cenchrus ciliaris,* dominates many areas and has spread into adjacent native pastureland in northwestern Australia (37, 76). *Pennisetum polystachyon* is invading native savanna regions in northern Australia (59, 152), and two species of *Ehrharta* (African veldt grass) have invaded coastal communities in southwestern Australia (20). Relatively undisturbed shrublands and woodlands in this region are also invaded by the African species, *Ehrharta longiflora, Rhyncoletrum repens* (Natal red top), and *Eragrostis curvula* (weeping lovegrass) (20).

Aquatic grasses have invaded perennial and seasonally moist habitats throughout Australia and Tasmania (76). Reed sweetgrass, *Glyceria maxima,* and para grass, *Brachiaria mutica,* form large floating mats in ponds and other wet areas (76, 104), and the Eurasian saltgrass *Spartina townsendii* has invaded coastal salt marshes (76).

Eurasia and Africa

Much of tropical Asia and Africa is covered by derived grasslands and savannas (defined as once-forested regions in which grasses now dominate as a consequence of human activity) (146). However, relatively few examples exist of large-scale invasions by alien grasses from other continents or other areas within these continents. One example is the invasion of coastal salt marshes in Britain by the North American species *Spartina alterniflora* and the hybrid species *S. anglica* which formed through hybridization between *S.*

alterniflora and the British native *S. maritima* (154). Other examples include the establishment of several European grasses in Mediterranean-climate regions of South Africa (27, 172), the spread of the South American species *Stipa trichotomea* and *Cortaderia jubata* in southern Africa (130, 179), and the naturalization of African *Panicum maximum* in India (136). Other African grasses are present in Asia, but they are less invasive (17). There are also examples of internal invasions, in which grasses from one region of a continent invade other areas: these include *Cenchrus ciliaris, Pennisetum clandestinum,* and *Eragrostis curvula* within Africa (37, 91).

In contrast to the rarity of grass invasions, several American shrubs and vines are aggressive invaders of Eurasian and African ecosystems (44, 92, 127).

EFFECTS OF GRASS INVASIONS

Framework

Grass invasions can have effects at multiple levels of ecological organization from population to the ecosystem. Table 1 suggests a framework for examining these effects and their interactions. The categories therein are not exclusive; competitive effects operating at the population level can also have ecosystem-level consequences. For example, where light absorption in a grass canopy prevents the establishment of tree seedlings, the interaction is competitive but the resultant ecosystem has the microclimate and flammability of a stable grass-dominated system. Conversely, ecosystem effects such as altered fire regimes will also alter competitive interactions by causing changes in resource availability.

Resource Competition

Where the canopy of an alien grass absorbs or intercepts incident light and thereby limits the establishment or growth of other species, this represents a

Table 1 A classification of possible population and ecosystem-level effects of grass invasions.

Resource competition (exploitation)	Effects on resource supply	Other ecosystem effects
Light absorption		Geomorphological effects
Water uptake	Altered water holding capacity	Microclimate effects
Nutrient uptake	Altered rates of mineralization and immobilization	Disturbance effects (e.g. fire)

competitive rather than an ecosystem effect. Similarly, a reduction in water or nutrient availability to other species that results from utilization of those resources by grasses constitutes a competitive effect. In contrast, where grasses alter water or nutrient availability by altering boundary-layer humidity or rates of nutrient mineralization, this represents an ecosystem effect.

Grasses have long been recognized as good competitors against herbaceous and woody species (e.g. 22, 33, 84, 88, 141, 159, 161). In numerous studies, the establishment of large seeded and woody perennials has been found to be limited in the presence of dense grasses or grass litter. As a result, the invasion of grasslands by other perennial species often requires soil disturbance such as gopher mounds (41, 52, 63, 66, 120, 162, 163).

Rapidly growing grasses can reduce light at the soil surface and thereby reduce the photosynthetic ability of competitors (153, 155). For example, the grass *Miscanthus sinensis* colonizes abandoned fields in Japan, reducing light availability and daily carbon gain of oak seedlings and thus slowing the rate of encroachment of oak trees into grassland (153). Similarly, alien grasses in Texas reduce growth rates of seedlings of woody species such as *Baccharis neglecta* and *Prosopis glandulosa* by reducing light availability (21, 162).

Grasses are also effective competitors for water and nutrients. Alien grasses can interrupt succession through competition for water with native perennials (40, 42, 43, 49, 64, 101, 141). In California grasslands, European annual grasses are considered one of the major causes of poor oak recruitment (40, 64). Oak seedlings are extremely sensitive to soil water, and alien annual grasses rapidly draw down soil moisture and suppress oak seedling growth more dramatically than do native perennial grasses (40). Growth of seedlings of the native shrub *Baccharis pilularis* in California is also reduced by competition for soil water with European annual grasses (42, 43). The exotic perennial grasses *Agropyron desertorum* and *Dactylis glomerata* suppress pine reestablishment after wildfires in the southwestern United States, and at least part of this effect appears to be due to competition for water (49). Alien grasses have been used to reduce shrub seedling survival in the conversion of chaparral to grassland in the Mediterranean climate region of western North America, and their success has been attributed in large part to their ability to rapidly draw down soil moisture (141).

Efficient use of water is also a means by which alien grasses can outcompete native grasses. Eissenstat (47) and Eissenstat & Caldwell (48) found that the alien grass *Agropyron desertorum* drew down soil moisture more rapidly and to lower levels than its native congener *A. spicatum*. *Agropyron desertorum* has replaced *A. spicatum* in portions of the Great Basin (75) and competes more effectively for water with the dominant native shrub *Artemesia tridentata* than does *A. spicatum* (48). The alien grass *Bromus tectorum* is more abundant in stands of *A. spicatum* than in stands of *A. desertorum*, and this difference may occur because *B. tectorum* begins root growth at a time when established *A. desertorum* are rapidly drawing down soil moisture (48).

Alien grasses also have been shown to compete effectively with native species for soil nutrients. Elliott & White (49) observed more rapid disappearance of nitrate from soil in plots planted with an alien grass than in those with native grasses; they hypothesized that competition for nitrogen was one of the factors responsible for poor pine seedling growth in the presence of the aliens. Seedlings of native shrubs have reduced growth and lowered tissue nitrogen when grown in the presence (compared to the absence) of alien grasses after wildfire in a Hawaiian woodland (73). The abundant alien bunchgrass *Agropyron desertorum* is an effective competitor for phosphorus with the native shrub *Artemesia tridentata,* and it is more effective at phosphorus extraction from the soil than its native congener *A. spicatum* (25, 26).

The effective uptake of water and nutrients by grasses is likely the result of their dense shallow root systems (43, 119, 133, 175). The root systems of most woody species are deeper and less dense than those of grasses. Once individuals are large, woody species are generally thought to have access to moisture and nutrients from portions of the soil profile below grass roots (141, 150). Grasses may therefore be most effective as competitors against seedlings rather than saplings or adults of woody species. However, Knoop & Walker (84) demonstrated that grasses can also reduce water availability in the subsoil (30–130 cm) where shrub roots are common.

Replacement of native species and dominance by alien grasses may also result from demographic differences between native and alien species. For example, *Agropyron desertorum* appears to be able to replace *A. spicatum* in sites dominated by the latter in part because of higher seed output, lower seed predation, and the buildup of a large seedbank in *A. desertorum* (124).

Numerous studies have reported negative correlations between alien grass cover and diversity or growth of native species without elucidation of the mechanisms leading to these patterns (14, 16, 23, 33, 74, 110, 183, 184, 187). For example, Bock et al (16) found 10 common native plant species to be significantly reduced in the presence of the African grass, *Eragrostis lehmanniana* in Arizona, and Billings (14) found substantially reduced diversity of native herbs in dense stands of *B. tectorum.*

The elimination of native plant species through competition with alien grasses in turn affects the diversity and persistence of animal populations that rely on grasses for food or habitat (16, 100, 147, 184). For example, the elimination of native dune species by the aggressive alien beachgrass *Ammophila arenaria* has resulted in a dramatic decline in native insect species, including the elimination of several rare species (147). The replacement of native herbs and shrubs in the Sonoran desert by *Eragrostis lehmanniana* has resulted in a local decline in native bird and insect species (16, 100). The presence of introduced grasses in disturbed prairie in Canada has caused simplification of habitat structure and a shift in species composition of birds (184).

A major consequence of the competitive success of alien grasses is the slowing or alteration of succession. This interacts with increasing rates of human disturbance to increase the proportion of the Earth's surface that is successional. On a local and regional scale this can result in the loss of both plant and animal diversity and the fragmentation of natural systems which has both genetic and population consequences. Even where alien grasses eventually are replaced by woody species, the competitive success of grass prolongs the period during which successional systems are susceptible to fire, the consequences of which are discussed later.

Rates of Resource Supply

For soil resources, individual species can affect rates of resource supply as well as the amount of resource that is available (Table 1). Nitrogen fixation is an obvious example, but there are also more subtle pathways. For example, the litter of different species can differ in rates of decomposition and nutrient immobilization or release (52). These differences can establish feedbacks that affect both litter quality and the rates at which soil nutrients are released from organic matter into inorganic forms (96, 117).

In one study involving grasses, several native grass species grown in experimental monocultures were shown to produce nutrient-poor litter that led to reduced soil nutrient supply (176). In contrast, two Eurasian grasses, (*Poa pratensis* and *Agropyron repens*), produced nitrogen-rich litter, and their soil had higher rates of nitrogen transformations (176).

The possibility of allelopathic effects of alien grasses fits uncertainly between competitive and ecosystem effects. To the extent that allelopathy alters nutrient dynamics (i.e. by suppressing nitrifying bacteria, 129) it should be considered an ecosystem effect. A number of common grass invaders (e.g. *Cynodon dactylon, Hyparrhenia* spp., and *Lolium* spp.) may be allelopathic (18, 53, 110, 148, 178), but the strength of these effects and their importance in field situations are uncertain.

Other Ecosystem Effects

Alien grasses can alter ecosystems through a number of pathways that are not obviously related to resource use or supply. Among these are: (i) geomorphological effects, (ii) microclimate effects, and (iii) disturbance effects, in particular fire regimes (which will be considered separately).

GEOMORPHOLOGICAL PROCESSES *Ammophila arenaria* (European beachgrass) alters dune formation patterns where it is planted or has invaded in North America, New Zealand, and Australia. Its ability to bind sand is greater than that of native species, and dunes formed by *Ammophila* tend to be steeper and taller than those formed by native species (8, 34, 69, 177). These changes may influence beach size, erosional patterns, and plant and animal diversity (8, 45, 60, 147).

Cynodon dactylon (Bermuda grass) invades stream courses in Arizona and appears to affect community development by increasing substrate stability during floods (46). Sites heavily dominated by *C. dactylon* retained more substrate during floods, including basal fragments of native aquatic macrophytes. Post-flood development of the aquatic macrophyte community proceeded more quickly in these sites than in those lacking *C. dactylon*.

The North American perennial bunchgrass *Andropogon virginicus* alters drainage patterns where it has invaded disturbed montane rainforest in Hawai'i (107). Its phenology, dense litter production, and low transpiration rates relative to forest species result in the accumulation of standing water and the formation of swampy areas. A number of introduced grasses in Australia have the opposite effect, colonizing ponds with water up to 2 m deep and thereby converting open water systems to wet grasslands (76).

MICROCLIMATE EFFECTS Grass invasion could alter microclimate on several scales. Grass litter can affect soil surface temperature and moisture and thereby influence seed germination, seedling growth, and nutrient transformations (52). For example, the buildup of litter of the alien grass *Bromus japonicus* in South Dakota decreases evaporation from the soil surface and favors further germination and establishment of *B. japonicus* (182). Similarly, litter of *Bromus tectorum* enhances seed germination of several alien species in desert shrublands because of improved water availability associated with the litter cover (51), and dense litter accumulations associated with the Eurasian grass *Lolium multiflorum*, reduced native species diversity in an Argentine grassland (53).

On a coarser scale, grass canopies are shallow and aerodynamically smooth in comparison to forest or woodland canopies. Where grass invasion leads to the replacement of woody vegetation by grassland, the pathway of energy partitioning leads to higher canopy and surface temperatures and lower relative humidities in grass-dominated systems (87, 143). These changes favor the growth of species with the C_4 photosynthetic pathway (mainly grasses), and also favor fire (160).

GRASS INVASION AND FIRE REGIMES

General

The most significant effects of alien grasses on ecosystems result from interactions between grass invasion and fire. A number of features make grasses and grass-dominated systems both relatively flammable and able to recover relatively rapidly following fire, in comparison to forests. First, the grass life form supports standing dead material that burns readily. Second, grass tissues have large surface/volume ratios and can dry out quickly. The flammability of biological materials is determined primarily by their surface/volume ratio and moisture content and secondarily by mineral content and tissue chemistry (132, 165, 174). The finest size classes of material (mainly grasses) ignite and spread fires under a broader range of conditions than do woody fuels or even surface leaf litter (80). Third, the grass life form allows rapid recovery following fire: there is little above-ground structural tissue, and so almost all new tissue fixes carbon and contributes to growth. Finally, grass canopies support a microclimate in which surface temperatures are hotter, vapor pressure deficits are larger, and the drying of tissues more rapid than in forests or woodlands. Thus, conditions that favor fire are much more frequent in grasslands. It is therefore reasonable to consider grasslands and fire as an "identity" (173) or to discuss the "pyrophytic grass life form" (108, 109). Indeed, human suppression of fire in natural grasslands is a disruption with major biological consequences.

Invasion can set in motion a grass/fire cycle where an alien grass colonizes an area and provides the fine fuel necessary for the initiation and propagation of fire. Fires then increase in frequency, area, and perhaps intensity. Following these grass-fueled fires, alien grasses recover more rapidly than native species and cause a further increase in susceptibility to fire. In fact land managers have seeded alien grasses for the purpose of increasing fire frequency and intensity in order to suppress woody species (141, 189).

Alteration of fire regimes clearly represents an ecosystem-level change caused by invasion. Fires themselves alter nutrient budgets profoundly; they volatilize some elements (notably carbon and nitrogen) while converting others into biologically more available, mobile forms for at least a short time (125). The selective loss of nitrogen in particular drives ecosystems toward nitrogen limitation (142, 169). Nutrient losses to streamwater, groundwater, and the atmosphere are also enhanced following fire (1, 137), and these can have significant effects on the chemistry of the atmosphere regionally and globally (38, 81).

Field Studies

A number of examples of effects of alien grasses on fire regimes have been described. These include:

HAWAI'I Several alien grasses have been implicated in increasing fire frequency and/or intensity in Hawai'i (148, 149). The best documented case involves invasion of the C_4 perennial grasses *Schizachyrium condensatum* and *Melinis minutiflora* in areas of seasonal submontane woodland in Hawaii

Volcanoes National Park (74, 149). Invasion by *S. condensatum* took place in the late 1960s, and prior to that time grass cover in these woodlands was sparse. Areas that have never burned now support 80% cover of alien grasses, mainly *Schizachyrium*. The grass thoroughly fills the interstices of the native shrubs and grows into their canopies, providing continuous layers of fine fuel. Prior to the invasion, 27 fires burned an average of 4 ha/fire in 48 years, but in the 20 years following invasion, 58 fires have burned an average of 205 ha/fire (149).

In Hawaii, a single grass-fueled fire can kill most native trees and shrubs. *S. condensatum* recovers rapidly, however, and the alien grass *Melinis minutiflora* also invades the burned area (74). Grass cover and fuel loading in the resultant community are greater than before fire. *Melinis* is highly resinous and more flammable than *Schizachyrium;* its green leaves burn vigorously and can burn in 95% relative humidity (111). Subsequent fires are more likely, and when they occur they cause further increases in *Melinis* cover. Overall, grass invasion sets in motion a positive feedback cycle that leads from a nonflammable, mostly native-dominated woodland to a highly flammable, low-diversity, alien-dominated grassland. Most of the dominant native species and at least one candidate endangered plant species are eliminated by this fire cycle (74).

Similar processes have been observed elsewhere in Hawai'i. In dry lowland areas and other seasonal submontane sites, the alien grasses *Andropogon virginicus, Hyparrhenia rufa, Pennisetum setaceum,* and *Cenchrus ciliaris* are abundant, enhance fire, and grow rapidly in response to it. In subalpine areas, the C_3 alien grasses *Holcus lanatus* and *Anthoxanthum odoratum* both add fuel and respond more rapidly to fire than do native species (149).

WESTERN NORTH AMERICA The best documented example of ecosystem effects of alien grasses in North America is a *Bromus tectorum* invasion into the intermountain west. Historical evidence suggests that much of this region was dominated by perennial grasses when Europeans arrived and that fires were not common (82, 94, 95, 181). *Bromus tectorum* invaded the region as the habitat was degraded by livestock (14, 94, 180, 187). It is a highly flammable winter annual that dies and dries out in the spring, and spreads fires rapidly. Grass-fueled summer fires can sweep through *B. tectorum*-colonized shrubland, killing or damaging shrubs and perennial grasses (82, 151). *B. tectorum* recovers rapidly following fire and can suppress the growth of native species (101).

On a regional level, *B. tectorum* increases both the size and number of fires. Whisenant (181) estimates that the fire return interval in Idaho shrublands before *B. tectorum* invasion was 60–110 years. Since invasion, sites burn every 3–5 years. Earlier, sites in eastern Oregon dominated by *B. tectorum* were considered 500 times more likely to burn than those under other cover (121, 151). The net effect of *B. tectorum* invasion is thus a positive feedback from initial colonization in the interstices of shrubs, followed by fire, to dominance by *B. tectorum* and more frequent fire (151, 181).

More recently, *B. tectorum* has been spreading to higher elevation sites, with a consequent increase in grass-fueled fires and a reduction in area of some pinyon-juniper woodlands (14). In other sites, the increase in fires has led to increased flooding and erosion (82). Overall, invasion by *B. tectorum* and the attendant fires affect at least 40 million hectares, making this perhaps the most significant plant invasion in North America (181). Presently, fire resistant plants are being seeded in portions of the Great Basin in an attempt to interrupt this fire cycle (118).

Several other alien grasses appear to alter fire regimes in North America. *Tainiatherum asperum* is colonizing portions of *B. tectorum*'s range (186); it increases under frequent fire at the expense of *B. tectorum* and also appears to be fire-enhancing (102, 188). Zedler et al (189) demonstrated that seeding burned California chaparral with the alien annual *Lolium perenne* as an erosion control measure fueled a second fire in an area that had burned less than 1 year previously. Nadkarni & Odion (110) observed a similar result and found that only the seeded portion of the original tract of burned shrubland was consumed in a second blaze two years after the original fire. Normal fire intervals in chaparral are considerably longer (19, 67). Schmid & Rogers (139) report increased fire frequencies in the Sonoran desert over the last 30 years and attribute at least some of this increase to the accumulation of fuels by exotic grasses.

TROPICAL AMERICA Central and particularly South America supported extensive savannas when Europeans arrived, but the dominant native grasses were unable to support intensive ungulate grazing. A number of C_4 African grasses, most importantly, *Hyparrhenia rufa, Melinis minutiflora, Panicum maximum,* and *Brachiaria* spp., were brought in to support grazing in savanna regions and in cleared forests (115). All four of these grasses burn readily and resprout rapidly following fire; the consequent grass-fire interaction is thereby capable of maintaining cleared forest land as a derived savanna or grassland (15, 99), preventing succession back to forest. In addition, *H. rufa* and *M. minutiflora* are able to invade otherwise intact native-dominated savanna ecosystems (11).

In Central America, *Hyparrhenia* has received the most attention from ecologists, since when it is not heavily grazed it forms tall, dense stands that burn readily and intensely (77, 85, 122). In contrast, fires in comparable sites dominated by native grasses are patchy and less intense (85). *H. rufa*-fueled fires can burn into successional and even intact tropical dry forest and represent a serious threat to preservation of this ecosystem in Guanacaste National Park in Costa Rica and elsewhere (78). Bilbao & Medina (13) and Coutinho (35) reported, surprisingly, that long-term fire suppression leads to colonization by *H. rufa* and *M. minutiflora* in some Brazilian and Venezuelan savannas. The difference is due to the very low soil fertility of the natural savannas; these alien grasses require more nutrients than native *Trachypogon* species (11), and fire suppression over time leads to increased nutrient availability (99). Cleared forest sites support much greater nutrient availability (13).

Nepstad et al (113) describe more complex interactions between alien grasses and fire near Paragominas, Brazil. Pastures there are cleared from forest, burned, and planted to the alien grass *Panicum maximum*. As the initial pulse of soil fertility declines, other species including *H. rufa* colonize the pastures. Grass/fire interactions can lead to arrested succession at this stage, or succession can proceed to a depauperate forest.

AUSTRALIA Several alien grasses have been identified as participating in a positive feedback with fire in Australia. The most significant of these is *Cenchrus ciliaris,* which produces 2–3 times as much flammable material as native grasses in central Australian watercourses (86). Colonization by *C. ciliaris* alters these watercourses from their historical role as a barrier to fire to one of being a "wick" for fire (76). Fire frequency in the region has increased as a consequence (86).

In northern Australia, the African perennial grass *Pennisetum polystachyon* maintains more litter and carries more intense fires than the native annual *Sorghum intrans,* which it is replacing in portions of Kakadu National Park (59, 152). There is concern that this invasion will allow fires to penetrate into and thereby alter areas of hitherto nonflammable monsoon vine forests (92).

The now familiar *Melinis minutiflora* also invades disturbed areas within the moist tropical zone near Cairns, Australia. In this region, cyclones frequently destroy intact native forest, thereby producing enough fuel to carry fires. Native grasses colonize the burned sites but now are being replaced by *Melinis*. The flammability and large litter mass of *Melinis* supports a grass/fire cycle that arrests succession at the grass stage (152). In southwestern Australia, *Ehrharta calycina* invasion initiates a grass-fire positive feedback cycle (6, 29, 30).

OTHER AREAS We did not locate any clear examples of alien grass/fire cycles in Eurasia or Africa, with the exception of the Mediterranean climate region of South Africa (27). These continents have supported human populations and anthropogenic fires for much longer than others, and they contain extensive, frequently burned areas dominated by native grasses. Many of these areas result from human clearing of forested areas (i.e. the extensive, economically destructive *Imperata* grasslands of Asia), and they are maintained by a positive feedback between grasses and fire.

The long history of human presence in Africa and Asia probably has had substantial effects on the grasses themselves. African grasses in particular have evolved with hominids for millions of years. They have also evolved with intense ungulate grazing and show adaptations to grazing that also confer adaptation to fire. These include perennating organs near or below the ground and a rapid growth response to defoliation (31, 72). It is therefore not surprising that Africa and to a lesser extent Asia are donors rather than recipients of fire-adapted alien grasses. Interestingly, the New World vine *Chromolaena odorata* was introduced to Asia and Africa specifically to interrupt the native grass/fire cycle and has become an aggressive invader on both continents (44, 50, 136).

LAND USE CHANGE, INVASION, AND FIRE: ARE THERE GLOBAL CONSEQUENCES?

Globally, the effects of alien grasses on fire and ecosystems must be viewed within the context of human-caused land use change (103), which has increased the importance of both fire and grasses world-wide (109). Humans often clear wooded lands in order to create grassland for domestic animals; fire is used as a tool in land clearing and the maintenance of cleared areas. Even where human land use merely involves selective logging, it increases both the amount and the flammability of combustible material on sites (62, 160), thereby greatly increasing the probability of fire and further land conversion (Figure 1). By increasing fire and grasses separately, both are increased synergistically through the grass-fire positive feedback cycle.

Land use change in the Americas, Australia, and Oceania is increasing fire and grazing to historically unprecedented levels and thereby selecting for grasses (many of them African and Eurasian) that are able to tolerate fire and grazing. In some cases (i.e. Hawai'i), grass invasion alone is sufficient to initiate grass/fire cycles that convert woodland to grassland (Figure 1). In a greater number of cases, the aliens increase the amount of biomass consumed during fire and/or the length of time flammable grassland persists before woody plant encroachment. The availability of alien grasses can even contribute significantly to human decisions to initiate land use change. For example, intensive cattle grazing could not be supported without African grasses in many areas of tropical America (115, 116).

Overall, there is abundant evidence that grass invasions and the grass-fire cycle arrest or alter succession in many regions, leading to substantial changes

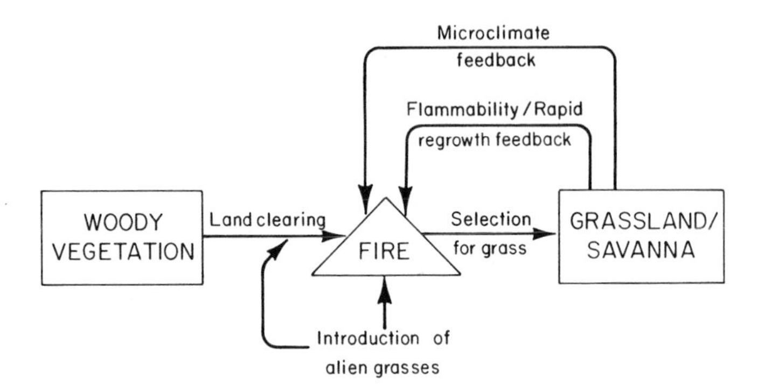

Figure 1 Conceptual diagram of land clearing and the grass-fire cycle (modified from Fosberg et al {54}), to illustrate the influence of alien grass invasion. In some cases grass invasion itself is sufficient to initiate grass-fire positive feedbacks: more often, it interacts with human-caused land use change.

in regional population, species, and landscape diversity in the Americas, Australia, and Oceania. These changes represent a significant challenge to conservation biology. How important are alien grass invasions as agents of regional and global functional change? Several of the studies reviewed here demonstrate that their effects on local ecosystem function can be profound. Is the overall effect of many local changes in ecosystem function sufficient to affect climate, water quality, or the composition of the atmosphere regionally or globally? We cannot answer that question unequivocally, but there are several plausible ways that grass invasions could have or could come to have regional or global consequences.

Two possible regional effects are alterations to climate caused by conversion of forest to grassland, and alterations to the oxidant chemistry of the atmosphere caused by fires. Using a simulation model of regional climate, Lean & Warrilow (87) and Shukla et al (143) suggested that complete conversion of the Amazon Basin forest to grassland would alter regional climate by increasing temperature and decreasing humidity and precipitation. These changes would also increase fire frequency and favor the persistence of grasses with the C_4 photosynthetic pathway. The simulations they created represent an extreme case, but partial conversion would still have significant climatic effects. To the extent that African grasses are essential to the initiation or maintenance of Amazonian forest clearing, grass invasion would contribute to any regional change. Similar regional climatic changes would not necessarily be expected elsewhere in the Americas, because much more of the precipitation is recycled through the biota within the Amazon basin than elsewhere (134).

Recent measurements have demonstrated that biomass burning has significant regional effects on the chemistry of the atmosphere and of precipitation during the dry season in west Africa and South America (3, 4, 55). Ozone levels in particular approach those in eastern North America and northern Europe, where significant effects on regional vegetation and agricultural yields have been demonstrated (128). Grass invasions contribute to regional atmospheric change in tropical America to the extent that burning reflects grasslands dominated by African grasses or forests being cleared to plant African grasses.

On a global level, grass invasions could contribute to functional change if alien grass-fueled fires added significantly to the increasing concentrations of radiatively active (greenhouse) or stratospheric ozone destroying gases in the atmosphere. Andreae (2) estimated that biomass fires globally contribute 6–8% of the total sources of the relatively stable gases methane, nitrous oxide, and carbonyl sulfide. The proportional contribution of the human-caused increase in fires to the human-caused increase in gas concentrations may be somewhat greater, but it probably does not exceed 15% for any of these gases. (We do not discuss CO_2 here because grass regrowth following fire is so rapid that grassland or savanna fires, unlike land clearing fires, do not represent a significant net source of CO_2).

Andreae (2) estimates that tropical forests, savannas, and grasslands now account for about 40% of the biomass burned globally; about half of that is now in Africa and so does not involve alien grasses. If half of the remainder involves alien grasses (probably an overestimate), then their overall contribution to increased gas concentrations could be .15 (maximum proportion of the increase in concentrations of stable gases that result from fire) $\times$.4 (proportion of global biomass combustion that occurs in tropical forests, savannas, and pastures) $\times$.25 (proportion of tropical biomass combustion that involves aliens) or approximately 1.5% of the total increase. It therefore appears likely that while human-altered grass/fire cycles may be a significant source of gases that drive climate change, the contribution of alien grasses to the total is relatively small.

CONCLUSIONS

The effects of alien grasses on ecosystem function (fire, nutrient loss, altered local microclimate, prevention of succession) are significant on the local scale and are becoming increasingly important on regional and global scales. Moreover, the interaction of competition with alien grasses, fire, and the prevention of succession now represents a substantial global threat to biological diversity on the genetic, population, and species levels. However, the number of cases in which ecosystem effects of grass invasions have been intensively studied (as opposed to described or speculated about) is small. A thorough understanding of additional cases, leading to a better overall understanding of the process, would be useful to the development of basic ecological principles as well as to the management of these invasions.

ACKNOWLEDGMENTS

We thank Z. Baruch, L. Henderson, S. Humphries, I. A. W. MacDonald, and S. Whisenant for information and references. R. Freifelder, D. Hooper, A. Townsend made comments on the manuscript, and T. Dudley assisted in tracking down references. Manuscript preparation was supported by a Pew Foundation Fellowship and NSF grant BSR-8918382 to Stanford University.

Literature Cited

1. Anderson, I. C., Levine, J. S., Poth, M. A., Riggan, P. J. 1988. Enhanced biogenic emissions of nitric oxide and nitrous oxide following surface biomass burning. *J. Geophys. Res.* 93:3893–98
2. Andreae, M. O. 1992. The influence of tropical biomass burning on climate and the atmospheric environment. In Proc. 10th Int. Symp. Environ. Biogeochem. In press
3. Andreae, M. O., Browell, E. V., Garstang, M., Gregory, G. L., Hariss, R. C., et al. 1988. Biomass burning emissions and associated haze layers over Amazonia. *J. Geophys. Res.* 93:1509–27
4. Andreae, M. O., Talbot, R. W., Berresheim, H., Beecher, K. M. 1990. Precipitation chemistry in central Amazonia. *J. Geophys. Res.* 95:987–99
5. Avon, W. I., Smith, S. H. 1971. Ship canals and aquatic ecosystems. *Science* 174:13–20
6. Baird, A. 1977. Regeneration after fire in Kings Park, Perth, Western Australia. *J. R. Soc. West. Aust.* 60:1–22
7. Barbour, M. G., Major, J., eds. 1977. *Terrestrial Vegetation of California*. New York: Wiley
8. Barbour, M. G., Johnson, A. F. 1977. Beach and dune. See Ref. 7, pp. 223–61
9. Barden, L. S. 1987. Invasion of *Microstegium vimineum* (Poaceae), an exotic, annual, shade-tolerant, C4 grass, into a North Carolina floodplain. *Am. Midl. Nat.* 118(1):40–45
10. Bartolome, J. W., Klukkert, S. E., Barry, W. J. 1986. Opal phytoliths as evidence for displacement of native California grassland. *Madrono* 33:217–22
11. Baruch, Z., Ludlow, M. M., Davis, R. 1985. Photosynthetic responses of native and introduced C4 grasses from Venezuelan savannas. *Oecologia* 67: 388–93
12. Baruch, Z., Hernandez, A. B., Montilla, M. G. 1989. Growth dynamics, phenology and biomass allocation in native and introduced grasses in a neotropical savanna. *Ecotropicos* 2(1):1–13
13. Bilbao, B., Medina, E. 1990. Nitrogen-use efficiency for growth in a cultivated African grass and a native South American pasture grass. *J. Biogeogr.* 17:421–25
14. Billings, W. D. 1990. *Bromus tectorum*, a biotic cause of ecosystem impoverishment in the Great Basin. In *The Earth in Transition: Patterns and Processes of Biotic Impoverishment*, ed. G. M. Woodell, pp. 301–22. Cambridge: Cambridge Univ. Press
15. Blydenstein, J. 1967. Tropical savanna vegetation of the llanos of Colombia. *Ecology* 48:1–15
16. Bock, C. E., Bock, J. H., Jepson, K. L., Ortega, J. C. 1986. Ecological effects of planting African lovegrasses in Arizona. *Nat. Geogr. Res.* 2(4):456–63
17. Bor, N. L. 1960. *The Grasses of Burma, Ceylon, India and Pakistan excluding Bambuseae*. Oxford: Pergamon
18. Boughey, A. S., Munro, P. E., Meiklejohn, J., Strang, R. M., Swift, M. J. 1964. Antibiotic relations between African savanna species. *Nature* 203: 1302–3
19. Bradbury, D. E. 1978. The evolution and persistence of a local sage/chamise community pattern in Southern California. *Yearbk. Assoc. Pacific Coast Geographers* 40:39–56
20. Bridgewater, P. B., Backshall, D. J. 1981. Dynamics of some Western Australian ligneous formations with special reference to the invasion of exotic species. *Vegetatio* 46:141–48
21. Bush, J. K., Van Auken, O. W. 1987. Light requirements for the growth of

Prosopis glandulosa seedlings. *South. Nat.* 32:469–73
22. Bush, J. K., Van Auken, O. W. 1989. Soil resource levels and competition between a woody and herbaceous species. *Torrey Bot. Club Bull.* 116:22–30
23. Cable, D. R. 1971. Lehmann Lovegrass on the Santa Rita experimental range, 1937–1968. *J. Range Manage.* 24:17–21
24. Caldwell, M. M., Richards, J. H., Johnson, D. A., Nowak, R. S., Dzurec, R. S. 1981. Coping with herbivory: photosynthetic capacity and resource allocation in two semiarid Agropyron bunchgrasses. *Oecologia* 50:14–24
25. Caldwell, M. M., Eissenstat, D. M., Richards, J. H., Allen, M. F. 1985. Competition for phosphorus: differential uptake from dual isotope-labeled soil interspaces between shrub and grass. *Science* 229:384–86
26. Caldwell, M. M., Richards, J. H., Manwaring, J. H., Eissenstat, D. M. 1987. Rapid shifts in phosphate acquisition show direct competition between neighboring plants. *Nature* 327:615–16
27. Campbell, B., Gubb, A., Moll, E. 1980. The vegetation of the Edith Stephens Cape Flats Flora Reserve. *J. So. Afr. Bot.* 46(4):435–44
28. Carpenter, S. R., Kitchell, J. F., Hodgson, J. R., Cochran, P. A., Elser, J. J., et al. 1987. Regulation of lake primary productivity by food web structure. *Ecology* 68:1863–76
29. Christenson, P., Abbott, I. 1989. Impact of fire in the eucalypt forest ecosystems of southern West Australia: A critical review. *Aust. For.* 52:103–21
30. Christenson, P. E., Burrows, N. D. 1986. Fire: an old tool with new uses. In *Ecology of Biological Invasions: An Australian Perspective,* ed. R.H. Groves, J. J. Burdon, pp. 97–105, Canberra: Aust. Acad. Sci.
31. Christie, E. K., Moorby, J. 1975. Physiological responses of semiarid grasses I. The influence of phosphorus supply on growth and phosphorus absorption. *Aust. J. Agr. Res.* 26:423–36
32. Coblentz, B. E. 1990. Exotic organisms: a dilemma for conservation biology. *Conserv. Biol.* 4:261–65
33. Cohn, E. J., Van Auken, O. W., Bush, J. K. 1989. Competitive interactions between *Cynodon dactylon* and *Acacia smallii* seedlings at different nutrient levels. *Am. Midl. Nat.* 121:265–72
34. Cooper, W. S. 1967. Coastal dunes of California. *Geol. Survey Am. Memoir 104*. Denver, Colo.
35. Coutinho, L. M. 1982. Aspectos ecologicos da sauva no cerrado: Os murundus de terra, as caracteristicas psamofiticas das especies de sua vegetacao e a sua invasao pelo capim gordura. *Revista Brasileira de Biol.* 42:147–53
36. Coutinho, L. M. 1990. Fire in the ecology of the Brazilian cerrado. See Ref. 61, pp. 82–105
37. Cox, J. R., Martin, M. H., Ibarra, F. A., Fourie, J. H., Rathman, N. F., et al. 1988. The influence of climate and soils on the distribution of four African grasses. *J. Range Manage.* 41(2):127–38
38. Crutzen, P. J., Andreae, M. O. 1990. Biomass burning in the tropics: impact on atmospheric chemistry and biogeochemical cycles. *Science* 250:1669–78
39. D'Angela, E. R., Leon, J. C., Facelli, J. M. 1986. Pioneer stages in a secondary succession of a Pampean subhumid grassland. *Flora* 178:261–70
40. Danielson, K. C., Halvorson, W. L. 1990. Valley oak seedling growth associated with selected grass species. In *Proceedings of the Symposium on Oak Woodlands and Hardwood Rangeland Management,* pp. 9–13. *Gen Tech. Report PSW-12*
41. D'Antonio, C. M. 1990. *Invasion of coastal plant communities by the introduced succulent,* Carpobrotus edulis *(Aizoaceae).* PhD thesis, Univ. California, Santa Barbara, Calif.
42. Da Silva, P. G., Bartolome, J. W. 1984. Interaction between a shrub, *Baccharis pilularis* subsp. *consanguinea* (Asteraceae), and an annual grass, *Bromus mollis* (Poaceae), in coastal California. *Madrono* 31(2):93–101
43. Davis, S. D., Mooney, H. A. 1985. Comparative water relations of adjacent California shrub and grassland communities. *Oecolgia* 66:522–29
44. de Rouw, A. 1991. The invasion of *Chromolaena odorata* (L.) King & Robinson (ex. *Eupatorium odoratum),* and competition with the native flora in a rainforest zone, southwest Cote d'Ivoire. *J. Biogeogr.* 18:13–23
45. Dolan, R., Godfrey, P. J., Odum, W. E. 1973. Man's impact on the barrier islands of North Carolina. *Am. Sci.* 61:152–62
46. Dudley, T. D., Grimm, N., Fisher, S. 1992. Modification of macrophyte resistance to disturbance by an exotic grass, and implications for desert stream community structure. *Verhandlungen Vereinung Fur Limnologie* In press
47. Eissenstat, D. M. 1986. *Belowground*

resource exploitation in semi-arid plants. A comparative study using two tussock grass species that differ in competitive ability. PhD thesis. Utah State Univ., Logan, Utah
48. Eissenstat, D. M., Caldwell, M. M. 1988. Competitive ability is linked to rates of water extraction. *Oecologia* 75:1–7
49. Elliott, K. J., White, A. S. 1989. Competitive effects of various grasses and forbs on ponderosa pine seedlings. *For. Sci.* 33(2):356–66
50. Eussen, J. H., deGroot, W. 1974. Control of *Imperata cylindrica* (L.) beauv. in Indonesia. *Mededelingen Fakulteit Landbouw-wetenschappen Gent* 39:451–64
51. Evans, R. A., Young, J. 1970. Plant litter and establishment of alien annual weed species in rangeland communities. *Weed Sci.* 18:697–703
52. Facelli, J. M., Pickett, S. T. A. 1991. Plant litter: its dynamics and effects on plant community structure. *Bot. Rev.* 57:1–32
53. Facelli, J. M., Montero, C. M., Leon, J. C. 1988. Effect of different disturbance regimen on seminatural grasslands from the subhumid Pampa. *Flora* 180:241–49
54. Farinas, M. R., San Jose, J. J. 1985. Cambios en el estrato herbaceo de una parcela de sabana protegida del fuego y del pastoreo durante 20 anos. *Acta Cientifica Venezolana* 36:199–200
55. Fishman, J., Fakhruzzaman, K., Cros, B., Nganga, D. 1991. Identification of widespread pollution in the southern hemisphere deduced from satellite analyses. *Science* 252:1693–96
56. Fosberg, M. A., Goldammer, J. G., Rind, D., Price, C. 1990. Global change: effects of forest ecosystems and wild fire severity. See Ref. 61, pp. 463–85
57. Gadgil, R. L., Knowles, A. L., Zabkiewisz, J. A. 1984. Pampas grass-A new forest weed problem. In *Proc. 37th New Zeal. Weed and Pest Control Conf.*, pp. 187–90
58. Gill, A. M., Groves, R. H., Clark, R., eds. 1981. *Fire and the Australian Biota*. Canberra: Aust. Acad. Sci.
59. Gill, A. M., Hoare, J. R. L., Cheney, N. P. 1990. Fires and their effects in the wet-dry tropics of Australia. See Ref. 61, pp. 159–78
60. Godfrey, P. J., Godfrey, M. M. 1974. An ecological approach to dune management in the National Recreation Areas of the United States East Coast. *Int. J. Biometer.* 18:101–10
61. Goldammer, J. G., ed. 1990. *Fire in the Tropical Biota: Ecosystem Processes and Global Challenges*. Berlin: Springer-Verlag
62. Goldammer, J. G., Siebert, B. 1990. Impact of droughts and forest fires on tropical lowland rainforest of east Kalimintan. See Ref. 61, pp. 11–31
63. Goldberg, D., Werner, P. A. 1983. The effects of size of opening in vegetation and litter cover on seedling establishment of goldenrods (Solidago spp.). *Oecologia* 60:149–55
64. Gordon, D., Welker, J. M., Menke, J. W., Rice, K. J. 1989. Competition for soil water between annual plants and blue oak (*Quercus douglasii*) seedlings. *Oecologia* 79:533–41
65. Griffin, G. F., Stafford-Smith, D. M., Morton, S. R., Allan, G. E., Masters, K. A. et al. 1989. Status and implications of the invasion of tamarisk (*Tamarix aphylla*) on the Finke River, Northern Territory, Australia. *J. Environ. Manage.* 29:297–315
66. Gross, K., Werner, P. A. 1982. Colonizing ability of biennial plant species in relation to ground cover: Implications for their distributions in a successional sere. *Ecology* 63:921–31
67. Hanes, T. 1977. Chaparral. See Ref. 7, pp. 417–70
68. Heady, H. F. 1977. Valley grassland. See Ref. 7, pp. 491–514
69. Heyligers, P. C. 1985. Impact of introduced grasses on foredunes in southeastern Australia. *Proc. Ecol. Soc. Aust.* 14:23–41
70. Hobbs, R. J., Atkins, L. 1988. Effect of disturbance and nutrient addition on native and introduced annuals in plant communities in the Western Australian wheatbelt. *Aust. J. Ecol.* 13:171–79
71. Hobbs, R. J., Atkins, L. 1990. Fire-related dynamics of a Banksia woodland in south-western Western Australia. *Aust. J. Bot.* 38:97–110
72. Hodkinson, K. C., Ludlow, M. M., Mott, J. J., Baruch, Z. 1989. Comparative responses of the savanna grasses *Cenchrus ciliaris* and *Themeda triandra* to defoliation. *Oecologia* 79: 45–52
73. Hughes, R. F. 1991. *Obstacles to native shrub colonization after fire in the seasonal submontane zone of Hawai'i*. MA thesis, Stanford Univ.
74. Hughes, R. F., Vitousek, P. M., Tunison, T. 1991. Alien grass invasion and fire

in the seasonal submontane zone of Hawai'i. *Ecology* 72:743–46

75. Hull, A. C., Klomp, G. J. 1967. Thickening and spread of crested wheatgrass stands on southern Idaho ranges. *J. Range Manage.* 20:222–27
76. Humphries, S. E., Groves, R. H., Mitchell, D. S. 1992. Plant invasions of Australian ecosystems: A status review and management directions. In *Plant Invasions: The Incidence of Environmental Weeds in Australia. Kowari 2,* ed. R. Longmore, pp. 1–127. Canberra: Austr. Natl. Parks Wildlife Service
77. Janzen, D. 1988. Tropical dry forests: the most endangered major tropical ecosystem. In *Biodiversity,* ed. E.O. Wilson, pp. 130–37. Washington, DC: Natl. Acad. Press
78. Janzen, D. 1988. Management of habitat fragments in a tropical dry forest. *Ann. Miss. Bot. Gar.* 75:105–16
79. Johnson, P. N. 1982. Naturalized plants in southwest South Island, New Zealand. *New Zeal. J. Bot.* 20:131–42
80. Kauffman, J. B., Uhl, C. 1990. Interactions of anthropogenic activities, fire and rain forests in the Amazon basin. See Ref. 61, pp. 117–34
81. Keller, M., Jacob, D. J., Wofsy, S. C., Harriss, R. C. 1991. Effects of tropical deforestation on global and regional atmospheric chemistry. *Clim. Change* 19:145–58
82. Klemmendson, J. O., Smith, J. G. 1964. Cheatgrass (*Bromus tectorum* L.). *Bot. Rev.* 30:226–61
83. Kloot, P. M. 1983. The role of common iceplant (*Mesembryanthemum crystallinum)* in the deterioration of medic pastures. *Aust. J. Ecol.* 8:301–6
84. Knoop, W. T., Walker, B. H. 1985. Interactions of woody and herbaceous vegetation in a southern African savanna. *J. Ecol.* 73:235–53
85. Koonce, A. L., Gonzales-Caban, A. 1990. Social and ecological aspects of fire in central America. See Ref. 61, pp. 135–58
86. Latz, P. K. 1991. Buffel and couch grass in Central Australian creeks and rivers. *Newslett. Cent. Aust. Conserv. Council, Inc.* April 1991:5
87. Lean, J., Warrilow, D. A. 1989. Simulation of the regional climatic impact of Amazon deforestation. *Nature* 342: 411–13
88. Litav, M., Kupernik, G., Orshan, G. 1963. The role of competition as a factor in determining the distribution of dwarf shrub communities in the Mediterranean territory of Israel. *J. Ecol.* 51:467–80
89. Loope, L. L., Hamman, O., Stone, C. P. 1988a. Comparative conservation biology of oceanic archipelagoes: Hawaii and the Galapagos. *Bioscience* 38:272–82
90. Loope, L. L., Sanchez, P. G., Tarr, P. W., Loope, W. L. Anderson, R. L. 1988. Biological invasions of arid land nature reserves. *Biol. Conserv.* 44:95–118
91. MacDonald, I. A. W. 1987. Invasive alien plants and their control in Southern African nature reserves. In *Management of Exotic Species in Natural Communities,* Vol. 5, ed. L. K. Thomas, pp. 63–79. Fort Collins: Colo. State Univ.
92. MacDonald, I. A. W., Frame, G. W. 1988. The invasion of introduced species into nature reserves in tropical savannas and dry woodlands. *Biol. Conserv.* 44:67–93
93. MacDonald, I. A. W., Thebaud, C., Strahm, W. A., Strasberg, D. 1991. Effects of alien plant invasions on native vegetation remnants on La Reunion (Mascarene Islands, Indian Ocean). *Enviro. Conserv.* 18:51–61
94. Mack, R. N. 1981. Invasion of *Bromus tectorum* L. into western North America: an ecological chronicle. *Agroecosystems* 7:145–65
95. Mack, R. N. 1986. Alien plant invasion into the intermountain west: A case history. See Ref. 106, pp. 191–213
96. Matson, P. A. 1990. Plant-soil interactions during primary succession in Hawaii Volcanoes National Park. *Oecologia* 65:241–46
97. McClintock, E. 1985. Escaped exotic weeds in California. *Fremontia* 12(4): 3–6
98. McClaran, M. P., Anable, M. E. 1992. Spread of introduced Lehman lovegrass along a grazing intensity gradient. *J. Appl. Ecol.* In press
99. Medina, E. 1987. Nutrients. Requirements, conservation, and cycles of nutrients in the herbaceous layer. In *Determinants of Tropical Savannas,* ed. B. H. Walker, pp. 39–65, Paris: IUBS Monograph 3
100. Medina, A. L. 1988. Diets of scaled quail in southern Arizona. *J. Wildl. Manage.* 52:753–57
101. Melgoza, G., Nowak, R. S., Tausch, R. J. 1990. Soil water exploitation after fire: competition between *Bromus*

tectorum (cheatgrass) and two native species. *Oecologia* 83:7–13

102. Menke, J. W. 1989. Management controls on productivity. In *Grassland Structure and Function: California Annual Grassland,* ed. L. F. Huenneke, H. A. Mooney, pp. 173–99. Dordrecht: Kluwer
103. Meyer, W. B., Turner, B. L. 1992. Human population growth and global land-use cover change. *Annu. Rev. Ecol. Syst.* 23: In press
104. Mitchell, D. S. 1978. *Aquatic Weeds in Australian Inland Waters.* Canberra: Aust. Govt. Publ. Serv.
105. Mooney, H. A., Bonnicksen, T. M., Christensen, N. L., Lotan, J. E., Reiners, W. A., eds. 1981. Fire regimes and ecosystem properties. *USDA For. Ser. Gen. Tech. Rep. WO-28* Washington, DC
106. Mooney, H. A., Drake, J. A. 1986. *Ecology of Biological Invasions of North America and Hawaii.* New York: Springer-Verlag
107. Mueller-Dombois, D. 1973. A non-adapted vegetation interferes with water removal in a tropical rain forest area in Hawaii. Trop. Ecol. 14:1-16
108. Mueller-Dombois, D. 1981. Fire in tropical ecosystems. See Ref. 105, pp. 137–76
109. Mueller-Dombois, D., Goldammer, J.G. 1990. Fire in tropical ecosystems and global environmental change: an introduction. See Ref. 61, pp. 1–10
110. Nadkarni, N., Odion, D. 1986. Effects of seeding exotic grass, *Lolium multiflorum* on native seedling regeneration following fire in a chaparral community. In *Proceedings of the Chaparral Ecosystems Res. Conf., Calif. Water Resources Cent. Rep. 62,* ed. J. J. DeVries, pp. 115–22. Davis, Calif.
111. National Park Service. 1990. *Fire Management Plan: Hawaii Volcanoes National Park.* Washington, DC: US Dept. Interior. 54p
112. Neill, W.M. 1983. The tamarisk invasion of desert riparian areas. *Educ. Bull. 83–4*. Spring Valley, Calif: Desert Protective Council
113. Nepstad, D.C., Uhl, C., Serrao, E. A. S. 1991. Recuperation of a degraded Amazonian landscape: forest recovery and agricultural restoration. *Ambio* 20:248-55
114. Paine, R.T. 1966. Food web complexity and species diversity. *Am. Nat.* 100:65–75
115. Parsons, J. 1972. Spread of African pasture grasses to the American Tropics. *J. Range Manage.* 25:12–7
116. Parsons, J. 1980. Europeanization of the savanna lands of northern South America. In *The Human Ecology of Savanna Environments,* ed. D. R. Harris, pp. 267–89. New York: Academic
117. Pastor, J., Gardner, R. H., Dale, V. H., Post, W. M. 1987. Successional changes in nitrogen availability as a potential factor contributing to spruce decline in boreal North America. *Can. J. For. Res.* 17:1394–1400
118. Pellant, M. 1990. The Cheatgrass-wildfire cycle—are there any solutions? In *Proceedings from the Symposium on Cheatgrass Invasion, Shrub Dieoff and Other Aspects of Shrub Biology and Management.* pp. 11–18. *USFS Gen. Tech. Rep. INT-276*
119. Phillips, W. S. 1963. Depth of roots in soil. *Ecology* 44:424–29
120. Platt, W. J. 1975. The colonization and formation of equilibrium plant species associations on badger disturbances in a tall-grass prairie. *Ecol. Monogr.* 445:285–305
121. Platt, K., Jackman, E. R. 1946. *The Cheatgrass Problem.* In *Ore. State Coll. Extension Serv. Bull. 668* Corvallis, Ore.
122. Pohl, R. W. 1983. *Hyparrhenia rufa* (Jaragua). In *Costa Rican Natural History,* ed. D. H. Janzen, p. 256. Chicago: Univ. Chicago Press
123. Purdie, R. W., Slatyer, R. O. 1976. Vegetation succession after fire in sclerophyll woodland communities in south-eastern Australia. *Aust. J. Ecol.* 1:223–36
124. Pyke, D. A. 1990. Comparative demography of co-occurring introduced and native tussock grasses: persistence and potential expansion. *Oecologia* 82: 537–43
125. Raison, R. J. 1979. Modification of the soil environment by vegetation fires with particular reference to nitrogen transformations: A review. *Plant Soil* 51:73–108
126. Ramakrishnan, P. S. 1991. *Ecology of Biological Invasions in the Tropics.* New Delhi: Int. Sci. Publ.
127. Ramakrishnan, P. S. 1991. Biological invasion in the tropics: an overview. See Ref. 126, pp. 1–19
128. Reich, P. B., Amundson, R. G. 1985. Ambient levels of ozone reduce net photosynthesis in tree and crop species. *Science* 230:566–70

129. Rice, E. L. 1984. *Allelopathy*. New York: Academic
130. Robinson, E. R. 1984. Naturalized species of Cortaderia (Poaceae) in southern Africa. *So. Afr. J. Bot.* 3:343–6
131. Rogler, G. A., Lorenz, R.L. 1983. Crested wheatgrass—early history in the United States. *J. Range Manage.* 36: 91–93
132. Rundel, P. W. 1981. Structural and chemical components of flammability. See Ref. 105, pp. 183–207
133. Russell, S. 1963. *Plant Root Systems*. New York: McGraw-Hill
134. Salati, E., Vose, P. B. 1984. Amazon Basin: A system in equilibrium. *Science* 225:129–38
135. Savidge, J. A. 1987. Extinction of an island forest avifauna by an introduced snake. *Ecology* 68:660–68
136. Saxena, K. G. 1991. Biological invasions in the Indian subcontinent: review of invasions by plants. See Ref. 126, pp. 53–73
137. Schindler, D. W., Newburn, R. W., Beaty, K. G., Prokopowich, J., Ruszczynski, T., et al. 1980. Effects of a windstorm and forest fire on chemical losses from forested watersheds and on the quality of receiving streams. *Can. J. Fish. Aqua. Sci.* 37:328–34
138. Schmalzer, P. A., Hinkle, C. R. 1987. Species biology and potential for controlling four exotic plants (*Ammophila arenaria, Carpobrotus edulis, Cortaderia jubata* and *Gasoul crystallinum)* on Vandenberg Air Force Base, Calif. *NASA Tech. Memorandum 100980*
139. Schmid, M. K., Rogers, G. F. 1988. Trends in fire occurrence in the Arizona upland subdivision of the Sonoran desert, 1955 to 1983. *Southwest. Nat.* 33: 437–44
140. Schofield, E. K. 1989. Effects of introduced plants and animals on island vegetation: examples from the Galapagos Archipelago. *Conserv. Biol.* 3: 227–38
141. Schultz, A. M., Launchbaugh, J. L., Biswell, H. H. 1955. Relationship between grass density and brush seedling survival. *Ecology* 33:226–38
142. Seastedt, T. R., Briggs, J. M., Gibson, D.J. 1991. Controls of nitrogen limitation in tallgrass prairie. *Oecologia* 87: 72–9
143. Shukla, J., Nobre, C., Sellers, P. 1990. Amazonian deforestation and climate change. *Science* 247:1322–4
144. Simoes, M., Baruch, Z. 1992. Responses to simulated herbivory and water stress in two tropical C4 grasses. *Oecologia*. In press
145. Singer, F.J., Swank, W. T., Clebsch, E. E. C. 1984. Effects of wild pig rooting in a deciduous forest. *J. Wildl. Manage.* 48:464–73
146. Singh, J.S., Hanxi, Y., Sajise, P. E. 1985. Structural and functional aspects of Indian and Southeast Asian savanna ecosystems. See Ref. 158, pp. 39–51
147. Slobodchikoff, C. F., Doyen, J. T. 1977. Effects of *Ammophila arenaria* on sand dune arthropod communities. *Ecology* 58:1171–5
148. Smith, C. W. 1985. Impact of alien plants on Hawaii's native biota. In *Hawaii's Terrestrial Ecosystems: Preservation and Management,* ed. C. P. Stone, J. M. Scott, pp. 180–250. Honolulu: Cooperative Natl. Park Resources Study Unit. Univ. Hawaii
149. Smith, C. W., Tunison, T. 1992. Fire and alien plants in Hawai'i: research and management implications for native ecosystems. In *Alien Plant Invasion in Hawaii: Management and Research in Native Ecosystems,* ed. C. P. Stone, C. W. Smith, J. T. Tunison, pp.394–408. Honolulu: Univ. Hawaii Press
150. Soriano, A., Sala, O. 1983. Ecological strategies in a Patagonian arid steppe. *Vegetatio* 56:9–15
151. Stewart, G., Hull, A. C. 1949. Cheatgrass (*Bromus tectorum)* L.—an ecological intruder in southern Idaho. *Ecology* 30:58–74
152. Stocker, G. C., Mott, J. J. 1981. Fire in the tropical forests and woodlands of northern Australia, See Ref. 58, pp. 427–42
153. Tang, Y., Washitani, I., Tsuchiya, T., Iwaki, H. 1988. Fluctuations of photosynthetic photon flux density within a *Miscanthus sinensis* canopy. *Ecol. Res.* 3:253–66
154. Thompson, J. 1991. The biology of an invasive plant. *Bioscience* 41:393–401
155. Thompson, L., Harper, J. L. 1988. The effect of grasses on the quality of transmitted radiation and its influence on the growth of white clover *Trifolium repens*. *Oecologia* 75:343–47
156. Tilman, D. 1987. Secondary succession and the pattern of plant dominance along experimental nitrogen gradients. *Ecol. Monogr.* 57:189–214
157. Tothill, J. 1985. American Savanna ecosystems. See Ref. 158, pp. 52–64

158. Tothill, J., Mott, J. J. eds. 1985. Ecology and management of the world's savannas, Canberra: Commonwealth Agricultural Bureaux
159. Ueckert, D. N., Smith, L. L., Allen, B. L. 1979. Emergence and survival of honey mesquite seedlings on several soils in west Texas. *J. Range Manage.* 32:284-87
160. Uhl, C., Kauffman, J. B. 1990. Deforestation, fire susceptibility and potential tree responses to fire in the eastern Amazon. *Ecology* 71:437–49
161. Van Auken, O. W., Bush, J. K. 1988. Competition between *Schizachyrium scoparium* and *Prosopis glandulosa*. *Am. J. Bot.* 75:782–89
162. Van Auken, O. W., Bush, J. K. 1990. Influence of light levels, soil nutrients, and competition on seedling growth of *Baccharis neglecta* (Asteraceae). *Bull. Torr. Bot. Club* 117:438–44
163. Van Auken, O. W., Bush, J. K. 1990. Importance of grass density and time of planting on *Prosopis glandulosa* seedling growth. *Southwest. Nat.* 35: 411–15
164. van Wilgen, B. W., Richardson, D. M. 1985. The effects of alien shrub invasions on vegetation structure and fire behavior in South African fynbos shrublands: a simulation study. *J. Appl. Ecol.* 22:955–66
165. Vines, R. G. 1981. Physics and chemistry of rural fires. See Ref. 58, pp. 151–176
166. Vitousek, P. M. 1986. Biological invasions and ecosystem properties: Can species make a difference? See Ref. 106, pp. 163–176
167. Vitousek, P. M. 1990. Biological invasions and ecosystem processes: towards an integration of population biology and ecosystem studies. *Oikos* 57:7–13
168. Vitousek, P. M., Walker, L. R., Whiteaker, L. D., Mueller-Dombois, D., Matson, P. A. 1987. Biological invasion by *Myrica faya* alters ecosystem development in Hawaii. *Science* 238:802–4
169. Vitousek, P. M., Howarth, R. W. 1991. Nitrogen limitation on land and in the sea: how can it occur? *Biogeochemistry* 13:87–115
170. Vitousek, P. M., Walker, L. R. 1989. Biological invasion by *Myrica faya* in Hawai'i: Plant demography, nitrogen fixation, ecosystem effects. *Ecol. Monogr.* 59: 247–65
171. Vivrette, N. J., Muller, C. H. 1977. Mechanism of invasion and dominance of coastal grassland by *Mesembryanthemum crystallinum*. *Ecol. Monogr.* 47:301–18
172. Vlok, J. H. J. 1988. Alpha diversity of lowland fynbos herbs at various levels of infestation by alien annuals. *So. Afr. J. Bot.* 54:623–27
173. Vogl, R. J. 1974. Effects of fire on grasslands. In *Fire and Ecosystems*, ed. T. T. Kozlowski, C. E. Ahlgren, pp.139–94. New York: Academic
174. Ward, D. E. 1990. Factors influencing the emissions of gases and particulate matter from biomass burning. See Ref. 61, pp. 418–36
175. Weaver, J. E. 1968. *Prairie Plants and Their Environment*. Lincoln: Univ. Neb. Press
176. Wedin, D. A., Tilman, D. 1990. Species effects on nitrogen cycling: A test with perennial grasses. *Oecologia* 84: 433–41
177. Weidemann, A. M. 1984. The ecology of Pacific Northwest coastal sand dunes: a community profile. *US Fish Wildlife Serv. FWS/OBS-84/04*. 130 pp
178. Weller, S., Skroch, W. A., Monaco, T. J. 1985. Common bermudagrass (*Cynodon dactylon*) interference in newly planted peach (*Prunus persica*) trees. *Weed Sci.* 33:50–56
179. Wells, M. J. 1978. Nasella tussock. In *Plant Invaders*, ed. C. H. Stirton, pp. 140–43. Capetown: Dep. Nat. Envir. Conserv.
180. West, N. E. 1988. Intermountain deserts, shrub steppes, and woodlands. In *North American Terrestrial Vegetation*, ed. M. G. Barbour, W. D. Billings, pp. 209–30, New York: Cambridge Univ. Press
181. Whisenant, S. 1990. Changing fire frequencies on Idaho's Snake River plains: ecological and management implications. From *Proceedings from the Symposium on Cheatgrass Invasion, Shrub Dieoff and Other Aspects of Shrub Biology and Management. USFS Gen. Tech. Rep. INT -276*, pp. 4–10
182. Whisenant, S. 1990. Postfire population dynamics of *Bromus japonicus*. *Am. Midl. Nat.* 123:301–8
183. Wilson, S. D. 1989. The suppression of native prairie by alien species introduced for revegetation. *Land. Urb. Plan.* 17:113–19
184. Wilson, S. D., Belcher, J. W. 1989. Plant and bird communities of native

prairie and introduced Eurasian vegetation in Manitoba, Canada. *Conserv. Biol.* 3:39–44
185. Witkoswski, E. T. F. 1991. Effects of invasive alien acacias on nutrient cycling in the coastal lowlands of the Cape fynbos. *J. Appl. Ecol.* 28:1–15
186. Young, J. A., Evans, R. A. 1971. Invasion of medusahead into the Great Basin. *Weed Sci.* 18:89–97
187. Young, J. A., Evans, R. A. 1973. Downy brome: intruder in the plant succession of big sagebrush communities in the Great Basin. *J. Range Manage.* 26:410–15
188. Young, J. A., Evans, R. A., Robison, J. 1972. Influence of repeated annual burning on a medusahead community. *J. Range Manage.* 25:372–75
189. Zedler, P. H., Gautier, C. R., McMaster, G. S. 1983. Vegetation change in response to extreme events: The effect of a short interval between fires in California chaparral and coastal scrub. *Ecology* 64:809–18

Annu. Rev. Ecol. Syst. 1992. 23:89–118

GLOBAL CHANGE AND CORAL REEF ECOSYSTEMS

S. V. Smith
Department of Oceanography, University of Hawaii, Honolulu, Hawaii 96822

R. W. Buddemeier
Geohydrology Section, Kansas Geological Survey, University of Kansas, Lawrence, Kansas 66047

KEYWORDS: global change, coral reef systems

INTRODUCTION

This paper reviews known or probable responses of coral reef ecosystems to "global change." Our focus is on the coral reef ecosystem, and on its defining organism, the symbiotic stony (scleractinian or milleporid) coral. One of the unique features of coral reefs is that the communities are commonly supported by biogenic geologic structures of their own creation. This creation of substratum by a taxonomically circumscribed group of organisms provides a dimension of interaction between community and environment not found in most other ecosystems. Other groups of organisms are also important to the structure and function of coral reefs; however, in most cases either functional roles are fulfilled by so many different taxa that it is impossible to generalize about responses to environmental variables, or there are very few data on which to base assessments of effects on them.

Individual organisms and ecosystems respond to their local environment without regard for the ultimate cause of environmental conditions. It is often not possible in practice to distinguish between the effects of climate change, of natural environmental variability, or of non-climatic anthropogenic alteration; effects may be interactive, or a single stress may have multiple sources. We use the term "global" in a broad sense, to include global climate change (the so-called "greenhouse effect"), non-climatic but global environmental changes such as the degradation of stratospheric ozone, and changes that can be called global in the sense that they are widespread, sustained, and likely to increase in distribution and intensity—the many effects of population growth and economic development. We emphasize that "global" does not mean

0066-4162/92/1120-0089$02.00

uniform, monotonic, or ubiquitous. We focus primarily on the nature and directions of present or reliably anticipated environmental change—warming and sea level rise rather than the onset of an ice age, increases rather than decreases in ultraviolet light, and so forth. Finally we focus on changes to be expected within the next century, rather than on longer-term or even less predictable effects.

This review shares with most reef-oriented research an intrinsic bias toward negative effects and stresses. If one starts with a "healthy" coral reef, there are many ways things can go wrong, but only a limited spectrum of "improvements" can be envisioned. Future reef communities may propagate to and thrive at sites not presently classified as coral reefs, but direct observations or specific predictions of such occurrences are—and seem likely to remain—restricted.

Global Change Scenarios

A brief summary of the status of global climate change predictions provides a context for subsequent discussions. We rely primarily on the work of the Intergovernmental Panel on Climate Change (78), and on MacCracken et al (106), who combine discussions of model-based predictions and paleoclimate records. This is a rapidly developing field. Although the detailed predictions derived from general circulation models are uncertain and subject to revision, there can be little doubt in a qualitative sense that the increased and still-increasing concentrations of radiatively active gases in the atmosphere will result in significant climate change of some sort.

Under the IPCC "Business as Usual" scenario (i.e. no substantial changes in present trends in greenhouse gas emissions), global mean temperatures are predicted to increase during the next century by about 0.3°C per decade (range: 0.2–0.5°). The net increase will amount to about 1° by 2030 and 3° by 2100. Land surfaces will warm faster than oceans, and high northern latitudes will warm more and faster than the global mean, especially in winter.

Present confidence in regional climate change predictions is low. In the oceanic tropics, the area of most interest to this review, the predictive ability of the general circulation models is highly questionable; both between-model agreement and calibration against present conditions are poor. Some models predict tropical sea-surface temperature increases of 1–3°C, but there is widespread debate about possible feedback mechanisms that might either stabilize values in the vicinity of 30–31°C (73, 123), or produce positive temperature feedbacks over the warmest part of the ocean (54). Although paleoclimatic conditions are not generally considered reliable predictors of future climate patterns, it may be relevant that during the Eemian warm period (125,000 BP) most northern hemisphere land areas were significantly warmer than at present, but tropical regions were not detectably warmer (106, 127).

Also under the IPCC Business-as-Usual Scenario, global sea level rise is predicted to average about 6 cm/decade over the next century (range: 3–10 cm/decade); this value compares with recently observed values of 1–2 cm/decade, and with maximum sustained rates of sea level rise during the Holocene transgression in excess of 20 cm/decade (7, 51).

Changes in the frequency and intensity of extreme events are probably more ecologically significant than moderate changes in the mean values of environmental factors. In addition to a probable increase in high-temperature events, two possible changes relevant to local coral reef environments are worthy of note (110). One is a shift in precipitation patterns so that more of the total precipitation falls during heavy storms; the other is a possible change in the frequency, magnitude, or geographic distribution of major tropical storms.

ECOSYSTEM ROLES AND RESPONSES

Ecosystems and their responses to environmental change may play various roles in a climate and global change context. Ecological control of reservoir size or rates of flux for climate-influencing materials such as CO_2 is a potentially important factor. In relation to human society, ecosystem sensitivity—or vulnerability—is an important issue; ecosystem collapse in response to environmental change may result in loss of resources; degradation could, in principle, serve as an early warning of increasing stress. From a scientific standpoint, large-scale environmental changes represent a natural experiment that may permit investigation and understanding of ecosystem structure and function not possible on a laboratory scale or in a stable environment. In this section we discuss these three issues before turning to a more detailed inventory of coral reef responses to individual environmental forcing functions.

Geochemical Reservoirs and Fluxes

Some systems, such as tropical rain forests, represent substantial global reservoirs of carbon. Temporal changes in the mass of such systems may be significant in terms of the atmospheric CO_2 reservoir, and such changes can provide feedbacks that may enhance or retard climate change forcing by atmospheric CO_2 or other biogenic gases. The $CaCO_3$ of coral reefs and other oceanic ecosystems represents another large carbon reservoir. Calcification represents the largest net carbon flux term associated with coral reefs (29, 94, 155). We therefore use reef calcification in order to assess reef significance to the global carbon cycle.

There is a widespread conception that coral reefs and other aquatic ecosystems dominated by calcifying organisms are sinks for atmospheric CO_2

(22, 95, 148). On time scales of human interest, $CaCO_3$ precipitation by aquatic ecosystems approximately in equilibrium with atmospheric gases is actually a source of CO_2 release to the atmosphere, rather than being an atmospheric CO_2 sink (155). This counter-intuitive result arises because $CaCO_3$ precipitation lowers pH and converts bicarbonate to CO_2 gas. If the water comes to gaseous equilibrium with the atmosphere, then CO_2 gas must escape from the water to the atmosphere. For coral reefs, the effect of calcification on the overlying atmospheric composition is small (59, 94, 155) and to some extent offset by CO_2 uptake from the atmosphere as a result of net organic carbon production (139, 144, 155). Reefs at steady state are a "greenhouse CO_2 source" of about 4 × 1012 mol/yr, equivalent to about 1% of the CO_2 released to the atmosphere by anthropogenic activities. Although diminution of marine net calcification could increase the uptake of atmospheric CO_2 by the oceans, coral reef contributions to the atmospheric CO_2 flux are well below the uncertainty levels of the major sources and sinks.

Ecosystem Sensitivity

SENSITIVE VERSUS ROBUST ECOSYSTEMS Widespread observations of apparent coral reef degradation (32, 158) have led to an ongoing debate as to whether reefs are "fragile" or "robust" ecosystems (34, 43, 44, 66).

The argument for robustness stems from geological evidence and from certain contemporary theories and observations of reef community development. On evolutionary and geologic time scales, it is argued that physical conditions are primarily responsible for mass extinction events (124), and that the climatic factor most responsible for mass extinction is refrigeration (145). In spite of common perceptions that reef biota are more vulnerable to major extinction events than are other marine organisms (83), this view has been contested (124, 131).

Maximum rates of sea-level rise during the Holocene exceeded 20 cm/decade during two separate periods lasting for roughly 1000 years each (7, 51). The difference in global mean temperature between the last glacial maximum and the present interglacial has been estimated at about 4°C (106), while the comparable difference of maximum temperatures in tropical marine environments ranged from <2° to 4° (127).

Similar fluctuations have occurred repeatedly throughout the Pleistocene. The persistence of coral reef communities as an ecosystem type (although not as temporally continuous individual communities) across repeated changes of these magnitudes suggests that reef ecosystems as a global phenomenon are not likely to be threatened by the predicted greenhouse-effect. The Quaternary climate oscillations occurred, however, on time scales that are long compared to the lifetimes of reef organisms, while greenhouse-forced global temperature change is predicted to be at least an order of magnitude faster—placing it on the same decadal time scale as coral lifetimes and community turnover.

Paulay (119) summarizes work indicating that while Plio-Pleistocene total extinctions of both corals and mollusks in the Caribbean were in excess of 30%, western Indo-Pacific faunas experienced extinctions of only 10–15% of corals (118, 152) and 20% of mollusks (rates described as "background levels") . Paulay's studies of the responses of Pacific island marine biota to Late Pleistocene sea level changes indicate variable patterns of local extirpation and recolonization, but with little large-scale permanent loss of diversity. Potts (122) has addressed the related issue of coral speciation and argues that it may be controlled by the relationship between longevities or generation times and the time-scale of habitat destruction.

A number of descriptive and theoretical approaches to coral reef ecosystems have indicated that reef communities are adapted to intermittent disturbances (27, 42, 44, 65, 66, 146). Indeed, disturbance may actually be necessary to the maintenance of a widely distributed healthy and adaptive ecosystem, analogous to the function of forest and grass fires in some terrestrial ecosystems.

In our opinion, the "robust vs fragile" debate is primarily the result of failure to define terms and conditions, particularly with reference to scale. The coral reef ecosystem, on a global scale and as a class of communities, is clearly robust with respect to natural climate change and variability over periods of millions of years. This robustness, coupled with wide geographic distribution and the possibility of range extensions to currently unsuitable environments as a result of global warming and sea level rise, suggests that the reef ecosystem is unlikely to be extirpated on a global basis by global climate changes currently predicted.

Reef fragility, however, is real and is a legitimate concern on time and space scales significant for human society. The robustness of reefs as a class of communities on a large-scale, long-term basis does not address specific concerns about the future of existing reefs as they respond to present-day stresses. Past natural stresses may not be adequate proxies for future accelerated climate change, which will occur on time scales much more comparable to the generation times of major reef organisms than was the case for past natural climate fluctuations. An even greater concern is the increasing extent of non-climatic anthropogenic stresses that reduce available refugia (83), impose stresses to which reef communities are not particularly well adapted (e.g. nutrient loading and sedimentation), and may interact with or exacerbate the more survivable climatic stresses.

ECOSYSTEMS AS "EARLY WARNING BIOINDICATORS Sensitive ecosystems or organisms may be "early warning systems" of environmental or global climate change. Such a role has been suggested for the "bleaching" of corals on coral reefs.

"Bleaching" has long been recognized as a coral response to various forms of environmental stress (17). This process may involve the loss of pigmented symbiotic algae (zooxanthellae) from the host coral, or of pigmentation from the retained algae (76, 86, 149); the physiological and environmental significance of the two modes of bleaching are not fully understood, and they are not readily distinguished by field observations. Corals without pigmented zooxanthellae appear white, or "bleached." Bleaching can be induced by a variety of stresses, including high or low temperature, salinity depression (and probably elevation), and excessive solar radiation or excessive shading. Bleached corals exhibit reduced production: respiration (P:R) ratios, reduction or cessation of calcification (121), and interruption of reproductive activity (149). If mortality does not result, apparent color recovery typically occurs within a few months, but long-term physiological recovery may be delayed for over a year (53, 149). Repeated bleaching on short time scales may result in increased cumulative mortality (130).

Extensive observations suggest that thermally induced bleaching and mortality, and subsequent community responses, have resulted from the unusually strong 1982–1983 El Niño (19, 56, 57, 60, 157). Other bleaching events, especially in the Caribbean have also been ascribed to high temperatures (19). Bleaching may be an early response of coral reef ecosystems to climatic warming (64). The evidence for both the temperature correlation and the climatic interpretation has been reviewed (32, 58, 59). Although episodes of regional or local high surface temperatures have surely been significant factors in major bleaching episodes, it is not possible to ascribe bleaching (or changes in its frequency of occurrence) solely to temperature. Given the definition of climate and our present understanding of climate change, short-term regional trends or variability in temperature or other directly measured environmental variables cannot be reliably ascribed to climate change (161). Thus far no rigorous explanation has been offered of how putative climate trends inferred from qualitative observations of uncalibrated biosensors can be more informative or reliable than direct quantitative measurements of the relevant variables.

Although not reliably useful specifically as a climate indicator, bleaching is a stress response and its occurrence is indicative of some form of environmental degradation or unsuitability (59). With adequate attention to diagnosis, mechanisms, and background levels of occurrence, it can be an important environmental bioindicator.

The Biogeophysical Experiment

Change in climate, or other large-scale environmental alterations, may be regarded as an experimental probe of ecosystem response, capable of yielding information of both practical and theoretical importance. Buddemeier & Smith (21) and Hopley & Kinsey (77) have suggested possible reef responses to sea level rise that, either alone or in combination with other environmental changes, could provide tests of key ecological or geological hypotheses over the next century. On land, similar questions have been raised about how forest or agricultural ecosystems might respond to the effects of elevated CO_2 partial pressure, air temperature, and moisture. How does one perform or interpret an uncontrolled geophysical "experiment" on a time scale of decades?

A number of symptoms or indicators of change in coral reef environments may be identified as possible results of such an "experiment":

1. Incidences of disease, parasitism, morbidity, and mortality (from any or all causes) are direct measures of both organism and ecosystem health.
2. Growth, particularly the rates or patterns of skeletal growth of long-lived sessile organisms such as corals, is a useful indicator for which there is a background of knowledge about normal rates and environmental correlates that can be used for hypothesis testing in the event of significant change.
3. Shifts in community dominance or population structure are important indicators; of particular importance are shifts from calcifying to non-calcifying organisms, as in the case of algal invasions of coral communities in response to increased nutrient loading in previously oligotrophic systems. Other examples include changes in populations of indicator organisms such as filter-feeders or echinoderms.
4. Community metabolism is a measure of large-scale function that can be used to document functional changes over time or to determine whether the metabolism of a given community falls within normal limits. The characteristics amenable to such treatment include organic production, P:R ratio, and net calcification. The rates of these processes on unperturbed reefs appear to be remarkably consistent (90). Departures from the norm are therefore readily recognized.
5. The reproductive behavior of existing community members (spawning, etc) and the recruitment of new members (e.g. larval settlement) are involved in maintaining community structure and function and may be observed directly as indicators of environmental suitability or organism health.

The experimental utility of the above observable characteristics suffers from the drawback that reef systems tend to exhibit high natural variability in time and space, and we lack adequate baseline data. If we are to predict or learn from the effects of climate change, the research community will need to make a transition to a more quantitative large-scale perspective.

An important and closely related factor is that reef structures and organisms in reef systems are potentially very good recorders of change. For example, reef structures in the geological record have been used as indicators of sea level and of relatively warm seawater. On shorter time scales, coral skeletons, especially massive forms, have annual growth bands analogous to tree rings, with some situations of subannual banding (8, 20). The bands themselves record variations in growth rate—which are in part responses to environmental variables such as temperature and solar radiation (103, 104). Moreover, the chemical, structural, and isotopic characteristics of the coral skeletons can reflect a wide variety of environmental forcing, with the growth bands serving as chronometers. The skeletal characteristics clearly can provide high-precision, uniquely diagnostic retrospective records at sub-annual intervals for periods ranging up to several centuries. Examples include information on river flow (81), oceanic circulation and turnover (45, 46, 132), and local stress events (121). Effective exploitation of these records can establish the baseline data needed to determine variabilities in reef environments on climatic time scales. A promising new approach to retrospective studies is the combination of population genetics with organism life histories and reproduction strategies to yield inferences about population dynamics over time (97).

REEF RESPONSES TO ENVIRONMENTAL VARIABLES

The sensitivity of corals, reefs, and related environments to environmental change, and their potential utility as monitors or recorders of change, is addressed in this section summarizing present knowledge about possible responses to changes in environmental factors. The order in which the variables are addressed is based on the level of certainty of near-term global change based on climate change considerations. Eustatic sea level rise and CO_2 concentration increases are extremely likely and globally equilibrated, and temperature increases are extremely likely although with significant uncertainty about regional variability. Other factors are considered to have a high probability of change but are either manifestations of climate change that are not predictable on a regional or local basis (light, physical oceanography, salinity) or are likely to result primarily from nonclimatic anthropogenic impacts (nutrients, sedimentation, resource utilization).

Sea Level

The predicted eustatic sea level rise (about 6 cm/decade) will be reflected in local relative sea level changes along with the influences of tectonism, atmospheric pressure, ocean current changes, and so forth. Coral reef responses to sea level rise need to be considered on two different scales. The potential exists for reefs to "drown" (i.e., be covered with such a depth of water that they are below the photic zone or otherwise cannot calcify sufficiently to catch up with sea level), but this requires a protracted imbalance between reef accretion rates and sea level rise. Depending on water clarity and other environmental conditions, the depth range of maximum reef calcification may extend from several meters to more than 20 m. This depth range represents a safety factor; transient sea level rise may inundate oceanic reefs to a depth of meters or even tens of meters without terminating reef growth if sea level rise subsequently returns to a rate less than reef vertical accretion rates.

Reef vertical accretion rates calculated from community calcification rates, growth rates of calcifying organisms in reef communities, and radiometric dating of cores through reefs range from less than 1 mm/yr to a maximum slightly in excess of 10 mm/yr (21, 35, 36, 77, 92, 138). A rate of 10 mm/yr is commonly taken as the consensus value for maximum sustained reef vertical accretion rates (21, 36). The present best prediction for eustatic sea level rise over the next century is 6 mm/yr on average (156); this is well within the range of reef accretion rates, and even with no net accretion would submerge reefs by less than a meter by the year 2100.

On the shorter time scale of years to decades, sea level is a changing environmental variable that may interact with other changes and be reflected in organism and community response. Because sea level has been within 1–2 m of its present elevation for several thousand years, many present-day reefs have grown to an elevation where further upward growth is constrained by sea level. Sea level rise can be expected to remove this constraint and result in increases in successful recruitment and coral longevity on intertidal and shallow subtidal reef flats (77), with a consequent increase in reef flat calcification (94). If rising sea level creates more benign conditions on shallow reefs, diversity and community structure may change as species other than the extremely hardy are able to survive. Increases in coral community diversity and productivity can also be expected in enclosed lagoons where salinity extremes, nutrient depletion, or other aspects of restricted circulation have restricted reef development (119, 140, 141, 143), since the probable effect of rising sea level on circulation will be to maintain reef/lagoon water composition closer to that of the local oceanic water. On the other hand, if deepening water subjects currently sheltered communities to more physical (wave) stress, calcification and sediment accumulation may not increase (21).

Other possible effects associated with sea level rise may include shifts in zonation and community structure associated with the relationship between wave energy regimes and sea level in a given locale (31). (These issues are discussed in the Currents, Waves, and Storms section below.) Second-order interactions may also be considered. For example, coral bleaching and associated mortality may selectively remove faster growing taxa, resulting in less rapid $CaCO_3$ accretion and more rapid net removal of framework material by bioerosion (56, 59). In many locations, coral reefs act as barriers that protect coastlines from incident wave energy; a change in reef depth could alter the coastal energy regime and erosion patterns (116). Over the next several decades these possibilities are likely to be significant only in particularly vulnerable locations (e.g. low islands) or in association with extreme events such as storms.

Carbonate Mineral Saturation State

A major feature of coral reefs is the precipitation of substantial quantities of $CaCO_3$ from the overlying water, thus modifying the chemical balance of the local marine CO_2 system. One of the best characterized human perturbations of the global climate system is the rate of CO_2 emission to the atmosphere and consequent changes of atmospheric CO_2 content. Changes in atmospheric CO_2 are well documented over the past several decades, reasonably well estimated over at least the last 150 years, understood in broad outline over the past 160,000 years, and at least semi-quantitatively predictable over the next century (78, 136). Estimates of increased atmospheric CO_2 invasion into the upper ocean are much modeled and discussed. While there is uncertainty on the quantity of CO_2 invasion into the ocean, it is reasonably well understood in comparison to other aspects of the global carbon budget (16, 150).

It is well established that the average CO_2 partial pressure (pCO_2) of surface ocean waters has remained within 10–20 μatm of the pCO_2 of the atmosphere as a result of CO_2 mass transfer across the air-sea interface. It seems likely that, as atmospheric pCO_2 continues to rise above the present level of about 350 μatm, the average partial pressure difference between the air and surface ocean will remain small. This increase in pCO_2 will change carbonate mineral saturation states, as outlined in the following discussion.

Consider tropical seawater (salinity = 35‰, temperature = 25°, total alkalinity = 2.3 meq/liter, and present pCO_2 = 350 μatm). It can be calculated (111, 120, 134) that seawater of this composition is about 520% saturated with respect to the common $CaCO_3$ mineral calcite; 340% saturated with respect to aragonite (the most common form of carbonate on coral reefs, deposited by reef-building corals and calcareous green algae); and 260% saturated with respect to 15 mol% Mg-calcite (commonly precipitated by coralline red algae, echinoderms, and many other reef taxa). With increasing Mg substitution for Ca in the calcite, mineral solubility increases, and saturation state of the water with respect to that mineral decreases. A 1–2° temperature increase (about what is predicted for tropical surface seawater over the next century) would not shift saturation states much, but a pCO_2 increase to 600 μatm (again, about what is expected) would decrease the saturation states with respect to the above minerals to about 360%, 240%, and 190%, respectively.

Despite the obvious potential for a link between oceanic CO_2 composition and the process of calcification, there is very limited information about the responses of reef organisms to changes in carbonate saturation state. Studies by Agegian (1), with portions also summarized in Mackenzie & Agegian,

1989 (107), Borowitzka (13), and Smith & Roth (135) on coralline algal growth in response to changes in the aqueous CO_2 system give discrepant results. Data by both Agegian and Borowitzka tend to indicate inhibition of algal growth by high pCO_2, while data by Smith & Roth show accelerated growth at least to modest levels of CO_2 elevation. The former results seem to us to be more plausible as long-term sustained phenomena. Yamazato (166) found that the growth rate of the coral *Fungia scutaria* varied as a function of Ca^{++} concentration in short-term experiments; this effect could result from altered saturation state.

A related issue is potential CO_2 fertilization effects on marine organic productivity. A doubling of atmospheric CO_2 has the potential to increase the photosynthesis of some macroalgae and seagrasses by a factor of two, and that of entire marine communities by up to 20% (159). Coral reef turf algal assemblages have productivity responses to altered pH that are consistent with the predicted fertilization effects of dissolved CO_2 (67). The response of calcifying algae or symbiotic corals to increasing pCO_2 may depend on whether tissue growth or calcification is the rate-limiting step in a given location. At the community level, the combination of photosynthetic enhancement and possible calcification reduction could enhance the competitive advantage of algal communities relative to coral communities, an effect that would be reinforced by community responses to nutrient loading (discussed below).

If we assume that, over time, calcification of many reef organisms will be generally proportional to carbonate mineral saturation state, we can postulate how reef ecosystems might respond to decreasing $CaCO_3$ saturation state. As calcification decreases, the ability of reefs to keep up with rising sea level may diminish. Alternatively, skeletal density may decrease, leaving the organisms more vulnerable to physical damage or bioerosion. The mineralogy of calcifying organisms may shift toward the less soluble mineral phases. Community structure may shift as noncalcifying organisms and bioeroders outcompete calcifying organisms. Habitat provided by corals for other organisms may become restricted.

We clearly know too little about the links between atmospheric CO_2, marine benthic productivity, and marine calcification; and available information is partially contradictory. Growth-rate and metabolic experiments that carefully and explicitly define and control aqueous CO_2 chemistry are required, and the experimental data will then require thoughtful extrapolation to environmentally relevant time scales.

Temperature

Unusually elevated water temperature is a stressor that commonly leads to bleaching (see discussion above) of corals and some other zooxanthellate organisms, and it will result in death if the "dose" (excess temperature times duration of excursion) is great enough. This factor is of particular concern because it is one of the few climate-related variables that has the potential for extensive near-term damage to coral reef communities (17). Temperature sensitivities of Indo-Pacific corals have been reviewed (86).

Temperature sensitivity in corals is clearly adaptational; the vulnerability of corals in a given locale varies by taxon and is related to the long-term historical record of mean temperature of the warmest month at that site (86). Temperature-induced bleaching can occur in response to 1–2 day exposures to temperatures 3–4° above the normal maximum, or to several weeks of exposure to elevations of 1-2°. Mortality is $> 90\%$ for temperature elevations in excess of 4° for periods as short as even a few hours (26), while relatively prompt recovery is the norm for bleaching episodes induced by temperature excursions $< 2°$.

Temperature responses of some coral symbionts, parasites, and co-occurring organisms appear similar to those of the hosts (60, 86). Bleaching in other symbiotic reef taxa (Millepora, soft corals, zoanthids) (82) also appears to be induced by elevated temperatures, but symbiotic sponges show less propensity for thermal bleaching (153). Irradiance (either visible or UV) may be interactive with elevated temperature in producing stress (86, 99, 129).

Systematic studies of coral temperature responses have been carried out independent of the bleaching issue for corals from tropical and subtropical locations (86). Calcification, P:R values, reproduction, and recruitment all showed maximum values near the local average maximum temperature (about 27° for Hawaii and 31° at Enewetak), with significant reductions at temperatures as little as 1–2° higher. From the standpoint of community response to temperature, these data need to be considered in view of a number of other observations (86): (i) At a given location a significant range of temperature responses appears among different colonies of the same species, and there may be systematic differences between intraspecific types or clones, or between microenvironments (28); (ii) there are systematic differences between species—those with the highest respiratory rates are more sensitive to elevated temperature, and branching species (especially Acropora and Pocillopora) are more vulnerable than massive corals (e.g. Porites); (iii) individual colonies have not been observed to exhibit temperature acclimation on time scales of months to years (86); and (iv) temperatures of 33–34° are tolerated by apparently healthy coral communities in the Great Barrier Reef (89) and the Arabian Gulf (23), where some species survive temperatures of 36–38°. However, recent observations in the Arabian Gulf indicate that even corals adapted to unusually high ambient temperatures remain susceptible to bleaching and other lethal or sublethal responses to slight additional temperature elevations (59, 128).

These observations indicate that reef ecosystem responses to rising temperature will be very dependent on the time scales of change and on other local environmental conditions. An increase in sea surface temperature or its variability, characterized primarily by increasing frequency of sublethal stress or partial mortality and occurring on time scales of the same magnitude as coral generation times, will probably tend to shift coral communities toward heat-tolerant "ecotypes" or species. More extreme temperature excursions or more rapid increases in mean water temperature could result in rapid change of reef diversity and community structure through extensive mortality and species-selective inhibition of reproduction and recruitment. The probability of local extirpations or total extinctions (58, 61) or of failure of reef communities to recover after thermal mortality (56, 60) would depend not only on the rate of change of mean temperatures, but also on the geographic distributions of extreme events and on other environmental controls on reef development.

In view of the fundamental importance of temperature sensitivity to questions of reef response to global change, we suggest that priority should be given to further research into the rates and mechanisms of organism and community adaptation, into the genetic and physiologic bases for thermal sensitivity, and into synergism or interactions between elevated temperature and other potential stressors. A particularly important avenue of research may relate to the question of the nature and distribution of strains or species of zooxanthellae (125). If temperature or stress response of the host-symbiont pair is influenced by algal as well as by coral genetics, and if multiple strains of algae are present in a given locale, then bleaching and reinfection could provide a mechanism for rapid adaptive change.

Visible and Ultraviolet Light

Corals have light requirements and adaptive mechanisms such that maximum rates of calcification and photosynthesis can be sustained to depths of as much as 20 m in clear water (52). Symbiotic corals occur to depths in excess of 60 m, albeit with much reduced rates of growth (80). The depth distribution of light, interacting with other environmental variables such as wave energy, is believed to account for such important reef characteristics as a broad maximum in biodiversity between 10 and 30 m (80) and maximum rates of reef accretion between 5 and 15 m (77).

Although high light levels do not appear to induce stress in corals if other environmental factors are optimal, they enhance coral sensitivity to temperature- and salinity-induced stress (86) and have been implicated as a factor in some episodes of coral bleaching (see discussion above). Photoinhibition appears to occur only at very shallow depths (80).

There are no anticipated climate or environmental changes that would increase peak clear-day visible light irradiance values. Net annual or seasonal irradiance incident on a reef community, and therefore the incidence of light-related stress and the depth range of optimal growth, could change in response to systematic changes in sea state, cloud cover, or turbidity (55). Cloud cover is one of the most poorly understood and poorly modeled features of the Greenhouse climate (30), and changes in surface wind fields (which influence light penetration by controlling surface roughness) are also not predictable at present. Increases in turbidity can result from input of suspended terrigenous sediments, or from planktonic primary production stimulated by nutrient loading. Both sedimentation and nutrient loading (see discussions below) tend to be anthropogenic stresses resulting from land use, population density, and resource exploitation (49, 105, 109, 142). Extreme turbidity can also occur from resuspension of naturally produced and trapped carbonate particles in restricted reef lagoonal environments (126).

Although not climatic in the narrowest definition of the term, the expected increase in ultraviolet-B (UVB) radiation levels due to destruction of the stratospheric ozone layer by chlorofluorocarbons may be significant; the factors controlling UVB penetration in the water column are the same as for visible light. Detrimental effects on coral reef fauna have been documented (85, 88, 133), and some of the mechanisms identified (99). UV levels in shallow water are clearly able to inhibit photosynthesis (102, 137, 163, 164). The depth-dependent occurrence of UV-blocking pigments in corals has been demonstrated (87, 88), as have photoadaptation and UV tolerance in specific strains of zooxanthellae (87, 88, 98).

Reproduction of reef organisms may also be affected by increased UV levels, either directly or through mortality of planula larvae. There are no data on these specific effects on reef organisms, but studies of mature and larval zooplankton (33, 37) indicate a wide range of UVB sensitivities, with at least some organisms vulnerable to present or increased UV levels in the near-surface environment.

Concern about the possible effects of increased UVB exposure is somewhat tempered by the fact that predicted increases in the tropics are relatively small; average low-latitude UVB surface exposures are expected to increase by 1–10%. This estimate is based on average ozone column depletion estimates of 0.5–5.0% (depending on the trace gas scenario used), although with the potential for much larger local variations (48), and an approximate UVB surface exposure increase of twice the ozone column depletions (165). However, lack of knowledge about UVB exposure effects, especially related to reproduction, combined with recognition that some organisms are sensitive at present levels, indicates that this topic merits further research.

On an overall basis, the present-day distribution of coral reefs across a range of light and water-clarity regimes suggests that changes in light availability and UVB exposure will probably be secondary factors in determining the response of coral reef systems to regional climate or local anthropogenic change. We note that visible light and UVB may be coincident contributors to reef stress under conditions of clear skies, calm water, and elevated temperature (86, 99, 129), but that in general the factors (cloudiness, sea level rise, turbidity) that reduce levels of desirable light exposure also protect against the deleterious effects of UVB. Because of the characteristics of light extinction with depth and the nature of community and organism adaptation, reduction of exposure levels will have little negative effect on shallow communities but may be deleterious to deeper ones; increased exposures may increase stress levels on shallow reefs but improve conditions at depth.

Currents, Waves, and Storms

Circulation exerts primary control over the distributions of salinity, temperature, nutrient levels, etc, by water mass advection and by determining the intensity and locations of upwelling, and secondary control of nutrient, sediment, and contaminant distributions by interacting with local (often anthropogenic) sources. Distributions of reefs and related or competing communities are controlled to a significant degree at the large and intermediate spatial scales by advective transport of propagules (3). The propagation of pathogens at the regional level has also been identified with major circulation patterns (100). Zonation, reef morphology, and the depth distributions of reef corals, algae, and other key organisms are strongly influenced by physical factors, including wave action (41, 80). Storms, waves, and currents are also the driving forces for sediment transport, coastal dynamics and geomorphology, and local relative sea level. Local circulation effects will interact with those of sea level rise, and in some significant cases they may be controlled primarily by sea level rise (see discussion above).

As with other factors, generalization about the effects of climate change on physical oceanography is difficult because specific effects will be regional or local in nature and because we lack adequate predictions about probable changes at those scales. In addition to modifications to the patterns mentioned above, there are specific reef-related issues: (i) Upwelling is an important local circulation phenomenon on several reef systems (83, 96, 162). It has been suggested that in some locations both the frequency and the intensity of temperate-latitude coastal upwelling may be significantly altered by changing weather conditions (4). The regional resolution of general circulation models is insufficient to predict how upwelling might change in specific reef settings. (ii) Storms provide long-term episodic control of reef community development by catastrophic pruning and/or substrate renewal (40, 72). In some instances, such disruption aids in coral reproduction (74). Storms influence community succession and diversity directly at the local level (74). Storm frequencies and/or intensities may increase under warmer conditions, but geographic distributions may shift; the result will be long-term shifts in community succession and development patterns in reefs subject to changed disturbance regimes.

Sedimentation

Sedimentation is not primarily a climatic factor, although natural coastal erosion and sediment accumulation patterns may alter as a result of changing currents, sea level, wave energy, or storm patterns. More significant factors are likely to be land use (e.g. deforestation, agricultural practices, dredging, coastal construction—158). The stress-related aspects of sedimentation have been reviewed by Brown & Howard (18) and Grigg & Dollar (66).

Rapid sedimentation can smother corals and other sedentary reef organisms (49, 109); slower but significant rates of sedimentation as well as high rates of sediment resuspension are known to cause reductions in coral growth rate (2, 38, 39, 105, 108, 117). It should be kept in mind, however, that coral morphology as well as sediment grain size affects the ability of coral colonies to cleanse themselves of sediment (79, 168). As a result, community structure will be altered by changing sedimentation regimes. Occlusion of hard substrate with soft sediment can reduce coral recruitment (12, 75). Increased sedimentation is often, but not always, accompanied by elevated turbidity and nutrient levels (discussed below). Among the mechanisms for deleterious effects are choking of feeding mechanisms, sulfide generation, prolonged contact with toxic components of the sediment, and facilitation of bacterial "infections" (18, 66).

Salinity

Although most reef systems occur near the apparently optimal salinity value of about 35 ‰, many reefs can be found in localities that show significant variations in mean or extreme salinities. It is therefore possible to extract considerable information about reef responses to salinity from natural distribution patterns.

Coles & Jokiel (25) have reviewed field and laboratory data on the responses of corals and other reef organisms to salinity, and they report that natural reef communities seem to do well within a salinity range of about 25–40 ‰, with rapid loss of taxa at higher salinities. Salinity maintained below 20 ‰ for longer than 24 hours is lethal for corals and most other reef taxa (49, 151). The lethal response is more rapid at lower salinities. Sublethal responses to salinity excursions include expulsion of zooxanthellae (hence, "bleaching") and altered metabolic rates (24, 114).

Salinity is unlikely to show significant regional-scale shifts of magnitudes that will affect coral reefs during a period of changing climate, but there will be local alterations of salinity regimes, especially in terms of variability. Local salinity excursions that are primarily associated with slow water exchange will tend to be reduced as sea level rises. Except in very restricted shallow water bodies, evaporation is unlikely to proceed rapidly enough to cause short-term excursions to excessively high salinity values. Salinity depressions are much more probable short-term excursions. Local reef areas adjacent to large or "flashy" drainage basins will be vulnerable to increased flow that will chronically or episodically depress salinity to well below currently observed levels. Such occurrences will have potentially dramatic local impacts on corals and other reef organisms.

Climate-induced changes in local salinity stress may result from shifts in rainfall amount or intensity, but as with sedimentation, a more important factor in populated areas is likely to be changes in runoff pattern resulting from land-use patterns such as urbanization and deforestation.

Nutrients

Inorganic plant nutrients have long been implicated in damage to coral reef communities (71), and increasing human pressures are leading to elevated nutrient discharges from sewage, agricultural practices, and land use. These changes may be the most pervasive human-induced alterations of the coastal zone. Perhaps the most intensively studied example of nutrient effects on coral reefs has been Kaneohe Bay, Hawaii (6, 89, 91, 108, 142, 160). This system received substantial and increasing discharges of sewage-derived nutrients over a 20-year period; the discharge was abruptly terminated in 1977. A particularly noteworthy characteristic of Kaneohe Bay is the degree to which reef communities had recovered several years after termination of sewage discharge into this system (50).

Other major reef systems such as the Australian Great Barrier Reef (167) and the Florida Reef Tract (70) are increasingly discussed in terms of observed or perceived nutrient stress due to anthropogenic activity. Nutrient loading is certainly increasing on many aquatic communities adjacent to growing human populations, particularly in areas where water quality regulations are lax.

The major impact of nutrients on coral reefs usually appears to be indirect. Corals require very little external nutrient supply because of effective internal nutrient recycling between the coral animal host and the zooxanthellae plant symbiont (68, 113). Other taxa of common photosynthetic reef organisms also thrive at low nutrient concentrations. The primary production rate of coral reefs is very high (90) but is almost exactly balanced by the respiration rate (29, 90). Coral reef communities subjected to high nutrient levels appear to undergo deterioration due to overgrowths by filamentous algae, bryozoans, and barnacles, increased plankton-generated turbidity, increased bioerosion, and poor coral recruitment (9, 69, 101). Hallock (69) and Birkeland (11) have suggested that reef growth diminishes along a gradient from oligotrophic to eutrophic conditions.

However, Kinsey (92) has observed that it is incorrect to jump from the observation that coral reefs can do well under low nutrient conditions to the conclusion that coral reefs require such low nutrient environments. We agree, noting that some reefs look healthy and are apparently doing well in a milieu of naturally high nutrient levels (62, 140). At the organism level, increased nutrients cause increases in both the pigmentation per cell and the absolute number of cells of symbiotic zooxanthellae (47, 112). Although there is no evidence that increased nutrients are a physiological stress in themselves, studies are needed on whether elevated nutrients change the sensitivity of the symbiotic assembly to other stresses, particularly temperature and/or light-induced effects. We suggest that an important consideration for any particular reef community (which may be adapted to a wide range of nutrient regimes) is the rate of change in the nutrient loading into that system and its interaction with other environmental variables.

At the community level, rapid nutrient changes can cause shifts in structure and function. With an increase in nutrients, rapidly growing phytoplankton and benthic algae gain a competitive advantage over corals, often overgrowing and eventually smothering them (6, 11, 66, 69, 142). Turbidity of the water resulting from increased plankton biomass and detritus can also cause reef deterioration (142). (See also discussion of light effects, above). Perhaps an even more important effect is that suspension-feeding animals, especially those that bore into calcareous substrata, become abundant in high nutrient waters replete with particulate organic matter (10, 14, 66, 69, 142, 160). Many of these organisms damage the structural integrity of the reef, and the larvae of sponges, barnacles, ascidians, and other sedentary organisms can outcompete coral larvae for space. High organic loading, which may be associated with some types of nutrient loading, can also kill corals and other organisms because of elevated biochemical oxygen demand and hydrogen sulfide release (66). Nutrient loading may also be associated with reef stress induced by increased sedimentation.

Some evidence suggests direct damage to reef organisms from elevated nutrients. Yamazato (166) reported that high levels of phosphate apparently inhibited coral calcification during incubation experiments. Kinsey & Domm (93) inferred a similar effect for a reef community deliberately fertilized with inorganic phosphorus and nitrogen. Walker & Ormond (154) found localized mortality of corals in the vicinity of sewage discharge and spillage of phosphate dust during ship loading. One might also postulate such direct effects from locally high levels of ammonium. If such direct nutrient-induced mortality does occur, it seems most likely to be associated with very high nutrient concentrations. Such effects could usually be anticipated to be localized. A "halo" around a discharge site would be inside of a much broader zone, exhibiting the secondary effects we have noted.

Other Anthropogenic Stresses

Anthropogenic environmental alterations, especially those connected with land use and waste disposal, have been discussed in connection with many of the factors considered above. Coral reef stresses have been recently reviewed (18, 66) and cataloged (158). Two classes of anthropogenic stress not previously discussed merit mention.

Toxins introduced into the marine environment may include such materials as oil (18, 84), metals (18, 63, 147), and pesticides (63, 147). Although these have thus far been documented as having only local effects, their obvious association with human population and economic development suggests that they deserve continued attention.

Direct reef resource utilization (quarrying, dredging, harvesting of organisms, tourism) represents a significant threat and a confounding variable when assessing other forms of environmental change. In particular, many coral reefs, especially those adjacent to land masses supporting large human populations, receive fishing pressure sufficient to reduce the population of fishes dramatically (158). Fish grazing is known to be an important control on plant biomass on reefs. Large algal populations are often attributed to excessive nutrient loading; an important alternative interpretation is that fishing pressure has removed those herbivorous fishes that might have grazed on the algae and prevented their excessive growth.

The effects of resource utilization are important to understanding environmental change, both because they can mimic or obscure the effects of climate or other environmental change, and because the utilization tends to exacerbate the effects of the more directly environmental influences discussed above.

"PATHWAYS" OF CLIMATE FORCING

Table 1 presents a slightly modified list of the factors discussed above, subdivided according to whether environmental changes over the next few decades are more likely to result from changing climate or from other anthropogenic environmental alterations.

In order to assess relative importance or sensitivity, these same stressors may be categorized by their primary pathway or mechanism of delivery to the local reef environment. Those factors transmitted primarily through the atmosphere or via air-sea interaction are identified with the atmospheric pathway; the hydrographic pathway corresponds to those features normally associated with physical oceanography, and the hydrologic pathway involves effects transmitted or created through the mechanisms of evaporation,

Table 1 Dominant sources of near-term coral reef stresses

Climate	Other
Sea level rise**	Ultraviolet light**
CO_2 changes**	Nutrients (*?)
Temperature change*	Sedimentation
Visible light	Turbidity
Current/storm change	Toxics
(Fresh water)	Resource use
	(Fresh water)

** global, trends monotonic
* trends in global mean monotonic, significant spatial/temporal exceptions

precipitation, runoff, or groundwater flow. This list is given in Table 2; although not rigorous, it provides a basis for identifying signal amplification possibilities.

It seems clear that all three pathways by which changing climate influences aquatic ecosystems will have their major influence in shallow water. From the standpoints of sensitivity, vulnerability, and monitoring, marginal environments are particularly sensitive to change because of their steep gradients and physical constraints. Coastal zones are a particularly good example of such environments, and a large fraction of the world's coral reefs occur within coastal zones or their physical extensions, enclosed and/or shallow basins, or shelf environments. In such locales the hydrologic pathway has a particularly high signal amplification because effects may be integrated over very large land areas and focused (e.g. through runoff) onto the narrow coastal zone. It is important to note that the stressors delivered via the hydrologic pathway are generally those in the "other" (non-climate) source category.

This analysis leads to the conclusion that reefs in proximity to land masses or in relatively shallow or enclosed basins will be particularly sensitive (or vulnerable) to the effects of environmental change but will have a low specificity for climatic variables. In the near term, the suggestion is strong that local and regional effects of population growth, land use, etc, will have

Table 2 Dominant pathways of stress delivery

Atmospheric	Hydrographic	Hydrologic
Temperature	Sea level	Fresh water
		Nutrients
UV light	Currents	Sedimentation
Visible light	Storms/waves	Toxics
CO_2 change		Turbidity

more effect on these reefs than the more distributed and gradual climatic changes. Oceanic reefs are probably more sensitive to the truly climatic component of environmental change and could serve as control sites for studies in the more complex nearshore environment. Stress responses can be observed on a local or regional basis but may represent various categories of response: (i) Direct physiological response of an organism to a single environmental variable such as temperature or light; (ii) interactive responses of organisms to concurrent changes in multiple variables (e.g. temperature plus light, water quality plus light or temperature); and (iii) indirect effects on competitive relationships or community interactions, whether or not accompanied by overt evidence of physiologic stress at the organism level.

We note that chemical stresses are more likely to be observed at the community level, while physical stress is more readily observed in terms of organism response. Indirect effects will be situational and highly variable on a spatial scale. This highlights the problems of scale and variability associated with the use of local reef responses for detection of global signals, "average" ecosystem conditions, or responses to environmental forcing; extensive networks of long-term observations will be required.

DISCUSSION

Large-Scale Issues

On large scales, the environmental effects specifically attributable to climate change would suggest that reefs as a global biotic phenomenon are not seriously threatened. We know that reefs are structurally and functionally similar over a wide range of conditions and dominant species, indicating that the communities are not highly dependent on specific individual taxa and may be resilient to the loss of some more vulnerable species. When we consider individual climatic factors in isolation, we see that the effects of sea level rise on coral reefs over the next century are more likely to be positive than negative. Increases in maximum sea surface temperatures will increase the frequency of temperature-induced stress or mortality events, but increases in minimum and/or average values may extend the geographic range of conditions suitable for reef development on a longer time scale. Climatic changes in hydrologic or hydrographic factors may have positive or negative effects on a local scale. Although we may expect significant changes in the details of reef characteristics and distributions, there is at present no basis for predicting widespread deleterious effects on average. There are two important caveats to this conclusion:

One is that we know remarkably little about the probable effects of increasing UV exposure and of decreasing carbonate saturation state. Although there is no evidence that either one represents a potential near-term cataclysm, their effects are predominantly negative, their distribution is global, and the nature of change is reasonably predictable. Focussed research is clearly needed.

A more important caveat stems from the artificiality of considering climate-induced stresses by themselves. Reef damage from anthropogenic environmental degradation is widespread, represents a much greater threat than climate change in the near future, and can reinforce negative effects of climate change. In particular, we note that many anthropogenic changes represent chronic stresses that may interfere with reef recovery from the acute stress events that are likely to increase in frequency as a result of climate change. The combination poses threats on a regional level that we believe should be taken seriously. As an example, we note that the Caribbean fits the profile of a vulnerable region: biodiversity is far lower than in the Indo-Pacific; it has been more vulnerable than the Indo-Pacific during past climate fluctuations; it is a relatively enclosed basin with a growing human population in its drainage area and abundant evidence of anthropogenic effects and terrigenous (e.g. runoff-related) influences; there are no other large-scale reef communities in the tropical Atlantic that can serve as refugia or sources of recolonization; and evidence of widely distributed reef stress has already been noted.

Local Issues

At the local level, concerns are about the specific condition of specific reefs, and reassurances on the global average level offer little joy. We gain some additional levels of predictability and impact assessment with a local focus, whereas in a general treatment such as this review we are forced to fall back on generalities and uncertainties. When we look at a specific site we are dealing with known hydrologic, hydrographic, population, and development characteristics; and specific vulnerabilities and probabilities are much more identifiable. The effects of chronic stress and altered disturbance regimes may be effectively evaluated on the local level; Bales (5) points out in a discussion of altered hydrologic regimes that "Human-related disturbances tend to differ from natural disturbances in being more frequent and of different intensity. Systems dominated by human-related disturbances tend to be less diverse biologically. Physically they have less structure and complexity . . . we need to better understand the long-term impacts of altering natural disturbance regimes in order to adequately address such issues as biodiversity and cumulative impacts." However, for some types of acute stress, the scale mismatch becomes more critical—if the effects of climate change are felt (as seems probable) primarily through changes in the frequency and magnitude of extreme events, then effects at small spatial scales can be predicted only on a probabilistic basis. Overall, we suspect that many human populations will, over the next few decades, come to grips with the previously abstract geological truism that individual reefs are transient phenomena.

Scientific Issues

There are critical needs for data on the nature and variability of organism and symbiont responses to specific environmental factors (notably carbonate mineral saturation state, temperature, and UVB), and especially on the combined effects of potentially interacting stresses and environmental conditions such as nutrient loading. There are enormous challenges ahead in learning how to relate observations, theory, and predictions across differing time and space scales; this has been identified as one of the highest priorities for research in hydrology (115), which has been pointed out above as an extremely sensitive pathway for the effects of environmental or climate change.

The human race appears to be in the process of conducting an unplanned planetary experiment in what might be loosely termed accelerated evolution. If the scientific community has the resolve and insight necessary to discern and take advantage of the "design" inherent in this experiment, there is enormous understanding to be gained—much of it vitally useful to society in preparing for or responding to the unintended consequences of our experiment.

Institutional Issues

Addressing the various needs and opportunities will require significant changes in our present scientific culture and institutional structures. One of the greatest administrative challenges is the need for funding and research institutions to develop an effectively coordinated approach to problems that have time scales in excess of budget cycles and political administration lifetimes, and both spatial scales and ranges of implications that transcend academic disciplines, agency missions, and geopolitical boundaries (15, 21).

A specific focus of the administrative problem is the need for long-term, globally integrated, and stable yet technically evolving data acquisition programs. The traditional term is monitoring, and the traditional attitude among researchers is that "monitoring" is an activity of dubious merit, particularly if it competes with "research" for resources. At the time and space scales of global change, however, there is no basis for this dichotomy. Monitoring is an intrinsic and essential part of research (32), and its design and effective conduct should be one of the highest priorities of the research community, because our ability to formulate and test hypotheses will depend on the data bases available. Retrospective extension of monitoring observations by integrating paleoecological and neoecological studies will be an important component of efforts to extend our baseline data both temporally and spatially.

Finally, we note (not for the first time) that the scientific culture will need to find ways for better integration across disciplinary boundaries as well as across time and space scales within disciplines. Reef research has long been dominated by biologists and geologists, yet climate and stresses induced by climate change are defined and studied in physical and chemical terms. A concerted effort to match studies of physical/chemical forcing functions with the biological/geological responses appears to us to be the only way to synthesize an effective approach to global change issues.

ACKNOWLEDGMENTS

We gratefully acknowledge the assistance of many colleagues who have provided access to information, and particularly the participants in the Miami Workshop on Coral Bleaching, Coral Reef Ecosystems and Global Change (32). The manuscript has been improved by review and comments of D. G. Fautin, G. Paulay, R. Holt, P. Glynn, and J. Ware. M. Pecheux kindly provided many useful references. This is the School of Ocean and Earth Science and Technology contribution No. 2829.

Literature Cited

1. Agegian, C. R. 1985. *The biogeochemical ecology of Porolithon gardineri (Foslie)*. PhD thesis. Univ. Hawaii, 178 pp.
2. Aller, R. C., Dodge, R. E. 1974. Animal-sediment relations in a tropical lagoon—Discovery Bay, Jamaica. *J. Mar. Res.* 32:209–32
3. Andrews, J. C., Pickard, G. L. 1990. The physical oceanography of coral-reef systems. See Dubinsky 1990, pp. 11–48
4. Bakun, A. 1990. Global climate change and intensification of coastal ocean upwelling. *Science* 247:198–201
5. Bales, R. C. 1991. Hydrology sessions at the 1990 fall AGU meeting. EOS, Transactions, Am. Geophys. Union 72:243
6. Banner, A. H. 1975. Kaneohe Bay, Hawaii: Urban pollution in a coral reef ecosystem. *Proc. 2nd Int. Symp. Coral Reefs, Brisbane* 2:685–702
7. Bard, E., Hamelin, B., Fairbanks, R. G., Zindler, A. 1990. Calibration of the ^{14}C timescale over the past 30,000 years using mass spectrometric U-Th ages from Barbados corals. *Nature* 345: 405–10
8. Barnes, D. J., Lough, J. M. 1989. The nature of skeletal density banding in scleractinian corals: Fine banding and seasonal patterns. *J. Exp. Mar. Biol. Ecol.* 126:119–34
9. Birkeland, C. 1977. The importance of rate of biomass accumulation in early successional stages of benthic communities to the survival of coral recruits. *Proc. Third Int. Coral Reef Symp., Miami.* 1:16–21
10. Birkeland, C. 1987. Nutrient availability as a major determinant of differences among hard-substratum communities in different regions of the tropics. In *Differences between Atlantic and Pacific Tropical Marine Coastal Ecosystems: Community Structure, Ecological Processes, and Productivity*, ed. C. Birkeland, pp. 45–90. Paris: UNESCO
11. Birkeland, C. 1988. Geographic comparisons of coral-reef community processes. *Proc. 6th Coral Reef Symp., Townsville* 1:211–20
12. Birkeland, C., Rowley, D., Randall, R. H. 1982. Coral recruitment patterns at Guam. *Proc. Fourth Int. Coral Reef Symp., Manila* 2:339–44
13. Borowitzka, M. A. 1981. Photosynthesis and calcification in the articulated coralline alga *Amphiroa anceps* and *A. foliaceae*. *Mar. Biol.* 62:17–23
14. Brock, R. E., Smith, S. V. 1983. Response of coral reef cryptofaunal communities to food and space. *Coral Reefs* 1:179–93
15. Broecker, W. S. 1987. Unpleasant surprises in the greenhouse? *Nature* 328: 123–26
16. Broecker, W. S., Takahashi, T., Simpson, H. J., Peng, T.-H. 1979. Fate of fossil fuel carbon dioxide and the global carbon budget. *Science* 206:406–18
17. Brown, B. E. 1990. Coral bleaching: special issue. *Coral Reefs* 8:153–232
18. Brown, B. E., Howard, L. S. 1985. Assessing the effects of "stress" on reef corals. *Adv. Mar. Biol.* 22:1–63
19. Brown, B. E., Suharsono 1990. Damage and recovery of coral reefs affected by El Niño related seawater warming

in the Thousand Islands, Indonesia. *Coral Reefs* 8:163–70

20. Buddemeier, R. W., Kinzie, R. A. III 1976. Coral growth. *Oceanogr. Mar. Biol. Ann. Rev.* 14:183–225
21. Buddemeier, R. W., Smith, S. V. 1988. Coral reef growth in an era of rapidly rising sea level: Predictions and suggestions for long-term research. *Coral Reefs* 7:51–56
22. Bunkley-Williams, L., Williams, E. H. 1990. Global assault on coral reefs. *Nat. Hist.* 4:47–54
23. Coles, S. L. 1988. Limitations on reef coral development in the Arabian Gulf: temperature or algal competition. *Proc. 6th Int. Coral Reef. Symp., Townsville* 3:211–16
24. Coles, S. L., Jokiel, P. L. 1978. Synergistic effects of temperature, salinity and light on the hermatypic coral *Montipora verrucosa*. *Mar. Biol.* 49: 187–95
25. Coles, S. L., Jokiel, P. L. 1992. Effects of salinity on coral reefs. In *Pollution in Tropical Aquatic Systems,* ed. D. Connell, D. Hawker. Boca Raton, Fla: CRC. In press
26. Coles, S. L., Jokiel, P. L., Lewis, C. R. 1976. Thermal tolerance in tropical versus subtropical Pacific reef corals. *Pac. Sci.* 30:159–66
27. Connell, J. 1978. Diversity in tropical rain forests and coral reefs. *Science* 199:1302–10
28. Cook, C. B., Logan, A., Ward, J., Luckhurst, B., Berg, C. J. Jr. 1990. Elevated temperatures and bleaching on a high-latitude coral reef: the 1988 Bermuda event. *Coral Reefs* 9:45–49
29. Crossland, C. J., Hatcher, B. G., Smith, S. V. 1991. Role of coral reefs in global ocean production. *Coral Reefs* 10:55- 64
30. Cubasch, U., Cess, R. D. 1990. Processes and modeling. In *Climate Change: The IPCC Assessment,* ed. J. T. Houghton, G. J. Jenkins, J. J. Ephraums, pp. 69–92. Cambridge: Cambridge Univ. Press
31. Cubit, J. D. 1985. Possible effects of recent changes in sea level on the biota of a Caribbean reef flat and predicted effects of rising sea levels. *Proc. Fifth Int. Coral Reef Congr., Tahiti* 3:111–18
32. D'Elia, C. F., Buddemeier, R. W., Smith, S. V. 1991. *Workshop on Coral Bleaching, Coral Reef Ecosystems and Global Change: Report of Proceedings. Maryland Sea Grant Coll. Publ. No. UM-SG-TS-91–03*. College Park, Md: Univ. Maryland
33. Damkaer, D. M., Dey, D. B. 1983. UV damage and photo-reactivation potentials of larval shrimp, Pandalus platyceros, and adult euphausiids, *Thysanoessa raschii*. *Oecologia* 60:169–75
34. Davies, P. J. 1988. Evolution of the Great Barrier Reef - reductionist dream or expansionist vision. *Proc. Sixth Int. Coral Reef Symp., Townsville* 1:9–17
35. Davies, P. J., Hopley, D. 1983. Growth facies and growth rates of Holocene reefs in the Great Barrier Reef. Bur. Mineral Resources. *J. Aust. Geol. Geophys.* 8:237–51
36. Davies, P. J., Marshall, J. F., Hopley, D. 1985. Relationship between reef growth and sea level in the Great Barrier Reef. *Proc. Fifth Int. Coral Reef Congr., Tahiti*. 3:95–103
37. Dey, D. B., Damkaer, D. M., Heron, G. A. 1988. UV-B dose/dose-rate responses of seasonally abundant copepods of Puget Sound. *Oecologia* 76: 321–29
38. Dodge, R. E., Aller, R. C., Thomson, J. 1974. Coral growth related to resuspension of bottom sediments. *Nature* 247:547–77
39. Dodge, R. E., Vaisnys, J. R. 1977. Coral populations and growth patterns: responses to sedimentation and turbidity associated with dredging. *J. Mar. Res.* 35:715–30
40. Dollar, S. J. 1982. Wave stress and coral community structure in Hawaii. *Coral Reefs* 1:71–81
41. Done, T. J. 1983. Coral zonation: Its nature and significance. In *Perspectives on Coral Reefs,* ed. D. J. Barnes, pp. 107–47. Manuka, Aust: Brian Clouston
42. Done, T. J. 1987. Simulation of the effects of outbreaks of Crown-of-Thorns starfish on massive corals in the genus Porites: Evidence of population resilience? *Coral Reefs* 6:75–90
43. Done, T. J. 1991. The debate continues—robust versus fragile reefs. *Reef Encounter* 9:5–7
44. Done, T. J. 1992. Phase changes in coral reefs. *Hydrobiologia*. In press
45. Druffel, E. R. M. 1985. Detection of El Niño and decade time scale variations of sea surface temperature from banded coral records: Implications for the carbon dioxide cycle. In *The Carbon Cycle and Atmospheric CO_2*: Natural Variations Archean to Present, ed. E. T. Sundquist, W. S. Broecker. Washington, DC: Am. Geophys. Union
46. Druffel, E. R. M., Dunbar, R. B., Wellington, G. M., Minnis, S. A. 1990. Reef-building corals and identification of ENSO warming periods. In *Global*

Ecological Consequences of the 1982–83 El Niño-Southern Oscillation, ed. P. W. Glynn, pp. 233–53. New York: Elsevier Oceanography Series 52

46a. Dubinsky, Z. 1990. *Coral Reefs: Ecosystems of the World 25*. New York: Elsevier

47. Dubinsky, Z., Stembler, N., Ban-Zion, M., McCloskey, L. R., et al. 1990. The effect of external nutrient resources on the optical properties and photosynthetic efficiencies of *Stylophora pistilleta*. *Proc. R. Soc. Lond., Ser. B*239:231–46

48. Eckman, R. S., Pyle, J. A. 1989. Numerical modelling of ozone perturbations. In *Ozone Depletion: Health and Environmental Consequences,* ed. R. R. Jones, T. Wigley, pp. 27–42. New York: Wiley

49. Edmondson, C. H. 1928. The ecology of an Hawaiian coral reef. *Bernice P. Bishop Museum Bull.* 45:64

50. Evans, C. W., Maragos, J. E., Holthus, P. F. 1986. Reef corals in Kaneohe Bay; Six years before and after termination of sewage discharge (Oahu, Hawaii Archipelago). *Univ. Hawaii Inst. Mar. Biol., Tech. Rep. 37*. Kaneohe, Hawaii

51. Fairbanks, R. G. 1989. A 17,000 year glacio-eustatic sea level record. *Nature* 342:637–42

52. Falkowski, P. G., Jokiel, P. L., Kinzie, R. A. III 1990. Irradiance and corals. See Dubinsky, pp. 89–107

53. Fitt, W. K., Spero, H. J., Halas, J., White, M. W., Porter, J. W. 1992. Recovery patterns of the coral *Montastrea annularis* after the 1987 "bleaching event" in the Florida Keys. *Coral Reefs*. In press

54. Flohn, H., Kapala, A., Knoche, H. R., Machel, H. 1990. Recent changes of the tropical water and energy budget and of mid-latitude circulations. Abstracts. *Int. TOGA Sci. Conf.* Honolulu, Hawaii (unnumbered)

55. Glynn, P. W. 1977. Coral growth in upwelling and nonupwelling areas off the Pacific coast of Panama. *J. Mar. Res.*, 35:567–85

56. Glynn, P. W. 1988. El Niño-southern oscillation 1982–83: Nearshore population, community and ecosystem. *Annu. Rev. Ecol. Syst.* 19:309–45

57. Glynn, P. W. 1990. Coral mortality and disturbances to coral reefs in the tropical Eastern Pacific. In *Global Ecological Consequences of the 1982–1983 El Niño-Southern Oscillation,* ed. P. W. Glynn, pp. 55–126. New York: Elsevier Oceanography Ser.

58. Glynn, P. W. 1991. Coral reef bleaching in the 1980's and possible connections with global warming. *Trends Ecol. Evol.* 6:175–79

59. Glynn, P. W. 1992. Coral reef bleaching: ecological perspectives. *Coral Reefs*. In press

60. Glynn, P. W., D'Croz, L. 1990. Experimental evidence for high-temperature stress as the cause of El Niño–coincident coral mortality. *Coral Reefs* 8:181–92

61. Glynn, P. W., de Weert, W. H. 1991. Elimination of two reef-building hydrocorals following the 1982–83 El Niño warming event. *Science* 253:69–71

62. Glynn, P. W., Stewart, R. H. 1973. Distribution of coral reefs in the Pearl Islands (Gulf of Panama) in relation to thermal conditions. *Limnol. Oceanogr.* 18:367–79

63. Glynn, P. W., Szmant, A. M., Corcoran, E. F., Cofer-Shabica, S. V. 1989. Condition of coral reef cnidarians from the northern Florida reef tract: Pesticides, heavy metals, and histopathological examination. *Mar. Poll. Bull.* 20: 568–76

64. Goreau, T. J. 1990. Coral bleaching in Jamaica. *Nature* 343:417

65. Grassle, J. F. 1973. Variety in coral reef communities. In *Biology and Geology of Coral Reefs,* ed. O. A. Jones, R. Endean, pp. 247–68. New York: Academic Press

66. Grigg, R. W., Dollar, S. J. 1990. Natural and anthropogenic disturbance on coral reefs. See Dubinsky, pp. 439–52

67. Hackney, J. M., Sze, P. 1988. Photorespiration and productivity rates of a coral reef algal turf assemblage. *Mar. Biol.* 98:483–92

68. Hallock, P. 1981. Algal symbiosis: a mathematical analysis. *Mar. Biol.* 62: 249–55

69. Hallock, P. 1988. The role of nutrient availability in bioerosion: consequences to carbonate buildups. *Palaeogeogr., Palaeoclimat., Palaeoecol.* 63:275–91

70. Hallock, P. 1990. Coastal pollution and coral communities. *Bull. Am. Littoral Soc.* 19:15–18

71. Hallock, P., Schlager, W. 1986. Nutrient excess and the demise of coral reefs and carbonate platforms. *Palaios* 1:389–98

72. Harmelin-Vivien, M. L., Laboute, P. 1986. Catastrophic impact of hurricanes on outer reef slopes in the Tuamotu (French Polynesia). *Coral Reefs* 5:55–62

73. Heymsfield, A. J., Miloshevich, L. M.

1991. Limit to greenhouse warming? *Nature* 351(2 May 1991):14–15
74. Highsmith, R. C. 1982. Reproduction by fragmentation in corals. *Mar. Ecol. Prog. Ser.* 7:207–26
75. Hodgson, G. 1990. Sediment and the settlement of larvae of the reef coral *Pocillopora damicornis*. *Coral Reefs* 9:41–43
76. Hoegh-Guldberg, O., Smith, G. J. 1989. Light, salinity, and temperature and the population density, metabolism, and export of zooxanthellae from *Stylophora pistillata* and *Seriatopora hystrix*. *J. Exp. Mar. Biol. Ecol.* 129:279–303
77. Hopley, D., Kinsey, D. W. 1988. The effects of a rapid short-term sea level rise on the Great Barrier Reef. In *Greenhouse: Planning for Climate Change,* ed. G. I. Pearman, pp. 189–201. New York: Brill
78. Houghton, J. T., Jenkins, G. J., Ephraums, J. J., eds. 1990. *Climate Change: The IPCC Assessment*. Cambridge Univ. Press: Cambridge
79. Hubbard, J. A., Pocock, Y. P. 1972. Sediment rejection by recent scleractinian corals: a key to paleo-environmental reconstruction. *Geol. Rundsch. (Sonderdr.)* 61:598–626
80. Huston, M. A. 1985. Patterns of species diversity on coral reefs. *Annu. Rev. Ecol. Syst.* 6:149–77
81. Isdale, P. 1984. Fluorescent bands in massive corals record centuries of coastal rainfall. *Nature* 310:578–79
82. Jaap, W. C. 1988. The 1987 zooxanthellae expulsion event at Florida reefs. In *Mass Bleaching of Corals in the Caribbean: a Research Strategy. Research Report 88–2.*, ed. J. Ogden, R. Wicklund, pp. 24–29. Rockville, Md: NOAA Undersea Res. Prog.
83. Jablonski, D. 1991. Extinctions: A paleontological perspective. *Science* 253: 754–57
84. Jackson, J. B. C. 1989. Ecological effects of a major oil spill on Panamanian coastal marine communities. *Science,* 243:37–44
85. Jokiel, P. L. 1980. Solar ultraviolet radiation and coral reef epifauna.*Science* 207:1060–71
86. Jokiel, P. L., Coles, S. J. 1990. Response of Hawaiian and other Indo-Pacific reef corals to elevated temperature. *Coral Reefs* 8:155–62
87. Jokiel, P. L., York, R. H., Jr. 1984. Importance of ultraviolet radiation in photoinhibition of microalgal growth. *Limnol. Oceanogr.* 29:192–99
88. Jokiel, P. L., York, R. H. J. 1982. Solar ultraviolet photobiology of the reef coral *Pocillopora damicornis* and symbiotic zooxanthellae. *Bull. Mar. Sci.* 32:301–15
89. Kinsey, D. W. 1979. *Carbon turnover and accumulation by coral reefs*. PhD thesis. Univ. Hawaii. 284 pp.
90. Kinsey, D. W. 1985. Metabolism, calcification and carbon production: System level studies. *Proc. Fifth Int. Coral Reef Congr., Tahiti.* 4:505–26
91. Kinsey, D. W. 1988. Coral reef system responses to some natural and anthropogenic stresses. *Galaxea* 7:113–28
92. Kinsey, D. W. 1991. The coral reef: an owner-built, high-density, fully-serviced, self-sufficient housing estate in the desert—or is it? *Symbiosis* 10:1–22
93. Kinsey, D. W., Domm, A. 1974. Effects of fertilization on a coral reef environment—primary production studies. *Proc. Second Int. Coral Reef Symp., Brisbane.* 1:49–52
94. Kinsey, D. W., Hopley, D. 1991. The significance of coral reefs as global carbon sinks—response to greenhouse. *Paleogeogr., Paleoclimatol., Paleoecol.* (Global and Planetary Change Section) 89:363–77
95. Lapointe, B. E. 1989. Caribbean coral reefs: Are they becoming algal reefs? *Sea Frontiers* 35:82–91
96. Lapointe, B. E., Smith, N. P. 1987. A preliminary investigation of upwelling as a source of nutrients to Looe Key National Marine Sanctuary. *NOAA Technical Memorandum NOS MEND,* 9:53
97. Lasker, H. R., Coffroth, M. A. 1991. Population dynamics and genetic population structure of a Gorgonian coral: Do multi-year studies characterize longer time scale dynamics? *Am. Zool.* 31(5):48A
98. Lesser, M. P., Shick, J. M. 1989a. Effects of irradiance and ultraviolet radiation on photoadaptation in the zooxanthellae of *Aiptasia pallida:* primary production, photoinhibition, and enzymic defenses against oxygen toxicity. *Mar. Biol.* 134:129–41
99. Lesser, M. P., Stochaj, W. R., Tapley, D. W., Shick, J. M. 1990. Bleaching in coral reef anthozoans: Effects of irradiance, ultraviolet radiation, and temperature on the activities of protective enzymes against active oxygen. *Coral Reefs* 8:225–32
100. Lessios, H. A. 1988. Mass mortality of *Diadema antillarum* in the Caribbean: What have we learned? *Annu. Rev. Ecol. Syst.* 19:371–93
101. Littler, M. M., Littler, D. S. 1984. Models of tropical reef biogenesis: the

contribution of algae. *Prog. Phycol. Res.* 3:323–64

102. Lorenzen, D. J. 1979. Ultraviolet radiation and phytoplankton photosynthesis. *Limnol. Oceanogr.* 24:1117–20
103. Lough, J. M., Barnes, D. J. 1990a. Possible relationships between environmental variables and skeletal density in a coral colony from the central Great Barrier Reef. *J. Exp. Mar. Biol.* 134:221–41
104. Lough, J. M., Barnes, D. J. 1990b. Intra-annual timing of density band formation of Porites coral from the central Great Barrier Reef. *J. Exp. Mar. Biol. Ecol.* 135:35–37
105. Loya, Y. 1976. Effects of water turbidity and sedimentation on the community structure of Puerto Rican corals. *Bull. Mar. Sci.* 26:450–66
106. MacCracken, M. C., Budyko, M. I., Hecht, A. D., Izrael, Y. A., ed. 1990. *Prospects for Future Climate: A Special US/USSR Report on Climate and Climate Change*. Chelsea, MI: Lewis Publ. 270 pp.
107. Mackenzie, F. T., Agegian, C. R. 1989. Biomineralization and tentative links to plate tectonics. In *Origin, Evolution, and Modern Aspects of Biomineralization in Plants and Animals,* ed. R. E. Crick, pp. 11–27. New York: Plenum
108. Maragos, J. E. 1972. *A study of the ecology of Hawaii reef corals*. PhD thesis. Univ. Hawaii. 290 pp.
109. Marshall, S. M., Orr., A. P. 1931. Sedimentation on Low Isles Reef and its relation to coral growth. *Sci. Rep. Great Barrier Reef Expedition* 1:94–133
110. Mitchell, J. F. B., Manabe, S., Meleshko, V., Tokioka, T. 1990. Equilibrium climate change—and its implication for the future. In *Climate Change: The IPCC Assessment*. ed. J. T. Houghton, G. J. Jenkins, J. J. Ephraums, pp. 131–64. Cambridge: Cambridge Univ. Press
111. Morse, J. W., Mackenzie, F. T. 1990. *Geochemistry of Sedimentary Carbonates*. New York, Elsevier. 707 pp.
112. Muscatine, L., Falkowski, P. G., Dubinsky, Z., Cools, P., McCloskey, L. R. 1989. The effect of external nutrient resources on the population dynamics of zooxanthellae in a reef coral. *Proc. R. Soc. Lond. Ser. B* 236:311–24
113. Muscatine, L., Porter, J. W. 1977. Reef corals: mutualistic symbioses adapted to nutrient-poor environments. *Bioscience* 27:454–59
114. Muthiga, N. A., Szmant, A. 1987. The effects of salinity stress on the rates of aerobic respiration and photosynthesis in the hermatypic coral Siderastrea siderea. *Biol. Bull.* 173:539–51
115. National Research Council. 1991. *Opportunities in the Hydrologic Sciences*. Washington, DC: Natl. Acad. Press. 348 pp.
116. Neumann, A. C., Macintyre, I. 1985. Reef response to sea level rise: Keep up, catch up or give up. *Proc. Fifth Int. Coral Reef Cong., Tahiti*. 3:105–10
117. Ott, B. 1975. Community patterns on a submerged barrier reef at Barbados, West Indies. *Int. Revue Gesamten. Hydrobiol.* 60:719–36
118. Paulay, G. 1988. *Effects of glacio-eustatic sea level fluctuations and local tectonics on the marine fauna of oceanic islands*. PhD thesis. Univ. Wash., Seattle
119. Paulay, G. 1991. Late Cenozoic sea level fluctuations and the diversity and species composition of insular shallow water marine faunas. In *The Unity of Evolutionary Biology: Proc. Fourth Int. Congress of Systematic and Evolutionary Biology,* ed. E. C. Dudley. Portland: Dioscorides
120. Plath, D. C., Johnson, K. S., Pytkowicz, R. M. 1980. The solubility of calcite—probably containing magnesium—in seawater. *Mar. Chem* 10:9–29
121. Porter, J. W., Fitt, W. K., Spero, H. J., Rogers, C. S., White, M. W. 1989. Bleaching in reef corals: Physiological and stable isotopic responses. *Proc. Natl. Acad. Sci. USA* 86 (Ecology): 9342–46
122. Potts, D. C. 1991. Evolution of reef-building corals during periods of rapid global change. In *The Unity of Evolutionary Biology: Proc. Fourth Int. Congr. of Systematic and Evolutionary Biology,* ed. E. C. Dudley. Portland, Dioscorides
123. Ramanathan, V., Collins, W. 1991. Thermodynamic regulation of ocean warming by cirrus clouds deduced from observations of the 1987 El Niño. *Nature* 351:27–32
124. Raup, D. M., Boyajian, G. E. 1988. Patterns of generic extinction in the fossil record. *Paleobiology* 14:109–25
125. Rowan, R., Powers, D. A. 1991. A molecular genetic classification of zooxanthellae and the evolution of animal-algal symbiosis. *Science* 251: 1348–51
126. Roy, K. J., Smith, S. V. 1971. Sedi-

mentation and coral reef development in turbid water: Fanning lagoon. *Pac. Sci.* 25:234–48

127. Ruddiman, W. F. 1985. Climate studies in ocean cores. In *Paleoclimate Analysis and Modeling*, ed. A. D. Hecht, pp. 197–257. New York: Wiley
128. Salm, R. V. 1990. IUCN Coastal Zone Management Project 9571, Sultanate of Oman. *CZMP4:F6*
129. Sandeman, I. M. 1988. Coral bleaching at Discovery Bay, Jamaica: a possible mechanism for temperature-related bleaching. In *Mass Bleaching of Corals in the Caribbean: a Research Strategy. Research report 88–2.*, ed. J. Ogden, R. Wicklund. Rockville, Md: NOAA Undersea Res. Prog., pp. 46–48
130. Savina, L. A. 1991. Naturally occurring and laboratory induced bleaching in two Caribbean coral species. *Am. Zool.* 31(5):48A
131. Sheehan, P. M. 1985. Reefs are not so different—they follow the evolutionary pattern of level-bottom communities. *Geology* 13:46–49
132. Shen, G. T., Sanford, C. L. 1990. Trace-element indicators of climate change in annually bonded corals. In *Global Ecological Consequences of the 1982–83 El Niño*, ed. P. W. Glynn, pp. 255–83. New York: Elsevier
133. Siebeck, O. 1988. Experimental investigation of UV tolerance in hermatypic corals (Scleractinia). *Mar. Ecol. Prog. Ser.* 43:95–103
134. Skirrow, G. 1975. The dissolved gases—carbon dioxide. In *Chemical Oceanography* ed. J. P. Riley, G. Skirrow, pp. 1–192. London: Academic
135. Smith, A. D., Roth, A. A. 1979. Effect of carbon dioxide concentration on calcification in the red coralline alga *Bossiella orbigniana*. *Mar. Biol.* 52: 217–25
136. Smith, J. B., Tirpak, D., ed. 1989. *The Potential Effects of Global Climate Change on the United States. Policy, Planning, and Evaluation (PM-221), EPA-230–05–89–050.* Washington, DC: US Environ. Protection Agency. 412 pp.
137. Smith, R. C., Baker, K. S. 1980. Stratospheric ozone, middle ultraviolet radiation and carbon-14 measurements of marine productivity. *Science* 208:592–93
138. Smith, S. V. 1983. Coral reef calcification. In *Perspectives on Coral Reefs*, ed. D. G. Barnes, pp. 240–47. Manuka, Aust: Brian Clouston
139. Smith, S. V. 1985. Physical, chemical and biological characteristics of CO_2 gas flux across the air-water interface. *Plant, Cell Environ.* 8:387–98
140. Smith, S. V., Henderson, R. S. 1978. Phoenix Island report. I: An environmental survey of Canton atoll lagoon 1973. *Atoll Res. Bull.* 221:183
141. Smith, S. V., Jokiel, P. L. 1978. Water composition and biogeochemical gradients in the Canton Atoll lagoon. *Atoll Res. Bull.* 221:15–53
142. Smith, S. V., Kimmerer, W. J., Laws, E. A., Brock, R. E., Walsh, T. W. 1981. Kaneohe Bay sewage diversion experiment: perspectives on ecosystem responses to nutritional perturbation. *Pac. Sci.* 35:279–395
143. Smith, S. V., Pesret, F. 1974. Processes of carbon dioxide flux in the Fanning Island lagoon. *Pac. Sci.* 28:225–45
144. Smith, S. V., Veeh, H. H. 1989. Mass balance of biogeochemically active materials (C, N, P) in a hypersaline gulf. *Estuarine, Coastal Shelf Sci.* 29:195–215
145. Stanley, S. M. 1984. Temperature and biotic crises in the marine realm. *Geology* 12:205–8
146. Stoddart, D. R. 1969. Ecology and morphology of recent coral reefs. *Biol. Rev. Cambridge Philos. Soc.* 44:433–98
147. Strom, R. M., Braman, R. S., Jaap, W. C., Dolan, P., Donnelly, K. B., Martin, D. F. 1991. Analysis of selected trace metals and pesticides offshore of the Florida Keys. *Florida Scientist* 54 In press
148. Swinbanks, D. 1991. Algae to the rescue. *Nature* 350:267
149. Szmant, A. M., Gasser, M. P. 1990. The effects of prolonged "bleaching" on the tissue biomass and reproduction of the reef coral *Montastrea annularis*. *Coral Reefs* 8:217–24
150. Tans, P. P., Fung, I. Y., Takahashi, T. 1990. Observational constraints on the global atmospheric CO_2 budget. *Science* 247:1431–38
151. Vaughan, T. W. 1916. The results of investigations of the ecology of the Floridian and Bahamian shoal-water corals. *Proc. Natl. Acad. Sci. USA* 2:95–100
152. Veron, J. E. N., Kelley, R. 1988. Species stability in reef corals of Papua New Guinea and the Indo-Pacific. *Memoirs Aust. Paleontolog.* 6:1–69
153. Vicente, V. P. 1990. Response of sponges with autotrophic endosymbionts during the coral-bleaching episode in Puerto Rico. *Coral Reefs* 8: 199–202
154. Walker, K. I., Ormond, R. F. G. 1982.

Coral death from sewage and phosphate pollution at Aqaba, Red Sea. *Mar. Pollution Bull.* 13:21–25
155. Ware, J. R., Smith, S. V., Reaka-Kudla, M. L. 1992. Coral reefs: sources or sinks of atmospheric CO_2? *Coral Reefs*. In press
156. Warrick, R. A., Oerlemans, H. 1990. Sea level rise. In *Climate Change: The IPCC Assessment*, ed. J. T. Houghton, G. J. Jenkins, J. J. Ephraums, pp. 257–82. Cambridge, Cambridge Univ. Press
157. Warwick, R. M., Clarke, K. R., Suharsono 1990. A statistical analysis of coral community responses to the 1982–83 El Niño in the Thousand Islands, Indonesia. *Coral Reefs* 8:171–79
158. Wells, S. M., ed. 1988. *Coral Reefs of the World*. Vol. 1–3. Cambridge: IUCN
159. Wetzel, R. G., Grace, J. B. 1983. Aquatic plant communities. In *CO_2* and Plants: The Response of Plants to Rising Levels of Atmospheric Carbon Dioxide, ed. E. R. Lemon, pp. 223–280. *AAAS Selected Symposium 84* Boulder, Colo: Westview
160. White, J. K. F. 1980. *Distribution, recruitment, and development of the borer community on shallow Hawaiian reefs*. PhD thesis. Univ. Hawaii 223 pp.
161. Wigley, T. M. L., Barnett, T. P. 1990. Detection of the greenhouse effect in the observations. In *Climate Change: The IPCC Assessment*, ed. J. T. Houghton, G. J. Jenkins, J. J. Ephraums, pp. 239–56. Cambridge: Cambridge Univ. Press
162. Wolanski, E., Pickard, G. L. 1983. Upwelling by internal tides and Kelvin waves at the continental shelf break on the Great Barrier Reef. *Aust. Jour. Mar. Freshwater Res.* 34:65– 80
163. Worrest, R. C. 1982. Review of literature concerning the impact of UV-B radiation upon marine organisms. In *The Role of Solar Ultraviolet Radiation in Marine Ecosystems*, ed. J. Calkins, pp. 429–57. New York: Plenum
164. Worrest, R. C. 1983. Impact of solar ultraviolet-B radiation (290–320 nm) upon marine microalgae. *Physiol. Plant* 58:428–34
165. Worrest, R. C., Grant, L. D. 1989. Effects of ultraviolet-B radiation in terrestrial plants and marine organisms. In *Ozone Depletion: Health and Environmental Consequences*, ed. R. R. Jones, T. Wigley, pp. 198–206. New York: Wiley
166. Yamazato, K. 1970. Calcification in a solitary coral *Fungia scutaria* Lamark, in response to environmental factors. *Bull. Sci. Engineer Div., U. Ryukyus, Math. Nat. Sci.* 13:59–122
167. Yellowlees, D. 1991. Land use patterns and nutrient loading of the Great Barrier Reef region. *Proc. Workshop, Nov. 17–18, 1990, Townsville. Sir George Fisher Cent. Trop. Mar. Stud. Queensland, Aust.*
168. Yonge, C. M. 1931. The biology of reef building corals. *Sci. Rep. Br. Mus. (Nat. Hist.)* 1:353–91.

Annu. Rev. Ecol. Syst. 1992. 23:119–39

GLOBAL CHANGE AND FRESHWATER ECOSYSTEMS

Stephen R. Carpenter
Center for Limnology, University of Wisconsin, Madison, Wisconsin 53706

Stuart G. Fisher and Nancy B. Grimm
Department of Zoology, Arizona State University, Tempe, Arizona 85287

James F. Kitchell
Center for Limnology, University of Wisconsin, Madison, Wisconsin 53706

KEY WORDS: aquatic ecology, climate change, lakes, streams, wetlands

INTRODUCTION

The biosphere contains only 0.014% of Earth's water, distributed among lakes (0.008%), soils (0.005%), and the atmosphere, rivers, and biota (0.001%) (79). An additional 2.58% of Earth's water is fresh, occurring as ice (1.97%) or groundwater (0.61%). The remainder of Earth's water is saline. Climate affects key fluxes (evaporation, water vapor transport, and precipitation) that determine the amount and distribution of freshwater (79). The availability of freshwater to both ecosystems and humans is therefore sensitive to changes in climate.

The small amount of water found in the biosphere has large significance for ecosystems and society. The availability of water influences the distributions of major biome types and potential agricultural productivity (35). Humans depend directly on freshwater for drinking, irrigation, industry, transportation, recreation, and fisheries. Expanding human populations and changes in global climate will exacerbate already severe stress to freshwater resources in some regions (1, 79, 108, 163, 172). In others, increased availability of water will potentially mitigate stress (1, 15, 163, 172).

This review addresses the potential effects of global climate change on lake and stream ecosystems. Concern for brevity severely constrained the topics covered. Sustained shifts in global climate will have enormous effects on distributions and interactions of species (38, 71). We limit this review to

0066-4162/92/1120-0119$02.00

fishes, which interact strongly with lower trophic levels (18, 122), illustrate key considerations common to other aquatic organisms, and are significant to the public. Although wetlands interact intimately with streams and lakes (169), our consideration is limited to their interactions with atmospheric chemistry.

The major consequences of global change for freshwater ecosystems depend on the temporal and spatial scales at which effects are assessed. We have organized this review around two broad scales of change. Transitional scales pertain to landscape-level shifts in location, morphometry, and persistence of lakes and streams and their biotas over decades to centuries. Perturbational scales pertain to events with return times of years to decades that affect entire stream or lake ecosystems, such as floods, droughts, or fish recruitment events. We then turn to ecosystem metabolism as an integrated response to climate, and to feedbacks between freshwaters and regional climate; we close the review with salient uncertainties that should influence future research priorities.

FRESHWATERS IN TRANSITION

Future Distributions of Freshwaters

Over the past 25,000 years, freshwater ecosystems have undergone massive changes in spatial extent correlated with trends in regional climate (150). Future climate change is likely to produce comparable changes in the supply and distribution of freshwater.

Forecasts of future precipitation and evaporation are uncertain, but the consensus is that greenhouse warming will increase both (139). Some areas of the Earth will become wetter and others drier, thereby shifting the global distribution of streams, lakes, and wetlands (163). Predictions of variability are more uncertain than predictions of means (139). A recent analysis indicates that greenhouse warming may decrease variability of temperature and the diurnal range, while increasing the variability of precipitation (133).

The difficulty of forecasting future distributions of freshwaters is exacerbated by the effect of transpiration on runoff (64, 135, 170). Climate change is likely to alter terrestrial vegetation, soils, and soil moisture, which affect evapotranspiration (29, 35, 113).

Numerous hydrologic studies predict changes in runoff over the next century if global warming of 2–4°C occurs (Table 1). These regional studies of large basins apply water balance models (155) driven by precipitation, soil moisture, and potential evapotranspiration. Generalizations that emerge are: (i) the precipitation-runoff relationship is nonlinear (170); (ii) under reduced precipitation scenarios (predicted for continental interiors—171), reductions in streamflow are likely, particularly in western US basins (Table 1; see also 172); (iii) arid basins are more sensitive to precipitation changes than are humid basins (Table 1; see also 69, 170); (iv) seasonal shifts in streamflow distributions are probably more significant than changes in total annual runoff, particularly in cold or mountainous regions where winter precipitation now occurs as snow (Table 1; see also Ref. 81).

In arid regions such as the North American Great Basin, drier conditions may decrease water supply rates below current human demand (44). Lakes and wetlands are expected to contract in area. Projected water quality changes are negative (172).

In currently moist regions such as the Laurentian Great Lakes, drying will decrease stream flows, lake levels, wetland areas, and water supplies (24, 132). In heavily populated areas, the magnitude of change is uncertain because effects on water demand for irrigation, energy, and cooling are unknown (23). Drier conditions may affect transportation in regions that now have abundant water. In the Great Lakes, shippers will enjoy a longer ice-free season but will encounter lower lake levels (87).

Certain regions may become wetter as a result of global climate change (140). Increased stream flow, more frequent floods, and expansion of lakes and wetlands are likely (15). Reduced snow cover will reduce the need to salt roads, and water quality will benefit (15).

Fluvial Geomorphology

Changes in river channel form will be a consequence of global warming and associated precipitation changes. Erosion and sediment transport are linked to climate (68, 173), with highest rates in regions having the most variable precipitation and runoff (2, 54). Holocene arid phases of centuries duration were characterized by rare, high-magnitude flash floods, while floods in humid phases were more frequent but less intense (76). High magnitude floods cause channel widening, which is exacerbated by sparse riparian vegetation characteristic of drought phases. Periods of increased precipitation permit reestablishment of vegetation with attendant channel narrowing and flood flow moderation (89). Arid and semiarid regions experience, in comparison to humid regions, amplified runoff response to precipitation change (28, 170), higher streamflow variability (118), and more severe flash floods (2, 48). These facts lead several authors to suggest that these systems might provide early warnings of global change (28, 52, 118).

In forest streams, wood is an important geomorphic agent with a long residence time (61). Large woody debris in streams influences channel morphology (55, 70, 159), habitat, and stream ecosystem processes (7, 160) but can be removed by debris flows (4, 151). Frequency of debris torrents, landslides, and earthflows, which are triggered by extreme rainstorms, will no

Table 1 Results of hydrologic modeling and water balance studies predicting regional impacts of climate warming on runoff. Scenarios are: 2°C temperature increase (unless otherwise indicated) and −20% to +20% change in precipitation. Runoff values are expressed as per cent change and are annual averages unless otherwise indicated. Where 2 values are listed, they represent minimum and maximum estimates.

Location and Characteristics	Per cent change in precipitation					Comments	Reference
	−20	−10	0	+10	+20		
Western USA (7 basins)		−40					1
		−76					
Arid basin		−50		+50		1°C temperature increase	2
Humid basin		−25		+25			
Colorado River Basin		−40		−18			3
Sacramento River	−42	−32	−22	−12	−1	Summer runoff	
	−24	−9	+8	+25	+44	Winter runoff	4
Great Basin (4 basins)	−33	−17				Col 1 is −25% precipitation	5
	−51	−28					
Delaware River	−51	−32	−14	+8	+31	Seasonal runoff change, especially in cold northern portion of basin	6
Greece-(3 basins)	−27	−15	0	+12	+25	Seasonal runoff change, especially mountainous basins; humid basin less sensitive.	7
	−31	−17	−3	+13	+27		
Saskatchewan River	−36	−21	0	−3	+27	Semiarid sub-basins most sensitive (highest extremes)	8
	−100	−100	−100	+63	+286		

References: 1- Stockton & Boggess 1979; 2- Nemec & Schaake 1982; 3- Revelle & Waggoner 1983; 4- Gleick 1987; 5- Flaschka et al. 1987; 6- McCabe and Ayers 1989; 7- Mimikou et al. 1991; 8- Cohen 1991.)

doubt change as precipitation variability and timing change. Altered regimes of drought and moisture may also increase the incidence of fire in forest, scrub, and grassland (113), thereby changing the rate, input regime, and role of wood in streams (103).

Inputs from Terrestrial to Freshwater Ecosystems

The zone of intense biogeochemical activity at the land-water interface will expand and contract with fluctuations in the water supply (169). This interface is especially important in streams, temporary ponds, and small lakes because of their large perimeter-to-volume ratio. Stream ecologists have paid close attention to the influence of terrestrial vegetation on freshwaters (43, 63). Utility of terrestrial leaf litter to detritivores varies greatly among species (115) and in the combination of species within individual leaf packs (5, 152). Change in climate will likely alter the total biomass, productivity, and species composition of the riparian community and, in turn, the supply of organic matter and nutrients to freshwaters (99).

Climate change may alter the composition of riparian vegetation (99). Past vegetation changes that had substantial effects on stream ecosystems include the demise of American chestnut (143) and the spread of exotic saltcedar (*Tamarisk* spp.) through the Colorado basin (147). Transplant experiments indicate that aquatic consumers prefer detritus derived from local plants (91, 102). As high CO_2 levels cause N content of leaves to decline, grazing by terrestrial insects may decline (36), increasing the proportion of leaf production that enters streams and slowing in-stream decomposition (99). Such changes in the riparian zone will likely force marked alterations in aquatic communities and the nature and rates of ecosystem processes. While there is little question that adjustment of the aquatic sector can keep pace with riparian dynamics, consequences of vegetation shifts may also be translated far downstream through transport, altering longitudinal patterns described by the River Continuum Concept (162).

Distribution of Freshwater Fishes

Temperature tolerances often govern both the local and biogeographic distribution limits of freshwater fishes, which are generally grouped as coldwater (e.g. Salmonidae), coolwater (e.g. Percidae) or warmwater (e.g. Centrarchidae and Cyprinidae) forms (84). Thermal limits will be altered by global warming. Distributions of aquatic species will change as some species invade more high latitude habitats or disappear from the low latitude limits of their distribution (66, 142).

Small, shallow habitats (ponds, headwater streams, marshes, and small lakes) will first express effects of reduced precipitation (83). Consequences for aquatic organisms are obvious. Of greatest immediate concern are the severely limited desert pool and stream habitats now occupied by threatened and endangered fishes (90, 100). Similarly, the spawning habits of many species require small and shallow habitats as refuge and nursery for both gametes and early life history stages. Prospects for successful relocation of spawning activities exist but may be thwarted by the strong imprinting and homing behavior known for many species.

Warming will alter the stream habitat of coldwater fishes such as trout, charr, and salmon (94). Projected increases in air temperature will be transferred, with local modification, to groundwaters, resulting in elevated temperatures and reduced oxygen concentrations. At low latitudes and altitudes these changes may have immediate adverse effects on eggs and larvae, which are usually deposited at sites of groundwater discharge. They may also reduce suitable habitat during summer months. Conversely, the winter stressors of osmoregulation at extremely low temperatures and physical damage by ice will be diminished. At higher latitudes and altitudes, elevated groundwater temperatures will increase the duration and extent of optimal temperatures; all life history stages benefit accordingly (94).

Many major river systems have an east-west drainage pattern. Lacking the opportunity to move north within the river courses, many freshwater fishes will not have access to a thermal refuge. Some genetic adaptation is possible but the rates of mutation are low and the prospect for development of tolerant strains is an inverse function of the rate of warming (90). Strong selection pressure and low capacity for adaptation may lead to local extinctions for some species (125).

Warming of freshwater habitats at higher latitudes is more likely to open them to invasion. Shuter & Post (142) examined life histories and energy requirements for a variety of temperate, North American fishes. All aspects of the life history were accelerated by global warming. The constraint of size-dependent winter starvation would be relaxed and permit many species to expand their distributions.

By coupling output from physical models of thermal structure for the Great Lakes with a fish bioenergetics model (74), Magnuson et al (85) analyzed potential responses for guilds of coldwater, coolwater, and warmwater fishes. Results included greater total annual growth due to increases in duration of the growing season and volume of habitat offering preferred temperatures (60). Higher water temperatures and increased metabolic costs would lead to growth declines if prey production was insufficient.

An insightful analysis of prospective invasion of the Great Lakes was offered by Mandrak (86). After glaciation, the Great Lakes were colonized by at least 122 fish species from the Mississippi and Atlantic refugia. Fishes now present in those drainages but not recorded in the Great Lakes were evaluated as candidates for future invasion on the basis of habitat requirements (trophic interactions, reproductive biology, thermal niche). Of the 58 species considered, 27 (mostly Cyprinidae and Centrarchidae) were deemed likely to invade Great Lakes habitats if global warming released the temperature constraints to the limits of their northerly distribution.

Mandrak's (86) large-scale analysis raises issues pertinent at the local scale of other lakes. Omnivorous warmwater fishes that have short life spans are most likely to invade and flourish in habitats where impoundment, water quality degradation, and fishery exploitation have disturbed the environment (82, 83). Few lakes and streams have been spared these anthropogenic effects. As a result, few fish communities may be resistant to colonization as thermal barriers to migration are removed (93). Autecological constraints, the regional pool of potential invaders, and local community composition must be considered to assess prospects for invasion (156).

Fish species diversity, biological production rates, and fisheries yields are inversely related to latitude (22, 128). The majority of freshwater fish species occupy low-to-middle latitude environments. Most of the world's lakes are of glacial origin and occupy higher latitudes. Global warming makes the thousands of glacial lakes of North America, Europe, and Asia more open to colonization and thermal enhancement of production. In total, the volume of habitat available to fishes is likely to increase. Destabilization of current communities may occur as glacial relicts and coldwater species are displaced (138), but compensatory replacement will likely proceed as invasions by coolwater and warmwater species occur. As a result, the overall diversity and productivity of fishes in north temperate lakes will probably increase. Principles expressed in the thermal ecology of fishes likely extend to other aquatic fauna. Freshwater communities will reflect changes in autecological constraints and the ecological ramifications of changing food webs (38, 71).

RHYTHMS OF PERTURBATION

Analysts of climate change expect the frequency of surprises to increase (139). Surprises include floods, droughts, and thermal extremes that cause abrupt change in ecosystem process rates or populations of key species. Such changes have substantial ecological significance because rhythms of temporal variability in streams and lakes are entrained by cycles of hydrology and recruitment of keystone species (17, 52). Such cycles typically have periods of one to many years, measured at the spatial scale of a particular stream or lake ecosystem. Here we consider the consequences of changing perturbation cycles for riparian and floodplain ecosystems, hydrologic disturbances of streams, water renewal in lakes, and fish recruitment.

Perturbations of Riparian Zones and Floodplains

The riparian zone undergoes a cycle of establishment and destruction owing to patterns of hydrologic disturbance and recruitment characteristics of riparian trees (3, 31, 49). Rare catastrophic events such as mass wasting of hillslopes remove soil, debris, and vegetation from affected areas and expose bedrock surfaces. These events affect only a small fraction of the watershed but may influence up to 10% of the channel network as material enters and dams recipient streams (151). The geomorphic effect of slumps and earthflows can persist for thousands of years (34); however, biota of recipient streams may recover to predisturbance levels in 1–2 years (77). Rate of recovery of riparian vegetation is between these extremes.

In arid regions, riparian vegetation is very sensitive to the availability of water, which in turn depends upon amount and distribution of precipitation. Conversion of chaparral to grass in a small Arizona watershed increased water yield and converted a previously intermittent stream to permanent flow, increasing density and extent of riparian vegetation (65). Similar oscillations may occur in unmanipulated desert streams in response to climate change (52). Recovery of in-stream processes after flood or drought is rapid (40); however, the riparian zone is likely to respond more slowly. Because of the strong interaction of aquatic and riparian components (even in aridland streams where riparian vegetation is sparse), it is unlikely that either phase approaches equilibrium given that return times for effective disturbance events differ markedly and disturbance-recovery cycles are always out of phase.

The floodplain is the zone of maximum interaction between a river and its valley (20); the floodplain depends upon regular inundation to supply nutrients and organic matter supply and to provide a breeding ground for fishes and their food organisms. Especially during initial stages, inundation influences river water chemistry (168). Inundation extent and duration are sensitive to climate, so changes in precipitation and runoff amount and timing will influence floodplain-river interactions. The inundation both provides waters and stresses riparian plants, and it controls colonization by exotics (11). Thus climate change has a substantial potential for altering riparian vegetation in this and similar drainages.

In the Mississippi River, floods of less than annual return frequency may be significant disturbances which influence, through erosion and sedimentation, the distribution of macrophytes and primary production (145). The influence of floodplains in this system extends far downstream. Similarly, dependence upon allochthonous inputs increases longitudinally with river size in the blackwater Ogeechee River of Georgia owing to lateral floodplain contributions (98), counter to predictions of the river continuum concept (162). Floodplain contribution to the stream's energy budget is linked to seasonal flooding, which provides a bimodal pulse of organic matter to stream consumers (80, 97).

Hydrologic Disturbance of Streams

The most obvious and immediate effects of global climate change on stream ecosystems involve changes in hydrologic patterns. Complex relationships link climate to runoff (163). Changes in total precipitation, extreme rainfall events, and seasonality will affect the amount, timing, and variability of flow. Changes in total amount or timing of runoff may result in increased or decreased intermittency (118, 165), while altered amount, variability, and extremes of runoff may affect timing, frequency and severity of flash flooding (47). Such changes in magnitude and temporal distribution of extreme events may disrupt ecosystems more than changes in mean conditions (52, 118, 164). Extreme events comprise the disturbance regime of lotic ecosystems and are mediated by catchment geology, soil chemistry, and vegetation (101, 118, 165).

The role of disturbance in lotic communities and ecosystems is currently a major focus of research (111, 114, 119, 124, 131, 175). Despite the attention it has received, there is no consensus as to the importance of disturbance in determining structure and functioning (but see 131), although a strong case can be made for expecting regional variation (39, 119, 120). Poff (118) suggests that in a warmer, drier climate many perennial streams fed by runoff might become intermittent because of their high flow variability, while groundwater-fed streams would be buffered against such changes. Streams fed by snowmelt, which have highly predictable flood and flow regimes, might become less predictable with winter warming (increased rain-on-snow events). Increasing aridity may render flows of many more streams unpredictable (52). The non-linear runoff response of arid regions may be caused by development of a hydrophobic layer on soils during dry periods, which increases the likelihood of overland flow (28).

Altered seasonality of both rainfall and spates is a likely consequence of global warming that would alter temporal variation in controls of ecosystem processes (50, 52). Post-flood succession (i.e. disturbance control) occurs during approximately 30 days following spates in desert streams (40); thereafter biotic controls predominate. Comparison of years that differ in flood timing but that have similar total annual runoff revealed dramatic shifts in the balance of disturbance controlled vs biotically controlled states (50). Similar inferences could be made for many other streams in which disturbance and biotic controls alternate (see 37, 58, 59, 121, 123).

As streams dry, mobile organisms are concentrated and biotic interactions surely intensify (148, 153). These interactions or physiological stresses may differentially affect species or size classes of organisms (e.g. 130). Competitive interactions are normally important only in summer in a California stream, but they persisted year round during a drought year when caddisfly densities were not reduced by the usual high winter flows (37). Year-to-year variation in competition was attributed in another study to variation in timing and intensity of winter floods (58).

Reduced flows can concentrate pollutants (9, 78, 157). For this reason, assessment of the likelihood of extreme low flows in streams influenced by pollutant discharges is a high priority (see also 45, 99). Chessman & Robinson (21), for example, reported poor water quality associated with record low flows. These water quality changes included reduced oxygen concentration, a likely consequence of reduced or intermittent flow even in unimpacted streams (149).

During floods, the hydrologic linkage between streams and hyporheic or groundwater zones is accentuated (88). As surface flow declines during drying, the hyporheic zone assumes more importance in its effects on surface water chemistry and biota (149), including input of low-oxygen and high-nutrient water (161). Loss of hydrologic connection between these subsystems (149), however, can result in significant stresses to biota in underground habitats.

Water Renewal and Chemical Inputs to Lakes

Allochthonous inputs to lakes depend on the timing, frequency, and magnitude of runoff events (72, 163). These effects are amplified where human activities have disturbed soils or vegetation (108). Increases in precipitation and its variability would increase nutrient inputs to lakes, contributing to eutrophication (127, 163).

Conversely, increases in water renewal rate can decrease concentrations of nutrients and contaminants derived from point sources or sediments (62, 127, 136). The water renewal rate scales the rate of ecosystem response to chemical inputs (127). In lakes of western Ontario during 1969–1989, decreased precipitation and increased evaporation decreased the water renewal rate, increasing concentrations of chemicals in both lakes and streams (138). Increased nutrient concentrations lead to increases in primary production and reductions in hypolimnetic oxygen, potentially endangering some fish stocks (138). In softwater lakes susceptible to acidification by acid deposition, drought can cause both pH and acid neutralizing capacity to decline (166). Drought reduces groundwater inputs rich in acid neutralizing chemicals (166). Concentrations of contaminants such as organochlorine compounds and heavy metals can also be expected to increase as the water renewal rate declines, possibly increasing accumulation in the aquatic food chain and ingestion of the chemicals by wildlife and humans (126, 154). Theory predicts that all of these trends would be reversed in lakes that undergo increases in water renewal rate (127). Fluctuations in water renewal rate caused by global change therefore translate into fluctuations in primary production and cycling of nutrients and contaminants.

Variability in Fish Recruitment

If climate becomes more variable (139), local populations will more often experience extreme events such as those that produce lethal conditions for short periods of time. Deletion of adult stocks by short-lived stress can ramify throughout the trophic network for extended periods. Most vunerable are the populations of coldwater fishes in habitats near the low-latitude limits of their range. For example, a summerkill of the pelagic, planktivorous lake herring (*Coregonus artedii)* in Lake Mendota, Wisconsin, occurred during a period of high temperatures and low wind velocities. As a result, total predation on zooplankton was reduced by nearly 50%, large *Daphnia* increased, and intensified grazing by *Daphnia* caused substantial reduction in phytoplankton (72). Analogous examples can be drawn from river systems where effects of variable temperatures and flows affect the fishes and alter the network of interactions regulating both herbivores and the filamentous algae they consume (121, 123). Similarly, the synchrony of predator and prey life histories in reservoirs is affected by rates of warming (30).

The relative strengths of year classes in many fishes correlate with weather conditions during the spawning and early life history stages (158). Relatively modest changes in the weather can dramatically increase variability in recruitment (27). A cascade of food web effects can follow from the removal or enhancement of fishes whose effects as predators cause variability in the trophic structure, productivity, and water quality of lakes (17–19, 30, 53).

One of the immediate interactions between the public and the effects of global change will occur in recreational and commercial fisheries. Fish populations are notoriously variable, and fisheries yields are often heavily dependent on the occasional strong year class (117). Because survival during early life history stages is a key to recruitment success (158, 174), the increased variability in critical habitat will increase variability in the success of individual cohorts. The social and economic effects will be immediately apparent and, if history is a guide, the subject of policy to remedy the wrongs of nature (46).

ECOSYSTEM METABOLISM

Stream Metabolism

Temperature directly affects rate of photosynthesis and respiration in streams, but it does so differentially: as temperature rises, respiration tends to increase faster than primary production, and the P/R ratio falls (16), assuming ample organic substrates are present. Other factors being equal, a drop in P/R renders streams more of a sink than a source for organic matter and will decrease export downstream (16).

In addition to production and respiration, stream ecosystem metabolism depends on import, export, and intrasystem storage pools of organic matter (43). Litter input from the riparian zone may change in quantity and quality as temperature rises (99). Dissolved organic carbon entering streams via interflow may decrease as soil respiration rises (99). Rate of decomposition of organic matter in situ may change as well if shredder (leaf-consuming) macroinvertebrate populations are enhanced, or it may decline as a result of increased temperature. Changes in inputs of wood affect the capacity of streams to retain leaves and fine particulate organic matter, altering respiration associated with organic substrates (67, 144).

Floods affect primary production in some streams (6, 41) and may flush significant amounts of detritus downstream (57). In some cases, floods remove entire ecosystem components. Grimm & Fisher (51) showed that the hyporheic zone can be an important contributor to ecosystem respiration, reducing the P/R ratio in proportion to hyporheic volume. Storm events can alter hyporheic volume by export of sand and shift the P/R ratio upward (39, 42). In this manner, altered hydrology due to changing climate can convert a system that is largely a processor of organic matter ($P/R < 1$) to one that is primarily an exporter ($P/R > 1$).

Heterotrophic streams of coniferous forests are equally sensitive to the hydrologic regime and are far less efficient during wet than dry years (26). Spiralling distance, an index of processing efficiency (109), also increases with discharge. This affects spiralling not only of organic matter (104, 110) but also of inorganic substances, because organic substrates are especially active in nutrient retention (106).

Lake Productivity

In temperate lakes, global warming is expected to increase the heat content and the durations of the ice-free and stratified seasons (92). Analyses across a latitude gradient from the equator to 75° N showed that primary production was directly related to mean air temperature and length of the growing season (14). Trophic transfer efficiencies across the same gradient suggest that secondary and tertiary production are also related inversely to latitude (14). The seasonal variability of primary production increased with latitude across a similar gradient (95). An empirical analysis of responses of aquatic systems to temperature change suggested that global warming would lead to a general increase in lake productivity at all trophic levels (128). Increased production will exacerbate hypolimnetic oxygen depletion in productive lakes (8).

Changes in cloud cover are among the critical uncertainties in projections of future climate (33). Increased frequency of cloudy days has significant implications for deep-dwelling phytoplankton that account for substantial primary production in oligotrophic and mesotrophic lakes during summer stratification (105, 146). Deep algal layers exploit high rates of nutrient supply by eddy diffusion from the hypolimnion, but these algae can be limited by grazers, low temperatures, or low light (146). If global warming reduces vertical mixing (92), the metalimnion will become a more important habitat for phytoplankton. Increased cloud cover may reduce surface irradiance to the point that algae cannot maintain a positive carbon balance in deep, nutrient rich water. If photosynthesis were restricted to nutrient poor surface waters, primary production would decrease by factors of two to three (105, 146).

FEEDBACKS

Greenhouse Gases

Wetlands contribute 22% of the annual methane flux to the atmosphere (56). These ecosystems, which achieve their greatest extent in polar and tropical belts (56), are active sites because of waterlogged soils and anoxia (required for the anaerobic process of methanogenesis). Most of the methane production by wetlands occurs above 40°N latitude (137), where global warming may be greatest (140). Rates of methanogenesis are directly related to temperature (56). Increased temperature will thaw permafrost, which may promote both areal expansion of wetlands and drying of extant ones (134). The balance of these processes will determine whether the northern wetland contribution to increasing atmospheric methane increases or decreases.

Extensive wetlands are connected to large tropical rivers through seasonal cycles of flood and drought (167). Substantial methane derives from these systems (32, 137). Disruptions of the hydrologic cycle could profoundly influence seasonal patterns of inundation and thereby affect methanogenesis in tropical and subtropical wetlands (25, 99).

Carbon Storage

One of the largest uncertainties in global climate feedbacks concerns alteration of carbon storage. Thawing of permafrost in polar regions could have significant effects on carbon storage (137). Aquatic systems are particularly important in fluxes of peat-derived carbon (116), which is oxidized in aquatic ecosystems (75) and exported from arctic rivers (112).

Regional Climate

Large freshwater ecosystems moderate regional climates through effects on the hydrologic cycle and on ocean circulation. Water evaporated from extensive tropical wetlands falls as rain over a much wider area (e.g. all of South America plus the southwestern Atlantic Ocean for Amazon basin evaporation; 96); drying or expansion of these areas will thus influence climate over large regions. Changes in discharge from the world's large rivers may alter ocean circulation and thus influence world climate (12, 13). Melack (96) summarized paleooceanographic evidence for a major climatic impact on North Atlantic regions of cool, glacial meltwater discharged by the Mississippi River into the Gulf of Mexico. North-flowing rivers like the MacKenzie can move enough heat into arctic regions to moderate regional climate (137). Glacier melting due to global warming, in addition to having catastrophic effects on stream ecosystems (112), may also contribute to changes in oceanic circulation and thus climate.

Anthropogenic Effects

Abundant freshwaters attract human settlement, activating feedbacks that alter ecosystems and regional climate. In most developed nations, extensive draining and filling of wetlands (108) has implications for trace gas budgets. In North America, trapping reduced beaver populations to near extinction, which affected the extent of waterlogged, anoxic soils over vast regions and potentially altered global methane flux (107). Simplification of river channels has reduced extent of floodplains and associated wetlands (10, 141) with unknown consequences for trace gas flux. Abuses of water supply create the need for water development, water treatment, desalinization, fertilizer use, health care, flow diversion, and ecosystem restoration, all of which require combustion of fossil fuels and release of greenhouse gases.

We can expect humans to redistribute freshwater as an adaptive response to global climate change (1, 137). The direct effects on aquatic resources and indirect effects on climate of such policies are largely unknown.

PRIORITIES

Freshwaters have been neglected in research planning for global change (1). Yet, freshwaters are critical for sustainability of ecosystems and society and are tightly coupled to climate and land use (38, 73). Although much of the literature of aquatic ecology is relevant to global change, little work has focused explicitly on the consequences of global change for freshwater ecosystems. It is time to correct these oversights. We will be surprised and disappointed if this review is not outdated in a few years.

Aquatic ecologists and global climatologists need to adjust their research to a common scale. Aquatic ecologists must reinvigorate research at landscape and watershed scales, breaching the walls that divide hydrology, wetland ecology, stream ecology, and limnology. We need mesoscale climate models

(139) capable of translating global climate scenarios to ecosystem scales and accommodating the effects of freshwaters on climate. Models that account for effects of precipitation, plant responses to CO_2, vegetation change, and soil change on runoff are also needed.

To date, the most severe stresses on freshwater ecosystems have come from watershed modifications and use and contamination of aquatic resources by humans. Climate change adds to and interacts with substantial ongoing anthropogenic changes in ecosystems. It is impossible to study the effects of climate change in isolation from the effects of land use change and direct human use of freshwater resources. Humans are an interactive component of aquatic ecosystems, responding to changes in freshwater resources and thereby causing further change. That powerful feedback must be incorporated in our views of ecosystem dynamics.

ACKNOWLEDGMENTS

We thank Dave Hill for discussions of climate models and Linda Holthaus for organizing the references. We are grateful for the support of the National Science Foundation and the University of Wisconsin Sea Grant Institute.

Literature Cited

1. Ausubel, J. H. 1991. A second look at the impacts of climate change. *Am. Sci.* 79:210–21
2. Baker, V. R. 1977. Stream-channel response to floods, with examples from central Texas. *Geol. Soc. Am. Bull.* 88:1057–71
3. Baker, W. L. 1990. Climatic and hydrologic effects on the regeneration of *Populus angustifolia* James along the Animas River, Colorado. *J. Biogeogr.* 17:59–73
4. Benda, D. J. 1990. The influence of debris flows on channels and valley floors in the Oregon Coast Range, U.S.A. *Earth Surf. Proc. Landforms* 15:457–66
5. Benfield, E. F., Jones, D. S., Patterson, M. F. 1977. Leaf pack processing in a pastureland stream. *Oikos* 29:99–103
6. Biggs, B. J. F. 1988. Algal proliferations in New Zealand's shallow stony foothills-fed rivers: toward a predictive model. *Verh. Int. Verein. Limnol.* 23: 1405–11
7. Bilby, R. E., Likens, G. E. 1980. Importance of organic debris dams in the structure and function of stream ecosystems. *Ecology* 61:1107–13
8. Blumberg, A. F., Di Toro D. M. 1990. Effects of climate warming on dissolved oxygen concentrations in Lake Erie. *Trans. Am. Fish. Soc.* 119:210–23
9. Boulton, A. J., Lake, P. S. 1990. The ecology of 2 intermittent streams in Victoria, Australia. 1. Multivariate analyses of physicochemical features. *Freshwater Biol.* 24:123–41
10. Bravard, J. P., Amoros, C., Pautou, G. 1986. Impact of civil engineering works on the successions of communities in a fluvial system. *Oikos* 47:92–111
11. Bren, L. J. 1987. The duration of inundation in a flooding river red gum forest. *Austr. For. Res.* 17:191–202
12. Broecker, W. S., Denton, G. H. 1989. The role of ocean-atmosphere reorganizations in glacial cycles. *Geochim. Cosmochim. Acta* 53:2465–501
13. Broecker, W. S., Peng, T-H., Jouzel, J., Russell, G. 1990. The magnitude of global fresh-water transports of importance to ocean circulation. *Clim. Dynam.* 4:73–79
14. Brylinsky, M., Mann, K. H. 1973. An analysis of factors governing productivity in lakes and reservoirs. *Limnol. Oceanogr.* 18:1–14
15. Bultot, F., Coppens, A., Dupriez, G.

L., Gellens, D., Meulenberghs, F. 1988. Repercussions of a CO_2 doubling on the water cycle and on the water balance: a case study for Belgium. *J. Hydrol.* 99:319–47
16. Busch, D. E., Fisher, S. G. 1981. Metabolism of a desert stream. *Freshwater Biol.* 11:301–7
17. Carpenter, S. R., ed. 1988. *Complex Interactions in Lake Communities.* New York: Springer-Verlag. 283 pp.
18. Carpenter, S. R., Kitchell, J. F., eds. *The Trophic Cascade in Lake Ecosystems.* London: Cambridge Univ. Press. In press
19. Carpenter, S. R., Kitchell, J. F., Hodgson, J. R. 1985. Cascading trophic interactions and lake productivity. *BioScience* 35:634–39
20. Chauvet, E., DeCamps, H. 1989. Lateral interactions in a fluvial landscape: the River Garonne, France. *J. No. Am. Benthol. Soc.* 8:9–17
21. Chessman, B. C., Robinson, D. P. 1987. Some effects of the 1982–1983 drought on water quality and macroinvertebrate fauna in the lower LaTrobe River, Victoria. *Aust. J. Mar. Freshwater Res.* 38:288–99
22. Christie, G. C., Regier, H. A. 1988. Measures of optimal thermal habitat and their relationship to yields for four commercial fish species. *Can. J. Fish. Aquat. Sci.* 45:301–14
23. Cohen, S. J. 1986. Impacts of CO_2-induced climatic change on water resources in the Great Lakes basin. *Clim. Change* 8:135–53
24. Cohen, S. J. 1987. Influences of past and future climates on the Great Lakes region of North America. *Water Int.* 12:163–69
25. Cook, A. G., Janetos, A. C., Hinds, W. T. 1990. Global effects of tropical deforestation: towards an integrated perspective. *Environ. Conserv.* 17:201–12
26. Cummins, K. W., Sedell, J. R., Swanson, F. J., Minshall, G. W., Fisher, S. G., et al. 1983. Organic matter budgets for stream ecosystems: problems in their evaluation. In *Stream Ecology: Application and Testing of General Ecological Theory,* ed. J. R. Barnes, G. W. Minshall, pp. 299–353. New York: Plenum
27. Cushing, D.H. 1982. *Climate and Fisheries.* London: Academic
28. Dahm, C. N., Molles, M. C., Jr. 1991. Streams in semiarid regions as sensitive indicators of global climate change. See Ref. 38, pp. 250–60
29. Davis, M. B. 1988. Ecological systems and dynamics. In *Toward an Understanding of Global Change,* Natl. Res. Council. Washington DC: Natl. Acad. Press. 213 pp.
30. DeAngelis, D. L., Cushman, R. M. 1990. Potential applications of models in forecasting the effects of climate change on fisheries. *Trans. Am. Fish. Soc.* 119:224–39
31. DeCamps, H., Fortune, M., Gazelle, F., Pautou, G. 1988. Historical influence of man on the riparian dynamics of a fluvial landscape. *Land. Ecol.* 1:163–73
32. Devol, A. H. , Richey, J. E., Clark, W., King, S., Martinelli, L. 1988. Methane emissions to the troposphere from the Amazon floodplain. *J. Geophys. Res.* 93:1583–92
33. Dickinson, R. E. 1989. Uncertainties of estimates of climate change: a review. *Clim. Change* 15:5–14
34. Dietrich, W. E., Dunne, T., Humphrey, N. F., Reid, L. M. 1982. Construction of sediment budgets for drainage basins. In *Sediment Budgets and Routing in Forested Drainage Basins,* ed. F. J. Swanson, R. J. Janda, T. Dunne, D. N. Swantson, pp. 5–23. *USDA For. Serv. Tech. Rep. PNW-141,* Portland, Ore.
35. Emanuel, W. R., Shugart, H. H, Stevenson, M. P. 1985. Climate change and broad-scale distribution of ecosystem complexes. *Clim. Change* 7:29–43
36. Fajer, E. D., Bowers, M. D., Bazazz, F. A. 1989. The effects of enriched carbon dioxide atmospheres on plant-insect herbivore interactions. *Science* 243:1198–200.
37. Feminella, J. W., V. H. Resh. 1990. Hydrologic influences, disturbance, and intraspecific competition in a stream caddisfly population. *Ecology* 71:2083–94
38. Firth, P., Fisher, S. G., ed. 1991. *Climate Change and Freshwater Ecosystems.* New York: Springer-Verlag. 321 pp.
39. Fisher, S. G. 1990. Recovery processes in lotic ecosystems: limits of successional theory. *Environ. Manage.* 14: 725–36
40. Fisher, S. G., Gray, L. J., Grimm, N. B., Busch, D. E. 1982. Temporal succession in a desert stream ecosystem following flash flooding. *Ecol. Monogr.* 52:93–110
41. Fisher, S. G., Grimm, N. B. 1988. Disturbance as a determinant of structure in a Sonoran Desert stream ecosystem. *Verh. Int. Verein Limnol.* 23: 1183–89
42. Fisher, S. G., Grimm, N. B. 1991. Streams and disturbance: are cross-eco-

system comparisons useful? In *Comparative Analyses of Ecosystems: Patterns, Mechanisms, and Theories,* ed. J. Cole, S. Findlay, G. Lovett, pp. 196–221. New York: Springer-Verlag

43. Fisher, S. G., Likens, G. E. 1973. Energy flow in Bear Brook, New Hampshire: An integrative approach to stream ecosystem metabolism. *Ecol. Monogr.* 43:421–39
44. Flaschka, I. M., Stockton, C. W., Boggess, W. R. 1987. Climatic variation and surface water resources in the Great Basin region. *Water Resources Bull.* 23:47–57
45. Ford, D. E., Thornton, K. W. 1991. Water resources in a changing climate. See Ref. 38, pp. 26–47
46. Glantz, M. H. 1990. Does history have a future? Forecasting climate change effects on fisheries by analogy. *Fisheries* 15:39–44
47. Gleick, P. H. 1987. Regional hydrologic consequences of increases in atmospheric CO_2 and other trace gases. *Clim. Change* 10:127–61
48. Graf, W. L. 1988. *Fluvial Processes in Dryland Rivers.* New York: Springer-Verlag. 346 pp.
49. Gregory, S. V., Swanson, F. J., McKee, W. A., Cummins, K. W. 1991. An ecosystem perspective of riparian zones. *BioScience* 41:540–51
50. Grimm, N. B. 1992. Implications of climate change for stream communities. See Ref. 71. In press
51. Grimm, N. B., Fisher, S. G. 1984. Exchange between interstitial and surface water: implications for stream metabolism and nutrient cycling. *Hydrobiologia* 111:219–28
52. Grimm, N. B., Fisher, S. G. 1991. Responses of arid-land streams to changing climate. See Ref. 38, pp 211–33
53. Gulati, R. D., Lammens, E. H. R. R., Meijer, M.-L., van Donk, E., eds. 1990. *Biomanipulation—Tool for Water Management.* Dordrecht: Kluwer
54. Harlin, J. M. 1980. The effect of precipitation variability on drainage basin morphometry. *Am. J. Sci.* 280: 812–25
55. Harmon, M. E., Franklin, J. F., Swanson, F. J., Sollins, P., Gregory, S. V., et al. 1986. Ecology of coarse woody debris in temperate ecosystems. *Adv. Ecol. Res.* 15:133–302
56. Harriss, R. C., Frolking, S. E. 1991. The sensitivity of methane emissions from northern freshwater wetlands to global warming. See Ref. 38, pp. 48–67
57. Hedin, L. O., Mayer, M. S., Likens, G. E. 1988. The effect of deforestation on organic debris dams. *Verh. Int. Verein Limnol.* 23:1135–41
58. Hemphill, N. 1991. Disturbance and variation in competition between 2 stream insects. *Ecology* 72:864–72
59. Hemphill, N., Cooper, S. D. 1983. The effect of physical disturbance on the relative abundance of two filter-feeding insects in a small stream. *Oecologia* 58:378–82
60. Hill, D. K., Magnuson, J. J. 1990. Potential effects of global climate warming on the growth and prey consumption of Great Lakes fishes. *Trans. Am. Fish. Soc.* 119:265–75
61. Hupp, C. R. 1984. Dendrogeomorphic evidence of debris flow frequency and magnitude at Mount Shasta, California. *Environ. Geol. Wat. Sci.* 6:121–28
62. Hutchinson, N. J., Neary, B. P., Dillon, P. J. 1991. Validation and use of Ontario's Trophic Status Model for establishing lake development guidelines. *Lake Reserv. Manage.* 7:13–23
63. Hynes, H. B. N. 1975. The stream and its valley. *Verh. Int. Verein. Limnol.* 19:1–15
64. Idso, S. B., Brazel, A. J. 1984. Rising atmospheric carbon dioxide concentrations may increase streamflow. *Nature* 312:51–53
65. Ingebo, P. A. 1971. Suppression of channelside chaparral cover increases streamflow. *J. Soil Water Conserv.* 26:79–81
66. Johnson, T. B., Evans, D. O. 1990. Size-dependent mortality of young-of-the-year white perch: climate warming and invasion of the Laurentian Great Lakes. *Trans. Am. Fish. Soc.* 119:301–13
67. Jones, J. B., Smock, L. A. 1991. Transport and retention of particulate organic matter in 2 low-gradient headwater streams. *J. No. Am. Benthol. Soc.* 10:115–26
68. Judson, S., Ritter, D. F. 1964. Rates of regional denudation in the United States. *J. Geophys. Res.* 69:3395–401
69. Karl, T. R., Riebsame, W. E. 1989. The impact of decadal fluctuations in mean precipitation and temperature on runoff: a sensitivity study over the United States. *Clim. Change* 15:423–47
70. Keller, E. A., Swanson, F. J. 1979. Effects of large organic material on channel form and fluvial processes. *Earth Surf. Proc.* 4:361–80
71. Kingsolver, J., Kareiva, P., Huey, R., eds. 1992. *Biotic Interactions and Global Change.* Sunderland Mass: Sinauer. In press

72. Kitchell, J. F., ed. 1992. *Food Web Management: A Case Study of Lake Mendota*. New York: Springer-Verlag. In press
73. Kitchell, J. F., Carpenter, S. R., Bayley, S. W., Ewel, K. C., Howarth, R. W., et al. 1991. Aquatic ecosystem experiments in the context of global climate change: Working group report. In *Ecosystem Experiments*, ed. H. A. Mooney et al., pp. 229–35. England: Wiley
74. Kitchell, J. F., Stewart, D. J., Weininger, D. 1977. Applications of a bioenergetics model to perch *(Perca flavescens)* and walleye (*Stizostedion vitreum). J. Fish. Res. Board Can.* 34:1922–35
75. Kling, G. W., Kipphut, G. W., Miller, M. C. 1991. Arctic lakes and streams as gas conduits to the atmosphere: implications for tundra carbon budgets. *Science* 251:298–301
76. Kochel, R. C., Baker, V. R. 1982. Paleoflood hydrology. *Science* 215: 353–61
77. Lamberti, G. A., Gregory, S. V., Ashkenas, L. R., Wildman, R. C., Moore, K. M. S. 1991. Stream ecosystem recovery following a catastrophic debris flow. *Can. J. Fish. Aquat. Sci.* 48:196–208
78. Larimore, R. W., Childers, W. F., Heckrotte, C. 1959. Destruction and re-establishment of stream fish and invertebrates affected by drought. *Trans. Am. Fish. Soc.* 88:261–85
79. la Riviere, J. W. 1989. Threats to the world's water. *Sci. Am.*Sept. 1989:80–94
80. LeCren, E. D., Lowe-McConnell, R. H., eds. 1981. *Functioning of Freshwater Ecosystems*. London: Cambridge Univ. Press.
81. Lettenmaier, D. P., Gan, T. Y. 1990. Hydrologic sensitivities of the Sacramento-San Joaquin River basin, California, to global warming. Water Resour. Res. 26:69–86
82. Li, H. W., Moyle, P. B. 1981. Ecological analysis of species introductions into aquatic systems. *Trans. Am. Fish. Soc.* 110:772–82
83. Lodge, D. M. 1992. Species invasions and deletions: community effects and responses to climate and habitat change. See Ref. 71. In press
84. Magnuson, J. J., Crowder, L. B., Medvick, P. A. 1979. Temperature as an ecological resource. *Am. Zool.* 19: 331–43
85. Magnuson, J. J., Meisner, J. D., Hill, D. K. 1990. Potential changes in the thermal habitat of Great Lakes fishes after climate warming. *Trans. Am. Fish. Soc.* 119:254–64
86. Mandrak, N. E. 1989. Potential invasion of the Great Lakes by fish species associated with climatic warming. *J. Great Lakes Res.* 15:306–16
87. Marchand, D., Sanderson, M., Howe, D., Alpaugh, C. 1988. Climatic change and Great Lakes levels: the impact on shipping. *Clim. Change* 12:107–33
88. Marmonier, P., Chatelliers, M. C. D. 1991. Effects of spates on interstitial assemblages of the Rhone River: importance of spatial heterogeneity. *Hydrobiologia* 210:243–51
89. Martin, C. W., Johnson, W. C. 1987. Historical channel narrowing and riparian vegetation expansion in the Medicine Lodge River Basin, Kansas, 1871–1983. *Ann. Assoc. Am. Geog.* 77:436–49
90. Matthews, W. J., Zimmerman, E. G. 1990. Potential effects of global warming on native fishes of the southern Great Plains and the Southwest. *Fisheries* 15:26–32
91. McArthur, J. V., Marzolf, G. R., Urban, J. E. 1985. Response of bacteria isolated from a pristine prairie stream to concentration and source of soluble organic carbon. *Appl. Environ. Microbiol.* 499: 238–41
92. McCormick, M. J. 1990. Potential changes in thermal structure and cycle of Lake Michigan due to global warming. *Trans. Am. Fish. Soc.* 119:183–94
93. Meffe, G. K. 1991. Failed invasion of a Southeastern blackwater stream by bluegills: Implications for conservation of native communities. *Trans. Am. Fish. Soc.* 120:333–38
94. Meisner, J. D., Rosenfeld, J. S., Regier, H. A., 1988. The role of groundwater in the impact of climate warming on stream salmonines. *Fisheries* 13:2–8
95. Melack, J. M. 1979. Temporal variability of phytoplankton in tropical lakes. *Oecologia* 44:1–7
96. Melack, J. M. 1991. Reciprocal interactions among lakes, large rivers, and climate. See Ref. 38, pp. 68–87
97. Merritt, R. W., Lawson, D. L. 1979. Leaf litter processing in floodplain and stream communities. In *Strategies for Protection and Management of Floodplain Wetlands and Other Riparian Ecosystems*, ed. R. R. Johnson, J. F. McCormick, pp. 93–105. *Gen. Tech. Rpt. WO-a2*. Washington, DC: USDA For. Serv. 410 pp.
98. Meyer, J. L., Edwards, R. T. 1990. Ecosystem metabolism and turnover of

organic carbon along a blackwater river continuum. *Ecology* 71:668–77
99. Meyer, J. L., Pulliam, W. M. 1991. Modifications of terrestrial-aquatic interactions by a changing climate. See Ref. 38, pp. 177–91
100. Minckley, W. L., Hendrickson, D. A., Bond, C. E. 1986. Geography of western north American freshwater fishes: description and relationships to intracontinental tectonism. In *The Zoogeography of North American Freshwater Fishes,* ed. C. H. Hocutt, E. O. Wiley, pp. 519–613. New York: Wiley. 866 pp.
101. Minshall, G. W. 1988. Stream ecosystem theory: a global perspective. *J. No. Am. Benthol. Soc.* 7:263–88
102. Minshall, G. W., Brock, J. T., LaPoint, T. W. 1982. Characterization and dynamics of benthic organic matter and invertebrate functional feeding group relationships in the Upper Salmon River, Idaho (USA). *Int. Rev. Ges. Hydrobiol.* 67:793–820
103. Minshall, G.W., Brock, J.T., Varney, J.D. 1989. Wildfires and Yellowstone's stream ecosystems. *BioScience* 39:707–15
104. Minshall, G. W., Petersen, R. C., Cummins, K. W., Bott, T. L., Sedell, J. R., et al. 1983. Interbiome comparison of stream ecosystem dynamics. *Ecol. Monogr.* 53:1–25
105. Moll, R. A., Stoermer E. F. 1982. A hypothesis relating trophic status and subsurface chlorophyll maxima of lakes. *Arch. Hydrobiol.* 94:425–40
106. Munn, N. L., Meyer, J. L. 1990. Habitat-specific solute retention in two small streams: an intersite comparison. *Ecology* 71: 2069–82
107. Naiman, R. J., Manning, T., Johnston, C. A. 1991. Beaver population fluctuations and tropospheric methane emissions in boreal wetlands. *Biogeochemistry* 12:1–15
108. National Research Council. 1992. *Restoration of Aquatic Systems: Science, Technology, and Public Policy.* Washington, DC: Natl. Acad. Press. 485 pp.
109. Newbold, J. D., Elwood, J. W., O'Neill, R. V., Van Winkle, W. 1981. Measuring nutrient spiralling in streams. *Can. J. Fish. Aquat. Sci.* 38:860–3
110. Newbold, J. D., Mulholland, P. J., Elwood, J. W., O'Neill, R. V. 1982. Organic carbon spiralling in stream ecosystems. *Oikos* 38:266–72
111. Niemi, G. J., Devore, P., Detenbeck, N., Taylor, D., Lima, A., et al. 1990. Overview of case studies on recovery of aquatic systems from disturbance. *Environ. Manage.* 14:571–87
112. Oswood, M. W., Milner, A. M., Irons, J. G. III. 1991. Climate change and Alaskan rivers and streams. See Ref. 38, pp. 192–210
113. Overpeck, J. T., Rind, D., Goldberg, R. 1990. Climate-induced changes in forest disturbance and vegetation. *Nature* 343:51–53
114. Peckarsky, B. L. 1983. Biotic interactions or abiotic limitations? A model of lotic community structure. In *Dynamics of Lotic Ecosystems,* ed. T. D. Fontaine III, S. M. Bartell, pp. 303–24. Ann Arbor: Ann Arbor Sci.
115. Petersen, R. C., Cummins, K. W. 1974. Leaf processing in a woodland stream. *Freshwater Biol.* 4:343–68
116. Peterson, B. J., Hobbie, J. E., Corliss, T. L. 1986. Carbon flow in a tundra stream ecosystem. *Can. J. Fish. Aquat. Sci.* 43:1259–70
117. Pitcher, T. J, Hart, P. J. B. 1982. *Fisheries Ecology.* London: Croom Helm
118. Poff, N. L. 1991. Regional hydrologic response to climate change: an ecological perspective. See Ref. 38, pp. 88–115
119. Poff, N. L., Ward, J. V. 1989. Implications of streamflow variability and predictability for lotic community structure: a regional analysis of streamflow patterns. *Can. J. Fish. Aquat. Sci.* 46:1805–18
120. Poff, N. L., Ward, J. V. 1990. Physical habitat template of lotic systems: recovery in the context of historical pattern of spatiotemporal heterogeneity. *Environ. Manage.* 14:629–45
121. Power, M. E. 1990. Effects of fish in river food webs. *Science* 250:811–14
122. Power, M. E. 1992. Top down and bottom up forces in food webs: do plants have primacy?*Ecology.* In press
123. Power, M. E., Matthews, W. J., Stewart, A. J. 1985. Grazing minnows, piscivorous bass and stream algae: dynamics of a strong interaction. *Ecology* 66: 1448–56
124. Power, M. E., Stout, R. J., Cushing, C. E., Harper, P. P., Hauer, F. R., et al. 1988. Biotic and abiotic controls in river and stream communities. *J. No. Am. Benthol. Soc.* 7:456–79
125. Quattro, J. M., Vrijenhoek, R. C. 1989. Fitness differences among remnant populations of the endangered Sonoran topminnow. *Science* 245:976–78
126. Rasmussen, J. B., Rowan, D. J., Lean, D. R. S., Carey, J. H. 1990. Food

chain structure in Ontario lakes determines PCB levels in lake trout (*Salvelinus namaycush*) and other pelagic fish. *Can. J. Fish. Aquat. Sci.* 47: 2030–38

127. Reckhow, K., Chapra, S. C. 1983. *Engineering Approaches for Lake Management*. Vol. 1. *Data Analysis and Empirical Modeling*. Boston: Butterworth's. 340 pp.
128. Regier, H. A., Holmes, J. A., Pauly, D. 1990. Influence of temperature change on aquatic ecosystems: an interpretation of empirical data. *Trans. Am. Fish. Soc.* 119:374–89
129. Regier, H. A., Magnuson, J. J., Coutant, C. C. 1990. Introduction to proceedings: Symposium on effects of climate change on fish. *Trans. Am. Fish. Soc.* 119:173–75
130. Resh, V. H. 1982. Age structure alteration in a caddisfly population after habitat loss and recovery. *Oikos* 38: 280–84
131. Resh, V. H., Brown, A. V., Covich, A. P., Gurtz, M. E., Li, H. W., et al. 1988. The role of disturbance in stream ecology. *J. No. Am. Benthol. Soc.* 7:433–55
132. Richey, J. E., Nobre, C., Deser, C. 1989. Amazon River discharge and climate variability, 1903–1985. *Science* 246:101–3
133. Rind, D., Goldberg, R., Ruedy, R. 1989. Change in climate variability in the 21st century. *Clim. Change* 7:367–89
134. Roots, E. F. 1989. Climate change: high latitude regions. *Clim. Change* 15:223–53
135. Rosenberg, N. J., Kimball, B. A., Martin, P., Cooper, C. F. 1990. See Ref. 163, pp. 151–75
136. Schindler, D. W. 1978. Factors regulating phytoplankton production and standing crop in the world's freshwaters. *Limnol. Oceanogr.* 23:478–86
137. Schindler, D. W., Bayley, S. E. 1990. Fresh waters in cycle. In *Planet Under Stress,* ed. C. Mungall, D. J. McLaren, pp. 149–67. Toronto: Oxford Univ. Press
138. Schindler, D. W., Beaty, K. G., Fee, E. J ., Cruikshank, D. R., DeBruyn, E. R., et al. 1990. Effects of climatic warming on lakes of the central boreal forest. *Science* 250:967–70
139. Schneider, S. H. 1992. Scenarios of global warming. See Ref. 71. In press
140. Schneider, S. H., Gleick, P. H., Mearns, L. O. 1990. Prospects for climate change. See Ref. 163, pp. 41–73
141. Sedell, J. R., Froggatt, J. L. 1984. Importance of streamside forests to large rivers: the isolation of the Willamette River, Oregon, U.S.A., from its floodplain by snagging and streamside forest removal. *Verh. Int. Verein. Limnol.* 22:1828–34
142. Shuter, B. J., Post, J. R. 1990. Climate, population variability, and the zoogeography of temperate fishes. *Trans. Am. Fish. Soc.* 119:314–36
143. Smock, L. A., MacGregor, C. M. 1988. Impact of the American chestnut blight on aquatic shredding macroinvertebrates. *J. N. Am. Benthol. Soc.* 7:212–21
144. Smock, L. A., Metzler, G. M., Gladden, J. E. 1989. Role of debris dams in the structure and functioning of low-gradient headwater streams. *Ecology* 70: 764–75
145. Sparks, R. E., Bayley, P. B., Kohler, S. L., Osborne, L. L. 1990. Disturbance and recovery of large floodplain rivers. *Environ. Manage.* 14:699–709
146. St. Amand, A., Carpenter, S. R. 1993. Metalimnetic phytoplankton dynamics. In *Trophic Cascades in Lakes,* ed. S. R. Carpenter, J. F. Kitchell. London: Cambridge Univ. Press. In press
147. Stanford, J. A., Ward, J. V. 1986. The Colorado River system. In *The Ecology of River Systems,* ed. B. R. Davies, K. F. Walker, pp. 353–74. Dordrecht:Junk
148. Stanley, E. H., Fisher, S. G. 1992. Intermittency, disturbance, and stability in stream ecosystems. In *Aquatic Ecosystems in Semi-Arid Regions,* ed. R. Robarts. Saskatoon, Sask: Natl. Hydrology Res. Cent. In press
149. Stanley, E. H., Valett, H. M. 1991. Interactions between drying and the hyporheic zone in a desert stream. See Ref. 38, pp. 234–49
150. Street, F. A., Grove, A. T. 1979. Global maps of lake-level fluctuations since 30,000 years B.P. *Quat. Res.* 12:83–118
151. Swanson, F. J., Benda, L. E., Duncan, S. H., Grant, G. E., Megahan, W. F., et al. 1987. Mass failures and other processes of sediment production in Pacific Northwest forest landscapes. In *Streamside Management: Forestry and Fishery Interactions,* ed. E. O. Salo, T. W. Cundy, pp. 9–38. Seattle, Wash: Univ. Wash. Inst. For. Resources
152. Sweeney, B. W., Vannote, R. L., Dodds, P. J. 1986. Effects of temperature and food quality of growth and development of a mayfly, *Leptophlebia intermedia*. *Can. J. Fish. Aquat. Sci.* 43:12–18
153. Taylor, R. C. 1983. Drought-induced changes in crayfish populations along

a stream continuum. *Am. Midl. Natur.* 110:286–98

154. Thomann, R. V. 1989. Deterministic and statistical models of chemical fate in aquatic systems. In *Ecotoxicology: Problems and Approaches,* ed. S. A. Levin, M. A. Harwell, J. R. Kelly, K. D. Kimball, pp. 245–77. New York: Springer-Verlag. 547 pp.
155. Thornthwaite, C. W., Mather, J. R. 1955. *The Water Balance. Publications in Climatology,* Vol. 8, No. 1. Drexel Inst. Technol. Philadelphia, PA. 104 pp.
156. Tonn, W. M. 1990. Climate change and fish communities: a conceptual framework. *Trans. Am. Fish. Soc.* 119: 337–52
157. Towns, R. D. 1985. Limnological characteristics of a South Australian intermittent stream, Brown Hill Creek. *Aust. J. Mar. Freshwater Res.* 36:821–37
158. Trippel, E. A., Eckmann, R., Hartmann, J. 1991. Potential effects of global warming on whitefish in Lake Constance, Germany. *Ambio* 20:226–31
159. Triska, F. J. 1984. Role of wood debris in modifying channel geomorphology and riparian areas of a large lowland river under pristine conditions: a historical case study. *Verh. Int. Verein. Limnol.* 22:1876–92
160. Trotter, E. H. 1990. Woody debris, forest-stream succession, and catchment geomorphology. *J. No. Am. Benthol. Soc.* 9:141–56
161. Valett, H. M., Fisher, S. G., Stanley, E. H. 1990. Physical and chemical characteristics of the hyporheic zone of a Sonoran desert stream. *J. No. Am. Benthol. Soc.* 9:201–15
162. Vannote, R. L., Minshall, G. W., Cummins, K. W., Sedell, J. R., Cushing, C. E. 1980. The river continuum concept. *Can. J. Fish. Aquat. Sci.* 37:130–37
163. Waggoner, P. E., ed. 1990. *Climate Change and U.S. Water Resources.* New York: Wiley. 496 pp.
164. Walker, B. H. 1991. Ecological consequences of atmospheric and climate change. *Clim. Change* 18:301–16
165. Ward, A. K., Ward, G. M., Harlin, J., Donahoe, R. 1991. Geological mediation of stream flow and sediment and solute loading to stream ecosystems due to climate change. See Ref. 38, pp. 116–42
166. Webster, K. E., Newell, A. D., Baker, L. A., Brezonik, P. L. 1990. Climatically induced rapid acidification of a softwater seepage lake. *Nature* 347: 374–76
167. Welcomme, R. L. 1979. *Fisheries Ecology of Floodplain Rivers.* London: Longmanns
168. Welcomme, R. L. 1988. Concluding remarks I. On the nature of large tropical rivers, floodplains, and future research directions. *J. No. Am. Benthol. Soc.* 7:525–26
169. Wetzel, R. G. 1990. Land-water interfaces: Metabolic and limnological regulators. *Verh. Int. Verein. Limnol.* 24: 6–24
170. Wigley, T. M. L., Jones, P. D. 1985. Influences of precipitation changes and direct CO_2 effects on streamflow. *Nature* 314:149–52
171. Wigley, T. M. L., Jones, P. D., Kelly P. M. 1980. Scenario for a warm, high-CO_2 world. *Nature* 283:17–21
172. Williams, P. 1989. Adapting water resources management to global climate change. *Clim. Change* 15:83–93
173. Wilson, L. 1973. Variations in mean annual sediment yield as a function of mean annual precipitation. *Am. J. Sci.* 273:335–49
174. Wooton, R. J. 1990. *Ecology of Teleost Fishes.* New York: Chapman & Hall
175. Yount, J. D., Niemi, G. J. 1990. Recovery of lotic communities and ecosystems from disturbance: a narrative review of case studies. *Environ. Manage.* 14:547–69

Annu. Rev. Ecol. Syst. 1992. 23:141–173

GLOBAL CHANGES DURING THE LAST 3 MILLION YEARS: Climatic Controls and Biotic Responses

T. Webb III
Department of Geological Sciences, Brown University, Providence, Rhode Island 02912-1846

P. J. Bartlein
Department of Geography, University of Oregon, Eugene, Oregon 97403

KEYWORDS: global climate, Quaternary, vegetation dynamics

INTRODUCTION

Research during the last 20 years has led to a major expansion in knowledge about long-term climatic variability and dynamics. Two developments in particular have advanced the theoretical understanding of the major environmental changes that induce continuous changes in ecosystems. The first development was a recognition that the alternation of glacial and interglacial climates has been paced by the variations in solar radiation generated by periodic variations in the Earth's orbit. The second development involved an increased understanding of the hierarchical controls of regional climatic variations. Studies of marine plankton from deep-sea cores were critical to the first development, whereas the second development arose from regional to global syntheses of paleoclimatic data combined with analyses of paleoclimatic simulations from climate models. These two sets of information illustrate a theoretical framework for understanding temporal climatic variations at the subcontinental scale, and they lead to a recognition that elements of the biosphere such as vegetation and marine plankton have experienced large periodic variations in climate for millions of years.

In this paper, we describe the major global climatic variations for the last 20 kyr (kilo years, i.e. 20,000 years), the last 175 kyr, and the last 3 Myr (million years). We combine a map view of changes from the last 20 kyr

0066-4162/92/1120-0141$02.00

with a time-series perspective for the two longer time intervals. The order of our discussion of these time periods allows us to interpret the longer records in terms of lessons learned from studies of the last 20 kyr. Large spatial data sets and climate-modeling experiments for the last 20 kyr show how global and regional controls of climate govern the regional patterns of environmental change. They also illustrate the biospheric response (*a*) to the change from full-glacial to interglacial conditions and (*b*) to a full cycle of moisture-balance changes in the northern tropics from dry to wet to dry conditions. Records from the last 175 kyr, which span almost two full glacial/interglacial cycles, help us evaluate how vegetation and marine plankton populations track these long-term environmental changes. Records spanning the last 3 Myr show the antiquity of the large periodic variations and indicate how selected ecosystem and biogeographic changes are embedded within longer-term nonreversing trends in climate.

From studying these three periods of large climate changes we have concluded that major elements of the biosphere track long-term environmental changes fairly closely. In 1976, Davis (42) used maps of changing range boundaries for eastern North American tree taxa to illustrate the individualistic response of taxa to general climatic changes accompanying deglaciation. Since 1976, we have developed a better understanding of how the climate system operates and how global controls create regionally varying responses in both climate and the biosphere. These new insights have helped to clarify the role of climate in continuously forcing major rearrangements in assemblages of plants and animals (45, 56, 66, 118, 147). This understanding when combined with records of the last 3 Myr and longer is fundamental to understanding how climatically induced ecological changes affect evolutionary history.

BACKGROUND

Before reviewing the data and modeling results, we provide some background information on the major climatic and biospheric changes within each of the time periods. We also describe the paleoclimatic data and climate models used in studying the time periods and note some of the major global syntheses of the data and model results.

Time Scales

For each of the three time scales (the last 20 kyr, 175 kyr, and 3 Myr), we describe (*a*) variations in the controls of climate, (*b*) the nature of the climatic variations induced by the changes in controls, and (*c*) selected, observed biotic responses. For each time scale, many of the key controls of climate are understood, and for both the last 20 kyr and the last 175 kyr the climate models and paleoclimatic data have been used to test ideas about how the system works.

These analyses have allowed "historical" or geological data to be used in an experimental framework (nature having already performed the experiments) to test hypothetical climate patterns simulated from the theory incorporated in the climate models. The model results also help to explain how the global controls affect regional climates. The lessons learned from these experiments are useful for the interpretation of the 3 Myr climate records and for understanding how global climatic controls affect regional climates.

Of the three time scales, the last 20 kyr contains the most information about climatic and biospheric changes. The well-dated large data sets from this period have allowed us to map the responses of vegetation and lake-levels to the large environmental changes from full-glacial to full-interglacial conditions (34). Time sequences of maps thus provide a spacetime view of the changes. Simulations from climate models are also available and allow testing of specific ideas about why climate changed. Studies of the last 175 kyr expand the view of responses to the whole range of variations from glacial to interglacial conditions and the reverse. Time series from this period show the response of the climate system to a variety of settings of the controls and show the behavior of biological systems under different climatic conditions. The records from the past 3 Myr show that periodic, large swings in the environment predate the Quaternary and that the biotic response is consistent with these changes.

Our focus on global controls for climate has precluded discussion of the Little Ice Age (ca. 1550 to 1850 AD) and other short-term climate variations relevant to many ecological issues. Climate changes at these scales have produced large regional variations in temperature and precipitation that have affected vegetation and patterns of succession (18, 20, 53, 57, 151), but none have large globally averaged signals nor are they well understood in terms of what is forcing the climatic changes. Grove (58) and Bradley (23) provide excellent reviews of these climate variations.

Paleoclimate Data

The primary sources of paleoclimatic data are quantitative, dated time series of ecological, geochemical, or geophysical phenomena that are sensitive to climate. For the shorter time scales, spatial networks of such time series exist. To illustrate variations in various aspects of the climate system, we use stable isotopes from marine cores to infer global ice volume, analyses of ice cores to estimate atmospheric composition, pollen and plant macrofossils to infer vegetation and climate, foraminifera to represent aspects of the marine biosphere and to estimate sea-surface temperatures, and geomorphic and paleolimnological data to record variations in lake levels and surface hydrology (34). Birks & Birks (19), Bradley (21, 22), Berglund (17), Crowley & North (38), Hecht (60), and Huntley & Webb (68) describe the general characteristics of the data, dating methods, and the methods for data collection, analysis, and display. Methods for the paleoclimatic calibration and interpretation of the data appear in Bartlein et al (11), Hecht (60), Webb et al (149), and Prentice et al (118).

Climate Controls

A distinction can be made between the ultimate controls of variations in the whole climate system and the proximal controls of climate at a location. The ultimate controls include factors external to the climate system as well as components within the system that vary slowly at the time scale under consideration (70, 132). External factors include insolation (incoming solar radiation at the top of the atmosphere); the arrangement of the continents, mountain chains, and oceanic gateways; and long-term changes in the composition of the atmosphere. The regular and predictable variations in the earth's orbital elements (the eccentricity of the orbit, the tilt of the axis, and the time of year of perihelion—also referred to as "precession" or "precession of the equinoxes") alter the latitudinal and seasonal distribution of insolation (14). Each of these orbital elements has a characteristic period of variation: 100 and 400 kyr (eccentricity), 41 kyr (tilt) and 23 and 19 kyr (precession). The variations of the orbital elements can be reliably determined back to 5 Myr (16), and the latitudinal and seasonal distribution of insolation can be determined throughout this interval.

At time scales of 10^4 years and longer, these external controls govern the growth and decay of the ice sheets and changes in sea surface temperatures (SSTs). At shorter timescales, however, the ice sheets change so slowly that they can be considered external controls. The specific combinations of the external controls and slowly varying internal components of the climate system that exist at any time are sometimes referred to as the "boundary conditions" of the climate system, particularly in the context of climate simulation modeling (86).

The proximal controls of climate include variations of atmospheric circulation that determine the location of storm tracks and the distribution of airmasses. The direct effects of insolation and clouds on net radiation (and hence on sensible heating and evapotranspiration) at a place may also be considered a proximal control of climatic variations.

Climate Models

The development of general-circulation models for the atmosphere and oceans has permitted quantitative paleoclimate modeling (38, 81). These mathematical models are based on nonlinear flow equations and the principles of mass and energy conservation (61, 145). Advances in computer design and computational power allow development of new models with increasing realism and complexity. In paleoclimatic experiments, the models simulate past geographic patterns for surface and upper-air temperature, winds, and pressure as well as for surface precipitation and moisture balance. Analysis of these data reveal the position of the jet stream and storm tracks (34). With appropriate analysis and empirical modeling, model simulations can be converted into estimates of past distributions of plant or plankton taxa (101, 149).

Syntheses

Extensive research and data generation within the Quaternary paleoclimatic and paleoecological community have been critical to the theoretical advances and improved understanding of the long-term global climate changes. But the spearhead for much of the advance has come from three international interdisciplinary research groups (CLIMAP, SPECMAP, AND COHMAP) that have focused on global syntheses of Quaternary data in map and time-series formats. CLIMAP (Climate: Mapping and Prediction) was active from 1971 to 1981, and involved mostly paleoceanographers studying deep-sea cores with marine plankton dating back 300 kyr or longer. CLIMAP produced global maps for sea-surface conditions for the last glacial maximum (18 to 21 ka) (ka = 1000 years ago) (32, 47) and the last interglacial (ca. 125 ± 5 ka) (33). From long time-series and power-spectrum analyses, CLIMAP demonstrated that orbital forcings were the pacemaker of the glacial/interglacial oscillations in climate (59). CLIMAP also initiated close collaboration between climate modelers and data generators in running and analyzing paleoclimatic experiments on the climate models (54, 94).

SPECMAP followed CLIMAP in focusing on long marine cores and time series analysis and has generated several syntheses (15, 71, 73) illustrating the leads and lags among components of the climate system over the main orbital frequency bands. COHMAP (Cooperative Holocene Mapping Project), which began in 1977, has focused on the climate, vegetation, and surface-hydrological history of the last 20 kyr. COHMAP researchers compiled a global data base and maps of terrestrial data for each 3 kyr interval from 18 ka to present, calibrated some of the data sets in climatic terms, produced a time series of climate simulations with a general circulation model, and compared the data and model results to illustrate the impact of global climate controls on regional climates (34, 153).

THE LAST 20,000 YEARS

During the last 20 kyr, the Earth's climate has undergone a transition from glacial to interglacial conditions, a change as large as any during the past 3 Myr. These climatic variations resulted in large biotic responses, including migrations of individual taxa and rearrangements of vegetation and planktonic associations. The immediate controls of the climatic variations during this interval are well known, and the syntheses by CLIMAP (32) and COHMAP (34, 153) have allowed the spatial and temporal patterns of the biotic response to be explored in some detail.

Variations in Boundary Conditions

Insolation varied from a latitudinal and seasonal distribution at 20 ka which resembles that at present (Figure 1) to one at 10 ka in which summer insolation in the northern hemisphere was about 8% greater than at present, with winter insolation about 8% less (13). This exaggeration of the seasonal cycle of insolation resulted from the occurrence of perihelion during the northern hemisphere summer then (now it occurs on January 4th), and from an increase in the tilt of the earth's rotational axis (to 24.5° as compared with the present 23.5°). Deglaciation was favored by the increased summer isolation. The ice sheets decreased in height and area from 15 to 6 ka (47, 49), and sea levels correspondingly rose (51). Sea surface temperatures, particularly in the high northern latitudes, increased from 15 to 9 ka (32, 129). Reduced SSTs act to lower temperatures in adjacent land areas, particularly in downwind land areas (to the east in the midlatitudes). Over the same period, atmospheric carbon dioxide concentrations also increased (5, 106), while atmospheric dust loadings decreased (46, 111). Carbon dioxide, one of the greenhouse gases, absorbs and re-emits outgoing terrestrial radiation, thus raising the equilibrium surface temperature of the globe, while dust reflects incoming solar radiation, acting to lower the equilibrium surface temperature. Most of the change in the boundary conditions occurred during the interval between 15 and 8 ka, with changes following that interval representing a gradual relaxation toward present (34, 86).

Regional Climate Changes—Model Results

Climate modeling results demonstrate major global responses to the changing seasonal cycle of insolation and to the changing size and area of the ice sheets (Table 1). These two major controls interacted to produce a rich variety of climatic patterns (34). The model results show that when the ice sheets were fully extended at 18 kyr, temperatures were more than 20°C lower than they are today over and at the edge of ice sheets (84). Sea ice limits were extended, and temperatures were 4–10°C lower than they are today in the North Atlantic and off eastern Asia (32). The model results also show that at their maximum size the height and high reflectivity of the Northern Hemisphere ice sheets acted in concert with lower sea-surface temperatures and lower concentrations of CO_2 to depress northern hemisphere annual temperatures by more than 5°C (25, 148). As the ice sheets retreated, regional and global temperatures increased.

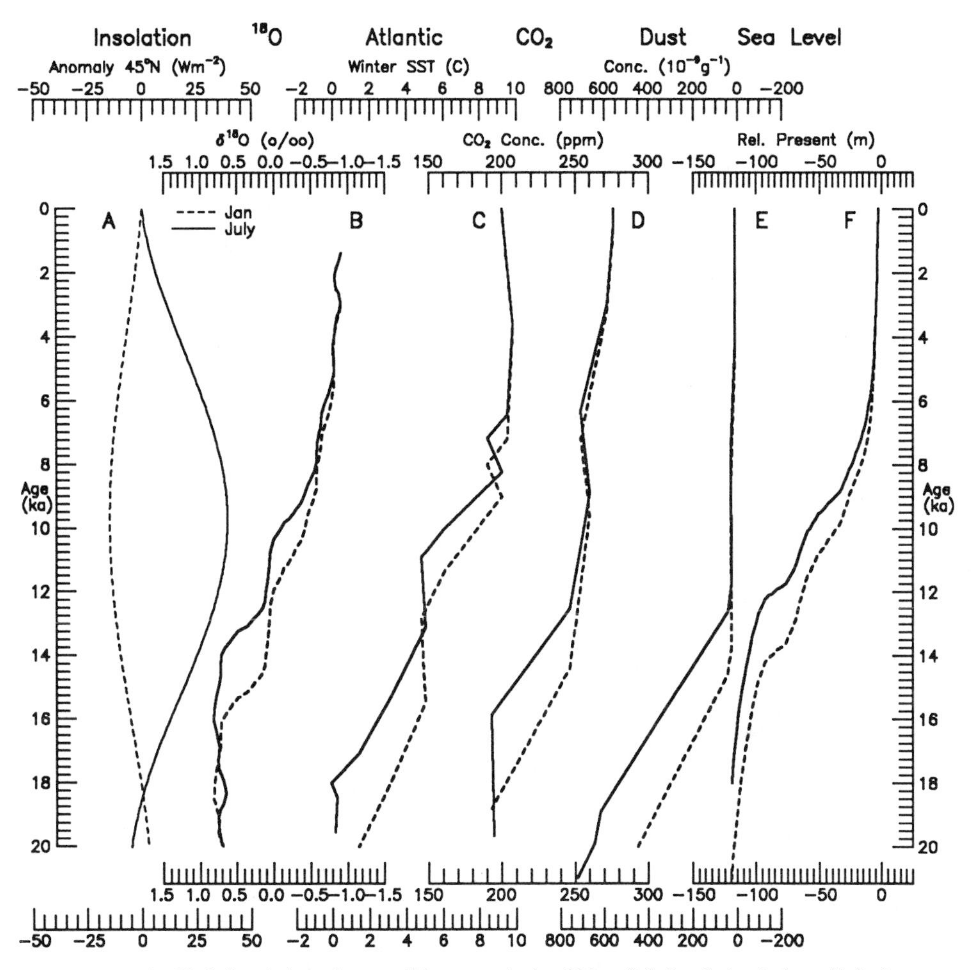

Figure 1 Variations in boundary conditions over the last 20 kyr (34). Insolation (solar radiation) is shown as anomalies, or differences from present, for July and January (*curve A.;* from 13) for 45° N. Global ice volume is represented by oxygen isotope (^{18}O) variations in foraminifera (*curve B;* from 99a); sea-surface temperatures by winter SST estimates from foraminifera in the North Atlantic (*curve C;* from 128). Carbon dioxide concentrations (*curve D*) and dust (*curve E*) are from the Vostok ice core (5, 111), and relative sea level estimates (*curve F)* are from cores off Barbados (51). For curves B–F, the solid lines show the data as plotted assuming that the published dates pertain to the radiocarbon time scale, whereas the dashed lines show the same data as plotted using the Uranium/Thorium calibration from Bard et al (Ref. 4a).

The Northern Hemisphere ice sheets also had a major influence on atmospheric circulation. In the simulations, the Laurentide ice sheet acted as a major mountain range, somewhat like the Himalayas, that was big enough to reorganize the hemispheric circulation patterns. The ice sheet split the jet

Table 1 Effects of large-scale controls on paleoclimates as simulated by general circulation model experiments

Controlling factor and effects	Global response (relative to present)	Northern Hemisphere mid-to-high latitude response (relative to present)	Key references
Ice sheet temperature effects ($\delta^{18}O$ > 12 ka level)	colder ($\Delta T_{Ann} \sim -3°C$)	colder ($\Delta T_{Ann} \sim -4$ to $-10°C$)	25, 84
Ice sheet circulation effects ($\delta^{18}O$ = 18 ka level)	stronger westerlies/increased storminess	split jet/glacial anticyclone over North America and Europe; advection of cold air over North Atlantic	25, 84, 122
(Reduced) sea-surface temperature effects (SSTs = 18 ka values)	cooler land surface temperatures downwind of oceans ($\Delta T_{Ann} \sim -2°C$)	cooler and drier in Europe ($\Delta T_{Ann} \sim -5°C$)	122, 123, 84
Direct insolation effects (K_{Jul} > 6 ka levels)	increased land-sea temperature contrast and seasonality in Northern Hemisphere; decreased in Southern Hemisphere; warmer oceans (Sept. Oct. Nov.)	warmer continental interiors ($\Delta T_{Jul} \sim +2$ to $+4°C$); reduced P-E and soil moisture	99, 83, 84
Indirect insolation effects (K_{Jul} > 6 ka levels)	stronger Northern Hemisphere monsoons; weaker Southern Hemisphere monsoons	greater precipitation in monsoon regions ($\Delta P_{Jul} \sim +1.5 mmd^{-1}$); reduced Arctic sea ice and snow cover	99, 83, 84
(Reduced) CO_2 temperature effects (CO_2 conc. $\leq$ 200ppm)	cooler than otherwise ($\Delta T_{Ann} \sim -1°C$)	cooler than otherwise ($\Delta T_{Ann} \sim -1.5°C$)	25, 84

stream in winter with the southern branch crossing the west coast of North America farther south than at present, bringing abundant precipitation to the arid Southwest (34, 89). The position of the jet stream was also controlled by the southern border of sea ice across the North Atlantic.

While the ice sheets were both at maximum extent and retreating, their direct and indirect effects combined to create climates with different combinations of temperature and precipitation than are found today. For example, from 18 to 12 ka, when cold summer temperatures were close to the key source of moisture in the Gulf of Mexico, and the associated patterns of precipitation and moisture balance differed from those of 9 ka and 6 ka as well as from those of today. Many of the simulated climates from 18 to 12 ka have no close analogs among modern climates.

As the influence of the ice sheet on climate diminished, the influence of insolation grew with the amplification of its seasonal cycle. The mechanisms through which insolation governs regional climates (independent of its influence on the ice sheets) are also both direct and indirect: through greater heating and evapotranspiration in summer (and correspondingly less in winter) and through changes in atmospheric circulation caused by the increased land/sea contrast in both seasons (88).

Because the oceans have a greater thermal inertia (i.e. heat less quickly) than the continents, the northern hemisphere surface heating effect was greatest in the continental interiors, and the land/sea temperature contrast increased. (The remnant Laurentide ice sheet at 9 ka, however, diminished the response over eastern North America.) Heating of the center of the continents resulted in the lowering of surface pressure relative to the surrounding oceans, particularly across northern Africa and Asia (87, 99). This increased pressure differential caused stronger monsoonal circulation, and consequently greater summer precipitation in the monsoon regions of the northern hemisphere and an increase in the area affected by the monsoons. In the southern hemisphere, the simulations show the opposite: lower summer (January) temperatures, lower precipitation, and reduced monsoons. These simulated patterns of climatic changes represent immediate (not ice-lagged) responses to insolation variations. The simulations suggest that tropical climates undergo regular periodic variations, particularly in effective moisture, in response to the insolation variations.

Many aspects of the simulated climatic changes over the last 20 kyr are evident in the mapped records of paleoclimatic indicators (34). For example, in the northern hemisphere tropics, lakes rose to levels much higher than those at present during the interval of maximum summer insolation from 12 to 6 ka, the interval for which the simulations project a pronounced increase in monsoonal precipitation (87, 99). Similarly, many features of the simulated midlatitude climates of the northern hemisphere are found in paleoclimatic data syntheses (34, 153).

Biotic Responses

The principal biotic response to the climatic changes accompanying deglaciation was migration (69). Throughout the last 20 kyr, migration allowed individual species to track the climates favorable to this growth. In Europe, for example (Figure 2), spruce (*Picea*) trees and most arboreal vegetation were at very low abundances 18 ka, in response to the cold, dry climate that prevailed then (34, 65). As temperatures increased, spruce populations increased in abundance in the Alps and northeastern Europe. During the last 9 ka, the abundance maximum of spruce populations in Europe shifted westward, as spruce migrated into Scandinavia from the east after 6 ka. This latter shift fits well with the development of cooler winters in Scandinavia (83), and the overall long-term, large-scale change for spruce appears to be mainly a response to climatic forcing (69). Huntley (65) and Huntley & Webb (69) have developed similar interpretations for the other major tree taxa in Europe. A key to understanding the individualistic behavior of taxa in Europe as well as elsewhere (see below) is an understanding of the regional climatic patterns and independent behavior of the different climate variables in response to the change in global controls.

In eastern North America, spruce populations show a similar type of behavior. At 18 ka, spruce populations were largest in the Midwest and their abundance then increased eastward south of the ice sheet by 15 ka before moving northward as the ice sheet retreated (Figure 2). Between 9 and 7 ka, spruce populations were at a minimum south of the remnant ice sheet. After the ice sheet disappeared at 6 ka, spruce populations expanded in the region of the modern boreal forest and then moved southward in southern Canada and the bordering regions of the United States after 2 ka (90). Again the scale and trends fit well with our understanding of the major climate changes, but the best evidence for climate forcing comes from the ability of the series of climate model simulations to reproduce the changes evident in the spruce maps (Figure 2).

Similar time series of maps for other eastern North American plant taxa show the individualistic changes among taxa in terms of movement in both their abundance maxima and range boundaries (43, 95, 149). Jacobson et al (74) have used color maps to illustrate the resultant changes in plant association from 18 ka to present. The maps show that different combinations of taxa form and later dissociate. The broad association of spruce and sedge (Cyperaceae) pollen from 18 to 12 ka indicates growth of a spruce parkland that disappeared before spruce and birch (*Betula*) combined to form of the modern boreal forest at 6 ka. Other series of color maps show similar types

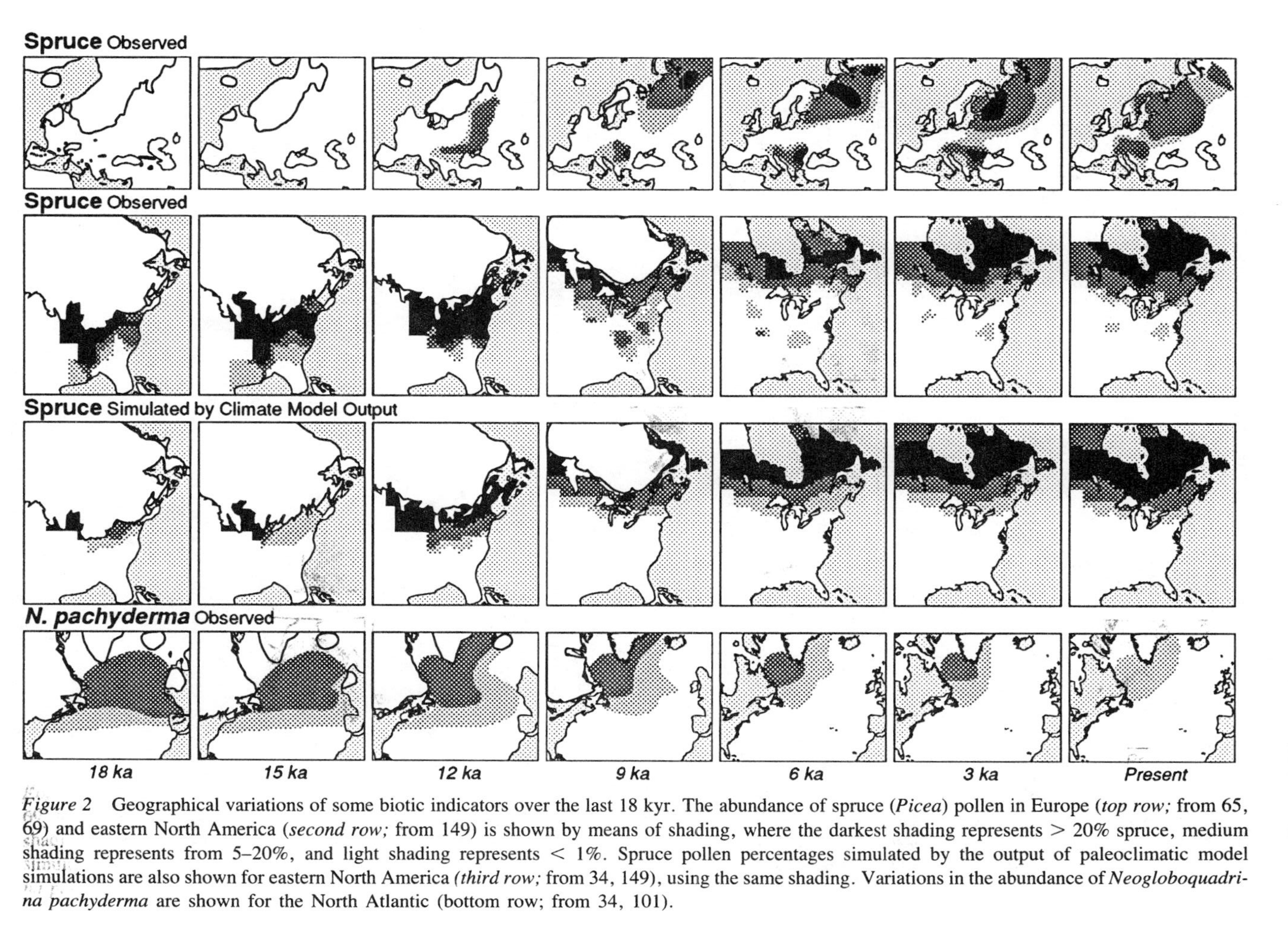

Figure 2 Geographical variations of some biotic indicators over the last 18 kyr. The abundance of spruce (*Picea*) pollen in Europe (*top row;* from 65, 69) and eastern North America (*second row;* from 149) is shown by means of shading, where the darkest shading represents > 20% spruce, medium shading represents from 5–20%, and light shading represents < 1%. Spruce pollen percentages simulated by the output of paleoclimatic model simulations are also shown for eastern North America (*third row;* from 34, 149), using the same shading. Variations in the abundance of *Neogloboquadrina pachyderma* are shown for the North Atlantic (bottom row; from 34, 101).

of changes for deciduous and mixed forest taxa. Recent work by Prentice et al (118) supports the interpretation that climate played a strong role in governing vegetation changes. They inferred a set of climate estimates from the changing distribution of six pollen types and then were able to simulate correctly the major changes in distribution of seven other major taxa that had not been used to estimate the climate values. They were even able to show that the inferred climate values led hickory (*Carya*, one of the seven independent taxa) to move into New England after oak (*Quercus*). Their study showed that there are plausible climate values that can account for the long-noted migration lag between oak and hickory in New England (42, 45).

Similar types and scales of spatial and temporal changes are evident among foraminifora taxa in the North Atlantic over the last 18 kyr (Figure 2). Maps for several taxa show that the foraminifera assemblages change in composition as well as location (101). Since individual forams are free to move, we infer that they closely tracked the environment most favorable for their growth. What is remarkable is that their scale of motion is similar to that for plant taxa on land. Therefore, for the time scales mapped, the populations of sessile trees on land behave like populations of mobile foraminifera in the ocean (48).

The tracking of climate we show here for plant taxa and foraminifera is mirrored by similar kinds of changes elsewhere, especially where the biological data have been mapped, e.g. the western United States, New Zealand, and Japan (8, 96, 138, 140, 141). The paleoecological data therefore reveal two common reactions among taxa to large climatic variation—individualistic response and migration (69, 75)—a situation that is not unique to the last 20 kyr.

THE PAST 175,000 YEARS

The climatic changes of the last 20 kyr represent one kind of variation of the climate system—a transition from full-glacial to full-interglacial conditions. Over the last 175 kyr, the climate system underwent two such transitions, along with a more protracted transition into the last full glaciation. Throughout this interval, the biosphere was challenged by a continuously varying combination of climatic controls, to which it responded using the same mechanisms it employed during the last 20,000 years.

Changes in Boundary Conditions and the Controls of Regional Climatic Variations

Insolation variations over the last 175 kyr clearly show the influence of the regular variations of the tilt of the earth's axis (at a 41 kyr period) and of the precession of the equinoxes or the time of year of perihelion (at a 23 kyr period) (Figure 3A). The most striking feature of the insolation record is the prominence of the precessional variations, with the influence of the tilt (and eccentricity) variations superimposed on the amplitude of these variations. Each precessional cycle is unique. The relative maximum of summer insolation in the Northern Hemisphere around 10 kyr is one of eight such maxima during the last 175 kyr. It is smaller than the maximum during the previous interglacial (around 125 ka), but larger than the two preceding ones (at 35 and 55 kyr). At 125 ka, the July insolation anomaly at 65°N reached values about 50% greater than the peak anomaly during the present interglacial at 10 ka.

Over intervals longer than the response time of the ice sheets (> 10–20 kyr), global ice volume and sea surface temperatures (SSTs) can be regarded as internal variables within the climate system that are interrelated and dependent on the insolation inputs (70, 73). These relationships are most clearly displayed by spectral analyses of insolation, oxygen-isotope ($\delta^{18}O$) and SST time series (Figures 3B–D), which show the strong expression of the orbital variations in these data (59, 72, 73). The specific physical mechanisms that link ice volume, sea surface temperatures, atmospheric composition, and insolation on the Quaternary time scales still remain to be explicated in detail, however (71, 73).

For climate modeling experiments of particular times, ice volume and SSTs can be regarded as boundary conditions that contribute to the variations of climate at the regional scale. Oxygen isotope records of global ice volumes show a characteristic "saw-tooth" pattern over the late Quaternary (27, 72), with rapid transitions from glacial to interglacial conditions (such as between 134 and 122 ka and 16 and 8 ka), and more gradual transitions into glacial conditions (Figure 3). Although broadly similar to one another, the "full-glacial" (deglacial or interglacial) intervals differ in detail from one another. For example, the transition into the previous interglacial period was more abrupt than the transition into the current interglacial. If the 14 ka ice volume is used as an index of the size of the ice sheet that significantly depressed global temperatures and may have also influenced atmospheric circulation, then the duration of such influence was longer during the previous full-glacial interval than during the last (about 28 kyr as opposed to 14 kyr). The oxygen-isotope record also reveals a variety of different ice-volume levels during the interval between the previous interglacial and the last glacial maximum, with ice-volume minima occurring after the summer insolation maxima. In general, the interval between 125 ka and 75 ka featured more extreme insolation maxima and ice-volume minima than those during the interval from 75 ka to the beginning of the present interglacial. Most of the time during the Quaternary, ice volume was at a level intermediate between the extremes of glacial maxima and interglacials (Figure 3; see 113).

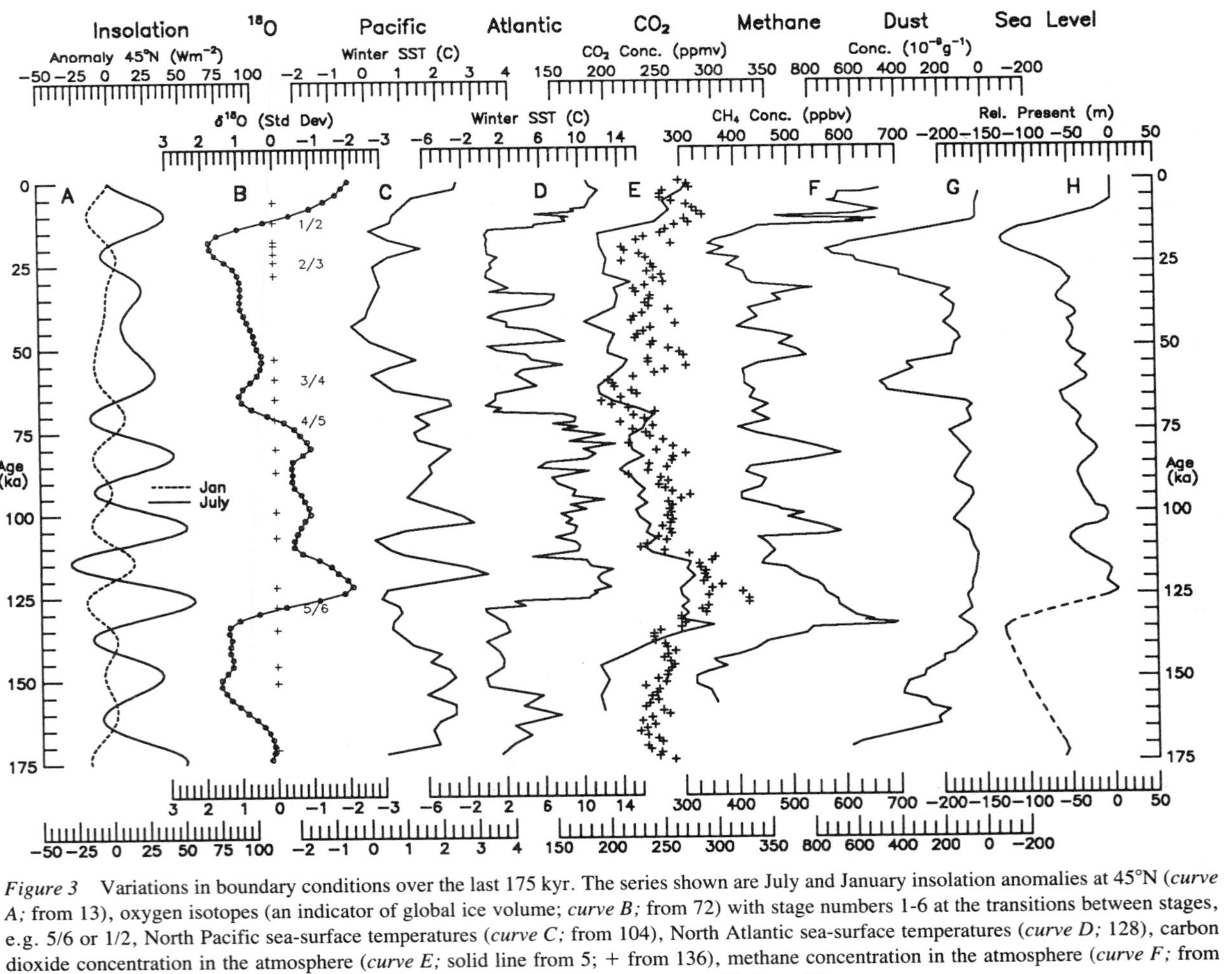

Figure 3 Variations in boundary conditions over the last 175 kyr. The series shown are July and January insolation anomalies at 45°N (*curve A;* from 13), oxygen isotopes (an indicator of global ice volume; *curve B;* from 72) with stage numbers 1-6 at the transitions between stages, e.g. 5/6 or 1/2, North Pacific sea-surface temperatures (*curve C;* from 104), North Atlantic sea-surface temperatures (*curve D;* 128), carbon dioxide concentration in the atmosphere (*curve E;* solid line from 5; + from 136), methane concentration in the atmosphere (*curve F;* from 30), dust loading (*curve G;* from 111), and sea level (relative to present; *curve H;* from 29).

Sea surface temperatures (SSTs) reconstructed from foraminiferal (Atlantic, Figure 3D) and radiolarian (Pacific, Figure 3C) fossils in ocean sediment cores show variations that are broadly similar to the ice-volume records. The interglacial warm intervals are evident, as are the glacial maxima, and the contrast between the first and second halves of the last interval of glaciation is also clearly marked. Both records suggest that the relatively low SSTs that characterized the last glacial maximum persisted for much of the time after 75 ka. Consequently, the influence of these cold oceans on adjacent land areas must also have persisted. Latitudinal transects of estimated SST in the Atlantic show the greater importance of the 100-kyr and 41-kyr period variations in records at higher latitudes, and the greater importance of the 23-kyr period variations in the records from the tropics (73, 128). This pattern likely reflects the pronounced influence of the Laurentide ice sheet on the northern oceans (127), coupled with the influence on the tropical oceans of insolation-driven variations in atmospheric circulation (97). Another feature evident in the record of North Atlantic SSTs (Figure 3) is the abrupt reversal of the deglacial warming trend about 10 kyr, marking the Younger Dryas climate reversal in the North Atlantic region (26, 123, 152).

Detailed records from the Vostok ice core from Antarctica reveal the history of variations in atmospheric composition including such greenhouse gases as carbon dioxide and methane over the last complete glacial/interglacial cycle (76). Carbon dioxide (Figure 3E) levels generally inversely parallel the ice-volume variations through time (5). They are highest during the interglacial periods and lowest during the full-glacial intervals. Recent work (137) suggests that CO_2 variations lead $\delta^{18}O$ by about 3 kyr, a result consistent with earlier analyses of atmospheric carbon dioxide concentrations as inferred from $\delta^{13}C$ records from benthic foraminifera (73, 136).

Methane levels (Figure 3F) reveal variations that seem to reflect the insolation variations to a greater degree than they do the ice-volume variations. This linkage may be a result of methane production from increased tropical precipitation, which in turn is related to the enhancement of the monsoons during Northern Hemisphere insolation maxima (30, 112, 139). It may also result from increased development of methane-producing northern wetlands (37).

Atmospheric dust and aerosol records from the Vostok core (111) also reveal a characteristic pattern of glacial/interglacial variations (Figure 3G). Dust concentrations are highest during the glacial maxima, consistent with exposure then of continental shelves (by lowered sea levels), and greater windiness and dryness of continental-interior source regions (76). Variation in the amount of ice on the continents has a direct effect on sea level, and relative sea-level variations (Figure 3H) closely parallel the oxygen-isotope record of ice volume.

The specific details of the interrelationship between SSTs and ice volume, and their joint dependence on insolation and atmospheric composition, are the subject of continuing research in paleoclimatology (71, 128). The specific mechanisms involved are incompletely known, but a general research strategy involving the retrieval and interpretation of paleoclimatic indicators, the development of models—both conceptual and numerical, and the testing of those models with the paleoclimatic data will help to determine the specific cause-and-effect and feedback mechanisms (34, 73). In general terms, insolation variations drive the global climatic variations on the time scale represented by the last 175 kyr (and longer), by governing global ice volume, and indirectly, SSTs and CO_2. Variations in atmospheric composition apparently amplify the influence of insolation and may provide a mechanism for synchronizing the climatic variations of the northern and southern hemispheres (25, 26, 107). Insolation variations directly control terrestrial climates by warming and drying the continental interiors during times of summer insolation maxima, and the variations indirectly control both terrestrial climates and SSTs through variations in the strength of the monsoons (103, 114).

Regional Climatic Patterns

Fewer examples of regional-scale syntheses of paleoclimatic records exist for the last 175 kyr than for the last 20 kyr, and these pertain mainly to the previous interglacial (e.g. 33, 91). The paleoclimatic records are based primarily on biotic indicators, and we are therefore faced with the prospect of using them to document both the climatic changes and the biospheric responses to those changes. We are encouraged to do this because of our experience with the last 20 kyr, for which independent paleoclimatic records are available in the form of the climate-model simulations. Also, some studies (114, 124) have attempted to extend modeling results over time and to test those simulations (31, 143). These efforts have been largely successful. Because the temporal variations in boundary conditions are known in detail for the last 175 kyr, the record of known variations can be combined with insights gained from the results of climate simulations over the last 20 kyr in order to estimate the climate variations in a particular region (9). We therefore expect that regional variations of climate over the last 175 kyr will become better known, and ultimately, independent climatic records, to which the biospheric response can be compared, will become available.

Biotic Responses

A selection of paleoecological records can be used to illustrate the nature of the response of the biosphere to the climatic forcing during the past 175 kyr (Figure 4). Two principal kinds of response of marine ecosystems are documented by the fossil record: (i) changes in species distributions and abundances, and (ii) changes in productivity. In both the North Pacific (Figure 4C; 100) and North Atlantic (Figure 4D; 128), for example, subtropical assemblages increase at the expense of colder water assemblages during the interglacial periods, and at times of the precessional insolation maxima and ice-volume minima during the glaciations. Similar changes in species distributions and abundances occur throughout the world ocean (32).

Productivity in marine ecosystems also varies through time, at the whole-ocean scale as well as at the regional scale (100, 105). Productivity changes are large in those regions where oceanographic changes are governed by atmospheric circulation controls that themselves are highly variable. For example, the opal flux in sediments from the Arabian Sea reflects productivity in a region of upwelling induced by monsoon winds; its variations are coherent with the insolational forcing of the monsoons (31). Other indicators of various aspects of marine ecosystems also vary coherently with climatic forcing. For example, variations in nutricline depth in the equatorial Atlantic can be linked to variations in the strength of tropical easterlies (Figure 4E; 103), which, like the monsoons, are governed by insolational forcing.

Terrestrial paleoecological records also reveal sweeping reorganizations of plant associations at the vegetation formation or biome levels, and, as in the oceans, these reorganizations proceed in a coherent fashion with the climatic forcing. The record of oak (*Quercus*) pollen abundances at Clear Lake (California) reflect variations in the vegetation between montane coniferous forests and oak woodland, with the latter best expressed during the interglacials (Figure 4F; 1). A second mid-northern-latitude record from Grande Pile (France) shows the alternation between a forested landscape (during warm intervals) and nonforested landscape (during cold) (Figure 4G; 102, 150). Similar variations, although of different amplitude, can be noted in a record from the Valle di Castiglione (Italy) (Figure 4H; 52) In each of these records, the variations evident in the single series shown in Figure 4 are not the product of simple replacements of one vegetation assemblage by another but instead reflect the individualistic variation of different taxa.

Large variations in species abundance are not restricted to the mid-latitudes. Variations in pollen types indicative of moist conditions in East Africa vary in phase with the strength of the monsoon (Figure 4I; see 114, 115). Similarly, variations in the abundance of sclerophyll taxa in tropical Queensland (indicative of dry conditions) vary in concert with changes in SSTs (and hence the moisture flux from tropical oceans) (Figure 4J; 78). Although few continuous paleoecological records for the last 175 kyr exist for the tropics as a whole, the records that do exist suggest that biotic variation in the tropics is considerable on this time scale (28, 35, 93, 144). In the arctic, too, pollen records reveal fluctuations between forest and tundra on this time scale (see reviews in 3, 125; see also 4, 50, 62).

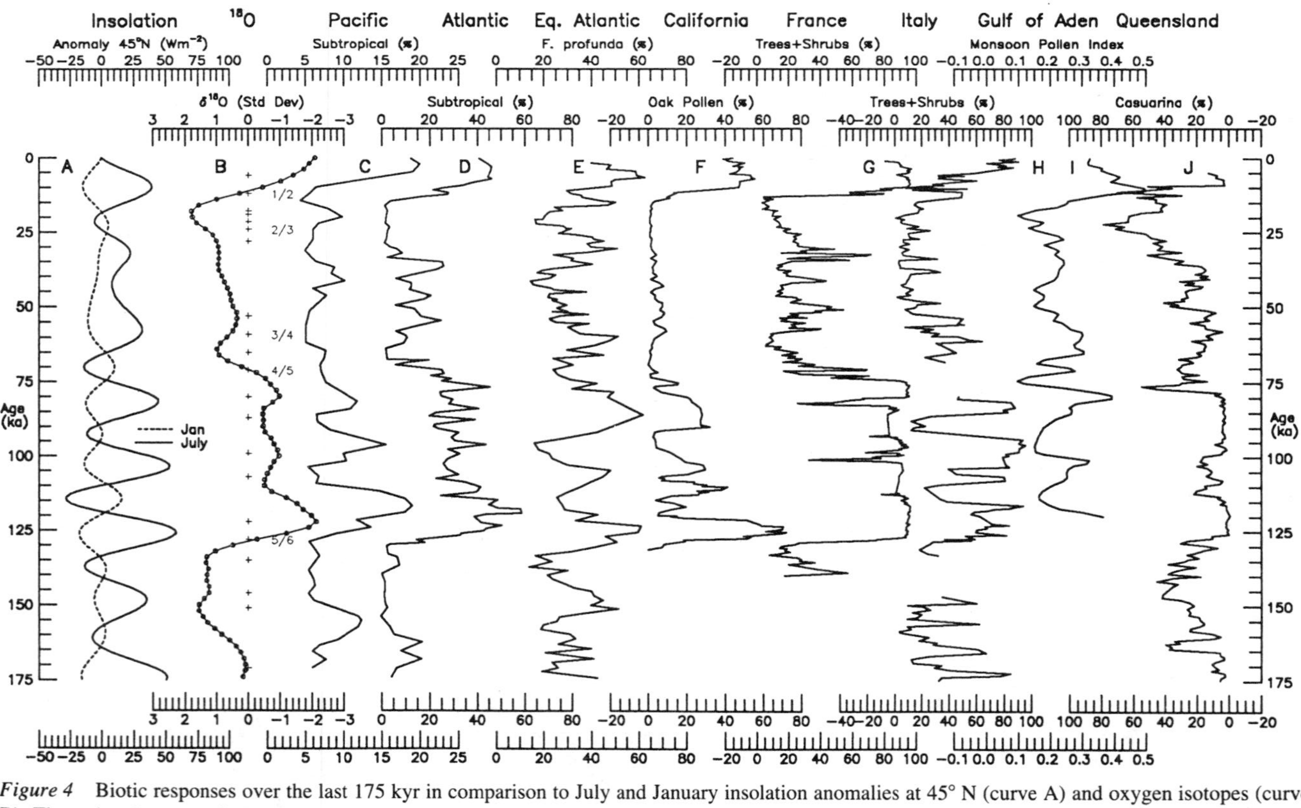

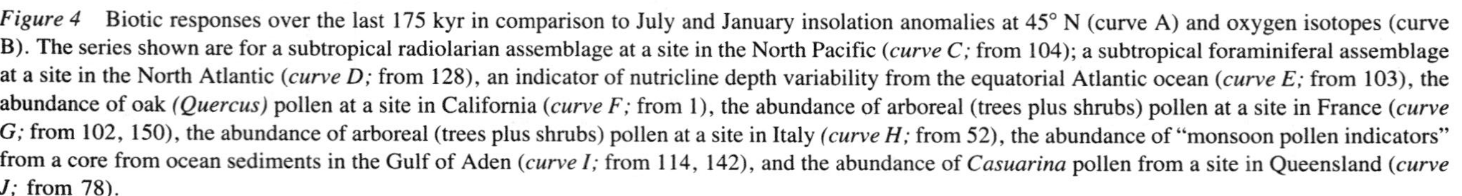
Figure 4 Biotic responses over the last 175 kyr in comparison to July and January insolation anomalies at 45° N (curve A) and oxygen isotopes (curve B). The series shown are for a subtropical radiolarian assemblage at a site in the North Pacific (*curve C;* from 104); a subtropical foraminiferal assemblage at a site in the North Atlantic (*curve D;* from 128), an indicator of nutricline depth variability from the equatorial Atlantic ocean (*curve E;* from 103), the abundance of oak (*Quercus*) pollen at a site in California (*curve F;* from 1), the abundance of arboreal (trees plus shrubs) pollen at a site in France (*curve G;* from 102, 150), the abundance of arboreal (trees plus shrubs) pollen at a site in Italy (*curve H;* from 52), the abundance of "monsoon pollen indicators" from a core from ocean sediments in the Gulf of Aden (*curve I;* from 114, 142), and the abundance of *Casuarina* pollen from a site in Queensland (*curve J;* from 78).

The paleoecological records clearly register the overall state of the climate system, as represented by the oxygen-isotope record of global ice volume, as well as the influence of insolation variations. The control of regional climates by insolation may be expressed either directly, as in the case of midcontinental temperature or moisture, or indirectly, as for atmospheric circulation, monsoonal precipitation, or ice volume. Individual records thus show a mixture of effects of various controls of regional climatic variations. For example, as has already been discussed, the marine record shows distinct latitudinal variations in the amplitudes of the periodic components of variation, with sites from higher latitudes exhibiting greater variability at the 100-kyr and 41-kyr periods, while lower latitude sites show greater variability at the 23-kyr period (see transects of data in 73, 128). Similarly, the terrestrial records also show different mixtures. The record of oak pollen percentages from Clear Lake (Figure 4F) exhibits an overall pattern that resembles a blend of the insolation and ice-volume records (Figures 4A and 4B), while the record from Grande Pile (Figure 4G) resembles a blend of ice-volume and North Atlantic SSTs (Figure 3D). Most of the variation in the individual biotic records can therefore be related to either the ultimate controls of regional climatic variations (i.e. the external boundary conditions along with ice volume and atmospheric composition) or the proximate controls (i.e. SSTs and circulation features in adjacent oceans).

Even though age determinations are more uncertain than those for the last 20 kyr, the biotic responses occur with little discernable delay, or with lag times on the order of, or shorter than, those between the ice sheets and insolation. This inference does not depend on the assumptions made in assigning ages to the various records. For example, Adam (1) and Robinson et al (126) describe how the age of the earliest part of the Clear Lake record was assigned by correlation with the oxygen-isotope record. While this strategy would necessarily bring the "oak maximum" of the previous interglacial in line with the insolation maximum and ice-volume minimum about 125 ka, it would not necessarily produce the good alignment between subsequent peaks evident in the figures. Apparent discrepancies in the timing of the extrema of the different records may therefore be as much a manifestation of the inherent time lags among components of the climate system as they are of any "inertia" in the biotic indicators. The close coupling of the biotic responses with the climatic controls implies that the specific mechanisms involved in the response allow the components of terrestrial and marine ecosystems to remain in a dynamic equilibrium with climate (see also 118). We believe that the same kinds of mechanisms employed during the last 20 kyr were also used by the biosphere to track climatic variations during the last 175 kyr.

THE PAST 3 MILLION YEARS

When the time span over which we view the record of past climatic changes is expanded to the last 3 Myr, the periodic variations that were evident in series for the last 175 kyr are supplemented by longer-term variations and trends and by changes in the pattern of variability. The biotic responses again track these changes but show the additional impact of aperiodic, nonreversing trends in Earth history, in particular, those related to continental drift and mountain building. These processes erect or remove barriers to migration or dispersal, and directly affect both terrestrial and marine ecosystems over the long run. These slowly varying processes may also have an indirect effect on the distribution of plants and animals by governing the evolution of climate on the 3 Myr time scale.

Variations in Insolation and Oxygen Isotopes

During the last 3 Myr, the variations of insolation show the expected periodic variations at the 41-kyr and 23-kyr time scales of tilt and precession, and the influence of the 100-kyr and 400-kyr eccentricity cycles on the envelope of the insolation variations (Figure 5A; 16). This latter pattern results from the influence of eccentricity on the effectiveness of the variations in the time of year of perihelion in influencing the seasonal cycle of insolation at a particular latitude. The insolation variations give the overall impression of regularity, with little abrupt change in pattern. Through time, unique sequences of oscillations do occur, however. Groups of several large maxima occur regularly, including, for example, the three at 125, 105, and 85 ka (Figure 3A). Other groups of oscillations are also evident. Conversely, there are intervals when the extrema of insolation are relatively small, as from 450 to 300 ka. In the context of the changes during last 3 Myr, the early Holocene and previous interglacial insolation values are exceeded 9% and 1% of the time, respectively (16). One hundred forty local maxima in the July insolation record occurred during the last 3 Myr (about every 20 kyr). Of these local maxima, that in the early Holocene is the fifty-fourth largest, whereas that for the previous interglacial is the eleventh largest. Maxima in July insolation values are similar to those for the last 20 and 175 kyr, respectively; maxima of these magnitudes are therefore not uncommon in the longer record.

In contrast to the regularity of the insolation record, the oxygen-isotope record shows a striking change in the pattern of its quasiperiodic variations, beginning about 700 ka (Figure 5B; 120, 131). This change is the superimposition of strong 100-kyr variations on the 41-kyr period variations that dominate most of this record. A second change, less evident in the series plotted in Figure 5B, is the increase in the overall variability of record at the onset of more extensive glaciation in the circum-North Atlantic region about 2.5 Myr (as recorded by the input of glacial detritus into the North Atlantic

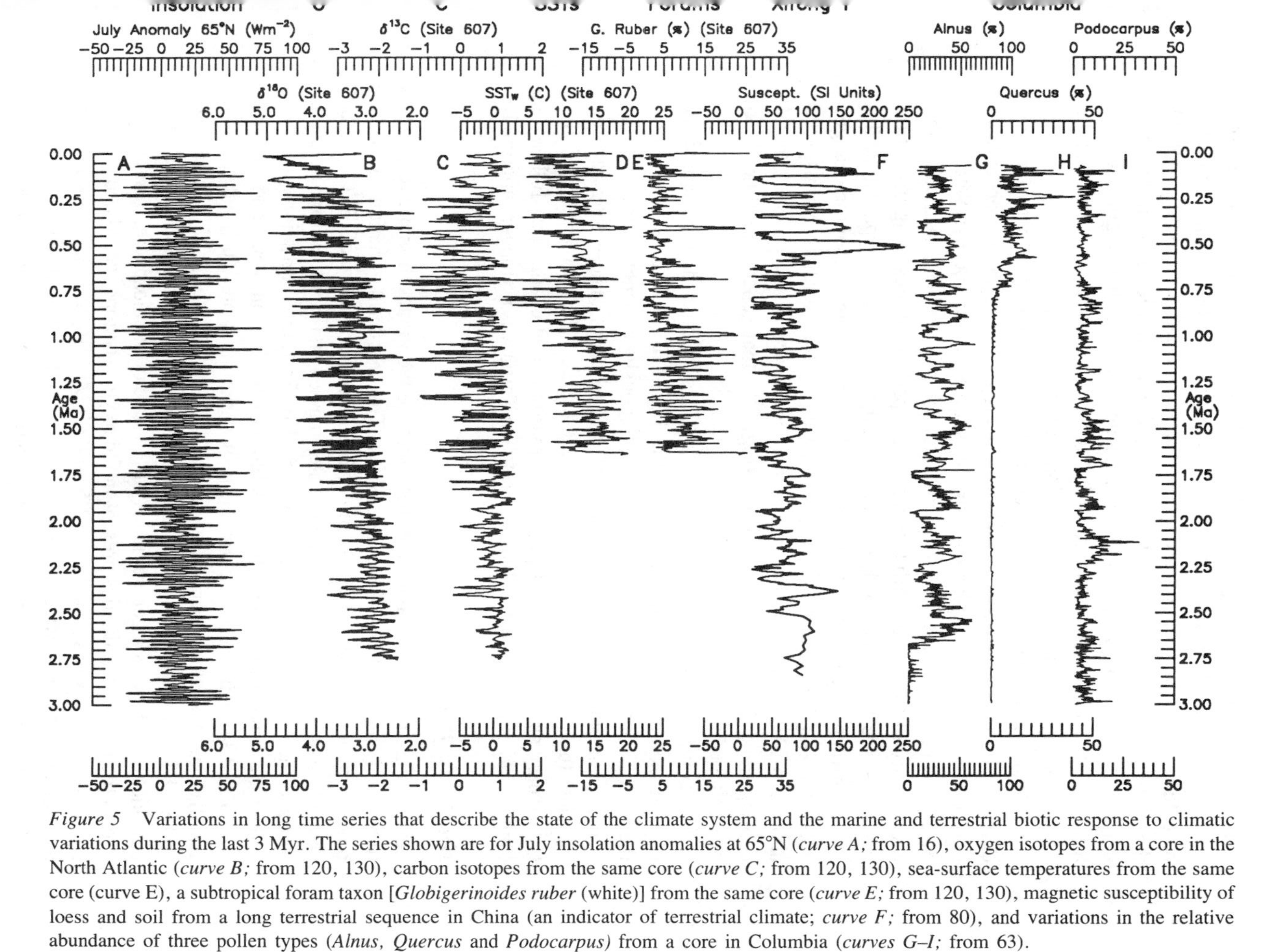

Figure 5 Variations in long time series that describe the state of the climate system and the marine and terrestrial biotic response to climatic variations during the last 3 Myr. The series shown are for July insolation anomalies at 65°N (*curve A;* from 16), oxygen isotopes from a core in the North Atlantic (*curve B;* from 120, 130), carbon isotopes from the same core (*curve C;* from 120, 130), sea-surface temperatures from the same core (curve E), a subtropical foram taxon [*Globigerinoides ruber* (white)] from the same core (*curve E;* from 120, 130), magnetic susceptibility of loess and soil from a long terrestrial sequence in China (an indicator of terrestrial climate; *curve F;* from 80), and variations in the relative abundance of three pollen types (*Alnus, Quercus* and *Podocarpus)* from a core in Columbia (*curves G–I;* from 63).

—e.g. 130, 135). This change is clearly expressed in other long oxygen-isotope series (77, 133). The general patterns in the oxygen-isotope record, including both the general periodicity throughout the record and the amplification of the 100-kyr cycle within the last million years are also present in a record of magnetic susceptibility in long loess sequences from China (Figure 5F; 79, 80). Variations in the terrestrial record of paleoenvironments confirm that the oxygen isotopic record represents a global climatic signal.

Comparison of the insolation record (Figure 5A) with the two paleoclimatic series (e.g. Figure 5B and 5F) illustrates one of the major tasks in paleoclimatology: the explanation of the source of the 100-kyr variability evident during the last 700 kyr (and its absence earlier). Although the eccentricity of the Earth's orbit varies at this period, the resulting insolation variations are too small to drive the ice-volume variations apparent in the oxygen-isotope record. A second major task is the explanation of the changes accompanying the growth of larger ice sheets in the Northern Hemisphere about 2.5 Myr. Although this change can be viewed as part of the general cooling trend throughout the Cenozoic (e.g. 98), that general trend needs to be explained as well.

The combined variations of insolation and ice volume (as represented by oxygen isotopes in foraminifera), and the regional climatic variations that these two controls govern, provide a continuously varying environment to which plant and animal taxa must have had to adapt. These long records indicate that environmental variation is the rule. The perception of the Quaternary as an exceptional period of environmental change that follows a long period of more gradual change during the Tertiary is no longer supported.

Biotic Responses

Several long records are available that illustrate the modes of biotic response to the climatic variations of the last 3 Myr. The long-term variations of $\delta^{13}C$ have two components (121; Figure 5C). Specific values of $\delta^{13}C$ are a characteristic property of individual water masses and can therefore be used to trace their mixing and circulation (see 40). Secondly, variations in $\delta^{13}C$ reflect the amount of carbon stored as biomass on the continents and shelves (see also 134). This second component of variation is present in all records equally, and visual comparison of several long records (e.g. 121; Figures 2 and 3) shows substantial common variation among those records. We therefore consider $\delta^{13}C$ here as a general index of the state of the biosphere. The record shows the major periodicities present in the oxygen-isotope record and clearly shows the changes in variability at 2.5 and 0.7 Myr (Figure 5). Viewed either as a tracer of different water masses or as a general terrestrial-marine biomass index, the variability in this $\delta^{13}C$ record suggests that the biotic tracking of environmental change that we noted for the late-Quaternary also prevails on longer time scales.

More specific indicators of biotic response can also be examined. A 1.6-Myr record of the subtropical foram *Globigeriniodes ruber* (white) (131) shows the characteristic orbital periodicities and changes of variance (Figure 5E). Two long terrestrial pollen records are available that show the biotic response to regional climatic changes over the last 3 Myr. A record from Tulelake (Northern California) does not have sufficient temporal resolution to record all of the orbital periods but does show some generalized changes corresponding to times of global climatic reorganization (2). The longer and more detailed record from the Sabana de Bogota (Columbia) has greater temporal resolution and shows several striking patterns (63) that are illustrated by time series for (*Alnus, Quercus* and *Podocarpus)* pollen (Figure 5G-I). All three pollen types show quasi-periodic variations with most major and minor glacial/interglacial cycles recorded. As striking is the first appearance in this record of *Alnus* (about 2.65 Myr) and *Quercus* (about 0.8 Myr). Both of these reflect the continuing reaction of the vegetation to the elevation of the isthmus of Panama (about 4 Myr). These large changes in the composition of the vegetation are superimposed on a continuous background of periodic variation (represented here by the *Podocarpus* record) that persisted throughout this time span. This pattern suggests that although large reorganizations of the biota can result from plate tectonic–driven processes, these reorganizations are carried out in the context of a continuously varying environment. For an account of the mammal exchange between North and South America since 2.5 Mya that recognizes the interplay of orbitally and tectonically induced climate changes, see Vrba (144a).

DISCUSSION

In our review of the data, we have emphasized three key observations: (i) the evidence on all three time scales for orbitally paced changes in climate, (ii) the current understanding of how the changing set of global controls leads to different climatic responses in each geographic region, and (iii) the major responses of the vegetation and marine plankton to these changes. The key results encourage us to project our understanding of climatic and biospheric changes during the Quaternary back into the Tertiary and earlier geological periods and thus onto evolutionary time scales. As Bennett (12) pointed out, such a projection helps explain Gould's (55) paradox of the first tier, i.e. "our failure to find any clear vector of fitfully accumulating progress," within species in the fossil record "when conventional [evolutionary] theory . . . expects it as a consequence of competition . . . ."

An understanding of the mechanisms and effects of orbital forcing on regional climates and biotic systems is key to this explanation. Since 1976, when Hays et al (59) used data from the last 400 kyr to present convincing evidence for orbital pacing of global climates, paleoclimatologists have used climate models and data to discover some of the mechanisms by which orbital variations have affected the climate system during both the Quaternary and earlier periods, (34, 73, 114, 115). They also obtained data to show the existence of orbitally paced changes from pre-Quaternary periods (i.e. older than 1.8 Myr).

Data and modeling results for the last 20 kyr show the important role of the contrast between land and sea in translating the changing seasonal intensity of insolation into stronger and weaker monsoons and thus in producing periodic large changes in moisture balance in tropical climates. This mechanism for direct climatic response to orbital forcing has been present throughout the geological record. For example, stratigraphic evidence exists for tropical lake level variations from a 20 Myr interval in the late Triassic (15, 108). Modeling studies also support the importance of this mechanism in the past by illustrating how continental drift, continental size, and mountain uplift modulate the magnitude of monsoonal oscillations (39, 85). In previous geological periods, therefore, ice sheets and ice-age climates are not required for orbitally induced changes in insolation to cause significant climate changes on the time scales of orbital variations, particularly for large continental areas like Afro-Eurasia today.

Data and modeling results from the last 20 kyr also show how global and regional climate controls produce a set of regional mid- to high-latitude climatic patterns that represent a complex but predictable response to relatively simple forcing (34). For example, the models and data illustrate how changes both in the seasonal cycle of insolation and in ice-sheet size can shift the position of the jet stream and produce glacial anticyclones. These in turn produce predictable changes in the seasonal patterns of temperature, precipitation, and moisture balance (8, 34, 149).

The changes in ice sheets and insolation produce changing seasonal patterns in the magnitude of different climate variables and thus lead to independent variations of individual climate variables. The changing controls, therefore, induce climates to change not just in location but also in character, and different past combinations of the controls therefore have created past climates unlike any at present (i.e. without modern analogs) (82). Among the biospheric responses to these climatic changes are time-transgressive changes in taxa that appear as differences in the patterns and timing of migration among taxa within regions (69, 75). The orbitally induced climate changes therefore have led to individualistic responses among plant and plankton taxa, and such responses, by definition, alter former plant and plankton assemblages and lead to the emergence of new associations and ecosystems (68, 95, 125, 147).

Oxygen isotope records for the last 3 Myr (Figure 5B) show that the recent large changes in climate like those of the last 20 kyr and 175 kyr extend back over 700 kyr with a dominant 100 kyr beat and that northern hemisphere ice sheets have varied significantly in size for at least the last 2.5 Myr. Conservative interpretations indicate that significant individualistic responses among climate variables and various taxa have at least this antiquity. In areas with significant monsoonal variations, the antiquity is greater. Recent studies present records for changes in global ice volume not only from the Pliocene (3 Myr ago) but also from much earlier in the Tertiary (119). The current ice age, i.e. this period in earth history with permanent ice sheets that vary in size, is, therefore, ancient. Tertiary paleoecological records predating those in Figure 5 show Quaternary-like variations (6, 7) whenever the temporal resolution of the sampling intervals is fine enough.

Modern species and many extinct species, therefore, evolved in the face of these regular changes in climate and the biosphere. Huntley & Webb (69) have even postulated that plant taxa developed an ability to migrate large distances relatively rapidly in order to cope with the orbitally induced variations in climate. Given that orbital forcing has a regularity similar to diurnal and annual forcing, we are not surprised that taxa are adapted to coping with the long-term orbital variations in much the same way that individual organisms are adapted to coping with the climate changes induced by diurnal and annual variations in insolation. The continuous changes in association among taxa have led to continuous changes in the competitive environment for each taxon. With favorable climate conditions appearing and disappearing and inducing large changes in species abundance, many currently abundant taxa have experienced long periods when they were rare and had fragmented distributions (10, 41). This latter condition should favor allopatric speciation. The problem is, however, that the fossil record indicates that speciation was relatively rare and that species have generally existed for 1 to 10 Myr (36). Aware of this latter observation, along with the evidence for little progressive change within species, Gould (55) posed the paradox "of the first tier" concerning the lack of evidence for fitfully accumulating progress. Bennett (12) responded by noting the importance of orbitally induced climate change and individualistic behavior. He proposed that Gould add the orbital time scale of 10 to 100 kyr to his three tier system of biologically significant time scales: (i) ecological time (< 1 kyr), (ii) geological time (1 to 10 Myr), and (iii) mass extinctions (ca. 26 Myr).

Equilibrium/Disequilibrium Responses and Lessons for the Future

Paleoecologists have long debated the relative roles of climate and biotic factors when interpreting both the causes for the changing associations among taxa and the evidence for past associations without modern analogs. This concern has led to a discussion of whether the vegetation is in equilibrium with climate (92, 116). We believe that the antiquity of the climate forcing implies that all modern species must have coping mechanisms that limit the degree of disequilibrium. We therefore believe that dynamic equilibrium theory must apply (117, 146), because climate changes on all time scales, making notions concerning steady-state equilibrium and climax theory inappropriate for time scales of 200 years or longer. Within the forcing-response perspective of a dependent variable (vegetation) always chasing an independent variable (climate), some lags or delays will exist between the climate and vegetation; but, as long as the lags are much shorter than the period of climate forcing, the vegetation is in dynamic equilibrium with climate. For the time scales described in this review, lags of 100 to 1000 years are part of an equilibrium response. Davis (45) has recently developed explanations that downplay the role of chance on continental scales, and she, like Prentice (117), notes the increasing role of biotic and edaphic factors as time and space scales decrease. We believe that this hierarchical scheme for attributing importance to the many factors controlling taxon location, abundance, and dispersal will ultimately lead to an adequate theory to unite explanation of continental migrations with gap-phase succession (117).

Recently the threat of global warming has raised concern and interest among many biologists about the impact of large rapid climate changes on species distributions, ecosystems integrity, and biological diversity (110). Paleoecological and paleoclimatic evidence has been much cited because the magnitude of global mean temperature changes during previous intervals (e.g. 15 to 6 ka) matches the predicted future increase in the global mean temperature. The rates of past global climate changes, however, appear to have been slower than those predicted (44, 148), and this fact raises concern about potential extinctions of species and disruptions of extant ecosystems. Several studies (44, 66, 67, 109, 148, 154) discuss these issues, and Hunter et al (64), Bradshaw & McNeilly (24), Huntley (67), and Peters & Lovejoy (110) propose some of the changes in planning that will be needed to minimize damage to biological diversity.

ACKNOWLEDGMENTS

NSF grants from the Climate Dynamics Program (ATM 91-03000 and ATM 91-07750) and a DOE grant from the Carbon Dioxide Research Office supported the writing of this review. We thank K. Anderson, B. Lipsitz, C. Mock, P. Newby, and L. Sheehan for technical assistance and G. MacDonald, R.S. Webb, and C. Whitlock for critical reviews of an early draft.

Literature Cited

1. Adam, D. P. 1988. Palynology of two upper Quaternary cores from Clear Lake, Lake County, California. : *US Geological Survey Professional Pap. 1363. 86 pp.*
2. Adam, D. P., Sarna-Wojcicki, A. M., Rieck, H. J., Bradbury, J. P., Dean, W. E., et al. 1989. Tulelake, California: the last 3 million years. *Palaeogeogr., Palaeoclimatol., Palaeoecol.* 72:89–103
3. Ager, T. A., Brubaker, L. 1985. Quaternary palynology and vegetational history of Alaska. In *Pollen Records of Late-Quaternary North American Sediments,* ed. V. M. Bryant Jr., R. G. Holloway, pp. 353–84. Dallas: Am. Assoc. Stratigraphic Palynologists
4. Anderson, P. M. 1985. Late Quaternary vegetational change in the Kotzebue Sound area, northwestern Alaska. *Quat. Res.* 24:307–21

4a. Bard, E., Hamelin, B., Fairbanks, R.G., Zindler, A. 1990. Calibration of the ^{14}C timescale over the past 30,000 years using mass spectrometric U-TH ages from Barbados corals. *Nature* 345:405–9

5. Barnola, J. M., Raynaud, D., Korotkevich, Y. S., Lorius, C. 1987. Vostok ice core provides 160,000-year record of atmospheric CO_2 *Nature* 329: 408–14
6. Barnosky, C. W. 1984. Late Miocene vegetational history and climatic variations inferred from a pollen record in northwestern Wyoming. *Science* 223: 49–51
7. Barnosky, C. W. 1987. Response of vegetation to climatic changes of different duration in the late Neogene. *TREE* 2:247–50
8. Barnosky, C. W., Anderson, P. M., Bartlein, P. J. 1987. The northwestern U.S. during deglaciation; Vegetational history and paleoclimatic implications. In North America and Adjacent Oceans During the Last Deglaciation, ed. W. F. Ruddiman, H. E. Wright Jr., pp. 289–321. Boulder: Geological Society of America.
9. Bartlein, P. J., Anderson, P. M., Edwards, M. E., McDowell, P. F. Prentice, I. C. 1992. A framework for interpreting paleoclimatic variations in eastern Beringia. *Quat. Int.* In press
10. Bartlein, P. J., Prentice, I. C. 1989. Orbital variations, climate and paleoecology. *TREE* 4:195–99
11. Bartlein, P. J., Webb, T. III, Fleri, E. 1984. Holocene climatic change in the northern Midwest: pollen-derived estimates. *Quat. Res.* 22:361–74
12. Bennett, K. D. 1990. Milankovitch cycles and their effects on species in ecological and evolutionary time. *Paleobiology* 16:11–21
13. Berger, A. 1978. Long-term variations of caloric insolation resulting from the earth's orbital elements. *Quat. Res.* 9:139–67
14. Berger, A. 1984. Accuracy and frequency stability of the Earth's orbital elements during the Quaternary. In *Milankovitch and Climate,* Part 1, ed. A. Berger, J. Imbrie, J. Hays, G. Kukla, B. Saltzman, pp. 527–37. Dordrecht: Reidel
15. Berger, A., Imbrie, J., Hays, J., Kukla, G., Saltzman, B. 1984. *Milankovitch and Climate*. Dordrecht, Netherlands: Reidel. 895 pp.
16. Berger, A., Loute, M. F. 1991. Insolation values for the climate of the last 10 million years. *Quat. Sci. Rev.* 10:297–317
17. Berglund, B. E. 1986. *Handbook of Holocene Paleoecology and Paleohydrology*. New York: Wiley. 869 pp.
18. Bernabo, J. C. 1981. Quantitative estimates of temperature changes over the last 2700 years in Michigan based on pollen data. *Quat. Res.* 15:143–59
19. Birks, H. J. B., Birks, H. H. 1980. *Quaternary Paleoecology*. London: Arnold. 289 pp.
20. Borchert, J. A. 1950. The climate of the central North American grassland. *Ann. Assoc. Am. Geogr.* 40:1–39
21. Bradley, R. S. 1985. *Quaternary Paleoclimatology*. Boston: Allen & Unwin. 472 pp.
22. Bradley, R. S. 1991. *Global Changes of the Past*. Boulder, Colo: Off. Interdiscipl. Earth Sci. 514 pp.
23. Bradley, R.S. 1992. *Climate Since A.D. 1500*. London: Routledge. 698 pp.
24. Bradshaw, A. D., McNeilly, T. 1991. Evolutionary response to global climatic changes. *Ann. Bot.* 67(Suppl. 1):5–14
25. Broccoli, A. J., Manabe, S. 1987. The influence of continental ice, atmospheric CO_2, and land albedo on the climate of the last glacial maximum. *Climate Dynamics* 1:87–99
26. Broecker, W. S., Denton, G. H. 1989.

The role of ocean-atmosphere reorganizations in glacial cycles. *Geochimica Cosmochimica Acta* 53:2465-501
27. Broecker, W. S., van Donk, J. 1970. Insolation changes, ice volumes, and the 0^{18} record in deep sea cores. *Rev. Geophys. Space* 8:169–98
28. Bush, M. B., Colinvaux, P. A. 1990. A pollen record of a complete glacial cycle from lowland Panama. *J. Vegetation Sci.* 1:105–18
29. Chappell, J., Shackleton, N. J. 1986. Oxygen isotopes and sea level. *Nature* 324:137–40
30. Chappellaz, J., Barnola, J. M., Raynaud, D., Korotkevich, Y. S., Lorius, C. 1990. Ice-core record of atmospheric methane over the past 160,000 years. *Nature* 345:127–31
31. Clemens, S., Prell, W., Murray, D., Shimmield, G., Weedon, G. 1991. Forcing mechanisms of the Indian Ocean monsoon. *Nature* 353:720–25
32. CLIMAP Project Members. 1981. *Seasonal Reconstruction of the Earth's Surface at the Last Glacial Maximum.*: Geological Society of America Map and Chart Series, Map MC. 36 pp.
33. CLIMAP Project Members. 1984. The last interglacial ocean. *Quat. Res.* 21: 123–224
34. COHMAP Members. 1988. Climatic changes of the last 18,000 years: observations and model simulations. *Science* 241:1043–52
35. Colinvaux, P. 1987. Amazon diversity in light of the paleoecological record. *Quat. Sci. Rev.* 6:93–114
36. Cronin, T.M. 1985. Speciation and stasis in marine Ostracoda: climatic modulation of evolution. *Science* 227:60–63
37. Crowley, T. J. 1991. Ice-age methane variations. *Nature* 353:122–23
38. Crowley, T. J., North, G. R. 1991. *Paleoclimatology*. New York: Oxford Univ. Press. 339 pp.
39. Crowley, T. J., Short, D. A., Mengel, J. G., North, G. R. 1986. Role of seasonality in the evolution of climate during the last 100 million years. *Science* 231:579-84
40. Curry, W. B., Duplessy, J. C., Labeyrie, L. D., Shackleton, N. J. 1988. Changes in the distribution of $\delta\ ^{13}C$ of deep water ΣCO_2 between the last glaciation and the Holocene. *Paleoceanography* 3:317–41
41. Cwynar,.L., MacDonald, G. 1987. Geographical variation of lodgepole pine in relation to population history. *Am. Nat.* 129:463–69
42. Davis, M. B. 1976. Pleistocene biogeography of temperate deciduous forests. *Geosci. & Man* 13:13–26
43. Davis, M. B. 1983. Holocene vegetational history of the eastern United States. In *Late-Quaternary Environments of the United States,* ed. H. E. Wright Jr., pp. 166–81. Minneapolis: Univ. Minn. Press
44. Davis, M. B. 1989. Lags in vegetation response to greenhouse warming. *Climatic Change* 15:75–82
45. Davis, M. B. 1991. Research questions posed by the paleoecological record of global change. In *Global Changes of the Past,* ed. R. S. Bradley, pp. 385–95. Boulder, Colo: UCAR/Off. Interdisc. Earth Stud.
46. De Angelis, M., Jouzel, J., Lorius, C., Merlivat, L., Petit, J. R., et al. 1984. Ice age data for climate modelling from an Antarctic (Dome C) ice core. In *New Perspectives in Climate Modelling,* ed. A. L. Berger, C. Nicolis, pp. 23–45. Amsterdam: Elsevier
47. Denton, G. H., Hughes, T. J. 1981. *The Last Great Ice Sheets*. New York: Wiley-Intersci. 484 pp.
48. Dexter, F., Banks, H. T., Webb, T. III. 1987. Modeling Holocene changes in the location and abundance of beech populations in eastern North America. *Rev. Palaeobot. Palynol.* 50:273–92
49. Dyke, A. S., Prest, V. K. 1987. Late Wisconsinan and Holocene history of the Laurentide ice sheet. *Geographie physique et Quaternaire* 41:237–63
50. Edwards, M. E., McDowell, P. F. 1991. Interglacial deposits at Birch Creek, Northeast Interior Alaska. *Quat. Res.* 35:41–52
51. Fairbanks, R. G. 1989. A 17,000-year glacio-eustatic sea level record: influence of glacial melting rates on the Younger Dryas event and deep-ocean circulation. *Nature* 342:637–42
52. Follieri, M., Magri, D., Sadori, L. 1989. Pollen stratigraphical synthesis from Valle di Castiglione (Roma). *Quat. Int.* 3/4:81–84
53. Gajewski, K. 1988. Late Holocene climate changes in eastern North America estimated from pollen data. *Quat. Res.* 29:255–62
54. Gates, W. L. 1976. The numerical simulation of ice-age climate with a global general circulation model. *J. Atmos. Sci.* 33:1844–73
55. Gould, S. J. 1985. The paradox of the first tier: an agenda for paleobiology. *Paleobiology* 11:2–12
56. Graham, R. W., Grimm, E. C. 1990. Effects of global climate change on

the patterns of terrestrial biological communities. *TREE* 5:289–92

57. Grimm, E. C. 1983. Chronology and dynamics of vegetation change in the prairie-woodland region of southern Minnesota. *New Phytologist* 93:311–50
58. Grove, J. M. 1988. *The Little Ice Age*. London: Methuen. 498 pp.
59. Hays, J. D., Imbrie, J., Shackleton, N. 1976. Variations in the earth's orbit: pacemaker of the ice age. *Science* 194:1121–32
60. Hecht, A. D. 1985. *Paleoclimate Analysis and Modeling*. New York: Wiley. 445 pp.
61. Henderson-Sellers, A., McGuffie, K. 1987. *A Climate Modelling Primer*. New York: Wiley. 217 pp.
62. Hillaire-Marcel, C., de Vernal, A. 1989. Isotopic and palynological records of the Late Pleistocene in eastern Canada and adjacent ocean basins. *Geographie physique et Quaternaire* 43:263–90
63. Hooghiemstra, H. 1989. Quaternary and Upper Pliocene glaciations and forest development in the tropical Andes: evidence from a long high-resolution pollen record from the sedimentary basin of Bogota, Colombia. *Palaeogeogr., Palaeoclimatol., Palaeoecol.* 72:11–26
64. Hunter, M. L., Jr., Jacobson, G. L., Jr., Webb, T. III. 1988. Paleoecology and the coarse-filter approach to maintaining biological diversity. *Conserv. Biol.* 2:375-85
65. Huntley, B. 1988. Europe. In *Vegetation History*, ed. B. Huntley, T. Webb III, pp. 341–83. Amsterdam: Kluwer Acad.
66. Huntley, B. 1990. Studying global change: the contribution of Quaternary palynology. *Palaeogeogr., Palaeoclimatol., Palaeoecol.* 82:53–61
67. Huntley, B. 1991. How plants respond to climate change: migration rates, individualism and the consequences for plant communities. *Ann.Bot.* 67(Suppl.) 1:15–22
68. Huntley, B., Webb, T., 111. 1988. *Vegetation History*. Dordrecht: Kluwer Acad. 803 pp.
69. Huntley, B., Webb, T. III. 1989. Migration: species' response to climatic variations caused by changes in the earth's orbit. *J. Biogeography* 16:5–19
70. Imbrie, J. 1985. A theoretical framework for the Pleistocene ice ages. *J. Geological Soc. London*. 142:417–32
71. Imbrie, J., Boyle, E. A., Clemens, S. C., Duffy, A., Howard, W. R., et al. 1992. The structure of major glaciation cycles. *Paleoceanography*. In Press
72. Imbrie, J., Hays, J. D., Martinson, D. G., McIntyre, A., Mix, A. C., et al. 1984. The orbital theory of Pleistocene climate: support from a revised chronology of the marine 1180 record. In *Milankovitch and Climate*, ed. A. Berger, J. Imbrie, J. Hays, G. Kukla, B. Saltzman, pp. 269–305. Dordrecht, Netherlands: Reidel
73. Imbrie, J., McIntyre, A., Mix, A. 1989. Oceanic response to orbital forcing in the Late Quaternary: observational and experimental strategies. In *Climate and Geo-Sciences*, ed. A. Berger, S. Schneider, J. C. Duplessy, pp. 121–64. Dordrecht: Kluwer Acad.
74. Jacobson, G. L., Webb, T. III, Grimm, E. C. 1987. Patterns and rates of vegetation change during the deglaciation of eastern North America. In *North America and Adjacent Oceans During the Last Deglaciation*, ed. W. F. Ruddiman, H. E. Wright, Jr., pp. 277–88. Boulder, Colo: Geological Soc. Am.
75. Johnson, W. C., Webb, T., 111. 1989. The role of bluejays (*Cyanocitta christata* L.) in the postglacial dispersion of Fagaceous trees in eastern North America. *J. Biogeogr.* 16:561–71
76. Jouzel, J., Barkov, N. I., Barnola, J. M., Genthon, C., Korotkevich, Y. S., et al. 1989. Global change over the last climatic cycle from the Vostok ice core record (Antarctica). *Quat. Int.* 2:15–24
77. Joyce, J. E., Tjalsma, L. R. C., Prutzman, J. M. 1990. High-resolution planktic stable isotope record and spectral analysis for the last 5.35 m.y.: ocean drilling program site 625 northeast Gulf of Mexico. *Paleoceanography* 5:507–29
78. Kershaw, A. P. 1986. Climatic change and Aboriginal burning in north-east Australia during the last two glacial/interglacial cycles. *Nature* 322:47–49
79. Kukla, G. 1990. Loess stratigraphy in central China and correlation with an extended oxygen isotope stage scale. *Quat. Sci. Rev.* 6:191–219
80. Kukla, G., An, Z. 1989. Loess stratigraphy in central China. *Palaeogeogr., Palaeoclimatol., Palaeoecol.* 72:203–25
81. Kutzbach, J. E. 1985. Modeling of paleoclimates. *Adv. Geophysics* 28A: 159–96
82. Kutzbach, J. E. 1987. Model simulations of the climatic patterns during

the deglaciation of North America. In *North America and Adjacent Oceans During the Last Deglaciation,* ed. W. F. Ruddiman, H. E. Wright Jr., pp. 425–46. Boulder, Colo: Geological Soc. Am.

83. Kutzbach, J. E., Gallimore, R. G. 1988. Sensitivity of a coupled atmosphere/ mixed layer ocean model to changes in orbital forcing at 9000 years BP. *J. Geophysical Res.* 93:803–21
84. Kutzbach, J. E., Guetter, P. J. 1986. The influence of changing orbital parameters and surface boundary conditions on climate simulations for the past 18,000 years. *J. Atmos. Sci. 43: 1726–59*
85. Kutzbach, J. E., Guetter, P. J., Ruddiman, W. F., Prell, W. L. 1989. Sensitivity of climate to Late Cenozoic uplift in southern Asia and the American West: numerical experiments. *J. Geophysical Res.* 94:18,393–407
86. Kutzbach, J. E., Ruddiman, W. F. 1992. Model description, external forcing, and surface boundary conditions. In *Global Climates Since the Last Glacial Maximum,* ed. H. E. Wright Jr., J. E. Kutzbach, T. Webb III, W. F. Ruddiman, F. A. Street-Perrott, P. J. Bartlein, Minneapolis: Univ. Minn. Press. In press
87. Kutzbach, J. E., Street-Perrott, F. A. 1985. Milankovitch forcing of fluctuations in the level of tropical lakes from 18 to 0 kyr BP. *Nature* 317:130–34
88. Kutzbach, J. E., Webb, T. III. 1991. Late Quaternary climatic and vegetational change in eastern North America: concepts, models, and data. In *Quaternary Landscapes,* ed. L. C. K. Shane, E. J. Cushing, pp. 175–217. Minneapolis: Univ. Minn. Press
89. Kutzbach, J. E., Wright, H. E. Jr. 1985. Simulation of the climate of 18,000 years BP: results for the North American/ North Atlantic/European sector and comparison with the geologic record of North America. *Quat. Sci. Rev.* 4:147–87
90. Lamb, H. F., Edwards, M. E. 1988. The Arctic. In *Vegetation History,* ed. B. Huntley, T. Webb III, pp. 519–55. Amsterdam: Kluwer Acad.
91. LIGA Members. 1992. The last interglacial in high latitudes of the Northern Hemisphere: terrestrial and marine evidence. *Quat. Int.* In press
92. MacDonald, G. M., Edwards, K. J. 1991. Holocene palynology: I. principles, population and community ecology, palaeoclimatology. *Prog. Phys. Geogr.* 15:261–89
93. Maley, J. 1991. The African rain forest vegetation and palaeoenvironments during the Late Quaternary. *Climatic Change* 19:79–98
94. Manabe, S., Hahn, D. G. 1977. Simulation of the tropical climate of an Ice Age. *J. Geophysical Res.* 82:3889–911
95. McDowell, P. F., Webb, T. III, Bartlein, P. J. 1991. Long-term environmental change. In *The Earth As Transformed By Human Action: Global and Regional Changes in the Biosphere Over the Past 300 Years,* ed. B. L. Turner, II, W. C. Clark, R. W. Kates, J. F. Richards, J. T. Mathews, W. B. Meyer, pp. 143–62. Cambridge: Cambridge Univ. Press
96. McGlone, M. S. 1988. New Zealand. In *Vegetation History,* ed. B. Huntley, T. Webb III, pp. 557–99. Amsterdam: Kluwer Acad.
97. McIntyre, A., Ruddiman, W. F., Karlin, K., Mix, A. C. 1989. Surface water response of the Equatorial Atlantic Ocean to orbital forcing. *Paleoceanography* 4:19-55
98. Miller, K. G., Fairbanks, R. G., Mountain, G. S. 1987. Tertiary oxygen isotope synthesis, sea level history, and continental margin erosion. *Paleoceanography* 2:1-19
99. Mitchell, J. F. B., Grahame, N. S., Needham, K. J. 1988. Climate simulations for 9000 years before present: seasonal variations and effect of the Laurentide ice sheet. *J. Geophysical Res.* 93:8283–8303

99a. Mix, A.C. 1987. The oxygen-isotope record of glaciation. In *North America and Adjacent Oceans During the Last Deglaciation,* ed. W. F. Ruddiman, H. E. Wright, Jr., pp. 111–35. Boulder: Geological Soc. Am.

100. Mix, A. C. 1989. Influence of productivity variations on long-term atmospheric CO_2. *Nature* 337:541–44
101. Molfino, B. 1992. Paleoecology of marine systems. In *Aquatic Ecology: Scale, Pattern, Process,* ed. P. Giller, D. Rafaelli. Oxford: Blackwell Sci. In press
102. Molfino, B., Heusser, L. H., Woillard, G. M. 1984. Frequency components of a Grande Pile pollen record: evidence of precessional orbital forcing. In *Milankovitch and Climate,* ed. A. Berger, J. Imbrie, J. Hays, G. Kukla, B. Saltzman, pp. 391–404. Dordrecht: D. Reidel

103. Molfino, B., McIntyre, A. 1990. Precessional forcing of nutricline dynamics in the equatorial Atlantic. *Science* 249:766–69
104. Morley, J. J., Pisias, N. G., Leinen, M. 1987. Late Pleistocene time series of atmospheric and oceanic variables recorded in sediments from the subarctic Pacific. *Paleoceanography* 2:49–62
105. Mortlock, R. A., Charles, C. D., Froelich, P. N., Zibello, M. A., Saltzman, J., et al. 1991. Evidence for lower productivity in the Antarctic Ocean during the last glaciation. *Nature* 351: 220–23
106. Neftel, A., Oeschger, H., Staffelbach, T., Stauffer, B. 1988. CO_2 record in the Byrd ice core 50,000–5,000 years BP. *Nature* 331:609–11
107. Nelson, C. S., Hendy, C. H., Jarrett, G. R., Cuthbertson, A. M. 1985. Near-synchroneity of New Zealand alpine glaciations and Northern Hemisphere continental glaciations during the past 750 ka. *Nature* 318:361–63
108. Olsen, P. E. 1986. A 40-million-year lake record of early Mesozoic orbital climatic forcing. *Science* 234:842–48
109. Overpeck, J. T., Bartlein, P. J., Webb, T. III. 1991. Potential magnitude of future vegetation change in eastern North America: comparisons with the past. *Science* 254:692–95
110. Peters, R., Lovejoy, T. E. 1992. *Global Warming and Biological Diversity*. New Haven: Yale Univ. Press. pp. 386
111. Petit, J. R., Mounier, L., Jouzel, J., Korotkevich, Y. S., Kotlyakov, V. I., et al. 1990. Palaeoclimatological and chronological implications of the Vostok core dust record. *Nature* 343: 56–58
112. Petit-Maire, N., Fontugne, M., Rouland, C. 1991. Atmospheric methane ratio and environmental changes in the Sahara and Sahel during the last 130 kyrs. *Palaeogeogr., Palaeoclimatol., Palaeoecol.* 86:197–204
113. Porter, S. C. 1989. Some geological implications of average quaternary glacial conditions. *Quat. Res.* 32:245–61
114. Prell, W. L., Kutzbach, J. E. 1987. Monsoon variability over the past 150,000 years. *J. Geophysical Res.* 92:8411–25
115. Prell, W. L., Van Campo, E. 1986. Coherent response of Arabian Sea upwelling and pollen transport to late Quaternary monsoonal winds. *Nature* 323:526–28
116. Prentice, I. C. 1983. Postglacial climatic change: vegetation dynamics and the pollen record. *Prog. Phys. Geogr.* 7:273–86
117. Prentice, I. C. 1986. Vegetation responses to past climatic changes. *Vegetatio* 67:131-41
118. Prentice, I. C., Bartlein, P. J., Webb, T. III. 1991. Vegetation and climate change in eastern North America since the Last Glacial Maximum. *Ecology* 72:2038–56
119. Prentice, M. L., Matthews, R. K. 1988. Cenozoic ice-volume history: Development of a composite oxygen isotope record. *Geology* 16:963–66
120. Raymo, M. E., Ruddiman, W. F., Beckman, J., Clement, B. M., Martinson, D. G. 1989. Late Pliocene variation in northern hemisphere ice sheets and North Atlantic deep water circulation. *Paleoceanography* 4:413–46
121. Raymo, M. E., Ruddiman, W. F., Shackleton, N. J., Oppo, D. W. 1990. Evolution of Atlantic-Pacific δ ^{13}C gradients over the last 2.5 m. y. *Earth Planetary Sci. Lett.* 97:353–68
122. Rind, D. 1987. Components of the ice age circulation. *J. Geophysical Res.* 92:4241–81
123. Rind, D., Peteet, D., Broecker, W., McIntyre, A., Ruddiman, W. 1986. The impact of cold North Atlantic sea surface temperatures on climate: implications for the Younger Dryas cooling (11–10 k). *Climate Dynamics* 1: 3–33
124. Rind, D., Peteet, D., Kukla, C. 1989. Can Milankovitch orbital variations initiate the growth of ice sheets in a general circulation model? *J. Geophysical Res.* 94:12,851–71
125. Ritchie, J. C. 1987. *The Postglacial Vegetation of Canada*. Cambridge: Cambridge Univ. Press. 187 pp.
126. Robinson, S. W., Adam, D. P., Sims, J. D. 1988. Radiocarbon content, sedimentation rates, and a time scale for core CL-73-4 from Clear Lake, California. In *Late Quaternary Climate, Tectonism, and Sedimentation in Clear Lake, Northern California Coast Ranges*, ed. J. D. Sims, pp. 151–60. Boulder,Colo: Geological Soc. Am. (Special Paper 214)
127. Ruddiman, W. F. 1987. Northern oceans. In *North America and Adjacent Oceans During the Last Deglaciation*, ed. W. F. Ruddiman, H. E. Wright, Jr., pp. 137–54. Boulder: Geological Soc. Am.
128. Ruddiman, W. F., McIntyre, A. 1984. Ice-age thermal response and climatic

role of the surface Atlantic Ocean, 40N to 63N. *Geol. Soc. Am. Bull.* 95:381–96

129. Ruddiman, W. F., Mix, A. C. 1992. The North and Equatorial Atlantic Ocean at 9000 and 6000 yr B.P. In *Global Climates Since the Last Glacial Maximum,* ed. H. E. Wright Jr., J. E. Kutzbach, T. Webb III, W. F. Ruddiman, F. A. Street-Perrott, P. J. Bartlein. Minneapolis: Univ. Minn. Press. In press
130. Ruddiman, W. F., Raymo, M. E. 1988. Northern Hemisphere climate regimes during the past 3 Ma: possible tectonic connections. *Philos. Trans R. Soc, London B* 318:411-30
131. Ruddiman, W. F., Raymo, M. E., Martinson, D. G., Clement, B. M., Beckman, J. 1989. Pleistocene evolution: northern hemisphere ice sheets and North Atlantic Ocean. *Paleoceanography* 4: 353–412
132. Saltzman, B. 1985. Paleoclimatic modeling. In *Paleoclimate Analysis and Modeling,* ed. A. D. Hecht, pp. 341–96. New York: Wiley
133. Sarnthein, M., Tiedemann, R. 1989. Toward a high-resolution stable isotope stratigraphy of the last 3.4 million years: sites 658 and 659 off northwest Africa. *Proceedings of the Ocean Drilling Program, Scientific Results* 108: 167-85
134. Shackleton, N. J. 1977. The carbon isotope record of the Cenozoic history of organic carbon burial and of oxygen in the ocean and atmosphere. In *Marine Petroleum Source Rocks,* ed. J. Brooks, A. J. Fleet, pp. 423–34. *Geological Soc. Special Publ. No. 26.*
135. Shackleton, N. J., Beckman, J., Zimmerman, H., Kent, D. V., Hall, M. A., et al. 1984. Oxygen isotope calibration of the onset of glaciation in the North Atlantic region. *Nature* 307:620–23
136. Shackleton, N. J., Pisias, N. G. 1985. Atmospheric carbon dioxide, orbital forcing and climate. In *The Carbon Cycle and Atmospheric CO_2 Natural Variations Archean to Present. (Geophysical Monograph, Vol. 32),* ed. E. T. Sundquist, W. S. Broecker, pp. 303–17. Washington, DC: Am. Geophysical Union
137. Sowers, T., Bender, M., Raynaud, D., Korotkevich, Y. S., Orchardo, J. 1991. The $\delta^{18}0$ of atmospheric 0_2 from air inclusions in the Vostok ice core: timing of CO_2 and ice volume changes during the penultimate deglaciation. *Paleoceanography* 6:679-96
138. Spaulding, W. G. 1990. Vegetational and climatic development of the Mojave Desert: the last glacial maximum to the present. In *Packrat Middens, The Last 40,000 Years of Biotic Change,* ed. J. L. Betancourt, T. R. Van Devender, P. S. Martin, pp. 166–99. Tucson: Univ. Ariz. Press
139. Street-Perrott, F. A. 1992. Tropical wetland sources. *Nature* 355:23–24
140. Thompson, R. S. 1990. Late Quaternary vegetation and climate in the Great Basin. In *Packrat Middens, The Last 40,000 Years of Biotic Change,* ed. J. L. Betancourt, T. R. Van Devender, P. S. Martin, pp. 200–39. Tucson: Univ. Ariz. Press
141. Tsukada, M. 1988. Japan. In *Vegetation History* ed. B. Huntley, T. Webb III, pp. 459–518. Amsterdam: Kluwer Acad.
142. Van Campo, E., Duplessy, J. C., Rossignol-Strick, M. 1982. Climatic conditions deduced from a 150-kyr oxygen isotope-pollen record from the Arabian Sea. *Nature* 296:56–59
143. Van Campo, E., Prell, W. L., Barratt, N., Sabatier, R. 1990. Comparison of terrestrial and marine temperature estimates for the past 135 ka off southeast Africa: a test for GCM simulation of palaeoclimate. *Nature* 348:209–12
144. Van Der Hammen, T. 1991. Palaeoecological background: neotropics. *Climatic Change* 19:37–47

144a. Vrba, E. S. 1992. Mammals as a key to evolutionary theory. *J. Mammal.* 73:1–28

145. Washington, W. M., Parkinson, C. L. 1986. *An Introduction to Three-Dimensional Climate Modeling.* Mill Valley: Univ. Science Books. 422 pp.
146. Webb, T. III. 1986. Is vegetation in equilibrium with climate? How to interpret late-Quaternary pollen data. *Vegetatio* 67:75–91
147. Webb, T. III. 1987. The appearance and disappearance of major vegetational assemblages: long-term vegetational dynamics in eastern North America. *Vegetatio* 69:177–87
148. Webb, T. III, 1992. Past Changes in vegetation and climate: lessons for the future. In *Global Warming and Biological Diversity,* ed. R. Peters, T. E. Lovejoy, pp. New Haven: Yale Univ. Press, pp. 59–75.
149. Webb, T. III, Bartlein, P. J., Kutzbach, J. E. 1987. Climatic change in eastern North America during the past 18,000 years; Comparisons of pollen data with model results. In *North America and*

Adjacent Oceans During the Last Deglaciation, ed. W. F. Ruddiman, H. E. Wright Jr., pp. 447–62. Boulder: Geological Soc. Am.
150. Woillard, G. M. 1978. Grande Pile peat bog: a continuous pollen record for the last 140,000 years. *Quat. Res.* 9:1–21
151. Woods, K. D., Davis, M. B. 1989. Paleoecology of range limits: beech in the Upper Peninsula of Michigan. *Ecology* 70:681–96
152. Wright, H. E. Jr. 1990. The amphi-Atlantic distribution of the Younger Dryas paleoclimatic oscillation. *Quat. Sci. Rev.* 8:295–306
153. Wright, H. E., Jr., Kutzbach, J. E., Webb, T. III, Ruddiman, W. F., Street-Perrott, F. A., et al. 1992. *Global Climates since the Last Glacial Maximum*. Minneapolis: Univ. Minn. Press. In press
154. Wyman, R. L. 1991. *Global Climate Change and Life on Earth*. New York: Routledge, Chapman & Hall. 282 pp.

Annu. Rev. Ecol. Syst. 1992. 23:175–200

SPATIAL SCALES AND GLOBAL CHANGE: Bridging the Gap from Plots to GCM Grid Cells

Carol A. Wessman
Cooperative Institute for Research in Environmental Sciences, and Department of Environmental, Population and Organismic Biology, University of Colorado, Boulder, Colorado 80309-0449

KEYWORDS: scaling, extrapolation, remote sensing, spatial modeling, atmosphere-biosphere interactions

INTRODUCTION

The complexity of earth system processes results from interactions among the physical, chemical, and biological subsystems that vary in both time and space. Gaining an understanding of these dynamics has taken on great importance in the context of current environmental change and the portent of even larger scale global change. Appreciation for the concept of "scaling" is increasing as we are challenged to integrate data and models from different disciplines and different time and space scales. In particular, biophysical and ecological information, intrinsically derived at the scale of the individual organism, must be extrapolated to the regional and global scales of climate models. Unfortunately, this may not always be a simple process due to complex spatial variations and nonlinearities in dynamics across landscapes. Bridging the gap between our site-level ecological understanding and global scale phenomena challenges our current disciplinary approach and requires new strategies for acquiring and interpreting information on large-scale earth system dynamics.

Research tools such as remote sensing and simulation modeling hold the potential for clarifying general ecological principles by expanding limitations inherent in site-level studies (81, 125). In combination, technologies of remote sensing, geographic information systems, and simulation modeling permit quantitative assessment of the consequences of heterogeneity in earth systems over a broad range of spatial and temporal scales. Remote sensing techniques extend measurements to scales over which biospheric processes operate and

0066-4162/92/1120-0175$02.00

provide the only practical means for consistent regional and global monitoring. Geographic information systems (GIS) allow a multivariate approach in the spatial and temporal domains, and simulation models aid in understanding and predicting long-term effects of environmental change. A combination of these technologies is instrumental for linking process-level studies with global scale interests (121, 154)

This paper reviews the experiences of recent attempts to extrapolate local measurements to regional scales through remote sensing and modeling. Issues in scaling and heterogeneity in ecosystems are introduced to provide a theoretical framework, but a thorough discussion of the ever-expanding literature on these topics is beyond the scope of this paper (e.g. see 151). Discussions focus on the roles remote sensing, simulation models, and geographic information systems offer together for the extrapolation of local measurements to regional scales. The discussion emphasizes those parameters that can be estimated remotely and that best assist in scaling of ecological information. Linking current and future remote sensing and GIS technologies with integrative, mechanistic models strengthens our investigations of responses and feedbacks among parameters operating at different scales, and improves predictions and understanding of biosphere/atmosphere interactions. Strategies for scaling process-level studies must be considered by the interdisciplinary scientific community if any progress is to be made in understanding large-scale phenomena.

PERSPECTIVES ON SCALING

Principal Considerations

Scientists have a natural tendency toward reductionism; we are intrigued by detail and we seek to explain through mechanisms. The key to scaling is determining what to ignore; what we can and cannot ignore affects how easily we move up or down in scale (see 36). Given that ecological systems are scaled in space and time (4), variables influencing a process may or may not change with scale (150). A shift in the relative importance of variables may result from the influence of additional factors and constraints. With our limited "resolving" power—i.e. limited sampling capabilities in both time and space of the frequencies of natural variance—our predictive powers are contingent on an understanding of the characteristic response of processes to changes in scale. In the broad context, scaling requires the identification of process nonlinearities with change in scale, the range in scales where linearity may hold, and the properties that may be coherent between scales. The object, then, is not to analyze all of the smaller-scale aspects of a process under observation, but to focus instead only on those that have direct importance to the scale under consideration (15).

Hierarchy theory serves as a framework to approach complex systems (4, 104) by considering only factors that are "operationally" significant. An important consequence of hierarchical structuring is embodied in the concept of constraint (103). The scale under observation is affected by the potential behaviors of its components and by the environmental constraints imposed by higher levels. Lower-level dynamics are too fast to be seen as variables; they are experienced as averages or integrated values and appear in patterns as a blend. These dynamics can often be ignored, given a relatively stable system (103). Reciprocally, higher-level dynamics are so slow that they are experienced as constants, and larger-scale spatial patterns are seen only as a uniform, local condition. However, when the system is disrupted, it is the response dynamics at the fine scales that break up the constraint system and move the system into a new configuration (103). Similarly, coupling large-scale models such as general circulation models (GCMS) with important subsystems such as biology is important to understanding the complex of regional and global process interactions and the integrity of their connections under directional environmental change (125).

Spatial heterogeneity constrains our ability to translate information from one scale to another. Scaling problems may not occur in spatially homogeneous systems because process measurements are likely to sum directly. However, in heterogeneous landscapes or aquatic systems, process measurements obtained at fine scales often cannot be summed directly to produce regional estimates because of the large number of interactions involved and the spatial heterogeneities that may influence processes in nonlinear ways (76, 116). Top-down approaches such as remote sensing must consider nonlinearities that may enter into coarse resolution measurements.

The response of a process to changes in scale can be defined as one of four conditions (43): (i) The process under consideration is invariant with scale, in which case no transformations are required. (ii) A process is similar in its effect across scales (e.g. a linear relationship) and scale transformations are simple. (iii) The process varies little and remains dominant across scales, but additional factors and constraints increase uncertainty of the prediction process. (iv) The importance of the process or the process constraints changes with scale. In both cases (iii) and (iv), inappropriate characterization of local scale processes may introduce substantial error in large-scale extrapolations (42, 75, 76, 116). Case (iv) may require the identification of critical thresholds below or above the observation scale for any sort of successful extrapolation (e.g. 43, 77).

In sum, to understand the response to change in scale we must acquire a quantitative understanding of: (i) the heterogeneity of the system under observation (frequencies of natural variation), and (ii) the linkage between spatial and temporal patterns and the processes that drive them. Given such knowledge, scaling from the plot to the GCM grid cell is a matter addressed (when extrapolations are complex) in the initial model formulation using hierarchy theory, or it is a problem in sampling strategy.

Integrating Perspectives: Bottom-Up and Top-Down

Scaling represents the transcending concepts that link processes at different levels of space and time (19). It implies a change in scale by identification of significant (often limiting) factors at each scale under consideration; the approach can be in an upward or downward direction. The bottom-up approach begins with individual or entity-based (e.g. communities, functional groups) measurements and adds appropriate constraints to explain phenomena observed at broader scales. The objective is to use information that is available at fine scales to predict phenomena at broader scales for which empirical data are lacking. Errors will occur if predictions do not take into account changes in process response or the appearance with change in scale of new constraining factors.

The top-down perspective begins with an observation, makes a generalization, and proceeds to describe the mechanism. In other words, this approach analyzes the pattern and infers the processes that generated the pattern. The concept of constraint can be used to predict phenomena at finer scales with the objective of identifying the factors that have the most control over variation at each scale. Error in this approach arises because of an inability to address heterogeneity or because of increases in variance coming from the bottom up; the question is to determine whether variance is subsumed or expressed at each higher scale.

Parallel approaches (bottom-up and top-down) to system processes may be taken by different sciences, but unless there is some awareness of the significance of scale and its consequences for interpretation, conflicts will occur with regard to extrapolation and prediction (68). While we can describe mechanisms using the scaling-up approach, we identify, from the top down, prevalent patterns which result from the feedbacks that may subsume much of the variation in those mechanisms (155). Coherent connections between small-scale site specific measurements and regional scale phenomena require the concurrent implementation of bottom-up and top-down approaches. Complex scale interactions will demand direct observations at both small and large scales; knowledge of small-scale dynamics will not be enough to predict large-scale processes (43).

Remote sensing involves both scaling-up and scaling-down processes. The synoptic nature of remotely sensed data presents the top-down modeler with large scale measurements of pattern. The resolution or grain size is immediately as large or significantly larger than the traditional field plot size, and it samples the landscape in a manner that may or may not represent natural frequencies of interest. Scaling-up becomes relevant in the attempt to understand the relationship of the measurement to surface conditions. This is a function of the surface reflectance properties and sensor optical characteristics; time-series measurements and the confounding influence of atmospheric conditions; and the influence of heterogeneity within the landscape relative to the pixel size.

ECOSYSTEM STRUCTURE, DYNAMICS, AND CHANGE

Ecosystem Processes over Time Scales of Global Change

The biotic response to and feedback on global environmental change is expected to vary on many scales. Recent research has emphasized short-term atmosphere-biosphere interactions at time scales of one to several years. There is general agreement that increasing global carbon dioxide concentrations will have direct physiological effects on plants, but the duration of these effects and their expression at the level of the population and ecosystem is relatively unknown (13, 36a, 50, 141). Biogeochemical and biogeographical relationships are likely to be affected over decades to centuries (125). Yet, while carbon and nutrient dynamics can be expected to be sensitive to changes in temperature and precipitation regimes, the complexity of response will vary between ecosystems (26, 45). Land management in agroecosystems may induce enough change in regional carbon balance to overshadow predicted climate change effects (17). Strong evidence suggests that land use changes that affect vegetation canopy characteristics and evapotranspiration will influence regional and perhaps global climate (34, 111, 134, 142). Alterations in disturbance regimes (e.g. severe storms, fires) may seriously influence ecosystem composition and functioning, and the prediction of potential species redistribution will likely be further confounded by the rate of climate change and spatial displacement of habitats (22, 27, 28, 107).

Efforts to predict the effects of global environmental change must consider the coupled nature of earth system components, namely, the hydrosphere, biosphere, and the atmosphere (125). Studying these interactions at landscape to global scales will require the integrated use of remote sensing, GIS, and simulation models. But because models have been parameterized with traditional environmental and biometeorological data, incorporation of spatial data bases and remotely sensed information requires reevaluation of both the biological/climatological processes and the qualitative and quantitative data types needed for large scale simulations. Successful modeling of biospheric-atmospheric interactions will need to consider the range of time steps at which interactions occur. These can be represented in essentially three characteristic time constants: (i) high frequency (seconds to days) physiological or biophysical processes, (ii) middle frequency (weeks to seasons) phenology and carbon and nutrient dynamics, and (iii) low frequency (annual to decadal) ecosystem compositional and edaphic changes (67). Data requirements at each of these scales have been recognized by the international community (67), and efforts to assemble relevant data bases are gaining momentum.

Spatial and Temporal Characterization

The current scientific data base is inadequate to address much of the spatial and temporal heterogeneity encountered at regional and global scales. Early efforts to assemble terrestrial vegetation and soil data sets at 1° latitude by 1° longitude resolution (87, 164) have contributed significantly to the incorporation of realistic biological regimes in modeling of climate processes (e.g. 40). Improved global land cover characterization and monitoring by remote sensing is feasible (144), but other global data sets such as land use, disturbance history, and physicochemical soil attributes will require substantial resource investment in the compilation of existing data and future ground-based measurements. These data sets must include mesoscale landscape characteristics and knowledge of their change over time for successful generalization to the global system (111).

Studies that link ground observations to regional and global scales are needed to take full advantage of the broad base of understanding at smaller scales. Research programs that coordinate simultaneous, integrated measurements across a range of scales are required to characterize spatial and temporal scales of variability (54). Several recent field programs have been specifically aimed at the scaling issues associated with a spatially heterogeneous environment. The First ISLSCP (International Satellite Land Surface Climatology Project) Field Experiment (FIFE) was designed to address, through simultaneous acquisition of satellite, atmospheric, and surface data, processes controlling surface energy and mass exchange over a range of scales, from those of individual plants to scales of GCM grid cells (53, 132). The Hydrologic Atmospheric Pilot Experiment (HAPEX) was a similarly intense measurement campaign for the study of the water budget and evaporation at climatic scales (5). Regional-scale studies of biosphere-atmosphere interactions on the chemistry of the troposphere over relatively undisturbed tropical forests and wetlands (Amazon Boundary Layer Experiment, ABLE) are part of a longer-term study of tropospheric chemistry supported by the Global Tropospheric Experiment (GTE) component of the US National Aeronautics and Space Administration (NASA) Tropospheric Chemistry Program (55, 56). These and other similarly designed experiments provide critical links between small-scale, process-oriented biogeochemical research and modeling at global scales.

Priorities for future comprehensive, multiscale studies need to be placed on regions expected to have the maximum potential for change, including tropical moist forest and savannah regions and northern-latitude tundra and boreal ecosystems (54). The second major ISLSCP experiment to be conducted in North America will focus on the latter. This Boreal Ecosystem-Atmosphere Study (BOREAS), planned for 1994–1996, will provide the integration in large-scale field experiments of biogeochemical objectives (e.g. GTE/ABLE) and biophysical objectives (e.g. FIFE) that has previously been lacking (54).

REMOTE SENSING-BASED EXTRAPOLATION MODELS

Extrapolation is the process of estimating unknown values from known conditions through the transfer of information from one scale to another or from one system to another at the same scale (150). Again, the heterogeneity of the system and the scale dependence of the process under question will dictate the ranges over which extrapolations may be made. Because of the wide variation in states and processes across temporal and spatial scales, extrapolation from local to regional and global scales appears an impossible task based on strict statistical requirements. Optimally, sampling techniques should involve (i) systematic and repetitive measurements to evaluate the variability of the data; and (ii) random sampling to provide a picture of the heterogeneity of the data (7, 70). For determination of regional processes, this would entail more ground sampling than is logistically possible for validating model predictions. Remote sensing can overcome these constraints to a certain extent by providing a means of sampling large areas repeatedly over time. The type of biophysical and geophysical attributes sensible from space will determine how widely ecological extrapolations can be implemented; and the calibration of the data between instruments and across time will affect how consistent those extrapolations will be.

The focus in research using remote sensing data has shifted over the last decade from empirically based classification and mapping procedures to more physically based characterization of the data with regard to radiative transfer and energy balance. Mapping and inventory remain important applications that are particularly relevant for tracking regional and global change. Through statistical classification procedures, remote sensing has been used to describe the spatial patterns in land cover types, their location, area, and change over time (e.g. 60, 62, 118). Modeling of structural characteristics of the canopy such as crown shape, size, and spacing by using radar (e.g. 89, 143) and optical remote sensing systems (e.g. 79, 80, 136, 137, 160) promises to both augment and refine classifications by providing more quantitative information on ecosystem structure. Applications to process-level questions will require explicit linkages between the process under study and spatial (and temporal) landscape patterns. Quantitative remote sensing that relates the biophysical and geophysical attributes of the surface (e.g. reflectance, phase, and backscatter) to physical units (e.g. biomass, absorbed photosynthetically active radiation (APAR), evapotranspiration) has already demonstrated potential to drive ecosystem and climate models (40, 123). Remote sensing can provide time-series data as additional input to models, allowing representation of pixel-to-regional scale variation with improved temporal accuracy.

Remote sensing data will be needed for extrapolations in three areas of relevance to global change research: (i) physical processes linking the biosphere to the atmosphere; (ii) biogeochemical cycling, including trace gas exchange; and (iii) ecosystem dynamics and change over time. Each of these will likely have profound influence on the others (e.g. ecosystem structure will influence rate and magnitude of biophysical processes). The following discussion reviews a number of parameters retrievable from remote sensing data that will be be important to scaling ground-level understanding in one or more of these three areas.

Biophysical Processes

Important interactions between the land surface and the atmosphere involve the exchanges of radiation, sensible heat, latent heat, and momentum. Until recently, these processes have been modeled at global scales in a very rudimentary fashion (see 33, 130). Incorporation of explicitly modeled vegetation effects in descriptions of land surface characteristics that govern albedo, roughness, and moisture availability has brought GCM simulations significantly closer in agreement with observations (34, 133, 142). Remote sensing can integrate the relatively small-scale understanding on which the modeling of these processes is based to the appropriate scales required for initialization of atmosphere-biosphere interaction models. There is promise that some of the components of the radiation budget and surface roughness can be estimated using a number of different sensors (69, 99). The following discussion focuses on attempts to remotely estimate the biophysical controls over evapotranspiration.

The idea of remote sensing as a fundamental tool for scaling biophysical rates of photosynthesis and transpiration has drawn on the simple thesis that plant growth is related to the fraction of incident radiation absorbed by the canopy and the dry matter:radiation quotient (an "efficiency" coefficient defining the carbon fixed per radiation intercepted) (96, 97). Chlorophyll's unique absorption of energy in the red (R) spectral region relative to the highly reflected near-infrared (NIR) region distinguishes live vegetation from soil and other nonphotosynthetic materials. Early field studies investigated the near-linear relationships between spectral reflectance indices composed of radiance measured in those regions (e.g. a simple ratio NIR/R or normalized difference vegetation index (NDVI = (NIR−R)/(NIR+R)) and standard measurements of the canopy properties of biomass, leaf area, and photosynthetically active radiation (PAR) absorbed by the canopy (8, 58, 59, 146, 147). Strong relationships were later demonstrated between time integrals of satellite-derived vegetation indices (VI) and net primary production (NPP) (40, 47), the geography and seasonality of vegetative cover (73, 149), and simulated photosynthesis and transpiration (122).

The amount of leaf surface area available for gas and moisture exchange (described by leaf area per ground area, the leaf area index—LAI) has been shown to relate linearly to NDVI (see 9, 109, 124), although changes in canopy closure, understory vegetation and background reflectance may affect broad-scale extrapolations (10, 138). VIs are, in fact, asymptotic in nature with respect to LAI, with linearity extending from LAIs of 2 to 6 for crop and grassland canopies (8, 119, 146, see also 131) and up to approximately 8 for coniferous forests (110, 123).

Theoretical analyses by Sellers (128) examined the links between spectral vegetation indices and canopy properties of LAI, absorbed PAR, photosynthetic capacity, and minimum canopy resistance. A mechanistic basis for the observed correlations (given a horizontally uniform canopy) was demonstrated with a two-stream approximation model of radiative transfer and simple leaf models of photosynthesis and stomatal resistance (129). The analysis suggested that VIs are indicative of instantaneous biophysical rates of photosynthesis and conductance but are not a reliable estimator for any state (leaf area, biomass) associated with vegetation. These arguments were a significant advance in theoretical understanding of remote sensing measurements, but problems associated with inadequate physiological models and optical contributions from variations in background and canopy architecture were admitted weaknesses.

The potential for scaling estimates of photosynthetic rates and conductances using VIs was defended more rigorously in work by Sellers et al (131). In the simpler model (129), it was assumed that all leaves in the canopy had the same light response curve, i.e. all leaves throughout the depth of the canopy responded identically to the flux of PAR. Accordingly, with increasing PAR flux above the canopy, the uppermost leaves would saturate and leaves lower down would remain below saturation, resulting in an increasingly nonlinear relationship between photosynthesis and the fraction of absorbed PAR. By considering that photosynthetic capacity changes in parallel with the depth-distribution of PAR (14), the bulk analytical canopy model of Sellers et al (131) closely reproduced an exact numerical integration of leaf models for normal environmental conditions. The modified model produced a stronger theoretical relationship between canopy biophysical rates (photosynthesis, conductance) and spectral VIs because the contributions of canopy structure and environmental forcings could be separated from those of leaf physiology and radiation flux. This argument supports the original hypothesis that area-averaged spectral vegetation indices do give good estimates of the area-integral of photosynthesis and conductance even for spatially heterogeneous (although physiologically uniform) vegetation covers.

Conditions still constraining the predictive powers of VIs include those that affect the photosynthesis/PAR relationship such as stressed vegetation and differing photosynthetic pathways (C_3, C_4), and conditions that may influence spectral estimates of absorbed PAR such as contributions from background soil and litter. The former can be addressed by considering the range of biological processes for different vegetation types and their respective sensitivity to VIs and environmental variables (12). For example, land cover can be stratified according to ecosystem or biome type before relationships are established between PAR and VIs. Fung et al (40) determined global net primary production from NDVI using an empirically derived scaling factor that essentially accounted for Monteith's conversion efficiency for each biome type. Prince (113) has cited efficiency factors converting annual APAR energy in megajoules (MJ) to NPP in grams for different biome types.

However, it is questionable whether a measurement in two spectral bands can provide an unambiguous measure of vegetation. Confounding influences from background variation, atmospheric attenuation, and off-nadir viewing cannot all be accounted for using a two-band ratio (10, 21, 46, 66, 93, 140). Huete (64) suggests a simple adjustment of VI to account for first-order soil-vegetation interactions (i.e. soil brightness effects); but secondary soil variations due to soil optical properties can only be addressed, using multiple spectral bands, through factor-analytic inversion models which allow composite plant-soil mixtures to be separated into component spectra (63, 65). In a similar approach, spectral mixture analysis systematically separates and quantifies vegetative and nonvegetative components at sub-pixel spatial resolution by identifying major sources of variance in remote sensing imagery (3, 135, 136, 137, 152, 153).

Understanding of the relationship of biophysical and biochemical processes to canopy reflectance is being extended by use of more defined measurements of spectral curvature available from high spectral resolution instruments. Variables of spectral shape such as width, depth, skewness, and symmetry of absorption features may be more directly indicative of biochemical state and canopy physiology than are broad-band measurements made with current operational sensors (see 156). Wavelength-specific absorption differences among photosynthetic pigments may permit quantification of their concentrations, and these may be related to photosynthetic activity (e.g. 30). In one case, a spectral change in green reflectance resulted from a light-induced change in a xanthophyll pigment that is closely linked to changes in photosynthetic capacity (41). Studies relating chlorophyll content to the location of the inflection point of the long wavelength edge of the feature have met with varied success (24, 94, 120, 126). Second derivatives of high spectral resolution reflectance data in the visible and near infrared regions appear to be strongly related to absorbed PAR and relatively insensitive to the reflectance of non-photosynthetically active materials such as litter and soils (52).

Biogeochemical Cycles

ELEMENTAL CYCLES Terrestrial carbon dynamics have presented significant challenges to global carbon cycle research. The magnitude of the terrestrial carbon pools varies in time and space, and the balance between the principal acting processes—namely, carbon assimilation by photosynthesis, respiration rates, and carbon turnover time—determine the net exchange between the biosphere and the atmosphere. Remote sensing can contribute information that can aid our understanding of these processes and their response to and feedback on the global system. Research linking reflectance measurements to photosynthesis and net primary production, reviewed in the above section, will be instrumental in the inventory of global terrestrial carbon. Some of the terms used to calculate carbon turnover time, nutrient availability and soil respiration may be provided by new techniques in imaging spectrometry. These processes are tightly linked with rates of decomposition, which are strongly regulated by the chemical quality of the organic matter (e.g. 91, 92). Remotely sensed estimations of lignin (the most recalcitrant material in litter), canopy nitrogen, or other constituents may serve to constrain decomposition submodels in ecosystem simulations, thus stabilizing model inversions (2, 125).

Quantitative determinations of vegetation biochemistry from reflectance data were initiated in the agriculture and the food industries (e.g. 102, 161) and later adopted by ecologists to replace standard laboratory wet chemistry analyses of lignin, cellulose, and nitrogen (90, 158). The physical basis for the extraction of biochemical information is the absorption of radiation by the molecular functional groups of C-H, O-H and N-H found within all foliar material. Overtone and combination bands specific to these compounds occur in the near infrared region and are particularly sensitive to changes in chemical concentrations. Principles of analytical spectroscopy concerned with separating individual component concentrations from organic mixtures (see 156) are currently being tested on reflectance data from field and aircraft spectrometers. An early application of NASA's Airborne Imaging Spectrometer (AIS) data successfully estimated canopy lignin concentrations in a series of northern temperate forest ecosystems and was subsequently used to derive images of annual nitrogen mineralization rates (157, 159). While the rate of nutrient cycling is influenced by a host of factors, foliar lignin is particularly dominant in modulating litter decomposition and therefore cycling rates in forest ecosystems. In this case, lignin provided the scaling factor and remote sensing the means to extrapolate beyond site measurements in order to observe variation in nitrogen availability at the scale of the landscape.

Further studies on the question of remote sensing of canopy chemistry are currently underway (20, 25, 44, 83). The nature of the remote sensing problem requires consideration of additional parameters beyond those used in the field of analytical chemistry. In the complete leaf or canopy condition, complications arise from the attenuation and multiple scattering of radiation along the path length, additional influences from soil, shadow and other background components, and atmospheric scattering and attenuation of the reflected signal. In the spectral region beyond 1 μm, the spectral reflectance of plant canopies is dominated by liquid water absorption features. The absorption intensity of the spectral features associated with canopy biochemical constituents can be quantified only after the effects of liquid water absorbance have been removed (44). Curve fitting techniques are used both to correct for the effects of liquid water (41a) and to solve for the canopy biochemical constituent concentrations. The accuracy of these techniques is dependent on the noise level of the spectral data. Thus, determinations of canopy biochemical constituent concentrations that are based on spectra derived from multiple pixels are more accurate than those performed on single pixels. The application of spectral unmixing techniques (e.g. 136, 137, 153) to derive end-member spectra and their spatial abundances can be used to provide low noise spectra for the curve fitting analysis. It also has the additional benefit of reducing the number of spectra for which the curve fitting analysis must be performed from a million or more (the number of pixels in a typical imaging spectrometer scene), to the number of spectral end-members.

The successful use of remote sensing in extrapolation models of biogeochemistry will depend on relating measurements to ecosystem properties indicative of underlying processes. This will, of course, require a better understanding of how those properties, such as plant physiology and biochemistry, reflect the balance between factors limiting to the system (125).

TRACE GAS EXCHANGE Considerable attention has been placed on predicting the effects on climate of increasing concentrations of radiatively active trace gases (see 95). Attempts to develop regional and global budgets of biogenically released trace gases have been made (see 74), but source/sink strengths and the processes that control flux rates remain areas of great uncertainty. This is largely due to the complexity of the biological, physical, and chemical systems that are involved and the difficulty of measuring exchange fluxes in the field (see 6). Significant effort has been given to understanding and quantifying flux processes within important ecosystems such as wetlands, grassland, tundra, and tropical forest (e.g. 11, 32, 56, 98, 127). However, these studies have disclosed great variability within and among vegetation types as well as with time at a given site.

Regional and global trace gas budgets have been largely based on area-weighted extrapolation of in situ flux measurements considered representative of a given land category. In other words, the region is first stratified into areas considered homogeneous and likely to have lower within-class variance in flux rates. Matthews & Fung (88) demonstrated the importance of wetlands to the global methane budget using a stratification of global wetlands based on environmental characteristics governing methane emissions. The original wetland data base at 1° resolution was developed from global digital data of vegetation, innundation characteristics, and soil properties. The five derived strata were then assigned typical methane fluxes integrated over the methane-production season as defined by latitude.

While useful for first-cut global and regional estimates and for targeting major contributors, area-weighted field estimates can incur significant error from the high spatial and temporal variability in point measurements and the subsequent process of aggregation. An assumption of "representativeness" for the land cover type may fail to account for sources of spatial variation. Matson & Vitousek (84) suggest that the high variability in tropical forests will result in a parallel variability in trace gas fluxes; i.e., there is no representative site for the general category of "tropical forest". Gradients of factors (e.g. soil fertility) that control both fluxes and ecosystem properties and processes in tropical forests may be more useful for extrapolating fluxes and for calculating budgets of nitrous oxide and other trace gases. In fact, stratification of tropical forest types that reflect soil characteristics reveal consistently higher nitrous oxide fluxes in forests on acid clay soils (terra firme) than other forest types within a Brazilian study area, even though within-type variation is significant (85). Estimates of nitrous oxide emission rates in moist tropical forests were extrapolated by areal estimates of forests stratified by soil fertility to derive a tropical contribution to the global nitrous oxide source (86). The total flux, calculated as 2.4 Tg/y, was considerably lower than that based on a general lumping of tropical and subtropical forests and woodlands (7.4 Tg/yr), but the total remains very significant in the global budget. Combination of these types of data with remote sensing-based areal extrapolations may be particularly useful in refining regional flux measurements or locating "hot spots" worthy of further investigation.

Single physical characteristics of the surface in wetlands (water and soil depth, soil temperature) do not appear to be quantitatively associated with the variability of methane flux rates within a single regional wetland system (11), nor have quantitative relationships been found to link wetlands in different physiographic and climatic regimes (88). In the absence of parameters that offer predictive relationships with fluxes, vegetation community distribution can provide a relevant stratification for area-weighted regional flux inventory. An emission inventory of the Everglades was substantially improved by using Landsat Thematic Mapper data to direct in situ sampling efforts in important habitats and by providing a means for calculating area-weighted mean fluxes for the system as a whole (11). Reiners et al (117) followed a similar strategy to predict nitrogen mineralization rates over sagebrush steppe landscape using an ecosystem simulation model parameterized for steppe ecosystem types defined by Landsat Thematic Mapper data.

Aircraft flux measurements provide independent data for testing regional and global extrapolations of trace gas fluxes (31). Regional tower and aircraft flux measurements which integrate gaseous exchange over large areas can provide impetus to relate ground-based measurements mechanistically to regional scales. The Amazon Boundary Layer Experiment (ABLE 2A and 2B) was designed to characterize quantitatively the spatial and temporal variability of trace gases and aerosols over the Brazilian Amazon during wet and dry season conditions (55, 57). Continuous ground-based sampling served to characterize temporal variability, with aircraft and satellite observations providing spatial sampling capabilities. The success of these experiments suggests the use of airborne eddy correlation flux surveys in future research programs to select representative ground sites for continuous tower measurements.

The great uncertainty in estimating global trace gas budgets arises from the heterogeneity of source distributions (56). Difficulties in producing consistent estimates result from (i) high variability in a single habitat, and (ii) uncertainty in geographical extent and seasonal variability in extent of wetland environments. Despite an extensive field sampling program, large sampling errors surrounding ground-based estimates of trace gas emissions occur if the stratification of the area does not consider the controlling factors. Combined remote sensing of landscape biophysical and ecological characteristics and trace gas measurements are needed to generalize to regional and global flux models. Approaches to extrapolations of local to regional measurements should include process-level modeling, stratified sampling by the most relevant landscape units, and synthesis with the aid of geographic information systems (54,56).

Patterns of Change in Terrestrial Ecosystems

The type and successional stage of ecosystems occurring within a landscape have profound implications for regional biogeochemical flux estimates and atmosphere-biosphere interactions. The rapid rate of land-use changes, particularly in the tropics, contribute directly to perturbations in those dynamics. Ecosystem successional patterns can indicate local variations in water availability and linked carbon and nitrogen cycles, which in turn may modify the effects of climate change or human disturbance (108). Yet the extrapolation of ecosystem research to regional and global scales has been hindered in the past by the difficulty of observing large-scale spatial heterogeneity and the long-term patterns of successional dynamics (51). Remote sensing and ground-based evaluations provide the most promising tools for compiling geographical information on the stage and condition of ecosystems over time.

The ability to detect long-term change in ecosystems requires that we are able to detect conditions in the static situation, e.g. health, structure, and seasonal productivity (60). Certainly, parameters discussed in this paper, such as seasonally integrated VIs and biochemistry, are variables that will be affected by and respond to environmental change. Image texture combined with spatial statistics provides a means to extract stand structure information from remotely sensed data (e.g. 38, 106). Textural analysis can be considered a quantitative measure of landscape heterogeneity and, when combined with VI data, it successfully documented tallgrass prairie response to burning and grazing treatments (16). Spectral mixture analysis also provides a means to estimate the spatial cover of vegetation in a sparse community, independent of the spectral characteristic of the substrate (136, 137, 152). Procedures have been developed for using the spatial variance in images to quantify the number and spacing of forest trees (37, 79, 80). A ten-year time-series Landsat Multispectral Scanner data was used to track changes in succession state, based on species composition and age structure, for northeastern Minnesota boreal forests (51). Once the images were rectified for changes in atmospheric conditions between years, it was possible to infer the rapid dynamics occurring in the spatial pattern of and the transition rates between forest ecological states.

Reliable information on global land cover/land use maps is a top priority for global change research (67). Existing maps of global vegetation are generally compiled from disparate sources at varying scales and contain many inconsistencies (144). Encouraging results with data from NOAA's Advanced Very High Resolution Radiometer (AVHRR) for monitoring regional land use change (72, 148) and classifying land cover at continental scales (47, 145, 149) suggest that global land cover classification by remote sensing is a real possibility (144). Operational provision of global land cover data will require systems with consistent internal (instrument) calibration, and appropriate temporal and spatial characteristics.

LINKING SIMULATION MODELS AND REMOTE SENSING

The use of remote sensing to drive models will require rethinking of traditional modeling approaches and the use of inverse modeling techniques. This, in turn, will be directed by the understanding of how remote measurements of plant physiology and biochemistry can reflect the balance between above and belowground limiting factors (125). As a consequence, ecosystem simulation models that recognize the relationship between fluxes and controlling factors will be best positioned to link to spatial data provided by remote sensing. While remote observations cannot simulate future change, they will be critical to describe the current state of the earth, monitor near-term change, and estimate initial conditions for predictive modeling.

Pilot efforts have been made to incorporate remote observations as drivers for flux rate calculations. Fung et al (40) integrated global NDVI values with field data on soil respiration and climate data to obtain global distributions of monthly atmosphere-biosphere exchange of CO_2. Satellite-derived LAI has been used to drive simulations of regional evapotranspiration and photosynthesis in a forest ecosystem using a model (Forest BGC) designed to be particularly sensitive to LAI, because LAI can be retrieved by satellite (123). This same model is being implemented as an integrative tool in NASA's Oregon Transect Ecosystem Research (OTTER) project. The goal of OTTER is to examine canopy and landscape characteristics along climate and fertility gradients that may be indicators of ecosystem processes and physiological dynamics. Data is currently being assimilated from the intensive field and airborne remote sensing experiment with the intention of interfacing with Forest BGC and providing a rigorous validation of the flux simulations (e.g. 71). Another NASA project of similar magnitude is the Forest Ecosystem Dynamics (FED) project located in Maine (e.g. 115, 163). This project, concerned with the scaling of ecosystem patterns and processes with northern forests, supports a hierarchy of submodels describing forest growth and development, soil processes, radiative transfer, and establishing functional linkages between them. Multiscale remote sensing acquisitions will serve as inputs to the models, as well as provide, through independent remote observations, validation for the submodels and integrated model.

Simulation models may be useful for developing methods to extrapolate across scales because they can test the implications of various scaling rules. However, ground validations for regional and global simulations are a fundamental problem. Remote sensing can act as an independent validation for spatial predictions from geographically based simulations, and it can test retrospective temporal predictions. Good correlation between time-integrated NDVI and simulated NPP for the central Great Plains indicates the value of remote sensing to verify simulations of important ecosystem dynamics such as productivity (17, 18, 125). As discussed earlier, nonlinear relationships among other processes such as trace gas evolution and soil carbon storage will require more attention to appropriate stratification schemes for spatial extrapolations and inference using remote sensing.

PROSPECTS FOR NEW TECHNOLOGY

Current satellite coverage is of limited use to international scientific needs because many of the measurements are only loosely coordinated in space and time (67). Measurement strategies must be long-term and global, at varying temporal and spatial scales. Remotely sensed data needed for large-scale monitoring and for integration with simulation models requires an observational system that offers: (i) moderate resolution, frequently repeated coverage sufficient to capture rapidly changing biophysical processes; and (ii) high resolution, periodic coverage to assure proper scaling procedures and necessary calibrations at subpixel scales.

Prospects for future remote sensing technology suggest greater capabilities for measuring and monitoring earth system dynamics (99, 105). The Mission to Planet Earth is an international plan to provide comprehensive, long-term, and continuous global observations for the development of quantitative earth system models on both regional and global scales (100). It incorporates both existing and new remote sensing instruments, including the Earth Observing System (EOS), a series of multipurpose polar-orbiting platforms to be initiated in the late 1990s. EOS will offer the first opportunity to obtain frequent (2–16 day repeat) coverage of the earth's surface with remote sensing data at optical, thermal, and microwave wavelengths for a 15-year period (see 162). A data and information system (EOS Data and Information System, EOSDIS) will be established to facilitate the archiving and analysis of raw, processed, and derived data products.

The combination of high spectral/low temporal resolution and low spectral/high temporal data provided by EOS surface imagers (High Resolution Imaging Spectrometer, HIRIS, Advanced Spaceborne Thermal Emission and Reflectance Radiometer (ASTER), Multi-Angle Imaging Spectra-Radiometer (MISR), and Moderate Resolution Imaging Spectrometer, MODIS) will supply several important variables required by ecosystem models (154). Rapid biophysical processes such as photosynthesis and evapotranspiration require data at daily if not diurnal intervals and at moderate resolution. MODIS will provide, at moderate resolutions (500 m), high temporal data that is currently provided at coarse spatial resolution (1–4 km) by NOAA's Advanced Very High Resolution Radiometer (AVHRR). Beyond its contributions to atmospheric calibrations, MISR will provide moderate resolution observations of multi-angle relectance properties of the surface. Applications of ASTER data will focus on high resolution measurements of land and water surface temperatures. Regional-scale variability of vegetation chemical quality is important for estimating decomposition rates and can be well represented by strategic sampling using HIRIS. Vegetation biochemistry important to decomposition is temporally less dynamic than biophysical processes. Sampling at two to three week intervals throughout the growing season and at time of senescence will be adequate to track seasonal dynamics that, when integrated over periods of years, will indicate chemical changes due to climate or rising CO_2. Techniques such as spectral mixture analysis, particularly suited to EOS' multiband image data sets, can provide a framework for systematically defining both large and small scale features in the image data.

CONCLUDING REMARKS

Advances in technology and integrative techniques are adding substantially to the scaling of ecological information from local to global systems. Remote sensing makes it possible to measure select variables at time and spatial scales consistent with our global interests. The link to geographic information systems and simulation models "animates" these data in a broader and more dynamic context. Use of these technologies is, however, predicated on the assumption that we recognize the appropriate variables to be scaled. Developments in theory and multiscaled, integrated measurement programs are necessary to provide the links to ground-based understanding. Approaches to extrapolation should consider the following:

1. *Stratified sampling procedures are required to address biospheric processes which vary in time and space.* The level of stratification will depend on the complexity of the process of interest and its response characteristics as the scale is changed. Accurate estimates of photosynthetic rates and conductances from vegetation indices, for example, will require stratification by biome type, at the very least, to account for conversion efficiencies and potential environmental limitations.
2. *Identification of controlling factors and the use of gradients sensible from space will enable regional extrapolations with remotely sensed data and will be amenable to simulation models.* Limiting or controlling factors will often govern variability in ecological processes. Remote observations can serve as state variables to drive rate calculations within models.
3. *Time-series of remote sensing data are central to studies in global change.* However, the utility of long-term data sets from current operational satellites is hindered by the lack of instrument calibrations to monitor changes in radiometric sensitivity over time. Future programs must consider this a priority.
4. *A hierarchy of spatial resolution measurements is required to address mesoscale heterogeneity if generalizations to global scales are to be successful.* Landscape and regional heterogeneity contribute significantly to the scaling properties of some processes and must be included in extrapolations. Remote sensing can be used to delineate habitats within

which process variation may be minimized. Moreover, characterization of sub-pixel variation for global-scale instruments such as AVHRR and MODIS is needed for scaling process measurements and calibrations.

5. *Global geographic data bases of parameters such as vegetation cover, land use, soil properties, etc will determine extrapolation and modeling capabilities at the global scale.* These data will be required at scales that reflect their effect on regional and global processes and which are relevant to modeling goals.
6. *Extrapolations of local to regional measurements must combine modeling at the process level, remote observations of biophysical and ecological characteristic relevant to landscape, regional and global dynamics, and synthesis with the aid of geographic information systems.*

ACKNOWLEDGMENTS

I thank D. Schimel for sharing his insights on scaling principles in many stimulating discussions. I am also grateful for input from P. Sellers and D. Graetz and for critical reviews of earlier versions of the manuscript from D. Schimel, M. Walker, D. Walker, and K. Wolter. Thanks also to B. Curtiss and S. Ustin for many interesting and relevant discussions; and NASA's Earth Observing System program for financial support.

Literature Cited

1. Aber, J. D., Nadelhoffer, K. J., Steudler, P., Melillo, J. M. 1989. Nitrogen saturation in northern forest ecosystems. *BioScience* 39(6):378–86
2. Aber, J. D., Wessman, C. A., Peterson, D. L., Melillo, J. M., Fownes, J. 1990. Remote sensing of litter and soil organic matter decomposition in forest ecosystems. In *Remote Sensing of Biosphere Functioning,* ed. R. Hobbs, H. Mooney, pp. 87–103. New York: Springer-Verlag. 312 pp.
3. Adams, J. B., Smith, M. O., Johnson, P. E. 1986. Spectral mixture modeling: a new analysis of rock and soil types at the Viking Lander I site. *J. Geophys. Res.* 91:8098–112
4. Allen, T. F. H., Starr, T. B. 1982. *Hierarchy: Perspectives for Ecological Diversity.* Chicago: Univ. Chicago Press. 310 pp.
5. Andre, J.-C., Goutorbe, J. P., Perrier, A. 1986. HAPEX-MOBILHY, a hydrologic atmospheric pilot experiment for the study of water budget and evaporation at the climatic scale. *Bull. Am. Meteorol. Soc.* 67:138–44
6. Andreae, M. O., Schimel, D. S., eds. 1989. *Exchange of Trace Gases between Terrestrial Ecosystems and the Atmosphere.* Chichester: Wiley. 346 pp.
7. Aselmann, I. 1989. Global-scale extrapolation: a critical assessment. Exchange of trace gases between terrestrial ecosystems and the atmosphere. In *Exchange of Trace Gases between Terrestrial Ecosystems and the Atmosphere,* ed. M. O. Andreae, D. S. Schimel, pp. 119–33. Chichester: Wiley. 347 pp.
8. Asrar, G., Fuchs, B. M., Kanemasu, E. T., Hatfield, J. L. 1984. Estimating absorbed photosynthetic radiation and leaf area index from spectral reflectance in wheat. *J. Agron.* 76:300–6
9. Asrar, G., Myneni, R. B., Kanemasu, E. T. 1989. Estimation of plant-canopy attributes from spectral reflectance measurements. In *Theory and Applications of Optical Remote Sensing,* ed. G. Asrar, pp. 252–96. New York: Wiley. 734 pp.
10. Baret, F., Guyot, G. 1991. Potentials and limits of vegetation indices for LAI and APAR assessment. *Remote Sensing Environ.* 35:161–73
11. Bartlett, D. S., Bartlett, K. B., Hartman,

J. M., Harriss, R. C., Sebacher, D. I. 1989. Methane emissions from the Florida Everglades: patterns of variability in a regional wetland ecosystem. *Global Biogeochem. Cycles* 3(4):363–74
12. Bartlett, D. S., Whiting, G. J., Hartman, J. M. 1990. Use of vegetation indices to estimate intercepted solar radiation and net carbon dioxide exchange of a grass canopy. *Remote Sensing Environ.* 30:115–28
13. Bazzaz, F. A. 1990. The response of natural ecosystems to the rising global CO_2 levels. *Annu. Rev. Ecol. Syst.* 21:167–96
14. Björkman, O. 1981. Responses to different quantum flux densities. In *Encyclopedia of Plant Physiology,* Vol. 12A. *Plant Physiological Ecology* I. ed. O. L. Lange, P. S. Nobel, C. B. Osmond, H. Ziegler, pp. 57–107. Berlin:Springer
15. Bolin, B. 1988. Linking terrestrial ecosystem process models to climate models. In *Scales and Global Change,* ed. T. Rosswall, R. G. Woodmansee, P. G. Risser, pp. 109–24. *SCOPE 35.* Chichester: Wiley. 355 pp.
16. Briggs, J. M., Nellis, M. D. 1991. Seasonal variation of heterogeneity in the tallgrass prairie: a quantitative measure using remote sensing. *Photogrammetric Engineer. Remote Sensing* 57(4):407–11
17. Burke, I. C., Kittel, T. G. F., Lauenroth, W. K., Snook, P., Yonker, C. M., Parton, W. J. 1991. Regional analysis of the central Great Plains. *BioScience* 41(10):685–92
18. Burke, I. C., Schimel, D. S., Yonker, C. M., Parton, W. J., Joyce, L. A., Lauenroth, W. K. 1990. Regional modeling of grassland biogeochemistry using GIS. *Landscape Ecol.* 4(1):45–54
19. Caldwell, M., Matson, P. A., Wessman, C. A., and Gamon, J. 1992. Prospects for scaling. In *Scaling Processes Between Leaf and Landscape Levels,* ed. J. Ehleringer, C. Field. San Diego: Academic Press. In press
20. Card, D. H., Peterson, D. L., Matson, P. A., Aber, J. D. 1988. Prediction of leaf chemistry by the use of visible and near infrared reflectance spectroscopy. *Remote Sensing Environ.* 26: 123–47
21. Choudhury, B. J. 1987. Relationships between vegetation indices, radiation absorption, and net photosynthesis evaluated by a sensitivity analysis. *Remote Sensing Environ.* 22:209–33
22. Clark, J. S. 1991. Ecosystem sensitivity to climate change and complex responses. In *Global Climate Change and Life on Earth,* ed. R. Wyman, pp. 65–98. New York: Routledge, Chapman & Hall
23. Collins, W., Chang, S. H., Canney, F., Ashley, F. 1983. Airborne biogeochemical mapping of hidden mineral deposits. *Econ. Geol.* 78:737–49
24. Curran, P. J., Dungan, J. L., Gholz, H. L. 1990. Exploring the relationship between reflectance red edge and chlorophyll content in slash pine. *Tree Physiol.* 7:33–48
25. Curran, P. J., Dungan, J. L., Macler, B. A., Plummer, S. E., Peterson, D. L. 1992. Reflectance spectroscopy of fresh whole leaves for the estimation of chemical composition. *Remote Sensing Environ.* 39(2):153–166
26. D'Arrigo, R., Jacoby, G. C., Fung, I. Y. 1987. Boreal forests and atmosphere-biosphere exchange of carbon dioxide. *Nature* 329:321–23
27. Davis, M. B. 1989. Lags in vegetation response to greenhouse warming. *Climatic Change* 15:75–82
28. Davis, M. B., Botkin, D. B. 1985. Sensitivity of cool-temperature forests and their fossil pollen record to rapid temperature change. Quat. Res. 23:327–40
29. Deleted in proof.
30. Demmig-Adams, B. 1990. Carotenoids and photoprotection in plants. A role for the xanthophyll zeaxanthin. Reviews on Bioenergetics. *Biochim. Biophys. Acta* 1020:1–24
31. Desjardins, R. L., MacPherson, J. I. 1989. Aircraft-based measurements of trace gas fluxes. In *Exchange of Trace Gases between Terrestrial Ecosystems and the Atmosphere,* ed. M. O. Andreae, D. S. Schimel, pp. 135–52. Chichester: Wiley. 347 pp.
32. Desjardins, R. L., MacPherson, J. I., Schuepp, P. H., Karanja, F. 1989. An evaluation of aircraft flux measurements CO_2, water vapor and sensible heat. *Boundary-Layer Meteorol.* 47:55–70
33. Dickinson, R. E. 1984. Modeling evapotranspiration for three-dimensional global climate models. Climate processes and climate sensitivity: *Am. Geophys. Union Monogr.* 29:58–72
34. Dickinson, R. E., Henderson-Sellers, A. 1988. Modelling tropical deforestation: a study of GCM land-surface parameterizations. *Q. J. R. Meteorol. Soc.* 114:439–62
35. Deleted in proof
36. Ehleringer, J., Field, C., eds. 1992.

Scaling Processes Between Leaf and Landscape Levels. San Diego: Academic Press. In press

36a. Field, C., Chapin, F. S. III, Matson, P. A., Mooney, H, A. 1992. Responses of terrestrial ecosystems to the changing atmosphere: a resource based approach. *Annu. Rev. Ecol. Syst.* 23: 201–36

37. Franklin, J., Strahler, A. H. 1988. Invertible canopy reflectance modeling of vegetation structure in semiarid woodland. *IEEE Transact. Geosci. Remote Sensing* 26:809–25

38. Franklin, S. E., Peddle, D. R. 1990. Classification of SPOT HRV imagery and texture features. *Int. J. Remote Sensing* 11(3):551–56

39. Fung, I., Lerner, J. J., Matthews, E., Prather, M., Steele, L. P., Fraser, P. J. 1992. Global budgets of atmospheric methane. *J. Geophys. Res.* In press

40. Fung, I. Y., Tucker, C. J., Prentice, K. C. 1987. Application of Advanced Very High Resolution Radiometer vegetation index to study atmosphere-biosphere exchange of CO_2. *J. Geophys. Res.* 92(D3):2999–3015

41. Gamon, J. A., Field , C. B., Bilger, W., Björkman, O., Fredeen, A. L., Penuelas, J. 1990. Remote sensing of the xanthophyll cycle and chlorophyll fluorescence in sunflower leaves and canopies. *Oecologia* 85:1–7

41a. Gao, B.-C., Goetz, A. F. H. 1990. Column atmospheric water vapor and vegetation liquid water retrievals from airborne imaging spectrometer data. *Ecology* 63(6): 1771–79

42. Gardner, R. H., Cale, W. G., O'Neill, R. V. 1982. Robust analysis of aggregation error. *Ecology* 63(6):1771–79

43. Gardner, R. H., O'Neill, R. V., Turner, M. G., Dale, V. H. 1989. Quantifying scale-dependent effects of animal movement with simple percolation models. *Landscape Ecol.* 3(3/4):217–227

44. Goetz, A. F. H., Gao, B. -C., Wessman, C. A., Bowman, W. D. 1990. Estimation of biochemical constituents from fresh, green leaves by spectrum matching techniques. *Proc. 10th Int. Geosci. Remote Sensing Symp.* 3:971–74. New York: IEEE

45. Gorham, E. 1991. Northern peatlands: Role in the carbon cycle and probable responses to climatic warming. *Ecol. Appl.* 1(2):182–95

46. Goward, S. N., Markham, B., Dye, D. G., Dulaney, W., Yang, J. 1991. Normalized difference vegetation index measurements from the Advanced Very High Resolution Radiometer. *Remote Sensing Environ.* 35:257–77

47. Goward, S. N., Tucker, C. J., Dye, D. G. 1985. North American vegetation patterns observed with the NOAA-7 advanced very high resolution radiometer. *Vegetatio* 64:3–14

48. Deleted in proof

49. Deleted in proof

50. Grulke, N. E., Riechers, G. H., Oechel, W. C., Hjelm, U. Jaeger, C. 1990. Carbon balance in tussock tundra under ambient and elevated atmospheric CO_2. *Oecologia* 83:485–94

51. Hall, F. G., Botkin, D. B., Strebel, D. E., Woods, K. D., Goetz, S. J. 1991. Large-scale patterns of forest succession as determined by remote sensing. *Ecology* 72(2):628–640

52. Hall, F. G., Huemmrich, K. F., Goward, S. N. 1990. Use of narrow-band spectra to estimate the fraction of absorbed photosynthetically active radiation. *Remote Sensing Environ.* 1:47–54

53. Hall, F. G., Sellers, P. J., Strebel, D. E., Kanemasu, E. T., Kelly, R. D., et al. 1991. Satellite remote sensing of surface energy and mass balance: results from FIFE. *Remote Sensing Environ.* 35:187–99

54. Harriss, R. C. 1989. Experimental design for studying atmosphere-biosphere interactions. In *Exchange of Trace Gases between Terrestrial Ecosystems and the Atmosphere*, ed. M. O. Andreae, D. S. Schimel, pp. 291–301. Chichester: Wiley. 347 pp.

55. Harriss, R. C., Garstang, M., Wofsy, S. C., Beck, S. M., Bendura, R. J., et al. 1990. The Amazon Boundary Layer Experiment: wet season 1987. *J. Geophys. Res.*, 95(D10):16,721–36

56. Harriss, R. C., Sebacher, C. D., Bartlett, K. B., Bartlett, D. S., Crill, P. M. 1988. Sources of atmospheric methane in the South Florida environment. *Global Biogeochem. Cycles* 2:231–43

57. Harriss, R. C., Wofsy, S. C., Garstang, M., Molion, L. C. B., McNeal, R. S., et al. 1988. The Amazon Boundary Layer Experiment (ABLE 2A): dry season 1985. *J. Geophys. Res.* 93: 1351–60

58. Hatfield, J. L., Kanemasu, E. T., Asrar, G., Jackson, R. D., Pinter, Jr., P. J., et al. 1984. Leaf area estimates from spectral reflectance measurements over various planting dates of wheat. *Int. J. Remote Sensing* 46:651–56

59. Hipps, L. E., Asrar, G., Kanemasu, E. T. 1983. Assessing the interception of photosynthetically active radiation

in winter wheat. *Agric. Meteorol.* 28: 253–59

60. Hobbs, R. J. 1990. Remote sensing of spatial and temporal dynamics of vegetation. In *Remote Sensing of Biosphere Functioning*, ed. R. Hobbs, H. Mooney, pp. 203–19. New York: Springer-Verlag. 312 pp.
61. Deleted in proof
62. Hope, A., Stow, D., Burns, B. 1989. Mapping arctic tundra vegetation types using digital SPOT/HRV-XS data: A preliminary assessment. *Int. J. Remote Sensing* 10(8):1451–57
63. Huete, A. R. 1986. Separation of soil-plant spectral mixtures by factor analysis. *Remote Sensing Environ.* 19: 237–51
64. Huete, A. R. 1988. A soil-adjusted vegetation index (SAVI). *Remote Sensing Environ.* 25:295–309
65. Huete, A. R., Escadafal, R. 1991. Assessment of biophysical soil properties through spectral decomposition techniques. *Remote Sensing Environ.* 35: 149–59
66. Huete, A. R., Jackson, R. D. 1988. Soil and atmospheric influences on the spectra of partial canopies. *Remote Sensing Environ.* 25:89–105
67. IGBP (International Geosphere-Biosphere Programme). 1990. *The International Geosphere-Biosphere Programme: A Study of Global Change Report No. 12. The Initial Core Projects*. Stockholm: IGBP Secretariat
68. Jarvis, P. G., McNaughton, K. G. 1986. Stomatal control of transpiration: scaling up from leaf to region. *Adv. Ecol. Res.*, 15:1–49
69. Jedlovec, G. J., Wilson, G. S., Dodge, J. C. 1989. A status report: NASA's plans for an earth science geostationary platform. *Proceedings of the GOES I-M Operational Satellite Conference*. Washington, DC: Natl. Oceanic Atmos. Admin.
70. Jeffers, J. N. R. 1988. Statistical and mathematical approaches to issues of scales in ecology. In *Scales and Global Change*, ed. T. Rosswall, R. G. Woodmansee, P. G. Risser, pp. 47–56. *SCOPE 35*. Chichester: Wiley. 355 pp.
71. Johnson, L. F., Peterson, D. L. 1991. AVIRIS observations of forest ecosystems along the Oregon Transect. *Proc. 3rd Airborne Visible/Infrared Imaging Spectrometer (AVIRIS) Workshop. JPL Pub. 91–28*. pp. 190–96. Pasadena: NASA
72. Justice, C. O., Hieurnaux, P. 1986. Monitoring the grasslands of the Sahel using NOAA/AVHRR data: Niger 1983. *Int. J. Remote Sensing* 7:1475–98
73. Justice, C. O., Townshend, J. R. G., Holben, B. N., Tucker, C. J. 1985. Analysis of the phenology of global vegetation using meteorological satellite data. *Int. J. Remote Sensing* 6(8): 1271–1381
74. Khalil, M. A. K., Rasmussen, R. A. 1990. Atmospheric methane: recent global trends. *Environ. Sci. Technol.* 24(4):549–53
75. King, A. W. 1991. Translating models across scales in the landscape. In *Quantitative Methods in Landscape Ecology*, ed. M. G. Turner, R. H. Gardner, pp. 479–517. New York: Springer-Verlag. 536 pp.
76. King, A. W., Johnson, A. R., O'Neill, R. V. 1991. Transmutation and functional representation of heterogeneous landscapes. *Landscape Ecol.* 5(4):239–53
77. Krummel, J. R., Garner, R. H., Sugihara, G., O'Neill, R. V., Coleman, P. R. 1987. Landscape patterns in a disturbed environment. *Oikos* 48:321–24
78. Deleted in proof
79. Li, X., Strahler, A. H. 1986. Geometric-optical bidirectional reflectance modeling of a coniferous forest canopy. *IEEE Transact. in Geoscience & Remote Sensing* GE-24:906–19
80. Li, X., Strahler, A. H. 1988. Modeling the gap probability of a discontinuous vegetation canopy. *IEEE Transact. Geosci. Remote Sensing* 26:161–70
81. Lubchenco, J., Olson, A. M., Brubaker, L. B., Carpenter, S. R., Holland, M. M., et al. 1991. The sustainable biosphere initiative: An ecological research agenda. *Ecology* 72(2):371–412
82. Deleted in proof
83. Martin, M. E., Aber, J. D. 1990. Effects of moisture content and chemical composition on the near infrared spectra of forest foliage. *Proc. SPIE Vol. 1298 Imaging Spectroscopy of the Terrestrial Environ.* Pp. 171–77
84. Matson, P. A., Vitousek, P. M. 1987. Cross-system comparisons of soil nitrogen transformations and nitrous oxide flux in tropical forest ecosystems. *Global Biogeochem. Cycles* 1:163–70
85. Matson, P. A., Vitousek, P. M., Livingston, G. P., Swanberg, N. A. 1990. Sources of variation in nitrous oxide flux from Amazonian ecosystems. *J. Geophys. Res.* 95(D10):16,789–98
86. Matson, P. A., Vitousek, P. M. 1990.

Ecosystem approach to a global nitrous oxide budget. *BioScience* 40(9):667–72
87. Matthews, E. 1983. Global vegetation and land use: New high-resolution data bases for climate studies. *J. Clim. Appl. Meteorol.* 22:474–87
88. Matthews, E., Fung, I. 1987. Methane emission from natural wetlands, global distribution, area, and environmental characteristics of sources. *Global Biogeochemical Cycles* 1(1):61–86
89. McDonald, K. C., Dobson, M. C., Ulaby, F. T. 1990. Using MIMICS to model L-band multiangle and multitemporal backscatter from a walnut orchard. *IEEE Transact. Geosci. Remote Sensing* 28:477–91
90. McLellan, T. M., Martin, M. E., Aber, J. D., Melillo, J. M., Nadelhoffer, K. J., Dewey, B. 1991. Comparison of wet chemistry and near infrared reflectance measurements of carbon-fraction chemistry and nitrogen concentration of forest foliage. *Can. J. For. Res.* 21:1689–93
91. Meentemeyer, V., Berg, B. 1986. Regional variation in rate of mass loss of *Pinus sylvestris* needle litter in Swedish pine forests as influenced by climate and litter quality. *Scand. J. For. Res.* 1:167–80
92. Melillo, J. M., Aber, J. D., Muratore, J. M. 1982. Nitrogen and lignin control of hardwood leaf litter decomposition dynamics. *Ecology* 63:621–26
93. Middleton, E. M. 1991. Solar zenith angle effects on vegetation indices in tallgrass prairie. *Remote Sensing Environ.* 38:45–62
94. Milton, N. M., Mouat, D. A. 1989. Remote sensing of vegetation responses to natural and cultural environmental conditions. *Photogrammetric Engin. Remote Sensing* 55:1167–74
95. Mitchell, J. F. B. 1989. The "greenhouse" effect and climate change. *Rev. Geophys.* 27:115–39
96. Monteith, J. L. 1972. Solar radiation and productivity in tropical ecosystems. *J. Appl. Ecol.* 9:747–66
97. Monteith, J. L. 1977. Climate and the efficiency of crop production in Britain. *Philos. Transact. R. Soc., London, Ser. B* 281:277–94
98. Mosier, A., Schimel, D., Valentine, D., Bronson, K., Parton, W. 1991. Methane and nitrous oxide fluxes in native, fertilized and cultivated grasslands. *Nature* 350(6316):330–32
99. NASA. 1991. *Earth Observing System 1991 Reference Handbook*. Washington, DC: Goddard Space Flight Center, NASA. 147 pp.
100. National Research Council. 1988. Space science in the twenty-first century: imperatives for the decades 1995–2015. Mission to planet earth. Washington, DC: Natl. Acad. Press
101. Norman, J. M., Cambell, G. S. 0000. Canopy structure. In *Plant Physiological Ecology: Field Methods and Instrumentation,* ed., R. W. Pearcy, J. R. Ehleringer, H. A. Mooney, P. W. Rundel, pp. 301–325. London: Chapman & Hall, 457 pp.
102. Norris, K. H., Barnes, R. F., Moore, J. E., Shenk, J. S. 1976. Predicting forage quality by infrared reflectance spectroscopy. *J. Anim. Sci.* 43:889–97
103. O'Neill, R. V. 1988. Hierarchy theory and global change. In *Scales and Global Change,* ed. T. Rosswall, R. G. Woodmansee, P. G. Risser, pp. 29–45. SCOPE 35. Chichester: Wiley. 355 pp.
104. O'Neill, R. V., DeAngelis, D. L., Waide, J. B., Allen, T. F. H. 1986. *A Hierarchical Concept of Ecosystems*. Princeton: Princeton Univ. Press. 253 pp.
105. Ormsby, J. P., Soffen, G. E. (eds.) 1989. Special Issue on the Earth Observing System (EOS). *IEEE Transact. Geosci. Remote Sensing* 27(2):106–242
106. Otterman, J. 1981. Satellite and field studies of man's impact on the surface in arid regions. *Tellus* 33:68–77
107. Overpeck, J. T., Rind, D., Goldberg, R. 1990. Climate-induced changes in forest disturbance and vegetation. *Nature* 343:51–53
108. Pastor, J., Post, W. M. 1986. Influence of climate, soil moisture, and succession on forest carbon and nitrogen cycles. *Biogeochemistry* 2:3–27
109. Peterson, D. L., Running, S. W. . 1989. Applications in forest science and management. In *Theory and Applications of Optical Remote Sensing,* ed. G. Asrar, pp. 429–73. New York: Wiley. 734 pp.
110. Peterson, D. L., Spanner, M. A., Running, S. W., Teuber, K. B. 1987. Relationship of Thematic Mapper simulator data to leaf area index of temperate coniferous forests. *Remote Sensing Environ.* 22:323–41
111. Pielke, R. A., Avissar, R. 1990. Influence of landscape structure on local and regional climate. *Landscape Ecol.* 4(2/3):133–55
112. Deleted in proof
113. Prince, S. D. 1991. A model of regional primary production for use with coarse

resolution satellite data. *Int. J. Remote Sensing* 12(6):1313

114. Deleted in proof
115. Ranson, K. J., Smith, J. A. 1990. Airborne SAR experiment for forest ecosystems research: Maine 1989 experiment. *Proc. 10th Annual Int. Geosci. Remote Sensing Symp,* Univ. Maryland. 1:861–64. New York: IEEE.
116. Rastetter, E. B., King, A. W., Cosby, B. J., Hornberger, G. M., O'Neill, R. V., Hobbie, J. E. 1992. Aggregating fine-scale ecological knowledge to model coarser-scale attributes of ecosystems. *Ecol. Appl.* 2(1):55–70
117. Reiners, W. A., Strong, L. L., Matson, P. A., Burke, I. C., Ojima, D. S. 1989. Estimating biogeochemical fluxes across sagebrush-steppe landscapes with Thematic Mapper imagery. *Remote Sensing Environ.* 28:121–29
118. Ringrose, S., Matheson, W., Boyle, T. 1988. Differentiation of ecological zones in the Okavango Delta, Botswana by classification and contextural analyses of Landsat MSS data. *Photogrammetric Engin. Remote Sensing* 54(5): 601–8
119. Ripple, W. J. 1985. Landsat Thematic Mapper bands for characterizing fescue grass vegetation. *Int. J. Remote Sensing* 6(8):1373–84
120. Rock, B. N., Hoshizaki, T., Miller, J. R. 1988. Comparison of in situ and airborne spectral measurements of the blue shift associated with forest decline. *Remote Sensing Environ.* 24: 109–27
121. Roughgarden, J., Running, S. W., Matson, P. A. 1991. What does remote sensing do for ecology? *Ecology* 72(6): 1918–22
122. Running, S. W., Nemani, R. R. 1988. Relating seasonal patterns of the AVHRR vegetation index to simulated photosynthesis and transpiration of forests in different climates. *Remote Sensing Environ.* 24: 347–67
123. Running, S. W., Nemani, R. R., Peterson, D. L., Band, L. E., Potts, D. F., et al. 1989. Mapping regional forest evapotranspiration and photosynthesis by coupling satellite data with ecosystem simulation. *Ecology* 70(4):1090–1101
124. Running, S. W., Peterson, D. L., Spanner, M. A., Teuber, K. B. 1986. Remote sensing of coniferous forest leaf area. *Ecology* 67:273–76
125. Schimel, D. S., Kittel, T. G. F., Parton, W. J. 1991. Terrestrial biogeochemical cycles: global interactions with atmosphere and hydrology. *Tellus* 43AB: 188–203
126. Schutt, J. B., Rowland, R. R., Heartly, W. H. 1984. A laboratory investigation of a physical mechanism for the extended infrared absorption ('red shift') in wheat. *Int. J. Remote Sensing* 5(1): 95–102
127. Sebacher, D. I., Harris, R. C., Bartlett, K. B., Sebacher, S. M., Grice, S. S. 1986. Atmospheric methane sources: Alaskan tundra bogs, an alpine fen, and a subarctic boreal marsh. *Tellus* 38B:1–10
128. Sellers, P. J. 1985. Canopy reflectance, photosynthesis and transpiration. *Int. J. Remote Sensing* 6(8):1335–1372
129. Sellers, P. J. 1987. Canopy reflectance, photosynthesis, and transpiration. II. The role of biophysics in the linearity of their interdependence. *Remote Sensing Environ.* 21:143–83
130. Sellers, P. J. 1991. Modeling and observing land-surface–atmosphere interactions on large scales. *Surv. Geophys.* 312:85–114
131. Sellers, P. J., Berry, J. A., Collatz, G. J., Field, C. B., Hall, F. G. 1992. Canopy reflectance, photosynthesis, and transpiration. III. A reanalysis using improved leaf models and a new canopy of integration scheme. *Remote Sensing Environ.* In press
132. Sellers, P. J., Hall, F. G., Markham, B. J., Wang, J. R., Strebel, D. E., et al. 1990. FIFE experiment design and operations. In *AMS Symp. on the First ISLSCP Field Experiment (FIFE),* Anaheim, CA, 7–9 February, pp. 1–5
133. Sellers, P. J., Mintz, Sud, Y. C., Dalcher, A. 1986. A simple biosphere model (SiB) for use within general circulation models. *J. Atmos. Sci.* 43(6):505–31
134. Shukla, J., Nobre, C., Sellers, P. 1990. Amazon deforestation and climatic change. *Science* 247:1322–25
135. Smith, M. O., Johnson, P. E., Adams, J. B. 1985. Quantitative determination of mineral types and abundances from reflectance spectra using principal components analysis. *J. Geophys Res.* 90:C797-C804
136. Smith, M. O., Ustin, S. L., Adams, J. B., Gillespie, A. F. 1990. Vegetation in deserts: I. A regional measure of abundance from multispectral images. *Remote Sensing Environ.* 31:1–26
137. Smith, M. O., Ustin, S. L., Adams, J. B., Gillespie, A. F. 1990. Vegetation in deserts: II. Environmental influences on regional abundance. *Remote Sensing Environ.* 31:27–52

138. Spanner, M. A., Pierce, L. L., Peterson, D. L., Running, S. W. 1990. Remote sensing of temperate coniferous forest leaf area index: the influence of canopy closure, understory vegetation and background reflectance. *Int. J. Remote Sensing* 11:95–111
139. Deleted in proof
140. Strahler, A. H., Jupp, D. L. B. 1990. Modeling bidirectional reflectance of forests and woodlands using boolean models and geometric optics. *Remote Sensing Environ.* 34:153–66
141. Strain, B. R. 1985. Physiological and ecological controls on carbon sequestering in ecosystems. *Biogeochemistry* 1:219–32
142. Sud, Y. C., Sellers, P. J., Mintz, Y., Chou, M. D., Walker, G. K., Smith, W. E. 1990. Influence of the biosphere on the global circulation and hydrologic cycle—a GCM simulation experiment. *Agric. For. Meteorol.* 52:133–80
143. Sun, G. Q., Simonett, D. S. 1988. Simulation of L-band HH microwave backscattering from coniferous forest stands: a comparison with SIR-B data. *Photogrammetric Engineering Remote Sensing* 54:1195–1201
144. Townshend, J., Justice, C., Li, W., Gurney, C., McManus, J. 1991. Global land cover classification by remote sensing: present capabilities and future possibilities. *Remote Sensing Environ.* 35:243–55
145. Townshend, J. R. G., Justice, C. O., Kalb, V. T. 1987. Characterization and classification of South American land cover types using satellite data. *Int. J. Remote Sensing* 8:1189–207
146. Tucker, C. J. 1977. Asymptotic nature of grass canopy reflectance. *Appl. Optics* 16(3):635–42
147. Tucker, C. J. 1979. Red and photographic infrared linear combinations for monitoring vegetation. *Remote Sensing Environ.* 8:127–50
148. Tucker, C. J., Holben, B. N., Goff, T. E. 1984. Intensive forest clearing in Rondonia, Brazil, as detected by satellite remote sensing. *Remote Sensing Environ.* 15:255–61
149. Tucker, C. J., Townshend, J. R. G., Goff, T. E. 1985. African land-cover classification using satellite data. *Science* 227:369–75
150. Turner, M. G., Dale, V. H., Gardner, R. H. 1989. Predicting across scales: theory development and testing. *Landscape Ecol.* 3(3/4):245–52
151. Turner, M. G., Gardner, R. H., eds. 1991. *Quantitative Methods in Landscape Ecol.*. New York: Springer-Verlag. 536 pp.
152. Ustin, S. L., Adams, J. B., Elvidge, C. D., Rejmanek, M., Rock, B. N., 1986. Thematic Mapper studies of semiarid shrub communities. *BioScience* 36: 446–52
153. Ustin, S. L., Smith, M. O., Adams, J. B. 1992. Remote sensing of ecological processes: A strategy for developing and testing ecological models using spectral mixture analysis. In *Scaling Ecological Processes between Leaf and Landscape,* ed. J. Ehlringer, C. Field. San Diego: Academic. In press
154. Ustin, S. L., Wessman, C. A., Curtiss, B., Kasischke, E., Way, J., Vanderbilt, V. C. 1991. Opportunities for using the EOS imaging spectrometers and synthetic aperture radar in ecological models. *Ecology* 72(6):1934–45
155. Vitousek, P. M. 1992. Concepts of scale at the landscape/global level. In *Scaling Ecological Processes between Leaf and Landscape,* ed. J. Ehlringer, C. Field. San Diego: Academic. In press
156. Wessman, C. A. 1990. Evaluation of canopy biochemistry using remote sensing. In *Remote Sensing of Biosphere Functioning,* ed. R. Hobbs, H. Mooney, pp. 135–56. New York: Springer-Verlag. 312 pp.
157. Wessman, C. A., Aber, J. D., Peterson, D. L. 1989. An evaluation of imaging spectrometry for estimating forest canopy chemistry. *Int. J. Remote Sensing* 10(8):1293–1316
158. Wessman, C. A., Aber, J. D., Peterson, D. L., Melillo, J. M. 1988. Foliar analysis using near infrared reflectance spectroscopy. *Can. J. For. Res.* 18:6–11
159. Wessman, C. A., Aber, J. D., Peterson, D. L., Melillo, J. M. 1988. Remote sensing of canopy chemistry and nitrogen cycling in temperate forest ecosystems. *Nature* 335:154–56
160. Wessman, C. A., Curtiss, B., Ustin, S. L. 1990. Large scale ecosystem modeling using parameters derived from imaging spectrometer data. *SPIE Tech. Symp. on Optics, Electro-Optics, and Sensors,* Orlando, Fla. 1298:164–70
161. Wetzel, D. L. 1983. Near-infrared reflectance analysis: sleeper among spectroscopic techniques. *Analyt. Chem.* 55:1165A-71A
162. Wickland, D. E. 1991. Mission to planet earth: the ecological perspective. *Ecology* 72(6): 1923–1933

163. Williams, D. L., Walthall, C. L. 1990. Helicopter-based multispectral data collection over the Northern Experimental Forest. Preliminary results from the 1989 field season. *Proc. 10th Annu. Int. Geosci. and Remote Sensing Symp, Univ. Md.* 1:875–78. New York: IEEE

164. Wilson, M. F., Henderson-Sellers, A. 1985. A global archive of land cover and soils data for use in general circulation climate models. *J. Climatol.* 5:119–43

Annu. Rev. Ecol. Syst. 1992. 23:201–35

RESPONSES OF TERRESTRIAL ECOSYSTEMS TO THE CHANGING ATMOSPHERE: A Resource-Based Approach*,**

Christopher B. Field
Department of Plant Biology, Carnegie Institution of Washington, Stanford, California 94305

F. Stuart Chapin III
Department of Integrative Biology, University of California, Berkeley, California 94720

Pamela A. Matson
Earth System Science Division, NASA-Ames Research Center, Moffett Field, California 94035-1000

Harold A. Mooney
Department of Biological Sciences, Stanford University, Stanford, California 94305-5020

KEYWORDS: global change, biogeochemistry, terrestrial biosphere, carbon cycle, nitrogen cycle

INTRODUCTION

Changes in the atmosphere impact a broad variety of ecosystem processes over a range of temporal and spatial scales (172). Several recent reviews provide detailed information on the observed and expected ecosystem-level responses to particular components of global change, or environmental forcing factors (EFFs), including increased atmospheric CO_2 (26, 77, 125, 170),

* CIWDPB Publication Number: 1129

increased temperature (159), acid precipitation (211, 212), nutrient deposition (3, 176), tropospheric ozone (19), UV-B (52), and sulfur dioxide (252).

Here, we focus on the effects of environmental forcing factors mediated through the atmosphere—changes in CO_2, temperature, pollutants, nutrient deposition, and radiation—on the function of terrestrial ecosystems, defining function as mass and energy exchange. To keep the topic manageable within the context of this brief review, we emphasize anthropogenic EFFs acting on the decade-to-century time scale, and we do not consider either direct effects on ecosystem structure and function from land use change (clearing, burning, fertilizer application, grazing, etc) or effects of nonanthropogenic EFFs (orbital variations, continental drift, etc). We consider the impacts of past and projected future changes on production, evapotranspiration, nutrient cycling, and trace gas exchange, as well as feedbacks among these ecosystem properties, the climate, and the composition of the atmosphere. We develop a conceptual framework for interpreting effects of EFFs, singly or in combination, across a broad range of ecosystem types. The conceptual framework is based on general mechanisms that operate on a range of spatial and temporal scales. Some of the mechanisms are purely physical (e.g. the feedback of the planetary boundary layer on ecosystem evapotranspiration); some are biophysical (e.g. the temperature dependence of decomposition); and still others—our primary focus here—may have been maintained by natural selection because they confer advantages in growth, survival, or reproduction (e.g. increased root allocation as plants become nutrient limited).

We begin with mechanisms that are organism-based, because direct effects of environmental forcing factors on organisms drive ecosystem-level responses. The direct effects may, however, be poor predictors of the full suite of responses at the ecosystem or regional scale. To account for indirect effects, positive and negative feedbacks, and external constraints, all of which have the potential to override the direct effects, we follow the discussion of organism-based mechanisms with a consideration of processes at the scale of the ecosystem and the biosphere. For this review, we emphasize broad mechanisms controlling the interactions among direct and indirect effects, feedbacks and constraints; we discuss few of the details, which are treated in an increasing body of quantitative literature (e.g. 8, 6, 101, 184, 192, 202, 208).

Even with our restricted focus on the response of terrestrial ecosystems to global change, the topic is still too broad to allow an exhaustive treatment. Both the processes and the literature we discuss are intended as examples. We discuss environmental forcing factors, organism responses, and ecosystem processes that are both quantitatively important and illustrate the kinds of factors that deserve consideration. The literature citations were selected to provide points of access to major topics.

A RESOURCE-BASED APPROACH

Direct and indirect effects of atmospheric environmental forcing factors can be analyzed in terms of their effects on the availability and the utilization of resources required by organisms. We define resources as the substances required for and consumed during growth. Utilization of resources by one organism makes them at least temporarily unavailable to others. For plants, the required resources are CO_2, light, water, mineral nutrients, and oxygen. For heterotrophs, the required resources are carbon compounds, oxidizing potential (usually oxygen), mineral nutrients, and water. Some of the atmospheric EFFs, for example, atmospheric CO_2, nutrient deposition, and precipitation, are resources. Others are resource modulators. They act on rates or efficiencies of resource acquisition and use. Modulators are either not consumed during growth (e.g. temperature) or not required (e.g. UV-B, ozone, SO_2).

Effects of environmental forcing factors are not fundamentally different from effects of "traditional" environmental factors considered outside the context of global change. The data base from studies of plant and ecosystem responses to, for example, soil fertility, light or water availability, or temperature can provide a starting point for initial assessments of ecosystem responses to global changes as well as a guide for designing future experiments. In addition, studies of ecosystem responses to global change can strengthen the foundations of ecology by testing basic principles across an expanded range of conditions.

Plants tend to respond predictably to resource variation (Figure 1). Plants of low resource environments have genetically determined low maximum growth potential, while plants of high resource environments have high maximum growth potential (54, 104, 237). The broad scaling of maximum growth potential to resource availability (54, 91, 105, 189) integrates over many processes, including photosynthetic capacity (90, 213), nutrient uptake capacity (62), tissue nutrient content (54, 189), and tissue longevity (54, 197). As described below, an individual plant responds to resources in a fashion that parallels these patterns for growth-form characteristics. This suggests that, if we can develop a rigorous framework for predicting the impacts of environmental forcing factors on resource availability and individual plants, we can extend these predictions to a broad range of species and growth forms.

Responses of heterotrophs to resource variation are controlled by a combination of direct effects and effects mediated through the chemical composition and abundance of the organic matter available in the environment (Figure 2). The relationships between temperature and litter decomposition (193), soil moisture and litter decomposition (156), litter chemistry and

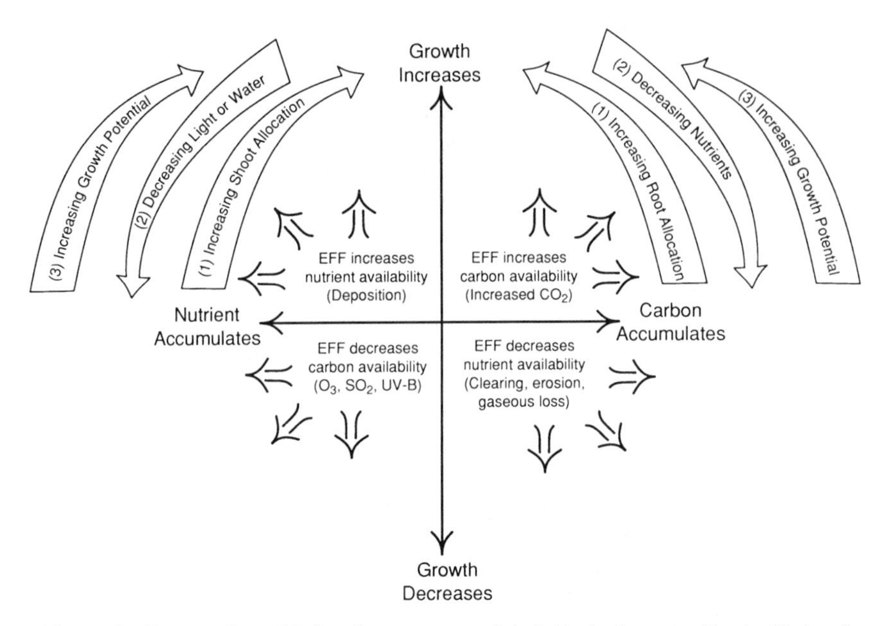

Figure 1 Conceptual model for the responses of individual plants to direct effects of environmental forcing factors. Effects of each EFF tend to push plants away from the center point, which represents the status of a plant prior to the imposition of an EFF. Effects of all EFFs fall into one of four quadrants, depending on whether they tend to increase or decrease the availability of carbon or one or more nutrients, relative to the other resources. Plant responses to any EFF, or combination of EFFs, range from effects only on growth (movement up or down the. vertical axis), with no change in tissue composition, to effects only on resource accumulation (movement left or right along the horizontal axis), with no effects on growth. The factors that determine where a plant moves within a quadrant are: (1) the extent to which changes in allocation compensate for the resource imbalance, (2) the availability of the resources not affected by an EFF, and (3) each species' maximum potential growth rate.

decomposition (1, 33, 158, 233), leaf chemistry and herbivore feeding (86, 143, 144), and soil solution chemistry and nitrogen loss (248) are all predictable enough to begin a whole-biosphere assessment.

ORGANISM RESPONSES

Organism responses to environmental forcing factors lie on a continuum from immediate biochemical responses to changes in the genetic composition of populations or identities of the species present. Some of the responses are essentially unregulated and occur independent of modifications to the biochemical machinery or the allocation program (61, 109). Examples include the response of photosynthesis to short-term changes in CO_2 concentration and the response of soil respiration to short-term changes in temperature.

Decreased plant growth in response to an EFF that decreases carbon or nutrient availability (Figure 1), carbon accumulation in leaves of plants exposed to elevated CO_2 (53), and nitrate accumulation in roots or leaves of plants exposed to elevated nitrate (130) all occur without active changes in the control of resource uptake and utilization.

Other phenotypic responses are highly regulated. They represent an active redistribution of carbon or nutrients deployed within an organism to facilitate acquisition or to decrease loss of a limiting resource. Either within a single species or across a range of species and growth forms, variation in biomass and nutrient allocation is a central mechanism through which plants regulate resource balance (Figure 1) (7, 44, 123, 169, 237), constrain losses to herbivores (66, 106), and manage investments in reproduction (27, 63). The mechanistic link between allocation and growth potential (164, 237) gives allocation a major role in the control of primary production. The role of root growth in nutrient (64) and water (50) uptake makes allocation an important regulator of nutrient cycling and evapotranspiration.

Changes in the abundance of any species or in the suite of species present in an ecosystem may be direct results of an environmental forcing factor, as when the EFF is a temperature lethal to one or more species. Most, however, are indirect and mediated by processes like changes in competitive balance that occur as EFFs differentially affect the ability of genotypes or species to acquire and use resources, or when effects of EFFs on factors like pollinators, dispersers, herbivores, pathogens, or microbial symbionts lead to altered competitive balances. The theory of resource competition (237, 238) and gap dynamics models (184, 222) can play a critcal role in providing concrete predictions of the outcome of competition, but with many species at several trophic levels, the potential for surprises is always present. Plants have evolved rapidly to EFFs like metal-contaminated soil (15), but the rapidity of some of the projected changes in EFFs may force evolutionary change to lag behind other response mechanisms, especially in organisms with long generation times (149). During past climate changes, plants have tended to respond by migration more strongly than by evolution (43).

Plant Gas Exchange

Most of the atmospheric environmental forcing factors have direct or indirect effects on plant photosynthesis and transpiration. These effects occur at several levels. Short-term effects of elevated CO_2 or temperature on the rate of photosynthesis or transpiration are very amenable to experimental study but may or may not be useful for predicting ecosystem responses to EFFs (26, 170). Effects of EFFs on stomatal conductance and the biochemical capacity for photosynthesis are also relatively straightforward to quantify under experimental conditions. Effects at the level of changing source-sink relations, canopy development, or the duration of the period of favorable conditions may, however, be equally important but less accessible to experimental study, especially if the ecosystem responses to the EFFs include changes in species composition.

EFFECTS OF RESOURCES CO_2 has three major, short-term, direct effects on plants. It increases photosynthesis in C_3 plants (230, 26, 146, 229, 170), decreases stomatal conductance, tending to decrease transpiration in C_3 and C_4 species (76), and it suppresses mitochondrial respiration (48).

The short-term stimulation of photosynthesis by elevated CO_2 in C_3 plants occurs because the primary carboxylating enzyme in C_3 photosynthesis, ribulose-1,5-bisphosphate carboxylase/oxygenase (Rubisco), is not CO_2-saturated under current ambient conditions. The slow saturation reflects Rubisco's low affinity for CO_2, competitive inhibition of CO_2 fixation by O_2, and CO_2 production in photorespiration (42). Photosynthesis in C_4 plants, which evidently evolved under low CO_2 (79), is generally CO_2-saturated under current ambient conditions.

The biophysical mechanism through which increased CO_2 reduces stomatal conductance and transpiration is unclear, but the functional implication may be substantial increases in photosynthetic water use efficiency (higher ratio of photosynthesis to transpiration) (76). For a broad range of C_3 and C_4 species exposed to short-term CO_2 enrichment, stomatal conductance at 2 × normal CO_2 is about 60% of the level at normal CO_2 (175). Often, the ratio of intercellular to ambient CO_2 remains approximately constant as CO_2 increases above the ambient level (147).

CO_2 has several effects on mitochondrial respiration. Respiration usually increases with plant or microbial growth rate, and CO_2-stimulation of plant growth can also increase respiration, especially in young tissues (13). Decreases in tissue nitrogen and in biomass allocation to leaves, which often occur in plants grown at elevated CO_2, tend to decrease the respiratory costs for growth and maintenance (13, 204), offsetting some or all of any increase in respiration associated with increased growth. Short-term exposure to elevated CO_2 may also depress plant respiration (48), though the mechanistic basis for this is unclear (88).

Indirect effects of elevated CO_2 on gas exchange generally tend to offset the increased availability of carbon relative to water and mineral nutrients. Decreased stomatal conductance in response to elevated CO_2 sacrifices uptake of CO_2, a relatively abundant resource, but conserves water, a resource that becomes relatively less available as carbon becomes more available. Decreased biochemical capacity for photosynthesis, which occurs frequently but not always in plants grown under elevated CO_2 (75, 140, 229), also limits the abundance of carbon relative to nutrients. This downregulation or acclimation of photosynthetic potential may reflect physical disruption of the chloroplast structure by excess starch reserves (53), sink feedback mediated through a shortage of inorganic phosphate in the chloroplast (229), or active regulation of biochemical potential (42, 108, 229). In N-limited ecosystems, acclimation may be an unavoidable consequence of allocating a relatively fixed pool of N over a larger canopy. Downregulation of photosynthetic capacity is greatest in species with low maximum potential growth rates (241) or that are grown under resource (especially nutrient) limitation (137). Growth under elevated CO_2 levels leads to unchanged or increased photosynthetic capacity in species with high maximum potential growth rates and in plants with active carbon sinks (17, 140, 205, 229) or abundant soil resources (18).

Environmental forcing factors that alter soil nutrients affect photosynthesis and transpiration through effects on canopy development and maximum photosynthetic capacity, with the balance between the two set by the growth potential of the species present and by the availability of other resources. Leaf-level photosynthetic capacity usually increases with leaf nitrogen (85, 90), but nitrogen levels appear to be regulated in relation to species' growth potential and to the availability of other limiting resources (54, 91, 189, 197). Thus, nitrogen deposition may or may not increase photosynthetic capacity. Especially in nutrient-poor sites, increases in photosynthesis may require species replacements (213). Responses of photosynthetic capacity to other nutrients, including phosphorus (178, 196) and potassium (84) may reflect increased sink strength as much as direct involvement of these elements in the photosynthetic machinery.

Mitochondrial respiration tends to increase with increasing leaf nutrients. Since plant growth requires respiration for biosynthesis, nutrient-induced increases in plant growth should also lead to increased respiration (12, 187). Mitochondrial respiration also provides energy for protein turnover and ion transport, two processes that should increase roughly in proportion to leaf nutrient content (11, 186; but see 134).

EFFECTS OF MODULATORS Effects of resource modulators on gas exchange are predictable from their mode of action. Over broad geographic ranges, the optimum temperature for photosynthesis tends to parallel local ambient temperatures, although the optimum temperature is 5–10°C above ambient in extremely cold environments and 5–10°C below ambient in extremely hot environments (35). Net photosynthesis in C_3 plants always decreases at temperatures above 35°, at least under current ambient CO_2. The temperature optimum for maximum photosynthesis increases with CO_2 concentration in C_3 plants (35). Many C_4 species have temperature optima for photosynthesis above 35°C (185).

Plant respiration generally increases exponentially with temperature, with a Q_{10} typically near 2 but ranging from 1.6 to 3 (11). The similarity of the temperature response of maintenance respiration among plants from a wide variety of habitats may indicate that plants generally operate near the fundamental biochemical constraints set by enzyme kinetics and membrane fluidity.

Temperature can affect transpiration through several pathways. Temperature effects on photosynthesis can alter stomatal conductance (67, 253), such that the temperature response of transpiration parallels that of photosynthesis. Effects of temperature on atmospheric humidity can affect transpiration both through effects on the driving gradient and effects on stomatal conductance (124). For photosynthesis, respiration, and transpiration, effects of altered temperature on the length of the growing season may be as important as instantaneous impacts (6).

Atmospheric pollutants may have direct effects on photosynthesis and stomatal conductance (163). Ozone exposure in the range common in urbanized areas generally reduces photosynthesis (195), while similarly defined SO_2 exposures reduce photosynthesis in some species but not others (163). In sensitive species, the negative effects of either gas depend on the amount of pollutant that actually enters leaves; plants that have low stomatal conductance or that close stomates in response to exposure are less sensitive (21, 194, 195). Environmental forcing factors that close stomates (elevated CO_2 and decreased soil moisture) should reduce damage from gaseous pollutants.

Effects of acid precipitation on gas exchange appear to involve two mechanisms—changes in nutrient availability and nutrient balance in the soil (211), and decreased leaf nutrients that result from increased leaching (161, 183, 212).

Although UV-B reduces photosynthesis under laboratory or greenhouse conditions, few studies have shown effects on photosynthesis or growth under field conditions. This difference may reflect the greater genetic variability utilized in field trials, or it may be due to the induction of protection against UV-B damage by exposure to high solar radiation (52). Genetic differences in sensitivity parallel exposure; races of plants from high elevations at low latitudes are more resistant than low elevation, high latitude races of the same species (25, 51, 201).

Growth

Plant growth is a key feature determining plant and ecosystem responses to environmental forcing factors because it governs rates of resource capture and availability of resources to other trophic levels. Production by individual plants is a product of their size and relative growth rate (RGR, growth rate per unit biomass). In dry or infertile soils, maximum plant biomass is low, and plants have an inherently low maximum potential RGR (RGR_{max}) (54, 103). Consequently, the potential growth response to EFFs of plants from low-resource environments is small and can be substantially increased only by changes in species composition (62). By contrast, plants from high-resource environments generally have higher RGR_{max} and respond more strongly to added N or water in terms of both photosynthesis and growth (213). The low RGR_{max} of plants adapted to low-resource environments is often associated with a high root:shoot ratio and long-lived leaves with low photosynthetic rate (54, 168, 197). Thus, species differences in RGR_{max} often parallel differences in photosynthetic capacity (but see (99, 190)), while truly reflecting a suite of correlated controls (57).

In general, impacts of environmental forcing factors on plant growth depend on potential RGR, the impact of the EFF on resource availability or efficiency of resource use, and the availability of other potentially growth-limiting resources (Figure 1). Plant growth responses should be large only when (i) the EFF increases the availability or efficiency of use of a limiting resource, (ii) other resources are not strongly limiting, and (iii) the plants are not growing at their maximum possible rates (Figure 1). If other resources are limiting, plants tend to accumulate the resource made more abundant by the EFF (luxury consumption), and the final growth response will depend on the extent to which the plant can shift allocation to decrease the limitation by other resources. Negative growth responses should be large when plants are operating in a resource range where growth is resource dependent and when the resouce diminished by the EFF is limiting.

How well do the observations match these predictions? The fit is good but not without exception. Growth of C_3 plants (whose photosynthesis is CO_2-limited) generally responds more strongly to elevated CO_2 than does that of C_4 plants (10, 26, 68), particularly when there are strong sinks (e.g. reproductive organs or belowground storage organs) for the added carbon (5, 182). However, growth of C_4 plants often responds to elevated CO_2 (227, 28), perhaps as a consequence of increased water use efficiency or decreased tissue nutrient levels. The prediction that plants with low intrinsic growth rates are less responsive to elevated CO_2 than are rapidly growing species is met in some studies (119), but not others (26, 144). Limited N and P availability generally restricts CO_2 responses (26). When nutrients and light are abundant, the stimulation of growth by elevated CO_2 may be larger under dry than under mesic conditions (120).

Water and nitrogen supply have the greatest effect on plant growth when availability of these resources is low (122), when other resources are nonlimiting (38), and for species with high growth potential (133, 54). Complications arise because drought reduces N availability (56) and low N availability alters water use efficiency (89, 198).

Modulators affect plant growth primarily through changes in resource gain and loss. For example, temperature has a greater effect on photosynthesis and maintenence respiration than on growth respiration (232). However, frost, pollutants and UV-B can reduce growth directly through damage of meristems and other tissues.

Tissue Chemistry

The maximum growth for any plant species requires an optimal balance of internal reserves of carbon and various nutrients (Figure 1) (58, 122). Whenever environmental forcing factors alter resource availability or efficiency of resource gain, nonlimiting resources tend to accumulate in plants, altering tissue chemistry (61). For example, elevated CO_2 causes increases in plant carbohydrate reserves (182, 26). This increased mass and associated increases in plant growth dilute nitrogen over a larger biomass, causing a decline in tissue N concentration. Similarly, increases in belowground resources (as caused by N deposition, increased soil moisture, or increased temperature) lead to increased plant nutrient concentration and a decline in carbohydrates, as carbon becomes the factor more limiting to growth.

The high carbohydrate status associated with a high C:N ratio often promotes synthesis of carbon-based secondary metabolites (i.e. those containing no nitrogen) such as lignin, terpenes, and condensed tannins. This response has not been consistent, however, in the few available studies of plants grown under elevated CO_2 (23, 26). Carbon-based secondary metabolites often deter herbivores and pathogens (23, 46, 106, 148, 199), and reduce the activity of soil microbes (111, 250). Those species that normally occupy low-resource sites are characterized by high concentrations of lignin, tannins, and terpenes, and therefore produce food and litter of low quality (46, 47, 93). High C:N ratios and high concentrations of carbon-based secondary compounds lay the foundation for a self-reinforcing loop in which the slow decomposition of litter with a high C:N ratio retards the release of nutrients and further decreases nutrient availability (246).

Trace Gas Emissions

Changes in the quantity of the two major nonmethane hydrocarbons emitted by plants—monoterpenes and isoprene—in response to environmental forcing factors include regulated and unregulated components. Isoprene emission tends to increase with photosynthesis but appears to be inhibited by elevated CO_2 (87, 167, 206, 218). The relative strength of these two responses is not well known. Monoterpenes, generally considered defensive compounds, should respond like the other carbon-based secondary compounds in high-CO_2 grown plants. The lack of consistent evidence for increased levels of carbon-based secondary compounds in these plants, however, prevents a simple extrapolation to monoterpenes.

Controls on isoprene and monoterpene emissions are still unclear, but emissions may respond differently to changes in soil resources. Because isoprene emission is positively related to net primary production (87, 167), we expect that changes in soil resources (including nitrogen added via anthropogenic deposition and fertilization) that increase production will probably lead to increases in isoprene emission; changes that reduce plant production will have the reverse effect. Increased soil nitrogen availability resulting from rising temperatures may also stimulate isoprene emission. Monoterpene production and emission may, on the other hand, be reduced by deposition or fertilization if allocation to secondary carbon compounds decreases as nutrient availability increases and carbon is allocated to growth (46, 199). Effects of continued sporadic fertilization are likely to represent a balance between the short-term effect—reduction in monoterpene pools and emission —and the long-term effect—an increase due to increased foliar biomass (142).

Emissions of monoterpenes and isoprene as well as ammonia and sulfur gases increase greatly with temperature. For monoterpenes, this temperature sensitivity (127, 135, 240) is partly due to the effect of temperature on monoterpene vapor pressure but may also involve changes in physical conductance through the mesophyll or changes in the monoterpene pools (240). Stomatal conductance does not exert strong control on monoterpene emissions. The temperature stimulation of isoprene emission (166, 218, 239) cannot be due to changes in isoprene vapor pressure since isoprene is not stored in the plant. Isoprene emission is not controlled by stomatal conductance, presumably because decreases in stomatal conductance are compensated by increased isoprene concentration (87, 167, 206, 218).

Allocation

Adjustments in allocation are the major mechanism by which plants compensate for changes in resource supply. Plants usually respond to altered resource balance by increasing allocation to organs that acquire the most strongly limiting resources (38, 251). Thus, increases in aboveground resources—light (168) and sometimes CO_2 (26, 88, 119, 182)—cause increases in the proportion of biomass allocated to roots, whereas increases in belowground resources (water and nutrients) lead to increased shoot allocation (54, 72, 123, 168). Allocation to the support of mycorrhizas and nitrogen fixing symbionts usually follows the same pattern—increasing investment with decreasing nutrient availability (36, 132), though investments in mycorrhizal associations are curtailed when P availability is extremely low (132). These compensatory changes in allocation are generally most pronounced in species from high-resource environments (103). Parallel changes in resource limitation and allocation may fail to occur when species run into the limits of their plasticity (62, 213), when foraging for one limiting resource is constrained by the availability of another—e.g. when the potential for N fixation is limited by the availability of P, Mo, or Co (247), or when a limiting nutrient is not a substrate for growth (e.g. potassium—123). The sensitivity of root:shoot ratio to elevated CO_2 appears to depend on nutrient availability and species characteristics (88).

Effects of modulators on allocation can be predicted from changes in plant carbon-nutrient balance. Modulators that directly reduce photosynthesis cause a decline in carbohydrate status and increased shoot allocation (173). Similarly, in cases where increased temperature leads to increased respiration and reduced carbohydrate status, we expect plants to allocate more biomass to shoots.

Nutrient Uptake, Use, and Loss

The effect of environmental forcing factors on patterns of nutrient uptake and use also can largely be predicted from changes in plant carbon/nutrient balance. Factors that enhance plant growth increase the "demand" of the plant for nutrients, leading to increased uptake through two mechanisms, increased root:shoot ratio and increased capacity for uptake per unit root (54, 65). Thus, uptake per plant increases with elevated CO_2 and more favorable growth temperatures, even though tissue N concentration declines (26). Similarly, declines in plant growth mediated by drought, ozone, and pollutants reduce nutrient uptake (56). N deposition directly enhances plant N uptake by increasing N availability in soil, despite compensatory declines in root:shoot ratio and N uptake capacity per unit root.

Combinations of environmental forcing factors can have more complex effects on nutrient uptake. For example, acid rain increases availability of both N and aluminum. The inhibitory effect of aluminum on growth of roots and mycorrhizae, combined with leaching loss of cations like magnesium, reduces magnesium uptake, inducing magnesium deficiency and declines in photosynthesis and growth (161, 183, 212).

Effects of environmental forcing factors on plant nutrient loss are also quite predictable from plant resource balance. On average, about 15% of plant aboveground nutrient loss occurs by leaching (55), 10% by herbivory (66), and the rest by tissue senescence. Leaching loss increases exponentially with increasing plant nutrient status (54, 243) and should, therefore, increase with N deposition and pollutants that inhibit photosynthesis; it should decline with elevated CO_2 and temperature regimes that reduce tissue N concentration through enhanced growth (54). Pollutants like SO_2 enhance leaching loss because hydrogen ions in the rain exchange with cations held on the cuticular exchange surface and because acidity alters cuticle chemistry (200, 221). The effect of plant N status on nutrient loss to herbivores is not clear. Losses to herbivores often increase with improved plant N status (152), but tissue consumption per herbivore increases in at least some insects fed leaves in which the C:N ratio was increased by growth at elevated CO_2 (86, 143).

Tissue turnover is the major avenue of nutrient loss in plants and is determined by both tissue longevity and the proportion of nutrients withdrawn from tissues at senescence. Both of these factors respond to plant resource balance and, therefore, to environmental forcing factors. Factors that enhance plant growth also increase the rates of tissue loss as lower leaves and branches become shaded, and as old roots senesce when an expanding root system explores new regions of soil (54, 219, 245). Thus, in general, we expect nutrient loss through tissue turnover to increase when elevated CO_2, N deposition, or temperature enhances growth and to decrease when pollutants or drought depresses growth. Some pollutants like SO_2 and ozone damage tissues directly, increasing the loss of tissues and the nutrients they contain (141).

Despite considerable study, there is conflicting evidence about whether plant nutrient status influences the proportion of nutrients that are resorbed from leaves at senescence. Low plant N status (as induced by elevated CO_2, altered temperature regime, etc) usually has either no effect (59, 37) or enhances N resorption efficiency (93, 228) and always results in litter with lower N concentration than that produced by high-N plants. However, in the single experiment examining resorption from CO_2-grown plants in the field, the low N content of CO_2-grown leaves was offset by a low efficiency of N resorption from senescing leaves (70). Other environmental forcing factors reduce resorption efficiency more directly. For example, drought reduces resorption efficiency (73, 191), as does acid rain (183).

COMMUNITY RESPONSES

The outcome of competition depends on resource supply (237) and can generally be predicted for major types of species. The observed vegetation change toward increased abundance of rapidly growing grasses and forbs and decreased abundance of evergreen shrubs in areas of heavy nitrogen deposition (31) is consistent with the hypothesis that rapidly growing species with high rates of resource acquisition should be competitively favored over slowly growing, nitrogen-efficient species under conditions of high nutrient availability (31, 54, 103, 238). Similarly, long droughts in semiarid grasslands increase the relative abundance of CAM and C_4 species over C_3 species (9, 139, 171, 235).

Elevated CO_2 is expected to favor C_3 more strongly than C_4 species, particularly under arid conditions where water use efficiency is a major factor governing competitive balance. Although this pattern has been observed in some studies (26), there are cases where elevated CO_2 favors C_4 species in competition with C_3 species, and where the changes in competitive balance among C_3 species are as large or larger than differences between C_3 and C_4 species (28). Because there have been few studies of effects of elevated CO_2 on competition among species under field conditions (241, 69), predictions about CO_2 effects on natural vegetation change remain tenuous. This is a critical gap in our ability to predict the impact of global change.

Effects of temperature, UV-B, and pollutants on competitive balance are highly species-specific and seldom lead to general predictions other than that species that occupy a particular thermal or pollution environment are favored in competition under conditions resembling those to which they are adapted. This would lead to predictions that C_4 species should expand northward with climatic warming, whereas CO_2 effects on water-use efficiency should allow C_3 species to expand into warmer, drier habitats, thus blurring the current geographic patterning of C_3 and C_4 species (235).

ECOSYSTEM RESPONSES

Trophic Dynamics

Contrary to predictions (46, 66, 106, 244), elevated CO_2 has no consistent effect on concentrations of carbon-based secondary metabolites in the few studies that have been done (23, 26). Therefore, most CO_2 effects on insect herbivory are mediated by the decline in tissue N concentration, which leads predictably to reduced digestion efficiency and growth rate, compensated by increases in the amount of plant material eaten and the time required to reach pupation (23, 86). From the plant perspective, the impact of CO_2 on herbivory depends on the relative magnitude of the enhanced tissue removal by an undisturbed larva versus the declines in herbivory due to increased predation and parasitism and decreased fecundity in the herbivores. We speculate that the latter factors will predominate and that herbivory will decline with increasing CO_2.

Although CO_2 effects on mammalian herbivores have not been investigated, we expect the same general responses as those observed in insects. Because nitrogen is cycled more rapidly through herbivore- than through detritus-based food chains (60), any changes in herbivory have important implications for nutrient cycling.

As with competition, the impact of global change on soil resources leads to consistent predictable changes in trophic dynamics. Nitrogen deposition increases herbivory through its combined effect on increased tissue N and reduced secondary metabolite concentrations (55). Moderate drought increases tissue carbohydrates and plant secondary metabolites because growth is reduced more strongly than photosynthesis, but severe drought further reduces

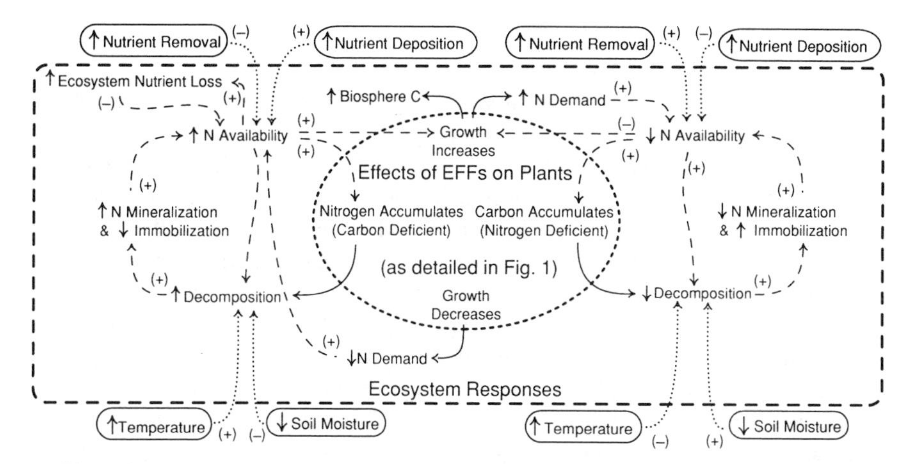

Figure 2. Conceptual model for the responses of ecosystem nitrogen status to environmental forcing factors. Direct effects of EFFs on plants, detailed in Figure 1, are summarized (central dashed oval) by their effects on growth or tissue composition. Plant responses that involve changes in both growth and tissue composition should affect ecosystem responses through both kinds of change. Effects of these plant responses on ecosystem nitrogen status are shown by solid arrows. Dashed arrows indicate feedbacks and indirect responses, with + indicating a positive effect and – indicating a negative effect. Direct effects of EFFs not initially mediated through changes in plants are indicated with dotted arrows. Note that changes in growth lead to changes in N availability that tend to limit further changes in growth, while changes in plant tissue composition lead to effects on N availability that tend to amplify the changes in tissue composition.

photosynthesis, carbohydrate status, and secondary metabolite levels (148), inducing major insect outbreaks (148, 153). When climate changes to the point that long-lived plants are maladapted to their environment, we expect reduced carbon balance and insect outbreaks to become more common phenomena. Pathogen attacks are often mediated by these same conditions (151) and are often spread by insect outbreaks.

Primary Production

Ecosystem-level feedbacks may either damp or amplify the production responses of individual plants to environmental forcing factors (Figure 2). Ecosystem feedbacks damp plant responses when a change in growth leads to an opposing change in resource availability, as when the rate of resource supply is fixed. Ecosystem feedbacks that lead to changes in rates of resource supply in response to changes in plant growth, allocation, or composition may amplify or damp effects of an EFF on plant growth.

How will resource availability change with plant growth? The pattern varies from resource to resource, but no resource is infinitely available. Local CO_2 concentration can be reduced slightly by local photosynthesis (256) or increased by local respiration (29), but atmospheric transport is sufficient that localized depressions in CO_2 within canopies are seldom large enough to have major effects on photosynthesis (24).

Supply rates of light are sensitive to ecosystem processes only indirectly and over large spatial scales, through ecosystem effects on cloudiness. On the local scale, the supply of light per unit of ground area is fixed. Once a canopy is closed, the upper boundary on light supply enforces strongly diminishing returns on ecosystem-level production. When all or nearly all of the light is intercepted, as in many reasonably fertile forest and grassland ecosystems, further production increases are limited to changes in light use efficiency, ϵ(dry matter produced per unit of light captured). Over a broad range of ecosystems, ϵ is quite conservative (approximately 0.7 g C MJ^{-1} of intercepted total solar radiation (203)). ϵ decreases slightly with nutrient limitation (98) and slowly applied drought (179). Nutrient additions tend to have much larger effects on the rate of canopy closure than on ϵ (98). Effects of elevated CO_2, O_3, SO_2, and UV-B on ϵ are unknown. In ecosystems that are rich in nutrients and water, light supply sets the upper limit to possible CO_2 responses, with the maximum set by the photon yield of photosynthesis. The sensitivity of the photon yield to an increase in intercellular CO_2 from 325 to 600 ppm increases with temperature, from an increase of about 30% at 10°C to as much as 100% at 38°C (78, 147). For ecosystems exposed to environmental forcing factors that decrease plant-level production, the loss of some leaf area will increase the mean light level through the canopy and may partially compensate for the loss of canopy.

Limited flexibility of supply rates of water also tends to counteract environmental forcing factor–driven changes in production. Ecosystem-level water storage capacity is sensitive to rooting depth, and canopy interception of precipitation increases with the leaf area index (LAI) (255), but the supply of precipitation is sensitive to ecosystem processes only over coarse spatial scales (74). In water-limited ecosystems, increases in production caused by environmental forcing factors should be no greater than their effects on water use efficiency. For elevated CO_2, increases in water use efficiency on the order of 30–40% (140) can account for similar increases in production. Since any decrease in transpiration tends to prolong the moisture availability in water-limited sites, EFFs that decrease photosynthesis or canopy development may result in longer growing seasons, and the increased season length will tend to offset the decrease in photosynthesis.

Unlike light and water, rates of nutrient supply are sensitive to a range of ecosystem-level processes on a range of time scales. In the short term, changes in plant uptake change availability in the opposite direction and damp production responses. For production decreases, however, this damping is limited by the tendency for ecosystems to lose excess nutrients through leaching and gaseous emission (233). On a longer time scale, the regulation of decomposition by tissue nitrogen tends to reinforce the low N status of sites that support plants with high C:N ratios and the high N status of sites supporting tissue with low C:N ratios (246). On a still longer time scale, total nitrogen stocks in an ecosystem reflect the balance between gains and losses (247).

Nutrient limitation is common in terrestrial ecosystems, with proximate limitation by nitrogen probably the most common (247). For ecosystems in which all or nearly all of the nitrogen mineralized is acquired by plants or microbes, production increases in response to environmental forcing factors should be possible only to the extent that the EFF increases nitrogen availability, shifts the competition for nutrients in favor of plants and away from soil microbes, or increases nitrogen use efficiency (growth per unit of nitrogen taken up—30).

Nutrient deposition has substantial potential to increase nitrogen availability, leading to increased production in nutrient-limited ecosystems, especially in combination with increased CO_2 (236).

At the ecosystem level, supply rates of light and water and, in the short term, nutrients are all constrained. As a result, environmental forcing factors that tend to increase production will also tend to increase the limitation by one or more of these resources. If none of the resources is limiting, as may be the case in some agricultural settings (e.g. 121), the production response to an EFF such as elevated CO_2 may be dramatic. Natural ecosystems with low disturbance and agricultural ecosystems with abundant nutrients and water tend to have well-developed plant canopies that capture all or most of the available light (207, 237). In these ecosystems, production increases in response to elevated CO_2 should be limited to effects due to the increased rate of canopy closure and the change in ϵ, the efficiency of light use in biomass production. In ecosystems with limiting nutrients or water, the response of production to elevated CO_2 should be limited by the effect of CO_2 on the efficiency of use of the limiting resource. This efficiency effect can be substantial for both nutrients and water, though feedbacks through decomposition should temper the effect of increased efficiency of nutrient use.

For environmental forcing factors that tend to decrease production, ecosystem-level feedbacks should, at least in some cases, act to offset some of the limitation by increasing the availability of previously limiting resources. This mechanism is probably more important for water and light than for nutrients, for which competition from microbes and the potential for loss from the ecosystem offset increases in availability.

Nutrient Cycling and Decomposition

Changes in plant production, allocation, and resorption resulting from environmental forcing factors all affect litter characteristics and thus have logical consequences for decomposition and nutrient cycling. These plant-mediated responses to EFFs affect litter both by altering its carbon and nutrient composition and by altering litter type and location.

Where changes in litter quality or decomposability (most commonly described by N and lignin concentrations) occur, ecosystem-level changes in nutrient cycling and carbon storage can be expected. Immobilization of N and other nutrients is inversely related to the content of N in fresh litter (34), and release of inorganic N (mineralization) is initially negatively correlated with the lignin content of the substrate (33). In later stages, decomposition rates are similar among litters of initially contrasting chemistries, and mineralization depends on nutrient content (2). Where litter lignin:N has been increased due to elevated CO_2 effects on plant chemistry, increased immobilization of N by microorganisms and decreased rates of decomposition and mineralization may increase C storage in the soil and reduce N availability to plants, thus limiting production and continued uptake of elevated CO_2 by plants. On the other hand, in areas where N availability is inherently high (e.g. agricultural systems, lowland tropical forests—248), increased immobilization and reduced mineralization may have only minor effects on production but may prevent leaching of inorganic N to ground water and aquatic ecosystems, or volatilization of N to the atmosphere.

Changes in plant allocation in response to environmental forcing factors also modify patterns of decomposition by changing the relative abundance of tissue types and the location of litter deposition. Litter in the soil decomposes and mineralizes N more rapidly than litter on the ground surface because of the more favorable moisture and nutrient regime below ground (110). High root allocation, as occurs with drought and elevated CO_2, increases N mineralization primarily because of the favorable nature of the belowground environment. Although roots often have a lower N content than leaves, the very limited studies to date show no N resorption from senescing roots (181, 220), so that root litter in some forests has an N concentration similar to, or even greater than, that of leaf litter. On the other hand, shifts in allocation to wood or changes in growth-form composition from herbaceous to woody species should reduce overall litter quality substantially.

While elevated CO_2 may slow decomposition and increase soil carbon storage through effects on litter quality, atmospheric deposition or fertilizer inputs of N cause increased litter N concentration (55, 162, 183), generally increasing rates of decomposition and N mineralization (32, 118, 157). Soil N additions can lead to increased rates of litter decomposition, reduced immobilization of nitrogen, increased rates of nitrogen mineralization and nitrification, and potentially increased losses of nitrogen via gas flux or leaching, on a time scale so rapid that changes in plant tissues and litter chemistry may not initially be involved (4, 138, 150, 180). In some systems, fertilizer inputs of P may affect cycling of nitrogen through their effect on free-living and symbiotic nitrogen fixing organisms (82, 247). Altogether, the effect of N additions should be to increase decomposition, N mineralization, and N availability.

Temperature and moisture are important direct limitations to decomposition in cold and dry environments, and strongly influence the seasonality of decomposition in all environments. Soil moisture affects both the rates and the endproducts of microbial transformations. The microbial processes that mediate the transfer of organically bound nutrient elements to inorganic forms generally respond positively and often exponentially to temperature. Therefore, the effect of increasing temperatures should be to increase decomposition rate and mineralization rates, regardless of initial litter quality.

The predictability of soil respiration from temperature and moisture is sufficiently strong to provide a basis for global estimates (210, 193). This could suggest that, at least on the global scale, variation in litter quality is relatively unimportant. On the other hand, litter quality may be a strong covariate with climate, at least in ecosystems that develop in equilibrium with climate. If changes in climate occur more rapidly than changes in ecosystem characteristics and if regional changes in deposition and fertilization drive changes in litter quality, then simple predictions of decomposition that have worked in the past may be less useful in the future.

Soil Trace Gas Exchange

For gases such as methane, nitrous oxide, nitric oxide, and sulfur gases, responses to environmental forcing factors are mediated by physical characteristics of the soil and atmosphere and by biotic controls on nutrient cycling. Biogenic trace gases are released as end products or byproducts of microbially mediated processes, with rates influenced by organic matter characteristics, trophic interactions, water and nutrient availability, and temperature.

Methane is a greenhouse gas produced primarily by methanogenic bacteria during decomposition of organic compounds under anaerobic conditions. A number of short-term, seasonal, and interannual studies of natural wetlands and rice paddies show strong positive response of methane emission to temperature (145, 214), but effects of future temperature changes on soil moisture, the depth of the water table, and organic matter production will also be important in controlling methane emissions. Changes in plant biomass and plant species composition in response to environmental forcing factors will also affect methane emissions, because wetland vegetation determines the amount and composition of the organic substrate for methane genesis and plays a major role in transport of methane out of soils and sediments (177, 214).

Emissions of nitrous oxide (N_2O) and nitric oxide (NO) are affected by a combination of microbial processes and physical charateristics of the soil, sediment, or water column. Both gases are lost from ecosystems as a consequence of nitrification (an aerobic process) and denitrification (an anaerobic process), with emissions sensitive to the microbially mediated partitioning of the gas production among N_2O, NO, and N_2 (92). Environmental forcing factors are likely to influence nitrogen trace gases in numerous, interacting ways (Figure 2). Fertilization leads to increased N_2O and NO flux in agricultural ecosystems and in forests (41, 81, 97); fluxes of nitrogen trace gases from forests impacted by high N deposition may also be very high. Increasing temperature should stimulate emission of both gases, but the stimulation of NO emission should be greater if soil drying concomitantly occurs—in natural situations, NO appears to dominate in dry or coarse textured soils, and N_2O dominates in wet, compacted, or fine textured soils (71). Direct effects of other resources and modulators are not known, but indirect effects mediated through plant processes are likely.

Evapotranspiration

At the ecosystem level, the stomatal conductance of individual leaves is only one of several factors with the potential to control evapotranspiration, complicating the extrapolation from effects of environmental forcing factors on leaf stomatal conductance to effects on ecosystem water use. Other potential controllers at the ecosystem level are leaf area index (LAI), canopy energy balance, dynamics of the planetary boundary layer, and the moisture-holding capacity of the soil.

In open canopies, transpiration varies strongly with leaf area index (225). In closed canopies, the change in transpiration resulting from altered LAI is usually small, proportionally less than the change in LAI. At the canopy level, transpiration responds nearly linearly to absorbed photosynthetically active radiation (APAR) (215, 216), but the exponential nature of light extinction in plant canopies (165) makes APAR a saturating function of LAI. Two factors explain the transpiration-APAR relationship. First, net radiation is the dominant energy source for evaporating water (126), and net radiation increases with APAR. Second, canopy stomatal conductance is often a linear or nearly linear function of incident photosynthetically active radiation, a major determinant of APAR. Thus, effects of environmental forcing factors on LAI will have potentially important consequences for evapotranspiration when LAI is less than one, but the potential for these effects decreases as LAI increases.

Effects of energy balance may have important implications for the realization of the potential increases in water use efficiency resulting from increased CO_2 or other environmental forcing factors that decrease stomatal conductance. When a canopy is aerodynamically rough—when the plants are tall or the structure is heterogenous—transpiration changes roughly in proportion to stomatal conductance, because the extra energy not dissipated as latent heat is transported as sensible heat. When a canopy is aerodynamically smooth, however, as for short crops and uniform grasslands, the small aerodynamic conductance prevents efficient transport by sensible heat. Under these conditions, a decrease in stomatal conductance causes an increase in canopy temperature, which increases the driving gradient for transpiration. In smooth canopies under typical conditions, a 10% change in stomatal conductance drives a change in transpiration of only 1–6% (126), with the sensitivities increasing as stomatal conductance decreases (155). Stomatal closure in response to the increased driving gradient for transpiration (16, 254), could lead to larger decreases in transpiration but only at the cost of large decreases in stomatal conductance and increases in leaf temperature, both of which may impact photosynthesis (67). Increased exchange of sensible heat also leads to increased growth and drying of the planetary boundary layer (22, 154) which can stimulate transpiration through increasing the driving gradient or decrease it through stimulating stomatal closure.

Evapotranspiration is also sensitive to changes in rooting depth resulting from changes in root allocation and community composition. Environmental forcing factors that lead to decreased root allocation (nutrient deposition, oxidants, and UV-B) have the potential to lead to a decrease in the total quantity of water accessible to plants. When an ecosystem changes from tree dominated to grass and forb dominated, the potential is even greater. Rooting depth may be a critical determinant of annual evapotranspiration (224).

Vegetation, especially over large regions, can have a major impact on atmospheric circulation, with dramatic effects on temperature and precipitation. Changing the surface of a continent, in an atmospheric general circulation model, from completely wet to completely dry can increase surface temperatures by up to 20°C and cause large decreases in precipitation (223). Even over regions of modest size, local evapotranspiration and precipitation are linked (22). In a region like the Amazon basin, which tends to develop internal weather systems, GCM simulations indicate that conversion of rainforest to grassland produces significant drying, with decreases in precipitation, evaporation, and convergence of moisture into the region (74).

Production and the Carbon Cycle

The major fluxes and pools in the carbon cycle of the terrestrial biosphere (photosynthesis, autotrophic respiration, heterotrophic respiration, carbon consumption in fires) are known only approximately, to an accuracy of about

± 020% (20, 116). Since annual global carbon fixation in terrestrial photosynthesis or loss in respiration is about 100 Pg, the uncertainty in the magnitude of these fluxes is several times the fossil fuel flux (5.7 ± 0.5 Pg in 1987—249) to the atmosphere. While the individual values are poorly known, the balance between fixation and release can be estimated from changes in atmospheric CO_2. Since, however, ocean-atmosphere carbon exchange is also poorly known but about the same in magnitude as terrestrial ecosystem-atmosphere exchange (39), it is difficult to use changes in the atmosphere to draw definitive conclusions about the carbon balance of terrestrial ecosystems.

From 1850 to 1986, cumulative anthropogenic carbon inputs into the atmosphere were 195 ± 20 Pg from fossil fuel combustion and 117 ± 35 Pg from deforestation and land use change (249). Of this, 41 ± 6% has remained in the atmosphere. The remaining 150–200 Pg carbon has been stored in the oceans, in the terrestrial biosphere, or both. For the 1980s, carbon inputs to the atmosphere from fossil fuel combustion and deforestation averaged 7.0 ± 1.5 Pg yr^{-1}. Accumulation in the atmosphere was 3.4 ± 0.2 Pg, and ocean uptake estimated from a number of models was 2.0 ± 0.8 Pg, leaving an unaccounted sink of 1.6 ± 1.4 Pg yr^{-1} (249).

Terrestrial ecosystems have clearly lost carbon as a result of changes in land use. Have these losses been partially offset by carbon accumulation in natural ecosystems, and if so, is the accumulation continuing? The answer to the first question appears to be yes. Deconvolution analysis is based on comparing the actual trajectory of atmospheric CO_2 against that expected from models that combine the historical trajectories of fossil fuel combustion and ocean exchange. Calculations based on the deconvolution approach for CO_2 concentration only (226) or CO_2 concentration and ^{13}C in CO_2 (128) indicate that, from about 1920 to the mid 1970s, carbon losses from the biosphere were consistently less than those expected from historical changes in land use (115–117). The deconvolutions yield estimates for the last 15 years that approach but do not rejoin the estimates based on land use (95, 128, 226). Inverse modeling approaches, which infer a distribution of surface sources and sinks from the spatial distribution of atmospheric CO_2 concentration (83, 234), also indicate the existence of a terrestrial carbon sink, with a magnitude from a few tenths to more than 3 Pg yr^{-1}.

Three mechansims may be driving carbon storage in the terrestrial biosphere. These are CO_2 fertilization, nutrient deposition, and twentieth-century warming (Figure 3).

The increase in atmospheric CO_2 during this century is sufficient, at least in short-term exposures, to substantially increase photosynthesis in C_3 plants. Of course, the persistence of the increase should depend on the availability of other resources (Figure 1). Based on chronosequences, Schlesinger (209) estimated that conversion of plant biomass into refractory humus substances in the soil has been only about 0.4 Pg yr^{-1}, although local rates shortly after a disturbance or vegetation change can be higher than the long-term average. Alternatively, elevated CO_2 could drive a faster increase in photosynthesis than in respiration and decomposition. Whether this occurs will depend on temperature, moisture, and the extent to which plants respond to elevated CO_2 by decreasing tissue nutrients and by increasing allocation to woody tissues resistant to decomposition. Since storing carbon as wood, surface litter, or new soil organic matter requires less nitrogen than storing it as leaves, roots, or old soil organic matter, the most likely locations for a CO_2-stimulated sink are in the high C:N pools. Estimates of forest biomass (40, 45), biomass changes from land use (107, 116, 113), and tree growth (102) do not indicate a major carbon sink in terrestrial ecosystems, although forests in Europe (127a) and many young forest ecosystems appear to be accumulating biomass (113).

Nitrogen limitation to production is a strong constraint on carbon storage in the biosphere (Figure 3). Carbon storage can be increased only if the carbon-to-nitrogen ratio of the stored material increases, a likely consequence of increased CO_2, or if nutrients are added to the ecosystem. Globally, anthropogenic nitrogen inputs into the terrestrial biosphere are approximately 135 Tg yr^{-1}, of which most is in fertilizer applied to agricultural systems (159). Wet and dry nitrogen deposition in the temperate and boreal regions of the northern hemisphere is up to 25 Tg yr^{-1} (160). Some of this nitrogen enters ecosystems that cannot store much carbon (e.g. annual agriculture—49), and some of it enters ecosystems that are nearly or completely nitrogen saturated (4). We can bound the implications with the following calculations. If half of the 25 Tg enters nitrogen-limited ecosystems and if the added N is all incorporated in wood with a C:N ratio of 200:1, then the potential carbon storage is 2.5 Pg yr^{-1}. But if the carbon were all stored as humus or passive soil organic matter with a C:N ratio of 12 or less, then the potential storage is only 0.3 Pg yr^{-1}. Modeling studies (188) and the limited evidence for increased wood production both indicate that the actual storage is probably closer to the smaller of these two values (but see 127a). The combination of increased CO_2 and nitrogen deposition acting together potentially account for a sink of up to 70 Pg since the start of the industrial revolution (236).

Increased temperature increases respiration, but it may also increase photosynthesis. Effects of temperature alone could lead to accumulation or loss of biosphere carbon, with the probability of loss an increasing function of temperature (242). Ecosystem-level nutrient limitation can increase the probability that climatic warming leads to temporary increases in biosphere

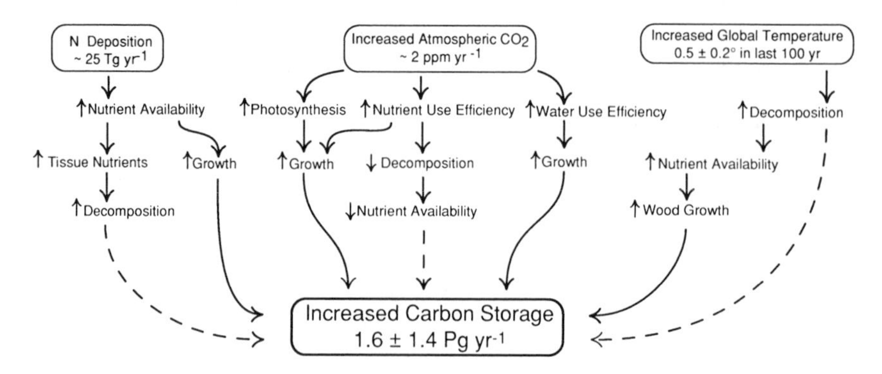

Figure 3 Schematic of three mechanisms potentially accounting for a biosphere carbon sink. The three driving EFFs are shown in ovals. Solid arrows indicate positive effects, and dashed arrows indicate negative effects. Note that this biosphere sink is not the net change in biosphere carbon. The net change in biosphere carbon is the sum of this quantity plus the loss from changing land use. Note that each of the mechanisms has both positive and negative effects on carbon storage.

carbon (Figure 3). Warming increases decomposition, tending to lead to a loss of soil organic matter (193), where nitrogen is stored at a C:N ratio of 10:20. If the nutrients released from the respiration of soil organic matter are incorporated in wood with a C:N ratio of 100:200, then the carbon stored per unit nitrogen immobilized increases by up to a factor of 20. This mechanism can persist only until the pool of soil organic matter reequilibrates at a new size. In rough terms, if the Q_{10} for decomposition averages 2.5 (193), then the 0.5° warming in the twentieth century (94) should have increased respiration by 5.3% of the decomposition total of 50–75 Pg yr^{-1} or 2.6–4 Pg yr^{-1}. The 0.1–0.4 Pg nitrogen released from organic matter could lead to storage of 10–80 Pg yr^{-1} carbon in wood, or much more than the missing sink. The evidence inconsistent with a major sink (40, 45, 102, 107, 116, 113) suggests that the sink is broadly distributed, that some carbon is stored through each of the three mechanisms, and that the increased storage includes above- and below-ground components.

The seasonal oscillation in atmospheric CO_2 is quite regular from year to year, but the amplitude of the oscillation is gradually increasing (128). This increase could reflect a more active biosphere (131). Identical annual increases in photosynthesis and respiration can drive an increase in the seasonal oscillation of atmospheric CO_2 as a consequence of the seasonal offset between the two processes (96). Alternatively, the increase in the seasonal oscillation could reflect a small, progressive change in the relative timing of production, decomposition, fossil fuel combustion, and land use change (114). The analysis of changes in the the seasonal oscillation is complicated by the broad temporal overlap between photosynthesis and respiration.

GENERALIZATIONS AND CONCLUSIONS

1. Direct and indirect impacts of atmospheric environmental forcing factors on terrestrial ecosystems can be interpreted on the basis of their effects on the availability and utilization of the resources required for ecosystem function.

2. Many of the effects of environmental forcing factors on plant and ecosystem processes are sufficiently general to support extrapolation across a broad range of terrestrial ecosystems. The quantitative ecosystem consequences of any particular combination of environmental forcing factors are influenced by interactions among the EFFs, the resource status of the site, and characteristics of the organisms present. The resource-based approach provides a foundation for quantitative models of these interactions.

3. Plant responses to resource imbalances caused by environmental forcing factors tend to compensate for the imbalance, at least partially. The dominant mechanisms of compensation are changes in the biochemical capacity for resource uptake, changes in biomass and nutrient allocation, and changes in rates of tissue loss. Compensation for changing resource balance allows plants to make growth a function of resource availability over a wide range of resource balances. The plasticity or compensation ability of any particular species is, however, limited, and effective compensation for large changes in resource balance caused by environmental forcing factors may require changes in species composition. When appropriate species are available, changes in species composition will not dramatically disrupt the response of growth to resource availability, but the availability of appropriate species may be compromised by EFFs, especially changing land use patterns.

4. Ecosystem responses to altered resource balance caused by environmental forcing factors are often constrained by limiting resources. The tendency for ecosystems to amplify nutrient deficiency or excess through the feedback of tissue chemistry on decomposition and nutrient availability can exaggerate carbon-nutrient imbalance. When nutrients are limiting, EFFs that increase carbon availability tend to further decrease nutrient availability through this feedback. This is very likely to make natural ecosystems less responsive to increased atmospheric CO_2 than many agricultural ecosystems or plants in controlled environments. The decomposition feedback also amplifies the effect of EFFs that directly increase nutrient availability.

5. Ecosystem and plant processes provide little buffering against the effects of environmental forcing factors that tend to decrease primary production. The tissue chemistry/decomposition feedback and the tendency for ecosystems to lose abundant nutrients and water further weakens the buffering. These mechanisms increase the likelihood of decreased primary production in ecosystems impacted by pollutants, UV-B, and loss of soil fertility.

6. Most ecosystems are exposed to a combination of atmospheric environmental forcing factors and increased intensity of human utilization. Accurate predictions of future responses will need to account for impacts of land use as well as impacts of atmospheric EFFs.

ACKNOWLEDGMENTS

Thanks to many colleagues with whom we have discussed these concepts over the last few months (including many who disagree with our interpretation). Comments from Peter Vitousek and David Schimel were particularly brutal but helpful. We acknowledge the support of NSF, NASA, and DOE.

Literature Cited

1. Aber, J. D., Melillo, J. M. 1982. Nitrogen immobilization in decaying hardwood leaf litter as a function of initial nitrogen and lignin content. *Can. J. Bot.* 60:2263–69
2. Aber, J. D., Melillo, J. M., McClaughtery, C. A. 1990. Predicting long-term patterns of mass loss, nitrogen dynamics, and soil organic matter formation from initial fine litter chemistry in temperate forest ecosystems. *Can. J. Bot.* 68:2201–8
3. Aber, J. D., Melillo, J. M., Nadelhoffer, K. J., Pastor, J., Boone, R. D. 1991. Factors controlling nitrogen cycling and nitrogen saturation in northern temperate forest ecosystems. *Ecol. Appl.* 1:303–15
4. Aber, J. D., Nadelhoffer, K. J., Steudler, P., Melillo, J. M. 1989. Nitrogen saturation in northern forest ecosystems. *BioScience* 39:378–86
5. Ackerson, R. C., Havelka, U. D., Boyle, M. G. 1984. CO_2-enrichment effects on soybean physiology. II. Effects of stage-specific CO_2 exposure. *Crop Sci.* 24:1150–54
6. Adams, R. M., Rosenzweig, C., Peart, R. M., Ritchie, J. T., McCarl, B. A., et al. 1990. Global climate change and US agriculture. *Nature* 345:219–24
7. Ågren, G. I., Ingestad, T. 1987. Root: shoot ratio as a balance between nitrogen productivity and photosynthesis. *Plant Cell Environ.* 10:579–86
8. Ågren, G. I., McMurtrie, R. E., Parton, W. J., Pastor, J., Shugart, H. H. 1991. State-of-the-art models of production-decomposition linkages in conifer and grassland ecosystems. *Ecol. Appl.* 2(1): 118–38
9. Albertson, F. W., Weaver, J. E. 1942. History of the native vegetation of western Kansas during seven years of continuous drought. *Ecol. Monogr.* 12: 23–51
10. Allen, L. H. Jr. 1990. Plant responses to rising carbon dioxide and potential interactions with air pollutants. *J. Environ. Qual.* 19:15–34
11. Amthor, J. S. 1984. The role of maintenance respiration in plant growth. *Plant Cell Environ.* 7:561–69
12. Amthor, J. S. 1989. *Respiration and Crop Productivity*. New York: Springer-Verlag. 215 pp.
13. Amthor, J. S. 1991. Respiration in a future, higher CO_2 world. *Plant Cell Environ.* 14:13–20
14. Andreae, M. O., Schimel, D. S., eds. 1989. *Exchange of Trace Gases between Terrestrial Ecosystems and the* Atmosphere. New York: Wiley. 347 pp.
15. Antonovics, J., Bradshaw, A. D., Turner, R. G. 1971. Heavy metal tolerance in plants. *Adv. Ecol. Res.* 7:1–85
16. Aphalo, P. J., Jarvis, P. G. 1991. Do stomata respond to relative humidity? *Plant Cell Environ.* 14:127–32
17. Arp, W. J. 1991. Effects of source-sink relations on photosynthetic acclimation to elevated CO_2. *Plant Cell Environ.* 14:869–76
18. Arp, W. J., Drake, B. G. 1991. Increased photosynthetic capacity of *Scirpus olneyi* afer 4 years of exposure to elevated CO_2. *Plant Cell Environ.* 14: 1003–6
19. Ashmore, M. R., Bell, J. N. B. 1991. The role of ozone in global change. *Ann. Bot.* 67(Suppl. 1):39–48
20. Atjay, G. L., Ketner, P., Duvigneaud, P. 1979. Terrestrial primary production

and phytomass. In *The Global Carbon Cycle, SCOPE 13*, ed. B. Bolin, E. Degens, S. Kempe, P. Ketner, pp. 129–82. Chichester: Wiley
21. Atkinson, C. J., Winner, W. E., Mooney, H. A. 1988. Gas exchange and SO_2 fumigation studies with irrigated and unirrigated field grown *Diplacus auriantiacus* and *Heteromeles arbutifolia*. *Oecologia* 75(3):386–93
22. Avissar, R., Pielke, R. A. 1991. The impact of plant stomatal control on mesoscale atmospheric circulations. *Agric. For. Meteorol.* 54:353–72
23. Ayres, M. P. 1992. Plant defense, herbivory, and climate change. See Ref. 129. In press
24. Baldocchi, D. D. 1993. Scaling water vapor and carbon dioxide exchange from leaves to a canopy: Rules and tools. See Ref. 80. In press
25. Barnes, P. W., Flint, S. D., Caldwell, M. M. 1987. Photosynthesis damage and protective pigments in plants from a latitudinal arctic/alpine gradient exposed to supplemental UV-B radiation in the field. *Arctic Alp. Res.* 19:21–27
26. Bazzaz, F. A. 1990. The response of natural ecosystems to the rising global CO_2 levels. *Annu. Rev. Ecol. Syst.* 21:167–96
27. Bazzaz, F. A., Chiariello, N. R., Coley, P. D., Pitelka, L. F. 1987. Allocation resources to reproduction and defense. *BioScience* 37:58–68
28. Bazzaz, F. A., Garbutt, K. 1988. The response of annuals in competitive neighborhoods: Effects of elevated CO_2. *Ecology* 69:937–46
29. Bazzaz, F. A., Williams, W. E. 1991. Atmospheric CO_2 concentrations within a mixed forest: Implications for seedling growth. *Ecology* 72(1):12–16
30. Berendse, F., Aerts, R. 1987. Nitrogen-use efficiency: A biologically meaningful definition? *Funct. Ecol.* 1: 293–96
31. Berendse, F., Elberse, W. T. 1990. Competition and nutrient availability in heathland and grassland ecosystems. In *Perspectives on Plant Competition*, ed. J. B. Grace, D. Tilman, pp. 93–116. San Diego: Academic Press. 484 pp.
32. Berg, B., McClaugherty, C. 1987. Nitrogen release from litter in relation to the disappearance of lignin. *Biogeochemistry* 4:219–24
33. Berg, B., McClaugherty, C. 1989. Nitrogen and phosphorus release from decomposing litter in relation to lignin. *Can. J. Bot.* 67:1148–56
34. Berg, B., Staaf, H. 1981. Leaching, accumulation, and release of nitrogen in decomposing forest litter. In *Terrestrial Nitrogen Cycles*, ed. F. E. Clark, T. Rosswall, pp. 163–78. Stockholm: Ecol. Bull.
35. Berry, J., Björkman, O. 1980. Photosynthetic response and adaptation to temperature in higher plants. *Annu. Rev. Plant Physiol.* 31:491–543
36. Bethlenfalvay, G. J., Abu-Shakara, S. S., Phillips, D. A. 1978. Interdependence of nitrogen nutrition and photosynthesis in *Pisum sativum* L. I. Effect of combined nitrogen on symbiotic nitrogen fixation and photosynthesis. *Plant Physiol.* 62:127–30
37. Birk, E. M., Vitousek, P. M. 1986. Nitrogen availability and nitrogen use efficiency in loblollo pine stands. *Ecology* 67:69–79
38. Bloom, A. J., Chapin, F. S. III, Mooney, H. A. 1985. Resource limitation in plants—an economic analogy. *Annu. Rev. Ecol. Syst.* 16:363–92
39. Bolin, B. 1986. How much CO_2 will remain in the atmosphere? In *The Greenhouse Effect, Climate Change, and Ecosystems, SCOPE 29*, ed. B. Bolin, B. Doos, J. Jager, R. Warrick, pp. 93–155. London: Wiley. 541 pp.
40. Botkin, D. B., Simpson, L. G. 1990. Biomass of North American boreal forest: a step toward accurate global measures. *Biogeochemistry* 9:161–74
41. Bowden, R. D., Melillo, J. M., Steudler, P. A. 1991. Effects of nitrogen fertilizer on annual nitrous oxide fluxes from temperate forest soils in the Northeastern United States. *J. Geophys. Res.* 96:9321–28
42. Bowes, G. 1991. Growth at elevated CO_2: photosynthetic responses mediated through RUBISCO: Commissioned review. *Plant Cell Environ.* 14:795–806
43. Bradshaw, A. D., McNeilly, T. 1991. Evolutionary response to global climate change. *Ann. Bot.* 67(Suppl.):5–14
44. Brouwer, R. 1962. Nutritive influences on the distribution of dry matter in the plant. *Neth. J. Agric. Sci.* 10:399–408
45. Brown, S., Gillespie, A. J. R., Lugo, A. E. 1991. Biomass of tropical forests in south and southeast Asia. *Can. J. For. Res.* 21:111–17
46. Bryant, J. P., Chapin, F. S. III, Klein, D. R. 1983. Carbon/nutrient balance of boreal plants in relation to vertebrate herbivory. *Oikos* 40:357–68
47. Bryant, J. P., Provenza, F. D., Pastor, J., Reichardt, P. B., Clausen, T. P., du Toit, J. T. 1991. Interactions between woody plants and browsing mammals

mediated by secondary metabolites. *Annu. Rev. Ecol. Syst.* 22:431–46
48. Bunce, J. A. 1990. Short- and long-term inhibition of respiratory carbon dioxide efflux by elevated carbon dioxide. *Ann. Bot.* 65:637–42
49. Burke, I. C., Yonker, C. M., Parton, W. J., Cole, C. V., Flach, K., et al. 1989. Texture, climate, and cultivation effects on soil organic matter content in U. S. grassland soils. *Soil Sci. Soc. Am. J.* 53:800–5
50. Caldwell, M. M., Richards, J. H. 1986. Competing root systems: morphology and models of absorption. See Ref. 100, pp. 251–73
51. Caldwell, M. M., Robberecht, R., Nowak, R. S. 1982. Differential photosynthetic inhibition by ultraviolet radiation in species from the arctic/alpine life zone. *Arctic Alp. Res.* 14:195–202
52. Caldwell, M. M., Teramura, A. H., Tevini, M. 1989. The changing solar ultraviolet climate and the ecological consequences for higher plants. *TREE* 4:363–67
53. Cave, G., Tolley, L. C., Strain, B. R. 1981. Effect of carbon dioxide enrichment on chlorophyll content, starch content and starch grain structure in *Trifolium subterraneum* leaves. *Physiol. Plant.* 51:171–74
54. Chapin, F. S. III. 1980. The mineral nutrition of wild plants. *Annu. Rev. Ecol. Syst.* 11:233–60
55. Chapin, F. S. III. 1991. Effects of multiple environmental stresses on nutrient availability and use. See Ref. 174, pp. 67–88
56. Chapin, F. S. III. 1991. Integrated responses of plants to stress. *BioScience* 41:29–36
57. Chapin, F. S. III. 1993. The functional role of growth forms in ecosystem and global processes. See Ref. 80. In press
58. Chapin, F. S. III, Bloom, A. J., Field, C. B., Waring, R. H. 1987. Plant responses to multiple environmental factors. *BioScience* 37(1):49–57
59. Chapin, F. S. III, Kedrowski, R. A. 1983. Seasonal changes in nitrogen and phosphorus fractions and autumn retranslocation in evergreen and deciduous taiga trees. *Ecology* 64:376–91
60. Chapin, F. S. III, McNaughton, S. J. 1989. Lack of compensatory growth under phosphorus deficiency in grazing adapted grasses from the Serengeti Plains. *Oecologia* 79:551–57
61. Chapin, F. S. III, Schulze, E.-D., Mooney, H. A. 1990. The ecology and economics of storage in plants. *Annu. Rev. Ecol. Syst.* 21:423–48
62. Chapin, F. S. III, Vitousek, P. M., Van Cleve, K. 1986. The nature of nutrient limitation in plant communities. *Am. Nat.* 127:48–58
63. Chiariello, N. R., Gulmon, S. L. 1991. Stress effects on plant reproduction. See Ref. 174, pp. 162–88
64. Clarkson, D. T. 1985. Factors affecting mineral nutrient acquisition by plants. *Annu. Rev. Plant Physiol.* 36:77–115
65. Clarkson, D. T., Hanson, J. B. 1980. The mineral nutrition of higher plants. *Annu. Rev. Plant Physiol.* 31:239–98
66. Coley, P. D., Bryant, J. P., Chapin, F. S. III. 1985. Resource availability and plant anti-herbivore defense. *Science* 230:895–99
67. Collatz, G. J., Ball, J. T., Grivet, C., Berry, J. A. 1991. Physiological and environmental regulation of stomatal conductance, photosynthesis and transpiration: a model that includes a laminar boundary layer. *Agric. For. Meteorol.* 54:107–36
68. Cure, J. D., Acock, B. 1986. Crop responses to CO_2 doubling: a literature survey. *Agric. For. Meteorol.* 38:127–45
69. Curtis, P. S., Drake, B. G., Leadley, P. W., Arp, W. J., Whigham, D. F. 1989. Growth and senescence in plant communities exposed to elevated CO_2 concentrations on an estuarine marsh. *Oecologia* 78(1):20–26
70. Curtis, P. S., Drake, B. G., Whigham, D. F. 1989. Nitrogen and carbon dynamics in C_3 and C_4 estuarine marsh plants grown under elevated CO_2 in situ. *Oecologia* 78(3):297–301
71. Davidson, E. A., Matson, P. A., Vitousek, P. M., Riley, R., Dunkin, K., et al. 1992. Process regulation of soil emissions of NO and N_2O in a seasonally dry tropical forest. *Ecology*. In press
72. Davidson, R. L. 1969. Effect of root/leaf temperature differentials on root/shoot ratios in some pasture grasses and clover. *Ann. Bot.* 67:1007–10
73. del Arco, J. M., Escudero, A., Vega Garrido, M. 1991. Effects of site characteristics on nitrogen retranslocation from senescing leaves. *Ecology* 72: 701–8
74. Dickenson, R. E. 1991. Global change and terrestrial hydrology—A review. *Tellus* 43AB:176–81
75. Drake, B. G., Leadley, P. W. 1991. Canopy photosynthesis of crops and native plants exposed to long-term el-

evated CO_2: Commissioned review. *Plant Cell Environ.* 14:853–60

76. Eamus, D. 1991. The interaction of rising CO_2 and temperatures with water use efficiency: Commissioned review. *Plant Cell Environ.* 14:843–52
77. Eamus, D., Jarvis, P. G. 1989. The direct effects of increase in the global atmospheric CO_2 concentration on natural and commercial temperate trees and forests. *Adv. Ecol. Res.* 19:1–55
78. Ehleringer, J., Björkman, O. 1977. Quantum yields for CO_2 uptake in C_3 and C_4 plants. *Plant Physiol.* 59:86–90
79. Ehleringer, J. R., Sage, R. F., Flanagan, L. B., Pearcy, R. W. 1991. Climate change and the evolution of C_4 photosynthesis under low atmospheric CO_2. *TREE* 6:95–99
80. Ehleringer, J. R., Field, C. B., eds. 1993. *Scaling Physiological Processes: Leaf to Globe.* San Diego: Academic Press. In press
81. Eichner, M. J. 1990. Nitrous oxide emissions from fertilized soils: summary of the available data. *J. Environ. Qual.* 19:272–80
82. Eisele, K. A., Schimel, D. S., Kapustka, L. A., Parton, W. J. 1989. Effects of available P and N:P ratios on nonsymbiotic dinitrogen fixation in tallgrass prairie soils. *Oecologia* 79:471–74
83. Enting, I. G., Mansbridge, J. V. 1991. Latitudinal distribution of sources and sinks of CO_2: results of an inversion study. *Tellus* 43B:156–70
84. Ericcson, T., Kahr, M. 1992. Growth and nutrition of birch seedlings in relation to potassium supply rate. *Trees.* In press
85. Evans, J. R. 1989. Photosynthesis and nitrogen relationships in leaves of C_3 plants. *Oecologia* 78(1):9–19
86. Fajer, E. D. 1989. The effects of enriched CO_2 on plant-insect herbivore interactions: growth responses of larvae of the specialist butterfly *Junconia coenia* (Lepidoptera: Nymphalidae). *Oecologia* 81:514–20
87. Fall, R. 1991. Isoprene emission from plants: summary and discussion. See Ref. 217, pp. 209–16
88. Farrar, J. F., Williams, M. L. 1991. The effects of increased atmospheric carbon dioxide and temperature on carbon partitioning, source-sink relations and respiration: commissioned review. *Plant Cell Environ.* 14:819–30
89. Field, C., Merino, J., Mooney, H. A. 1983. Compromises between water-use efficiency and nitrogen-use efficiency in five species of California evergreens. *Oecologia* 60:384–89
90. Field, C., Mooney, H. A. 1986. The photosynthesis-nitrogen relationship in wild plants. See Ref. 100, pp. 25–55
91. Field, C. B. 1991. Ecological scaling of carbon gain to stress and resource availability. See Ref. 174, pp. 35–65
92. Firestone, M. K., Davidson, E. A. 1989. Microbiological basis of NO and N_2O production and consumption in soil. See Ref. 14, pp. 7–24
93. Flanagan, P. W., Van Cleve, K. 1983. Nutrient cycling in relation to decomposition and organic matter quality in taiga ecosystems. *Can. J. For. Res.* 13:795–817
94. Folland, C. K., Karl, T. R., Vinnikov, K. Y. A. 1990. Observed climate variations and change. See Ref. 112, pp. 195–238
95. Fung, I. Y. 1992. Models of oceanic and terrestrial sinks of anthropogenic CO_2: A review of the contemporaty carbon cycle. In *The Biogeochemistry of Global Change: Radiative Trace Gases,* ed. R. S. Oremland. New York: Chapman & Hall. In press
96. Fung, I. Y., Tucker, C. J., Prentice, K. C. 1987. Application of advanced very high resolution radiometer vegetation index to study atmosphere-biosphere exchange of CO_2. *J. Geophys. Res.* 92:2999–3015
97. Galbally, I. E. 1989. Factors controlling NO_x emissions from soils. See Ref. 14, pp. 23–37
98. Garcia, R., Kanemasu, E. T., Blad, B. L., Bauer, A., Hatfield, J. L., et al. 1988. Interception and use efficiency of light in winter wheat under different nitrogen regimes. *Agric. For. Meteorol.* 44:175–86
99. Garnier, E. 1991. Resource capture, biomass allocation and growth in herbaceous plants. *TREE* 6:26–31
100. Givnish, T. J., ed. 1986. *On the Economy of Plant Form and Function.* Cambridge: Cambridge Univ. Press. 717 pp.
101. Goudriaan, J., van Laar, H. H., van Keulen, H., Louwerse, W. 1984. Simulation of the effect of increased atmospheric CO_2 on assimilation and transpiration of a closed crop canopy. *Wiss. Z. Humboldt.-Univ. Berlin Math.-Naturwiss. Reihe* 23:3252–3356
102. Graumlich, L. J. 1991. Subalpine tree growth, climate, and increasing CO_2: an assessment of recent growth trends. *Ecology* 72(1):1–11

103. Grime, J. P. 1977. Evidence for the existence of three primary strategies in plants and its relevance to ecological and evolutionary theory. *Am. Nat.* 111: 1169–94
104. Grime, J. P. 1979. *Plant Strategies and Vegetation Processes*. Chichester: Wiley. 222 pp.
105. Grime, J. P., Campbell, B. D. 1991. Growth rate, habitat productivity, and plant strategy as predictors of stress response. See Ref. 174, pp. 143–61
106. Gulmon, S. L., Mooney, H. A. 1986. Costs of defense on plant productivity. See Ref. 100, pp. 681–98
107. Hall, C. S., Uhlis, J. 1991. Refining estimates of carbon release from tropical land-use change. *Can. J. For. Res.* 21:118–31
108. Hilbert, D. W. 1990. Optimization of plant root:shoot ratios and internal nitrogen concentration. *Ann. Bot.* 66:91–99
109. Hobbie, S., Jensen, D., Chapin, F. S. III. 1991. Global changes in resources and disturbance: Predictions for Biodiversity. In *Ecosystem Function and Biodiversity*, ed. E.-D. Schulze, H. A. Mooney. Berlin: Springer-Verlag. In press
110. Holland, E. A., Coleman, D. C. 1987. Litter placement effects on microbial and organic matter dynamics in an agroecosystem. *Ecology* 68:425–33
111. Horner, J. D., Gosz, J. R., Cates, R. G. 1988. The role of carbon-based plant secondary metabolites in decomposition in terrestrial ecosystems. *Am. Nat.* 132:869–83
112. Houghton, J. T., Jenkins, G. J., Ephraums, J. J., eds. 1990. *Climate Change: The IPCC Scientific Assessment*. Cambridge: Cambridge Univ. Press. 365 pp.
113. Houghton, R. A. 1986. Estimating changes in carbon content of terrestrial ecosystems from historical data. In *The Changing Carbon Cycle: A Global Analysis*, ed. J. R. Trabalka, D. E. Reichle, pp. 175–93. New York: Springer-Verlag. 592 pp.
114. Houghton, R. A. 1987. Biotic changes consistent with the increased seasonal amplitude of atmospheric CO_2 concentrations. *J. Geophys. Res.* 92:4223–30
115. Houghton, R. A., Boone, R. D., Fruci, J. R., Hobbie, J. E., Melillo, J. M., et al. 1987. The flux of carbon from terrestrial ecosystems to the atmosphere in 1980 due to changes in land use: geographic distribution of the global flux. *Tellus* 39B:122–39
116. Houghton, R. A., Boone, R. D., Melillo, J. M., Palm, C. A., Woodwell, G. M., et al. 1985. Net flux of carbon dioxide from tropical forests in 1980. *Nature* 316:617–20
117. Houghton, R. A., Hobbie, J. E., Melillo, J. M., Moore, B., Peterson, B. J., et al. 1983. Changes in the carbon content of terrestrial biota and soils between 1860 and 1980: A net release of CO_2 to the atmosphere. *Ecol. Monogr.* 53: 235–62
118. Hunt, H. W., Ingham, E. R., Coleman, D. C., Elliott, E. T., Reid, C. P. P. 1988. Nitrogen limitation of production and decomposition in prairies, mountain meadow, and pine forest. *Ecology* 69: 1009–16
119. Hunt, R., Hand, D. W., Hannah, M. A., Neal, A. M. 1991. Response to CO_2 enrichment in 27 herbaceous species. *Funct. Ecol.* 5:410–21
120. Idso, S. B. 1988. Three phases of plant response to atmospheric CO_2 enrichment. *Plant Physiol.* 87:5–7
121. Idso, S. B., Kimball, B. A., Allen, S. G. 1991. CO_2 enrichment of sour orange trees: 2.5 years into a long-term experiment. *Plant Cell Environ.* 14: 351–53
122. Ingestad, T. 1982. Relative addition rate and external concentration; driving variables used in plant nutrition research. *Plant Cell Environ.* 5:443–53
123. Ingestad, T., Ågren, G. I. 1991. The influence of plant nutrition on biomass allocation. *Ecol. Appl.* 2(1):168–74
124. Jarvis, P. G. 1976. The interpretation of the variations in leaf water potential and stomatal conductance found in canopies in the field. *Philos. Trans. R. Soc. London Ser. B* 273:593–610
125. Jarvis, P. G., Dewar, R. C. 1992. Forests in the global carbon balance from stand to region. See Ref. 80. In press
126. Jarvis, P. G., McNaughton, K. G. 1986. Stomatal control of transpiration: Scaling up from leaf to region. *Adv. Ecol. Res.* 15:1–49
127. Juuti, S., Arey, J., Akinson, R. 1990. Monoterpene emission rate measurements from a Monterey pine. *J. Geophys. Res.* 95(?):7515–19
127a. Kauppi, P. E., Mielikäinen, K., Kuusela, K. 1992. Biomass and carbon budget of European forest, 1971 to 1990. *Science* 256:70–74
128. Keeling, C. D., Bacastrow, R. B., Carter, A. F., Piper, S. C., Whorf, T. P., et al. 1989. A three-dimensional model of atmospheric CO_2 transport based on observed winds: 1. Analysis of

observational data. In *Aspects of Climate Variability in the Pacific and the Western Americas,* ed. D. H. Peterson, pp. 165–236. Washington, DC: Am. Geophys. Union. 445 pp.

129. Kingsolver, J. G., Kareiva, P. M., Huey, R. B., eds. 1992. *Biotic Interactions and Global Change*. Sunderland, Mass: Sinauer. In press
130. Koch, G., Schulze, E.-D., Percival, F., Mooney, H. A., Chu, C. 1988. The nitrogen balance of *Raphanus sativus raphanistrum* plants. II. Growth, nitrogen redistribution and photosynthesis under NO_3- deprivation. *Plant Cell Environ.* 11:755–67
131. Kohlmaier, G. H., Siré, E., Janecec, A., Keeling, C. D., Piper, S. C., et al. 1989. Modelling the seasonal contribution of a CO_2 fertilization effect of the terrestrial vegetation to the amplitude increase in atmospheric CO_2 at Mauna Loa observatory. *Tellus* 41B: 487–510
132. Koide, R. T. 1991. Tansley Review No. 29: Nutrient supply, nutrient demand and plant response to mycorrhizal infection. *New Phytol.* 117:365–86
133. Kramer, P. J., Kozlowski, T. T. 1979. *Physiology of Woody Plants*. New York: Academic Press. 811 pp.
134. Kraus, E., Wilson, D., Robson, M. J., Pilbeam, C. J. 1990. Respiration: Correlation with growth rate and its quantitative significance for the net assimilation rate and biomass production. See Ref. 136, pp. 187–98
135. Lamb, B., Westberg, H., Allwine, G. 1985. Biogenic hydrocarbon emissions from deciduous and coniferous trees in the United States. *J. Geophys. Res.* 90:2380–90
136. Lambers, H., Cambridge, M. L., Konings, H., Pons, T. L., eds. 1990. *Causes and Consequences of Variation in Growth Rate and Productivity in Higher Plants*. The Hague: SPB Academic. 365 pp.
137. Larigauderie, A., Hilbert, D. W., Oechel, W. C. 1988. Effect of CO_2 enrichment and nitrogen availability on resource acquisition and resource allocation in a grass, *Bromus mollis*. *Oecologia* 77: 544–49
138. Larsson, S., Wiren, A., Ericsson, T. 1986. Effects of light and nutrient stress on leaf phenolic chemistry in *Salix dasyclados* and susceptibility to *Galerucella lineola*. *Oikos* 47:205–10
139. Lauenroth, W. K., Dodd, J. L., Simms, P. L. 1978. The effects of water- and nitrogen-induced stresses on plant community structure in a semiarid grassland. *Oecologia* 36:211–22
140. Lawlor, D. W., Mitchell, R. A. C. 1991. The effects of increasing CO_2 on crop photosynthesis and productivity: Commissioned review. *Plant Cell Environ.* 14:807–18
141. Lechowicz, M. J. 1987. Resource allocation by plants under air pollution stress: implications for plant-pest-pathogen interactions. *Bot. Rev.* 53: 281–300
142. Lerdau, M. T. 1991. Plant function and biogenic terpene emission. See Ref. 217, pp. 121–34
143. Lincoln, D. E., Couvet, D. 1989. The effect of carbon supply on allocation to allelochemicals and caterpillar consumption of peppermint. *Oecologia* 78: 112–14
144. Lindroth, R. L., Kinney, K. K., Platz, C. L. 1992. Responses of forest trees to elevated atmospheric CO_2: Productivity, phytochemistry and insect resistance.*Ecology*.In press
145. Livingston, G. P., Morrissey, L. A. 1991. Methane emission from Alaska arctic tundra in response to climate warming. In *Proc. Int. Conf. on the Role of the Polar Regions in Global Change,* Univ. Alaska, Fairbanks
146. Long, S. P. 1991. Modification of the response of photosynthetic productivity to rising temperature by atmospheric CO_2 concentrations: Has its importance been underestimated? Opinion. *Plant Cell Environ.* 14:729–40
147. Long, S. P., Hutchin, P. R. 1991. Primary production in grasslands and coniferous forests with climate change: An overview. *Ecol. Appl.* 1:139–56
148. Lorio, P. L. Jr. 1986. Growth-differentiation balance: a basis for understanding southern pine beetle-tree interactions. For. Ecol. Manage. 14: 259–73
149. Lynch, M. 1992. Evolution versus extinction in response to environmental change. See Ref. 129. In press
150. Matson, P. A., Hain, F. P., Mawby, W. 1987. Susceptibility indices in loblolly pine vary with silvicultural treatment. *For. Ecol. Manage.* 22:107–18
151. Matson, P. A., Waring, R. H. 1984. Effects of nutrient and light limitation on mountain hemlock: susceptibility to laminated root rot. *Ecology* 65:1517–24
152. Mattson, W. J. 1980. Herbivory in relation to plant nitrogen content. *Annu. Rev. Ecol. Syst.* 11:119–61
153. Mattson, W. J., Haack, R. A. 1987. The role of drought in outbreaks of

plant-eating insects. *BioScience* 37: 110–18
154. McNaughton, K., G. 1989. Regional intertactions between canopies and the atmosphere. In *Plant Canopies: Their Growth, Form and Function*, ed. G. Russell, B. Marshall, P. G. Jarvis, pp. 62–81. Cambridge: Cambridge Univ. Press. 178 pp.
155. McNaughton, K. G., Jarvis, P. G. 1991. Effects of spatial scale on stomatal control of transpiration. *Agric. For. Meteorol.* 54:279–302
156. Meetenmeyer, V. 1978. Climatic regulation of decomposition rates of organic matter in terrestrial ecosystems. In *Environmental Chemistry and Cycling Processes. CONF 760429*, ed. D. C. Adriano, I. L. Brisbin, pp. 779/-89. Springfield, Va: Natl. Tech. Info. Serv. 911 pp.
157. Meetenmeyer, V., Berg, B. 1986. Regional variation in rate of mass loss in *Pinus sylvestris* needle litter in Swedish pine forests as influenced by climate and litter quality. *Scand. J. For. Res.* 1:167–80
158. Melillo, J. M., Aber, J. D., Muratore, J. F. 1982. Nitrogen and lignin control of hardwood leaf litter decomposition dynamics. *Ecology* 63:621–26
159. Melillo, J. M., Callaghan, T. V., Woodward, F. I., Salati, E., Sinha, S. K. 1990. Effects on Ecosystems. See Ref. 112, pp. 283–310
160. Melillo, J. M., Steudler, P. A., Aber, J. D., Bowden, R. D. 1989. Atmospheric deposition and nutrient cycling. See Ref. 14, pp. 263–80
161. Meyer, J., Schneider, B. U., Werk, K. S., Oren, R., Schulze, E. 1988. Performance of two *Picea abies* (L.) Karst. stands at different stages of decline. V. Root tip and ectomycorrhiza development and their relationship to above ground and soil nutrients. *Oecologia* 77:7–13
162. Miller, H. G., Cooper, J. M., Miller, J. D. 1976. Effects of nitrogen supply on nutrients in litterfall and crown leaching in a stand of Corsican pines. *J. Appl. Ecol.* 13:233–48
163. Miller, J. E. 1989. Effects on photosynthesis, carbon allocation, and plant growth associated with air pollutant stress. In *Assessment of Crop Loss from Air Pollutants*, ed. W. W. Heck, O. C. Taylor, D. T. Tingey, pp. 287–314. London: Elsevier Appl. Sci. 552 pp.
164. Monsi, M. 1968. Mathematical models of plant communities. In *Functioning of Terrestrial Ecosystems at the Primary Production Level*, ed. F. Eckardt, pp. 131–49. Paris: UNESCO. 516 pp.
165. Monsi, M., Saeki, T. 1953. Über den Lichtfaktor in den Pflanzengesellschaften und seine Bedeutung für die Stoffproduktion. *Jpn. J. Bot.* 14:22–52
166. Monson, R. K., Fall, R. 1989. Isoprene emission from aspen leaves. The influence of environment and relation to photosynthesis and photorespiration. *Plant Physiol.* 90:267–74
167. Monson, R. K., Guenther, A. B., Fall, R. 1991. Physiological reality in relation to ecosystem- and global-level estimates of isoprene emission. See Ref. 217, pp. 185–207
168. Mooney, H. A. 1972. The carbon balance of plants. *Annu. Rev. Ecol. Syst.* 3:315–46
169. Mooney, H. A., Chiariello, N. R. 1984. The study of plant function: The plant as a balanced system. In *Perspectives on Plant Population Biology*, ed. R. Dirzo, J. Sarukhán, pp. 305–23. Sunderland, Mass: Sinauer. 478 pp.
170. Mooney, H. A., Drake, B. G., Luxmoore, R. J., Oechel, W. C., Pitelka, L. F. 1991. Predicting ecosystem responses to elevated CO_2 concentrations. *BioScience* 41:96–104
171. Mooney, H. A., Dunn, E. L. 1970. Convergent evolution of mediterranean-climate evergreen sclerophyll shrubs. *Evolution* 24:292–303
172. Mooney, H. A., Vitousek, P. M., Matson, P. A. 1987. Exchange of materials between terrestrial ecosystems and the atmosphere. *Science* 238:926–32
173. Mooney, H. A., Winner, W. E. 1991. Partitioning response of plants to stress. See Ref. 174, pp. 129–41
174. Mooney, H. A., Winner, W. E., Pell, E. J., eds. 1991. *Response of Plants to Multiple Stresses*. San Diego: Academic Press. 422 pp.
175. Morison, J. I. L. 1987. Intercellular CO_2 concentration and stomatal response to CO_2. In *Stomatal Function*, ed. E. Zeiger, G. D. Farquhar, I. R. Cowan, pp. 229–52. Stanford, Calif: Stanford Univ. Press. 503 pp.
176. Morris, J. T. 1991. Effects of nitrogen loading on wetland ecosystems with particular reference to atmospheric deposition. *Annu. Rev. Ecol. Syst.* 22: 257–80
177. Morrissey, L. A., Zobel, D. B., Livingston, G. P. 1992. Significance of stomatal control on methane release from Carex-dominated wetlands. *Chemosphere*. In press
178. Mulligan, D. R. 1989. Leaf phosphorus and nitrogen concentrations and net

photosynthesis in *Eucalyptus* seedlings. *Tree Physiol.* 5:149–57
179. Munchow, R. C. 1985. An analysis of the effects of water deficits on grain legumes grown in a semi-arid tropical environment in terms of radiation interception and its efficiency of use. *Field Crops Res.* 11:309–23
180. Muzika, R., Pregitzer, K., Hanover, J. 1989. Changes in terpene production following nitrogen fertilization of grand fir seedlings. *Oecologia* 80:485–89
181. Nambiar, E. K. S., Fife, D. N. 1987. Growth and nutrient retranslocation in needles of radiata pine in relation to nitrogen supply. *Ann. Bot.* 60:147–56
182. Oechel, W. C., Strain, B. R. 1985. Native species responses to increased atmospheric carbon dioxide concentrations. See Ref. 230, pp. 117–54
183. Oren, R., Schulze, E.-D., Werk, K. S., Meyer, J. 1988. Performance of two *Picea abies* (L.) Karst. stands at different stages of decline. VII. Nutrient relations and growth. *Oecologia* 77: 163–73
184. Pastor, J., Post, W. M. 1988. Responses of northern forests to CO_2-induced climate change. *Nature* 334:55–58
185. Pearcy, R. W., Ehleringer, J. 1984. Comparative ecophysiology of C_3 and C_4 plants. *Plant Cell Environ.* 7:1–13
186. Penning de Vries, F. W. T. 1975. The cost of maintenance processes in plant cells. *Ann. Bot.* 39:77–92
187. Penning de Vries, F. W. T., Brunsting, A. H. M., van Laar, H. H. 1974. Products, requirements, and efficiency of biosynthesis: A quantitative approach. *J. Theor. Biol.* 45:339–77
188. Peterson, B. J., Melillo, J. M. 1985. The potential storage of carbon caused by eutrophication of the biosphere. *Tellus* 37B:117–27
189. Poorter, H. 1990. Interspecific variation in relative growth rate: On ecological causes and physiological consequences. See Ref. 136, pp. 45–68
190. Poorter, H., Remkes, C. 1990. Leaf area ratio and net assimilation rate of 24 wild species differing in relative growth rate. *Oecologia* 83:553–59
191. Pugnaire, F. I., Chapin, F. S. III. 1992. Environmental and physiological factors governing nutrient resorption efficiency in barley. *Oecologia* 90: 120–26
192. Raich, J. W., Rastetter, E. B., Melillo, J. M., Kicklighter, D. W., Steudler, P. A., et al. 1991. Potential net primary production in South America. *Ecol. Appl.* 1(4):399–429
193. Raich, J. W., Schlesinger, W. H. 1992. The global carbon dioxide flux in soil respiration and its relationship to climate. *Tellus.* 44B:81–99
194. Reich, P. B. 1987. Quantifying plant response to ozone: A unifying theory. *Tree Physiol.* 3:63–91
195. Reich, P. B., Amundson, R. G. 1985. Ambient levels of ozone reduce net photosynthesis in tree and crop species. *Science* 230:566–70
196. Reich, P. B., Schoettle, A. W. 1988. Role of phosphorus and nitrogen in photosynthetic and whole plant carbon gain and nutrient use efficiency in eastern white pine. *Oecologia* 77:25–33
197. Reich, P. B., Walters, M. B., Ellsworth, D. S. 1992. Leaf lifespan in relation to leaf, plant and stand characteristics among diverse ecosystems. *Ecology.* In press
198. Reich, P. B., Walters, M. B., Tabone, T. J. 1989. Response of Ulmus americana seedlings to varying nitrogen and water status. 2. Water and nitrogen use efficiency. *Tree Physiol.* 5:173–84
199. Reichardt, P. B., Chapin, F. S. III, Bryant, J. P., Mattes, B. R., Clausen, T. P. 1991. Carbon/nutrient balance as a predictor of plant defense in Alaskan balsam poplar: potential importance of metabolite turnover. *Oecologia* 88: 401–6
200. Reuss, J. O., Johnson, D. W. 1986. *Acid Deposition and the Acidification of Soils and Waters.* New York: Springer-Verlag. 119 pp.
201. Robberecht, R., Caldwell, M. M. 1980. Leaf ultraviolet optical properties along a latitudinal gradient in the arctic-alpine life zone. *Ecology* 61:612–19
202. Running, S. W., Coughlan, J. C. 1988. A general model of forest ecosystem processes for regional applications. I. Hydrologic balance, canopy gas exchange and primary production processes. *Ecol. Model.* 42:125–54
203. Russell, G., Jarvis, P. G., Monteith, J. L. 1989. Absorption of radiation by canopies and stand growth. In *Plant Canopies: Their Growth, Form and Function,* ed. G. Russell, B. Marshall, P. G. Jarvis, pp. 21–39. Cambridge: Cambridge Univ. Press. 178 pp.
204. Ryan, M. G. 1991. Effects of climate change on plant respiration. *Ecol. Appl.* 1(2):157–67
205. Sage, R. F., Sharkey, T. D., Seemann, J. R. 1989. Acclimation of photosynthesis to elevated CO_2 in five C_3 species. *Plant Physiol.* 89:590–96
206. Sanadze, G. A. 1991. Isoprene effect-light dependent emission of isoprene

by green parts of plants. See Ref. 217, pp. 135–52
207. Schimel, D. S., Kittell, T. G. F., Knapp, A. K., Seastedt, T. R., Parton, W. J., et al. 1991. Physiological interactions along resource gradients in a tallgrass prairie. *Ecology* 72:672–84
208. Schimel, D. S., Parton, W. J., Kittel, T. G. F., Ojima, D. S., Cole, C. V. 1990. Grassland biogeochemistry: Links to atmospheric processes. *Clim. Change* 17:13–25
209. Schlesinger, W. H. 1990. Evidence from chronosequence studies for a low carbon-storage potential of soils. *Nature* 348:232–34
210. Schlesinger, W. H. 1991. *Biogeochemistry: An Analysis of Global Change*. San Diego: Academic Press. 443 pp.
211. Schulze, E.-D., Lange, O. L., Oren, R., eds. 1988. *Air Pollution and Forest Decline: A Study of Spruce on Acid Soils*. Berlin: Springer-Verlag
212. Schulze, E.-D. 1989. Air pollution and forest decline in a spruce (*Picea abies*) forest. *Science* 244:776–83
213. Schulze, E.-D., Chapin, F. S. III. 1987. Plant specialization to environments of different resource availability. In *Potentials and Limitations in Ecosystem Analysis*, ed. E.-D. Schulze, H. Zwolfer, pp. 120–48. Berlin: Springer-Verlag. 435 pp.
214. Schütz, H., Schroder, P., Rennenberg, H. 1991. Role of plants in regulating the methane flux to the atmosphere. See Ref. 217, pp. 29–63
215. Sellers, P. J. 1985. Canopy reflectance, photosynthesis and transpiration. *Int. J. Remote Sens.* 6:1335–72
216. Sellers, P. J., Berry, J. A., Collatz, G. J., Field, C. B., Hall, F. G. 1992. Canopy reflectance, photosynthesis and transpiration, III. A reanalysis using enzyme kinetics-electron transport models of leaf physiology. *Remote Sens. Environ.* In press
217. Sharkey, T. D., Holland, E. A., Mooney, H. A., eds. 1991. *Trace Gas Emissions by Plants*. San Diego: Academic Press. 365 pp.
218. Sharkey, T. D., Loreto, F., Delwiche, C. F. 1991. The biochemistry of isoprene emission from leaves during photosynthesis. See Ref. 217, pp. 153–84
219. Shaver, G. R. 1981. Mineral nutrition and leaf longevity in an evergreen shrub, *Ledum palustre* ssp. *decumbens*. *Oecologia* 49:362–65
220. Shaver, G. R., Chapin, F. S. III, Gartner, B. L. 1986. Factors limiting seasonal growth and peak biomass accumulation in *Eriophorum vaginatum* in Alaskan tussock tundra. *J. Ecol.* 74:257–78
221. Shriner, D. S., Johnston, J. W. Jr. 1985. Acid rain interactions with leaf surfaces: a review. In *Acid Deposition: Environmental, Economic, and Policy Issues*, ed. D. D. Adams, W. P. Page, pp. 241–53. New York: Plenum. 560 pp.
222. Shugart, H. H. 1992. Global models of change based on species and/or functional groups. *Annu. Rev. Ecol. Syst.* 23:15–38
223. Shukla, J., Mintz, Y. 1982. Influence of land-surface evapotranspiration on the earth's climate. *Science* 215:1498–1501
224. Shukla, J., Nobre, C., Sellers, P. 1990. Amazon deforestation and climate change. *Science* 247:1322–25
225. Shuttleworth, W. J., Wallace, J. S. 1985. Evaporation from sparce crops—an energy combination theory. *Q. J. R. Meteorol. Soc.* 111:839–55
226. Siegenthaler, U., Oeschger, H. 1987. Biospheric CO_2 emissions during the past 200 years reconstructed by deconvolution of ice core data. *Tellus* 39B: 140–54
227. Sionit, N., Patterson, D. T. 1984. Responses of C_4 grasses to atmospheric CO_2 enrichment. I. Effect of irradiance. *Oecologia* 65:30–34
228. Stachurski, A., Zimka, J. R. 1975. Methods of studying forest ecosystems: leaf area, leaf production and withdrawal of nutrients from leaves of trees. *Ekol. Pol.* 23:637–48
229. Stitt, M. 1991. Rising CO_2 levels and their potential significance for carbon flow in photosynthetic cells: Commissioned review. *Plant Cell Environ.* 14: 741–62
230. Strain, B. R., Cure, J. A., eds. 1985. *Direct Effects of Increasing Carbon Dioxide on Vegetation*. Washington, DC :US Dep. Energy. 286 pp.
231. Deleted in proof
232. Szaniawski, R. K., Kielkiewicz, M. 1982. Maintenance and growth respiration in shoots and roots of sunflower plants grown at different temperatures. *Physiol. Plant.* 54:500–4
233. Tamm, C. O. 1990. *Nitrogen in Terrestrial Ecosystems: Questions of Productivity, Vegetational Change, and Ecological Stability*. Berlin: Springer-Verlag. 116 pp.
234. Tans, P. P., Fung, I. Y., Takahashi, T. 1990. Observational constraints on the global CO_2 budget. *Science* 247:1431–38
235. Teeri, J. A., Stowe, L. G. 1976. Climatic

patterns and the distribution of C_4 grasses. *Oecologia* 23:1–12
236. Thornley, J. H. M., Fowler, D., Cannell, G. R. 1991. Terrestrial carbon storage resulting from CO_2 and nitrogen fertilization in temperate grasslands. *Plant Cell Environ.* 14:1007–11
237. Tilman, D. 1988. *Plant Strategies and the Dynamics and Function of Plant Communities, 26*. Princeton: Princeton Univ. Press. 360 pp.
238. Tilman, D. 1990. Constraints and tradeoffs: toward a predictive theory of competition and succession. *Oikos* 58: 3–15
239. Tingey, D. T., Manning, M., Grothaus, L. C., Burns, W. F. 1979. The influence of light and temperature on isoprene emission rates from live oak. *Physiol. Plant.* 47:112–18
240. Tingey, D. T., Turner, D. P., Weber, J. A. 1991. Factors controlling the emissions of monoterpenes and other volatile organic compounds. See Ref. 217, pp. 93–119
241. Tissue, D. T., Oechel, W. C. 1987. Response of *Eriophorum vaginatum* to elevated CO_2 and temperature in the Alaskan tussock tundra.*Ecology* 68(2): 401–10
242. Townsend, A. R., Vitousek, P. M., Holland, E. A. 1992. Tropical soils dominate the short-term carbon cycle feedbacks to atmospheric carbon dioxide. *Clim. Change*. In press
243. Tukey, H. B. Jr. 1970. The leaching of substances from plants. *Annu. Rev. Plant Physiol.* 21:305–24
244. Tuomi, J., Niemela, P., Haukioja, E., Neuvonen, S. 1984. Nutrient stress: an explanation for plant anti-herbivore responses to defoliation. *Oecologia* 61: 208–10
245. Turner, J. 1977. Effect of nitrogen availability on nitrogen cycling in a Douglas-fir stand. *For. Sci.* 23:307–16
246. Vitousek, P. 1982. Nutrient cycling and nutrient use efficiency. *Am. Nat.* 119:553–72
247. Vitousek, P. M., Howarth, R. W. 1991. Nitrogen limitation on land and in the sea: How can it occur? *Biogeochemistry* 13:87–115
248. Vitousek, P. M., Gosz, J. R., Grier, C. C., Melillo, J. M., Reiners, W. A. 1982. A comparative analysis of potential nitrification and nitrate mobility in forest ecosystems. *Ecol. Monogr.* 52:155–77
249. Watson, R. T., Rhodhe, H., Oeschger, H., Siegenthaler, U. 1990. Greenhouse gases and aerosols. See Ref. 112, pp. 1–40
250. White, C. S. 1988. Nitrification inhibition by monoterpenoids: Theoretical mode of action based on molecular structures. *Ecology* 69:1631–33
251. Wilson, J. B. 1988. A review of evidence on the control of shoot:root ratio, in relation to models. *Ann. Bot.* 61:433–49
252. Winner, W. E., Mooney, H. A., eds. 1985. *Sulfur Dioxide and Vegetation*. Stanford, Calif: Stanford Univ. Press. 597 pp.
253. Wong, S. C., Cowan, I. R., Farquhar, G. D. 1979. Stomatal conductance correlates with photosynthetic capacity. *Nature* 282:424–26
254. Woodrow, I. W., Ball, J. T., Berry, J. A. 1987. A general expression for the control of the rate of photosynthetic CO_2 fixation by stomata, the boundary layer and radiation exchange. In *Progress in Photosynthesis Research*, ed. J. Biggins, 4:225–28. The Netherlands: Martinus Nijhoff
255. Woodward, F. I. 1990. Global change: Translating plant ecophysiological responses to ecosystems. *TREE* 5:308–11
256. Wright, J. L., Lemon, E. R. 1966. Photosynthesis under field conditions. VIII. Analysis of windspeed fluctuation data to evaluate turbulent exchange within a corn crop. *Agron. J.* 58:265–68

Annu. Rev. Ecol. Syst. 1992. 23:237–261

NATURAL HYBRIDIZATION AS AN EVOLUTIONARY PROCESS

Michael L. Arnold
Department of Genetics, University of Georgia, Athens, Georgia, 30602

KEYWORDS: introgression, hybrid speciation, natural selection, adaptation

INTRODUCTION

Studies of natural hybridization have generally addressed evolutionary questions using one of the three following frameworks: (i) taxonomy or systematics (57, 59, 73, 79, 111, 172); (ii) mechanisms of reproductive isolation and speciation (21, 23, 24, 26, 32, 34, 47, 56, 60, 82, 91, 110, 132, 134, 136, 141, 143, 157, 158, 167, 169, 173); or (iii) natural hybridization as a fundamental evolutionary process that produces consequences that are significant in their own right (1–4, 6, 31, 62, 77, 90, 100, 116, 122, 128, 130, 162, 166, 174). This review emphasizes the conceptual and empirical basis for the last of these contexts. The evolutionary significance of natural hybridization is thus addressed through two general questions: "To what extent has this process been involved in the evolutionary history of plant and animal species?" and "How does the fitness of hybrid individuals compare with that of the parental taxa?"

Natural hybridization and introgression (the transfer of genetic material between the hybridizing taxa through backcrossing; 2, 3) have been ascribed varying levels of importance with regard to the genetic makeup of species and the evolutionary history of species complexes (2, 4, 6, 24, 61, 62, 72, 86, 89, 90, 97, 100, 116, 122, 124, 136, 156, 170, 172). In the extreme, introgression may lead to either the merging of the hybridizing forms (61, 178) or the reinforcement of reproductive barriers through selection for assortative (conspecific) mating (45, 76; but see 37, 136). Another potential consequence is the production of more or less fit introgressed genotypes (107; 87), allowing the expansion of the introgressed form into a novel habitat (89; but see 33, 55). Hybrid individuals may also act as a "hybrid sink" to which pest species are preferentially attracted, thus limiting the ability of pest species

0066-4162/92/1120-0237$02.00

to adapt to parental individuals (170). Finally, natural hybridization may lead to the formation of hybrid species (9, 41a, 120). These hybrid derivatives may involve polyploidy (58, 63, 94, 151, 158a, 180), parthenogenesis, hybridogenesis (36, 75, 105, 161, 169), or homoploid (diploid) speciation (9, 62, 127).

Hypotheses pertaining to the evolutionary significance of natural hybridization (2, 62, 90, 156, 163, 164) raise questions concerning both the phylogenetic distribution and the adaptive nature of this process. If natural hybridization has been a widespread evolutionary phenomenon (156), it should be possible to detect the genetic consequences in contemporary species. One prediction is that extant taxa of hybrid origin would possess a combination of genetic markers that are diagnostic for two or more parental taxa. Such a finding would support the hypothesis of past reticulate evolution (natural hybridization between different evolutionary lineages; 127, 128) and might be indicated by an incongruence between the phylogenetic histories of different genetic markers (124, 128, 168, 171). Furthermore, examinations of genetic variation in contemporary hybrid populations may lead to the definition of processes that have given rise to the pattern of reticulation. In particular, analyses of present-day gene flow between hybridizing taxa have identified both localized and dispersed introgression of various molecular, biochemical, chromosomal, and morphological characteristics (20, 22, 25, 35, 38, 42, 49, 54, 66, 67, 74, 83, 93, 109, 116, 127, 133, 135, 137, 155, 165, 175). Frequently, estimates of the magnitude of gene flow between the hybridizing taxa vary for different markers, and the barriers to reproduction have been referred to as "semipermeable" (69, 78). This aspect of introgression would seem of particular relevance in plant systems where the decoupling of male and female markers is due to the two avenues for gene flow, namely, seed ("female") and pollen ("male") dispersal (10, 11, 17).

Studies of the micro- and macroevolutionary consequences of natural hybridization can suggest the extent to which natural hybridization has been involved in the evolution of extant and extinct populations and taxa. However, neither phylogenetic analysis nor the pattern of genetic variation in contemporary hybrid associations can finally reveal the effect of natural hybridization on specific phenotypes. A determination of the relative fitness of parental and hybrid taxa in different experimental and natural environments is necessary for such inferences.

In this review, I first give an overview of studies that document the phylogenetic distribution of reticulate evolution. This treatment emphasizes examples from animal systems because recent reviews of the plant literature are already available (128, 130). Data from studies of contemporary hybridization are then used to illustrate the effect that differential gene flow may have on reticulate evolution. The remainder of the review examines evidence concerning the effects of natural selection on the outcome of natural hybridization. The fertility and viability of parental and hybrid classes are discussed. Examples are also provided where the relative fitness of hybrid and parental forms has been estimated for different environments.

RETICULATE EVOLUTION: PHYLOGENETIC DISTRIBUTION AND MICROEVOLUTIONARY PATTERNS

Phylogenetic Reconstruction and Reticulate Evolution

Recent reviews have discussed the use of phylogenetic analyses to infer reticulate evolution involving plant species (125, 128, 130). In addition, phylogenetic analyses of animal taxa have detected the influence of recent or ancient natural hybridization (14, 38, 46, 83, 153). One plant group apparently impacted by introgressive hybridization is the sunflower genus *Helianthus*. Analyses of morphological and chromosomal variation and experimental crossing successes have resulted in inferences concerning species relationships (40, 73). Recently, molecular data have also been used to construct phylogenies for the annual species of this genus (123, 124, 126, 127, 129). The phylogenetic histories derived from these genetic markers have suggested the occurrence of wide-scale reticulate evolution. The data suggest that these reticulate events included (i) a quantitative bias in the introgression of cytoplasmic genetic material between the progenitors of extant species, and (ii) the formation of diploid, hybrid species (123–125). Figure 1 summarizes the results from a chloroplast DNA (cpDNA) analysis of 26 *Helianthus* taxa (123). Numerous conspecific taxa are placed in disparate regions of the phylogenetic tree based upon cpDNA markers (Figure 1). In contrast, an analysis of phylogenetic relationships derived from ribosomal DNA (rDNA) variation shows considerable concordance with a morphologically based phylogeny (123). Rieseberg (123) has concluded that the nonconcordance between the rDNA and cpDNA phylogenies, and the distribution seen in the cpDNA phylogeny, is reflective of past introgressive hybridization involving the chloroplast genomes of several species. A number of other plant groups reflect the same pattern of chloroplast DNA introgression (128); Rieseberg & Soltis (128) recognized instances of cpDNA transfer in 16 different plant genera. This finding supports earlier conclusions concerning the widespread occurrence of reticulate evolution in plant species (4, 156).

Reticulate evolution involving animal taxa has been inferred from two patterns of nonconcordance for different phylogenetic histories: one is between a mitochondrial DNA (mtDNA) phylogeny and the species delimitations, and the other is a lack of resolution of defined clades when using data derived from mtDNA studies (46, 83, 153). The former pattern was detected

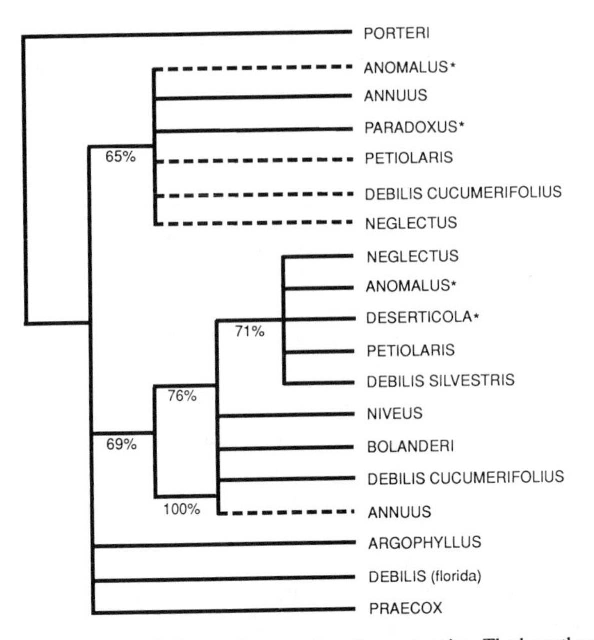

Figure 1 Chloroplast DNA phylogeny for annual sunflower species. The hypothesized instances of chloroplast DNA transfer are indicated by dashed lines leading to the species involved. The percentages below certain lineages indicate how often these lineages were supported in a bootstrap analysis. The asterisks (*) indicate putative stabilized hybrid derivatives (Redrawn from Figure 3, Ref. 128).

in an analysis of mtDNA phylogenies for three closely related species of *Drosophila; D. simulans, D. mauritiana* and *D. sechellia*. These species are chromosomally homosequential (identical banding patterns of polytene chromosomes; 39) and morphologically very similar (84), and they readily form experimental hybrids (85). A mtDNA phylogeny for these three species placed certain interspecific mtDNA types together (Figure 2). One clade in this phylogeny contained one mitochondrial type from each of the three species, and the second clade contained two mitochondrial types from *D. simulans*, grouped with one mtDNA type from *D. mauritiana* (Figure 2; 153). The conclusion drawn from the *Drosophila* result was that introgressive hybridization had taken place among these three species (153). This conclusion has since been supported through experimental crosses and determinations of the fitness of hybrid offspring (18). A second study has resolved a similar pattern of nonconcordance between a mtDNA-based phylogeny and the species identification; this analysis involved the examination of populations of gray wolves (*Canis lupus)* and coyotes (*C. latrans;* 83). The samples analyzed in this study represented allopatric and sympatric populations of both species. A

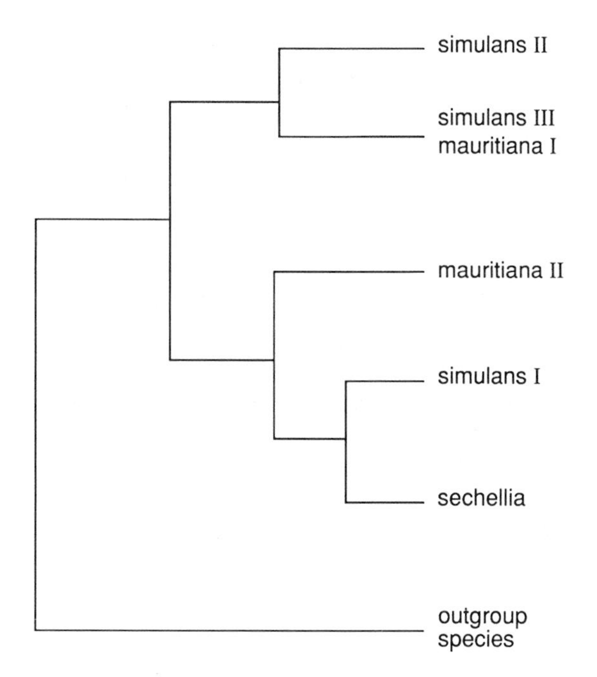

Figure 2 Mitochondrial DNA phylogeny for *D. simulans, D. mauritiana* and *D. sechellia*. Roman numerals following the species designations refer to mitochondrial DNA types (Redrawn from Figure 2, Ref. 153).

monophyletic assemblage was resolved for all of the coyote mtDNA types (83). In contrast, 7 of the 13 mtDNA types found in individuals identified as gray wolves fell within the coyote clade (83). Of these 7 mtDNA types, 4 were identical to a corresponding mtDNA type found in coyotes (83). The remaining 3 were more similar to coyote mtDNA than to other wolf types but possessed varying numbers of restriction site differences (83). The conclusions drawn from these observations were that there had been (i) recent contact between these two species and (ii) introgression of the coyote mtDNA into wolf populations (83). Although these authors argued for a recent event (due to human caused disturbance), they also recognized that the three divergent mtDNA types found in wolf populations could reflect the result of more ancient introgression and subsequent mtDNA evolution (83).

The second phylogenetic pattern that may result from reticulate evolution is exemplified by a study of minnow species from the genus *Notropis* (46). This study included an examination of both allozyme and mtDNA variation for four species that represented three subgenera; *N. chrysocephalus, N.*

cornutus (subgenus *Luxilus)* and *N. photogenis, N. rubellus* (subgenus *Notropis* and *Hydrophlox,* respectively; 46). A phylogenetic analysis of the allozyme data resolved the species into two monophyletic clades (46). However, a phylogenetic analysis using the mtDNA character set produced a phylogeny that did not resolve defined clades. The lack of phylogenetic resolution was reflected by a clade containing three of the four species that was not supported by a statistical analysis (46). This result was particularly surprising because the mitochondrial genome is expected to show a higher rate of evolution than the allozyme loci (46). Dowling & Brown (46) discussed in detail two explanations for this result; (i) heterogeneity in the rate of evolution for various mtDNA genes and lineages, and (ii) introgressive hybridization (46). Although it was not possible to decide between these two alternative explanations, the levels of mtDNA divergence between the four species were similar to numerous comparisons for other animal taxa (46). Thus, if the lack of resolution is due to rate heterogeneity it is not reflected in greater mtDNA divergences among these four species relative to other interspecific comparisons. In contrast, numerous species of *Notropis,* including three of the species examined in this analysis, are involved in contemporary, natural hybridization (46). These observations are consistent with the hypothesis that hybridization has resulted in the observed phylogenetic pattern (46).

Contemporary Hybridization and Ancient Introgression

The causal agents for evolutionary change include mutation, natural selection, genetic drift, and gene flow (43, 50, 53, 150, 176). Although these processes, either singly or in concert, may lead to genetic evolution, gene flow is thought to have the potential for affecting the likelihood of evolution (50, 95, 150, 176). As with gene flow per se, the process of introgressive hybridization may have either a restraining or a creative effect on the genetic makeup and evolution of species complexes. In particular, examples of contemporary hybridization often reflect a pattern of differential transfer of various genetic and morphological characters. The differential transfer of certain diagnostic markers through hybridization and introgression may be a quantitative or qualitative phenomenon (70). Quantitative differences are reflected by varying frequencies of the markers in introgressed populations or by the distance at which they are located from the source of contemporary hybridization (a hybrid zone or population). Qualitative differences in the pattern of introgression are evident when certain markers are transferred while others are not. Either of these patterns of introgression can give rise to introgressed populations characterized by different mixtures of the various markers. This is the case for numerous examples where there is ongoing hybridization and introgression (6, 10, 13, 31, 38, 52, 69, 71, 83, 93, 118, 127, 154, 159, 162, 175).

CALEDIA CAPTIVA The grasshopper species *Caledia captiva* possesses a number of chromosomal taxa that are distinguishable by centromere position, chromosomal distribution of highly repeated sequences (12, 140, 144, 147), and varying degrees of reproductive isolation (140). Furthermore, introgressive hybridization has been detected between two of the chromosomal taxa ("Moreton" and "Torresian") within this species complex (13, 41, 92, 93, 103, 104, 144, 145). These taxa demonstrate significant postmating reproductive isolation with 100% of the F_2 and 50% of the first generation backcross progeny being inviable (146, 148). Figure 3 illustrates the geographic distribution of rDNA, mtDNA, and allozyme variants that are characteristic for either the Moreton or Torresian form. Marchant et al (93) reported the presence of diagnostic Moreton mtDNA, rDNA, and allozyme markers in Torresian individuals collected up to 400 km north of the present-day hybrid zone (Figure 3). In contrast to the molecular markers, these populations were characterized as having the Torresian chromosomal type. Although each of the molecular markers was found at high frequencies north of the present-day area of contact, there were differences in the frequency and distance from the hybrid zone for the various introgressed markers. The rDNA and some of the allozyme markers were discovered farthest (ca. 400 km) from the contemporary contact point (Figure 3; 93, 146). The Moreton mtDNA markers were found a maximum of approximately 250 km from the hybrid zone (Figure 3). Finally, some of the diagnostic Moreton allozymes were discovered only 100 km from the present-day hybrid zone. Marchant et al (93) concluded that the introgression of the Moreton molecular markers, in the abscence of the Moreton chromosome rearrangements, might have resulted from the movement of the hybrid zone during past climatic changes and the action of normalizing selection against the disruption of *cis*-acting gene complexes. It has also been argued that the high frequency of the Moreton rDNA markers in this area of introgression was likely due to biased gene conversion in favor of the Moreton rDNA repeat unit (8). The introgression of the Moreton markers illustrates how stochastic and deterministic processes may lead to quantitative and qualitative differences in the extent of introgressive hybridization.

LOUISIANA IRIS SPECIES COMPLEX Wright (177) concluded that the differential transfer of "male" and "female" genetic components between populations should lead to genetic differentiation. The occasion for different levels of gene flow involving male and female components is most obvious in plant

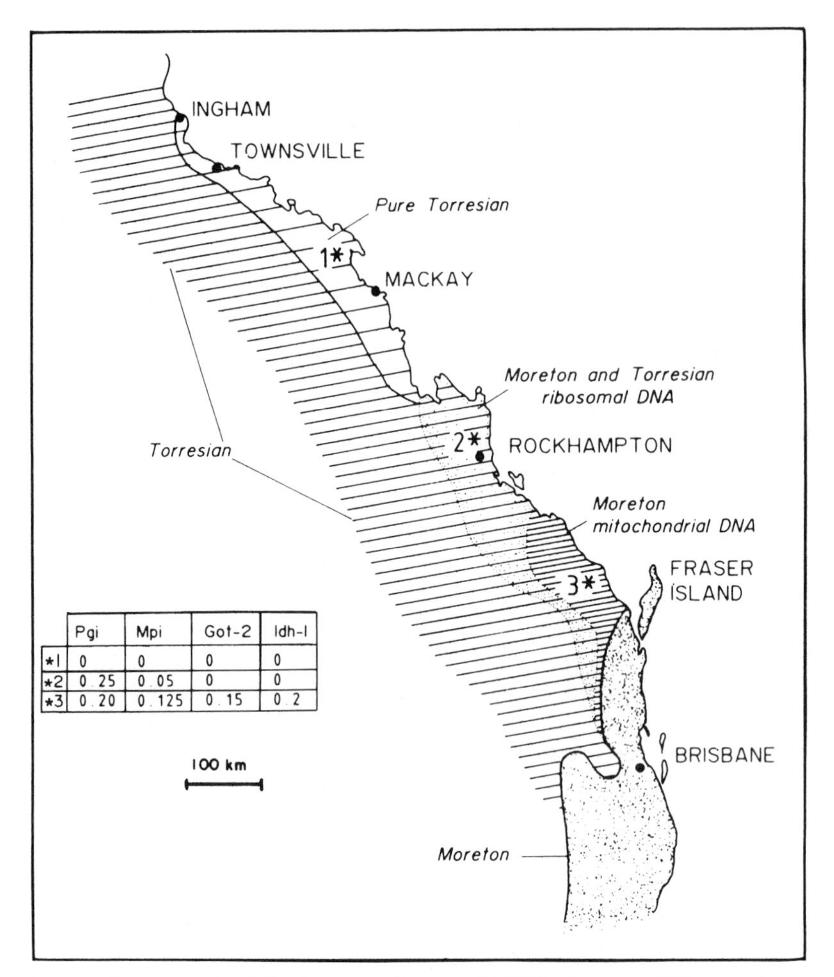

	Pgi	Mpi	Got-2	Idh-1
*1	0	0	0	0
*2	0.25	0.05	0	0
*3	0.20	0.125	0.15	0.2

Figure 3 The geographical distribution of mitochondrial DNA, ribosomal DNA, and allozyme markers characteristic for the Moreton subspecies of the grasshopper *C. captiva* in populations of the Torresian subspecies. The present-day area of parapatry is indicated by the intersection of the stippled and crosshatched areas west and northwest of Brisbane (Redrawn from Figure 4, Ref. 145).

systems due to the two avenues available for gamete/zygote dispersal; genetic material may be dispersed via pollen or seed movement. Differences in the magnitude of gene flow attributable to either pollen or seed dispersal may be quite large within a given species (152). This, in turn, may affect the amplitude of the "creative" or "restraining" effect of gene flow (150). As with intraspecific gene flow, introgressive hybridization may be accomplished by seed movement followed by pollen transfer or, alternatively, by the direct transfer of pollen. These two mechanisms for gene flow provide an opportunity for the differential transfer of cytoplasmic and nuclear genetic elements. If introgression occurs solely by the transfer of pollen, maternally inherited genetic elements (e.g. cpDNA) will not introgress. Alternatively, if pollen transfer occurs between plants produced from immigrant seeds and individuals from the "invaded" population, then there is the opportunity for introgression involving both maternal and paternal genetic elements.

Iris fulva and *Iris hexagona* The genetic interaction between the Louisiana Iris species, *Iris fulva* and *Iris hexagona,* has been used as a classic example of plant hybridization and introgression (2, 6, 7, 9, 10, 11, 28, 29, 120, 121, 131, 160). Although a morphological and cytological examination led to alternative conclusions (121), recent molecular analyses of these species and their hybrids have supported the original hypothesis of introgression (6, 7, 9–11). Alternate nuclear (rDNA, allozyme, and random DNA sequences) and cytoplasmic (cpDNA) markers have thus far been detected in areas of sympatry and in allopatric populations of both species. The genetic data indicate both localized and dispersed introgression (6, 7, 9–11); however, the chloroplast and nuclear markers demonstrate markedly different patterns of variation in the sympatric and allopatric populations. Figure 4 illustrates the distribution of nuclear and cpDNA markers in individuals from an area of overlap between *I. fulva* and *I. hexagona.* The array of multilocus, nuclear genotypes suggests the presence of advanced generation backcross individuals scattered throughout this population (7, 9, 10). Nason et al (108), using a maximum likelihood analysis, concluded that a large proportion of the genotypes found in this population probably resulted from repeated backcrossing into *I. hexagona.*

In contrast to the results from the nuclear analyses, the cpDNA markers reflect a much more limited area of overlap between the two species (Figure 4). The discrepancy between the two data sets is particularly striking for the area defined by the bayou channel and the roadway (Figure 4). In this region, there is a mixture of individuals that possess either *I. hexagona* or a hybrid genotype based upon the nuclear data. In contrast, all of these individuals are characterized by the *I. hexagona* cpDNA haplotype (Figure 4). Such an observation suggests that the introgressed individuals in this region are the result of pollen transfer from *I. fulva* individuals onto *I. hexagona* flowers. The resulting seeds would possess the nuclear, biparental markers of both *I. fulva* and *I. hexagona*. However, only the *I. hexagona* cpDNA markers would be reflected in the progeny since this cytoplasmic genome is apparently maternally inherited (10). A similar finding was made when allopatric populations of *I. fulva* and *I. hexagona* were analyzed for the nuclear and cpDNA markers (7, 9, 10). One allopatric population of each of these species

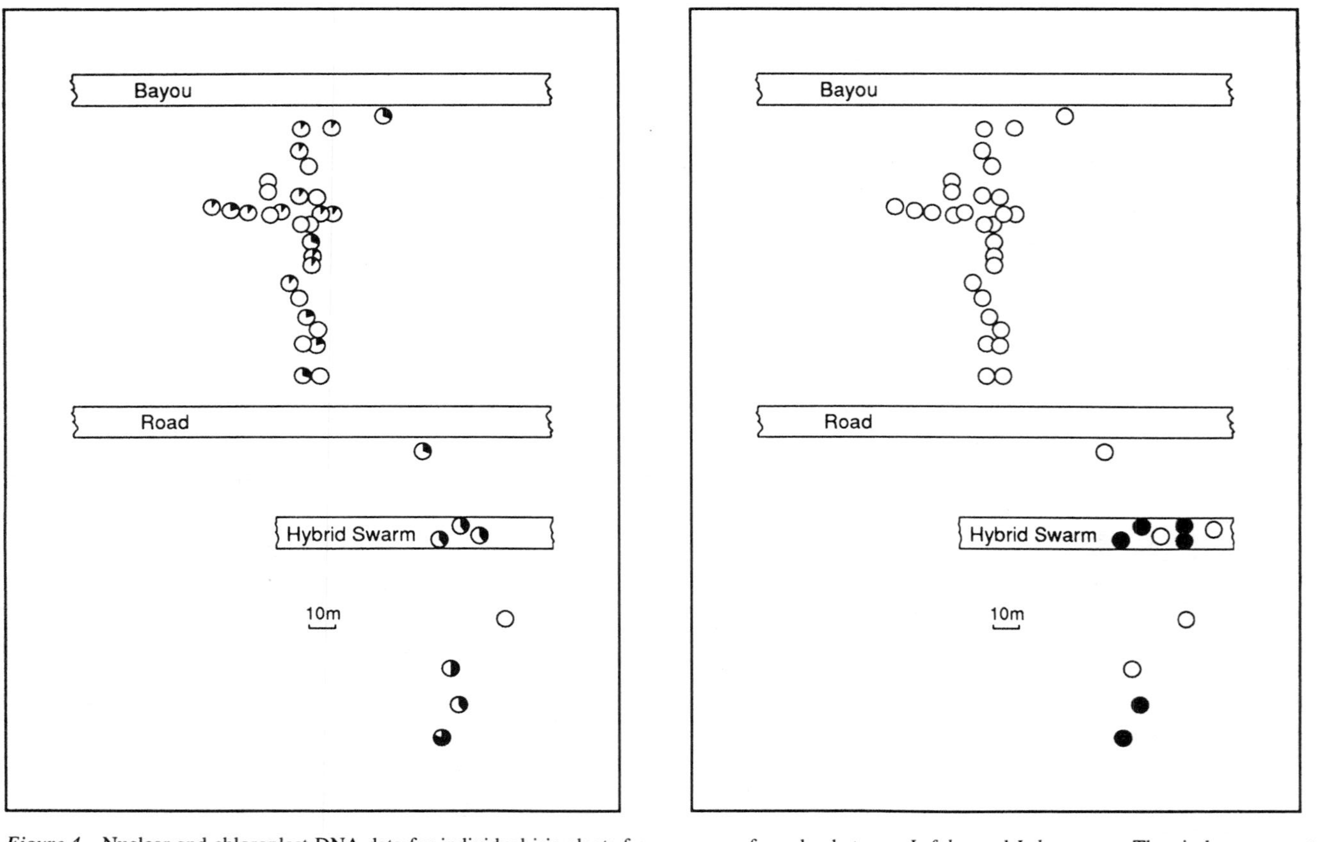

Figure 4 Nuclear and chloroplast DNA data for individual iris plants from an area of overlap between *I. fulva* and *I. hexagona*. The circles represent genetic data derived from single plants. The left-hand panel illustrates the relative proportion of *I. fulva* (shaded portion of circles) and *I. hexagona* (unshaded portion of circles) nuclear markers in a sample of 37 plants. The right-hand panel reflects the distribution of *I. fulva* (shaded circles) and *I. hexagona* (open circles) chloroplast DNA markers for these 37 plants and three additional plants (Figure 4, Ref. 11).

contained a low frequency of the diagnostic, nuclear markers from the alternate species (7, 9, 10); this indicates the occurrence of allopatric introgression. When these same populations were assayed for chloroplast DNA variation, the cpDNA characteristic for *I. fulva* and *I. hexagona* was found in the *I. fulva* and *I. hexagona* populations, respectively (10). This result is consistent with the pattern of variation seen in the area of overlap in that the nuclear markers are introgressing in the absence of cpDNA introgression. Arnold et al (10) have suggested that this pollen mediated introgression is due to the presence of pollinators capable of long-distance foraging and the absence of special adaptations for seed dispersal. As with the area of sympatry, the introduction of "foreign" pollen would result in hybrid offspring carrying the nuclear but not the cpDNA markers of the alternate species.

I. fulva, I. hexagona and I. brevicaulis Differential introgression due to a quantitative bias in the levels of pollen and seed dispersal has also been hypothesized for hybridization involving *I. fulva*, *I. hexagona* and *I. brevicaulis*. Combinations of diagnostic allozyme markers for each of these three species were detected in individuals from a single population (9). These data suggested that localized introgression had occurred between *I. fulva*, *I. hexagona* and *I. brevicaulis*. A recent analysis of the cpDNA from these same individuals allowed a test for the mode of gene flow in this population (11). The cpDNA haplotypes detected in this population were identical to those found in either *I. fulva* or *I. brevicaulis* individuals. Arnold et al (11) drew two conclusions from these data. First, the hybrid population reflects an area of sympatry and introgression between *I. fulva* and *I. brevicaulis*. This is indicated by the presence of individuals that are genetically *I. fulva* or *I. brevicaulis* and by the detection of recombinant individuals possessing various combinations of the *I. fulva* and *I. brevicaulis* nuclear and cpDNA markers. The second conclusion is that the presence of the *I. hexagona* nuclear markers is most likely due to pollen transfer from allopatric *I. hexagona* plants onto flowers of *I. brevicaulis*, *I. fulva* or hybrid individuals.

Hybrid speciation in Louisiana Irises: I. nelsonii Differential introgression in this species complex is also exemplified by the putative hybrid species, *I. nelsonii*. The formal designation of *I. nelsonii* was based upon unique morphological, chromosomal, and habitat characteristics and included the hypothesis that this species was a hybrid derivative of *I. fulva*, *I. hexagona* and possibly *I. brevicaulis* (120). Analyses involving allozyme, random nuclear, and cpDNA markers have supported the hypothesis that *I. nelsonii* is a hybrid derivative containing genetic material from each of the above three species (9, 10). Although this species has a combination of nuclear markers characteristic of *I. fulva*, *I. hexagona*, and *I. brevicaulis*, it possesses a cpDNA haplotype identical to that found in *I. fulva* (10). Arnold et al (10) concluded that *I. nelsonii* likely arose through the transfer of pollen from *I. hexagona* and possibly *I. brevicaulis* to *I. fulva* flowers. Since *I. nelsonii* has a preponderance of *I. fulva* markers (nuclear and cpDNA; 9, 10), it is most likely that the initial hybridization was followed by backcrossing into *I. fulva*.

In summary, the Louisiana Iris species complex exemplifies the effect that the unique mechanism of gene flow possessed by plants can have on the pattern and degree of introgression. Having two avenues for gene flow may act to decouple the nucleus and cytoplasm, resulting in a qualitative difference in the transfer of the various genomic elements.

NATURAL HYBRIDIZATION AND ADAPTATION

Evolutionary hypotheses have been proposed that predict various outcomes of hybrid zones between species (24, 25, 101). These have been differentiated into the "adaptive speciation" hypothesis, the "bounded hybrid superiority" hypothesis, and the "dynamic-equilibrium" hypothesis (102). The adaptive speciation hypothesis predicts that the level of reproductive isolation between hybridizing taxa will either decrease or increase depending upon the effectiveness of natural selection against hybrid types. The bounded hybrid superiority hypothesis suggests that hybrid zones are areas of ecological transition in which the hybrids are superior to the parental types. Therefore, the hybrid zone would be maintained because the recombinant individuals will outcompete the parental types in the unique habitat. Finally, the dynamic-equilibrium hypothesis assumes that hybrid zones are maintained by a balance between dispersal and hybrid inferiority (24).

Hybrid Zone Theory: A Synopsis

The relative fitness of hybrid and parental individuals may be estimated by fitting theoretical models to the genotypic data. Two theoretical frameworks have been the basis for mathematical models designed to predict hybrid zone structure and evolution. These two frameworks involve either cline theory or, alternatively, models based upon estimates of "cytonuclear disequilibrium."

CLINE THEORY AND HYBRID ZONES Until recently hybrid zone dynamics have been modeled using cline theory (24, 26, 51, 65, 100). Measurements of the slope, width, and concordance of clinal variation in hybrid populations have been used to infer the relative contribution of dispersal and natural selection to the structuring of hybrid zones (24, 100, 102). It is thus possible to infer whether the hybrid zone is maintained due to negative or positive selection on hybrids and whether the habitat is influencing this pattern of selection (26). Models of hybrid zone maintenance dependent upon hybrid superiority in an ecotonal habitat and models that assume the joint contributions of dispersal and hybrid inferiority differ in the "characteristic scale of selection," l (26). If a hybrid zone is maintained by dispersal and negative selection, l should be approximately the same as the width of the clinal variation (w = distance between the gene frequency of 0.2 and 0.8; 51). Alternatively, w will be much larger than l if a hybrid zone is maintained by positive selection for the hybrids (26). One study that exemplifies the use of these models to estimate dispersal and selection associated with a hybrid zone involved the fire-bellied toad species, *Bombina bombina* and *B. variegata* (157). This analysis examined genetic variation as measured using diagnostic allozyme markers. The results from this study indicated strongly concordant clines for each of the allozyme markers corresponding closely with clines for both morphological and mtDNA characters (157). These data, along with estimates of dispersal and fitness, suggested that the clines were probably maintained by strong selection against the hybrid individuals (157).

CYTONUCLEAR DISEQUILIBRIUM AND HYBRID ZONES Estimates of cytonuclear disequilibrium also allow a description of factors that may underlie the genetic structure and evolution of hybrid populations (5, 15). These estimates are made by comparing genotypic frequencies of diagnostic nuclear and cytoplasmic (mtDNA, cpDNA) markers. The cytonuclear models have been designed to test the genetic structure of hybrid populations for the effects of natural selection, nonrandom mating, migration and pollen versus seed dispersal (5, 15–17, 139). One example of the use of cytonuclear disequilibrium estimates involved a study of two species of North Atlantic eels, *Anguilla anguilla* ("European" eels) and *A. rostrata* ("American" eels; 19). Data from allozyme, mtDNA, and morphological characters were collected from populations of the two species and from a morphologically variable population in Iceland. These data suggested that there had been gene flow from the American populations into the European populations and, in particular, into the Iceland population. The cytonuclear disequilibria estimates allowed two additional conclusions. The estimates were consistent with the hypothesis that the Iceland population included pure *A. anguilla* and *A. rostrata* animals as well as a low frequency of individuals that were hybrids between these two species (19). Also, the best goodness-of-fit estimate was associated with a model in which there was a large quantitative bias in the migration of European eels into the Iceland population (19).

Habitat Associations and Adaptationist Inferences

A second approach for estimating the adaptation of the parental and hybrid forms to different ecological settings involves making observations concerning distinctive habitats or behavioral characteristics associated with parental and hybrid populations. This class of inference has been made for numerous instances of hybridization involving either plant or animal species (e.g. 27, 30, 35, 44, 74, 81, 96, 98, 99, 101, 102, 106, 149, 160, 179). Analyses of the grasshopper species group, *C. captiva* and the cricket species *Gryllus pennsylvanicus* and *G. firmus* exemplify two distinctive approaches for deriving such inferences (81, 119).

Kohlmann et al (81) used a bioclimate model to predict geographic distributions for the *C. captiva* races and species. These authors found the distributions to be correlated with gradients in both rainfall and temperature (81). In particular, this model predicted an area of parapatric association involving the Moreton and Torresian subspecies. The predicted area of association coincides closely with a present-day hybrid zone. The hybrid zone occurs in an area that is climatically marginal for both of these taxa (81). These authors also demonstrated a significant difference in the percentage ground cover between the Toressian and the Moreton sides of the hybrid zone, indicated by a clinal pattern of change in vegetation (81). This hybrid zone is also characterized by a pattern of clinal change in genetic markers that is explicable by severe hybrid inviabilities (148). However, the coincidence of a marginal climate and a clinal changeover in ground cover with the area of contemporary hybridization suggests an influence by environmental variables on the positioning and genetic structure of this hybrid zone (81).

Analyses of the hybrid zone between the cricket species *G. firmus* and *G. pennsylvanicus* have included an examination of the morphologies and genotypes associated with different soil types ("sand" and "loam"; 119). Experimental crosses were also undertaken to assay the levels of reproductive isolation between individuals from sand and loam habitats. The results from these analyses suggested an interaction between the genotypic constitution of individuals and their environment. Rand & Harrison (119) thus discovered little differentiation in either morphology or genotypic frequencies between populations occupying the same soil type, but they detected significant differences for these markers between spatially closely associated populations occupying different soil types (119). The experimental crosses defined a pattern of asymmetry, with females from sand habitats that were crossed with males from loam habitats producing many fewer hatchlings than the reciprocal cross (female/loam × male/sand); mean = 1.25 individuals and 28 individuals, respectively (119). A pattern of asymmetrical success is also apparent for crosses involving the parental taxa. Crosses involving *G. firmus* females and *G. pennsylvanicus* males are unsuccessful, while the reciprocal cross produces viable and fertile F_1 individuals (68). Significantly, *G. firmus* is associated with sandy soils and *G. pennsylvanicus* is found in loam soil habitats. As with the *C. captiva* study, the results from the analysis of the habitat association in *Gryllus* support the hypothesis that there is a causal relationship between the environment and the observed position and genetic structure of a hybrid zone.

Natural Hybridization, Introgression, and the Fitness of Hybrids

The most informative approach for estimating the adaptive effects of hybridization and introgression is to measure the relative fitness of different hybrid and parental genotypes under identical experimental or natural conditions. An ideal analysis should include the experimental manipulation of environmental factors thought to affect the fitness of various genetic classes. Observations also need to be made regarding the relative fertility and viability of different genotypes in habitats characteristic of the parental taxa and the naturally occurring hybrids. No study of natural hybridization has encompassed all of these parameters; however, analyses involving two species groups can be used to exemplify different portions of such an "ideal" study.

EUCALYPTUS RISDONII AND *E. AMYGDALINA* Although natural hybridization in the genus *Eucalyptus* has apparently been recognized for nearly two centuries, the frequency and taxonomic distribution of hybridization within this genus may be more restricted than previously postulated (64). Indeed, Griffin et al (64) concluded that the relative paucity of natural hybridization between species in this genus might reflect selection against hybrid individuals. However, fitness estimates for hybrids and parent species may be equivalent (48, 112, 115). Furthermore, introgressive hybridization followed by selection for phenotypes approaching the pollen parent may be an evolutionary mechanism for range expansions by the hybridizing taxa (113–117; but see 138).

Detailed experimentation and field studies have defined ecological associations and relative fitnesses for the species *E. risdonii* and *E. amygdalina* and for their natural hybrids (80, 112, 115). A progeny trial involving seeds from trees of the parental species and hybrid individuals allowed a measurement of seedling survivorship and vigor (115). The only significant difference was detected between one population of *E. amygdalina* and all the other parental and hybrid samples (115). It was concluded that the progenies of *E. risdonii*, *E. amygdalina* and hybrid individuals did not differ significantly in either survival or vigor (115). Analyses of regeneration following fire disturbance, involving these same species, allowed a test for the relative fitness of different phenotypes in both parental and hybrid habitats (112). The phenotypes that were scored in the parental and hybrid populations included *E. amygdalina*, *E. risdonii*, "Hybrid" (individuals intermediate in morphology between the species), "Backcross *E. risdonii*" (hybrid individuals more similar to *E.*

risdonii) and "Backcross *E. amygdalina*" (hybrid individuals more similar to *E. amygdalina*; 112).

The findings from this analysis indicated a consistent—and for many of the characteristics measured, a significant—difference in the fitness of the various classes (112). Within the hybrid populations, individuals scored as *E. risdonii* demonstrated a higher fitness than *E. amygdalina* in seed output, productivity over a 20-year period, mortality, vegetative recovery, and seedling establishment (112). The "backcross *E. amygdalina*" and "Hybrid" classes generally demonstrated an intermediate or the lowest fitness for all of the characteristics scored. The individuals identified as "backcross *E. risdonii*" were either intermediate between the two parental species or equivalent to *E. risdonii* for the various characteristics (112). Although the "backcross *E. risdonii*" class demonstrated equivalent fitness to *E. risdonii* for some characteristics, it was concluded that competition between these two classes would probably lead to replacement of the hybrid type by *E. risdonii* (112). There is an important caveat to the pattern described by Potts (112). Because the classes of individuals in this study were determined on the basis of morphological characters, the potential exists for the misidentification of advanced generation hybrids as pure parental individuals, due to the diluting effect of recombination on quantitative characters (3). In the *E. amygdalina* and *E. risdonii* example, the classes of individuals defined as one species or the other might actually contain a proportion of advanced generation hybrids. If this is the case, the relative fitness of the hybrid individuals might actually be higher (or lower) than estimated.

I. FULVA AND *I. HEXAGONA* Genetic and morphological studies have suggested that natural hybridization between *I. fulva* and *I. hexagona* has resulted in both introgression and hybrid speciation (2, 6, 7, 9–11, 120, 131). Analyses of pollen viability, experimental hybridization, germination, competitive ability, shade tolerance, and salt tolerance have yielded fitness estimates for parental and hybrid classes (6, 28, 29, 121).

A larger proportion of subfertile individuals have been detected in natural hybrid populations when compared to allopatric samples of *I. fulva* and *I. hexagona* (121). Furthermore, experimental crosses and germination studies have detected a lower success rate for inter- versus intra-specific crosses, as well as a lower germination frequency and the greatest mean number of days to germination associated with the hybrid seeds (6). Relative fitness estimates have also been determined for *I. fulva*, *I. hexagona*, and hybrid classes under various environmental conditions. An analysis of shade tolerance included the two species and two hybrid classes (29). One hybrid class was genetically more similar to *I. hexagona* ("Hybrid Purple") and one was genetically more similar to *I. fulva* ("Hybrid Red"; 7, 9–11). Both hybrid classes were intermediate in their responses relative to the parental taxa; the Hybrid Red individuals demonstrated a higher fitness (greater shade tolerance) than Hybrid Purple individuals in all but one comparison (29). The taxon with the lowest fitness in all of the comparisons was *I. hexagona*. The taxon showing the most shade tolerance and, therefore, the highest inferred fitness was *I. fulva* (29).

Experiments designed to measure the relative effect of varying levels of salt concentrations resolved the following fitness hierarchy: *I. hexagona* > Hybrid Purple = Hybrid Red > *I. fulva* (28). Finally, interspecific competition experiments determined that *I. hexagona* and the Hybrid Purple class were equivalent in their competitive abilities, but both were superior to *I. fulva* (28). The habitats occupied by *I. fulva*, *I. hexagona* and hybrid individuals seem to conform to expectations drawn from the experimental data. *I. fulva* occurs along the heavily shaded banks of bayous of the Mississippi drainage, *I. hexagona* occurs in open, freshwater marsh habitats, and the hybrid populations are found in apparent ecotonal areas (6, 28, 160). The patterns of fitness for the hybrid classes and parental species would suggest that the hybrids, and in particular the "Hybrid Purple" individuals, may have a higher fitness in an ecotonal habitat where they would be superior to *I. fulva* in competitive ability and superior to *I. hexagona* in shade tolerance (6).

Experimental and demographic data for the *Iris* and *Eucalyptus* species and hybrids reveal a range of fitness estimates that suggest selection against hybrids. However, the inferior nature of the various hybrids may not be qualitatively different from that seen for the parental forms. The parental species thus have varying levels of relative fitness, sometimes demonstrating a less fit response than certain hybrid classes. This in turn leads to the prediction that certain hybrid classes may have a competitive advantage over the less fit parental form leading to a displacement of the parent by the hybrid derivative.

CONCLUSIONS AND FUTURE DIRECTIONS

In his 1973 review of introgressive hybridization (72), Heiser asked the following questions: "Does it [introgressive hybridization] really occur? And, if so, does it have any evolutionary significance?" After considering a number of the best-known putative examples of introgressive hybridization, he was led to conclude that "It may play a very significant role; but it must be admitted, there is as yet no strong evidence to support such a claim." Given the inferential nature of evolutionary studies in general (88), any answer concerning the role that natural hybridization may have had on the genetic structure and patterns of speciation is always a qualified one. However, additional data, derived mainly from molecular and ecological analyses, are now available for assessment. These new findings suggest a prominent role for natural hybridization in numerous species complexes. Reticulate events are apparent in the phylogenies for a diverse set of plant and animal species complexes. Furthermore, it is possible to identify processes from contemporary hybridization that explain these phylogenetic patterns. Ecological and demographic studies have revealed the association of hybrids and parental types with specific and different habitats. More importantly, the ecological approaches allow predictions concerning where hybrid zones should occur and indicate that the hybrid and parental individuals display varying adaptive responses to different environments.

What future studies would underlie a more detailed understanding of the evolutionary significance of natural hybridization? Determining the evolutionary significance of natural hybridization will rely on an understanding of its phylogenetic distribution and the processes that determine its effect. At the present time, the greatest lack in analyses of natural hybridization, as with most areas of evolutionary biology, involves experimentation. Reciprocal transplants, life history tables, experimental hybridization, and gene flow analyses in natural and experimental hybrid populations are necessary to measure interactions between the environment and genotype. Such studies will allow an assessment of the relative roles of deterministic and stochastic processes in the evolution of hybrid populations and thus lead to predictions concerning the evolutionary consequences of natural hybridization.

ACKNOWLEDGEMENTS

I want to thank the following persons for reviewing the manuscript: J. C. Avise, R. G. Harrison, L. H. Rieseberg, and P. W. Sattler. This work was supported by National Science Foundation grants BSR-9004242 (M. L. Arnold, J. L. Hamrick, and B. D. Bennett) and BSR-9106666 (M. L. Arnold and J. L. Hamrick), and a National Science Foundation REU supplement, and a University of Georgia Faculty Research Grant to M. L. Arnold.

Literature Cited

1. Anderson, E. 1948. Hybridization of the habitat. *Evolution* 2:1–9
2. Anderson, E. 1949. *Introgressive Hybridization*. New York: Wiley. 109 pp.
3. Anderson, E., Hubricht, L. 1938. Hybridization in *Tradescantia*. III. The evidence for introgressive hybridization. *Am. J. Bot.* 25:396–402
4. Anderson, E., Stebbins, G. L. Jr. 1954. Hybridization as an evolutionary stimulus. *Evolution* 8:378–88
5. Arnold, J., Asmussen, M. A., Avise, J. C. 1988. An epistatic mating system model can produce permanent cytonuclear disequilibria in a hybrid zone. *Proc. Natl. Acad. Sci. USA* 85:1893–96
6. Arnold, M. L., Bennett, B. D. 1992. Natural hybridization in Louisiana irises: genetic variation and ecological determinants. In *Hybrid Zones and the Evolutionary Process,* ed. R. G. Harrison. In press. Oxford: Oxford Univ. Press
7. Arnold, M. L., Bennett, B. D., Zimmer, E. A. 1990. Natural hybridization between *Iris fulva* and *I. hexagona:* pattern of ribosomal DNA variation. *Evolution* 44:1512–21

8. Arnold, M. L., Contreras, N., Shaw, D. D. 1988. Biased gene conversion and asymmetrical introgression between subspecies. *Chromosoma* 96:368–71
9. Arnold, M. L., Hamrick, J. L., Bennett, B. D. 1990. Allozyme variation in Louisiana irises: a test for introgression and hybrid speciation. *Heredity* 65:297–306
10. Arnold, M. L., Buckner, C. M., Robinson, J. J. 1991. Pollen mediated introgression and hybrid speciation in Louisiana irises. *Proc. Natl. Acad. Sci. USA* 88:1398–1402
11. Arnold, M. L., Robinson, J. J., Buckner, C. M., Bennett, B. D. 1992. Pollen dispersal and interspecific gene flow in Louisiana Irises. *Heredity* 68:399–404
12. Arnold, M. L., Shaw, D. D. 1985. The heterochromatin of grasshoppers from the *Caledia captiva* species complex. II. Cytological organisation of tandemly repeated DNA sequences. *Chromosoma* 93:183–90
13. Arnold, M. L., Shaw, D. D., Contreras, N. 1987. Ribosomal RNA-encoding DNA introgression across a narrow hybrid zone between two subspecies of grasshopper. *Proc. Natl. Acad. Sci. USA* 84:3946–50
14. Arnold, M. L., Wilkinson, P., Shaw, D. D., Marchant, A. D., Contreras, N. 1987. Highly repeated DNA and allozyme variation between sibling species: evidence for introgression. *Genome* 29:272–79
15. Asmussen, M. A., Arnold, J., Avise, J. C. 1987. Definition and properties of disequilibrium statistics for associations between nuclear and cytoplasmic genotypes. *Genetics* 115:755–68
16. Asmussen, M. A., Arnold, J., Avise, J. C. 1989. The effects of assortative mating and migration on cytonuclear associations in hybrid zones. *Genetics* 122:923–34
17. Asmussen, M. A., Schnabel, A. 1991. Comparative effects of pollen and seed migration on the cytonuclear structure of plant populations I. Maternal cytoplasmic inheritance. *Genetics* 128:639–54
18. Aubert, J., Solignac, M. 1990. Experimental evidence for mitochondrial DNA introgression between *Drosophila* species. *Evolution* 44:1272–82
19. Avise, J. C., Nelson, W. S., Arnold, J., Koehn, R. K., Williams, G. C., Thorsteinsson, V. 1990. The evolutionary genetic status of Icelandic eels. *Evolution* 44:1254–62
20. Avise, J. C., Saunders, N. C. 1984. Hybridization and introgression among species of sunfish *(Lepomis):* analysis by mitochondrial DNA and allozyme markers. *Genetics* 108: 237–55
21. Baker, M. C., Baker, A. E. M. 1990. Reproductive behavior of female buntings: isolating mechanisms in a hybridizing pair of species. *Evolution* 44:332–38
22. Baker, R. J. 1981. Chromosome flow between chromosomally characterized taxa of a volant mammal, *Uroderma bilobatum* (Chiroptera: Phyllostomatidae). *Evolution* 35:296–305
23. Baker, R. J., Davis, S. K., Bradley, R. D., Hamilton, M. J., Van Den Bussche, R. A. 1989. Ribosomal-DNA, mitochondrial-DNA, chromosomal, and allozymic studies on a contact zone in the pocket gopher, *Geomys*. *Evolution* 43:63–75
24. Barton, N. H. 1979. The dynamics of hybrid zones. *Heredity* 43:341–59
25. Barton, N. H., Hewitt, G. M. 1981. A chromosomal cline in the grasshopper *Podisma pedestris*. *Evolution* 35: 1008–1018
26. Barton, N. H., Hewitt, G. M. 1985. Analysis of hybrid zones. *Annu. Rev. Ecol. Syst.* 16:113–48
27. Bell, N. B., Lester, L. J. 1978. Genetic and morphological detection of introgression in a clinal population of *Sabatia* section *Campestria* (Gentianaceae). *Syst. Bot.* 3:87–104
28. Bennett, B. D. 1989. *Habitat differentiation of* Iris fulva Ker Gawler, Iris hexagona Walter, *and their hybrids*. PhD thesis. Louisiana State Univ. 117 pp.
29. Bennett, B. D., Grace, J. B. 1990. Shade tolerance and its effect on the segregation of two species of Louisiana *Iris* and their hybrids. *Am. J. Bot.* 77:100–7
30. Benson, L., Phillips, E. A., Wilder, P. A. et al. 1967. Evolutionary sorting of characters in a hybrid swarm. I: Direction of slope. *Am. J. Bot.* 54: 1017–26
31. Bert, T. M., Harrison, R. G. 1988. Hybridization in western Atlantic stone crabs (Genus *Menippe):* evolutionary history and ecological context influence species interactions. *Evolution* 42:528–44
32. Bigelow, R. S. 1965. Hybrid zones and reproductive isolation. *Evolution* 19:449–58
33. Birch, L. C., Vogt, W. G. 1970. Plasticity of taxonomic characters of the Queensland fruit flies *Dacus tryoni* and

Dacus neohumeralis (Tephritidae). *Evolution* 24:320–43

34. Bradley, R. D., Davis, S. K., Baker, R. J. 1991. Genetic control of premating-isolating behavior: Kaneshiro's hypothesis and asymmetrical sexual selection in pocket gophers. *J. Hered.* 82:192–96
35. Briggs, B. G. 1962. Interspecific hybridization in the *Ranunculus lappaceus* group. *Evolution* 16:372–90
36. Brown, W. M., Wright, J. W. 1979. Mitochondrial DNA analyses and the origin and relative age of parthenogenetic lizards (genus *Cnemidophorus*). *Science* 203:1247–49
37. Butlin, R. 1989. Reinforcement of premating isolation. In *Speciation and its Consequences,* ed. D. Otte, J. A. Endler, pp. 158–79. Sunderland: Sinauer. 679 pp.
38. Carr, S. M., Ballinger, S. W., Deer, J. N., Blankenship, L. H., Bickham, J. W. 1986. Mitochondrial DNA analysis of hybridization between sympatric white-tailed deer and mule deer in west Texas. *Proc. Natl. Acad. Sci. USA* 83:9576–80
39. Carson, H. L., Clayton, F. E., Stalker, H. D. 1967. Karyotypic stability and speciation in Hawaiian *Drosophila. Proc. Natl. Acad. Sci. USA* 57:1280–85
40. Chandler, J. M., Jan, C. C., Beard, B. H. 1986. Chromosomal differentiation among the annual *Helianthus* species. *Syst. Bot.* 11:354–71
41. Daly, J. C., Wilkinson, P., Shaw, D. D. 1981. Reproductive isolation in relation to allozymic and chromosomal differentiation in the grasshopper *Caledia captiva. Evolution* 35:1164–79

41a. DeMarais, B. D., Dowling, T. E., Douglas, M. E., Minckley, W. L., Marsh, P. C. 1992. Hybrid origin of *Gila seminuda* (Pisces: Cyprinidae): Implications for evolution and conservation. *Proc. Natl. Acad. Sci. USA* 89:2747–51

42. dePamphilis, C. W., Wyatt, R. 1989. Hybridization and introgression in Buckeyes (*Aesculus:* Hippocastanaceae): a review of the evidence and a hypothesis to explain long-distance gene flow. *Syst. Bot.* 14:593–611
43. Dobzhansky, Th. 1937. *Genetics and the Origin of Species.* New York: Columbia Univ. Press. 364 pp.
44. Dobzhansky, Th. 1953. Natural hybrids of two species of *Arctostaphylos* in the Yosemite region of California. *Heredity* 7:73–79
45. Dobzhansky, Th. 1970. *Genetics of the Evolutionary Process.* New York: Columbia Univ. Press. 505 pp.
46. Dowling, T. E., Brown, W. M. 1989. Allozymes, mitochondrial DNA, and levels of phylogenetic resolution among four minnow species (*Notropis:* Cyprinidae). *Syst. Zool.* 38:126–43
47. Dowling, T. E., Moore, W. S. 1985. Evidence for selection against hybrids in the family Cyprinidae (Genus *Notropis*). *Evolution* 39:152–58
48. Drake, D. W. 1981. Reproductive success of two *Eucalyptus* hybrid populations. I. generalized seed output model and comparison of fruit parameters. *Aust. J. Bot.* 29:25–35
49. Echelle, A. A., Connor, P. J. 1989. Rapid, geographically extensive genetic introgression after secondary contact between two pupfish species (*Cyprinodon,* Cyprinodontidae). *Evolution* 43:717–27
50. Endler, J. A. 1973. Gene flow and population differentiation. *Science* 179: 243–50
51. Endler, J. A. 1977. *Geographic Variation, Speciation and Clines.* Princeton, NJ: Princeton Univ. Press. 246 pp.
52. Ferris, S. D., Sage, R. D., Huang, C., Nielsen, J. T., Ritte, U., Wilson, A. C. 1983. Flow of mitochondrial DNA across a species boundary. *Proc. Natl. Acad. Sci. USA* 80:2290–94
53. Fisher, R. A. 1930. *The Genetical Theory of Natural Selection.* Oxford: Clarendon Press. 272 pp.
54. Fleischer, R. C., Rothstein, S. I. 1988. Known secondary contact and rapid gene flow among subspecies and dialects in the brown-headed cowbird. *Evolution* 42:1146–58
55. Gibbs, G. W. 1968. The frequency of interbreeding between two sibling species of *Dacus* (Diptera) in wild populations. *Evolution* 22:667–83
56. Gill, F. B., Murray, B. G. Jr. 1972. Discrimination behavior and hybridization of the blue-winged and golden-winged warblers. *Evolution* 26:282–93
57. Gillett, G. W. 1966. Hybridization and its taxonomic implications in the *Scaevola gaudichaudiana* complex of the Hawaiian islands. *Evolution* 20: 506–16
58. Goldman, M. A., LoVerde, P. T., Chrisman, C. L. 1983. Hybrid origin of polyploidy in freshwater snails of the genus *Bulinus* (Mollusca: Planorbidae). *Evolution* 37:592–600
59. Good, D. A. 1989. Hybridization and cryptic species in *Dicamptodon* (Caudata: Dicamptodontidae). *Evolution* 43:728–44

60. Gore, P. L., Potts, B. M., Volker, P. W., Megalos, J. 1990. Unilateral cross-incompatibility in *Eucalyptus:* the case of hybridisation between *E. globulus* and *E. nitens*. *Aust. J. Bot.* 38:383–94
61. Grant, V. 1963. *The Origin of Adaptations*. New York: Columbia Univ. Press. 606 pp.
62. Grant, V. 1981. *Plant Speciation*. New York: Columbia Univ. Press. 563 pp.
63. Grant, V., Grant, K. A. 1971. Dynamics of clonal microspecies in cholla cactus. *Evolution* 25:144–55
64. Griffin, A. R., Burgess, I. P., Wolf, L. 1988. Patterns of natural and manipulated hybridisation in the genus *Eucalyptus* L'Herit.-a review. *Aust. J. Bot.* 36:41–66
65. Haldane, J. B. S. 1948. The theory of a cline. *J. Genet.* 48:277–84
66. Hall, H. G. 1990. Parental analysis of introgressive hybridization between African and European honeybees using nuclear DNA RFLPs. *Genetics* 125: 611–21
67. Hardin, J. W. 1975. Hybridization and introgression in *Quercus alba*. *J. Arnold Arb.* 56:336–63
68. Harrison, R. G. 1983. Barriers to gene exchange between closely related cricket species. I. Laboratory hybridization studies. *Evolution* 37:245–51
69. Harrison, R. G. 1986. Pattern and process in a narrow hybrid zone. *Heredity* 56:337–49
70. Harrison, R. G. 1990. Hybrid zones: windows on evolutionary process. In *Oxford Surveys in Evolutionary Biology,* ed. D. Futuyma, J. Antonovics, 7:69–128. Oxford: Oxford Univ. Press.
71. Harrison, R. G., Rand, D. M., Wheeler, W. C. 1987. Mitochondrial DNA variation in field crickets across a narrow hybrid zone. *Mol. Biol. Evol.* 4:144–58
72. Heiser, C. B. 1973. Introgression reexamined. *Bot. Rev.* 39:347–66
73. Heiser, C. B. Jr., Smith, D. M., Clevenger, S. B., Martin, W. C. Jr. 1969. The North American Sunflowers (*Helianthus). Mem. Torrey Bot. Club* 22:1–218
74. Heywood, J. S. 1986. Clinal variation associated with edaphic ecotones in hybrid populations of *Gaillardia pulchella*. *Evolution* 40:1132–40
75. Honeycutt, R. L., Wilkinson, P. 1989. Electrophoretic variation in the parthenogenetic grasshopper *Warramaba virgo* and its sexual relatives. *Evolution* 43:1027–44
76. Howard, D. J. 1986. A zone of overlap and hybridization between two ground cricket species. *Evolution* 40:34–43
77. Hubbs, C. L. 1955. Hybridization between fish species in nature. *Syst. Zool.* 4:1–20
78. Key, K. H. L. 1968. The concept of stasipatric speciation. *Syst. Zool.* 17: 14–22
79. Key, K. H. L. 1981. Species, parapatry, and the morabine grasshoppers. *Syst. Zool.* 30:425–58
80. Kirkpatrick, J. B., Nunez, M. 1980. Vegetation-radiation relationships in mountainous terrain: eucalypt-dominated vegetation in the Risdon Hills, Tasmania. *J. Biogeogr.* 7:197–208
81. Kohlmann, B., Nix, H., Shaw, D. D. 1988. Environmental predictions and distributional limits of chromosomal taxa in the Australian grasshopper *Caledia captiva* (F.). *Oecologia* 75:483–93
82. Lamb, T., Avise, J. C. 1986. Directional introgression of mitochondrial DNA in a hybrid population of tree frogs: the influence of mating behavior. *Proc. Natl. Acad. Sci. USA* 83:2526–30
83. Lehman, N., Eisenhawer, A., Hansen, K., Mech, L. D., Peterson, R. O., et al. 1991. Introgression of Coyote mitochondrial DNA into sympatric North American gray wolf populations. *Evolution* 45:104–19
84. Lemeunier, F., Ashburner, M. 1984. Relationships within the melanogaster species subgroup of the genus *Drosophila (Sophophora)*. IV. The chromosomes of two new species. *Chromosoma* 89:343–51
85. Lemeunier, F., David, J. R., Tsacas, L., Ashburner, M. 1986. The melanogaster species group. In *The Genetics and Biology of Drosophila,* Vol 3, ed. M. Ashburner, H. L. Carson, J. N. Thompson, pp. 147–256. London: Academic Press
86. Levin, D. A. 1969. The challenge from a related species: a stimulus for saltational speciation. *Am. Nat.* 103:316–22
87. Levin, D. A., Bulinska-Radomska, Z. 1988. Effects of hybridization and inbreeding in Phlox. *Am. J. Bot.* 75: 1632–39
88. Lewontin, R. C. 1974. *The Genetic Basis of Evolutionary Change*. New York: Columbia Univ. Press. 346 pp.
89. Lewontin, R. C., Birch, L. C. 1966. Hybridization as a source of variation for adaptation to new environments. *Evolution* 20:315–36
90. Lotsy, J. P. 1916. *Evolution by Means of Hybridization*. The Hague: M. Nijhoff. 166 pp.
91. Mallet, J. 1986. Hybrid zones of *Heliconius* butterflies in Panama and

the stability and movement of warning colour clines. *Heredity* 56: 191–202
92. Marchant, A. D. 1988. Apparent introgression of mitochondrial DNA across a narrow hybrid zone in the *Caledia captiva* species-complex. *Heredity* 60:39–46
93. Marchant, A. D., Arnold, M. L., Wilkinson, P. 1988. Gene flow across a chromosomal tension zone. I. Relicts of ancient hybridization. *Heredity* 61: 321–28
94. Matfield, B., Ellis, J. R. 1972. The allopolyploid origin and genomic constitution of *Potentilla anglica*. *Heredity* 29:315–27
95. Mayr, E. 1942. *Systematics and the Origin of Species*. New York: Columbia Univ. Press. 334 pp.
96. McArthur, E. D., Welch, B. L., Sanderson, S. C. 1988. Natural and artificial hybridization between big sagebrush (*Artemisia tridentata)* subspecies. *J. Hered.* 79:268–76
97. Mecham, J. S. 1960. Introgressive hybridization between two southeastern treefrogs. *Evolution* 14:445–57
98. Meyn, O., Emboden, W. A. 1987. Parameters and consequences of introgression in *Salvia apiana* × *S. mellifera* (Lamiaceae). *Syst. Bot.* 12:390–99
99. Montanucci, R. R. 1983. Natural hybridization between two species of collared lizards (*Crotaphytus*). *Copeia* 1983:1–11
100. Moore, W. S. 1977. An evaluation of narrow hybrid zones in vertebrates. *Q. Rev. Biol.* 52:263–77
101. Moore, W. S. 1987. Random mating in the northern flicker hybrid zone: implications for the evolution of bright and contrasting plumage patterns in birds. *Evolution* 41:539–46
102. Moore, W. S., Buchanan, D. B. 1985. Stability of the northern flicker hybrid zone in historical times: implications for adaptive speciation theory. *Evolution* 39:135–51
103. Moran, C. 1979. The structure of the hybrid zone in Caledia. *Heredity* 42:13–32
104. Moran, C., Shaw, D. D. 1977. Population cytogenetics of the genus *Caledia* (Orthoptera: Acridinae). III. Chromosomal polymorphism, racial parapatry and introgression. *Chromosoma* 63: 181–204
105. Moritz, C. 1983. Parthenogenesis in the endemic Australian lizard *Heteronotia binoei* (Gekkonidae). *Science* 220:735–37
106. Muller, C. H. 1952. Ecological control of hybridization in Quercus: a factor in the mechanism of evolution. *Evolution* 6:147–61
107. Nagle, J. J., Mettler, L. E. 1969. Relative fitness of introgressed and parental populations of *Drosophila mojavensis* and *D. arizonensis*. *Evolution* 23:519–24
108. Nason, J. D., Ellstrand, N. C., Arnold, M. L. 1992. Patterns of hybridization and introgression in populations of oaks, manzanitas, and irises. *Am. J. Bot.* 79:101–111
109. Nelson, K., Baker, R. J., Honeycutt, R. L. 1987. Mitochondrial DNA and protein differentiation between hybridizing cytotypes of the white-footed mouse, *Peromyscus leucopus*. *Evolution* 41:864–72
110. Nevo, E., Bar-El, H. 1976. Hybridization and speciation in fossorial mole rats. *Evolution* 30:831–40
111. Porter, A. H. 1990. Testing nominal species boundaries using gene flow statistics: the taxonomy of two hybridizing admiral butterflies (*Limenitis:* Nymphalidae). *Syst. Zool.* 39:131–47
112. Potts, B. M. 1986. Population dynamics and regeneration of a hybrid zone between *Eucalyptus risdonii* Hook. f. and *E. amygdalina* Labill. *Aust. J. Bot.* 34:305–29
113. Potts, B. M., Reid, J. B. 1983. Hybridization between *Eucalyptus obliqua* L'Herit. and *E. pulchella* Desf. *Aust. J. Bot.* 31:211–29
114. Potts, B. M., Reid, J. B. 1985. Variation in the *Eucalyptus gunnii-archeri* complex. II. The origin of variation. *Aust. J. Bot.* 33:519–41
115. Potts, B. M., Reid, J. B. 1985. Analysis of a hybrid swarm between *Eucalyptus risdonii* Hook.f. and *E. amygdalina* Labill. *Aust. J. Bot.* 33:543–62
116. Potts, B. M., Reid, J. B. 1988. Hybridization as a dispersal mechanism. *Evolution* 42:1245–55
117. Potts, B. M., Reid, J. B. 1990. The evolutionary significance of hybridization in *Eucalyptus*. *Evolution* 44:2151–52
118. Powell, J. R. 1983. Interspecific cytoplasmic gene flow in the absence of nuclear gene flow: evidence from *Drosophila*. *Proc. Natl. Acad. Sci. USA* 80:492–95
119. Rand, D. M., Harrison, R. G. 1989. Ecological genetics of a mosaic hybrid zone: mitochondrial, nuclear, and reproductive differentiation of crickets by soil type. *Evolution* 43:432–49
120. Randolph, L. F. 1966. *Iris nelsonii,* a new species of Louisiana iris of hybrid origin. *Baileya* 14:143–69

121. Randolph, L. F., Nelson, I. S., Plaisted, R. L. 1967. Negative evidence of introgression affecting the stability of Louisiana Iris species. *Cornell Univ. Agr. Expt. Stat. Memoir* 398:1–56
122. Rattenbury, J. A. 1962. Cyclic hybridization as a survival mechanism in the New Zealand forest flora. *Evolution* 16:348–63
123. Rieseberg, L. H. 1991. Homoploid reticulate evolution in *Helianthus* (Asteraceae) evidence from ribosomal genes. *Am. J. Bot.* 78:1218–37
124. Rieseberg, L. H., Beckstrom-Sternberg, S. M., Liston, A., Arias, D. M. 1991. Phylogenetic and systematic inferences from chloroplast DNA and isozyme variation in *Helianthus* sect. *Helianthus* (Asteraceae). *Syst. Bot.* 16:50–76
125. Rieseberg, L. H., Beckstrom-Sternberg, S., Doan, K. 1990. *Helianthus annuus* ssp. *texanus* has chloroplast DNA and nuclear ribosomal RNA genes of *Helianthus debilis* ssp. *cucumerifolius*. *Proc. Natl. Acad. Sci. USA* 87:593–97
126. Rieseberg, L. H., Brunsfeld, S. 1992. Molecular evidence and plant introgression. In *Plant Molecular Systematics*, ed. D. E. Soltis, P. S. Soltis, J. J. Doyle, pp. 151–76. New York: Chapman & Hall
127. Rieseberg, L. H., Carter, R., Zona, S. 1990. Molecular tests of the hypothesized hybrid origin of two diploid *Helianthus* species (Asteraceae). *Evolution* 44:1498–1511
128. Rieseberg, L. H., Soltis, D. E. 1991. Phylogenetic consequences of cytoplasmic gene flow in plants. *Evol. Trends Plants* 5:65–84
129. Rieseberg, L. H., Soltis, D. E., Palmer, J. d. 1988. A molecular re-examination of introgression between *Helianthus annuus* and *H. bolanderi* (Compositae). *Evolution* 42:227–38
130. Rieseberg, L. H., Wendel, J. F. 1992. Introgression and its consequences in plants. In *Hybrid Zones and the Evolutionary Process*, ed. R. G. Harrison. Oxford: Oxford Univ. Press. In press
131. Riley, H. P. 1938. A character analysis of colonies of *Iris fulva, Iris hexagona* var. *giganticaerulea* and natural hybrids. *Am. J. Bot.* 25:727–38
132. Ritchie, M. G., Butlin, R. K., Hewitt, G. M. 1989. Assortative mating across a hybrid zone in *Chorthippus parallelus* (Orthoptera: Acrididae). *J. Evol. Biol.* 2:339–52
133. Ross, K. G., Vander Meer, R. K., Fletcher, D. J. C., Vargo, E. L. 1987. Biochemical phenotypic and genetic studies of two introduced fire ants and their hybrid (Hymenoptera: Formicidae). *Evolution* 41:280–93
134. Sage, R. D., Heyneman, D., Lim, K-C., Wilson, A. C. 1986. Wormy mice in a hybrid zone. *Nature* 324:60–63
135. Sage, R. D., Selander, R. K. 1979. Hybridization between species of the *Rana pipiens* complex in central Texas. *Evolution* 33:1069–88
136. Sanderson, N. 1989. Can gene flow prevent reinforcement? *Evolution* 43: 1223–35
137. Sattler, P. W. 1985. Introgressive hybridization between the Spadefoot toads *Scaphiopus bombifrons* and *S. multiplicatus* (Salientia: Pelobatidae). *Copeia* 1985:324–32
138. Schemske, D. W., Morgan, M. T. 1990. The evolutionary significance of hybridization in *Eucalyptus*. *Evolution* 44: 2150–51
139. Schnabel, A., Asmussen, M. A. 1989. Definition and properties of disequilibria within nuclear-mitochondrial-chloroplast and other nuclear-dicytoplasmic systems. *Genetics* 123:199–215
140. Shaw, D. D. 1976. Population cytogenetics of the genus *Caledia* (Orthoptera: Acridinae). I. Inter-and intra- specific karyotype diversity. *Chromosoma* 54:221–43
141. Shaw, D. D. 1981. Chromosomal hybrid zones in orthopteroid insects. In *Evolution and Speciation: Essays in Honour of M. J. D. White*, ed. W. R. Atchley, D. S. Woodruff, pp. 146–70. Cambridge: Cambridge Univ. Press
142. Shaw, D. D., Coates, D. J., Arnold, M. L. 1988. Complex patterns of chromosomal variation along a latitudinal cline in the grasshopper *Caledia captiva*. *Genome* 30:108–17
143. Shaw, D. D., Coates, D. J., Arnold, M. L., Wilkinson, P. 1985. Temporal variation in the chromosomal structure of a hybrid zone and its relationship to karyotypic repatterning. *Heredity* 55: 293–306
144. Shaw, D. D., Marchant, A. D., Arnold, M. L., Contreras, N. 1988. Chromosomal rearrangements, ribosomal genes and mitochondrial DNA: introgression across a narrow hybrid zone. In *Kew Chromosome Conference III*, ed. P. E. Brandham, M. D. Bennett, pp. 121–129. London: Allen & Unwin
145. Shaw, D. D., Marchant, A. D., Arnold, M. L., Contreras, N., Kohlmann, B. 1990. The control of gene flow across a narrow hybrid zone: a selective role for chromosomal rearrangement? *Can. J. Zool.* 68:1761–69

146. Shaw, D. D., Moran, C., Wilkinson, P. 1980. Chromosomal reorganisation, geographic differentiation and the mechanism of speciation in the genus *Caledia*. In *Insect Cytogenetics,* ed. R. L. Blackman, G. M. Hewitt, and M. Ashburner, pp. 171–94. Oxford: Blackwell Scientific
147. Shaw, D. D., Webb, G. C., Wilkinson, P. 1976. Population cytogenetics of the genus *Caledia* (Orthoptera: Acridinae). II. Variation in the pattern of C-banding. *Chromosoma* 56:169–90
148. Shaw, D. D., Wilkinson, P. 1980. Chromosome differentiation, hybrid breakdown and the maintenance of a narrow hybrid zone in Caledia. *Chromosoma* 80:1–31
149. Sibley, C. G. 1954. Hybridization in the red-eyed towhees of Mexico. *Evolution* 8:252–90
150. Slatkin, M. 1987. Gene flow and the geographic structure of natural populations. *Science* 236:787–92
151. Smith-Huerta, N. L. 1986. Isozymic diversity in three allotetraploid *Clarkia* species and their putative diploid progenitors. *J. Hered.* 77:349–54
152. Smyth, C. A., Hamrick, J. L. 1987. Realized gene flow via pollen in artificial populations of Musk Thistle, *Carduus nutans* L. *Evolution* 41:613–19
153. Solignac, M., Monnerot, M. 1986. Race formation, speciation, and introgression within *Drosophila simulans, D. mauritiana,* and *D. sechellia* inferred from mitochondrial DNA analysis. *Evolution* 40:531–39
154. Spolsky, C., Uzzell, T. 1984. Natural interspecies transfer of mitochondrial DNA in amphibians. *Proc. Natl. Acad. Sci. USA* 81:5802–5
155. Stangl, F. B. Jr. 1986. Aspects of a contact zone between two chromosomal races of *Peromyscus leucopus* (Rodentia: Cricetidae). *J. Mamm.* 67: 465–73
156. Stebbins, G. L. 1959. The role of hybridization in evolution. *Am. Philos. Soc.* 103:231–51
157. Szymura, J. M., Barton, N. H. 1986. Genetic analysis of a hybrid zone between the fire-bellied toads, *Bombina bombina* and *B. variegata,* near Cracow in southern Poland. *Evolution* 40:1141–59
158. Tegelstrom, H., Gelter, H. P. 1990. Haldane's rule and sex biased gene flow between two hybridizing flycatcher species (*Ficedula albicollis* and *F. hypoleuca,* Aves: Muscicapidae). *Evolution* 44:2012–21
158a. Thompson, J. D. 1991. The biology of an invasive plant. *BioScience* 41: 393–401
159. Vanlerberghe, F., Dod, B., Boursot, P., Bellis, M., Bonhomme, F. 1986. Absence of *Y*-chromosome introgression across the hybrid zone between *Mus musculus domesticus* and *Mus musculus*. *Genet. Res.* 48:191–97
160. Viosca, P. Jr. 1935. The irises of southeastern Louisiana: a taxonomic and ecological interpretation. *J. Am. Iris Soc.* 57:3–56
161. Vrijenhoek, R. C., Angus, R. A., Schultz, R. J. 1977. Variation and heterozygosity in sexually vs. clonally reproducing populations of *Poeciliopsis*. *Evolution* 31:767–81
162. Wagner, D. B., Furnier, G. R., Saghai-Maroof, M. A., Williams, S. M., Dancik, B. P., Allard, R. W. 1987. Chloroplast DNA polymorphisms in lodgepole and jack pines and their hybrids. *Proc. Natl. Acad. Sci. USA* 84:2097–100
163. Wagner, W. H. Jr. 1969. The role and taxonomic treatment of hybrids. *BioScience* 19:785–789
164. Wagner, W. H. Jr. 1970. Biosystematics and evolutionary noise. *Taxon* 19:146–51
165. Wagner, W. H. Jr., Schoen, D. J. 1976. Shingle oak (*Quercus imbricaria)* and its hybrids in Michigan. *Mich. Bot.* 15:141–55
166. Warwick, S. I., Bain, J. F., Wheatcroft, R., Thompson, B. K. 1989. Hybridization and introgression in *Carduus nutans* and *C. acanthoides* reexamined. *Syst. Bot.* 14:476–94
167. Wasserman, A. O. 1957. Factors affecting interbreeding in sympatric species of spadefoots (Genus *Scaphiopus*). *Evolution* 11:320–38
168. Wendel, J. F. 1989. New World tetraploid cottons contain Old World cytoplasm. *Proc. Natl. Acad. Sci. USA* 86:4132–36
169. White, M. J. D. 1978. *Modes of Speciation*. San Francisco: Freeman. 455 pp.
170. Whitham, T. G. 1989. Plant hybrid zones as sinks for pests. *Science* 244: 1490–93
171. Whittemore, A. T., Schaal, B. A. 1991. Interspecific gene flow in sympatric oaks. *Proc. Natl. Acad. Sci. USA* 88: 2540–44
172. Wiegand, K. M. 1935. A taxonomist's experience with hybrids in the wild. *Science* 81:161–66
173. Williams, E. G., Rouse, J. L. 1988. Disparate style lengths contribute to

isolation of species in *Rhododendron*. *Aust. J. Bot.* 36:183–91
174. Woodruff, D. S. 1973. Natural hybridization and hybrid zones. *Syst. Zool.* 22:213–18
175. Woodruff, D. S., Gould, S. J. 1987. Fifty years of interspecific hybridization: genetics and morphometrics of a controlled experiment on the land snail *Cerion* in the Florida keys. *Evolution* 41:1022–45
176. Wright, S. 1931. Evolution in mendelian populations. *Genetics* 16:97–159
177. Wright, S. 1969. *Evolution and the Genetics of Populations*, Vol. 2, *The Theory of Gene Frequencies*. Chicago: Univ. Chicago Press. 511 pp.
178. Wynn, A. H. 1986. Linkage disequilibrium and a contact zone in *Plethodon cinereus* on the del-mar-va peninsula. *Evolution* 40:44–54
179. Yang, S. Y., Selander, R. K. 1968. Hybridization in the grackle *Quiscalus quiscula* in Louisiana. *Syst. Zool.* 17: 107–44
180. Zohary, D., Feldman, M. 1962. Hybridization between amphidiploids and the evolution of polyploids in the wheat (*Aegilops-Triticum*) group. *Evolution* 16:44-61

Annu. Rev. Ecol. Syst. 1992. 23:263–86

THE NEARLY NEUTRAL THEORY OF MOLECULAR EVOLUTION

Tomoko Ohta
National Institute of Genetics, Mishima 411, Japan

KEY WORDS: molecular evolution and polymorphism, random drift, weak selection

INTRODUCTION

For a long time the study of evolution has been based on morphology; the long neck of a giraffe, the human brain, a bird's wing, and so on. Morphological change in evolution is explained by Darwin's theory of natural selection, but this theory is largely qualitative rather than quantitative. Population genetics started more than half a century ago as an attempt to understand evolutionary change quantitatively. Because evolution must take place in all individuals of a species, the change of gene frequency in the population has been analyzed. However, so long as the facts of evolution are based on morphological traits, evolutionary change is very difficult to connect with gene frequency change except in relatively few circumstances.

The remarkable progress of molecular biology has made it possible to apply population genetics theory to real data. We now know that genetic information is stored in linear sequences of DNA which are stably transmitted from generation to generation, and we can compare the linear sequences of DNA and amino acids among species. It is also possible to compare secondary and tertiary structures of proteins and nucleic acids from various sources.

Because of such progress, some aspects of traditional neo-Darwinism are beginning to need revision. The first step in such a revision is the neutral mutation-random drift hypothesis put forward by Kimura (47) in 1968. In the next year, King & Jukes (53a) published a similar idea, though from a more biochemical point of view than that of Kimura. This theory states that most evolutionary changes at the molecular level are caused by random genetic drift of selectively neutral or nearly neutral mutations rather than by natural selection. Because this theory was contrary to the neo-Darwinian view at that time, it provoked much controversy. A complete review of the theory is found

0066-4162/92/1120-0263$02.00

in Kimura's book (49), so I shall give only a brief outline here, together with the basic concepts of the stochastic theory of population genetics.

The second step in the revision is to clarify the interaction of natural selection and random drift at the molecular level. Natural selection cannot be so simple as to be "all or nothing." There are numerous types of mutations, whose behavior is influenced by both selection and random drift. In this article, theoretical studies of such "nearly neutral" mutations are reviewed, together with some recent findings on DNA sequence variation. In addition, some possible ways of connecting molecular and phenotypic evolution are discussed with special reference to the versatile nature of DNA, also clarified in recent years.

Population Genetics

The study of population genetics was started by a trio of pioneers, Fisher (23), Haldane (35), and Wright (112), who formulated the basic framework. Gene frequency changes in the population, caused by Darwinian natural selection, are essential in this theory. Through successive investigations, various detailed theories on the effects not only of selection but also of mutation, migration, and random sampling drift at reproduction have been worked out.

The results have been used for establishing the modern synthetic theory of evolution (20). In addition, human population studies (12) as well as plant and animal breeding (22) depend heavily on population genetics theory.

In the period from 1966 to 1980, many studies on genetic variation at protein producing loci within species were published. Lewontin & Hubby (60) and Harris (36) were among the first to use electrophoresis for studying protein variation from the standpoint of population genetics. It turned out that biochemical variation is rather abundant in human and in *Drosophila* populations. The typical individual is heterozygous at 5–15% of its protein loci. At about the same time, data on sequence comparisons of hemoglobins and cytochrome *c* between mammalian species became available (116). By examining such data, Kimura (47), who was an authority on stochastic processes of population genetics, proposed the neutral mutation–random drift hypothesis of molecular evolution. This hypothesis was based on the stochastic theory of population genetics and has been most stimulating to the further development of both experimental and theoretical population genetics at the molecular level.

Rate of Molecular Evolution

Since the advent of rapid DNA sequencing techniques, data on the primary structure of genes are accumulating amazingly fast, and now statistical studies of DNA sequences are quite popular. However, only 20 or so years ago, most of the available data on molecular evolution were in the form of amino acid sequences (59, 4). Hemoglobin α of mammals consists of 141 amino acids, and it is one of the best-studied molecules. If one compares human hemoglobin α with that of the gorilla, all amino acids are identical except one, but 18 amino acids differ between human and horse. Such data on sequence divergence faithfully reflect the phylogenetic relationship. Since we know the approximate time of divergence of mammalian species, it is possible to estimate the rate of amino acid substitution. From a comparison of various species the rate of amino acid substitution of hemoglobins α and β is about 10^{-9} per amino acid site per year. It is quite impressive to find that almost the same value is obtained from any two species. Especially noteworthy is the fact that the rate seems to be almost the same in the line to living fossils whose morphological characters have hardly changed for tens of millions of years, and in the line leading to humans.

The apparent uniformity of evolutionary rates, as compared with phenotypic evolution, is a most remarkable characteristic of protein evolution, (116). By applying similar analysis to cytochrome *c* data, the rate seems again to be almost uniform for diverse organisms including plants, fungi, and mammals. However, the rate is much lower in cytochrome *c* than in hemoglobins. This is thought to be caused by stronger structural constraints on cytochrome *c* than on hemoglobins. This is another characteristic of molecular evolution, i.e. the stronger the constraint on the molecule, the lower is its rate of evolution. For some examples of the rates of protein evolution, see (49). It is now well known that fibrinopeptides have evolved rapidly with little constraint whereas histone IV has been evolving extremely slowly.

We attempted to estimate the variance of the evolutionary rate of a particular protein, in order to test the uniformity of the rate, and we found that the variance of amino acid substitution is larger than that expected if a simple Poisson process is assumed (96). This analysis may be flawed because it is based on paleontological estimates of divergence time. But the conclusion has been confirmed by Langley & Fitch (58), who used more data and statistics less dependent on paleontological estimates. Kimura (49) used data from mammalian orders that are believed to form a star phylogeny, i.e. simultaneous divergence of many mammals about 80 million years ago (198a). However, see Ref. 21a on some problems of this approach. For hemoglobins, myoglobin, cytochrome *c* and ribonuclease, the ratio, R, of the variance to the mean divergence turned out to be from 1.3 to 3.3, again confirming the previous result. Gillespie (28) extended such analyses for more proteins; hemoglobins, cytochrome *c*, insulin, prolactin, ribonuclease, LHP, albumin, cytochrome oxidase, ATPase 6 and cytochrome *b*, and he found that R takes values between 0.2 and 34.1. The extreme values are thought to be due to small sample size (three species with star phylogenies were examined).

Gillespie (28) argues that this pattern fits the "episodic" process in which a burst of amino acid substitutions is followed by a static phase.

Since methods of DNA sequencing have become available, more and more data on DNA sequences are accumulating and molecular evolutionary studies have shifted from analysis of amino acid sequences to that of DNA sequences. Several remarkable features of DNA evolution have emerged. The majority of genomic DNA of higher organisms evolves more rapidly than protein coding regions, i.e. those DNA regions that apparently do not carry genetic information in their primary structure are evolving rapidly. In mammals, the rate of nucleotide substitution in these regions is roughly 5×10^{-9} per site per year (e.g. see 49, 76, 63). The rate of synonymous substitutions in coding regions is slightly lower than this but is rather uniform among various genes, whereas the rate of amino acid replacement substitutions differs greatly from gene to gene. These values agree with the results of DNA hybridization studies (10, 104).

An as yet unsettled problem is whether the rate of DNA evolution depends upon generation length. An examination of amino acid sequences revealed little effect of generation time (55, 57). DNA hybridization studies, however, indicate that the longer the generation, the lower the DNA evolution rate (57, 55, 10, 104). This is still a controversial problem and is discussed later.

Another topic of interest is the isochore concept, the differentiation of chromosomal regions of warm blooded vertebrates into GC- and AT-rich segments (6, 41). Such segments are called "isochores" and have an average size well above 200 kilobases (6). A noteworthy fact is that most housekeeping genes locate in GC-rich isochores, whereas many tissue-specific genes are in AT-rich ones. In addition, codon usage is different in these two groups of genes (41). Thus, a codon bias in mammals appears to be largely determined by mutation pressure, but the bias in lower organisms with intermediate GC content such as in *E. coli* is influenced by selection because of the efficient selection in very large populations (102). Such selection must be nearly neutral, i.e. at the border between neutrality and selection.

Population Dynamics of Mutant Genes

A basic requirement for understanding the mechanisms of nucleotide or amino acid substitutions in evolution is to distinguish mutations from evolutionary substitutions. Numerous mutations appear in Mendelian populations in every generation, but the majority will be lost within a few generations. Thus, those mutations that contribute to evolution are a very small minority of all mutations. It is also necessary to understand the process of frequency increase of mutants in the population in the course of their substitution.

For neutral mutants, the process has been theoretically analyzed using the diffusion equation method (51, 53). A neutral mutant, if it is ultimately fixed in the population, takes on the average 4N generations until this occurs.

$$t_1 = 4N, \qquad 1.$$

where N is the effective population size. If N is large, the time is very long. The rate of molecular evolution is measured by averaging the number of substitutions over very long period of time. It may be expressed as follows,

$$k = \lim_{T \to \infty} \frac{n(T)}{T} \qquad 2.$$

where T is the period, and n(T) is the number of mutant substitutions in this period. Obviously, for k to be measured accurately, $T >> N$.

Now consider a locus encoding a protein. Let the rate of occurrence of base substitutions in this DNA region be v_g per generation, and let u be the probability of fixation of a mutant. Then in a population of N individuals, the total number of mutations appearing in the population is $2Nv_g$ per generation, and a fraction u of them spread through the population, so the rate of substitution per generation becomes,

$$k_g = 2Nv_g u. \qquad 3.$$

Here u generally depends upon the magnitude of natural selection. It should be remembered that, at the molecular level, the number of nucleotide sites of a locus is so large that the probability of having identical mutations more than once is almost nil. Also, the probability of back mutation is negligibly small.

Let us now examine how natural selection influences the rate. If most substitutions are caused by Darwinian natural selection, and the average selective advantage of such substitutions is s with no dominance, the fixation probability is roughly twice the selective advantage (23, 35), and we have,

$$k_g = 4Nsv_g. \qquad 4.$$

Hence k_g depends on the product of three parameters, N, s and v_g. But when most substitutions are selectively neutral, the fixation probability is equal to the initial frequency, ½ N (113, 46), and we have,

$$k_g = v_g. \qquad 5.$$

In other words, the evolutionary rate is simply equal to the mutation rate, and it is independent of population size (47).

The actual rate of molecular evolution seems to be roughly constant per year for each protein as reviewed in the previous section. In order to explain this fact by a selection model (Eq. 4), one has to assume that parameters like N and s have been nearly equal in various lineages. Such a situation could hardly hold in very different environments, particularly in the lines leading to living fossils and in those with rapid phenotypic evolution. On the other hand, if the majority of substitutions are selectively neutral as was first proposed by Kimura (47), from formula (5), the observed pattern may be explained if the neutral mutation rate is constant per year. Here the possible problems are: (i) how the generation length affects the evolution rate, (ii) how fluctuation in the evolutionary rate is related to natural selection, and (iii) how selective constraints influence the rate. These problems are discussed later in relation to weak natural selection, i.e. the interaction of random genetic drift and selection.

Selective Constraints and Evolutionary Rate

The rate of evolution is different from protein to protein, and the difference reflects the degree of constraint as explained before. This constraint is directly connected to the function and structure of protein or RNA molecules, i.e. the more rigid their function and structure are, the lower is the rate. This point has been beautifully shown for fibrinopeptides, hemoglobin, cytochrome *c* and histone IV by Dickerson (19).

Such evolutionary features may be explained by the neutral mutation random drift hypothesis as follows. As in Kimura (49), v_T designates the rate of occurrence of new mutations in terms of nucleotide substitutions. Let f_0 be the fraction of such mutations that are selectively neutral, and the remaining $(1 - f_0)$ be the fraction of mutations that have a deleterious effect. Then the rate of neutral evolution becomes,

$$k = f_0 v_T, \qquad 6.$$

because deleterious mutations do not contribute to evolution. Favorable mutations are assumed to be too rare to have any statistical influence. Various degrees of selective constraint may be taken into account by f_0, e.g. for pseudogenes without constraint, $f_0 = 1$, whereas for amino acid replacement substitutions of histone IV (see Table 1), $f_0 = 0$. Here the question is whether or not mutations may be simply divided into neutral and deleterious classes. This problem is discussed later.

Polymorphisms at the Molecular Level

Elucidating the mechanism for maintaining genetic variability has been one of the most important problems of population genetics. Particularly at the molecular level, stimulated by the neutral mutation theory, numerous attempts were made around the 1970s to clarify the roles of selection and drift (59). According to the neutral theory, molecular polymorphisms are a phase of evolution (52), whereas to selectionists, they are actively maintained by balancing selection and independent of mutant substitutions.

The pioneering work on protein polymorphisms was published by Lewontin

& Hubby (60), who found that heterozygosity at enzyme loci measured by electrophoresis is about 12%, and that about one third of the loci are polymorphic in *D. melanogaster*. At about the same time, Harris (36) reported that heterozygosity by electrophoresis is about 6% in human. Since then, many investigators examined the level of protein polymorphisms of many species (5, 8, 100, 77).

If protein polymorphisms are mostly neutral, they represent an intermediate stage of neutral gene substitution. Let us ask how much genetic variation is expected in a population when an equilibrium is reached between mutational input and random extinction in this process. A most convenient measure of genetic variation at a locus is a quantity called "virtual heterozygosity" (49), which is the probability that two randomly chosen alleles from the population will be different. In a randomly mating diploid population, heterozygosity and virtual heterozygosity are the same. The virtual heterozygosity at equilibrium has been shown to be,

$$\hat{H} = \frac{4Nv}{1 + 4Nv}, \qquad 7.$$

where v is the mutation rate (50, 69). This is remarkably simple in that only one product, Nv, comes into the formula. Nevertheless the formula is robust and applicable to many cases. As was pointed out before, the observed values of heterozygosity at enzyme loci are often around 12% in *Drosophila* species, and 5–6% in mammalian species. Then, Nv is predicted to be roughly 0.035 for *Drosophila,* and 0.015 in mammals, provided that most polymorphisms are selectively neutral and that the species are in equilibrium between random drift and mutation.

Of course, various species have greater or less heterozygosity. So far, *E. coli* is the species with highest value—47% (101). In several large animal species such as cheetah (78), polar bear (1), and elephant seal (8), the heterozygosity is practically nil. These species are thought to have had small population size in the past so that Nv is extremely small (49). However, if the neutral theory is correct for existing enzyme polymorphisms, the problem is still not settled. This is discussed further in the next section.

Recently, data on DNA sequence polymorphisms are becoming available. As expected from the neutral theory, polymorphisms in noncoding regions or at synonymous sites are much higher than at amino acid–replacing sites. The percentage of nucleotide differences among sampled alleles at the Adh locus in *D. melanogaster* was 0.13–1.22%, with an average value of 0.65% (56). Since an ordinary protein coding region consists of several hundred nucleotide sites, a heterozygosity of 0.65% per site is much higher than the 12% at the enzyme loci, even if the latter is measured by electrophoresis, which detects only a part of the total nucleotide change. In a way similar to the difference of evolutionary rates between protein and noncoding DNA regions, polymorphisms are different between the two regions, i.e. the stronger the constraints, the fewer polymorphisms are observed.

A notable finding on DNA polymorphism is the contrasting patterns at the rosy region in *D. melanogaster* and *D. simulans* reported by Aquadro et al (3). These authors found that, unlike protein polymorphisms, DNA sequence variation in this region of *D. simulans* is estimated to be several times as great as that of *D. melanogaster*. Note that the average heterozygosity over many protein loci is almost the same in the two species but that geographic differentiation is more pronounced in *D. melanogaster* than in *D. simulans* (106). Aquadro et al (3) suggested that the effective population size of *D. simulans* is larger than that of *D. melanogaster,* and that the differences in species effective population size may be responsible for the pattern, i.e. weak selection would be effective on protein variation, but not on DNA variation. This fact has an important bearing on the nearly neutral theory and is discussed in the next section.

Meaning of Near Neutrality

In the previous sections, I have repeatedly emphasized that the rate of molecular evolution is strongly dependent upon selective constraints of proteins or nucleic acids. Under the neutral mutation–random drift theory, it is assumed that a certain fraction of new mutations are free of constraint or are selectively neutral, while the rest have deleterious effects and are selectively eliminated. An important question is how the two classes are distinguished by natural selection. Let us now examine theoretically the behavior of the mutants belonging to the borderline class. The critical quantity is the fixation probability of mutant genes. I examine the simplest case of a semidominant gene with selective advantage, s. Fixation probability in finite populations has been shown to become the function of the product, Ns, as follows (46, 69),

$$u = \frac{1 - e^{-4Nsp}}{1 - e^{-4Ns}}, \qquad 8.$$

where p is the initial frequency in the population, and is assumed to be much less than unity. Figure 1 shows the fixation probability of mutants as a function of Ns relative to the completely neutral ones ($Ns = 0$). As can be seen from the figure, the fixation probability is a continuous monotone function of Ns. Thus, when discussing molecular evolution by nearly neutral mutants, one has to consider all mutants around $Ns = 0$.

Let us examine the borderline mutations in some detail. Figure 2 shows classification of new mutations. The upper part shows the simple neutral model, and the lower part, the nearly neutral model. In the figure, neutral

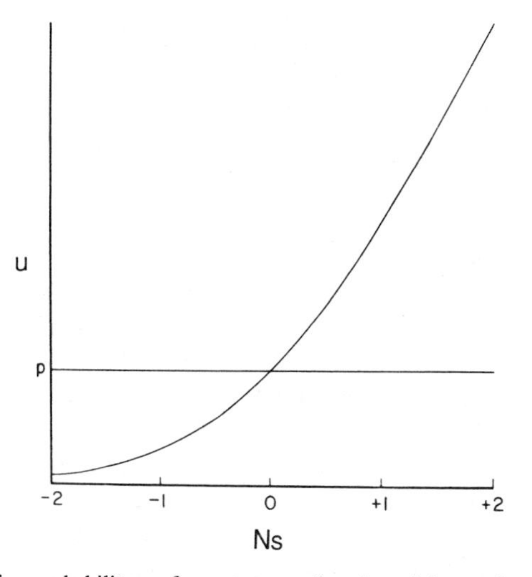

Figure 1 The fixation probability u of a mutant as a function of the product of population size and selection coefficient, Ns, relative to the completely neutral case. p is the initial frequency and the line, $u = p$, is the value for a neutral mutant.

means strictly neutral, deleterious means definitely deleterious, and nearly neutral means intermediate between neutral and selected. The behavior of mutations of the nearly neutral class is affected by both selection and drift.

As shown in Figure 1, the effectiveness of selection is determined by the product of the effective population size and the selection coefficient, Ns. Actual species have various population sizes from very small to very large, and therefore the effectiveness of selection will differ among species. In addition, physiological conditions may influence weak selection, e.g. a constraint on an enzyme function may differ between homeotherms and poikilotherms. The simple separation of new mutations into the deleterious and neutral classes will then not be satisfactory. There should be a substantial fraction of mutations that belong to the nearly neutral class.

Especially in view of the importance of negative selection caused by constraints, it is likely that many nearly neutral mutants are very slightly

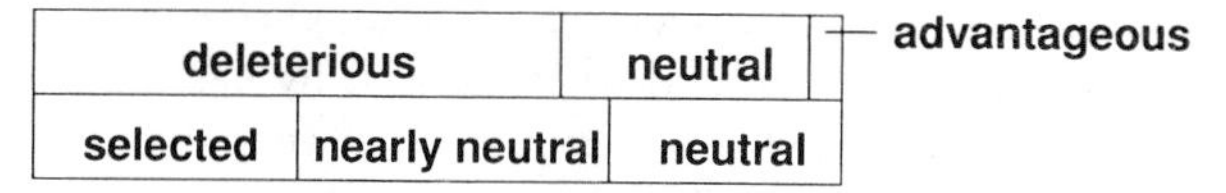

Figure 2 A schematic diagram shows the proportion of various classes of mutations. Deleterious mutants are definitely deleterious and neutral mutants are strictly neutral, while most selected mutants are deleterious, but the group also includes advantageous alleles. Nearly neutral mutants comprise a class intermediate between neutral and selected mutants.

deleterious, i.e. on the left side of $Ns = 0$ in Figure 1. My hypothesis of slightly deleterious mutants is based on this view and regards the near neutrality as the limit when the selection coefficient approaches zero (80–84). In other words, I propose that a substantial fraction of mutant substitutions at the molecular level are caused by random fixation of very slightly deleterious mutations.

This hypothesis seems realistic when one realizes that the effect of an amino acid substitution in a protein often produces only a minor modification of a reaction coefficient (45). It is also in accord with a recent finding that many enzyme variants in *E. coli* are likely to be mildly deleterious based on statistical analyses of their frequency distribution (99). Molecular variants that disturb very slightly the secondary structure of molecules, e.g. by opening the stem region of a clover leaf structure of tRNA, may represent a mutant class with mild deleterious effects (82, 83). In this case, it is likely that a slightly deleterious base substitution is followed by a slightly advantageous (compensatory) substitution. This is supported by the observation that G.U or U.G pairs that represent intermediate steps account for about two thirds of all non-Watson-Crick pairs in the helical regions of analyzed tRNAs (38).

The model of very slightly deleterious mutation is related to the molecular clock in an important way. One problem with the molecular clock is that the observed substitution rate is constant per year rather than per generation. The generation-time effect is particularly evident when divergence is measured by DNA hybridization (57, 55, 10, 105). According to these studies, the rate of divergence of single-copy genomic DNA varies from 10^{-9} to 10^{-8} per base pair per year depending upon the taxon, and there appears to be a high negative correlation between the rate and generation length. Similar effects have been found when synonymous and other unimportant DNA divergences are examined (13, 34, 37, 66, 71, 109, 114). Thus, it seems that the rate of evolution of the majority of genomic DNA depends upon generation length, whereas the divergence rate of amino acid sequences is relatively insensitive to it.

In the early 1970s, I tried to explain this seemingly contradictory observation by using population genetics theory. The fundamental idea is that most genomic DNA of higher organisms can freely accumulate base substitutions (80), i.e. most new mutations are selectively neutral. On the other hand, amino acid substitutions are more likely to be influenced by natural selection, i.e. many of them may be regarded as nearly neutral, or very slightly deleterious. Let us now examine how slightly deleterious mutations are related to the generation-time effect. The left side of $Ns = 0$ in Figure 1 shows that there is a negative correlation between the fixation probability and population size, provided that the selection coefficient is unchanged by population size. In other words, the chance of spreading by random drift is much higher in a small population than in a large population. To be more quantitative, several models of nearly neutral but slightly deleterious mutations have been studied by assuming some distribution of the selection coefficients for new mutations. I assumed an exponential function (85), Kimura (48), a gamma function, and Li (61), several classes of selection coefficients of new mutations. Again, all these models predict that the rate of mutant substitution is negatively correlated with the species population size, although the magnitude of the correlation depends on the model.

This prediction has an important bearing on the generation-time effect. In general, large organisms have a long generation length and small population size and vice versa. Thus there is a negative correlation between population size and generation length. On the other hand, the mutation rate per year will be lower in organisms with longer generation times as it is ordinarily believed (110, p. 254), and DNA divergence reflects such an effect. With selection this effect is expected to be partially cancelled by the negative correlation between the substitution rate and population size. Thus the amino acid substitution rate is expected to be relatively insensitive to generation length as compared to DNA divergence because amino acids are more likely to be subject to negative selection than are noncoding regions.

Actually, the small selection coefficients are not likely to remain unchanged in the long course of their substitutions (e.g. 25, 75). When varying selection intensity is examined, one notices that there are two quite different approaches. One is the so-called variable selection model in which selection coefficients fluctuate randomly around zero (e.g. see 26, 27, 30, 108). Here the selection coefficient varies from generation to generation or with some correlation between near generations, i.e. the variation is short term.

Another model considers long-term variation of the selection coefficient (81, 90, 70). In this model, the coefficients remain stable for a certain period of time or within a certain local area, but may vary in another period or area through a change of environmental conditions, which include both the internal genetic background and external ecological factors. Here the coefficient varies on a longer time scale than in the previous model. This model is difficult to formulate mathematically, and no precise analysis has yet been worked out.

We have performed extensive Monte Carlo simulations (97, 107). In the simulations, the selection coefficient of a new mutant in a local colony is assumed to follow the normal distribution with mean μ_s and variance σ_s^2. It is assumed that there are l habitats in the total population, and that each colony occupies a habitat. Let $s_{k,i}$ be the selection coefficient for the k-th mutant in the i-th colony and assume they are independent, so that the selection intensity fluctuates independently over space. This model is different from the landscape model of Gillespie (27) because we are concerned with the interaction effect of random drift and natural selection, whereas Gillespie treated the situation of strong selection and weak mutation. Let N be the effective population size of a local colony, and $N_T = lN$ be the total size. The most important parameter is twice the product of the population size and the standard deviation of the selection coefficient, $2N\sigma_s$, because σ_s is the measure of selection intensity in this model.

It should be noted that the present model is different from those of Ohta (85), Kimura (48) and others, because here the distribution of selection coefficients is fixed irrespective of mutant substitutions and is the same as Kingman's (54) house-of-cards model. Let us call the present one the fixed model, and the previous one the shift model. Note that in the shift model, the population mean is reset to zero after each substitution. What do these models imply about the real process? Consider the structure of a protein. In the shift model, an amino acid substitution has no effect on the other amino acid substitutions in the protein. Therefore each substitution is independent. On the other hand, in the fixed model, the effect of each substitution remains and affects subsequent substitutions by changing the mean fitness. Therefore substitutions are interrelated in their effects on fitness. In terms of the higher order structure of a protein, substitutions that occur at many amino acid sites do not behave independently, and the fixed model I believe is a better description of nature. The shift model also implies, unrealistically, that proteins can improve or deteriorate indefinitely by successive substitutions, whereas this process may stop under the fixed model. In a study of hemoglobin genes, Goodman (32) suggested that evolution had accelerated after gene duplication but had slowed down in later phylogeny, and he argued that this reflects improvement of gene functions. The fixed model is consistent with this scenario.

Simulations were performed by choosing realistic values for the products of $2N$ and other parameters. The mean fitness of the population fluctuates with time because of random drift. Figure 3 shows the distribution of the selection coefficient of new mutations around the population mean. If selection is strong enough, the mean moves towards the right without fluctuation. For nearly neutral mutations, it moves erratically but tends to increase. When the mean is positive, the average selection coefficient of new mutations becomes negative, i.e. new mutations are slightly deleterious on the average.

The results of our simulations show the properties of the model in detail. A panmictic population is easiest to treat, so consider the case of a single colony. Tachida (107) examined various quantities of biological interest. When $4N\sigma_s$ is larger than 3, several advantageous mutants are quickly fixed, and thereafter almost all new mutations are deleterious. Then the fitness of the population changes very slowly, and substitution almost stops. However, the system takes a very long time to reach true equilibrium. When $4N\sigma_s$ is in

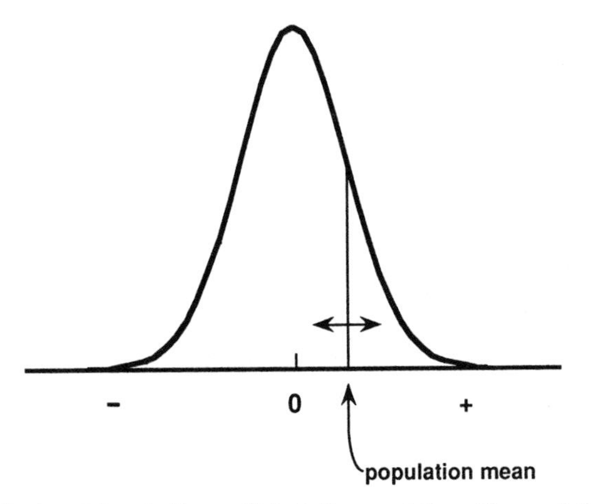

Figure 3 Distribution of the selection coefficient of new mutations. The population mean moves to the left or to the right by selection and drift.

the range 0.2 to 3, both random drift and selection affect the population fitness, which very slowly increases and reaches an equilibrium value in a period of the order of number of generations of the reciprocal of the mutation rate. Even after reaching equilibrium, substitution does not stop but is less frequent than in the completely neutral case. When $4N\sigma_s$ is less than 0.2, the mutants' behavior is almost the same as a completely neutral one. Tachida (107) called these three classes of mutations, selected ($4N\sigma_s > 3$), nearly neutral ($3 \geq 4N\sigma_s \leq 0.2$) and effectively neutral ($4N\sigma_s < 0.2$).

When the population is subdivided, and the selection coefficient fluctuates spatially, the situation is not so simple. Firstly, the selection intensity of the total population becomes the average of selection coefficients of the local colonies. If there are l habitats (colonies) and the selection coefficient of a colony is independently distributed as we assumed, the selection coefficient of the total population is normally distributed with the same mean, μ_s but with a smaller variance, σ_s^2/l, as expected. If the migration rate is high enough, the mutants' behavior becomes similar to that in panmictic populations with the variance of selection coefficient, σ_s^2/l.

Let us now examine the results for the cases where migration is less. Wright's island model was used, incorporating local extinction and recolonization, in the belief that natural populations often pass through bottlenecks, so that current populations are descended from a small number of founders (11). A most interesting question is: How does the total population size affect evolution by nearly neutral mutations? Recall that one problem of the nearly neutral theory is its prediction of a reduction of the evolutionary rate when population size becomes large. As mentioned before, this is related to the constancy of the evolutionary rate, particularly the effect of generation length which is thought to be partially cancelled by the population-size effect. In general, the nearly neutral model predicts that the stronger the selection is, the larger the population-size effect becomes. Under the shift model with constant selection coefficients, the transition from the neutral to the selected class is quite steep as the population size increases (see Figure 1). Under the fixed model, the transition becomes more gentle, as Figure 4 shows. The transition from the neutral to the selected class of mutations is quite gentle especially when the migration rate is low.

The above study is a quantitative evaluation of my intuitive argument on fluctuating weak selection (81). The idea is that, in a stable environment, a random mutant needs to be beneficial only under restricted circumstances, whereas in a more variable environment, a mutant must be beneficial in many circumstances. Usually, the smaller the population, the more restricted the environmental variability. This idea is somewhat similar to Mani's (70) model, but the selection coefficient is assumed to be large in his study. In any case, the nearly neutral model now incorporates very slightly advantageous mutations for the region $s > 0$, and the concept of near neutrality approaches that of selectionists (e.g. 117).

What kind of evidence do we have to support the nearly neutral theory? Recent studies on DNA polymorphism in natural populations of *Drosophila*

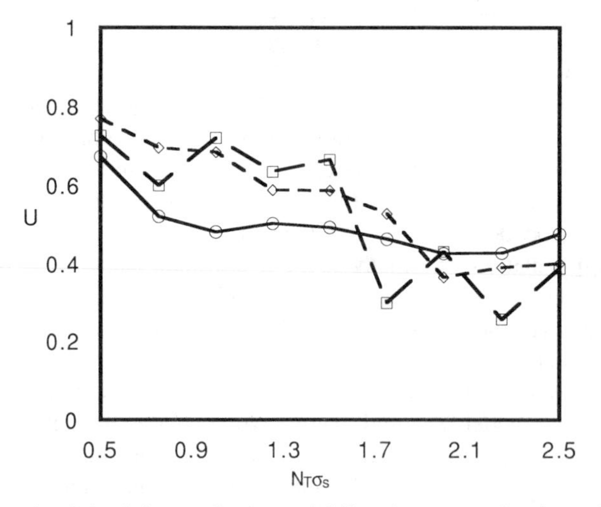

Figure 4 Results of simulations on fixation probability of mutants as functions of the product of total population size and selection intensity, $N_T\sigma_s$, relative to the completely neutral case. Each point is obtained by a simulation continued for $800/v$ generations, where v is the mutation rate. Three lines are for different levels of migration rate (m), (solid line) : $2Nm = 0.1$;(bold broken line) : $2Nm = 0.5$; (broken line): $2Nm = 4.0$.

reveal several facts that are better explained by the nearly neutral theory than by the strictly neutral one. As mentioned before, the pattern of DNA polymorphism at the rosy region in *Drosophila melanogaster* and *D. simulans* is quite different, i.e. unlike protein polymorphisms, DNA sequence heterozygosity in this region of *D. simulans* is estimated to be several times as great as that of *D. melanogaster* (3). Note that the average heterozygosity over many protein loci is almost the same in the two species but that geographic differentiation is more pronounced in *D. melanogaster* than in *D. simulans* (15). Aquadro et al (3) suggested that the differences in the effective population sizes of species may be responsible for the pattern, such that slightly deleterious mutations are common in proteins but not in DNA. Aquadro (2) studied more loci and found a similar pattern in all three loci examined; *per* in the X chromosome, Adh in the second chromosome, in addition to *rosy* in the third chromosome. By simulation studies, I have shown that the data fit the present model not only for heterozygosity but also for a proportion of polymorphism and the fixation index (95).

Another important subject related to the nearly neutral model is the rate of molecular evolution. For nearly neutral mutations, the effect of total population size on the substitution rate becomes larger as the mutant's effect gets stronger. Then the cancellation between generation length effect and negative selection (see above) would be more pronounced for amino acid substitutions than for synonymous changes. The data of Li et al (66) suggest that the generation-time effect is stronger on synonymous substitutions than on amino acid replacement substitutions, agreeing with the above prediction. By using the data of Li et al (66), Gillespie (31) obtained weighting factors. The factors were obtained such that the average is one, and the relative values remain the same. These are the ratios of the numbers of substitutions of the three branches for each of the mammalian orders, artiodactyls, rodents, and primates. Let w_a, w_r, and w_p be the weights for the above three orders respectively. For replacement substitutions, they turned out to be, $w_a = 0.885$, $w_r = 1.279$, $w_p = 0.836$. For synonymous substitutions, $w_a = 0.762$, $w_r = 1.611$, $w_p = 0.627$. Thus, amino acid replacement substitutions are less dependent on generation length. As I have repeatedly emphasized, the generation time effect is coupled with the population size effect. An interesting example is the differentiation of mitochondrial genome, among Hawaiian *Drosophila*(18). Mitochondrial evolution in two closely related lineages of Hawaiian *Drosophila* that have different histories shows that the rate is three times higher in lineages with repeated founder events than in lineages without bottlenecks.

Let us now turn our attention to the variance of the evolutionary rate. As discussed before, the variance is often larger than that under the simple Poisson process. Possible causes that inflate the variance are: (i) a difference of mutation rate among the lineages, (ii) weak selection that results in a difference of intensity due to the change of population size, and environment, and (iii) other factors. The total variance would be determined by the sum of the several causes. The generation-time effect on mutation rate is partially cancelled by the population size effect, but the cancellation is measurable only when a large number of loci are examined, as the above weight analysis shows, and the variance is inflated by the difference of mutation rate. As to the second cause, the larger the mutants' effect becomes, the more inflation of the variance is expected. The comparison of the variance of the amino acid replacement substitution with that of synonymous substitution does not show clearly that the former variance is larger than the latter (28). One needs more data and analysis for clarification.

The final problem on near-neutrality concerns the assumption of fluctuating selection coefficients. Results from Dykhuizen & Hartl (21) suggest that many naturally occurring enzyme polymorphisms in *E. coli* are neutral or nearly neutral, but a latent potential for selection can be observed in the polymorphism at the 6-phosphogluconate dehydrogenase locus of *E. coli*. Sawyer et al (99) estimated that the average selection coefficient for that locus is approximately -1.6×10^{-7}. This estimate corresponds to our μ_s, and local values of the selection coefficient may be larger. Dean et al (17) examined the fitness effect of newly occurring amino acid substitutions at the β-galactosidase locus of *E. coli*, and again found that the majority of amino acid changes have minor effects on fitness. This would imply that the "Dykhuizen-Hartl" effect applies not only to naturally occurring polymorphic alleles but also to many newly arisen mutant alleles.

The model considered here incorporates spatial fluctuation of selection intensity but not a temporal fluctuation. It is therefore not quite realistic. However, for such large-scale fluctuation as is considered here, temporal changes of fitness would not usually occur while a mutant is on its way to fixation. The effect of temporal variation would be on changing the mean fitness of the population such that the value of μ_s is simply lowered on rare occasions. Then the results of the simulation studies would hold in a more general situation that incorporates temporal fluctuation of selection intensity.

Adaptation at the Molecular Level

So far we are mainly concerned with selection for keeping the gene function status quo. Once gene function attains a state sufficiently near an optimum, genes are expected to evolve by nearly neutral mutant substitutions. As long as fixed loci are considered, gene functions are usually kept as they are, and on rare occasions an environmental change causes a shift of gene function with the appropriate mutant substitutions. It may occur in protein coding regions as well as in regulatory regions such as a promoter or an enhancer. Such substitutions may constitute a major process of adaptation in bacteria.

Adaptation at the molecular level of higher organisms like mammals appears to include minor chromosomal changes such as duplication and illegitimate crossing-over. These changes would be acceptable under a genome structure with a large noninformational part. This is because noninformational part would provide flexibility on gene organization. The *Drosophila* genome seems to be the intermediate between mammalian and bacterial ones. It is interesting to find that the detailed analyses of DNA polymorphisms in *Drosophila* species reveal many different patterns among loci suggesting various ways of adaptation (2, 56). In many cases, the interaction of random drift and selection would be important.

In vertebrates, it is now known that multigene families are common. Organismal development is governed by spatially and temporally regulated expression of various gene families that are the products of hundreds of millions of years of evolution. Thus, the reorganization of genes by duplication or illegitimate recombination is very important for organismal evolution. Such a process is occurring much more frequently than was imagined by leading geneticists, Bridges (9), Muller (72) and Ohno (79), who thought of gene duplication as the major way to acquire new genes. Indeed, comparative studies of gene families show that the genetic material is more versatile than was previously thought, i.e. various illegitimate recombinational processes must have been rather common in evolution. In general, it has been thought that natural selection works to keep genes in status quo for those gene families that were established a long time ago. On the other hand, incipient gene families may be on the way to further progress, in the sense of acquiring more diverse function while positive Darwinian selection may be operating.

Population genetic models of established multigene families have been extensively analyzed by incorporating gene conversion (87, 88, 73, 74), unequal crossing-over (86), and duplicative transposition (14, 89). The results can satisfactorily explain the so-called concerted evolution of multigene families. In other words, each member of a gene family does not differentiate independently but evolves in concert with other members because of their functional interrelationship and because of the homogenizing effect of the above processes. The study of copy number regulation has just started (67). The origin of gene families with diverse function has also been studied by population genetics (91, 92). Starting from a single copy, my simulations have shown how beneficial genes may accumulate on the chromosomes under various conditions. With realistic values of deleterious mutation rate, positive selection is needed for acquiring gene families with desirable functions.

Is it possible to find incipient gene families in the process of acquiring new functions? There are now several examples of duplicated genes that show accelerated amino acid substitution relative to synonymous substitution as surveyed in (62, 93). The first example is the emergence of fetal hemoglobin from embryonic hemoglobin in primates (33, 24). Higher primates have two duplicated hemoglobin γ genes, and other primates have only one. The γ genes of higher primates are turned on in fetal life, whereas the single gene of other primates is turned on in embryonic life and turned off at the beginning of fetal life. Examination of the DNA sequences of the γ gene family suggests that the amino acid substitution was accelerated in the duplicated genes in the period when the switch occurred from the embryonic gene to the fetal one phylogenetically (33, 24). The substitution rate is about 3.5 times the standard rate of hemoglobin genes. The next example is the β globin gene cluster of goat and sheep (64). However, in this case, it is difficult to judge whether the acceleration was caused by positive selection or simply by relaxation of selective constraint.

The evolution of the stomach lysozyme of ruminants is another interesting example. An amino acid sequence study has shown that the gene for stomach lysozyme arose by duplication of the gene for nonstomach lysozyme at the time when ruminants diverged from other mammals (42–44). It has been shown that, about 30 million years after divergence of ruminants from other mammals, the rate of amino acid substitution was three times as high as that of ordinary nonstomach lysozyme (42–44). Other examples surveyed in (93) include genes of the visual pigment (115), histocompatibility antigen (39, 40), immunoglobulin constant region (103), and protease inhibitor (16). For genes of the immune function, there seems to be no clear distinction between incipient and established gene families (94), and some multigene families belonging to the immunoglobulin superfamily are undergoing continuous reorganization via unequal crossing-over and gene conversion (98). Such examples reveal remarkable strategies to acquire enormous diversity of immune reaction, and positive selection must have operated for their origin, even if selective force may have been very weak at the level of an individual amino acid site (86).

Prospect

As explained in the previous section, recent studies of genetic variation of *Drosophila* species at the DNA level have revealed a number of unexpected properties such as the contrasting pattern of polymorphisms at the DNA and at the protein levels among the closely related species (2). The pattern now extends to the human population, as the nucleotide diversity is much less in human than in *Drosophila* populations in spite of a similar level of protein diversity (65). Also, it appears that each locus has its own characteristic pattern of polymorphism. In some loci, protein diversity appears to be very low as compared with DNA diversity within species, but not so in others (2, 56, 68). Thus the relative importance of drift and selection may differ from locus to locus, and such data provide excellent material for further study on the mechanisms of maintaining polymorphism. For example, the population-size effect of the nearly neutral model can be used to discriminate it from the simple neutral model. Here the relative numbers of the amino acid altering and the synonymous or other unimportant substitutions are most convenient—a point discussed in the previous section. It may be difficult to discriminate the nearly neutral model from the selection model, in which most amino acid substitutions are caused by positive Darwinian selection (29, 30, 68). However, I would like to point out that the classical model of selection predicts that evolution is more rapid in large populations than in small ones (23; see page 102). Thus species with large population size accumulate more mutant substitutions under the selection model. This prediction is contrary to that of the nearly neutral model and will be useful for discriminating the two.

As was emphasized before, the genetic material is more versatile than previously thought, and such versatility has been used for organismal evolution. In particular, the complexity of higher organisms has been attained through numerous trials and errors of gene duplication and other illegitimate recombination of DNA. Various simulation studies have attempted to understand how Darwinian selection has been responsible for the origin of complicated genetic systems (91, 92). The results indicate that interaction between random genetic drift and selection is important, i.e. nearly neutral illegitimate recombination and nucleotide substitution are thought to be raw materials for organismal evolution.

My model of simulation studies on the origin of gene families incorporates gene duplication by unequal crossing-over and mutation (91, 92). The evolution of actual gene families seems to be more complicated than such a simple model, i.e. gene conversion occurs continuously in addition to unequal crossing-over and nucleotide substitution. For example, Irwin & Wilson (43) found that two different trees are obtained for the coding region and the 3′ untranslated region of stomach lysozyme genes of ruminant. Gene conversion had been more frequent in the coding region than in the untranslated region, because coding sequences are more conservative. There are other examples that show a similar pattern. Fitch et al (24) examined the emergence of fetal γ gene of the β gene cluster of primates, and again found that two γ genes show concerted evolution via frequent gene conversion. In this case, gene conversion had occurred not only in the coding region but also in the 5′ regulatory sequences, and the authors suggest that concerted evolution plus selection for favorable mutations would have been responsible for the emergence of the fetal gene. Another example is the evolution of the rat kallikrein gene family. Wine et al (111) suggest that gene conversion has played an important role in the evolution of the functional diversity of the duplicated genes after comparing gene sequences in detail. These three examples indicate that the frequent occurrence of gene conversion at the time of evolution of a new function seems to be the rule rather than the exception. Basten & Ohta (5a) performed simulations to show that gene conversion is indeed effective in accelerating evolution by compensatory advantageous mutations. In the coming years, the combined study of experimental and theoretical analyses of various gene families will be a fascinating research project.

ACKNOWLEDGMENTS

I thank Dr. J. F. Crow, Dr. C. Aquadro, and Dr. C. Basten for going over the manuscript and offering numerous valuable comments. This is contribution No. 1915 from the National Institute of Genetics, Japan.

Literature Cited

1. Allendorf, F. W., Christiansen, F. B., Dobson, T., Eanes, W. F., Frydenberg, O. 1979. Electrophoretic variation in large mammals. I. The polar bear, *Thalarctos maritimus*. *Hereditas* 91: 19–22
2. Aquadro, C. F. 1990. Contrasting levels of DNA sequence variation in *Drosophila* a species revealed by "six-cutter" restriction map surveys. In *Molecular Evolution,* ed. M. Clegg, S. O'Brien, pp. 179–89. New York: Liss
3. Aquadro, C. F., Lado, K. M., Noon, W. A. 1988. The rosy region of *Drosophila melanogaster* and *Drosophila simulans*. I. Contrasting levels of naturally occurring DNA restriction map variation and divergence. *Genetics* 119: 875–88
4. Ayala, F. J., ed. 1976. *Molecular Evolution*. Sunderland, Mass: Sinauer
5. Ayala, F. J., Powell, J. R., Dobzhansky, Th. 1971. Polymorphisms in continental and island populations of *Drosophila willistoni*. *Proc. Nat. Acad. Sci. USA* 68:2480–83
5a. Basten, J. B., Ohta, T. 1992. Simulation study of a multigene family, with special reference to the evolution of compensatory advantageous mutations. *Genetics*. In press
6. Bernardi, G., Olofsson, B., Filipski, J., Zerial, M., Salinas, J., Cuny, G., et al. 1985. The mosaic genome of warm-blooded vertebrates. *Science* 228:953–58
7. Bonnell, M. L., Selander, R. K. 1974. Elephant seals: genetic variation and near extinction. *Science* 184:908–9
8. Boyer, S. H., Crosby, E. F., Noyes, A. N., Fuller, G. F., Leslie, S. E., et al. 1971. Primate hemoglobins: Some sequences and some proposals concerning the character of evolution and mutation. *Biochem. Genet.* 5:405–48
9. Bridges, C. B. 1935. Salivary chromosome maps. *J. Hered.* 26:60–64
10. Britten, R. J. 1986. Rates of DNA sequence evolution differ between taxonomic groups. *Science* 231:1393–98
11. Carson, H. L. 1976. The population flush and its genetic consequences. In *Population Biology and Evolution,* ed. R. C. Lewontin, pp. 123–37. Syracuse, NY: Syracuse Univ. Press
12. Cavalli-Sforza, L. L., Bodmer, W. F. 1971. *The Genetics of Human Populations*. San Francisco: W. H. Freeman
13. Chang, L-Y. E., Slightom, J. L. 1984. Isolation and nucleotide sequence analysis of the β-type globin pseudogene from human, gorilla and chimpanzee. *J. Mol. Biol.* 180:767–84
14. Charlesworth, B., Charlesworth, D. 1983. The population dynamics of transposable elements. *Genet. Res.* 42: 1–28
15. Choudhary, M., Singh, R. S. 1987. A comprehensive study of genetic variation in natural populations of *Drosophila melanogaster*. III. Variations in genetic structure and their causes between *Drosophila melanogaster* and its sibling species *Drosophila* simulans. *Genetics* 117:697–710
16. Creighton, T. E., Darby, N. J. 1989. Functional evolutionary divergence of

proteolytic enzymes and their inhibitors. *Trends Biochem. Sci.* 14:319–24

17. Dean, A. M., Dykhuizen, D. E., Hartl, D. L. 1988. Fitness effects of amino acid replacements in the β-galactosidase of *Escherichia coli*. *Mol. Biol. Evol.* 5:469–85
18. DeSalle, R.,Templeton, A. R. 1988. Founder effects and the rate of mitochondrial DNA evolution in Hawaiian *Drosophila*. *Evolution* 42:1076–84
19. Dickerson, R. E. 1971. The structure of cytochrome *c* and the rate of molecular evolution. *J. Mol. Evol.* 1:26–45
20. Dobzhansky, Th. 1937. *Genetics and the Origin of Species*. New York: Columbia Univ. Press. 1st ed.
21. Dykhuizen, D. E., Hartl, D. L. 1980. Selective neutrality of 6PGD allozymes in *E. coli* and the effects of genetic background. *Genetics* 96:801–17

21a. Easteal, S. 1985. Generation time and the rate of molecular evolution. *Mol. Biol. Evol.* 2:450–53

22. Falconer, D. S. 1960. *Introduction to Quantitative Genetics*. London, New York: Longmans
23. Fisher, R. A. 1930. *The Genetical Theory of Natural Selection*. Oxford: Clarendon
24. Fitch, D. H. A., Bailey, W. J., Tagle, D. A., Goodman, M., Sieu, L., Slightom, J. L. 1991. Duplication of the γ-globin gene mediated by repetitive L1 LINE sequences in an early ancestor of simian primates. *Proc. Natl. Acad. Sci. USA* 88:7396–7400
25. Gillespie, J. H. 1974. Polymorphism in patchy environments. *Am. Nat.* 108: 145–51
26. Gillespie, J. H. 1978. A general model to account for enzyme variation in natural populations. V. The SAS-CFF model. *Theor. Popul. Biol.* 14:1–45
27. Gillespie, J. H. 1983. A simple stochastic gene substitution model. *Theor. Popul. Biol.* 23:202–15
28. Gillespie, J. H. 1986. Variability of evolutionary rates of DNA. *Genetics* 113:1077–91
29. Gillespie, J. H. 1986. Natural selection and the molecular clock. *Mol. Biol. Evol.* 3:138–55
30. Gillespie, J. H. 1987. Molecular evolution and the neutral allele theory. *Oxford Surveys Evol. Biol.* 4:10–37
31. Gillespie, J. H. 1989. Lineage effects and the index of dispersion of molecular evolution. *Mol. Biol. Evol.* 6:636–47
32. Goodman, M. 1976. Protein sequences in phylogeny. In *Molecular Evolution*, ed. F. J. Ayala, pp. 141–59. Sunderland, Mass: Sinauer
33. Goodman, M., Czelusniak, J., Koop, B. F., Tagle, D. A., Slightom, J. L. 1987. Globins: A case study in molecular phylogeny. *Proc. Cold Spring Harbor Symp. Quant. Biol.* 52:875–90
34. Goodman, M., Koop, B. F., Czelusniak, J., Weiss, M. L., Slightom, J. L. 1984. The η-globin gene; its long evolutionary history in the β-globin gene family of mammals. *J. Mol. Biol.* 180:803–23
35. Haldane, J. B. S. 1932. *The Causes of Evolution*. New York: Harper & Row
36. Harris, H. 1966. Enzyme polymorphisms in man. *Proc. R. Soc. London, Ser. B*, 164:298–310
37. Harris, S., Barrie, P. A., Weiss, M. L., Jeffreys, A. J. 1984. The primate ψβ1 gene; An ancient β-globin pseudogene. *J. Mol. Biol.* 180:785–801
38. Holmquist, R., Jukes, T. H., Pangburn, S. 1973. Evolution of transfer RMA. *J. Mol. Biol.* 78:91–116
39. Hughes, A. L., Nei, M. 1988. Pattern of nucleotide substitution at major histocompatibility complex loci reveals overdominant selection. *Nature* 335: 167–70
40. Hughes, A. L., Nei, M. 1989. Nucleotide substitution at major histocompatibility complex class II loci: Evidence for overdominant selection. *Proc. Natl. Acad. Sci. USA* 86:958–62
41. Ikemura, T. 1985. Codon usage and tRNA content in unicellular and multicellular organisms. *Mol. Biol. Evol.* 2:13–34
42. Irwin, D. M., Wilson, A. C. 1989. Multiple cDNA sequences and the evolution of bovine stomach lysozyme. *J. Biol. Chem.* 264:11387–93
43. Irwin, D.M., Wilson, A. C. 1990. Concerted evolution of ruminant stomach lysozymes. *J. Biol. Chem.* 265:4944–52
44. Jollés, J., Pollés, P., Bowman, B. H., Prager, E. M., Stewart, C. -B., Wilson, A. C. 1989. Episodic evolution in the stomach lysozymes of ruminants. *J. Mol. Evol.* 28:528–35
45. Kacser, H., Burns, J. A. 1981. The molecular basis of dominance. *Genetics* 97:639–66
46. Kimura, M. 1962. On the probability of fixation of mutant genes in a population. *Genetics* 47:713–19
47. Kimura, M. 1968. Evolutionary rate at the molecular level. *Nature* 217:624–26

48. Kimura, M. 1979. A model of effectively neutral mutations in which selective constraint is incorporated. *Proc. Natl. Acad. Sci. USA* 76:3440–44
49. Kimura, M. 1983. *The Neutral Theory of Molecular Evolution*. Cambridge: Cambridge Univ. Press
50. Kimura, M., Crow, J. F. 1964. The number of alleles that can be maintained in a finite population. *Genetics* 49:725–38
51. Kimura, M., Ohta, T. 1969. The average number of generations until fixation of a mutant gene in a finite population. *Genetics* 61:763–71
52. Kimura, M., Ohta, T. 1971. Protein polymorphism as a phase of molecular evolution. *Nature* 229:467–69
53. Kimura, M., Ohta, T. 1971. *Theoretical Aspects of Population Genetics*. Princeton: Princeton Univ. Press
53a. King, J. L., Jukes, T. H. 1969. Non-Darwinian evolution. *Science* 164:788–98
54. Kingman, J. F. C. 1978. A simple model for the balance between selection and mutation. *J. Appl. Probab.* 15:1–12
55. Kohne, D. E. 1970. Evolution of higher-organism DNA. *Q. Rev. Biophysics* 3(3):327–75
56. Kreitman, M. 1987. Molecular population genetics. *Oxford Surv. Evol. Biol.* 4:38–60
57. Laird, C. D., McConaughy, B. L., McCarthy B. J. 1969. Rate of fixation of nucleotide substitutions in evolution. *Nature* 224:149–54
58. Langley, C. H., Fitch, W. M. 1974. An examination of the constancy of the rate of molecular evolution. *J. Mol. Evol.* 3:161–77
59. Lewontin, R. C. 1974. *The Genetic Basis of Evolutionary Change*. New York, London: Columbia Univ. Press
60. Lewontin, R. C., Hubby, J. L. 1966. A molecular approach to the study of genic heterozygosity in natural populations. II. Amount of variation and degree of heterozygosity in natural populations of *Drosophila pseudoobscura*. *Genetics* 54:595–609
61. Li, W-H. 1979. Maintenance of genetic variability under the pressure of neutral and deleterious mutations in a finite population. *Genetics* 92:647–67
62. Li, W.-H. 1985. Accelerated evolution following gene duplication and its implication for the neutralist-selectionist controversy. In *Population Genetics and Molecular Evolution*, ed. T. Ohta, K. Aoki, pp. 333–52. Tokyo: Jpn. Sci. Soc. Press
63. Li, W-H., Graur, D. 1991. *Fundamentals of Molecular Evolution*. Sunderland, Mass: Sinauer
64. Li, W-H., Gojobori, T. 1983. Rapid evolution of goat and sheep globin genes following gene duplication. *Mol. Biol. Evol.* 1:94–108
65. Li, W-H., Sadler, L. A. 1991. Low nucleotide diversity in man. *Genetics* 129:513–23
66. Li, W-H., Tanimura, M., Sharp, P. M. 1987. An evaluation of the molecular clock hypothesis using mammalian DNA sequences. *J. Mol. Evol.* 25:330–42
67. Lyckegaard, E. M. S., Clark, A. G. 1991. Evolution of ribosomal RNA gene copy number on the sex chromosomes of *Drosophila melanogaster*. *Mol. Biol. Evol.* 8:458–74
68. McDonald, J. H., Kreitman, M. 1991. Adaptive protein evolution at the *Adh* locus in *Drosophila*. *Nature* 351:652–54
69. Malécot, G. 1948. *Les mathématiques de l'hérédité*. Paris: Masson
70. Mani, G. S. 1984. A Darwinian theory of enzyme polymorphism. In *Evolutionary Dynamics of Genetic Diversity*, ed. G. S. Mani, pp. 242–98. Berlin: Springer
71. Moriyama, E. N. 1987. Higher rates of nucleotide substitution in *Drosophila* than in mammals. *Jpn. J. Genet.* 62: 139–47
72. Muller, H. J. 1936. Bar duplication. *Science* 83:528–30
73. Nagylaki, T. 1984. The evolution of multigene families under intrachromosomal gene conversion. *Genetics* 106: 529–48
74. Nagylaki, T. 1984. Evolution of multigene families under interchromosomal gene conversion. *Proc. Natl. Acad. Sci. USA* 81:3796–800
75. Nei, M. 1983. Genetic polymorphism and the role of mutation in evolution. In *Evolution of Genes and Proteins*, ed. M. Nei, R. K. Koehn, pp. 165–90. Sunderland, Mass: Sinauer
76. Nei, M. 1987. *Molecular Evolutionary Genetics*. New York: Columbia Univ. Press
77. Nevo, E., Kim, Y. J., Shaw, C. R., Thaeler, C. S. Jr. 1974. Genetic variation, selection and speciation in *Thomomyo talpoides* pocket gophers. *Evolution* 28:1–23
78. O'Brien, S. J., Wildt, D. E., Bush, M.

1986. The cheetah in genetic peril. *Sci. Am.* 254(5):68–76
79. Ohno, S. 1970. *Evolution by Gene Duplication*. Berlin: Springer-Verlag
80. Ohta, T. 1972. Evolutionary rate of cistrons and DNA divergence. *J. Mol. Evol.* 1:150–57
81. Ohta, T. 1972. Population size and rate of evolution. *J. Mol. Evol.* 1:305–14
82. Ohta, T. 1973. Slightly deleterious mutant substitutions in evolution. *Nature* 246:96–98
83. Ohta, T. 1974. Mutational pressure as the main cause of molecular evolution and polymorphisms. *Nature* 252:351–54
84. Ohta, T. 1976. *Role of very slightly deleterious mutations in molecular evolution and polymorphism. Theor. Popul. Biol.* 10:254–75
85. Ohta, T. 1977. Extension to the neutral mutation random drift hypothesis. In *Molecular Evolution and Polymorphism,* ed. M. Kimura, pp. 148–67. Mishima: Natl. Inst. Genet.
86. Ohta, T. 1980. *Evolution and Variation of Multigene Families*. New York: Springer-Verlag
87. Ohta, T. 1982. Allelic and non-allelic homology of a supergene family. *Proc. Natl. Acad. Sci. USA* 79:3251–54
88. Ohta, T. 1983. On the evolution of multigene families. *Theor. Popul. Biol.* 23:216–40
89. Ohta, T. 1986. Population genetics of an expanding family of mobile genetic elements. *Genetics* 113:145–59
90. Ohta, T. 1987. Very slightly deleterious mutations and the molecular clock. *J. Mol. Evol.* 26:1–6
91. Ohta, T. 1987. Simulating evolution by gene duplication. *Genetics* 115: 207–13
92. Ohta, T. 1988. Evolution by gene duplication and compensatory advantageous mutations. *Genetics* 120:841–47
93. Ohta, T. 1991. Multigene families and the evolution of complexity. *J. Mol. Evol.* 33:34–41
94. Ohta, T. 1991. Evolution of the multigene family: A case of dynamically evolving genes at major histocompatibility complex. In *Evolution of Life,* ed. S. Osawa, T. Honjo, pp. 145–59. Berlin: Springer
95. Ohta, T. 1992. Theoretical study of near neutrality. II. Effect of subdivided population structure with local extinction and recolonization. *Genetics* 130: 917–23
96. Ohta, T., Kimura, M. 1971. On the constancy of the evolutionary rate of cistrons. *J. Mol. Evol.* 1:18–25
97. Ohta, T., Tachida, H. 1990. Theoretical study of near neutrality. I. Heterozygosity and rate of mutant substitution. *Genetics* 126:219–29
98. Parham, P. 1989. Getting into the groove. *Nature* 342:617–18
98a. Romer, A. S. 1968. *The Procession of Life*. London: Weidenfeld & Nicolson
99. Sawyer, S. A., Dykhuizen, D E., Hartl, D. L. 1987. Confidence interval for the number of selectively neutral amino acid polymorphisms. *Proc Natl. Acad. Sci. USA* 84:6225–28
100. Selander, R. K. 1976. Genic variation in natural populations. In *Molecular Evolution,* ed. F. J. Ayala. Sunderland, Mass: Sinauer
101. Selander, R. K., Levin, B. R. 1980. Genetic diversity and structure in *Escherichia coli* populations. *Science* 210: 545–47
102. Sharp, P. M. 1989. Evolution at 'silent' sites in DNA. In *Evolution and Animal Breeding,* ed. W. G. Hill, T. F. C. Mackey, pp. 23–32. Wallingford: C. A. B. Int.
103. Sheppard, H. W., Gutman, G. A. 1981. Allelic forms of rat Kappa chain genes: evidence for strong selection at the level of nucleotide suquence. *Proc. Natl. Acad. Sci. USA* 78:7064–68
104. Sibley, C. G., Ahlquist, J. E. 1984. The phylogeny of the hominoid primates, as indicated by DNA-DNA hybridization. *J. Mol. Evol.* 20:2–15
105. Sibley, C. G., Ahlquist, J. E. 1987. DNA hybridization evidence of hominoid phylogeny: Results from an expanded data set. *J. Mol. Evol.* 26: 99–121
106. Singh, R. S. 1989. Population genetics and evolution of species related to *Drosophila melanogaster*. *Annu. Rev. Genet.* 23:425–53
107. Tachida, H. 1991. A study on a nearly neutral mutation model in finite populations. *Genetics:* 128:183–92
108. Takahata, N., Kimura, M. 1979. Genetic variability maintained in a finite population under mutation and antocorrelated random fluctuation of selection intensity. *Proc. Natl. Acad. Sci. USA* 76:5813–17
109. Templeton, A. 1985. The phylogeny of the hominoid primates: A statistical analysis of the DNA-DNA hybridization data. *Mol. Biol. Evol.* 2:420–33
110. Watson, J. D. 1976. *Molecular Biology of the Gene*. Menlo Park, Calif: Benjamin. 3rd ed.

111. Wines, D. R., Brady, J. M., Southard, E. M., MacDonald, J. 1991. Evolution of the rat kallikrein gene family: gene conversion leads to functional diversity. *J. Mol. Evol.* 32:476–92
112. Wright, S. 1931. Evolution in Mendelian populations. *Genetics* 16:97–159
113. Wright, S. 1942. Statistical genetics and evolution. *Bull. Am. Math. Soc.* 48:223–46
114. Wu, C-I., Li, W-H. 1985. Evidence for higher rates of nucleotide substitution in rodents than in man. *Proc. Natl. Acad. Sci. USA* 82:1741–45
115. Yokoyama, S., Yokoyama, R. 1990. Molecular evolution of visual pigment genes and other G-protein-coupled genes. In *Population Biology of Genes and Molecules,* ed. N. Takahata, J. F. Crow. Tokyo: Baifukan
116. Zuckerkandl, E., Pauling, L. 1965. Evolutionary divergence and convergence in proteins. In *Evolving Genes and Proteins,* ed. V. Bryson, H. J. Vogel, pp. 97–166. New York: Academic Press
117. Zuckerkandl, E. 1987. On the molecular evolutionary clock. *J. Mol. Evol.* 26: 34–46

Annu. Rev. Ecol. Syst. 1992. 23:287–310

ARTIFICIAL SELECTION EXPERIMENTS

William G. Hill and Armando Caballero
Institute of Cell, Animal and Population Biology, University of Edinburgh, West Mains Road, Edinburgh EH9 3JT, Scotland.

KEY WORDS: quantitative genetics, heritability, mutation, genetic variation, evolution

INTRODUCTION

A knowledge of the magnitude of genetic variability and covariability of quantitative traits in natural populations and an understanding of the action of forces that maintain variation and those that lead to change are fundamental to the study of evolutionary biology. Artificial selection experiments, in which known strong forces are applied to laboratory or field populations over a greatly curtailed evolutionary timescale, have been and continue to be an important source of information on the genetics of quantitative characters and their effects on fitness.

Historical Perspective

Although artificial selection has been applied in the breeding of animals and plants since their domestication, adequately recorded selection experiments date only roughly from the Mendelian rediscovery at the start of this century; the great majority are post 1945. The earliest selection experiments were important in showing that almost any quantitative trait could be permanently altered, that responses (mostly) occurred as a consequence of changes in the frequencies of genes affecting the traits and not from mutations in the genes, and that, as responses could continue well outside the range of the initial population, many genes must be involved (for reviews see 38, 54, 99, 138). By 1955 artificial selection experiments were becoming a standard tool in quantitative genetic research, as may be seen in that year's Cold Spring Harbor Symposium (22).

The longevity of research using artificial selection is exemplified by the Illinois Corn Experiment, started in 1896, which continues to show responses after 90 generations of selection (28) and remains a model for testing ideas and theory.

0066-4162/92/1120-0287$02.00

More recent experiments have been established to evaluate the theory of the mean and variance of short- and long-term response, from both original and de novo variation. By their number these experiments provide replication, because one selection line is merely one realization of a Markov process. Selection has been practiced on a wide range of traits in populations brought from the wild, in populations of laboratory animals, and in agricultural species as a standard quantitative genetic procedure. Detailed studies at the molecular, cellular, and physiological level have been undertaken on the divergent selected lines to understand the basis for the differences observed. Until quite recently, artificial selection experiments have been directed more vigorously to problems in animal and plant improvement than to evolutionary biology and ecology. Yet they offer the critical benefit that, in contrast with analysis of natural evolution (71), the selection pressures are largely known and do not have to be inferred from the data.

Why Do Selection Experiments?

Short-term experiments, lasting say five generations or less, enable an estimate to be obtained of the genetic variance and heritability of a trait in the base population. This can be used, for example, to test if there is any appreciable variance, or if the estimate agrees with that obtained from covariance of collateral relatives such as sibs as predicted by theory, or if there is symmetry of response to high and low selection. Similarly the selection experiment can be used to estimate genetic covariances, correlations, and regressions among traits or of the same trait across different environments. Selection can be conducted at the group rather than individual or family level, but this is not a main concern of this review.

Initial rates of response to selection depend on summary statistical parameters such as heritability, whereas changes in rates depend on the number, effects, and frequencies of the genes influencing the trait. Changes in rates of response to selection are unlikely to be detected in a few generations, so longer term experiments are necessary to make more fundamental inferences about the inheritance of quantitative traits. Attenuation of response and selection limits can be used to check theory and to make inferences about gene numbers, effects, and frequencies. Manipulation of plateaued populations can give information on the genetic basis of the limit, including, for example, interactions between artificial and natural selection. Highly divergent lines and their crosses are a valuable resource for study of the biochemical and physiological basis of the traits under study, and for identifying and mapping the genes affecting the trait. Selection from an inbred base enables the rate of mutation to the quantitative trait and the magnitude and pleiotropic effects of mutant genes to be estimated, as an aid to understanding the basis of quantitative genetic variation.

The aim is to review aspects of the design and analysis of selection experiments and of the inferences that can be drawn from them so as to provide a background, rather than to come to definitive conclusions. The relation between theory and observation is discussed, both as a check on theory and to gain information from their differences. The role of selection experiments in estimating mutation rates is reviewed. In the final section the information gained from selection experiments about the nature of quantitative genetic variation is assessed. Additional references to selection experiments are given in conference proceedings (e.g. 58, 98, 106, 135) and in recent reviews (30, 41, 56, 83, 90, 91, 110, 114).

ESTIMATION OF QUANTITATIVE GENETIC PARAMETERS

Realized Heritability Estimation

A selection experiment of only one generation is essentially a study of regression of offspring on parent in which only some of the potential parents are allowed to breed and both parent and progeny means are computed to provide single values on each. Some information is thereby wasted, but with the benefit that pedigrees (i.e. distinct families) do not have to be kept. By selecting for more generations, the range of the independent variate, the cumulative selection differential, is increased and the sampling error of regression estimates is reduced, providing that changes in variance produced by selection and inbreeding can be accommodated in the analysis (50, 52). In populations in which generations overlap, care has to be taken to compute an unbiased estimate of the selection differential (63, 127).

Without knowledge of the distribution of gene effects and frequencies, changes in variance resulting from change in gene frequency cannot be predicted; but unless there are genes with large effects, say well over one tenth of a phenotypic standard deviation, such changes are likely to be small (13, 35) and inferences back to the base population not severely affected. As variation among selected parents is reduced by truncation selection, a consequent reduction in genetic variance occurs between but not within families (the "Bulmer effect"). This change in variance, which is due to gametic association (linkage disequilibrium) among all pairs of loci (linked or not), is not negligible, particularly when heritability is high and selection is intense. For the infinitesimal model of very many additive unlinked genes, each of very small effect, such that gene frequency changes can be assumed to be negligible, precise predictions of changes in genetic variance can be made (12, 13).

The classical estimator of heritability from selection experiments is Falconer's realized heritability, usually obtained from the regression of response on selection differential, each accumulated over generations (34, 35). Because of the Bulmer effect, this a somewhat biased estimator of heritability in the base population if mass or other selection utilizing between family differences is practiced, but not if selection is practiced within families. To obtain an unbiased estimate, a simple approach is to correct the realized heritability to allow for the selection (e.g. 4). Such a correction is all that can be done if records of individual families are not kept. Changes in response between the first and later generations, however, largely went unnoticed in selection experiments, presumably because of the high sampling errors of response.

Variance of Response and Parameter Estimates

The variance among (conceptual) replicates of a selection experiment increases with generation number, as does the covariance of means in successive generations. The standard error of an estimate of realized heritability is therefore substantially biased downwards if a simple regression model assuming independent identically distributed errors is used (50, 51). The genetic variance within a replicate line can change as a result of the Bulmer effect, changes in gene frequency, and sampling. Consequently many assumptions have to be made in order to predict the sampling variance of selection responses. As a first approximation the same variance as for an unselected line has been suggested (52); this is a gross overestimate if, for example, gene effects are so large that all lines become fixed for the same alleles. A comprehensive analysis of the observed response among replicates of selection lines has yet to be undertaken and is long overdue.

Sampling variances of responses or realized heritabilities can be estimated from replicated experiments without making assumptions about the genetic model. While replication is the obvious ideal, the practical problem is that, if resources are limited, few replicates can be kept unless each line is maintained with few breeding individuals, so the sampling error of the mean response itself has a high standard error. It is probably best to have as many replicates as is feasible without lines becoming too inbred, but there may be no alternative to constructing some estimate of sampling error for single replicates, for example, by correcting the standard error of a heritability estimated by regression to account for drift using simple assumptions (51). An alternative approach (W. M. Muir, personal communication) would be to deduce the necessary variance and covariances from the variance of responses in successive generations.

The sampling properties of response have, of course, to be taken into account in designing and evaluating selection experiments (50, 52, 93). Providing symmetry of response can be assumed, resources are best used if divergent high and low selected lines are started from the same base. Otherwise, some form of genetically constant control population has to be maintained to eliminate the effects of environmental change. The control may, in addition to drift, show genotype × environmental interaction with respect to the selected line, and there are various ways of using the control efficiently (92).

Genetic Correlations

In analysis of a selection experiment where the interest is in several traits, it is easy both to draw incorrect inferences and to miss others that may be useful. Let us consider first which parameters of a base population that are informative about the relationships among traits can be estimated from selection experiments, assuming families are not identified and that only selection differentials and generation means are available (35, 52, 55).

From a single line (or high-low divergent lines) selected for trait X , its additive genetic variance and heritability can be estimated. By taking observations on some other trait Y, even if only in the last generation, estimates can be made of the genetic regression of Y on X and of their genetic covariance, but not their genetic correlation. The regression or ratio of correlated response in Y to its secondary selection differential does not, however, estimate its heritability but only the ratio of genetic to phenotypic covariances (Cov_A / Cov_P). If selection is practiced in one of a pair of lines for X and in the other for Y, or if two different indices are used, the genetic variances and heritabilities of both traits and their genetic correlation can be estimated (35). Their sampling errors can also be estimated by making simple assumptions about the genetic model, even without replication (50).

Experimental designs for estimation of genetic correlations and related parameters have had limited discussion (17, 50, 52, 55); the range of possible questions precludes much generality. Usually, however, experiments efficient for estimating heritability are also efficient for correlations. Only if X and Y have a low correlation is the best estimate of their genetic correlation obtained by selecting on X alone in one line and Y alone in the other; otherwise two indices should be used (17).

Gains in efficiency are likely if recording of traits which are difficult or expensive to measure can be deferred until selection has generated wide differences. An example is when the aim is to identify the relation between component traits such as specific hormone levels or enzyme activities in analysis of composite traits such as growth or reproductive rate. This may be either to evaluate genetic associations or, in the context of animal breeding, to find juvenile predictors of economic traits of mature animals. Thus the genetic covariance between the primary selected trait and any other trait can be estimated from results of a single line. From these covariances the correlated response in the primary trait from selection on each of the secondary traits can be predicted (55). In effect, the experiment is done "backwards."

Mixed Model Analysis

The realized heritability, even if generation means are weighted to take account of their sampling correlations, does not utilize all the information available in a pedigreed population because there are within-generation covariances, for example, among sibs and between offspring and parent. These covariances are affected by the selection, however, and can only be used with appropriate statistical methods. A full analysis of the data is now feasible with the use of mixed model statistical methods using the so-called animal model. In this model, the phenotypic value of each individual is expressed as a linear function of defined environmental fixed effects such as age or season at recording, and of random effects including its genotype (breeding value), common litter effects, and environmental error (7, 66, 88, 126). A relationship matrix is used to specify the covariances among individuals. The methodology has its origins in Henderson's (49) best linear unbiased prediction (BLUP) procedure for evaluating farm animals in a breeding program, and in showing how multiple generation selection experiments fit into this structure (118). More generally, the methodology drew on the development of restricted (residual) maximum likelihood (REML) to obtain efficient and unbiased estimates of variance components (96), on increases in computer power, and on the development of computer programs to undertake REML analysis (87).

The animal model incorporates the effects of selection and of finite population size, provided all the data on which selection is practiced are included (130), so that unbiased estimates of the heritability in the base population can be obtained. In an application of mixed model methods to analysis of selection experiments, a BLUP procedure was used (8), but the estimates of heritability depended on the prior values assumed in the analysis (125). REML or other iterative methods are necessary to avoid such dependence.

Maximum likelihood with the animal model for pedigreed populations enables sampling properties of estimates of parameters of the base population, such as genetic variances and heritabilities to be obtained. Also, hypothesis tests can be undertaken using the likelihood surface and likelihood ratio test. The analysis is dependent on the appropriateness of the infinitesimal model, but results so far indicate that the variance among actual replicates in early generations accords with the expectation from the likelihood curves (7, 88). Also, multivariate mixed model methods enable the genetic variances and correlations of traits Y and Z and their sampling errors to be estimated from a single line in which selection is practiced for X . Computational requirements for such multivariate REML analyses are very large, however, and the estimate of genetic variance on the correlated trait is likely to be much poorer than that of the selected trait.

The increase in precision using mixed model analysis compared to a realized heritability estimate depends on the structure of the experiment and on the true heritability. The increase in precision is greatest when the heritability is low (126). If no control population is maintained and environmental change cannot be eliminated, there is no information in the realized heritability; and though estimates can be obtained by mixed model methods from within generation comparisons (8, 118), the selection renders these inefficient. Information on the design and efficiency of experiments using mixed model methods of analysis is still limited; it needs to be developed to include more detailed partition of variance into additive and environmental components, and to estimate changes in genetic parameters.

The ability to incorporate more sophisticated statistical procedures and to control family size are benefits deriving from pedigreed populations, which may be sufficient to justify the extra labor involved.

EXPERIMENTAL CHECKS ON THEORY

Short-Term Response

Many selection experiments have provided an opportunity to test quantitative genetic theory, in particular to check whether short-term response agrees with predictions from the base population. Clayton et al (19) found satisfactory agreement in the first such check. Sheridan (114) has recently reviewed laboratory and farm animal experiments and compared estimates of heritability from the base population, obtained by standard methods using sib correlations or offspring-parent regressions, with estimates of realized heritability from selection (Table 1). In *Drosophila* , for example, the two values differed by less than 10% of the realized heritability estimate in only 29 of 66 comparisons. Where standard errors of both estimates were available and their difference could be tested, 14 differed significantly at the 5% level, 47 did not. All 11 selection lines of laboratory animals that were started from populations at plateaux (*Pl*) differed by more than 50%: i.e. sib correlation estimates differed from zero whereas the realized estimates did not.

Although these results at first seem very disturbing, there are caveats: (i) No account was taken of the Bulmer effect. (ii) For low heritabilities, a large percentage difference is small both in absolute terms and compared to its standard error. (iii) Although predictions of selection response from base population parameters would be expected to worsen the longer selection proceeded, agreement actually improved, perhaps because the realized heritability was estimated more precisely. (iv) Although 25% of the statistical tests were apparently significant at the 5% level, proper sampling errors of the realized estimates are unlikely to have been used in all experiments, and the distributional properties are not well known and certainly are not normal as was assumed. (v) The analyses of populations at a plateau are clearly

Table 1 Comparison of realized and estimated heritabilities. Based on Tables 1 and 3 of Sheridan (114). (Printed with permission.)

Species	Level of agreement (%)[a]					Total number of comparisons	Tested comparisons[b]	
	0–10	10–20	20–30	30–50	>50		$p < 0.05$	NS
Drosophila	29	7	3	5	22	66	14	47
Tribolium	8	2	3	10	5	28	7	19
Mice & rats	9	0	5	9	19	42	6	28
Poultry & quail	8	2	7	2	22	41	5	6
Pigs & sheep	4	1	4	5	7	21	6	13
All species	58	12	22	31	75	198	38	113
(%)	(29)	(6)	(11)	(16)	(38)		(25)	(75)
Pl[c]	0	0	0	0	11	11		

[a] 100x|Estimated-Realized|/Realized
[b] Test of estimated and realized heritability, incorporating standard errors of both
[c] *Pl*: selection from plateaud populations, 11 comparisons in the total of 198

confounded both by mutation and by segregation of genes deleterious to fitness. Even so, the results should not be dismissed.

Although artificial selection experiments are usually directed toward a change in the mean of the traits, they are sometimes specifically designed to change their variances. Stabilizing selection (selection of the individuals with intermediate score) is expected to reduce the genetic variance (12), and this has been observed experimentally (see 65, 83). Disruptive selection (selection of the most extreme in both directions) is conversely expected to increase genetic variance, caused by a build up of linkage disequilibrium (13, 117). This increase in variance has been observed in many selection experiments (83, 117, 123). If selection is intense, disruptive selection simply becomes a two-way selection experiment with limited migration during the first few generations (104).

Long-Term Responses and Plateaux

Absolute predictions of long-term response and limits cannot be made from parameters of the base population because they are dependent on gene effects and frequencies, but there is theory for predicting, for example, the relative effects of population size and selection intensity on short- and long-term response, based on Robertson's theory (102). Results have been reviewed by López-Fanjul (74). Reasonable agreement was obtained between predicted and observed comparisons of selection intensity. High values gave greater short-term but lesser long-term response, although only for experiments in which neither the duration nor the total population size was very large. Effects of complete suppression of recombination on limits were well predicted (86), whereas within-family selection did not lead to higher limits than mass selection (45), contrary to expectations from the infinitesimal model (24).

The selection limit is expected to increase in proportion to effective population size (N) under the infinitesimal model, but taking $1.4N$ generations to get halfway to the limit (102). With genes of larger effect, the limit is expected to show an asymptotic relation to N and to be reached earlier. In both cases, mutation is likely to contribute variance before a limit is reached (53). Many reviews have focussed on long-term experiments and selection limits (e.g. 3, 29, 69, 134) showing, in general, a qualitative but not quantitative agreement with theory. There is good experimental evidence of increased long-term responses and limits with increased population size in experiments in *Drosophila* (23, 64). Even with populations of only 10–40 breeding pairs, replicated long-term experiments designed to compare population sizes are highly labor intensive. Weber has recently applied automatic scoring methods to the traits of wing-tip height (133) and ethanol tolerance (134) in *Drosophila* and used these in populations with up to 1600 parents, obtaining increased responses with increased population sizes in what are, relative to N, "short" experiments. Somewhat surprisingly, comparisons of various selection experiments with mice showed long-term responses (probably near limits) in proportion to population size, even though selection was not of duration greatly in excess of N generations (29). The effect of an initial population size bottleneck on subsequent response to selection (e.g. 40) has been in good agreement with theory (62).

Selection experiments have been established to compare alternative breeding schemes (for review see 5, 74). Motivated by Wright's shifting balance theory, comparisons have been made of selection in single populations with cycles of subdivision into small lines and subsequent crossing of selected lines. All these experiments have, however, been conducted using a trait such as bristle number in *Drosophila* with apparently additive gene action and, as predicted, the subdivision generally gave lower responses. As López-Fanjul (74) remarks "it is desirable that experiments of this kind be carried out for epistatic traits in order to test Wright's hypothesis."

Some evidence in support of the shifting balance theory has been obtained by interdemic selection on large groups of laboratory populations of *Tribolium castaneum* (132). When the population was subdivided into demes of constant size and selection made by differential migration of individuals from demes of high into demes of low productivity, there was an increase in average fitness; but the response was not proportional to the selection differential applied in different treatments, which suggests a nonadditive genetic basis for fitness.

The use of inbreeding may also increase selection responses (25), particularly for recessive genes (15). Experiments involving alternate generations of full sib mating and random mating for a fitness associated trait have, however, shown little effect on rates of response (26, 80).

It is perhaps best to view such experiments not as tests of alternative breeding programs but as ways of elucidating the genetic basis of traits of interest, for example, the extent of nonadditive gene action. If this is known, prediction of response and the merits of alternative schemes by mathematical or simulation methods are straightforward. Indeed, selection experiments to compare alternative schemes may be no more than poorly replicated simulations unless the basic questions are about the genetic model. What defeats the theoreticians is lack of information on which to base models.

Correlated Responses

Genetic correlations among traits have been estimated in many experiments, notably for assessment of components of traits such as litter size, growth, and body composition of laboratory animals (for reviews see, e.g., 30, 35). There have not been extensive comparisons of realized correlated responses with base population predictions. It became clear long ago, however, that there were asymmetries of correlated responses—selection on trait *X* did not produce the responses expected from selection on *Y* (9). In modern jargon, this was attributed to a breakdown of the infinitesimal model, but an explanation solely in terms of sampling or linkage, after allowance is made for changes in correlations expected from the Bulmer effect (131), cannot be ruled out. Multitrait selection indexes used in animal (and less often plant) breeding are constructed using estimated variances and covariances of component traits. Caballero (14) reviewed experiments and problems due to poor estimates and concluded that the main cause of inefficiency was temporal change in the parameters as selection proceeded.

Selection experiments to investigate correlated responses for fitness components have been undertaken to understand the evolution of life-history characters. For example, *Drosophila* lines have been selected for high fitness early or late in life (e.g 75, 111). Although the results support the hypothesis that genes increasing fitness early in the lifespan do so at the expense of reducing fitness later in life (136), alternative explanations have been given (95).

MUTATION RATES AND SELECTION RESPONSE

Spontaneous Mutations

Spontaneous mutations influencing selection response have occurred in many selection lines, and the evidence has been extensively reviewed (39, 59, 78). One clear example is the occurrence of the bobbed gene (with different number of DNA sequence repeats) in selection lines of *D. melanogaster* (42). While most examples come from *Drosophila*, others have been revealed in selection lines of mice, mostly decreasing, but also increasing (10), body size. Unless the mutants have very large effects or they have some pleiotropic (visible) effect, they may not be noticed. Further, unless the base population is small, it is difficult to exclude the possibility that the gene, particularly if recessive, was segregating in it (105).

The magnitude of the variance in quantitative traits from mutation is important both for an understanding of the levels of variance maintained in natural populations and for predictions of selection response. Selection experiments started from inbred base populations, either immediately or after many generations in which mutations have been able to accumulate, can be used to obtain estimates of the mutational variance (20). Alternative methods to selection from an inbred base include selection from crosses of inbred sublines, and accumulation of mutations in unselected replicate inbred lines or on replicate chromosomes maintained against a balancer. These last two methods may not give the precision of a selection experiment for the trait of particular interest, but they have the benefit that estimates of the mutational variance can be obtained for any other trait. Where families are kept distinct and the infinitesimal model can be assumed, mixed model methods using maximum likelihood can be employed (66, 137). Estimates of the mutational heritability, h_m^2, the increment in variance from mutation relative to the environmental variances from different studies have been reviewed (53, 70, 76), and others have been obtained in recent selection experiments (16, 33, 66, 134). The consensus value using different methods is 0.1% per generation from spontaneous mutations for bristle number in Drosophila and pupal weight in Tribolium ; for other traits and species the estimates differ.

Estimates of h_m^2 from selection are highly dependent on the size of mutant effects and are based on the assumption that mutants are neutral with respect to fitness. Estimates are therefore biased if mutants are deleterious, particularly if mutations are allowed to accumulate before selection is practiced. Spontaneous mutants, however, have been found both to have large effects on the trait and to affect fitness (16).

Whilst selection from an inbred base gives a clearer experiment, estimates of mutational variance can, in principle, also be obtained by selection from a noninbred base if it is continued for sufficiently long in a small population that the original additive genetic variance is exhausted or if maximum likelihood methods are used to estimate the two sources of variation simultaneously (7). Such estimates depend strongly on the assumption that natural selection, either at the level of individual genes in natural/artificial selection balance or on overall phenotype, is unimportant, and so these estimates are likely to be unreliable. Nevertheless, even though Yoo's lines showed strong regression of response on relaxation of selection and had segregating lethal genes (139, 140), responses after 50 generations accord with estimates of h_m^2 obtained by other methods (53).

Induced Mutations

Following the realization that X-irradiation induced mutations and the early successes in accelerating responses to selection for bristle number in *Drosophila* (112), experiments in laboratory and domestic animals (reviewed in 48) and in plants (reviewed in 47) have been conducted using irradiation with a view to increasing response. In artificial selection experiments in animals, usually initiated from inbred base populations, there is some evidence of increases in genetic variance induced by X-rays, but successes have been limited. A contributing, perhaps crucial, factor has been that the mutations are often associated with chromosome damage and are deleterious with respect to fitness. Chemical mutagenesis may have less deleterious side effects (48).

Striking responses to artificial selection from populations founded from crosses of *Drosophila melanogaster* which led to the mobilization of P transposable elements were obtained by Mackay (77). In the first experiments the dysgenic cross of M and P lines yielded much higher responses in abdominal bristle number than the nondysgenic reciprocal cross, used as a control because the elements were expected not to be mobilized. In subsequent studies, while evidence of mutations with large effects was obtained, responses were smaller and little difference was found between dysgenic and nondysgenic crosses, presumably because elements are mobilized several generations after the nondysgenic cross (97, 128). To overcome the lack of satisfactory control in such experiments started from a crossbred base in which many loci segregate, Moran (89) injected P elements into an inbred M base stock and selected for abdominal bristle number. Estimates of h_m^2 averaged 4.5%, 45 times higher than the spontaneous standard referred to above.

Nevertheless, the way ahead for experiments using transposable element–induced quantitative genetic variation does not seem to be via selection experiments. Mutant genes have a molecular tag for the element, and this enables individual genes affecting quantitative traits to be identified and cloned providing there are sufficiently few inserts in a line. Selected populations from a cross-bred base have too many inserts for this purpose (115), and selection of lines with injected P may likewise be difficult to use. A more powerful approach is to use crosses (or injection) of defective P elements such that isogenic lines of *Drosophila* can be established with a few, preferably only one, stable insert and then to select among these lines to find mutations of interest (79).

INFERENCES ON THE NATURE OF QUANTITATIVE VARIATION

It is a major challenge to elucidate the number of genes influencing quantitative traits, their effects, their frequencies, and the forces that maintain their segregation in natural populations or in laboratory populations founded from the wild. A more restricted objective is to describe with few parameters the joint distribution of gene effects on the quantitative trait and on fitness, degrees of dominance, and frequencies. Selection experiments have contributed to what is still limited knowledge. Tentative conclusions can usually be drawn, but as Robertson (103) stressed, "we are always dealing with the variation that we observe in a particular measurement made in a certain way on a particular population," and we should not "pretend to ourselves that we are dealing with a general property of the individual rather than a very specific observation of that property."

Gene Number and Effects

Information on the number and size of effects can, in principle, be obtained from selection experiments because the rate of attenuation of response with successive generations and the relation of the divergence, at the limit, to the initial variation are both dependent on numbers of genes involved. For example, Falconer (35) gives values of half-lives for four examples of lines selected to limits, all of which are much less than the $1.4N$ generations expected for the infinitesimal model (102). These results point to a small number of genes controlling the character. There are major difficulties with such inferences, however. Sometimes the genetic variance is exhausted when the plateau is reached (e.g. 101), but very often it is not, indicating nonadditive variance associated with heterozygote superiority or opposing fitness effects; the limit may be delayed or never attained because of new mutations. Yoo's results illustrate these complications: almost linear responses for 80 generations, reversion on relaxation of selection, and large levels of residual genetic variability in the lines at the limit (139).

Extensive application has been made of the Wright-Castle index (18), $\textit{number of genes} = (\textit{range})^2/(8 \times \textit{variance})$, for pairs of high and low selected lines. This requires many assumptions: All increasing alleles are in one line, and all decreasing in the other; gene effects are equal; gene action is additive; and there is no linkage. Violations of these assumptions bias estimates, usually downwards. Estimates of the number of genes for the four examples referred to above range from 2 to 98 (35). The weakness of the method is illustrated by comparison of the estimates of the number of genes for oil percentage in the Illinois corn experiment, which increased from 20 to 40 in 1934 (121), to 54 in 1977 (27), and to 69 in 1992 (28). Theoretical analyses have been undertaken recently with the aim of improving the index and interpreting more clearly the estimate obtained from divergent selected lines; but Zeng et al (141) concluded that "even in the best of circumstances, information from [the index] is very limited and can be misleading".

Perhaps the most unrealistic assumption underlying these methods is equality of gene effects. Robertson (103) suggested that the distribution of gene effects is highly leptokurtic, with most genes having little or no effect on the trait and a few having large effect. Artificial selection experiments are one source of evidence that genes of large effect are segregating in natural populations and occur as new mutants; genes of large effect were found in many selection experiments and ascribed to specific loci (78). There are also indirect sources of evidence. Variation among replicates may be greater than expected from the population size and may be caused by genes with large effects. There may be sudden jumps in response followed by periods of stasis or small response, a phenomenon typical of many selection experiments (e.g. 21, 84, 113, 124, 139), though rare recombinants or particular epistatic effects cannot be ruled out in some cases (84, 108, 124).

Analyses of selection experiments using REML with the animal model can be used to check the validity of the infinitesimal model by estimating base population parameters from data for different numbers of generations, for example, 0–5 and 5–10, but with full pedigrees fitted to generation 0 in both cases. Differences among such estimates imply a breakdown of the model, for both the Bulmer effect and inbreeding are allowed for (88, 130). For example, estimates from selected lines of additive genetic variance (in arbitrary units) of lean mass of mice using data from generations 0–4, 5–7, 8–14, and 15–20 were 60, 40, 31, and 14, respectively (7).

When selection is practiced immediately from an inbred base, the pattern of response is highly dependent on the size of mutant gene effects. If effects are small, variance accumulates approximately linearly and cumulative response quadratically. If mutant genes have large effects and are therefore rapidly fixed by selection, the cumulative response is expected to be nearly linear, but subject to large variation between generations and replicate lines (53). This latter variability is usually observed from both spontaneous (16, 78, 85) and induced mutation (21, 60, 68).

Divergent selection from isogenic or inbred populations has the advantage of allowing direct analysis of individual mutations (16) and may give information about the symmetry of the distribution of mutant effects. Most information in this respect refers to bristle traits in *Drosophila* , pointing to strong asymmetry of the distributions, with genes of very large effect predominantly in one tail of the distribution (16, 78). The information drawn from these experiments on distribution of gene effects must be carefully interpreted, however, because the minimum detectable gene effect depends on the size of the experiment and the distribution of mutant effects may differ from that of genes segregating in wild populations where natural selection is continuously screening deleterious genes.

A more practical approach than to infer general properties of the distribution of gene effects may be to locate and identify the genes themselves, using crosses of highly divergent selected lines so that all high genes are on one chromosome and all low genes are on the other. Markers with identifiable phenotypes in *Drosophila* were proposed for defining recombinant chromosomes (122). This laborious technique has been used to identify small chromosome regions containing genes with substantial effect on bristle traits (116, 119). Now that large numbers of polymorphic molecular markers are available, more direct approaches using F2 or backcrosses of selected lines can be used in which effects of chromosome regions are associated with RFLP markers (72).

The chromosome walk from the RFLP marker to the actual trait gene may be a long one. Alternative approaches for directly identifying genes associated with the differences between selection lines are possible. Insertional mutagenesis enables straightforward cloning, but the effects may not be typical of variants in natural populations. The dissection of commercial traits into their physiological components in order to locate the products of genes through their action, and the display and assessing of expression of a large number of genes or gene products at random, have been suggested (11).

Indeed it seems likely (at least to us) that in the future a main use of artificial selection experiments will be to produce lines for subsequent mapping analysis. The theoretical problem will then be to use the results of this analysis to infer the distribution of effects and frequencies in the source population and to ascertain what forces maintain variation in populations. In any case, gene mapping and transgenesis will not substitute but will enhance the practice of artificial selection for animal and plant improvement.

Gene Frequency

Very little can be inferred from selection experiments about the shape of the distribution of gene frequencies of quantitative traits because it does not strongly affect the selection limit (57). A rough idea can be obtained by comparing the limiting response to selection (with all the problems discussed previously) with that in a line selected from the same base following a bottleneck in population size. If gene frequencies are extreme and gene effects are large, a restriction in population size will have a greater effect than if frequencies are intermediate or effects are small. A bottleneck to a single pair reduced response by only about 30% (40, 103), as might be expected by infinitesimal theory (62), indicating the absence of genes of large effect at extreme frequencies.

Inferences about average gene frequency can also be made from the response of selected lines after a reversal of the direction of selection. A short period of reverse selection will have the greatest influence—as does a bottleneck—on response from genes of large effect at low initial frequency (2). Thus, it was inferred that the average frequency of genes for high bristle number in a cage population of *Drosophila* was in the region 0.3 to 0.4 (1).

Some information about the mean gene frequency can also be obtained from the asymmetry of response to divergent selection. For example, if alleles increasing the trait have high average frequency, greater responses are expected to selection downwards. This result was obtained in many selection experiments, for example, in selection for wing-length by Robertson & Reeve (108). The same pattern of asymmetry could, however, have other causes (35) such as directional dominance or natural selection opposing the response in the high line. In fact, the high line in this experiment regressed partially when selection was ceased, while the low line did not, pointing to an asymmetry due to opposing natural selection. Analyses of asymmetry are also complicated by scale effects. For example in the Illinois corn oil experiment, symmetry would now require negative oil contents in the low line, yet, based on the asymmetry, Dudley (27, 28) inferred that initial gene frequencies were low. Asymmetrical short-term responses to selection can also be interpreted as a nonlinear offspring-parent regression in the base population due to skewness of the distributions of gene and environmental effects in opposite directions (36).

Deleterious mutations are predominantly recessive, or nearly so, and are likely to be found at low frequency in natural populations. Frankham's (41) review bears out this conclusion: genes that increase traits associated with fitness are at high frequencies in founder stocks.

Gene Action

Directional dominance, i.e. the dominant effects of genes being preponderantly in one direction, may cause asymmetry in the response to divergent selection as fixation of recessive genes occurs more quickly. In general, substantial evidence suggests directional dominance for higher reproductive rate, where asymmetry of response is usually observed (41). Unfortunately, this asymmetry is not easily distinguishable from that caused by asymmetrical gene frequencies, though directional dominance can be unequivocally diagnosed by inbreeding depression.

Directional dominance could be responsible for the asymmetry in the limits achieved in selection lines for thorax length in *Drosophila* (107). Three different wild populations were selected high and low; eventually an apparent limit was achieved in all lines, the response being greater in the small ones. Reversed selection was only effective in the three large lines, but relaxation had little effect, precluding natural selection as the main cause of the limit. This is consistent with the theory that most of the genes reducing the trait are recessive, though an alternative possibility would be overdominance (selection favoring heterozygotes) in the direction of large thorax length. Other experiments where asymmetrical responses could be due to directional dominance are discussed by Falconer (36).

Analysis of diallel crosses among replicated selected lines might shed some light on dominant (e.g. 27) or epistatic (e.g. 31) effects. For example, the cross of two highly selected lines for pupal weight in *Tribolium* showed negative heterosis, suggesting that specific sets of genes had been selected in the two lines and that epistatic interactions could be important (31).

A method available in *Drosophila* for studying interactions among homologous and nonhomologous chromosomes from selected lines is to cross these lines to others containing inversions on each of the chromosomes, and then to score the offspring. Widespread interactions in body size among nonhomologous chromosomes and nonadditive relations between homologues have been found (e.g. 109). A recent assay using this method was undertaken on chromosomes of lines of *Drosophila* which Hirsch selected divergently and intermittently for geotaxis for over 600 generations (probably the longest recorded artificial selection experiment in eukaryotes in terms of number of generations !) (100). The results were compared with a similar analysis of the same lines made about 500 generations earlier. In the early study extreme positive geotaxis seemed to be disfavored by natural selection (relaxation of selection repeatedly moved the mean toward the value in the base population), while in the recent analysis the opposite seemed to happen (reverse selection moved the mean away from the extreme but subsequent relaxation returned it back to the extreme). This, along with the observed increase in chromosomal interactions, indicated a possible genomic coadaptation in the line.

In quantitative genetic terms, local adaptation to the environment is described as genotype × environment interaction (G × E). The problem in designing or describing experiments to investigate G × E is to generalize in any way outside the specific set of environments tested. For example, does a diet giving reduced animal growth due to low protein content reflect changes that would occur with any other diet or environmental manipulation giving similar poor growth? Falconer (37) recently reviewed such selection (and related) experiments, and he distinguishes between "antagonistic" selection, downwards in a good or upwards in a bad environment, and "synergistic" selection, the converse. He found that in practice antagonistic selection was significantly better than synergistic for changing mean performance over two environments; in other words, better overall adaptation occurred with selection in the poorer environment, but there was no quantitative genetic theoretical basis for this finding.

Relation to Fitness

Artificial selection experiments have contributed a substantial proportion of the estimates of heritability for a wide range of traits available from populations founded from the wild (90, 110). Traits associated with fitness usually show low heritability, but the additive genetic variance may not be negligible because the phenotypic variance is typically very large. Let us consider what selection experiments can tell us about the forces that maintain this variance.

The most direct evidence of the effect of natural selection on a quantitative trait is that lethal genes with pleiotropic effects on the trait and other highly deleterious genes are a common feature of selection experiments (e.g. 21, 43, 59, 81, 140). Some consequences of this are a reduction in the selective pressure on the other loci causing a reduction in the selection response, and an accumulation of other lethals in the same chromosome region (46, 81).

A simple drift-mutation equilibrium is not an adequate model because heritabilities of bristle number are insufficiently large to conform with the values of h_m^2 observed (67), mean scores are similar in different populations, and lines selected for bristle number (and other traits) typically show decreased fitness (73).

Models of either heterozygote superiority or stabilizing selection (i.e. intermediate scoring individuals more fit) predict the half-life of response to be less than predicted by the infinitesimal neutral model (61, 94) and that selection followed by relaxation would lead to a return towards the base value. Such a reversion is typically observed, sometimes leading the mean back to its initial value (e.g. 113); but it is usually much smaller than the initial response even when substantial genetic variation is still present (43, 73). Indeed, when only small initial displacements have been made such that fixation has not had time to occur, the regression is very small (120; see also 103). Furthermore, when only one chromosome is allowed to segregate in a line in which other chromosomes have been fixed at extreme values of the trait, there is not a rapid return toward the base population value (120); this is evidence against substantial stabilizing selection. Even in cases where relaxation of selection has shown strong effects of natural selection, this could not be ascribed to stabilizing selection. For example, in an experiment for increasing pupal weight in *Tribolium* (32), relaxation of a highly selected line for heavy pupa led to a considerable reversion to the original mean and restoration of reproductive performance, but another heavy line did not show such reversion, indicating specific genes that affect pupa weight and that have a negative impact on fitness.

Response to selection that is resumed after a period of relaxation is usually faster than before it (e.g. 84); sometimes it also reaches a higher limit (113). This could be explained by the release of variation by recombination (82, 83), by the elimination of deleterious genes during relaxation, or by the transition between interactive adaptive peaks, following Wright's shifting balance theory (138).

It is a prediction of the mutation-stabilizing selection "house-of-cards" model that segregating genes will be at extreme frequencies, and that response to selection would increase in rate after early generations (129). There is no good evidence of this, but further analysis of Sheridan's (114) summary is necessary. In any case, variance in response among replicate lines under this model is likely to be large. Furthermore, results from selection experiments with population bottlenecks (40) do not seem to support this prediction, though their interpretation is very subjective (62). A range of models cannot be eliminated by the selection experiments.

Apparent stabilizing selection on a quantitative trait can result from genes that have pleiotropic effects on the trait and on fitness and that segregate either with heterozygote superiority (103) or as deleterious mutations. In the latter case the apparent selection is unlikely to be strong (6, 67). This model is not supported by recent data where the fitness of selected populations was clearly improved without reducing selection response by culling the least fit lines (44).

Asymmetries in divergent selection experiments are most likely to be found for traits closely associated with fitness, for example, litter size or mating speed. Smaller responses appear in the direction of increased fitness (41). In 24 out of 30 studies with divergent selection reviewed by Frankham (41), response to downward (reduced fitness) selection exceeded that to upward selection, and the mean realized heritabilities were 0.26 downwards and 0.17 upwards. Inbreeding depression was not an adequate explanation, as control populations similar in size were kept in most cases. Asymmetry due to scale, a problem in assessing symmetry when responses are large, would act in different ways for different traits (e.g. mating speed vs litter size). Thus these results point most clearly to the presence of genes, perhaps recessives, of sufficiently large effect to cause departure from the predictions of the infinitesimal model in the expected direction.

CONCLUSIONS

Artificial selection experiments have a long history, but improved methods for their analysis have been developed recently. They offer the opportunity to study a range of genetic properties of a population and are a resource for analysis of individual genes. Extensive information on the nature of quantitative genetic variation has been obtained from selection experiments, but definitive conclusions about the forces maintaining variation in natural populations are still sought. Further experiments will be needed to check on alternative models proposed by theoreticians.

Acknowledgments

We are grateful to the Agricultural and Food Research Council for financial support, and to N. H. Barton, D. S. Falconer, I. M. Hastings, and P. D. Keightley for helpful comments.

Literature Cited

1. Allan, J. S. 1963. A study of variability in selected populations of *Drosophila melanogaster*. PhD thesis. Univ. Edinburgh, Scotland. 56 pp.
2. Allan, J. S., Robertson, A. 1964. The effect of initial reverse selection upon total selection response. *Genet. Res.* 5:68–79
3. Al-Murrani, W. K. 1974. The limits to artificial selection. *Anim. Breed. Abstr.* 42:587–92
4. Atkins, K. D., Thompson, R. 1986. Predicted and realized responses to selection for an index of bone length and body weight in Scottish Blackface sheep. I. Responses in the index and component traits. *Anim. Prod.* 43:421–35
5. Barker, J. S. F. 1989. Population structure. See Ref. 56, pp. 75–80
6. Barton, N. H. 1990. Pleiotropic models of quantitative variation. *Genetics* 124:773–82
7. Beniwal, B. K., Hastings, I. M., Thompson, R., Hill, W. G. 1992. Estimation of changes in genetic parameters in selected lines of mice using REML with an animal model. 1. Lean mass. *Heredity*. In press
8. Blair, H. T., Pollak, E. J. 1984. Estimation of genetic trend in a selected population with and without the use of a control population. *J. Anim. Sci.* 58:878–86
9. Bohren, B. B., Hill, W. G., Robertson, A. 1966. Some observations on asymmetrical correlated responses to selection. *Genet. Res.* 7:44–57
10. Bradford, G. E., Famula, T. R. 1984. Evidence for a major gene for rapid postweaning gain in mice. *Genet. Res.* 44:293–308
11. Bulfield, G. 1989. The biochemical control of quantitative traits. See Ref. 56, pp. 227–31
12. Bulmer, M. G. 1971. The effect of selection on genetic variability. *Am. Nat.* 105:201–11
13. Bulmer, M. G. 1980. *The Mathematical Theory of Quantitative Genetics*. Oxford: Clarendon. 254 pp.
14. Caballero, A. 1989. Efficiency in prediction of response in selection index experiments. *J. Anim. Breed. Genet.* 106:187–94
15. Caballero, A., Keightley, P. D., Hill, W. G. 1991. Strategies for increasing fixation probabilities of recessive mutations. *Genet. Res.* 58:129–38
16. Caballero, A., Toro, M. A., López-Fanjul, C. 1991. The response to artificial selection from new mutations in *Drosophila melanogaster*. *Genetics* 127:89–102
17. Cameron, N. D., Thompson, R. 1986. Design of multivariate selection experiments to estimate genetic parameters. *Theor. Appl. Genet.* 72:466–76
18. Castle, W. E. 1921. An improved method of estimating the number of genetic factors concerned in cases of blending inheritance. *Science* 54:223
19. Clayton, G. A., Morris, J. A., Robertson, A. 1957. An experimental check on quantitative genetical theory. I. Short-term responses to selection. *J. Genet.* 55:131–51
20. Clayton, G. A., Robertson, A. 1955. Mutation and quantitative variation. *Am. Nat.* 89:151–58
21. Clayton, G. A., Robertson, A. 1957. An experimental check on quantitative genetical theory. II. The long-term effects of selection. *J. Genet.* 55:152–70
22. Cold Spring Harbor Symposia on Quantitative Biology. 1955. *Population Genetics: The Nature and Causes of Genetic Variability in Populations*. Vol. XX. Cold Spring Harbor, NY: Biol. Lab.
23. da Silva, J. M. P. 1961. *Limits of response to selection*. PhD thesis. Univ. Edinburgh, Scotland. 87 pp.
24. Dempfle, L. 1974. A note on increasing the limit of selection through selection within families. *Genet. Res.* 24:127–35
25. Dickerson, G. E., Lindhé, N. B. H. 1977. Potential uses of inbreeding to increase selection response. See Ref. 98, pp. 323–42
26. Dion, N., Minvielle, F. 1985. The effects of alternated cycles of full-sib and random mating on selection for pupa weight in *Tribolium castaneum*. *Can. J. Genet. Cytol.* 27:251–54
27. Dudley, J. W. 1977. 76 generations of selection for oil and protein percentage in maize. See Ref. 98, pp. 459–73
28. Dudley. J. W., Lambert, R. J. 1992. Ninety generations of selection for oil and protein in maize. *Maydica* 37:1–7
29. Eisen, E. J. 1980. Conclusions from long-term selection experiments with mice. *Z. Tierzuchtg. Zuchtgsbiol.* 97:305–19
30. Eisen, E. J. 1989. Selection experiments for body composition in mice and rats:

A review. *Livestock Prod. Sci.* 23:17–32
31. Enfield, F. D. 1977. Selection experiments in *Tribolium* designed to look at gene action issues. See Ref. 98, pp. 177–90
32. Enfield, F. D. 1980. Long-term effects of selection; the limits to response. See Ref. 106, pp. 69–86
33. Enfield, F. D., Braskerud, O. 1989. Mutational variance for pupa weight in *Tribolium castaneum*. *Theor. Appl. Genet.* 77:416–20
34. Falconer, D. S. 1954. Asymmetrical responses in selection experiments. In *Symp. Genet. Popul. Struc. Int. Union Biol. Sci. Naples, Series B* 15:16–41 X
35. Falconer, D. S. 1989. *Introduction to Quantitative Genetics*. Harlow, Essex: Longman. 438 pp. 3rd. ed
36. Falconer, D. S. 1989. Selection experiments and the nature of quantitative variation. See Ref. 56, pp. 121–27
37. Falconer, D. S. 1990. Selection in different environments: effects on environmental sensitivity (reaction norm) and on mean performance. *Genet. Res.* 56:57–70
38. Falconer, D. S. 1992. Early selection experiments. *Annu. Rev. Genet.* In press
39. Frankham, R. 1980. Origin of genetic variation in selection lines. See Ref. 106, pp. 56–68
40. Frankham, R. 1980. The founder effect and response to artificial selection in *Drosophila*. See Ref. 106, 87–90
41. Frankham, R. 1990. Are responses to artificial selection for reproductive fitness characters consistently asymmetrical? *Genet. Res.* 56:35–42
42. Frankham, R., Briscoe, D. A. Nurthen, R. K. 1980. Unequal crossing over at the rRNA tandon as a source of quantitative genetic variation in *Drosophila*. *Genetics* 95:727–42
43. Frankam, R., Jones, L. P., Barker, J. S. F. 1968. The effects of population size and selection intensity in selection for a quantitative character in *Drosophila*. III. Analyses of the lines. *Genet. Res.* 12:267–83
44. Frankham, R., Yoo, B. H., Sheldon, B. L. 1990. Reproductive fitness and artificial selection in animal breeding: culling on fitness prevents a decline in reproductive fitness in lines of *Drosophila melanogaster* selected for increased inebriation time. *Theor. Appl. Genet.* 76:909–14
45. Gallego, A., López-Fanjul, C. 1983. The number of loci affecting a quantitative trait in *Drosophila melanogaster* revealed by artificial selection. *Genet. Res.* 42:137–49
46. García-Dorado, A., López-Fanjul, C. 1983. Accumulation of lethals in highly selected lines of *Drosophila melanogaster*. *Theor. Appl. Genet.* 66:221–23
47. Gottschalk, W., Wolff, G. 1983. *Induced Mutations in Plant Breeding*. Berlin: Springer. 238 pp.
48. Gründl, E., Dempfle, L. 1990. Effects of spontaneous and induced mutations on selection response. See Ref. 58, 13:177–84
49. Henderson, C. R. 1984. *Application of Linear Models in Animal Breeding*. Guelph, Canada: Univ. Guelph. 462 pp.
50. Hill, W. G. 1971. Design and efficiency of selection experiments for estimating genetic parameters. *Biometrics* 27:293–311
51. Hill, W. G. 1972. Estimation of realised heritabilities from selection experiments. I. Divergent selection. *Biometrics* 28:747–65
52. Hill, W. G. 1980. Design of quantitative genetic selection experiments. See Ref. 106, pp. 1–13
53. Hill, W. G. 1982. Predictions of response to artificial selection from new mutations. *Genet. Res.* 40:255–78
54. Hill, W. G. 1984. *Quantitative Genetics*. Part II. *Selection. Benchmark Papers in Genetics*, Vol.15 . New York: Van Nostrand Reinhold. 347 pp.
55. Hill, W. G. 1985. Detection and genetic assessment of physiological criteria of merit. In *Genetics of Reproduction in Sheep*, ed. R. B. Land, D. W. Robinson. pp. 319–31. London: Butterworths. 427 pp.
56. Hill, W. G., Mackay, T. F. C., eds. 1989. *Evolution and Animal Breeding*. Wallingford, UK: C.A.B. Int. 313 pp.
57. Hill, W. G., Rasbash, J. 1986. Models of long-term artificial selection in finite populations. *Genet. Res.* 48:41–50
58. Hill, W. G., Thompson, R., Woolliams, J. A., eds. 1990. *Proceedings of the 4th World Congress on Genetics Applied to Livestock Production*. Vols. 13–16. Edinburgh: Organising Committee, 4th World Congress on Genetics Appl. to Livestock Production
59. Hollingdale, B. 1971. Analysis of some genes from abdominal bristle number selection lines in *Drosophila melanogaster*. *Theor. Appl. Genet.* 41:292–301
60. Hollingdale, B., Barker, J. S. F. 1971. Selection for increased abdominal bris-

tle number in *Drosophila melanogaster* with concurrent irradiation. *Theor. Appl. Genet* . 41:208–15
61. James, J. W. 1962. Conflict between directional and centripetal selection. *Heredity* 17:487–99
62. James, J. W. 1971. The founder effect and response to artificial selection. *Genet. Res.* 16:241–50
63. James, J. W. 1986. Cumulative selection differentials and realized heritabilities with overlapping generations. *Anim. Prod.* 42:411–15
64. Jones, L. P. Frankham, R., Barker, J. S. F. 1968. The effects of population size and selection intensity in selection for a quantitative character in *Drosophila*. III. Long-term response to selection. *Genet. Res.* 12:249–66
65. Kaufman, P., Enfield, F. D., Comstock, R. E. 1977. Stabilizing selection for pupal weight in *Tribolium castaneum*. *Genetics* 87:327–41
66. Keightley, P. D., Hill, W. G. 1990. Estimating new mutational variation in growth rate of mice. See Ref. 58, 13:325–28
67. Keightley, P. D., Hill, W. G. 1990. Variation maintained in quantitative traits with mutation-selection balance: pleiotropic side-effects on fitness traits. *Proc. R. Soc. Lond. B* 242:95–100
68. Kitagawa, O. 1967. The effects of X-irradiation on selection response in *Drosophila melanogaster*. Jpn. J. Genet. 42:121–37
69. Kress, D. D. 1975. Results from long-term selection experiments relative to selection limits. *Genet. Lect.* 4:253–71
70. Lande, R. 1975. The maintenance of genetic variability by mutation in a polygenic character with linked loci. *Genet. Res.* 26:221–34
71. Lande, R., Arnold, S. J. 1983. The measurement of selection on correlated characters. *Evolution* 37:1210–26
72. Lander, E. S., Botstein, D. 1989. Mapping Mendelian factors underlying quantitative traits using RFLP linkage maps. *Genetics* 121:185–99
73. Latter, B. D. H., Robertson, A. 1962. The effects of inbreeding and artificial selection on reproductive fitness. *Genet. Res.* 3:110–38
74. López-Fanjul, C. 1989. Tests of theory by selection experiments. See Ref. 56, pp. 129–33
75. Luckinbill, L. S., Arking, R., Clare, M. J., Cirocco, W. C., Buck, S. A. 1984. Selection for delayed senescence in *Drosophila melanogaster*. *Evolution* 38:996–1003
76. Lynch, M. 1988. The rate of polygenic mutation. *Genet. Res.* 51:137–48
77. Mackay, T. F. C. 1987. Transposable element-induced polygenic mutations in *Drosophila melanogaster*. *Genet. Res.* 49:225–33
78. Mackay, T. F. C. 1990. Distribution of effects of new mutations affecting quantitative traits. See Ref. 58, 13:219–28
79. Mackay, T. F. C., Lyman, R. F., Jackson, M. S. 1992. Effects of P-element insertions on quantitative traits in *Drosophila melanogaster*. *Genetics* 130: 315–32
80. MacNeil, M. D., Kress, D. D., Flower, A. E., Blackwell, R. L. 1984. Effects of mating system in Japanese Quail. 2. Genetic parameters, response and correlated response to selection. *Theor. Appl. Genet.* 67:407–12
81. Madalena, F. E., Robertson, A. 1975. Population structure in artificial selection: Studies with *Drosophila melanogaster*. *Genet. Res.* 24:113–26
82. Mather, K. 1941. Variation and selection of polygenic characters. *J. Genet.* 41:159–93
83. Mather, K. 1983. Response to selection. In *The Genetics and Biology of Drosophila, Vol. 3, ed. M. Ashburner, H. L. Carson, J. N. Thompson, Jr. New York: Academic Press. 305 pp.*
84. Mather, K., Harrison, B. J. 1949. The manifold effects of selection. *Heredity* 3:1–52
85. Mather, K., Wigan, L. G. 1942. The selection of invisible mutations. *Proc. R. Soc. Lond. B* 131:50–64
86. McPhee, C. P., Robertson, A. 1970. The effect of suppressing crossing-over on the response to selection in *Drosophila melanogaster*. *Genet. Res.* 16:1–16
87. Meyer, K. 1989. Restricted maximum likelihood to estimate variance components for animal models with several random effects using a derivative-free algorithm. *Genet. Selection Evol.* 21: 317–40
88. Meyer, K., Hill, W. G. 1991. Mixed model analysis of a selection experiment for food intake in mice. *Genet. Res.* 57:71–81
89. Moran, C. 1990. The role of transposable elements in generating quantitative genetic variation. See Ref. 58, 13:229–37
90. Mousseau, T. A., Roff, D. A. 1987. Natural selection and heritability of fitness components. *Heredity* 59:181–97
91. Mrode, R. A. 1988. Selection experi-

ments in beef cattle. Part 2. A review of responses and correlated responses. *Anim. Breed. Abstr.* 56:155–67

92. Muir, W. M. 1986. Estimation of response to selection and utilization of control populations for additional information and accuracy. *Biometrics* 42:381
93. Nicholas, F. W. 1980. Size of population required for artificial selection. *Genet. Res.* 35:85–105
94. Nicholas, F. W., Robertson, A. 1980. The conflict between natural and artificial selection in finite populations. *Theor. Appl. Genet.* 56:57–64
95. Partridge, L., Fowler, K. 1992. Direct and correlated responses to selection on age at reproduction in *Drosophila melanogaster*. *Evolution* 46:76–91
96. Patterson, H. D., Thompson, R. 1971. Recovery of inter-block information when block sizes are unequal. *Biometrika* 58:545–54
97. Pignatelli, P. M., Mackay, T. F. C. 1989. Hybrid dysgenesis-induced response to selection in *Drosophila melanogaster*. *Genet. Res.* 54:183–95
98. Pollak, E., Kempthorne, O., Bailey, T. B. Jr., eds. 1977. *Proceedings of the International Conference on Quantitative Genetics*. Ames: Iowa State Univ. Press. 872 pp.
99. Provine, W. B. 1971. *The Origins of Theoretical Population Genetics* . Chicago: Univ. Chicago Press
100. Ricker, J. P., Hirsch, J. 1988. Genetic changes occurring over 500 generations in lines of *Drosophila melanogaster* selected divergently for geotaxis. *Behav. Genet.* 18:13–25
101. Roberts, R. C. 1966. The limits to artificial selection for body weight in the mouse. II. The genetic nature of the limits. *Genet. Res.* 8:361–75
102. Robertson, A. 1960. A theory of limits in artificial selection. *Proc. R. Soc. Lond. B* 153:234–49
103. Robertson, A. 1967. The nature of quantitative genetic variation. In *Heritage from Mendel* , ed. R. A. Brink, pp. 265–80. Madison, Wisc: Univ. Wisc. Press
104. Robertson, A. 1970. A note on disruptive selection experiments in *Drosophila*. *Am. Nat.* 104:561–69
105. Robertson, A. 1978. The time of detection of recessive visible genes in small populations. *Genet. Res.* 31:255–64
106. Robertson, A., ed. 1980. *Selection Experiments in Laboratory and Domestic Animals*. Slough, UK: Commonwealth Agric. Bur. 245 pp.
107. Robertson, F. W. 1955. Selection response and the properties of genetic variation. See Ref. 22, pp. 166–77
108. Robertson, F. W., Reeve, E. C. R. 1952. Studies in quantitative inheritance. I. The effects of selection on wing and thorax length in *Drosophila melanogaster*. *J. Genet.* 50:414–48
109. Robertson, F. W., Reeve, E. C. R. 1953. Studies in quantitative inheritance. IV. The effects of substituting chromosomes from selected strains in different genetic backgrounds. *J. Genet.* 51:586–610
110. Roff, D. A., Mousseau, T. A. 1987. Quantitative genetics and fitness: lessons from *Drosophila*. *Heredity* 58: 103–18
111. Rose, M. R., Charlesworth, B. 1981. Genetics of life history in *Drosophila melanogaster*. II. Exploratory selection experiments. *Genetics* 97:187–96
112. Scossiroli, R. E. 1954. Effectiveness of artificial selection under irradiation of plateaud populations of *Drosophila melanogaster*. In *Symp. Genet. Popul. Struc. Inter. Union Biol. Sci. Naples, Series B* 15:42–66
113. Sheldon, B. L., Milton, M. K. 1972. Studies on the scutellar bristles of *Drosophila melanogaster*. II. Long-term selection for high bristle number in the Oregon RC strain and correlated responses in abdominal chaetae. *Genetics* 71:567–95
114. Sheridan, A. K. 1988. Agreement between estimated and realised genetic parameters. *Anim. Breed. Abstr.* 56: 877–89
115. Shrimpton, A. E., Mackay, T. F. C., Leigh Brown, A.J. 1990. Transposable element-induced response to artificial selection in *Drosophila melanogaster*: molecular analysis of selection lines. *Genetics* 125:803–11
116. Shrimpton, A. E., Robertson, A. 1988. Isolation of polygenic factors controlling bristle score in *Drosophila melanogaster*. 2. Distribution of third chromosome bristle effects within chromosome sections. *Genetics* 118:445–59
117. Sørensen, D. A., Hill, W. G. 1983. Effects of disruptive selection on genetic variance. *Theor. Appl. Genet.* 65:173–80
118. Sørensen, D. A., Kennedy, B. W. 1984. Estimation of response to selection using least-squares and mixed model methodology. *J. Anim. Sci.* 58:1097–106
119. Spickett, S. G., Thoday, J. M. 1966. Regular responses to selection. 3. In-

teraction between located polygenes. *Genet. Res.* 7:96–121
120. Spiers, J. G. D. 1974. *The effects of larval competition on a quantitative character in Drosophila melanogaster*. PhD thesis. Univ. Edinburgh, Scotland
121. "Student." 1934. A calculation of the minimum number of genes in Winter's selection experiment. *Ann. Eugen.* 6: 77–82
122. Thoday, J. M. 1961. Location of polygenes. *Nature* 191:368–70
123. Thoday, J. M. 1972. Review lecture: disruptive selection. *Proc. R. Soc. Lond. B* 182:109–43
124. Thoday, J. M., Boam, T. B. 1961. Regular responses to selection. I. Description of responses. *Genet. Res.* 2:161–76
125. Thompson, R. 1986. Estimation of realized heritability in a selected population using mixed model methods. *Genet. Selection Evol.* 18:475–83
126. Thompson, R., Atkins, K. D. 1990. Estimation of heritability in selection experiments. See Ref. 58, 13:257–60
127. Thompson, R., Juga, J. 1989. Cumulative selection differentials and realized heritabilities. *Anim. Prod.* 49:203–8
128. Torkamanzehi, A., Moran, C., Nicholas, F. W. 1988. P-element-induced mutation and quantitative variation in *Drosophila melanogaster* : lack of enhanced response to selection in lines derived from dysgenic crosses. *Genet. Res.* 51:231–38
129. Turelli, M. 1984. Heritable genetic variation via mutation-selection balance: Lerch's zeta meets the abdominal bristle. *Theor. Popul. Biol.* 25:138–93
130. Van der Werf, J. H. J., de Boer, I. J. M. 1990. Estimation of additive genetic variance when base populations are selected. *J. Anim. Sci.* 68:3124–32
131. Villanueva, B., Kennedy, B. W. 1992. Asymmetrical correlated responses to selection under an infinitesimal genetic model. *Theor. Appl. Genet.* In press
132. Wade, M. J., Goodnight, C. J. 1991. Wright's shifting balance theory: an experimental study. *Science* 253:1015–18
133. Weber, K. E. 1990. Increased selection response in larger populations. I. Selection for wing-tip height in *Drosophila melanogaster* at three population sizes. *Genetics* 125:579–84
134. Weber, K. E., Diggins, L. T. 1990. Increased selection response in larger populations. II. Selection for ethanol vapor resistance in *Drosophila melanogaster* at two population sizes. *Genetics* 125:585–97
135. Weir, B. S., Eisen, E. J., Goodman, M. M., Namkoong, G., eds. 1988. *Proceedings of the Second International Conference on Quantitative Genetics*. Sunderland, Mass: Sinauer. 724 pp.
136. Williams, G. C. 1957. Pleiotropy, natural selection, and the evolution of senescence. *Evolution* 11:398–411
137. Wray, N. R. 1990. Accounting for mutation effects in the additive genetic variance-covariance matrix and its inverse. *Biometrics* 46:177–86
138. Wright, S. 1977. *Evolution and the Genetics of Populations*. Vol.3. *Experimental Results and Evolutionary Deductions*. Chicago: Univ. Chicago Press. 613 pp.
139. Yoo, B. H. 1980a. Long-term selection for a quantitative character in large replicate populations of *Drosophila melanogaster*. I. Responses to selection. *Genet. Res.* 35:1–17
140. Yoo, B. H. 1980b. Long-term selection for a quantitative character in large replicate populations of *Drosophila melanogaster*. II. Lethals and visible mutants with large effects. *Genet. Res.* 35:19–31
141. Zeng, Z.-B., Houle, D., Cockerham, C. C. 1990. How informative is Wright's estimator of the number of genes affecting a quantitative character? *Genetics* 126:235–47

Annu. Rev. Ecol. Syst. 1992. 23:311–38

GASTROPOD PHYLOGENY AND SYSTEMATICS

Rüdiger Bieler
Department of Zoology, Field Museum of Natural History, Chicago, Illinois 60605

KEY WORDS: evolution, biology, taxonomy, cladistics, diversity, Mollusca, snails

> As every specialist of gastropod morphology will know, we are still far from having a natural system of the gastropods, and extensive studies will be necessary to develop one . . . . With adequate knowledge of the phylogeny, one rule would suffice; "to place in a systematic group always all and only such forms which are derived from a common ancestor."
>
> Adolf Naef, 1911 (p. 152, trans. from original German)

INTRODUCTION

Gastropod phylogenetic systematics has seen a recent boost, prompted by the discovery of new taxonomic groups (especially in the hydrothermal-vent faunas), the development of new and refined morphological and molecular techniques, and the application of new analytical methods of phylogenetic systematics. The class Gastropoda ("snails and slugs") is the largest group of mollusks in terms of species and one of the few animal groups successfully to inhabit marine, freshwater, and terrestrial biotopes. The enormous morphological and trophic diversity as well as other aspects of gastropod biology has been reviewed elsewhere (e.g. 7, 18, 78, 80, 90, 95, 99, 103, 154, 172, and numerous articles in 190). For a general overview see Cox (24) and Solem (152); comprehensive systematic references include the works of Thiele (164), Wenz (188), Zilch (194), Knight et al (94), Franc (34), and Boss (17). I report and comment on the status of phylogenetic investigations in the Gastropoda and give a review of attempted classifications.[1]

Estimates for extant gastropod species range from 40,000 (16) to over 100,000 (51), comprising about 80% of all extant molluscan taxa. In the traditional division of subclasses, an estimated 53% of the recognized species are prosobranchs (largely marine but with terrestrial and freshwater representatives), 4% opisthobranchs (marine), and 43% pulmonates (terrestrial and

[1]For the sake of brevity, most author-date references of mentioned taxa had to be omitted. No new names are introduced.

0066-4162/92/1120-0311$02.00

fresh water, few marine littoral) (16). Almost 13,000 genus-group names have been introduced for Recent and fossil gastropods, with an annual increase of about 83 over the last 30 years (20). Many groups are in need of taxonomic attention at the alpha-level. For example, in the prosobranch family Turridae alone, nearly 700 genus-group taxa and an estimated 10,000 Recent and fossil nominal species have been described (19). At least 20% of the recently introduced taxa escape the abstracting service of the *Zoological Record*, with Russian and paleontological literature especially underrepresented (20). Few comprehensive taxonomic listings and catalogs are available (e.g. 135, 171), and those few usually are not equipped with supporting bibliographies; others are restricted to particular geographic areas and smaller taxonomic groups.

MONOPHYLY, THE FOSSIL RECORD, AND THE SISTER-GROUP QUESTION

Several grand traditional controversies among malacologists have not been resolved. One involves the debate over molluscan ancestry and molluscan sister-group relationships (for recent reviews of the variously modified flatworm and annelid-arthropod theories, see 42, 64, 192). Recent molecular data support the hypothesis of molluscan evolution from a coelomate metameric lineage (32, 42, 98).

The second controversy concerns the Gastropoda more directly and lies in the interpretation of the phenomenon of torsion (e.g. 3, 28, 41, 53, 64, 65). Torsion, which occurs in the ontogeny of all gastropods, is the counterclockwise rotation of the visceral mass and mantle 180° relative to the anteroposterior axis of the head-foot complex. With its associated anatomical conditions (anterior position of mantle cavity, asymmetrical arrangement of pallial organs, streptoneury, twisted esophagus, spiral operculum, and, possibly, loss of left gonad), torsion is the only undebated synapomorphy for gastropods. Many features commonly associated with gastropods, such as the coiled shell, radula, and clearly demarcated head, are symplesiomorphies shared with other molluscan groups.

Torsion is thus a key issue in phylogenetic studies of gastropod ancestry and the sister-group relations of the class. Gastropods have an excellent fossil record from the earliest Paleozoic. However, the morphological data that can be derived from fossils are largely restricted to shell form, sculpture, and microstructure. Torsion or its absence in a fossil must be deduced from controversial circumstantial evidence such as muscle scars and shell slits. In particular, the interpretation of planispirally coiled Paleozoic bellerophontiform shells as gastropods (25, 57, 109, 121) or untorted organisms (15, 144, 177–179) is still a matter of debate (see also 119, 128a, 149, 181a). Several fossil groups previously classified as gastropods have been removed from the class to the Monoplacophora (in its various definitions; see 100, 177, 192) or are placed in preliminary groupings like the Paragastropoda (105).

Phylogenetic investigations of gastropods often rely on fossils to establish or corroborate polarity (11, 87, 102, 104, 157), and the interpretation of one particular shell may add or subtract millions of years of perceived fossil history and strongly influence ancestor-descendant hypotheses. How problematic this may be has recently been shown by Bandel (5), who dissolved the much-cited Mesozoic "Euomphaloidea" into members of numerous unrelated clades.

The available fossil record cannot resolve the question of sister-group relationship among the molluscan classes. Fossils classified as Monoplacophora, Rostroconchia, and Gastropoda appear suddenly and concurrently in the earliest Cambrian deposits (143). Based on the comparative anatomy and ontogeny of extant mollusks, several evolutionary scenarios have been presented. Intrigued by the discovery of (apomorphic) bivalved gastropods, some authors hypothesized a derivation of bivalves from within gastropods (e.g. 55). Others (e.g. 52, 101) supported a sister-group hypothesis for the gastropods with Monoplacophora, largely on the basis of the original description of a spiral larval shell in the first-discovered Recent monoplacophoran *Neopilina galatheae*, an interpretation now known to be in error (192). Most authors currently consider as a sister-group pair Gastropoda-Cephalopoda (a clade named Rhacopoda—77), for which numerous shared derived characters have been identified: epipodial tentacles, cerebrally innervated eyes, dorsal body elongation, restriction of mantle cavity to (pre-torsional) posterior end of body, free head and restriction of mantle/shell to visceral part of body, and body appendages extended by a muscle system (e.g. 64, 77, 100, 178, 192).

Von Ihering's (175) hypothesis of diphyletic gastropod derivation from annelids (prosobranchs) and flatworms (opisthobranchs and pulmonates) has never found subsequent proponents, but the monophyly of extant gastropods is not unquestioned (159). Some authors view the bellerophonts as a grade from which the recent gastropods derived in several lines (28). The great difference between RNA sequences of Docoglossa (= Patellogastropoda, see Table 1) and other Gastropoda led Tillier et al (168) to suggest further testing of monophyly.

TRADITIONAL ARRANGEMENTS AND NEW APPROACHES

Renaissance of Comparative Studies

When gastropod systematics began in pre-Linnéan times, it usually focused on shell shape, sculpture, and coloration. Comparative anatomical and histological investigations reached a first peak in the late nineteenth century

Table 1 Sequential arrangement of 'Prosobranchia' groups (Docoglossa to Caenogastropoda)[a]

(Prosobranchia Milne-Edwards, 1848 [= Streptoneura Spengel, 1881])
  Docoglossa Troschel, 1861 [= Patellogastropoda Lindberg, 1986][b]
  Archaeogastropoda Thiele, 1925
    unnamed [comprising most of the recently described hot-vent and deep sea groups]
      Neomphaloidei Sitnikova & Starobogatov, 1983[c]
      Lepetelloidea Dall, 1882 [Lepetellida Moskalev, 1971][d]
    Vetigastropoda von Salvini-Plawen, 1980[e]
  Neritopsina Cox & Knight, 1960 [= Neritimorpha Golikov & Starobogatov, 1975][f]
  Cocculinoidea Dall, 1882 [Cocculinida Moskalev, 1971 = Cocculiniformia Haszprunar, 1987][g]
  Caenogastropoda Cox, 1960 [= Mesogastropoda Thiele, 1925 + Neogastropoda Wenz, 1938]
    Cyclophoroidea Gray, 1847
    Ampullarioidea Guilding, 1828[h]
    Cerithioidea Férussac, 1819 [Cerithiimorpha Golikov & Starobogatov, 1975]
    Campanilidae Douvillé, 1904 [Campanilimorpha Haszpunar, 1988]
    unnamed [remaining caenogastropods][i]
  Heterobranchia

[a] illustrating recent hypotheses of relationship, naming and major synonyms; compare to Figures 1, 2 (based on sources cited in text and table notes; and R. Bieler, unpubl.). Original, rank-indicating, endings of suprafamilial taxa not adjusted.

[b] comprising Patellina von Ihering, 1876 [including Nacellina Lindberg, 1988] (with Patelloidea Rafinesque, 1815; Nacelloidea Thiele, 1891; Lottioidea Gray, 1840); McLean (112) recently added Lepetopsina McLean, 1990 [= Hot-vent group-C limpets], with Lepetopsoidea McLean, 1990; (38, 102, 112).

[c] comprising Neomphaloidea McLean, 1981 (with Neomphalidae McLean, 1981; Cyathermiidae McLean, 1990); *Melanodrymia aurantiaca* Hickman, 1984; Peltospiroidea McLean, 1989 [= Hot-vent group-A, tapersnout limpets]; (37, 66, 113, 185). Sitnikova & Starobogatov cited after (46).

[d] recently split from Cocculiniformia sensu Haszprunar (64) by Lindberg & Ponder (104 and pers. comm.) based on new phylogenetic analysis; see also (46).

[e] comprising Lepetodriloidea McLean, 1988; Fissurelloidea Fleming, 1822; Scissurelloidea Gray, 1847; Haliotoidea Rafinesque, 1815; Pleurotomarioidea Swainson, 1840; Trochoidea Rafinesque, 1815; Seguenzioidea Verrill, 1884 [Seguenziida von Salvini-Plawen & Haszprunar, 1987]; (64, 110, 137, 181; Lindberg & Ponder, pers. comm.).

[f] comprising Neritoidea Rafinesque, 1815; Titiscanoidea Bergh, 1890; Hydrocenoidea Troschel, 1857; Helicinoidea Latreille, 1825; (35, 104, 135).

[g] (64; Lindberg & Ponder, pers. comm.).

[h] Cyclophoroidea and Ampullarioidea often combined as Architaenioglossa Haller, 1890; but not supported by synapomophies (135; Lindberg & Ponder, pers. comm.).

[i] comprising Neotaenioglossa Haller, 1892 *auct.* (probably paraphyletic; numerous superfamilies); Ptenoglossa Gray, 1853 [= Ctenoglossa Gray, 1853] (with Triphoroidea Gray, 1847; Epitonioidea Berry, 1910; Eulimoidea H. & A. Adams, 1853), monophyly questioned (135); Neogastropoda Wenz, 1938 [= Stenoglossa Bouvier, 1887] (with Muricoidea Rafinesque, 1815; Cancellarioidea Forbes & Hanley, 1851; Conoidea Rafinesque, 1815); (134; Houbrick, pers. comm.; Lindberg & Ponder, pers. comm.).

with numerous works mostly by French and German workers; a second peak was achieved with the publication of Fretter & Graham's milestone work *British Prosobranch Molluscs* (39). With the advent of new and refined technologies, such as the transmission electron microscope combined with thin-sectioning techniques, comparative anatomical work is undergoing a renaissance. In addition to traditional anatomical studies, now often employed across traditional group limits, excellent comparative research on osphradial, ctenidial, excretory system, and sperm ultrastructure (e.g. 1, 44, 58, 59, 63, 70, 96) has provided much new data. With their numerous hard-parts—the shells, radulae, jaws, and opercula—gastropods lend themselves to studies with the scanning electron microscope (151), and now organ surfaces and other parts of soft anatomy are also being investigated (e.g. 162). Gastropods were among the first organisms to which electrophoretic methods were applied to address taxonomic questions (26, 27), and such investigations continue, especially at lower taxonomic levels (e.g. 2, 83, 85). At higher levels, the study of cellular DNA content (84), chromosome numbers (123, 128), shell matrix proteins (31, 43), and nucleic acid sequences (see below) have provided additional support for phylogenetic hypotheses.

New Data and New Questions

In addition to improving the resolution of gastropod phylogenies, new data have added to the mosaic picture, giving rise to questions about presumed homologies and character polarity. Recent anatomical studies have shown that some features previously employed as characters to separate larger groups need critical reevaluation. A few examples follow. Sperm dimorphism, previously thought restricted to "higher" gastropods (75), has been reported from archaeogastropod Trochoidea (72). Protandrous hermaphroditism has been documented in several groups previously regarded as exclusively gonochoristic (13, 139); pollution-induced pseudohermaphroditism, especially widespread in prosobranch neogastropods (33), demonstrates the lability of this feature. The position of the salivary glands relative to the circumesophageal nerve ring, an often-employed taxonomic character in caenogastropod systematics (87, 131), was found to vary widely within a single population of the vitrinellid *Cyclostremiscus* (13). Intraspecific and ontogenetic variation has also been documented for radular characters (10, 183), which traditionally provide a very important taxobase in gastropod systematics.

Single-Organ-Systems: Dealing With Homoplasy

The long evolutionary history, the often rapid radiations, and the adaptation to many habitats by members of the same evolutionary line and to the same habitat by distantly related forms make phylogenetic studies of gastropods challenging. Especially problematic are groups such as terrestrial and marine slugs in which numerous reductions and losses of structures occur (49, 166, 189), the stylommatophoran land snails (29, 155, 167), and the large groups of small caenogastropod prosobranchs that display very similar and relatively simple morphology (e.g. Rissooidea; 131).

The history of phylogenetic studies and classifications of gastropods is a record of single-character or at least single-organ system approaches. In the gastropod literature examples cover nearly every organ system or structure from the shell to the ultrastructure of spermatozoa. While some authors claimed that their system was finally the breakthrough to resolve questions of relationship, others were well aware of the shortcomings of such approaches

(40). A modified route has been to change the taxobase categorically for every hierarchical level, coarsely weighting for perceived relative classificatory importance of organ systems, e.g., Monotocardia (heart morphology), Androgyna (reproductive morphology), Gymnobranchia (gill morphology), Pleurognatha (buccal morphology) (118).

A holistic approach to resolve conflict between single-organ-system hypotheses in gastropod research (87, 116, 120, 129) faces several major problems. Most studies rely on single species (if not single specimens) as representatives for larger groups, and the actual taxonomic distribution of the character states throughout the group is often unknown or undocumented. If studies must rely on representatives, the sensible approach (for subsequent taxonomic decisions) is to employ the type species of a group. A danger exists, however, when the type species is atypical of the larger group. A form is often initially recognized and described as something new because of its conspicuousness, i.e. its set of uniquely derived characters. An example is *Architectonica,* whose members' "ptenoglossate" radulae grouped the Architectonicidae with other "ptenoglossate" families, although the feature is not shared with other architectonicid genera and is now viewed as a convergently acquired autapomorphy (11). Comprehensive comparative works such as Bouvier's (22) on the nervous system and Bandel's (4) on radulae are rare, and much of the published data has been generated outside any comparative context or taxonomic framework.[2] Thus, we may have detailed knowledge of the nervous system of species (or, by inference, of family) A, B, and C, the osphradial fine morphology of C, D, and E, but the reproductive anatomy of only A and E. At the higher level, this scenario is often a result of different traditions in approaches and available material (e.g. study of shells in prosobranchs, and of reproductive anatomy in pulmonates).

Trees and Rhizomes

Most published gastropod classifications adhere to the Prosobranchia-Opisthobranchia-Pulmonata scheme of Milne-Edwards (1848) or the Streptoneura-Euthyneura division by Spengel (1881), based on the overall organization of respiratory or nervous systems, respectively (for historic reviews, see 25, 150). Although some groups with strongly mosaic character suites were recognized early (106), the former systems proved to be workable and largely undebated arrangements, with the Prosobranchia (or Streptoneura) usually presented as a linear *scala naturae,* from the lowliest archaeogastropod via mesogastropods to the most advanced neogastropods. Similar linearity was also implied in the phylogeny of Pulmonata and Opisthobranchia (separately,

[2]Especially studies published in nonsystematics journals often neglect to deposit voucher material in permanent systematic collections, and thus part of the data cannot reliably be linked to particular taxa.

or united as Euthyneura). The necessarily serial arrangement of taxa in the major handbooks such as Thiele's (164) perpetuated gradistic thinking. Recent approaches still attempt to resolve the Gastropoda into such rhizome-like "trees" with regular, equally ranked offshoots (e.g. 64).

Several early statements stress the necessity of basing gastropod classifications on monophyletic groups (e.g. Naef—122, as quoted above). In the mid-1950s, Morton (120) presented a "cladistic" data matrix with primitive and advanced character states as part of a genus-level analysis of the Ellobiidae. Cladistic methodology (often supplemented by multivariate or phenetic approaches) has become a standard tool in gastropod systematics, allowing *testable* hypotheses of phylogeny and relationships at the species, genus, and family levels, or for analyses involving only a few characters at higher levels.

However, the relationships among many of the gastropod families and orders, and between Gastropoda and other groups, remain elusive. The wish for a resolved phylogenetic tree is strong and spawns ad-hoc constructions such as so-called "retrospective cladograms," created from noncladistically derived classifications (82). Several papers offering analyses and providing phylogenetic trees at higher systematic levels have appeared, but few document their data in formats allowing the presented hypotheses to be tested. Problems encountered in applying parsimony interpretation to their data led Gosliner and Ghiselin (49, 50) to apply (informal) character-weighting to overrule the homoplasious noise caused by extensive parallelism in the data set. Similar disagreement with cladistic parsimony approaches led to Haszprunar's (62) "clado-evolutionary methodology," which promised to combine the advantages of phylogenetic and evolutionary approaches. This was used in recent major gastropod reclassifications (e.g. 64) and involved undocumented character weighting leading to a preferred classification "primarily based on the nervous system" (64, p. 428; for discussion see 12, 67).

Molecular Phylogeny Reconstructions

Tools have recently become available to compare morphology-based reconstructions of gastropod phylogenies with molecular data. As with morphological investigations of mollusks, they involve various approaches depending on questions asked and technologies available. Studies in previous years concentrated on the higher-level relationships of mollusks with other invertebrate groups (32, 42, 47, 98), or they addressed questions at population and species levels. Now the early results of studies concerning higher-level relationships within the Gastropoda have become available. Emberton et al (30) employed large-subunit ribosomal RNA sequencing (D6 region and 5′ end) to compare single representatives of "Archaeogastropoda" and "Mesogastropoda," and several of Pulmonata. Tillier et al (168), also using 28S L*r*RNA partial sequencing (D1 and adjacent regions), presented a broader (ingroup) analysis across the Gastropoda. The results and implications of both studies are addressed below in the discussion of Major Groups.

Naming and Ranking of Clades, Grades, and Polyphyla

At the species-to-family levels, recent naming of taxa has resulted from (i) standard taxonomic work (e.g. in monographic group revisions); (ii) increased studies in previously unworked geographic areas, where occasionally 50–90% of the studied genera and species are recognized as new (107, 156); and (iii) the discovery of new family groups, especially in the hydrothermal-vent faunas (e.g. 108, 111, 112, 127, 186, 187).

However, the recent deluge of naming and renaming at higher, ICZN-unregulated (91), levels is of different origin. New naming and renaming has occurred because of perceived change in rank or group composition, or because the previous name was deemed insufficiently descriptive. This formal naming, often in the excitement of new phylogenetic hypotheses, especially by Austrian and Russian schools, has caused much of the current taxonomic confusion. One recent work reclassified the Gastropoda into an inflated system of 8 subclasses and 216 ordinal taxa (46). Not only does this defeat the main purpose of taxonomic classification, namely information retrieval, but it often also hampers future phylogenetic studies by immediately freezing hypotheses of relationship into an internested "higher" framework. Additional confusion in gastropod systematics is caused by a technicality: The name endings "-acea" and "-oidea" are used interchangeably at both ordinal and superfamilial levels. Depending on individual interpretation, ICZN's (91) name priority rule (which does not cover the ordinal level) either is or is not applied.

Attempts to align the preliminary research results in phylogenetic systematics with user-friendliness and the bookkeeping needs of collection management (usually based on the traditionally accepted classificatory schemes of Thiele, Wenz, and Zilch) have resulted in unproductive discussions about the relative ranks of supraspecific taxa, a confusion of clades and grades, and the retention of the latter for the sake of familiarity (64). Possibilities only beginning to be explored in gastropod systematics are informal groupings (82), unranked classifications, and accepting clades with different traditional hierarchies as sister-groups (without loading monotypic taxa with redundant names—12).

MAJOR GROUPS (Figures 1–4)

The subdivision of Gastropoda into the "subclasses" Prosobranchia, Opisthobranchia, and Pulmonata has become such textbook knowledge that the "naturalness" of these taxa is usually assumed. Some groups, however, were long known to have mosaic sets of traditional prosobranch, opisthobranch, or pulmonate features, and they were repeatedly reclassified (Figures 2, 4). Much of the published discussion of ancestry, relationship, and fossil record of the larger groups was thus influenced by the (expressed or implied) inclusion or exclusion of these misfits. Gradistic terminology has made it easy to deal with this situation, for instance, by viewing the extant Rissoelloidea as "a connecting link between marine taenioglossan prosobranchs and the heterobranchs" (60, p. 15), or by seeing in the (extant) genus *Architectonica* an "excellent precursor for the opisthobranchs and pulmonates" (43, p. 13). However, attempting to define the larger gastropod groups as monophyletic clades based on synapomorphic character suites should remain the goal.

Prosobranchia (Streptoneura) (Table 1, Figures 1, 2)

The Prosobranchia is the largest of the traditional subclasses and represents most marine shelled gastropods, as well as numerous land and freshwater groups. Prosobranchia is certainly paraphyletic in the sense that one of its subgroups (currently assumed to be the Caenogastropoda) stands in sister-group relationship with a clade comprising the remaining gastropods (Heterobranchia, see below). Prosobranchia is traditionally divided into Archaeogastropoda, Mesogastropoda, and Neogastropoda; the last two more recently have been combined as Caenogastropoda. When Thiele (164 and earlier papers) introduced the tripartite classification, he merged some previously distinguished groups and created others that are now no longer considered monophyletic (Figures 1, 2; see Table 1 for names used in the following discussion).

Among the traditional "archaeogastropods," the Neritoidea and related groups have long been known to have strong affinities to caenogastropods (122), and their removal from the archaeogastropods has been repeatedly advocated. They form a clearly defined clade with numerous apomorphies in anatomy, shell morphology, embryonic development, sperm ultrastructure, and sex chromosomes (3, 35, 39, 69, 96, 123, 177).

However, the breakup of the remaining "archaeogastropods" is a recent development. When it became clear that this group was not monophyletic, two routes were taken. One was to raise the various subgroups to equal rank of more-or-less uncertain interrelationship, and to use the name "archaeogastropod" to indicate a diotocardian, aspidobranch grade of anatomical organization with two sets of gills, osphradia, auricles, and kidneys, in contrast to a meso- or caenogastropod (= monotocardian, pectinibranch) level with only one such set (37, 54, 64). The other route was to restrict the use of the name Archaeogastropoda to a monophyletic unit by excluding certain groups (12, 81, 82, 104, 137; Table 1).

Docoglossa (= Patellogastropoda) is clearly recognized as a distinct

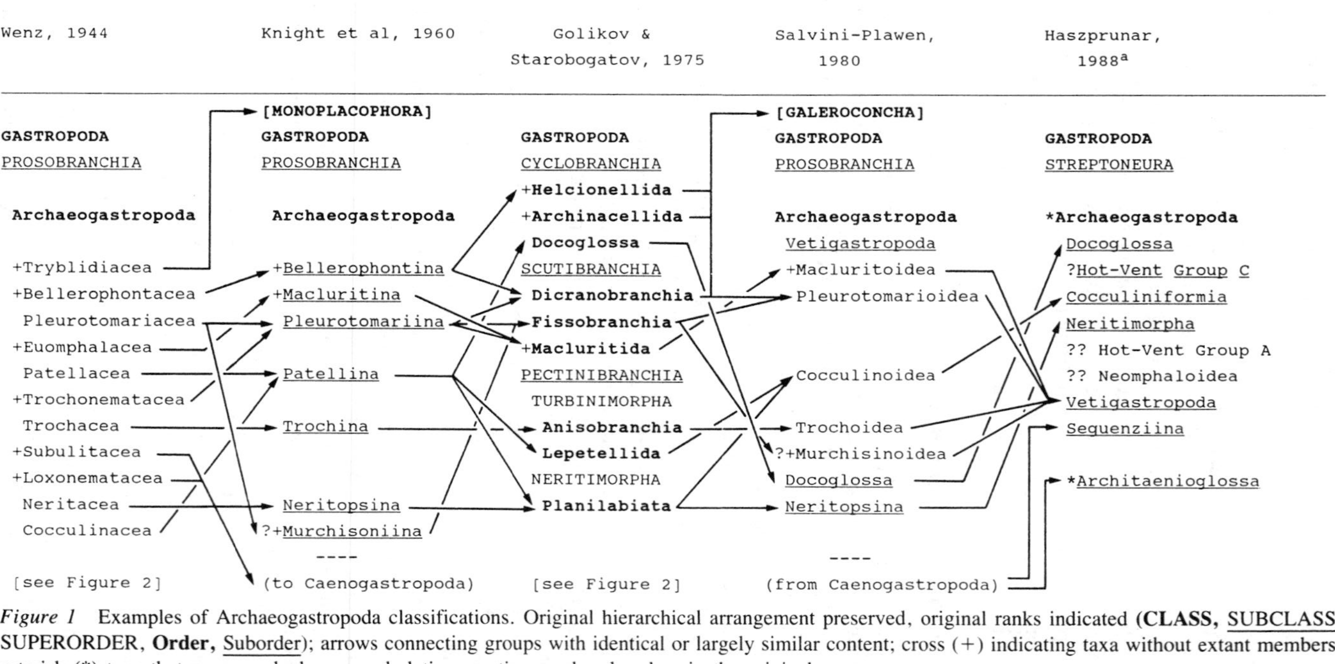

Figure 1 Examples of Archaeogastropoda classifications. Original hierarchical arrangement preserved, original ranks indicated (**CLASS**, SUBCLASS, SUPERORDER, **Order**, Suborder); arrows connecting groups with identical or largely similar content; cross (+) indicating taxa without extant members, asterisk (*) taxa that were marked as paraphyletic, question marks placed as in the original.
[a]based on Haszprunar 1988 (64), p. 428, table 5(a).

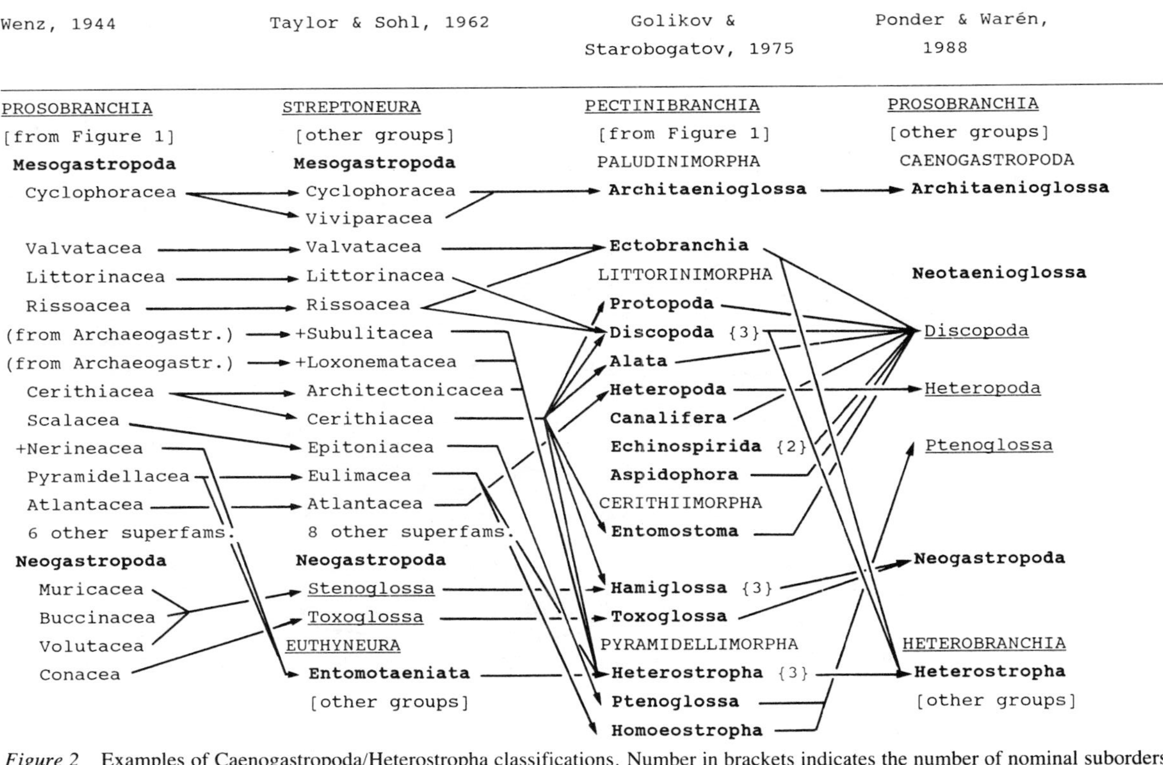

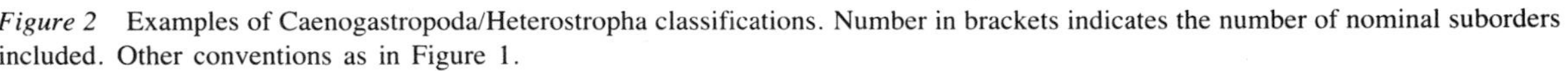
Figure 2 Examples of Caenogastropoda/Heterostropha classifications. Number in brackets indicates the number of nominal suborders included. Other conventions as in Figure 1.

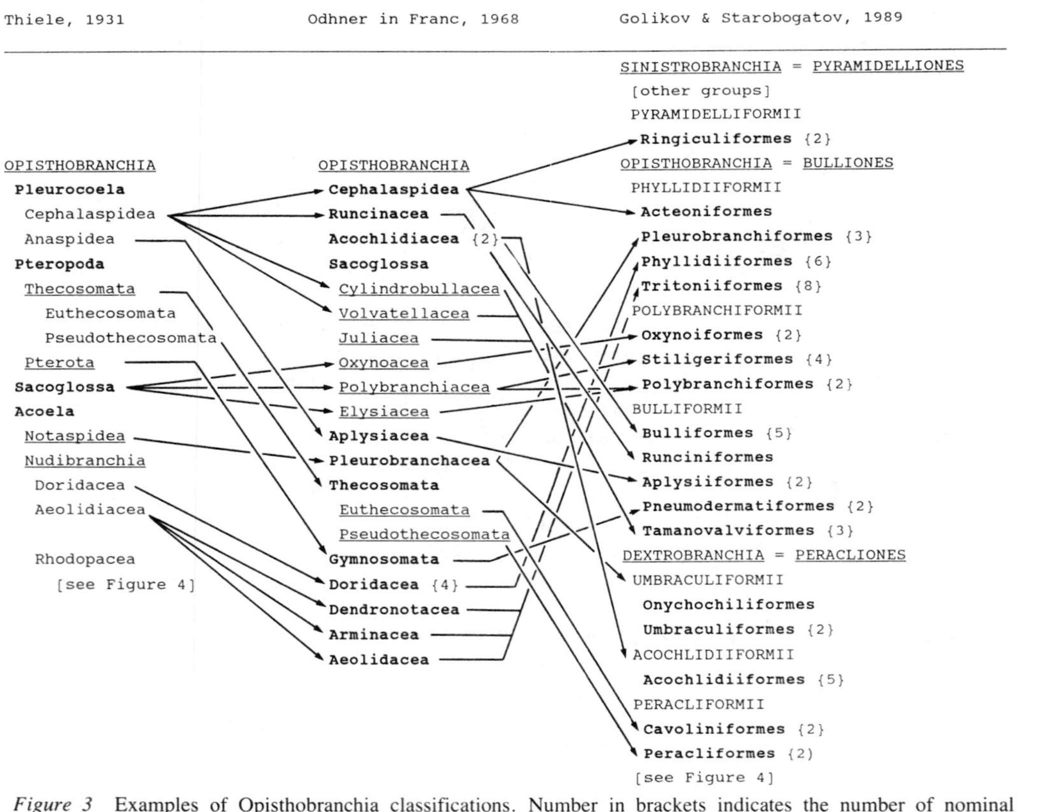

Figure 3 Examples of Opisthobranchia classifications. Number in brackets indicates the number of nominal suborders included. Other conventions as in Figure 1.

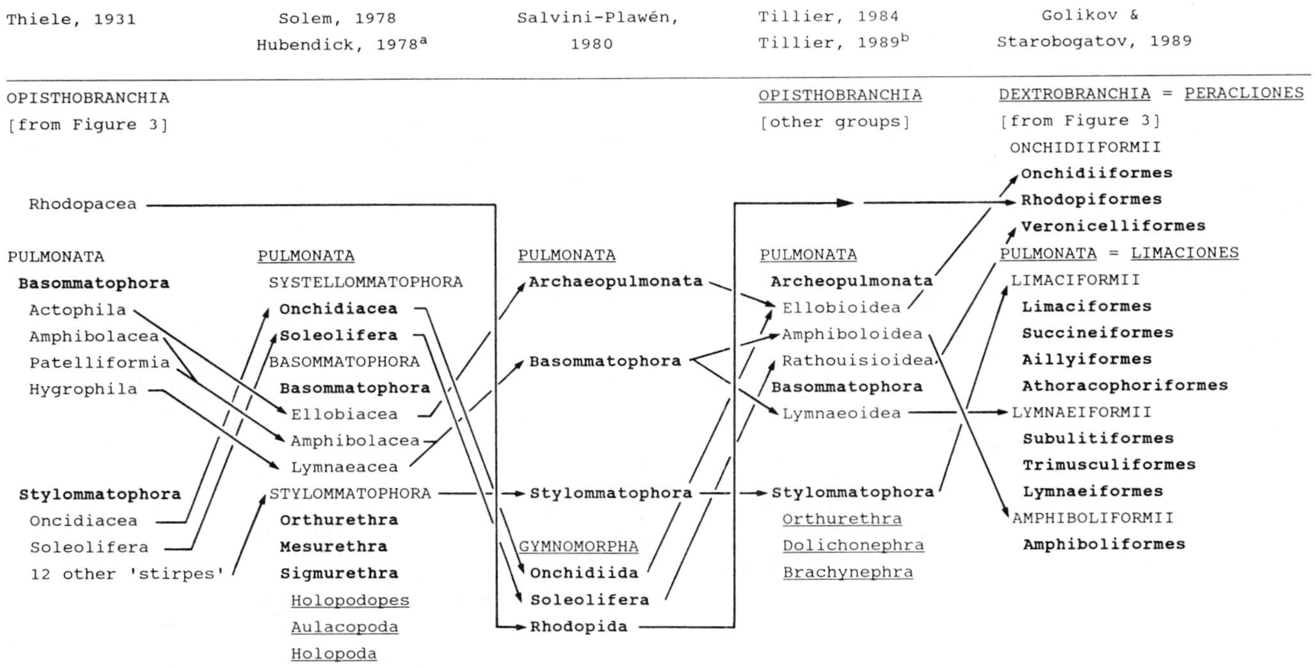

Thiele, 1931

OPISTHOBRANCHIA
[from Figure 3]
Rhodopacea
PULMONATA
Basommatophora
Actophila
Amphibolacea
Patelliformia
Hygrophila
Stylommatophora
Oncidiacea
Soleolifera
12 other 'stirpes'

Solem, 1978
Hubendick, 1978[a]

PULMONATA
SYSTELLOMMATOPHORA
Onchidiacea
Soleolifera
BASOMMATOPHORA
Basommatophora
Ellobiacea
Amphibolacea
Lymnaeacea
STYLOMMATOPHORA
Orthurethra
Mesurethra
Sigmurethra
Holopodopes
Aulacopoda
Holopoda

Salvini-Plawén, 1980

PULMONATA
Archaeopulmonata
Basommatophora
Stylommatophora
GYMNOMORPHA
Onchidiida
Soleolifera
Rhodopida

Tillier, 1984
Tillier, 1989[b]

OPISTHOBRANCHIA
[other groups]
PULMONATA
Archeopulmonata
Ellobioidea
Amphiboloidea
Rathouisioidea
Basommatophora
Lymnaeoidea
Stylommatophora
Orthurethra
Dolichonephra
Brachynephra

Golikov & Starobogatov, 1989

DEXTROBRANCHIA = PERACLIONES
[from Figure 3]
ONCHIDIIFORMII
Onchidiiformes
Rhodopiformes
Veronicelliformes
PULMONATA = LIMACIONES
LIMACIFORMII
Limaciformes
Succineiformes
Aillyiformes
Athoracophoriformes
LYMNAEIFORMII
Subulitiformes
Trimusculiformes
Lymnaeiformes
AMPHIBOLIFORMII
Amphiboliformes

Figure 4 Examples of Pulmonata classifications. Conventions as in Figure 1.
[a] Basommatophora classification after Hubendick 1978 (89)
[b] Stylommatophora classification after Tillier 1989 (167)

evolutionary line, currently placed as sister-group to all other gastropods. The group has retained numerous symplesiomorphies shared with other molluscan classes (e.g. in radula, nervous system, and gill morphology) and is supported by several autapomorphies, e.g. characters of shell structure and alimentary system, and the presence of so-called wart-organs (1, 3, 64, 96, 102, 174). This clade differs considerably from other gastropods in low chromosome number (see 128), low cellular DNA content (84), and L*r*RNA sequence (168), indicating early divergence.

Vetigastropoda was originally based on the supposed synapomorphic predominance of a post-torsional *right* shell muscle (177), which was later shown to be in error (61, 181). However, other apomorphies for the group are recognized, including sensory pockets (bursicles) on the gill leaflets (63, 160, 181). Preliminary molecular data, including representatives of Haliotidae, Fissurellidae, and Trochidae, support the hypothesis of monophyly for this group (168).

Caenogastropoda, which is considered monophyletic based on anatomical characters (58, 64, 104), comprises several groups. The largest, containing most of the traditional "Mesogastropoda" and in recent literature often termed Neotaenioglossa (58, 64, 131, 135), is an assemblage of numerous superfamilies whose sister-group relationships have not been resolved. A neotaenioglossate subgroup consisting of Rissooidea (= Truncatelloidea), Cingulopsoidea, and Littorinoidea appears to be a distinct clade (36, 131); Bandel's name "Metamesogastropoda," casually introduced for "Littorinoidea or Rissooidea" (8, p. 17), could be applied to this group. The Cerithioidea have been the subject of recent revisions by R. S. Houbrick (87, 88, personal communication), who transferred many previously assigned taxa to other gastropod groups. With the latest removal of Vermetidae, Campanilidae (70, 135), and Melanopsidae (R. S. Houbrick, personal communication quoting unpublished dissertation by Altaba), the superfamily with its now 18 families appears monophyletic, based on characters of euspermatozoa (70). The five caenogastropod species employed in Tillier et al's (168) molecular analysis (representatives of Littorinidae, Buccinidae, Calyptraeidae, and Muricidae) form a monophyletic clade. Also within the Caenogastropoda, the Neogastropoda forms a monophyletic group defined by anatomical synapomorphies such as gland and valve of Leiblein, anal gland, and osphradium structure (130, 163). The clade is of fairly recent origin (late Cretaceous radiation) and almost uniformly displays high chromosome numbers (see 128) and large cellular DNA content compared to other gastropods (84), which led to the suggestion of polyploidy (84, 130). Sister-group relationships of the Neogastropoda to other prosobranch groups is not clear, but evidence from sperm and osphradial data indicates close relationship to members of the "Neotaenioglossa" (64, 163).

Newly discovered hydrothermal-vent and deep-water gastropods, mostly interpreted as archaeogastropod (grade or clade) taxa, have prompted the erection of numerous higher groups (e.g. 79, 108, 111, 112, 127, 186, 187). Many more taxonomic changes can be expected as more material and comparative data become available for these groups. In contrast, extensive taxonomic work on other prosobranchs has led to considerable consolidation. Examples are the supraspecific revision of the Trochoidea by Hickman & McLean (82), and the work of Warén (182) who merged 14 familial and ordinal taxa into a revised prosobranch family Eulimidae.

The recently published RNA sequence study by Tillier et al (168) supports the general pattern of the anatomy-derived hypotheses of relationships among the larger clades, but that study did not address most of the groups with currently debated position. Using the nomenclature of Table 1, these cladograms resolved to (Docoglossa (Vetigastropoda (Caenogastropoda (Heterobranchia)))). Emberton et al's (30) molecular analysis, which included prosobranchs as outgroups, resolved to (Neritopsida (Caenogastropoda (Heterobranchia))). The fossil record at present cannot resolve the interrelationships of the larger groups. While much of the caenogastropod diversity most likely dates from the mid-Mesozoic, all major clades (Table 1) probably have Paleozoic origin (104, 188).

Several extant prosobranch taxa with unusual character sets have been discussed as direct links to groups with no previously recognized extant representatives. The Galapagos rift limpet *Neomphalus* was initially viewed as a relict member of the Euomphaloidea (108). The Mesozoic "Euomphaloidea" have recently been shown to be a polyphyletic assemblage of several unrelated groups (5). In a discussion of *Abyssochrysos,* Houbrick (86) addressed the similarity in shell characters with members of the fossil Loxonematidae. This caused the tentative placement of the Abyssochrysidae in the Loxonematoidea (135), a group thought to have become extinct in the Jurassic (188). Recent anatomical and spermatological studies led to the description of Provannidae as closely related to the Abyssochrysidae (21, 71, 187). Evidence from sperm studies suggests affinity of the thus "modernized" Loxonematoidea (71, 73) and of the Heteropoda (= Carinarioidea; 135) (92) to the Rissooidea-Littorinoidea. Like the Euomphaloidea, Loxonematoidea is a problematic group that has been used as a catch-all in the paleontological literature (6, 193), and its interpretation as a member of the Recent fauna may prove to be premature.

Heterobranchia (Heterostropha and Euthyneura) (Figures 3, 4)

The other two traditional subclasses of Gastropoda are the Opisthobranchia and Pulmonata. The name Euthyneura Spengel, 1881, combines these two groups whose members secondarily reduce or revert ("detort") the effects of torsion on the nervous system and other organ systems (15). Euthyneury itself is convergently arrived at by detorsion, nerve concentration, or a combination of both (15, 49, 60, 114). But there are several synapomorphies supporting the monophyly of Euthyneura, e.g. the presence of two additional "parietal" ganglia on the visceral loop (60, 68).

Families such as Pyramidellidae and Architectonicidae, displaying a mosaic of "typically prosobranch" and "typically opisthobranch" characters (140, 141), have puzzled malacologists for over a century. After a complicated taxonomic history (11, 36, 60, 74 and earlier papers, 97, 141; see Figure 2), these families form a group now referred to as Heterostropha Fischer, 1885 (45, 135) or Allogastropoda Haszprunar, 1985 (60, 64); numerous other synonyms are available (135). Several characters for members of this group have been described, including hyperstrophically coiled larval shells (the larval protoconch positioned at an angle to the post-larval teleoconch = heterostrophy) (141), a distinct osphradial type (59), and unique sperm morphology (74). Several other families initially grouped here have been removed as caenogastropod prosobranchs, and the Heterostropha now comprises the extant Architectonicoidea, Pyramidelloidea, Rissoellidae, Omalogyridae, Glacidorboidea, and Valvatoidea, plus the extinct Nerineoidea (11, 60, 64, 73, 132–134, 138, 158). Some of the currently included families (Cornirostridae, Amathinidae, and the Jurassic Provalvatidae) were described recently, and many are still insufficiently known. The composition of this group and its identified synapomorphies are changing rapidly, and this assemblage is a paraphyletic holding vessel for problematic taxa.

Heterostropha (= Triganglionata Haszprunar, 1985) and Euthyneura (= Pentaganglionata Haszprunar, 1985) share a number of presumed synapomorphic characters such as sperm type, heterostrophy[3], egg chalazae, and a pallially-located and supplied-kidney (60, 74, 75). The distinction between the two appears increasingly problematic, exemplified by the recently described Tjaernoeidae, a group sharing characters with Rissoellidae and Cornirostridae as well as "opisthobranch" Cephalaspidea (184). The name Heterobranchia is currently employed for the Heterostropha + Euthyneura clade (60, 74, 132–135, 184). Somewhat confusingly, the name Heterostropha has also been used in this wide sense in several recent publications (7, 8).

[3]The recently reported heterostrophy in trochoid archaeogastropods (56) appears to be a phenomenon of convergence. The heterostrophy of "higher" gastropods refers to a relationship between the larval shell and the teleoconch, with a change in direction of shell coiling caused by a reorientation of the mantle. The coiling of the embryonic shell in trochoidean archaeogastropods is a result of muscle activity (3) on the organic shell before calcification

Opisthobranchia (Figure 3)

The unclear phylogenetic relationships of Opisthobranchia and its subgroups caused many authors to rely on "evolutionary trends" in their definitions. The statement "Opisthobranchia are not defined by a set of features common to all members of the subclass but by certain marked tendencies, one of which is the tendency to lose or to reduce the shell in the course of evolution" (136, p. 223) stands as an example. Few modern analyses of larger groups such as Willan's (191) on the Notaspidea are available. Problems with encountered parallelism (claimed to reach 60–80% in many taxa—49) have led to the rejection of parsimony approaches in this group by some workers (50). Paleontological data are restricted to shell-bearing groups or the occasional find of a larval shell of an otherwise "naked" group (e.g. the cephalaspid acteonids and ringiculids can be traced into the Triassic, some into the Carboniferous—6). The L*r*RNA study by Tillier et al (168) included only two opisthobranch taxa, which formed a paraphyletic assemblage and grouped with the pulmonates in a Heterobranchia clade.

Boettger (15, and earlier papers) abandoned the traditional serial arrangement of the opisthobranch and pulmonate groups and developed a concept of several separate lineages, originating from the "order" Cephalaspidea presented as a polyphyletic stem group. The Opisthobranchia, with its subclades in sister-group relationships with the pulmonate group(s), thus became paraphyletic. The extant genus *Acteon,* whose members display many plesiomorphic characters (streptoneury, spiral adult shell, adult operculum, anteriorly directed mantle cavity), is often represented as being "close" to the stem form of the Euthyneura (15, 148). Ghiselin demonstrated the mosaic nature of Acteonidae characters (40), and Gosliner (48) suggested that members of Ringiculidae more closely approach the ancestral form.

The key to a better understanding of opisthobranch ancestry and the relationships between its subgroups appears to lie in the currently unresolved assemblage of the previously mentioned "heterostroph" and the "primitive cephalaspidean" groups. Haszprunar (60) formally separated the "conservative group" (180)—Diaphanoidea, Ringiculoidea, and Acteonoidea—from the Cephalaspidea s.l., recognized the assemblage as paraphyletic, and named it Architectibranchia. Von Salvini-Plawen (180) saw this group as the stem from which all other opisthobranchs derived. In the Russian literature, the Cephalaspidea s.l. were likewise sharply divided into several groups (46, 117; Figure 3). However, even more than is true for prosobranch and pulmonate work, Russian and "western" opisthobranch systematics appear to have developed with little interaction.

Depending on the author, the opisthobranchs are usually grouped in 8–12 orders (e.g. 148, 165, 194; Figure 3) which are often formally or informally arranged as tectibranchs, nudibranchs s.l., and (planktonic) pteropods. Currently, the relationships between the larger opisthobranch subgroups are largely uncertain, and character polarity is often problematic (40, 49). Rudman (142) hypothesized that an oodiaulic reproductive system is plesiomorphic in the Opisthobranchia, but this has been discounted by subsequent authors (48, 147). Schmekel (147) and von Salvini-Plawen (180) based most of their hypotheses about opisthobranch phylogeny on the assumption that there are two major lineages, but Gosliner (49) found difficulties with this: androdiauly evolved at least four times within the opisthobranchs, and fusion of cerebral and pleural ganglia also has occurred repeatedly within the group.

In addition to the problematic Cephalaspidea (Bullomorpha), the following opisthobranch "orders" are currently recognized:

(i) Anaspidea (= Aplysiomorpha), a group that includes the long-disputed Akeridae (40, 49), is united by synapomorphies such as "migration" of the intestinal ganglia posteriorly and fusion with the visceral ones (40, 147).
(ii) Sacoglossa (= Ascoglossa = Monostichoglossa), aside from the problematic Cylindrobullidae, this group is well-defined by several anatomical apomorphies (e.g. uniseriate radula with unicuspid teeth, ascus retaining discarded teeth, nervous system with 2–3 ganglia on visceral loop) (93, 145, 147).
(iii) Acochlidiacea is probably monophyletic and the sister-group of Sacoglossa (50).
(iv) Thecosomata and Gymnosomata are pelagic snails (with the gymnosomes ecologically dependent on the thecosomes as prey). The monophyly or separate origin of these groups has long been a subject of discussion (summarized in 99). Several models have been proposed for deriving pteropods neotenously from Cephalaspidea (40) or "primitive mesogastropod prosobranchs" (9).
(v) Nudibranchia is often subdivided into four (Doridacea, Dendronotacea, Arminacea, and Aeolidacea) or two (Anthobranchia, Cladobranchia) orders (Figure 3). Minichev (115) argued for an independent origin of two groups, while Schmekel (147) suggested a common origin of nudibranchs based on several synapomorphies (e.g. 13 chromosome pairs). Current data (summarized in 76) support the hypothesis that Nudibranchia and Notaspidea are closely allied; Willan (191) concluded that they probably shared a common cephalaspidean ancestor.
(vi) Notaspidea (= Umbraculomorpha + Pleurobranchomorpha) consists of two major clades (191). The Umbraculoidea have been variously removed from the remaining "pleurobranch" notaspids (Umbraculida, 116; Umbraculomorpha, 146, 180); recent Russian works even group

them in separate gastropod subclasses (Figure 3). However, new phylogenetic analyses advocated monophyly, based on several presumed anatomical synapomorphies (e.g. the longitudinally slit rhinophores) (49, 191).

Rhodopidae, a small group based on enigmatic, turbellarian-like species, was variously placed in the Opisthobranchia (34), Pulmonata (60), or other formal groups (46, 177) (see Figure 4). It is now regarded as a taxon of uncertain systematic rank in the Opisthobranchia (68, 166, 180).

Pulmonata

Numerous character suites have been employed to infer phylogenetic relationships within and between the larger pulmonate groups (see 124, 167, 169, 173). Among characters of the central nervous system have been the concentration of the visceral loop, the elaboration of the cerebral complex, the cell types of the procerebrum, and the position of the pleural ganglia (14, 68, 170). However, a recent analysis by Emberton (29) indicates that ganglionic fusion in pulmonates may not be irreversible.

Two major areas of discussion in pulmonate phylogenetic systematics are (i) the composition and relationships of the nonstylommatophoran groups assigned to Pulmonata, and (ii) the internal structure and interrelationships of Stylommatophora (Figure 4).

The current differences of opinion and resulting classifications of the groups traditionally referred to as Basommatophora, Systellommatophora (Figure 4), the genus *Otina,* and the families Succineidae and Trimusculidae are so great (and the comparative data base so insufficient) that a concise summary of "current knowledge" is impossible. An example will suffice: The opisthobranch versus pulmonate nature of the slugs belonging to the three families Onchidiidae, Veronicellidae, and Rathouisiidae has been the object of long debate, and conclusions often depended on whether the author studied the marine Onchidiidae or the two terrestrial families (see reviews in 68, 155, 166). This small group is known under numerous names in the literature, including Systellommatophora, Soleolifera, Ditremata, Teletremata, Gymnophila, and Opisthopneumona. Also placing the enigmatic *Rhodope* (see above) here, von Salvini-Plawen (176, 177) elevated the group as subclass Gymnomorpha (Figure 4), to which a monospecific order Smeagolida was later added (23). Haszprunar (60) accepted the Gymnomorpha as clearly separated from Pulmonata by their secondary lung, while Tillier (166, see Figure 4), using only characters unlinked to slug morphology, returned the original systellommatophoran groups to the Archaeopulmonata. Haszprunar & Huber (68) then redistributed the groups comprising Archaeopulmonata sensu Tillier (166) over Basommatophora, Systellommatophora, and the newly named "Eupulmonata," based on characters of the central nervous system.

It is clear that character suites and hypotheses of synapomorphies are strongly influenced by ongoing changes of group composition. As long as the various nonstylommatophoran pulmonate groups have not stabilized as definable clades, hypotheses about sister-group relationships of "higher taxa" must remain premature.

Stylommatophora is the largest pulmonate group, with 71–92 families (30). Despite recent advances in pulmonate research, the resolution of phylogenetic relationship among stylommatophorans remains low (29, 30, 167); the most stable resolution is that between orthurethrans and nonorthurethrans (167; K. C. Emberton, personal communication). Five recently presented morphology-based classifications of this group (17, 125, 126, 146, 153, 167), employing various character suites (excretory, locomotory, digestive, nervous, reproductive, shell) and analytical approaches, were found to differ up to 52%, 79%, and 74% at the ordinal, subordinal, and superfamilial levels (30). Traditional anatomical characters were evaluated and supplemented by Emberton's (1991) cladistic analysis of 17 stylommatophoran subfamilies, the results of which nearly matched Nordsieck's (125, 126). Emberton et al's (30) L*r*RNA sequence analysis of ten species (including outgroups but without systellommatophoran or archaeopulmonate representatives) resulted in a topology for the pulmonates of (*Biomphalaria* (Polygyridae (*Haplotrema* (Zonitidae)))), representing (Basommotophora (Holopoda (Holopodopes (Aulacopoda)))), with nonsignificant resolution between *Haplotrema* and the zonitids. Tillier et al's (168) sequence analysis also could not reliably resolve clades within the Pulmonata but recognized a distinct clade (Elasmognatha) formed by *Aneita* and *Succinea,* of uncertain position within or in sister-group relationship with the remaining Pulmonata.

CONCLUDING REMARKS

Gastropods have remained surprisingly underutilized as models for and objects of evolutionary studies. No other animal group offers an equal opportunity to combine the findings of comparative morphological and molecular studies on the diverse extant fauna with data derived from the extensive fossil record.

The frequency of new-and-improved higher level phylogenetic reconstructions and classifications in recent literature is not so much a reflection of major breakthroughs in phylogenetic research as it is an indication of a new exciting phase of data gathering and discussion after a long period of relative stasis. Despite the many new discoveries and advances in methodology, the group is far from having a sound taxonomic framework. There is a definite need for more broadly based comparative study and monographic work, and there is a clear indication that the expansion of molecular research will contribute greatly to our understanding of gastropod phylogeny. Recent research has resulted in a better definition of monophyletic groups; however, any attempt to present *the classification* of Gastropoda at this point would be premature. Naef's statement, quoted at the beginning of this paper, is as valid now as it was 80 years ago.

ACKNOWLEDGMENTS

I wish to thank the many colleagues who provided the publications on which this review is based; page-limit constraints allowed citation of only a few. I am particularly grateful to those who shared unpublished data and manuscripts with me: Thierry Backeljau, Philippe Bouchet, Kenneth J. Boss, Kenneth C. Emberton, Richard S. Houbrick, David Lindberg, Gary Rosenberg, Winston F. Ponder, James F. Quinn, and Simon Tillier. For helpful comments on an earlier draft of the manuscript, I thank K.J. Boss, T. Collins, K.C. Emberton, M.G. Harasewych, R.S. Houbrick, D. Lindberg, P.M. Mikkelsen, and G. Rosenberg.

Literature Cited

1. Andrews, E. B. 1985. Structure and function in the excretory system of archaeogastropods and their significance in the evolution of gastropods. *Philos. Trans. R. Soc. London Ser. B* 310:383–406
2. Backeljau, T. 1989. Electrophoresis of albumen gland proteins as a tool to elucidate taxonomic problems in the genus *Arion* (Gastropoda, Pulmonata). *J. Med. Appl. Malac.* 1:29–41
3. Bandel, K. 1982. Morphologie und Bildung der frühontogenetischen Gehäuse bei conchiferen Mollusken. *Facies* 7:1–198
4. Bandel, K. 1984. The radulae of Caribbean and other Mesogastropoda and Neogastropoda. *Zool. Verh.* 214:1–188
5. Bandel, K. 1988. Repräsentieren die Euomphaloidea eine natürliche Einheit der Gastropoden? *Mitt. Geol.-Paläontol. Inst. Univ. Hamburg* 67:1–33
6. Bandel, K. 1988. Early ontogenetic shell and shell structure as aids to unravel gastropod phylogeny and evolution. In *Prosobranch Phylogeny,* ed. W. F. Ponder. *Proc. Symp. 9th Int. Malac. Congr., Edinburgh, 1986. Malac. Rev. Suppl.* 4:267–72
7. Bandel, K. 1990. Shell structure of the Gastropoda excluding Archaeogastropoda. In *Skeletal Biomineralization: Patterns, Processes and Evolutionary Trends,* Vol. 1, ed. J. G. Carter, pp. 117–34. New York: Van Nostrand Reinhold
8. Bandel, K. 1991. Gastropods from brackish and fresh water of the Jurassic-Cretaceous transition (a systematic reevaluation). *Berl. geowiss. Abh.* (A)134:9–55
9. Bandel, K., Almogi-Labin, A., Hemleben, C., Deuser, W. G. 1984. The conch of *Limacina* and *Peraclis* (Pteropoda) and a model for the evolution of planktonic gastropods. *Neues Jahrb. Geol. Paläontol. Abh.* 168:87–107
10. Bertsch, H. 1976. Intraspecific and ontogenetic radular variation in opisthobranch systematics (Mollusca: Gastropoda). *Syst. Zool.* 25:117–22
11. Bieler, R. 1988. Phylogenetic relationships in the gastropod family Architectonicidae, with notes on the family Mathildidae (Allogastropoda). See Ref. 6, pp. 205–40
12. Bieler, R. 1990. Haszprunar's "clado-evolutionary" classification of the Gastropoda—a critique. *Malacologia* 31: 371–80
13. Bieler, R., Mikkelsen, P. M. 1988. Anatomy and reproductive biology of

two western Atlantic species of Vitrinellidae, with a case of protandrous hermaphroditism in the Rissoacea. *Nautilus* 102:1–29

14. Bishop, M. J. 1978. The value of the pulmonate central nervous system in phylogenetic hypothesis. *J. Molluscan Stud.* 44:116–19
15. Boettger, C. R. 1955. Die Systematik der euthyneuren Schnecken. *Zool. Anz. Suppl.* 18:253–80
16. Boss, K. J. 1971. Critical estimate of the number of Recent Mollusca. *Occas. Pap. Mollusks* 3:81–135
17. Boss, K. J. 1982. Mollusca [and] classification of Mollusca. In *Synopsis and Classification of Living Organisms,* ed. S. P. Parker, 1:945–1166; 2:1092–96. New York: McGraw-Hill
18. Bouchet, P. 1989. A review of poecilogeny in gastropods. *J. Molluscan Stud.* 55:67–78
19. Bouchet, P. 1990. Turrid genera and mode of development: the use and abuse of protoconch morphology. *Malacologia* 32:69–77
20. Bouchet, P., Rocroi, J-P. 1992. Supraspecific names of molluscs: a quantitative review. *Malacologia.* 34:75–86
21. Bouchet, P., Warén, A. 1991. *Ifremeria nautilei,* nouveau gastéropode d'évents hydrothermaux, probablement associé à des bactéries symbiotiques. *C. R. Acad. Sci. Paris Ser. III* 312:495–501
22. Bouvier, E. L. 1887. Système nerveux, morphologie générale et classification des Gastéropodes prosobranches. *Ann. Sci. Nat. Zool.* (7)3:1–510
23. Climo, F. M. 1980. Smeagolida, a new order of gymnomorph molluscs from New Zealand based on a new genus and species. *NZ J. Zool.* 7:513–22
24. Cox, L. R. 1960. Gastropoda. General characteristics of Gastropoda. In *Treatise on Invertebrate Paleontology,* Part I. *Mollusca* 1, ed. R. C. Moore. pp. 84–169. Lawrence: Univ. Kansas
25. Cox, L. R. 1960. Thoughts on the classification of the Gastropoda. *Proc. Malac. Soc. London* 33:239–61
26. Davis, G. M. 1978. Experimental methods in molluscan systematics. In *Pulmonates,* Vol. 2A, *Systematics, Evolution and Ecology*, ed. V. Fretter, J. Peake, pp. 99–169. London: Academic
27. Davis, G. M., Lindsay, G. K. 1967. Disc electrophoretic analysis of molluscan individuals and populations. *Malacologia* 5:311–34
28. Edlinger, K. 1988. Beiträge zur Torsion und Frühevolution der Gastopoden. *Z. Zool. Syst. Evolutionsforsch.* 26:27–50
29. Emberton, K. C. 1991. Polygyrid relations: a phylogenetic analysis of 17 subfamilies of land snails (Mollusca: Gastropoda: Stylommatophora). *Zool. J. Linn. Soc.* 103:207–24
30. Emberton, K. C., Kuncio, G. S., Davis, G. M., Phillips, S. M., Monderewicz, K. M., et al. 1990. Comparison of Recent classifications of stylommatophoran land-snail families, and evaluation of large-ribosomal-RNA sequencing for their phylogenetics. *Malacologia* 31:327–52
31. Erben, H. K., Krampitz, G. 1972. Ultrastruktur und Aminosäuren-Verhältnisse in den Schalen der rezenten Pleurotomariidae (Gastropoda). *Biominer. Forschungsber.* 6:12–31
32. Field, K. G., Olsen, G. J., Lane, D. J., Giovannoni, S. J., Ghiselin, M. T., et al. 1988. Molecular phylogeny of the animal kingdom. *Science* 239:748–53
33. Fioroni, P., Oehlmann, J., Stroben, E. 1991. The pseudohermaphroditism of prosobranchs; morphological aspects. *Zool. Anz.* 226:1–26
34. Franc, A. 1968. Classe des Gastéropodes. In *Traité de Zoologie: anatomie, systématique, biologie,* ed. P. Grassé, 5(3):1–893 ["Sous-classe des Opisthobranches. Systématique," pp. 834–88, posthumously compiled from N. H. Odhner's manuscript]. Paris: Masson
35. Fretter, V. 1965. Functional studies of the anatomy of some neritid prosobranchs. *J. Zool. London* 147:46–74
36. Fretter, V. 1980. The evolution of some higher taxa in gastropods. *J. Malac. Soc. Aust.* 4:226–27
37. Fretter, V. 1989. The anatomy of some new archaeogastropod limpets (Superfamily Peltospiracea) from hydrothermal vents. *J. Zool. London* 218:123–69
38. Fretter, V. 1990. The anatomy of some new archaeogastropod limpets (Order Patellogastropoda, Suborder Lepetopsina) from hydrothermal vents. *J. Zool. London* 222:529–55
39. Fretter, V., Graham, A. 1962. *British Prosobranch Molluscs; Their Functional Anatomy and Ecology.* London: Ray Soc. 755 pp.
40. Ghiselin, M. 1966. Reproductive function and the phylogeny of opisthobranch gastropods. *Malacologia* 3:327–78
41. Ghiselin, M. 1966. The adaptive significance of gastropod torsion. *Evolution* 20:337–48
42. Ghiselin, M. T. 1988. The origin of molluscs in the light of molecular evidence. *Oxford Surv. Evol. Biol.* 5: 66–95
43. Ghiselin, M. T., Degens, E. T., Spencer, D. W., Parker, R. H. 1967. A phylo-

genetic survey of molluscan shell matrix proteins. *Breviora* 262:1–35
44. Giusti, F. 1971. L'ultrastruttura dello spermatozoo nella filogenesi e nella sistematica dei mollusci gasteropodi. *Atti Soc. Ital. Sci. Nat. Mus. Civ. Stor. Nat. Milano* 112:381–402
45. Golikov, A. N., Starobogatov, Y. I. 1975. Systematics of prosobranch gastropods. *Malacologia* 15:185–232
46. Golikov, A. N., Starobogatov, Y. I. 1989. Problems of phylogeny and system of the prosobranchiate gastropods. In *Systematics and Fauna of Gastropoda, Bivalvia and Cephalopoda*, ed. Y. I. Starobogatov. *USSR Acad. Sci. Proc. Zool. Inst., Leningrad* 187(1988):4–77 (Russian with English abstr.; English transl. K. J. Boss, Spec. Occas. Publ., Dept. Mollusks, Mus. Comp. Zool., Harvard Univ. In press
47. Goodman, M., Pedwaydon, J., Czelusniak, J., Suzuki, T., Gotoh, T., et al. 1988. An evolutionary tree for invertebrate globin sequences. *J. Mol. Evol.* 27:236–49
48. Gosliner, T. M. 1981. Origins and relationships of primitive members of the Opisthobranchia (Mollusca: Gastropoda). *Biol. J. Linn. Soc.* 16:197–225
49. Gosliner, T. M. 1991. Morphological parallelism in opisthobranch gastropods. *Malacologia* 32:313–27
50. Gosliner, T. M., Ghiselin, M. T. 1984. Parallel evolution in opisthobranch gastropods and its implication for phylogenetic methodology. *Syst. Zool.* 33:255–74
51. Götting, K. -J. 1974. *Malakozoologie: Grundriss der Weichtierkunde*. Stuttgart: Gustav Fischer. 320 pp.
52. Götting, K. -J. 1980. Origin and relationships of the Mollusca. *Z. Zool. Syst. Evolutionsforsch.* 18:24–27
53. Graham, A. 1979. Gastropoda. In *The Origin of Major Invertebrate Groups*, ed. M. R. House. *Syst. Assoc. Spec. Vol.* 12:359–65. London: Academic
54. Graham, A. 1985. Evolution within the Gastropoda: Prosobranchia. In *The Mollusca*, Vol. 10, *Evolution*, ed. E. R. Trueman, M. R. Clarke, pp. 151–86. Orlando: Academic
55. Gutmann, W. F. 1974. Die Evolution der Mollusken-Konstruktion: ein phylogenetisches Modell. *Aufsätze Reden Senckenb. Naturforsch. Ges.* 25:1–84
56. Hadfield, M. G., Strathmann, M. F. 1990. Heterostrophic shells and pelagic development in trochoideans: implications for classification, phylogeny and palaeoecology. *J. Molluscan Stud.* 56:239–56
57. Harper, J. A., Rollins, H. B. 1982. Recognition of Monoplacophora and Gastropoda in the fossil record: a functional morphological look at the bellerophont controversy. *Proc. 3rd N. Am. Paleontol. Conv.* 1:227–32
58. Haszprunar, G. 1985. The fine morphology of the osphradial sense organs of the Mollusca. I. Gastropoda, Prosobranchia. *Philos. Trans. R. Soc. London Ser. B* 307:457–96
59. Haszprunar, G. 1985. The fine morphology of the osphradial sense organs of the Mollusca. II. Allogastropoda (Architectonicidae, Pyramidellidae). *Philos. Trans. R. Soc. London Ser. B* 307:497–505
60. Haszprunar, G. 1985. The Heterobranchia—a new concept of the phylogeny of the higher Gastropoda. *Z. zool. Syst. Evolutionsforsch.* 23:15–37
61. Haszprunar, G. 1985. On the innervation of gastropod shell muscles. *J. Molluscan Stud.* 51:309–14
62. Haszprunar, G. 1986. Die kladoevolutionäre Klassifikation—Versuch einer Synthese. *Z. Zool. Syst. Evolutionsforsch.* 24:89–109
63. Haszprunar, G. 1987. The fine structure of the ctenidial sense organs (bursicles) of Vetigastropoda (Zeugobranchia, Trochoidea) and their functional and phylogenetic significance. *J. Molluscan Stud.* 53:46–51
64. Haszprunar, G. 1988. On the origin and evolution of major gastropod groups, with special reference to the Streptoneura. *J. Molluscan Stud.* 54:367–441
65. Haszprunar, G. 1989. Die Torsion der Gastropoda—ein biomechanischer Prozess?. *Z. zool. Syst. Evolutionsforsch.* 27:1–7
66. Haszprunar, G. 1989. The anatomy of *Melanodrymia aurantiaca* Hickman, a coiled archaeogastropod from the East Pacific hydrothermal vents (Mollusca, Gastropoda). *Acta Zool.* 70:175–86
67. Haszprunar, G. 1990. Towards a phylogenetic system of Gastropoda Part I: traditional methodology—a reply [to Bieler's (1990) critique]. *Malacologia* 32:195–202
68. Haszprunar, G., Huber, G. 1990. On the central nervous system of Smeagolidae and Rhodopidae, two families questionably allied with the Gymnomorpha (Gastropoda: Euthyneura). *J. Zool. London* 220:185–99
69. Healy, J. M. 1988. The ultrastructure of spermatozoa and spermiogenesis in

pyramidellid gastropods, and its systematic importance. *Helgol. Meeresunters.* 42:303–18

70. Healy, J. M. 1988. Sperm morphology and its systematic importance in the Gastropoda. See Ref. 6, pp. 251–66
71. Healy, J. 1989. Spermatozeugmata of *Abyssochrysos:* ultrastructure, development and relevance to the systematic position of the Abyssochrysidae (Prosobranchia, Caenogastropoda). *Bull. Mus. natn. Hist. nat. Paris* (Ser. 4)11A:509–33
72. Healy, J. M. 1990. Euspermatozoa and paraspermatozoa in the trochoid gastropod *Zalipais laseroni* (Trochoidea: Skeneidae). *Mar. Biol.* 105:497–507
73. Healy, J. M. 1990b. Spermatozoa and spermiogenesis of *Cornirostra, Valvata* and *Orbitestella* (Gastropoda: Heterobranchia) with a discussion of valvatoidean sperm morphology. *J. Molluscan Stud.* 56:557–66
74. Healy, J. M. 1991. Sperm morphology in the marine gastropod *Architectonica perspectiva* (Mollusca): unique features and systematic relevance. *Mar. Biol.* 109:59–65
75. Healy, J. M., Willan, R. C. 1984. Ultrastructure and phylogenetic significance of notaspidean spermatozoa (Mollusca, Gastropoda, Opisthobranchia). *Zool. Scr.* 13:107–20
76. Healy, J. M., Willan, R. C. 1991. Nudibranch spermatozoa: comparative ultrastructure and systematic importance. *Veliger* 34:134–65
77. Hennig, W. 1979. *Taschenbuch der speziellen Zoologie*. Teil 1, *Wirbellose* I. 4. Aufl. Jena: Gustav Fischer. 392 pp.
78. Hickman, C. S. 1980. Gastropod radulae and the assessment of form in evolutionary paleontology. *Paleobiology* 6: 276–94
79. Hickman, C. S. 1984. A new archaeogastropod (Rhipidoglossa, Trochacea) from hydrothermal vents on the East Pacific Rise. *Zool. Scr.* 13:19–25
80. Hickman, C. S. 1985. Gastropod morphology and function. In *Mollusks, Notes for a Short Course,* ed. T. W. Broadhead. *Univ. Tenn. Stud. Geol.* 13:138–56
81. Hickman, C. S. 1988. Archaeogastropod evolution, phylogeny and systematics; a re-evaluation. See Ref. 6, pp. 17–34
82. Hickman, C. S., McLean, J. H. 1990. Systematic revision and suprageneric classification of trochacean gastropods. *Nat. Hist. Mus. Los Angeles County Sci. Ser. 35.* 169 pp.
83. Hillis, D. M., Dixon, M. T., Jones, A. L. 1991. Minimal genetic variation in a morphologically diverse species (Florida Tree Snail, *Liguus fasciatus). J. Hered.* 82:282–86
84. Hinegardner, R. 1974. Cellular DNA content of the Mollusca. *Comp. Biochem. Physiol.* 47A:447–60
85. Hoagland, K. E. 1984. Use of molecular genetics to distinguish species of the genus *Crepidula* (Prosobranchia: Calyptraeidae). *Malacologia* 25:607–28
86. Houbrick, R. S. 1979. Classification and systematic relationships of the Abyssochrysidae, a relict family of bathyal snails (Prosobranchia, Gastropoda). *Smithson. Contrib. Zool.* 290:1–21
87. Houbrick, R. S. 1988. Cerithiodean phylogeny. See Ref. 6, pp. 88–128
88. Houbrick, R. S. 1989. Campanile revisited: implications for cerithioidean phylogeny. *Am. Malac. Bull.* 7:1–6
89. Hubendick, B. 1978. Systematics and comparative morphology of the Basommatophora. See Ref. 26, pp. 1–47
90. Hughes, R. N. 1986. *A Functional Biology of Marine Gastropods*. Baltimore: Johns Hopkins Univ. Press. 245 pp.
91. International Commission on Zoological Nomenclature. 1985. *International Code of Zoological Nomenclature*. London: Int. Trust Zool. Nomencl. 338 pp. 3rd ed.
92. Jamieson, B. G. M., Newman, L. J. 1989. The phylogenetic position of the heteropod *Atlanta gaudichaudi* Souleyet (Mollusca, Gastropoda), a spermatological investigation. *Zool. Scr.* 18:269–78
93. Jensen, K. R. 1991. Comparison of alimentary systems in shelled and nonshelled Sacoglossa (Mollusca, Opisthobranchia). *Acta Zool.* 72:143–50
94. Knight, J. B., Cox, L. R., Keen, A. M., Batten, R. L., Yochelson, E. L., et al. 1960. Systematic descriptions. See Ref. 24, pp. 169–331
95. Kohn, A. J. 1985. Gastropod paleoecology. See Ref. 80, pp. 74–89
96. Koike, K. 1985. Comparative ultrastructural studies on the spermatozoa of the Prosobranchia (Mollusca: Gastropoda). *Sci. Rep. Fac. Educ. Gunma Univ.* 34:33–153
97. Kosuge, S. 1966. The family Triphoridae and its systematic position. *Malacologia* 4:297–324
98. Lake, J. A. 1990. Origin of the Metazoa. *Proc. Natl. Acad. Sci. USA* 87:763–66
99. Lalli, C. M., Gilmer, R. W. 1989.

Pelagic Snails: the Biology of Holoplanktonic Gastropod Mollusks. Stanford, Calif: Stanford Univ. Press. 259 pp.

100. Lauterbach, K. -E. 1984. Das phylogenetische System der Mollusca. *Mitt. dtsch. Malakozool. Ges.* 37:66–81
101. Lemche, H., Wingstrand, K. G. 1959. The anatomy of *Neopilina galatheae* Lemche, 1957. *Galathea Rept.*3:9–71
102. Lindberg, D. R. 1988. The Patellogastropoda. See Ref. 6, pp. 35–63
103. Lindberg, D. R. 1988. Heterochrony in gastropods: a neontological view. In *Heterochrony in Evolution,* ed. M. L. McKinney, pp. 197–216. New York: Plenum
104. Lindberg, R., Ponder, W. F. 1991. A phylogenetic analysis of the Gastropoda. Geol. Soc. Am. Abstr. with Programs 1991:A160–61
105. Linsley, R. M., Kier, W. M. 1984. The Paragastropoda: a proposal for a new class of Paleozoic Mollusca. *Malacologia* 25:241–54
106. MacDonald, J. D. 1860. Further observations on the metamorphosis of Gasteropoda, and the affinities of certain genera, with an attempted natural distribution of the principal families of the order. *Trans. Linn. Soc. London* 23:69–81
107. Marshall, B. A. 1991. Mollusca Gastropoda: Seguenziidae from New Caledonia and the Loyalty Islands. In *Résultats des Campagnes MUSORSTOM,* Vol. 7, ed. A. Crosnier, P. Bouchet. *Mém. Mus. natn. Hist. nat.* (A)150:41–109
108. McLean, J. H. 1981. The Galapagos rift limpet *Neomphalus:* relevance to understanding the evolution of a major Paleozoic-Mesozoic radiation. *Malacologia* 21:291–336
109. McLean, J. H. 1984. A case for derivation of the Fissurellidae from the Bellerophontacea. *Malacologia* 25:3–20
110. McLean, J. H. 1988. New archaeogastropod limpets from hydrothermal vents; superfamily Lepetodrilacea. I. Systematic descriptions. *Philos. Trans. R. Soc. London Ser. B* 319:1–32
111. McLean, J. H. 1989. New archaeogastropod limpets from hydrothermal vents: new family Peltospiridae, new superfamily Peltospiracea. *Zool. Scr.* 18:49–66
112. McLean, J. H. 1990. Neolepetopsidae, a new docoglossate limpet family from hydrothermal vents and its relevance to patellogastropod evolution. *J. Zool. London* 222:485–528
113. McLean, J. H. 1990. A new genus and species of neomphalid limpet from the Mariana Vents with a review of current understanding of relationships among Neomphalacea and Peltospiracea. *Nautilus* 104:77–86
114. Merker, E. 1913. Nervenkreuzungen als Folgen einer ehemaligen Chiastoneurie bei den pulmonaten Gastropoden und die zweifache Art ihrer Rückbildung. *Zool. Anz.* 41:337–54
115. Minichev, Y. 1970. On the origin and system of nudibranchiate molluscs (Gastropoda Opisthobranchia). *Monit. Zool. Ital.* (NS) 4:162–82
116. Minichev, Y. S., Starobogatov, Y. I. 1975. On the systematization of euthyneuran snails. In *Molluscs, Their System, Evolution and Signification in the Nature,* ed. I. M. Likharev, pp. 8–10. USSR Acad. Sci. Zool. Inst. (Russian with English title; English version publ. 1978 in *Malac. Rev.* 11:67–68)
117. Minichev, Y. S., Starobogatov, Y. I. 1979. Peculiarities of the evolution of reproductive system of the Opisthobranchia and their taxonomy. In *Molluscs, Main Results of Their Study*, ed. I. M. Likharev, pp. 16–20. USSR Acad. Sci. Zool. Inst. (Russian with English title; English version publ. 1984 in *Malac. Rev.* 17:112–14)
118. Mörch, O. A. L. 1865. On the systematic value of the organs which have been employed as fundamental characters in the classification of Mollusca. *Ann. Mag. Nat. Hist.* (Ser. 3)16:385–97
119. Morris, N. J. 1990. Early radiation of the Mollusca. In *Major Evolutionary Radiations,* ed. P. D. Taylor, G. P. Larwood. *Syst. Assoc. Spec. Vol.* 42: 73–90
120. Morton, J. E. 1955. The evolution of the Ellobiidae with a discussion on the origin of the Pulmonata. *Proc. Zool. Soc. London* 125:127–68
121. Morton, J. E. 1988. The pallial cavity. In *The Mollusca,* Vol. 11, *Form and Function,* ed. E. R. Trueman, M. R. Clarke, pp. 253–86. San Diego: Academic
122. Naef, A. 1911. Studien zur generellen Morphologie der Mollusken. 1. Teil: Uber Torsion und Asymmetrie der Gastropoden. *Ergeb. Fortschr. Zool.* 3:73–164
123. Nakamura, H. K. 1986. Chromosomes of Archaeogastropoda (Mollusca: Prosobranchia), with some remarks on

their cytotaxonomy and phylogeny. *Publ. Seto Mar. Biol. Lab.* 31:191–267
124. Nordsieck, H. 1966. Grundzüge zur vergleichenden Morphologie des Genitalsystems der Schnecken, unter besonderer Berücksichtigung der Stylommatophora. *Arch. Molluskenk.* 95:123–42
125. Nordsieck, H. 1985. The system of the Stylommatophora (Gastropoda), with special regard to the systematic position of the Clausiliidae, I. Importance of the excretory and genital systems. *Arch. Molluskenk.* 116:1–24
126. Nordsieck, H. 1986. The system of the Stylommatophora (Gastropoda), with special regard to the systematic position of the Clausiliidae. II. Importance of the shell and distribution. *Arch. Molluskenk.* 117:93–116
127. Okutani, T., Saito, H., Hashimoto, J. 1989. A new neritacean limpet from a hydrothermal vent site near Ogasawara Islands, Japan. *Venus* 48: 223–30
128. Pattersen, C. M., Burch, J. B. 1978. Chromosomes of pulmonate molluscs. See Ref. 26, pp. 171–217
128a. Peel, J. S. 1991. Functional morphology of the class Helcionelloida nov., and the early evolution of the Mollusca. In *The Early Evolution of the Metazoa and the Significance of Problematic Taxa,* ed. A. Cimonetta, S. Conway Morris, pp. 157–77. Cambridge: Cambridge Univ.
129. Pilsbry, H. A. 1893–94. Guide to the study of Helices. *Manual Conch.* (Ser. 2)9:1–366
130. Ponder, W. F. 1973. The origin and evolution of the Neogastropoda. *Malacologia* 12:295–338
131. Ponder, W. F. 1988. The truncatelloidean (=rissoacean) radiation—a preliminary phylogeny. See Ref. 6, pp. 129–66
132. Ponder, W. F. 1990. The anatomy and relationships of the Orbitestellidae (Gastropoda: Heterobranchia). *J. Molluscan Stud.* 56:515–32
133. Ponder, W. F. 1990. The anatomy and relationships of a marine valvatoidean (Gastropoda: Heterobranchia). *J. Molluscan Stud.* 56:533–55
134. Ponder, W. F. 1991. Marine valvatoideans gastropods—implications for early heterobranch phylogeny. *J. Molluscan Stud.* 57:21–32
135. Ponder, W. F., Warén, A. 1988. Classification of the Caenogastropoda and Heterostropha—a list of the family-group names and higher taxa. *Malac. Rev. Suppl.* 4:288–328
136. Poulicek, M., Voss-Foucart, M-F., Jeuniaux, C. 1991. Regressive shell evolution among opisthobranch gastropods. *Malacologia* 32:223–32
137. Quinn, J. F. 1992. Systematic position of *Basilissopsis* and *Guttula,* and a discussion of the phylogeny of the Seguenzioidea (Gastropoda: Prosobranchia). *Bull. Mar. Sci.* 49(1–2):575–98
138. Rath, E. 1988. Organization and systematic position of the Valvatidae. See Ref. 6, pp. 194–204
139. Reid, D. G. 1986. *Mainwaringia* Nevill, 1885, a littorinid genus from Asiatic mangrove forests, and a case of protandrous hermaphroditism. *J. Molluscan Stud.* 52:225–42
140. Robertson, R. 1974. The biology of the Architectonicidae, gastropods combining prosobranch and opisthobranch traits. *Malacologia* 14(1973):215–20
141. Robertson, R. 1985. Four characters and the higher category systematics of gastropods. *Am. Malac. Bull. Spec. Ed.* 1:1–22
142. Rudman, W. B. 1978. A new species and genus of the Aglajidae and the evolution of the philinacean opisthobranch molluscs. *Zool. J. Linn. Soc.* 62:89–107
143. Runnegar, B. 1983. Molluscan phylogeny revisited. *Mem. Assoc. Australas. Paleontol.* 1:121–44
144. Runnegar, B., Pojeta, J. 1985. Origin and diversification of the Mollusca. See Ref. 54, pp. 1–57
145. Sanders-Esser, B. 1984. Vergleichende Untersuchungen zur Anatomie und Histologie der vorderen Genitalorgane der Ascoglossa (Gastropoda, Euthyneura). *Zool. Jahrb. Anat.* 111:195–243
146. Schileyko, A. A. 1979. Sistema Otryada Geophila (= Helicida) (Gastropoda Pulmonata). Morfologiya, Sistematika i Filogeniya Mollyuskov. *Trudy Zool. Inst. Akad. Nauk SSR* 80:44–69 [English transl., *The System of the Order Geophila (= Helicida) (Gastropoda Pulmonata),* ed. K. J. Boss, M. K. Jacobson. Spec. Occas. Publ., Dept. Mollusks, Mus. Comp. Zool., Harvard Univ. 6:1–45 (+10)]
147. Schmekel, L. 1985. Aspects of evolution within the opisthobranchs. See Ref. 54, pp. 221–67
148. Schmekel, L., Portmann, A. 1982. *Opisthobranchia des Mittelmeeres. Nudibranchia und Saccoglossa.* Berlin: Springer. 410 pp.
149. Signor, P. W. 1985. Gastropod evolutionary history. See Ref. 80, pp. 157–73

150. Simroth, H. 1896–1907. *Dr. H. G. Bronn's Klassen und Ordnungen des Tier-Reichs, wissenschaftlich dargestellt in Wort und Bild. Dritter Band. Mollusca. II. Abteilung: Gastropoda prosobranchia*. Leipzig: C. F. Winter. 1056 pp.
151. Solem, A. 1970. Malacological applications of scanning electron microscopy. I. Introduction and shell surface features. *Veliger* 12:394–400
152. Solem, A. 1974. *The Shell Makers: Introducing Mollusks*. New York: Wiley. 289 pp.
153. Solem, A. 1978. Classification of the land Mollusca. See Ref. 26, pp. 49–97
154. Solem, A. 1984. A world model of land snail diversity and abundance. In *World-Wide Snails: Biogeographical Studies on Non-Marine Mollusca,* ed. A. Solem, A. C. van Bruggen, pp. 6–22. Leiden: Brill
155. Solem, A. 1985. Origin and diversification of pulmonate land snails. See Ref. 54, pp. 269–93
156. Solem, 1985. Camaenid land snails from Western and central Australia (Mollusca: Pulmonata: Camaenidae). V. Remaining Kimberley genera and addenda to the Kimberley. *Rec. W. Aust. Mus. Suppl.* 20:707–981
157. Solem, A., Yochelson, E. L. 1979. North American Paleozoic land snails, with a summary of other Paleozoic nonmarine snails. *Geol. Surv. Prof. Pap.* 1072:1–42
158. Starobogatov, Y. I. 1989. On the systematic position of the genus *Glacidorbis* (Gastropoda incertae sedis). In *Systematics and Fauna of Gastropoda, Bivalvia and Cephalopoda,* ed. Y. I. Starobogatov. USSR Acad. Sci. Proc. Zool. Inst. 187(1988): 78-84 (Russian with English abstr.)
159. Stasek, C. R. 1972. The molluscan framework. In *Chemical Zoology,* Vol. VII, *Mollusca,* ed. M. Florkin, B. T. Scheer, pp. 1–44. New York: Academic
160. Szal, R. 1971. "New" sense organ of primitive gastropods. *Nature* 229:490–92
161. Taylor, D. W., Sohl, N. F. 1962. An outline of gastropod classification. *Malacologia* 1:7–32
162. Taylor, J. D., Miller, J. A. 1989. The morphology of the osphradium in relation to feeding habits in meso- and neogastropods. *J. Molluscan Stud.* 55: 227–37
163. Taylor, J. D., Morris, N. J. 1988. Relationships of neogastropods. See Ref. 6, pp. 67–79
164. Thiele, J. 1929–35. *Handbuch der systematischen Weichtierkunde*. Jena: Gustav Fischer. 1154 pp. (English transl., *Handbook of Systematic Malacology,* ed. R. Bieler, P. M. Mikkelsen. Washington: Smithson. Inst. Libr. & Natl. Sci. Found. 1992 and in press
165. Thompson, T. E. 1976. *Biology of Opisthobranch Molluscs*. Vol. I. London: Ray Soc. 206 pp.
166. Tillier, S. 1984. Relationships of gymnomorph gastropods (Mollusca: Gastropoda). *Zool. J. Linn. Soc.* 82: 345–62
167. Tillier, S. 1989. Comparative morphology, phylogeny and classification of land snails and slugs (Gastropoda: Pulmonata: Stylommatophora). *Malacologia* 30:1–303
168. Tillier, S., Masselot, M., Philippe, H., Tillier, A. 1992. Phylogénie moléculaire des Gastropoda (Mollusca) fondée sur le séquençage partiel de l'ARN ribosomique 28 S. *C. R. Acad. Sci. Paris*. 314:(Ser. 3):79–85
169. Tompa, A. S. 1984. Land snails (Stylommatophora). In *The Mollusca,* Vol. 7, *Reproduction,* ed. A. S. Tompa N. H. Verdonk, J. a. M. van den Biggelaar, pp. 47–140. Orlando: Academic
170. Van Mol, J. -J. 1974. Evolution phylogénétique du ganglion cérébroïdes des Gastéropodes Pulmonés. *Haliotis* 4:77–86
171. Vaught, K. C. 1989. *A Classification of the Living Mollusca*. Melbourne: Am. Malac. Inc. 195 pp.
172. Vermeij, G. J. 1978. *Biogeography and Adaptation; Patterns of Marine Life*. Cambridge: Harvard Univ. Press. 332 pp.
173. Visser, M. H. C. 1988. The significance of terminal duct structures and the role of neoteny in the evolution of the reproductive system of Pulmonata. *Zool. Scr.* 17:239–52
174. Voltzow, J. 1988. The organization of limpet pedal musculature and its evolutionary implications for the Gastropoda. See Ref. 6, pp. 273–83
175. von Ihering, H. 1876. Versuch eines natürlichen Systemes der Mollusken. *Jahrb. dtsch. Malakolozool. Ges.* 3:97–148
176. von Salvini-Plawen, L. 1970. Zur systematischen Stellung von Soleolifera and *Rhodope* (Gastropoda, Euthyneura). *Zool. Jahrb. Syst.* 97:285–99
177. von Salvini-Plawen, L. 1980. A reconsideration of systematics in the Mollusca (phylogeny and higher classification). *Malacologia* 19:249–78
178. von Salvini-Plawen, L. 1984. Die

Cladogenese der Mollusca. *Mitt. dtsch. malak. Ges.* 37:89–118

179. von Salvini-Plawen, L. 1985. Early evolution and the primitive groups. See Ref. 54, pp. 59–150
180. von Salvini-Plawen, L. 1991. The status of the Rhodopidae (Gastropoda: Euthyneura). *Malacologia* 32:301–11
181. von Salvini-Plawen, L., Haszprunar, G. 1987. The Vetigastropoda and the systematics of streptoneurous gastropods (Mollusca). *J. Zool. London* A211: 747–70

181a. Wahlman, G. P. 1992. Middle and Upper Ordovician symmetrical univalved mollusks (Monoplacophora and Bellerophontina) of the Cincinnati arch region. *US Geol. Surv. Prof. Pap.* 1066-0:i–vi, 01–0213

182. Warén, A. 1984. A generic revision of the family Eulimidae (Gastropoda, Prosobranchia). *J. Molluscan Stud. Suppl.* 13(1983):1–96
183. Warén, A. 1990. Ontogenetic changes in the trochoidean (Archaeogastropoda) radula, with some phylogenetic interpretations. *Zool. Scr.* 19:179–87
184. Warén, A. 1991. New and little known Mollusca from Iceland and Scandinavia. *Sarsia* 76:53–124
185. Warén, A., Bouchet, P. 1989. New gastropods from east Pacific hydrothermal vents. *Zool. Scr.* 18:67–102
186. Warén, A., Bouchet, P. 1991. Mollusca Gastropoda: systematic position and revision of *Haloceras* Dall, 1889 (Caenogastropoda, Haloceratidae fam. nov.). See Ref. 107, pp. 111–61
187. Warén, A., Ponder, W. F. 1991. New species, anatomy, and systematic position of the hydrothermal vent and hydrocarbon seep gastropod family Provannidae fam. n. (Caenogastropoda). *Zool. Scr.* 20:27–56
188. Wenz, W. 1938–1944. Gastropoda. Teil 1. Allgemeiner Teil und Prosobranchia. In *Handbuch der Paläozoolgie,* Vol. 6, ed. O. H. Schindewolf. Berlin: Borntraeger. 1639 pp.
189. Wiktor, A., Likharev, I. M. 1979. Phylogenetische Probleme bei Nacktschnecken aus den Familien Limacidae und Milacidae (Gastropoda, Pulmonata). *Malacologia* 18:123–31
190. Wilbur, K. M., ed. 1983–88. *The Mollusca*. Vols. 1–11. New York: Academic
191. Willan, R. C. 1987. Phylogenetic systematics of the Notaspidea (Opisthobranchia) with reappraisal of families and genera. *Am. Malac. Bull.* 5:215–41
192. Wingstrand, K. G. 1985. On the anatomy and relationships of recent Monoplacophora. *Galathea Rept.* 16:1–94
193. Yin Hong-F., Yochelson, E. L. 1983. Middle Triassic Gastropoda from Qingyan, Ghizhou Province, China: 3 — Euomphalacea and Loxonematacea. *J. Paleontol.* 57:1098–1127
194. Zilch, A. 1959–60. Gastropoda Euthyneura. Teil 2. In *Handbuch der Paläozoologie,* Vol. 6, ed. O. H. Schindewolf. Berlin: Borntraeger. 834 pp.

Annu. Rev. Ecol. Syst. 1992. 23:339–60

TRACE FOSSILS AND BIOTURBATION: THE OTHER FOSSIL RECORD

T. Peter Crimes
Department of Earth Sciences, University of Liverpool, Liverpool L69 3BX, United Kingdom

Mary L. Droser
Department of Earth Sciences, University of California, Riverside, California 92521

KEYWORDS: trace fossils, ichnofabric, early Metazoans, tiering

INTRODUCTION

Most of the evidence for marine metazoan life during the past 570 million years (m. y.) comes from two sources. The diversity and abundance of shelly organisms is best known from body fossils. The activity of both shelly and soft-bodied organisms in sediment can be understood by studying the burrows, tracks, and trails which they create. When preserved in the fossil record these features are called trace fossils. Because preservation of soft-bodied organisms is rare, evidence of behavior is particularly important for understanding the history of soft-bodied marine life. Thus, while the fossil record is generally viewed as consisting of body fossils, the record of trace fossils offers an alternative and independent data source. In the many environmental settings where body fossils are poorly preserved, trace fossils may provide the only evidence of past life.

The study of trace fossils and the record of bioturbation (the process by which animals rework the sediment) is a subdiscipline of paleontology called ichnology. Ichnological data is being used as critical evidence for such questions as the timing of the evolution of the Metazoa, changes of life habits, and the invasion of new habitats.

This review of trace fossils concentrates on several aspects of the nature of the marine ichnological record. Specifically, after introducing several concepts of ichnology, we discuss three broad areas within ichnology where trace fossils and the record of bioturbation are providing paleobiological data

0066-4162/92/1120-0339$02.00

otherwise unavailable: (i) the early history of metazoans, (ii) tiering of the infauna, and (iii) behavior of extinct and extant arthropods.

What Is a Trace Fossil?

At its simplest, a trace is a structure left in or on a soft sediment or in a hard substrate by a living organism. When preserved it becomes a trace fossil, regardless of whether the sediment is hardened by cementation. Trace fossils therefore include footprints, tracks, trails, and burrows formed initially in soft sediments as well as raspings, borings, and etchings in rigid cemented substrates. Coprolites and fecal pellets, since they too are left on or in the sediment by an animal are included. Some workers also include plant traces such as root penetration structures, algal mats, and stromatolites.

Trace fossils are normally discrete and readily identifiable structures which can be classified morphologically under the International Code of Zoological Nomenclature.

Trace fossils are distinct from and should not be confused with body fossils which are biological skeletal structures or other remains of the actual bodies of animals. Body fossils, therefore, are structures in their own right, whereas trace fossils (except coprolites and fecal pellets) exist only by virtue of the soft or hard substrate on or in which they formed and which preserves them.

This essential difference provides one of the most useful features of trace fossils. They are an integral part of the substrate and cannot readily be transported. Therefore, with few (and obvious) exceptions, trace fossils were made at the location where they are found. This contrasts with body fossils which, being discrete objects, may be transported by ocean currents. After death an animal's body may be moved a long distance into a totally different environment than that in which the animal lived. It is not always possible to ascertain whether body fossils have been transported, and hence their usefulness in environmental interpretation may be questioned, whereas trace fossils are virtually always preserved in situ.

Infaunal burrowing animals commonly mix the sediment, and in addition, a burrowing animal may cross-cut a structure previously constructed by another animal. Thus, trace fossil associations can be a composite accumulation over time that results in a complex sedimentary fabric. All aspects of sedimentary rock fabric caused by bioturbation are called ichnofabric (36). Thus, ichnofabric includes individual discrete trace fossils and composite associations of trace fossils as well as mottled bedding produced by bioturbation where discrete trace fossils are not readily identifiable. Most studies have traditionally concentrated on individual discrete trace fossils. However, as discussed below, studies of associations of trace fossils and their relationships as well as of overall ichnofabric yield significant paleontological information.

The Ichnofacies Concept

Seilacher (68, 69) first suggested that trace fossils do not occur in the oceans in random combinations; rather, they appear in distinct communities, the distribution of which he related to water depth. More recent studies have shown that many other factors control trace fossil distribution, including energy conditions at the depositional interface, substrate type, food availability, salinity, and oxygen (see discussion by Ekdale—35—for review). However, the broad applicability of Seilacher's model to Phanerozoic sequences has been repeatedly demonstrated. According to this model, each of the four main soft-substrate marine communities is named after a characteristic trace fossil, and all are broadly characterized by behavior/feeding types. They are, in order of increasing water depth: *Skolithos* (littoral zone), *Cruziana* (littoral zone to wave base), *Zoophycos* (wave base to deep ocean), and *Nereites* (deep ocean). The shallow water *Skolithos* ichnofacies is the domain of traces such as simple vertical burrows (*Skolithos*), U-burrows (*Arenicolites*), vertical spreite burrows (*Diplocraterion*) and coelenterate resting traces (*Bergaueria*). In Lower Paleozoic sequences, the slightly deeper water *Cruziana* ichnofacies is dominated by trilobite traces, including furrows, (*Cruziana*), resting impression (*Rusophycus*) and walking traces (*Diplichnites*). The *Zoophycos* ichnofacies typically has complex feeding burrows, whereas the deepest water *Nereites* ichnofacies is particularly rich in complex meandering traces (e.g. *Cosmorhaphe, Helminthoida crassa*), patterned traces (e.g. *Paleodictyon, Squamodictyon*) and intricate spirals (e.g. *Spirophycus, Spirorhaphe*). Firmground (*Glossifungities*), woodground (*Teredolites*), and hardground (*Trypanites*) ichnofacies have also been established. The latter two are dominated by boring organisms.

The ichnofacies model provides a conceptual framework for analyzing the distribution of trace fossils and thus the distribution of behavior. Exceptions to this model provide significant signals of local or large-scale varying physical or biological conditions.

THE EARLY RECORD OF TRACE FOSSILS AND BIOTURBATION

One of the most important contributions of trace fossils and the ichnological record to our understanding of the history of life has been documenting the early history of metazoans and supplementing our body fossil record from this time. Trace fossils have proven to be critical to our understanding of the timing, rate, diversity, and nature of the first appearances and subsequent radiations of metazoans.

Trace Fossils as the Record of the First Metazoans

Paleontologists have generally considered that the earliest record of metazoans would be found through trace fossil evidence because it is likely that the earliest metazoans were soft-bodied. The earliest widely accepted occurrence of metazoan body fossils is that of the globally distributed Ediacaran faunas of latest Precambrian age (see review by Conway Morris—19). Workers have examined older sedimentary rocks for evidence of metazoan activity.

There have been claims of trace fossils from rocks more than 2000 m.y. old. For example, Hofmann (46) described as *Rhysonetron lahtii* and *R. byei* structures that he regarded as probably organic from the 2200 m.y. old Upper Huronian of Ontario, Canada. However, he subsequently recognized that they could no longer be considered as biogenic, and so he reinterpreted them as mud cracks (47 p. 38), which is clearly what they represent. No undoubted examples of trace fossils of this age exist.

As far as traces of Metazoa are concerned, such early examples are implausible because the development of eukaryotic cells would be a necessary precursor and it is now widely considered that this evolutionary advance did not take place until 1500–1650 million years ago (m.y.a.) at most (44, 65–67, 83, 84). Even after that, a long time may have been required for metazoans to evolve (4) from eukaryotes.

Most records of presumed metazoan burrows in Riphaen rocks about 1,000 m.y. old are inadequately or inaccurately described, or they are from sequences whose age is not known with any certainty. For example, Clemmey (16) described what he believed to be burrows from the Middle Roan Formation of Zambia (about 1,000 m.y. old), but Cloud et al (17) reinterpreted them as Recent termite burrows. Similarly, Squire (81) described trace fossil structures from the slightly younger Briovarian of the Channel Islands, believed to be about 750 m.y. old, but these have been discounted as markings produced in old rocks by Recent annelids (8). Palij (55) described as *Rugoinfractus ovruchensis* structures from 1000–1400 m. y. old strata in the Ukraine, but these appear to be desiccation cracks (see also Ref. 45, p. 26). Similar examples in such old rocks from the former USSR are mostly poorly stratigraphically constrained, so even when their organic nature is apparent, their age is uncertain.

Undoubted trace fossils occur in the late Precambrian, particularly after the Vendian Varanger glaciation (c. 610 m.y.a.). Crimes (24) summarized the worldwide occurrence of trace fossils in the Vendian and Lower Cambrian. Some forms such as *Harlaniella, Nenoxites,* and *Palaeopascichnus* are apparently restricted to the Upper Vendian, but others extend into, and in some cases continue right through, the Phanerozoic (e.g. *Arenicolites, Cochlichnus, Didymaulichnus, Neonereites)* (Figures 1–3). The first Vendian traces are

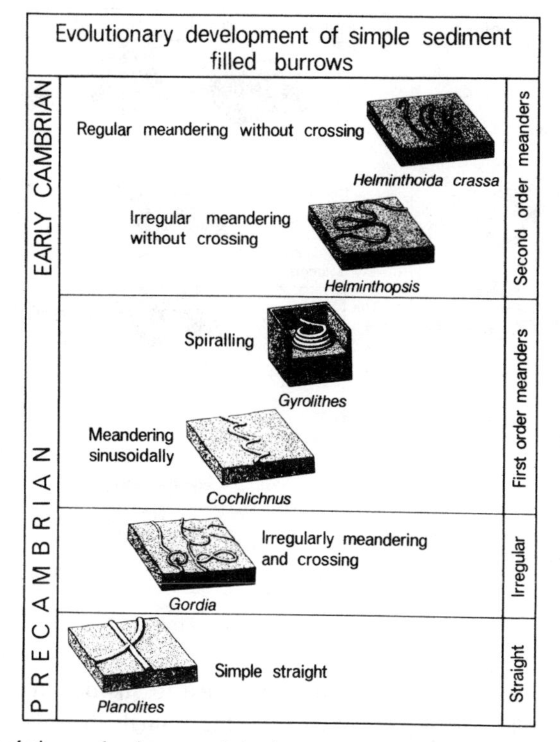

Figure 1 Evolutionary development of simple sediment filled burrows. From Crimes (26).

surface trails or horizontal burrows; vertical burrows become common only near the Vendian-Cambrian boundary (see also 4, 18, 60, 61).

We must therefore conclude with Bergstrom (4, p. 1) and Signor (79, p. 512) that there are few, if any, reliable reports of metazoan trace fossils significantly pre-dating the Vendian, a view previously expressed by Banks (3).

Increase in Trace Fossil Diversity

Trace fossil diversity is a reflection of the diversity of patterns of animal behavior. One animal may, within minutes, produce several different trace fossils that are classified as belonging to different genera. Equally, the same trace fossil may be produced by several separate animals. The latter point applies particularly to morphologically simple trace fossils, such as *Planolites* (Figure 1), which is only a sediment filled tube.

It is important to realize that trace fossil diversity is not the same as either

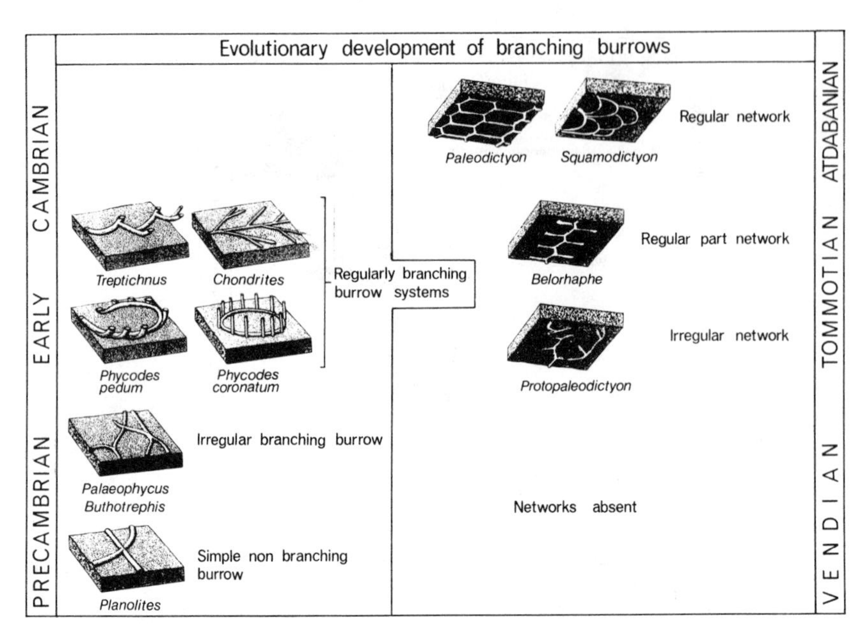

Figure 2 Evolutionary development of complex burrows. From Crimes (26).

animal or body fossil diversity. However, trace fossils do, to some extent, reflect the morphology of parts of or all of the animals that made them, and it is reasonable to expect that significant variations in animal diversity might be reflected by similar changes in trace fossil diversity. Furthermore, whereas almost all body fossils are produced from hard-bodied animals, trace fossils can equally be produced by soft-bodied forms. Trace fossils may therefore reflect initial animal diversity more accurately than do body fossils, which are heavily biased in favor of hard-bodied forms. Trace fossils should nevertheless be considered entirely separately from body fossils, and the two should never be counted together in diversity studies.

Studies of both trace fossil and body fossil diversity are plagued by variations in the area of exposed strata, quality of outcrop, lengths of geological periods, and uneven spatially and stratigraphically distribution of research. Nevertheless, recent research has tended to show that significant changes in body and trace fossil diversity do reflect a primary signal and cannot be dismissed as a result of secondary factors (75, 68, 69).

The increase in diversity of trace fossils from the Precambrian through the Phanerozoic has been considered by numerous authors (e.g. 22, 25, 26, 28, 57, 71, 75). Crimes (25) shows that the early Vendian (Redkino Horizon) is the first to have yielded a significant ichnofauna, with some 29 ichnogenera

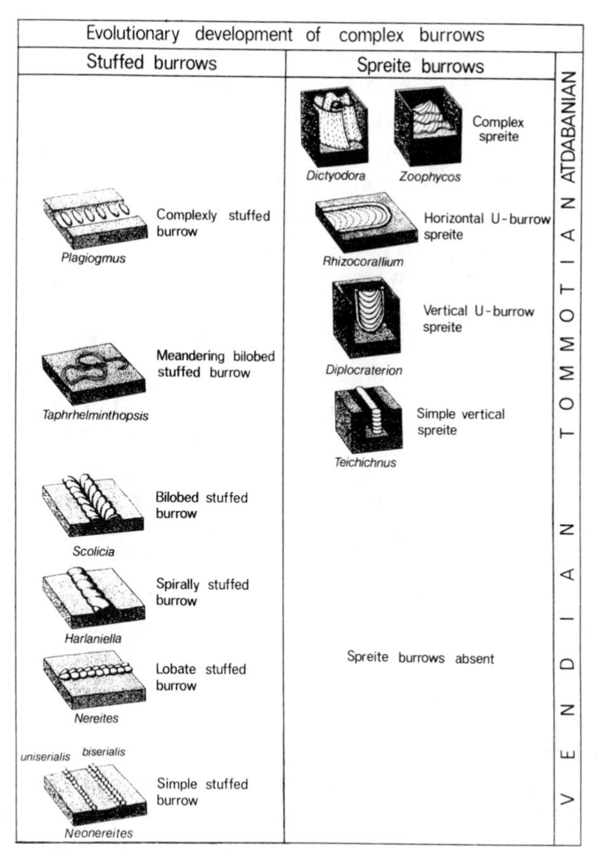

Figure 3 Evolutionary development of branching burrows. From Crimes (26).

present. There is little change in the middle Vendian, but a further burst in diversity occurs at about the level of the Precambrian-Cambrian boundary, in the late Vendian Rovno Horizon and the overlying Tommotian strata. Fedonkin (37, 38) considered the evolution of behavior patterns reflected in the increased diversity across the Precambrian-Cambrian boundary. The main part of the Vendian has only simple feeding traces, including open meandering simple feeding traces, the tight meandering *Palaeopascichnus,* and corkscrewing *Harlaniella* (Figure 3). More complex traces first appear around the Precambrian-Cambrian boundary, including complex branching burrows (*Chondrites, Phycodes, Treptichnus*), spiralling traces (*Gyrolithes)*, and spreite burrows (*Teichichnus)* (Figures 2-3). A complex evolutionary sequence of trace fossils was suggested (37; Figure 3). Similar changes were also inferred by Paczesna (54) from her investigation of Vendian and Lower

Cambrian strata in southeastern Poland. She concluded that Vendian traces are simple and consist mainly of primitive feeding burrows (e.g. *Torrawangea)*, with a few regarded as crawling traces (e.g. *Gordia)* and grazing traces (*Palaeopascichnus)*. All were considered of one trophic type: deposit feeders grazing on or within the sediment. Vertical traces, such as *Skolithos*, were absent. The Lower Cambrian trace fossils showed a rapid increase in ethological diversity with dwelling structures (e.g. *Bergaueria, Monocraterion)*, feeding traces (e.g. *Treptichnus, Phycodes)*, and crawling traces (e.g. *Bilinichnus)* all common. Two trophic groups were represented: suspension feeders and deposit feeders.

Crimes (26) discussed the genesis and evolutionary development of the main Phanerozoic trace fossil lineages. For example, the evolutionary development of simple sediment filled burrows is illustrated in Figure 1. The very simple, more or less straight burrow *Planolites* occurs first, low in the Vendian or earlier. Irregularly meandering burrows with crossings (e.g. *Gordia)* occur later in the Vendian and represent slightly more complex behavior. *Cochlichnus* has regular sinusoidal first order meanders and typically occurs first at about the Precambrian-Cambrian boundary, to be followed by tighter, vertically spiralling, first order meanders of *Gyrolithes*. *Helminthopsis*, which consists of irregular but non-crossing first and second order meanders, normally occurs later and is followed by the similar, but more regular, guided meanders of *Helminthoida crassa*. This evolutionary progression can be followed in one sequence in the Burin Peninsula, Newfoundland (27, 51) where *Planolites* occurs first, in the transitional beds between the Rencontre and Chapel Island Formation. *Gordia* is in the overlying Member 1 of the Chapel Island Formation, with *Cochlichnus* at the base of Member 2 and *Gyrolithes* slightly higher. *Helminthopsis* and the carefully programmed *Helminthoida crassa* are present only in Member 3. The evolutionary progression therefore seems to be from more or less straight burrows, through irregular meanders with crossings, to first order meanders and second order meanders, which represent the highest behavioral development.

Similar evolutionary sequences are shown by Crimes (26) for other trace fossil lineages such as branching burrows, spreite burrows, trails and resting traces (Figures 2 & 3). Each lineage shows the development of increasing complexity with time and more advanced behavioral programming. However, all the main lineages are well established, with the development of complex, sophisticated traces, as early as the Lower Cambrian.

Changes in trace fossil diversity through Phanerozoic time were first considered by Crimes (22) and Seilacher (71), with much more detail provided for the Lower Paleozoic by Crimes & Crossley (28) and Crimes (25, 26).

The data for the late Precambrian and Lower Paleozoic are summarized in

Figure 4, which shows that overall diversity reached a very high level in the Lower Cambrian. Then, after a slight fall (possibly because of inadequate study of Middle and Upper Cambrian strata), there is a steady rise through to the Upper Ordovician and Lower Silurian. The next major diversity increase is in the Cretaceous (22, 71, 74). These changes in trace fossil diversity essentially mirror those of body fossils (74, 79).

Many hypotheses have been advanced to explain the initial sharp diversity increase in the Vendian and Lower Cambrian, including build-up of atmospheric oxygen above a threshold value, increase in organic content of ocean floor muds, and temperature and sea level changes (22, 79). The burst in diversity in the Cretaceous and Cenozoic is mainly due to increased variety of trace fossils in the deep oceans; this diversity might reflect an increase in foraminiferal and calcareous cellulose debris reaching the ocean floor (73).

Trace Fossil Colonization of the Deep Oceans

While the Seilacher (68, 69) ichnofacies model has generally held up for most of the Phanerozoic, it was later shown that Precambrian and Cambrian deep water strata contain very few trace fossils and almost no examples of typical deep water *Nereites* ichnofacies types (22, 27, 29, 71). In the last few

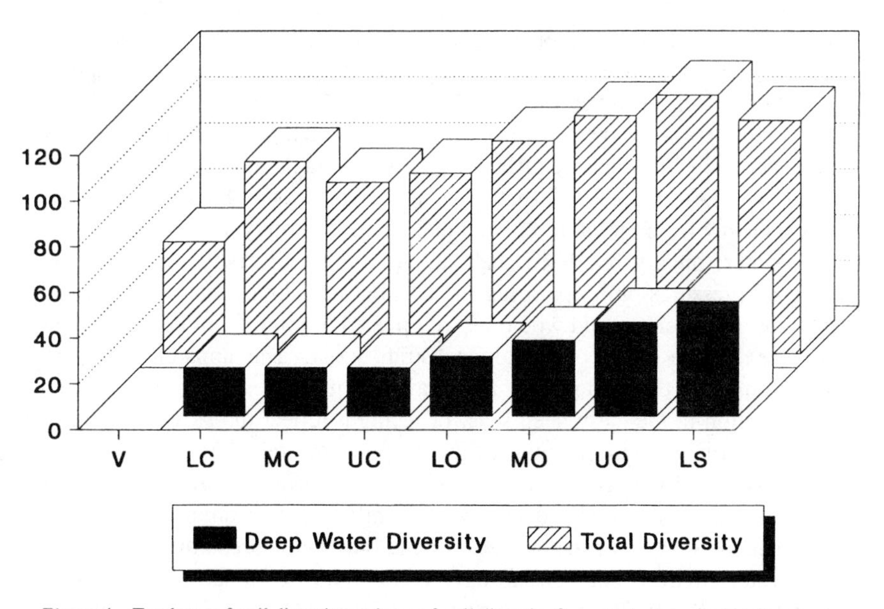

Figure 4 Total trace fossil diversity and trace fossil diversity from strata deposited in deep-water settings from the Vendian through the Lower Silurian. LC is Lower Cambrian; MC Middle Cambrian; UC Upper Cambrian; LO Lower Orodivican; MO Middle Ordovician, UO Upper Ordovician, and LS Lower Silurian. From Crimes et al (29).

years there have, however, been descriptions of what would be regarded in Seilacher's model as "deep water" trace fossils, from sequences of unquestionably shallow water sediments of Precambrian and Lower Cambrian age. Fedonkin (38–40) figured meandering traces from the late Precambrian of the former USSR. Crimes & Anderson (27) showed that the early Cambrian strata of southeastern Newfoundland contain complex, closely programmed, meandering traces (*Helminthoida crassa)* as well as patterned forms (*Protopaleodictyon, Paleodictyon, Squamodictyon). Paleodictyon* has also been recorded from a Cambrian shallow water succession in Poland by Paczesna (53). Regularly meandering trace fossils were described, as *Taphrhelminthoida dailyi,* by Hofmann & Patel (48) from the Lower Cambrian of New Brunswick, Canada. Finally, simple spiral traces (*Planispiralichnus, Protospiralichnus)* have been recorded from late Precambrian (Vendian) rocks of the former USSR (80).

Most meandering, patterned and spiral traces first appear in deep water successions during the Ordovician and Silurian (29). This suggests that these behavioral patterns evolved in shallow water at about the Precambrian-Cambrian boundary and only slowly dispersed into the deep ocean. The alternative hypothesis, that they evolved independently in the deep water, can be rejected because their arrival in the deep sea is followed by an almost total absence from shallow water successions. Crimes (25) also points out that it is difficult to understand why such closely programmed, efficient grazing patterns should be developed in the deep sea, where rich organic mud covers extensive areas. In shallow water, however, organically rich, muddy habitats occupy restricted areas, and there would be a distinct advantage to grazing the habitat efficiently and hence producing close regular, meanders. The rapid increase in trace fossil diversity in Lower Cambrian shallow water clastic seas probably forced some animals away from the more favored and extensive sandy substrates and into quiet water, areally restricted, habitat. As animal diversity increased and dispersal pressures built up during the Cambrian, animals that were adapted to living within or on mud would be well placed to migrate into deeper water where extensive areas of muddy substrate exist. The apparent delay in migration until the Ordovician may have been the result of low oxygen concentration in the early deep seas, or it might reflect the slow build up of organic detritus within the muds of the deep ocean floor (25).

Analysis of macroinvertebrate assemblages of various geological ages has shown that some relatively advanced or recently evolved communities appear first in shallow water nearshore settings and only later spread offshore, while deep water environments tend to have communities of long ranging archaic types (5–7, 49, 50, 76, 77). Bottjer & Jablonski (11) and subsequent papers have documented that, at least for individual post-Paleozoic clades, apparent community patterns are underlain by individualist clade histories. It has been inferred (11) that these migrations are not in response to any physical parameters such as sea level rise, but are biologically driven. The migration offshore can be accompanied by a loss of onshore representatives (termed "retreat") or retention of some onshore representatives (termed "expansion"). Mechanisms for "retreat" include (i) competitive exclusion by superior innovations that have originated onshore, and (ii) passive replacement, because speciation in stabler, offshore environments may be more probable than speciation back in less stable onshore environments. "Expansion" might be due to (i) the competitive superiority of onshore novelties relative to offshore taxa and (ii) simple diffusion, with offshore expansion the inevitable outcome of preferential seeding of innovations onshore (11).

The initial migration of trace fossils into the deep sea, accompanied as it is by a concomitant absence from shallow water, is an example of retreat. Bottjer et al (10) have presented a detailed analysis regarding changes in environmental preference with time for the common trace fossils *Zoophycos* and *Ophiomorpha*. They concluded that *Zoophycos* may show expansion in the Paleozoic and then retreat in the Mesozoic and Cenozoic, while *Ophiomorpha* originated in nearshore environments in the late Paleozoic and then "expanded" through the Mesozoic. Environmental migrations, as expressed both by body fossils and trace fossils, are likely to be a fruitful subject for continuing research: the process and driving mechanisms are still far from fully understood.

The Timing of the Evolution of Behavioral Patterns

For many years it was considered that there was a gradual improvement in animal behavioral programming through the Phanerozoic, so that behavioral patterns and resultant trace fossils gradually became more complex and "perfect." Seilacher (69, p. 8) suggested that in Cambrian times none of the animals on the ocean floor had evolved the meandering method. He considered that simple "scribbles" occurred first in the Ordovician but that complex meanders and dense double spirals did not appear until Cretaceous times and indicated significant progress in the foraging efficiency of sediment feeders.

In a more extensive summary, Seilacher (71, p. 238–239) concluded that, in a general way, scribbles, spirals, simple meanders, and complex meanders may be considered as a sequence of increasing behavioral complexity, as well as increasingly efficient in exploiting a given area. He claimed that scribbles are found only in Ordovician and Silurian deep sea deposits, while complex meanders did not appear before the later part of the Mesozoic. This was illustrated with the development of *Nereites* (Ref. 71, Figure 2), which was shown with somewhat irregular meanders in the Devonian, with very close meanders appearing only in the Cretaceous. Similar views and examples were later also given in Frey & Seilacher (43).

Recent investigations of ancient deep sea sediments, however, have revealed complex, very carefully programmed, behavioral patterns preserved as trace fossils in rocks as old as the Lower Paleozoic. Crimes & Crossley (28, Figure 9b) illustrate an example of *Nereites* from Lower Silurian deep sea sediments that is as tightly meandering as anything from the Mesozoic or Tertiary strata. They also showed examples of *Helminthoida crassa* with relatively tight meanders (Ref. 29, Figure 10 a-f). Pickerill (57, Figure 2b, 3e) recorded a double spiral *(Spirorhaphe)* and tight meanders *(Helminthoida)* in deep water deposits as far back as the early to Middle Ordovician. Clearly, these complex forms occurred in the deep ocean much earlier than Seilacher envisaged, although they may not have been abundant.

Perhaps even more surprising is the discovery of closely programmed meanders in the Lower Cambrian, but in shallow water deposits. Crimes & Anderson (27) found meanders of *Helminthoida crassa* near the base of the Lower Cambrian in Newfoundland, while Hofmann & Patel (48) recorded very carefully programmed meanders *(Taphrhelminthopsis dailyi)* from rocks of similar age in New Brunswick. Spirals have been reported from shelf sequences that date as far back as the late Precambrian (80).

The remarkable conclusion is that animals had evolved the ability to produce complex and closely programmed responses in shallow water environments by the time of Precambrian-Cambrian transition, about 570 m.y.a.

Ecospace Utilization in the Early Paleozoic

Many of the changes in the history of Earth's biota have been associated with the colonization of previously empty or underutilized ecospace. Because fossils from early Paleozoic marine settings are mostly of epifaunal organisms (72–74), much is known about trends in ecospace utilization for early Paleozoic epifauna (1, 2, 9, 82) Discrete trace fossils provide evidence on the timing, behavior, diversity, and evolution of early metazoans, but they cannot be used to determine trends in amount of bioturbation—or to what extent the infaunal habitat was utilized (33).

Extent of bioturbation must be evaluated through analysis of ichnofabric. Droser & Bottjer (31–33) examined the extent of bioturbation in Cambrian through Ordovician strata using a semiqualitative scheme based on the amount of original physical sedimentary structures disrupted by bioturbation. These studies reveal a two-phase stepwise increase in extent of bioturbation. In the Precambrian very little of the infaunal habitat was utilized. Most trace fossils are surficial, and those that are vertical are a centimeter or two deep. The first increase in bioturbation corresponded with the advent of trilobites and other large benthic macroinvertebrates (Figure 5). This is after the initial diversification of discrete trace fossils recorded near the Precambrian-Cambrian boundary, discussed above. The second increase, during which the percentage of reworked beds increased by an order of magnitude, corresponded with the

Ordovician radiation (510–440 m.y.a.) of skeletonized metazoans. Thus, this step-wise pattern of increasing bioturbation is contemporaneous with the Cambrian and Ordovician radiations of skeletonized metazoans. However, much of this recorded bioturbation cannot be attributed to any known skeletonized metazoans (33). Thus, these increases in extent of bioturbation were likely due to a soft-bodied infauna. These stepwise increases in extent of bioturbation, which are also accompanied by increases in bioturbation depth, are likely to have been due to increased ecospace utilization in the early Paleozoic. This could simply have been due to expansion of bioturbating organisms into deeper infaunal tiers, which may or may not have been accompanied by the development of new life habits at shallow depths. These data provide evidence independent of the body fossil record showing that the infaunal habitat, colonized primarily by soft-bodied organisms, shows a similar pattern of development to the two Paleozoic diversity increases of skeletonized metazoans.

TIERING

Animals and even plants commonly are distributed vertically at various levels, primarily for resource partitioning. In ecology, this vertical stacking has been referred to as "stratification" (e.g. 52). In part, because the term stratification

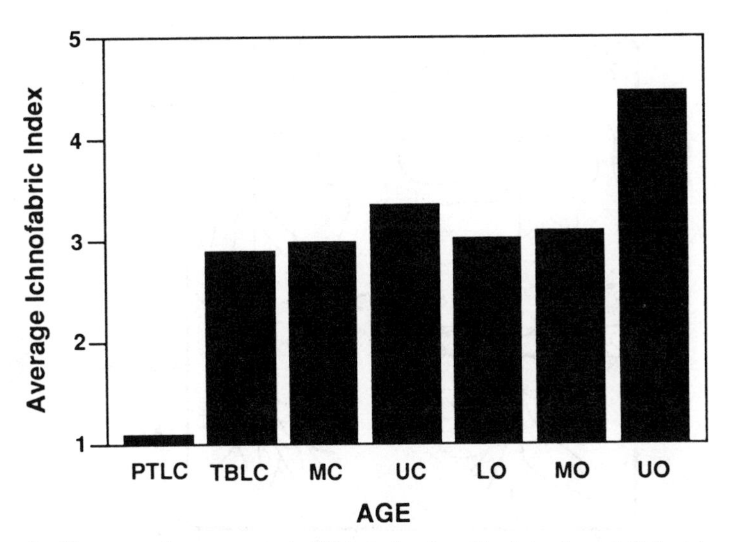

Figure 5 Histogram of average extent of bioturbation from Cambrian through Ordovician strata deposited in a carbonate shallow shelf settings. Ichnofabric index 1 is no bioturbation. Index 2 is up to 10% bioturbation. Index 3 is 10%–40%. Index 4 is 40%–60%. Index 5 is 60%–90%. PTLC is Pre-trilobite Lower Cambrian. TBLC is trilobite-bearing Lower Cambrian. All others as listed above.

has a well-defined geologic meaning, Ausich & Bottjer (1) coined the term "tiering" for the establishment of a community structure in which different organisms are distributed vertically both above and below the sediment-water interface. Ausich & Bottjer (1) documented a Phanerozoic history of tiering by suspension-feeding benthic macroinvertebrates in soft substrate communities, from both the body and trace fossil record.

Trace Fossils as Evidence of Tiering

Trace fossils preserve, in situ, the activity of animals. Thus, for the infaunal realm, trace fossils can provide data on depth of bioturbation and vertical distribution of animals (or their activity) in the sediment. Therefore study of trace fossils has significance for our understanding of tiering. First, trace fossils provide direct evidence of tiering. Tiering in the infaunal realm results in the juxtaposition of several trace fossils as animals burrow to different depths. This produces an ichnofabric composed of cross-cutting burrows (Figure 6). Extensive studies (13, 14, 36, 63, 64) of Cretaceous chalks and strata deposited under fluctuating oxygen conditions have demonstrated that cross-cutting relationships among trace fossils that reflect tiering patterns are common in the stratigraphic record. By examining these cross-cutting relationships, the sequence of emplacement can be discerned.

In studies of tiering relationships in chalk, Bromley & Ekdale (14)

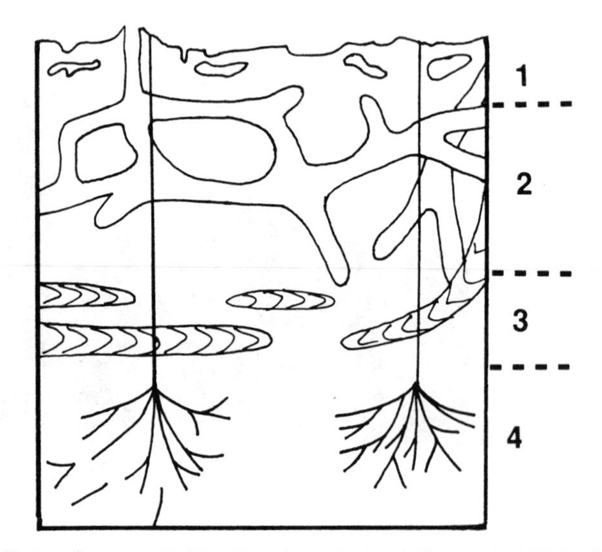

Figure 6 Tiering diagram with Planolites shown in the shallowest tier. Thalassinoides in the second tier, Zoophycos in the third and Chondrites in the deepest tier. After Bromley (13) and Savrda and Bottjer (62).

recognized several tiering levels. For example, in a study of Danish chalk, trace fossil suites represent numerous tiers, although, in any given sample, not all of the tiers will be represented. The shallowest tier is apparently represented by burrowing bivalves and echinoids, but this trace fossil suite is rarely preserved because it is destroyed by subsequent deeper-tier burrowers. The second tier contains densely packed *Planolites*. The third, fourth, fifth, and sixth tiers are occupied by *Thalasssinoides, Zoophycos,* and large and small *Chondrites*, respectively. Presumably, the deeper tier burrowers were in stiffer sediment beneath the superficial uncompacted sediments. In the chalks trace fossil cross-cutting relationships and tiering patterns remain consistent throughout many studied units even though the identical suite of trace fossils may not be present in every situation (13, 14).

Tiering patterns have also been extensively studied in strata deposited under oxygen-deficient conditions (62–64). Cross-cutting relationships of trace fossils demonstrate a predictable pattern dependent on factors such as the level of oxygen, length of anoxic/oxic event, and rate of reoxygenation. Data collected from strata of various ages have been used to establish a powerful model of tiering in low-oxygen strata which is utilized to reconstruct paleo-oxygen curves (62–64).

Phanerozoic Patterns of Tiering

Ausich & Bottjer (1) documented tiering in suspension-feeders through the Phanerozoic. But evidence from chalks and other strata suggest that tiering is common among suspension-feeding as well as deposit-feeding infauna. Thus, a second major contribution that trace fossils make to our understanding of tiering is that they can be used to develop further a model of Phanerozoic tiering in the infaunal realm.

In a study of tiering and guild structure of a Silurian community in a shallow carbonate setting, Watkins (85) recognized two tiers present throughout the examined unit. A shallow tier—approximately 2 cm below the sediment surface—is composed of the burrow *Palaeophycus*, which has been interpreted as the mucous-lined dwelling burrow of a suspension-feeding worm (56). These burrows are consistently cross-cut by *Chondrites* and *Planolites*, which are interpreted to represent deposit-feeding at depths of up to 10 cm. In older Cambrian strata deposited under similar conditions, however, deeper-tiered burrows are absent (32). Tiering patterns vary among sedimentary environments depending on the physical, biological, and chemical environments. Analysis of tiering relationships among trace fossils, particularly in similar environments through time, represents a relatively new approach in ichnology. Future studies will improve our understanding of Phanerozoic tiering patterns as well as allowing us to discern other tiering patterns characterizing specific substrate types and environmental conditions.

BEHAVIOR OF THE EXTINCT AND EXTANT

As discussed above, specific discrete trace fossils provide valuable information about the evolution of behavior. Commonly it is impossible to assign a trace maker to a given trace fossil, much to the dismay of many whose first question upon viewing a trace fossil is "what made it?" However, there are certain trace fossils of which we are relatively confident of the maker—or at least the clade of the maker. The benefits for understanding extinct animals are obvious, but trace fossils also allow us to document the development of behavior of extant animals, which is particularly useful in situations where animals may be poorly preserved. Two examples from the Arthropoda are discussed below.

The Life of Trilobites

The trilobites are a fascinating group of early arthropods that became extinct at about the end Permian, approximately 245 m.y.a. Fortunately, they left abundant evidence of their behavior patterns in the form of trace fossils (Figure 7), particularly in Lower Paleozoic nearshore and inner shelf successions. Consequently, despite a complete absence of modern examples, it is possible to study their behavior in detail.

The earliest trilobites are found in Lower Cambrian strata, but in numerous sections worldwide their trace fossils appear stratigraphically before their first preserved body fossils (24). This suggests that they initially developed as soft-bodied forms, which would not be preserved; only later did they develop hard parts. This idea has received recent support from cladistic analysis (59). The interval between the first appearance of their trace and body fossils is probably only a few million years. The earliest trilobite trace fossils are simple, shallowly impressed, structures that would be produced by slow movement of a simple muscle system of low efficiency and without the support of a rigid exoskeleton (23). Later examples are more complex and some suggest more rapid locomotion, changes which would follow the development of a hard sclerotized or mineralized integument, attached to which the muscles would work more efficiently.

The trace fossils show that trilobites indulged in several types of excavation and locomotion. The traces that show the closest resemblance to the form of trilobite exoskeletons are resting excavations (*Rusophycus*—Figure 7b). To produce these a trilobite dug in the mud at one location, so the shape and ornamentation of the trace may correspond closely to its undersurface. In some examples, the trilobite coxa, limbs, pleurae, headshield and genal spines have left their impressions in the sediment and form part of the *Rusophycus* (20, 58, 70). The trilobite behavior responsible for *Rusophycus* is unclear. Possibilities include that it was used for a brooding chamber, or as a place of protection either during moulting or from a predator, or that it represented a brief stop to scavenge into the sediment for food.

The best known trilobite trace is the very distinctive *Cruziana* (Figure 7c) which consists of a deeply impressed row of V-markings produced by the animal furrowing through the sediment. In some ichnospecies of *Cruziana* there is a parallel mark on either side of the Vs and this represents the trail of the genal spine. The Vs never extend beyond the genal spine markings, thus demonstrating that, as with some modern arthropods, trilobites dug only within the limits of their carapace.

When the trilobite walked, rather than furrowed, a much less deeply impressed series of "footprints" was produced, closely spaced during slow locomotion but more widely spaced as step distance was increased during more rapid movement. Trace fossils of this type are included within the ichnogenus *Diplichnites* (Figure 7a).

It appears that at times trilobites swam or drifted with the current and keeled over, so that the limbs of one side of the body would produce a simple set of scratch marks made by the animal's claws. *Monomorphichnus* is included under this trace, so named because the marking were of a single form (mono-morph) in contrast to *Dimorphichnus* which has a set of small deeply impressed marks on one side where the animal has dug in and longer, shallower, markings on the other side where the "free" limbs have scraped the mud, presumably in search for edible benthos.

These types of trilobite traces are found throughout the Paleozoic and testify to the stability of these behavior patterns through time.

In a study of over 500 trilobite trace fossils from the Upper Cambrian of Wales, Crimes (21) was able to demonstrate that for a given species, the smallest traces were much larger than the size at which the trilobites hatched, and the resting traces were, statistically, smaller than the walking traces, which in turn were smaller than the furrow. He therefore suggested that these trilobites changed their pattern of behavior during their life cycle. The youngest trilobites were planktonic; thus they left no trace of their early development. On becoming benthic they first excavated resting traces and

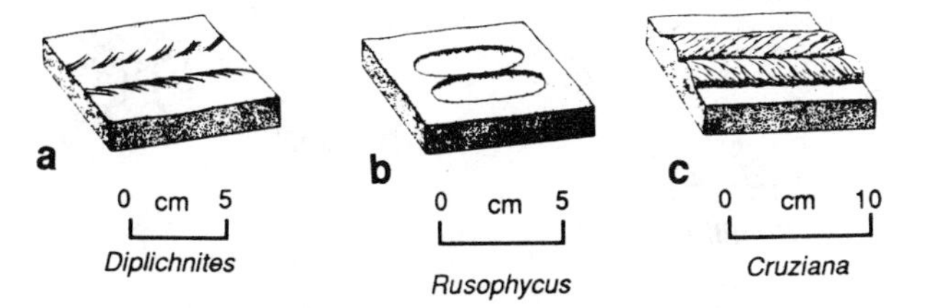

Figure 7 Common trilobite trace fossils. (a) Rusophycus, (b) Cruziana, and © Diplichnites. After Crimes (24).

may have used them for protection, especially during periods of molting. Later, as they matured, they took to walking and then to furrowing in search of food.

Life in Nearshore Sands

While other ichnofacies were not established until the Ordovician, in the nearshore Skolithos ichnofacies, deep and extensive bioturbation first appeared in the Early Cambrian and persisted for the rest of the Paleozoic. The *Skolithos* ichnofacies is characterized by shifting sandy substrates and high energy conditions (42). In these settings, where body fossils are poorly preserved, trace fossil history is well known. Dense assemblages of *Skolithos,* a simple vertical tube (Figure 8a), characterizes Paleozoic strata deposited under these conditions (30). Various modern marine animals make such a burrow including many sipunculans, polychaetes, phoronids, and actinian anemones (13). However, the animal(s) responsible for the abundant Paleozoic *Skolithos* remain(s) unknown. In the post-Paleozoic, dense assemblages of *Skolithos* are relatively uncommon and *Ophiomorpha* (Figure 8b) dominates the *Skolithos* ichnofacies. The appearance of *Ophiomorpha* represents the development of a behavioral program that allowed burrowing decapod crustaceans to colonize sandier, and thus less cohesive, substrates than muds. This innovative behavior is the ability to produce a pelleted lining for burrows, which prevents burrow collapse in substrates of relatively low cohesive strength. These burrow linings were as significant an evolutionary innovation for burrowing decapods as the development of siphons was for burrowing bivalves.

In a pivotal study relating recent biogenic structures and their makers to the trace fossil record, Weimer & Hoyt (86) demonstrated the similarities of burrows of *Callianassa major* to Cretaceous *Ophiomorpha.* Since then, several other species of *Callianassa,* as well as *Upogebia* and possibly Axius,

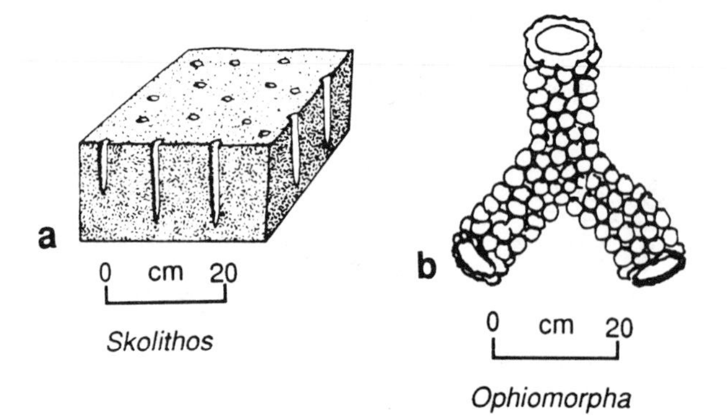

Figure 8 Characteristic burrows of nearshore sandstones. (a) Skolithos and (b) Ophiomorpha. After Crimes (24).

have been shown to produce a pelleted structure (41). There are several "species" of *Ophiomorpha*, which vary according to the nature of the pellets, thickness of lining, nature of branching and overall burrow morphology (41). *Ophiomorpha* burrows may even have double pelleted lining. *Ophiomorpha* grades into other trace fossils such as the spiral shaped *Gyrolithes* and the stacked *Teichichnus*. However, *Ophiomorpha* is a consistent feature of post-Paleozoic nearshore sandstones. In these strata, *Ophiomorpha* is commonly the only burrow present. Burrows are not crowded together, and most of the original physical sedimentary structures are usually preserved and have not been destroyed by these burrowing decapods (34)

Decapod crustaceans are not well preserved in the stratigraphic record. Shrimp-like malacostracan crustaceans first appeared in the Devonian and underwent a diversification in the Carboniferous (12). The earliest *Ophiomorpha* is reported from the Permian from several localities in shallow water strata (10, 15). There are no reported Triassic occurrences, but in the Jurassic, *Ophiomorpha* becomes common in shallow water sands. However, in the Cretaceous, the *Ophiomorpha*-producing animals expanded into the deep-sea realm, although still occurring in sandstones, generally fan and turbidite deposits (10).

SUMMARY

Although trace fossils have been recognized since the 1800s, the study of ichnology is still a relatively new field of paleontology, and came of age only in the 1970s. Taxonomic work provided the foundation of ichnology, but studies simply describing new trace fossils are no longer the norm. The ichnological record is becoming appreciated as a significant, independent, and alternative fossil record providing data otherwise unavailable. We have not attempted to summarize all of the contributions that have been made or are potentially significant to paleontology. Rather, we have presented three broad areas as examples where ichnological research has contributed to our understanding of the history of life. In recent years, paleontologists have begun to understand more fully the nature of the ichnological record, and we are now poised to contribute even more to knowledge of the evolution of behaviors, functional morphology, ecospace utilization, and the history of soft-bodied animals.

ACKNOWLEDGMENTS

Funding was provided by the Petroleum Research Fund of the American Chemical Society. Acknowledgment is also made to the White Mountain Research Station. This paper benefitted from discussions with and critical reviews by David Bottjer and Nigel Hughes.

Literature Cited

1. Ausich, W. I., Bottjer, D. J. 1982. Tiering in suspension-feeding communities on soft substrata throughout the Phanerozoic. *Science* 216:173–74
2. Bambach, R. K. 1983. Ecospace utilization and guilds in marine communities through the Phanerozoic. In *Biotic Interactions in Recent and Fossil Benthic Communities,* ed. M. J. S. Tevesz, P. L. McCall, pp. 719–46. New York: Plenum. 837 pp.
3. Banks, N. L. 1970. Trace fossils from the Late Precambrian and Lower Cambrian of Finnmark, Norway. See Ref. 28a, pp. 19–34
4. Bergstrom, J. 1990. Precambrian trace fossils and the rise of bilaterian animals. *Ichnos* 1:3–13
5. Berry, W. B. N. 1972. Early Ordovician bathyurid province lithofacies, biofacies, and correlations and their relationships to a proto-Atlantic Ocean. *Lethaia* 5:69–84
6. Berry, W. B. N. 1974. Types of early Paleozoic faunal replacements in North America: their relationships to environmental change. *J. Geol.* 82:371–82
7. Berry, W. B. N. 1977. Graptolite biostratigraphy: a welding of classical principles and current concepts. In *Concepts and Methods of Biostratigraphy,* ed. E. G. Kauffman, J. E. Hazel, pp. 321–38. Stroudsberg, Penn: Dowden, Hutchinson & Ross. 658 pp.
8. Bland, B. H., Evans, G., Goldring, R., Mourant, A. E., Renouf, J. T., Squire, A. D. 1987. Supposed Precambrian trace fossils from Jersey, Channel Islands. *Geol. Mag.* 124:123
9. Bottjer, D. J., Ausich, W. I. 1986. Phanerozoic development of tiering in soft substrata suspension-feeding communities. *Paleobiology* 12:400–20
10. Bottjer, D. J., Droser, M. L., Jablonski, D. 1988. Paleonvironmental trends in the history of trace fossils. *Nature* 333:252–55
11. Bottjer, D. J., Jablonski, D. 1988. Paleonvironmental patterns in the evolution of post-Paleozoic benthic marine invertebrates. *PALAIOS* 3:540–60
12. Briggs, D. E. G., Clarkson, E. N. K. 1989. Environmental controls on the taphonomy and distribution of Carboniferous malacostracan crustaceans. *Trans. R. Soc. Edinburgh Earth Sci.* 80:293–301
13. Bromley, R. G. 1990. *Trace Fossils: Biology and Taphonomy.* London: Unwin Hyman. 280 pp.
14. Bromley, R. G., Ekdale, A. A. 1986. Composite ichnofabrics and tiering of burrows. *Geol. Mag.* 123:59–65
15. Chamberlain, C. K., Baer, J. 1973. *Ophiomorpha* and a new Thallassinid burrow from the Permian of Utah. *Brigham Young Univ. Geol. Stud.* 20: 79–94
16. Clemmey, H. I. 1976. World's oldest animal traces. *Nature* 261:576–78
17. Cloud, P., Gustafson, L. B., Watson, J. A. L. 1980. The works of living social insects and the age of the oldest known Metazoa. *Science* 210:1013–15
18. Cloud, P. E. 1983. Are the Medicine Peak Quartzite "dubiofossils" fluid-evasion tracks? *Geology* 11:618–19
19. Conway Morris, S. 1990. Late Precambrian and Cambrian soft-bodied faunas. *Annu. Rev. Earth Planet. Sci.* 18:101–22
20. Crimes, T. P. 1970. Trilobite tracks and other trace fossils. *Geol. J.* 7:47–68
21. Crimes, T. P. 1970. The significance of trace fossils in sedimentology, stratigraphy and palaeoecology with examples from Lower Paleozoic strata. See Ref. 28a, pp. 101–26
22. Crimes, T. P. 1974. Colonisation of the early ocean floor. *Nature* 248:328–30
23. Crimes, T. P. 1975. The stratigraphical significance of trace fossils. In *The Study of Trace Fossils,* ed. R. W. Frey, pp. 109–30. Berlin: Springer-Verlag. 562 pp.
24. Crimes, T. P. 1987. Trace fossils and the correlation of Late Precambrian and Early Cambrian strata. *Geol. Mag.* 214:97–119
25. Crimes, T. P. 1992. Changes in the trace fossil biota across the Proterozoic-Phanerozoic boundary. *Geol. Soc. London Q. J.* In press
26. Crimes, T. P. 1992. The record of trace fossils across the Proterozoic-Cambrian boundary. In *Origins and Early Evolution of the Metazoa,* ed. J. H. Lipps, P. W. Signor. New York: Plenum. In press
27. Crimes, T. P., Anderson, M. M. 1985. Trace fossils from late Precambrian-Early Cambrian strata of southeastern Newfoundland (Canada): temporal and environmental implications. *J. Paleontol.* 59:310–43
28. Crimes, T. P., Crossley, J. P. 1991. A diverse ichnofauna from the Silurian flysch of the Aberystwyth Grits Formation, Wales. *Geol. J.* 26:27–62

28a. Crimes, T. P., Harper, J. C., eds. 1970.

Trace Fossils. Geol. J. Spec. Issue I, Vol. 3. Liverpool: Seel House
29. Crimes, T. P., Hidalgo, G. J. F., Poire, D. G. 1992. Trace fossils from Arenig flysch sediments of Eire and their bearing on the early colonisation of the deep sea. *Ichnos* 2:1–17
30. Droser, M. L. 1991. Ichnofabric of the Paleozoic Skolithos ichnofacies and the nature and distribution of Skolithos piperock. *PALAIOS* 6:316–25
31. Droser, M. L., Bottjer, D. J. 1986. A semiquantitative field classification of ichnofabric. *J. Sediment. Petrol.* 56: 558–59
32. Droser, M. L., Bottjer, D. J. 1988. Trends in depth and extent of bioturbation in Cambrian carbonate marine environments, western United States. *Geology* 16:233–36
33. Droser, M. L., Bottjer, D. J. 1989. Ordovician increase in extent and depth of bioturbation: Implications for understanding early Paleozoic ecospace utilization. *Geology* 17:850–52
34. Droser, M. L., Bottjer, D. J. 1990. Ichnofabric of sandstones deposited in high-energy nearshore environments: measurement and utillization. *PALAIOS* 4:598–604
35. Ekdale, A. A. 1988. Pitfalls of paleobathymetric interpretations based on trace fossil assemblages. *PALAIOS* 3: 464–72
36. Ekdale, A. A., Bromley, R. G. 1983. Trace fossils and ichnofabric in the Kjolby Baard Marl, Upper Cretaceous, Denmark. *Bull. Geol. Soc. Denmark* 31:107–19
37. Fedonkin, M. A. 1978. Ancient trace fossils and the ways of behavioral evolution of mud eaters. *Palaeontol. J.* 12:106–12
38. Fedonkin, M. A. 1980. Early stages of evolution of Metazoa on the basis of palaeoichnological data. *Izv. Akad. Nauka SSSR Ser. Geol.* 2:226–33
39. Fedonkin, M. A. 1980. Fossil traces of Precambrian Metazoa. *Izv. Akad. Nauka SSSR Ser. Geol.* 1:39–46
40. Fedonkin, M. A. 1981. Belomorskaya biota Venda (White Sea biota of the Vendian). *Trudy Geol. Inst. Akad. Nauka SSSR* 342:99 pp.
41. Frey, R. W., Howard, J. D., Pryor, W. A. 1978. *Ophiomorpha*: its morphologic, taxonomic, and environmental significance. *Palaeogeogr. Palaeoclimatol. Palaeoecol.* 23:199–229
42. Frey, R. W., Pemberton, S. G. 1984. Trace fossil facies models: In *Facies Models,* ed. R. G. Walker, *Geosci. Can., Rep. Ser.* 1:189–207
43. Frey, R. W., Seilacher, A. 1980. Uniformity in marine invertebrate ichnology. *Lethaia* 13:183–207
44. Glaessner, M. F. 1979. Precambrian. In *Treatise of Invertebrate Paleontology,* Part A, *Introduction*. ed. R. A. Robison, C. Teichert, pp. A79-A118. Lawrence, Kans: Geol. Soc. Am., Univ. Kans. Press
45. Glaessner, M. F. 1984. *The Dawn of Animal Life*. Cambridge: Cambridge Univ. Press. 244 pp.
46. Hofmann, H. J. 1967. Precambrian fossils (?) near Elliot Lake, Ontario. *Science* 156:500–4
47. Hofmann, H. J. 1971. Precambrian fossils, pseudofossils, and problematica in Canada. *Geol. Soc. Canada Bull.* 189:146 pp.
48. Hofmann, H. J., Patel, I. M. 1989. Trace fossils from the type 'Etcheminian Series' (Lower Cambrian Ratcliffe Brook Formation), Saint John area, New Brunswick, Canada. *Geol. Mag.* 126:139–57
49. Jablonski, D., Bottjer, D. J. 1983. Softbottom epifaunal suspension-feeding assemblages in the Late Cretaceous: implications for the evolution of benthic paleocommunities. In *Biotic Interactions in Recent and Fossil Benthic Communities,* ed. M. J. S. Tevesz, P. L. McCall, pp. 747–812. New York: Plenum. 837 pp.
50. Jablonski, D., Sepkoski, J. J. Jr., Bottjer, D. J., Sheehan, P. M. 1983. Onshore-offshore patterns in the evolution of Phanerozoic shelf communities. *Science* 222:1123–25
51. Narbonne, G., Myrow, P. M., Landing, E., Anderson, M. M. 1987. A candidate stratotype for the Precambrian-Cambrian boundary, Fortune Head, Barin Peninsula, southeast Newfoundland. *Can. J. Earth Sci.* 24:1277–93
52. Odum, E. P. 1983. *Basic Ecology*. Philadelphia, Penn: Saunders. 613 pp.
53. Paczesna, J. 1985. Ichnorodzaj Paleodictyon Meneghini z dolnego Kambru zbilutki (Gory Swietokrzyskie). *Kwart. Geol.* 29:589–96
54. Paczesna, J. 1986. Upper Vendian and Lower Cambrian ichnocoenoses of Lublin region. Biul.Inst. Geol. 355:32–47
55. Palij, V. M. 1974. On finding of the trace fossil in the Riphean deposits of the Orrutch Ridge. *Rep. Ukrainian Acad. Sci. SSSR Ser. B Geol. Geophys.* 36:34–37 (In Ukrainian)
56. Pemberton, S. G., Frey, R. W. 1982. Trace fossil nomenclature and the Planolites-Palaeophycus dilemma. *J. Paleontol.* 56:843–81
57. Pickerill, R. K. 1980. Phanerozoic trace fossil diversity—observations based on

an Ordovician flysch ichnofauna from the Aroostook–Metapedia Carbonate Belt of northern New Brunswick. *Can. J. Earth Sci.* 17:1259–70
58. Radwanski, A., Roniewicz, P. 1963. Upper Cambrian ichnocoensis from Wielsica Wisnikowa (Holy Cross Mountains) Poland. *Acta Paleontol.* 8: 259–81
59. Ramskold, L., Edgecombe, G. D. 1988. Trilobite monophyly revisited. *Hist. Biol.* 4:267–83
60. Runnegar, B. 1982. Oxygen requirements, biology, and phylogenetic significance of the Late Precambrian worm Dickinsonia and the evolution of the burrowing habit. *Alcheringa* 6: 223–39
61. Runnegar, B. 1982. The Cambrian explosion: animals or fossils? *J. Geol. Soc. Aust.* 29:395–411
62. Savrda, C. E., Bottjer, D. J. 1986. Trace fossil model for the reconstruction of paleo-oxygenatyion in bottom waters. *Geology* 14:3–6
63. Savrda, C. E., Bottjer, D. J. 1989. Trace fossil model for reconstructing oxygenation histories of ancient marine bottom waters: application to Upper Cretaceous Niobrara Formation, Colorado. *Palaeogeogr. Palaeoclimatol. Palaeoecol.* 74:49–74
64. Savrda, C. E., Bottjer, D. J. 1989. Anatomy and implications of bioturbated beds in "black shale" sequences: examples from the Jurassic Posidonienschiefer (southern Germany). *PALAIOS* 4:330–42
65. Schopf, J. W. 1972. Evolutionary significance of the Bitter Springs (late Precambrian) microflora. *Proc. Int. Geol. Congr., 24th, Montreal,* Sect. 1, pp. 68–77
66. Schopf, J. W. 1975. Precambrian paleobiology: problems and perspectives. *Annu. Rev. Earth Planet. Sci.* 3:213–49
67. Schopf, J. W., Oehler, O. Z. 1976. How old are the eukaryotes? *Science* 193:47–49
68. Seilacher, A. 1964. Biogenic sedimentary structures. In *Approaches to Paleoecology,* ed. J. Imbrie, N. O. Newell, pp. 296–316. New York: Wiley
69. Seilacher, A. 1967. Bathymetry of trace fossils. *Mar. Geol.* 5:189–200
70. Seilacher, A. 1970. *Cruziana,* stratigraphy of 'non-fossiliferous' Palaeozoic sandstones. See Ref. 28a, pp. 447–76
71. Seilacher, A. 1974. Flysch trace fossils: Evolution of behavioral diversity in the deep-sea. *Neues Jahrb. Geol. Palaeontol. Abh.* 4:233–51
72. Sepkoski, J. J. Jr. 1978. A kinetic model of Phanerozoic taxonomic diversity, I. Analysis of marine orders. *Paleobiology* 4:223–51
73. Sepkoski, J. J. Jr. 1979. A kinetic model of Phanerozoic taxonomy diversity. II. Early Phanerozoic families and multiple equilibrium. *Paleobiology* 5: 222–51
74. Sepkoski, J. J. Jr. 1981. A factor analytic description of the Phanerozoic marine fossil record. *Paleobiology* 7: 36–53
75. Sepkoski, J. J. Jr., Bambach, R. K., Raup, D. M., Valentine, J. W. 1981. Phanerozoic marine diversity and the fossil record. *Nature* 293:435–37
76. Sepkoski, J. J. Jr., Miller, A. I. 1985. Evolutionary faunas and the distribution of Paleozoic benthic communities in space and time. In *Phanerozoic Diversity Patterns,* ed. J. W. Valentine, pp. 159–90. Princeton, NJ: Princeton Univ. Press
77. Sepkoski, J. J. Jr., Sheehan, P. M. 1983. Diversification, faunal change and community replacement during the Ordovician radiation. In *Biotic Interactions in Recent and Fossil Benthic Communities,* ed. M. J. S. Tevesz, P. L. McCall, pp. 673–718. New York: Plenum
78. Signor, P. W. 1982. Species richness in the Phanerozoic: Compensating for sampling bias. *Geology* 10:625–28
79. Signor, P. W. 1990. The geological history of diversity. *Annu. Rev. Ecol. Syst.* 21:509–39
80. Sokolov, B. S., Ivanovski, A. B. 1985. *The Vendian System,* Vol. I. Moscow: Acad. Sci. USSR. 221 pp.
81. Squire, A. D. 1973. Discovery of late Precambrian trace fossils in Jersey, Channel Islands. *Geol. Mag.*110:223–26
82. Vermeij, G. J. 1987. *Evolution and Escalation.* Princeton, NJ: Princeton Univ. Press. 557 pp.
83. Vidal, G., Ford, T. D. 1985. Microbiotas from the late Proterozoic Chuar Group (northern Arizona) and Uinta Mountain Group (Utah) and their chronostratigraphic implications. *Precamb. Res.* 28:349–89
84. Vidal, G., Knoll, A. H. 1983. Proterozoic plankton. *Geol. Soc. Am. Mem.* 161:265–77
85. Watkins, R. 1991. Guild structure and tiering in a high-diversity Silurian community, Milwaukee County. *PALAIOS* 6:465–78
86. Weimer, R. J., Hoyt, J. H. 1964. Burrows of *Callianassa major* Say, Geologic indicators of littoral and shallow neritic environments. *J. Paleontol.* 38:761–67

Annu. Rev. Ecol. Syst. 1992. 23:361–81

BEHAVIORAL HOMOLOGY AND PHYLOGENY

John W. Wenzel
Department of Entomology, University of Georgia, Athens, Georgia 30602*

KEYWORDS: behavior, homology, phylogeny, ethology, systematics

INTRODUCTION

The sociobiology debates of the 1970s increased interest in the biology of behavior. At the same time, the growth of cladistics increased interest in how to do systematics and phylogenetic reconstruction. Yet, there are surprisingly few recent papers dealing explicitly with behavior from a phylogenetic perspective. Lack of communication between students of behavior and students of systematics is partly to blame. If one says to a behavioral ecologist, "Isn't it curious that there are white bears in the arctic?" he may say that there is nothing curious about it because they are white like all the other arctic mammals, and the fact that they are bears is irrelevant to the broad patterns of evolution. If one asks the same of a systematist he may reply that there is nothing curious about it because they are still bears like all the others, and the fact that they are white is irrelevant to the broad patterns of evolution. Both perspectives are partly right, and both are less than the whole story. Systematists tend to look for constraints of history, while behaviorists usually prefer to work with a warm ball of clay that lies ready to take on any shape the outside forces push upon it.

Some of what follows is review and some is more philosophical, but the point of the paper is simple. Determining homology among behaviors is no different than determining homology among morphological structures. Behavior is not special, it is only more difficult to characterize. Ethology (the study of behavior) is a relatively young science and does not yet have the benefit of centuries of debate and consensus, but that provides more reason for us to take up the challenge now. Ethology has made almost no advance with respect to a phylogenetic understanding of behavior since the late 1950s, and most

*Current address: Department of Entomology, American Museum of Natural History, Central Park West at 79th, New York, NY 10024-5192

0066-4162/92/1120-0361$02.00

modern ethologists simply do not work toward that goal. To honor the proud heritage of Lorenz and Tinbergen we need only to be brave and begin.

There is an immense literature dealing with "evolution of behavior," but only a tiny fraction of ethological efforts are relevant to the question of how one postulates homology among specific elements of an animal's behavioral repertoire. The majority of studies on behavioral evolution are related to theories of the process of evolution, and they therefore compare grades to illuminate the way in which analogous transitions occur in different groups (33). The focus of these studies is the transition itself, and homology of the steps is not an issue. Also the taxonomic literature is skewed toward finding species-specific behaviors that allow identification more easily than morphological variation allows (1). Such unique traits do not assist in reconstructing phylogeny; only shared traits are useful for finding a nested hierarchy of order. I have tried to include here both classical and more recent works that illustrate explicit postulates of homology, or cases where behavioral characters were critical for defining or supporting a phylogenetic scheme, but this discussion is streamlined to serve more as an introduction to a developing field than as the conclusive study of a mature body of science.

CRITERIA OF HOMOLOGY

The distinction between homology and analogy provided the foundation for systematic biology, and the terms have been the source of disagreement for close to two centuries. Evidently the discussion is not over yet, and if the reader chooses to pursue the philosophical issue, there is plenty to do (9, 24, 36, 46, 78, 79, 86, 98). Some of these papers generate more heat than light, and the entire debate is almost completely without reference to behavior (but see 3, 54). For the present purpose, homologous behaviors are defined as those that find their origin in the same ancestor and are similar because of descent from that common ancestor. Even if two behaviors satisfy all other criteria (below), they are not homologous if they are derived independently from different ancestors.

In the past, this sort of definition was regarded as partly circular because the postulates of homology are used to generate a phylogeny that in turn informs us of ancestry. Prior to the growth of Hennigian methods (10, 43, 112), most systematists based their classifications and evolutionary scenarios upon several critical characteristics that were believed a priori to have some special importance much greater than other characters. The amnion is not a particularly flamboyant structure, nor is it even present for most of an animal's life, but its influence in taxonomy is great because it is a "good" character. To maintain the postulate of homology among all amnia, we sacrifice any other character. This amounts to saying that there is a way to tell a "true" homology from mistaken homologies (analogies) in the absence of phylogenetic information. Although this view is represented here in an extreme that would apply better to scientists of a century ago, it is presented by Wagner (98, p. 62) who felt that "self-regulatory mechanisms of organ differentiation" were the key.

Congruence with Other Data

Opposing the traditional view, the Hennigian perspective dictates that because we cannot observe the relevant ancestor directly, there is no way to know what are the "true" homologies. Because we cannot confirm or refute any hypothesis of homology for a given character directly, we have no choice but to rely on some kind of indirect method to evaluate the postulate of homology for each character. If characters are allowed to weight themselves, "true" homologies (synapomorphies for relevant clades) will be concordant and will support each other, while analogies will not form a pattern (24, 79). The correct statements of homology will win if the characters are allowed to fight among themselves for the simplest resolution of discordant evidence. First, the Hennigian systematist uses the same methods systematists have always used to infer homology (below; "primary homology" sensu de Pinna, 24), then s/he observes the most parsimonious phylogenetic tree and deduces from it, post facto, which of the a priori determinations were correct and which were incorrect. Some methods allow characters to adopt the weights suggested by the phylogeny they produce (12). The advantage of this recursive procedure is that we simply do our best and see what happens.

This Hennigian perspective on homology is applied to behavioral data with increasing frequency, both for reconstructing phylogeny and for understanding the evolution of the behavioral characters themselves (10, 13, 14, 17, 18, 26, 39, 54, 67, 83, 106). Whether the characters are used to generate a cladogram or they are simply plotted on an accepted phylogeny, derivation of the same trait in separate lineages demonstrates a false postulate of homology. Casual critics often remark on the possibility for circular reasoning because the investigator may recode his data to get the result that suits his needs, but this hazard exists in every branch of science. Of course, character coding is by far the most difficult step in the process because it requires explicit statements of how many states there are for each character, whether these states are arranged linearly or in a hierarchical fashion, why the states are polarized as they are, whether all transitions are equally likely or some are favored, whether convergence is more likely than reversal, as well as other more subtle decisions (10). When done properly, the work is repeatable by others less familiar with the taxa and characters, or the points of contention are defined clearly. For this paper, congruence with other data is regarded as the ultimate arbiter of homology, and all other criteria are merely tools to assist in forming the original postulates.

The main disadvantage of this approach is that it operates nearly in reverse of how most ethologists prefer to work. As a young biologist steps into the field to do a comparative study of several species, he usually asks himself, "How do they differ?" For success in the Hennigian paradigm he must ask himself, "How are they the same?"

Morphological Criteria

Morphology may appear to provide a more stable reference than behavior itself, but morphological criteria for homology amount to little more than a case of Remane's (84) "special quality" (below). Neuroethology has progressed to the point where certain behaviors can be mapped onto anatomy (6, 47). Unfortunately, the output of a given system can vary a great deal due to hormonal (52) or other influences. Furthermore, the cascade of neural activities that leads to, or regulates, a behavior is often discovered indirectly so that statements of causality are more inductive than deductive. Bullock (11, p. 408) remarked that "innumerable instances in the literature testify to the temptation to conclude a causal relation which turns out later to be a parallel but distinct system." For the purposes of determining homology among behaviors, the field of neuroethology appears to be not yet mature enough to lend much assistance.

Taking a more macroscopic view, Jander (49) considered motions of legs and antennae to define traits in the grooming behavior of 45 families of insects. The breadth of her study required that the characters be defined broadly, such as "antennae rubbed between upraised forelegs." Whether one leg (Planipennia, Mecoptera, Diptera, Lepidoptera) or both (Megaloptera, Trichoptera) are used is regarded as variation among homologues. Although there is no obvious causal linkage between the morphological and behavioral evolution, the general pattern of evolution of grooming across insect orders corresponds well with phylogeny based upon morphological data. In some cases, behavioral data appear to be more stable than morphological data and show lower average homoplasy across the same taxa (18, 106; A. de Queiroz, P. Wimberger, unpublished information).

More strict application of morphological criteria require that the behaviors cannot be considered homologous unless the precise motions of the same body parts are used. Atz (2) took an extreme position, "the whole neural-behavioral organization of the bird is so unlike that of the mammal that any similarities in behavior between them must be attributed to convergence" (2, p. 65), concluding that "homology is a concept inappropriate for behavioral characters." He rejected behavioral products such as architecture despite its importance in species, genus, and family-level taxonomy of many speciose groups including birds (20), spiders (81), termites (32), caddisflies (85), bees (69), and social wasps (27). Hodos (45) chose to demonstrate his position by reviving Lorenz's classic example of the homology of scratching behavior in tetrapods (also discussed by Atz—2). Following Heinroth (42), Lorenz (60) pointed out that passerine birds do not scratch the back of their heads by doing anything that would seem easy and natural, but rather they dip one shoulder, extend a wing and pass the ipsilateral foot over the back and across the opened wing to reach the head in a most awkward fashion. He proposed that the deep phyletic root of scratching is evident from the fact that birds, dogs, and most other tetrapods pass the hind foot over a shoulder to scratch the ipsilateral ear, a primitive motion which persists in passerines despite the awkwardness. Hodos (45) claimed this example shows that scratching in primates is not homologous with other tetrapods because it is performed with the front foot rather than the hind foot. According to this position, substitution of serially homologous structures for the same function inspired by the same motivation constitutes a novel character (nonhomologous) rather than a derived state of the same old character (homologous). Yet, many morphologists see that substitution of elements does not invalidate a statement of homology regarding the composite structure (3, morphology and fish behavior; 66, insect thorax), so it would be extreme to disallow morphological modification or substitution when studying behavioral traits.

The strictest application of morphological criteria alone assumes that evolution does not take separate paths for morphology and behavior. When this assumption fails, two errors are possible. First, if morphological change occurs under the mantle of homologous behaviors, the behaviors may lose the status of homologues despite the fact that they may be conserved as much as possible. This is implicit in Hodos' (45) view of substitution discussed above. The morphological variation should be seen rather as alternative states of the same character rather than two characters that have no historical connection. When spiders try to localize the sticky spiral during web construction, or they need to support a combing leg, they may use either the outer or inner leg (relative to the hub), and Coddington (18; characters 55 and 61) treated this as alternative states of homologues, not as two different characters.

The second problem with strict morphological criteria is the corollary of the first: if separate and nonhomologous behaviors evolve to take advantage of the same morphological elements, they will lose their identity as independently derived traits. This may be common because behaviors serving other purposes will have to be laid over the anatomy and motor patterns defined for primary functions (5). Barlow (4, p. 228) characterized the problems of the morphological reductionist extreme by stating " . . . it reduces all to one. Thus in the end everything is locomotion, respiration, or feeding. Clearly the motivational world of the animal is more structured than this." The fundamental problem is that there is no assurance that postulates of homology for behavior will be at a level corresponding to the meaningful postulates of homology for morphology.

Remane's Criteria

Statements of morphological homology generally use Remane's (84) classical criteria: position, special quality, and connection by intermediates. Applying Remane's criteria directly to behaviors themselves provides just as accurate, if not more accurate, an assessment of the homology among behaviors than does relying upon Remane indirectly by way of morphology. Remane's three criteria all have ethological counterparts and are already used (perhaps unconsciously) by ethologists.

POSITION Baerends (3) referred to the behavioral analogue of Remane's similarity in position as "similarity in topography" (3, p. 408). He homologized behaviors by using position relative to others in the sequence to find broad patterns common to all species. One of his better examples is that the tail wagging movements of two species of *Tilapia* look somewhat different but are assumed to be homologous because they occur in exactly the same places in courtship ceremonies. Similarly, Tinbergen (96) divided the greeting ceremony of gulls and kittiwakes into arbitrary steps of a progression, homologizing the major phases despite variation across taxa. Simple and easy to understand, this is the weakest criterion of homology. This approach is usually used in combination with other criteria.

SPECIAL QUALITY This is the most useful criterion, though it is also the most difficult to define. If the complex movements of several species take the same distinct form in the same context and appear to be largely innate, they may be thought of as homologues. A morphological criterion for homology may be considered a special case of Remane's "special quality" criterion. The more complicated and distinctive the behavior is, the stronger the postulate of homology will be. Eberhard (28) and Coddington (18) divided web weaving into qualitatively different steps of frame, radius, hub, temporary spiral, and sticky spiral construction, etc, with variants in these categories then regarded as homologues, including taxa that are so derived as to produce a web no longer recognizable as an orb (16). Prum (83) used distinctive, arbitrary portions of displays (e.g. "wing-shiver twist display," or "double-snap jump display") of 21 species of manakins. Eberhard (28), Coddington (18), and Prum (83) homologized characters across taxa, but special quality can also be used to homologize elements within the repertoire of an animal. Tinbergen

(93, 95) and Daanje (23) homologized certain ritual behaviors of territorial and sexual displays with ordinary preening, feeding, and locomotory motions.

Motivation as quality Cichlids use identical components in such different behaviors as territorial display and calling fry, and Baerends' (3) explanation relied upon the idea that these contextually different situations share a homologous special quality—motivational conflict. Although the precise functions of defense or calling differ, the animal must resolve a conflict between advertising itself (to the intruder, or the young) and departing (to flee from combat, or to relocate the young). This emphasis on motivational factors and internal "drive" was common among early ethologists (23, 93) and is retained in some recent work. Mundinger's (75) "criterion 3" explicitly considered motivation to be separate from Remane's criterion of special quality.

Function as quality As with the morphological criteria (above), the exact motions do not always allow definition of homology, but "special quality" may be fulfilled partly through distinctive function. A broadly defined attribute such as host plant choice or the like (10, 99) may have some complex physiological foundation suggesting homology by special quality despite differences in motor patterns. If the same distinct function is served by different series of acts in different species, then the composite behaviors may be worth comparing as different states of the same trait. Such cases would include mating behaviors, territorial behaviors, and other ritualized behaviors that vary in the reduced components of the ritual but serve the same function. Tinbergen's (96) implicit assumption of homology between the parts of the greeting ceremony display seems plausible because the study was constrained to identical context in closely related taxa; therefore the different behaviors are homologous at the functional level of "greeting ceremony."

However, homologizing traits of "special quality" through function is more difficult than it first appears. Identical behaviors in identical circumstances pose no problem, but what of identical behaviors in very different contexts? Beer (5) showed a table of 15 display elements and their various combinations that give rise to 12 functional behaviors in laughing gulls, demonstrating that exclusive correspondence between display and function does not exist for these birds; Beer stated that as a practical matter "ethologists have used criteria of form rather than function to decide what is to count" (5, p. 49).

Many scientists equate function with adaptation, and the latter thereby serves to identify behavioral variation and its purpose: "diverse patterns of behavior, from foraging to mating, are being understood in terms of their individual selective benefits" (87, p. 3). People studying behavioral ecology and adaptation may benefit from studying traits defined by function, but students of phylogeny and homology should be wary of such traits. Certainly, the adaptational postulates of homology will be no less vague, axiomatic, and circular than the definitions of adaptation often are (17, 38). Furthermore, behaviors are specifically believed to be more plastic than morphology and to play an important role in evolution (101, 107), suggesting that functionally defined characters may appear repeatedly and are more likely to be convergent or broadly overlapping than are morphological adaptations (30, 105).

Functional definitions of characters are sometimes difficult to rationalize. Wiggins & Wichard (111) proposed a new classification of Trichoptera (caddisflies) based upon a functional and adaptive view of the cocoon or larval case, but this view is evidently not congruent with other data (102). It should be emphasized that while one should be cautious from the point of view of homology and phylogeny about adaptively defined traits, the general aversion to using phylogenetic perspectives in adaptive story-telling (25) is without foundation. Cladistic methodology itself has nothing to say about the presumed process by which characters became what they are today; it merely states the most parsimonious resolution of the observed facts and patterns (19).

The significance of functional and adaptive information for phylogenetic studies is at best indecisive. Special quality defined by function will be no better. Bock (8, p. 12) declared that "the morphological description of taxonomic characters cannot be done in the absence of proper functional understanding because no guide would exist to the functionally and adaptively significant aspects of these features." Cracraft (22) denied the idea that functional information contributes anything to either the postulate or evaluation of homology, and wrote, "that there may be one or more functions for a structure does not imply an associated stereotypic behavior pattern for each function; it is the behavior pattern, then, and not the function, which constitutes a systematic character" (22, p. 23). Beer (5, above) would likely agree. Lauder (55, p. 322, 323) took a position in the middle, stating that "functional characters are not different from any other attribute that might contribute to our understanding of genealogy", that "functional characters may be homologous or convergent (just like morphological features)", and that "function does not take primacy in homology decisions."

CONNECTION BY INTERMEDIATES Continuity of intermediates is used widely by ethologists, usually to build an evolutionary scenario that runs between two poles. When Tinbergen (93) and Daanje (23) homologized displacement and intention movements with preening, feeding, and locomotory motions, they supplied a number of species and contexts with intermediate degrees of ritualization to connect the highly ritualized forms with their putative origins. One hazard with this approach is that even if the postulate of homology is correct, polarity may be determined incorrectly because there is no assurance that the inferred "ethocline" actually matches the path of evolution (70). Lorenz (58, 62) reminded his readers that the "intermediates" are actually evolutionary terminals in their own right. This is demonstrated by Kessel's (50) logical spectrum explaining the bizarre development of worthless (but not meaningless) nuptial offerings in balloon flies. Recent phylogenetic study (14) shows that a branching pattern is more appropriate than an ethocline for these traits. The evolution of spider webs (15) presents a similar problem in which the so-called intermediates confuse the interpretation of polarity.

If homology is suggested by different species showing intermediate forms, then it might be rejected by a single species showing both forms together. This is Patterson's "conjunction test" (79, p. 38). The test may not provide unequivocal rejection of homology even for morphological situations (24), and I do not believe that the test will prove useful for behavioral problems. As mentioned previously for displacement behaviors (93) or intention movements (23), an animal may incorporate a behavior from one part of its repertoire into another suite of behaviors. Motions derived from feeding, preening, or locomotory functions become established as part of a territorial or mating ritual, and the same animal displays both the original behavior and the behavior derived from it. Behavioral evolution does not require complete replacement.

Relative Utility of the Criteria

Those who live in the realm of pure theory often disagree with those who work with the practical problems of simply getting the job done. An example of this schism is the dispute about what criteria serve to establish homology in behavioral traits. Table 1 shows a brief list of some of the major contributions in which behavioral variation is discussed in a phylogenetic context. Although the list is not exhaustive, such papers are surprisingly rare, and one sees that the "general discussion" papers all rely primarily on morphology while those dealing with the phylogeny of a particular group use other criteria. Chronology shows no evidence of resolution of this problem except perhaps in that the most modern papers show a Hennigian focus in which the final arbiter of homology is congruence with the other data at hand.

UNITS OF BEHAVIOR

Recent developments in biology have led to a revolution in our ability to examine DNA directly and to make use of this molecular information. Because we can now read the genetic code itself, why shouldn't we focus all our attention on the molecular record of evolution? The reason we should not is that biology is full of properties that emerge in an unforeseeable manner from a combination of lesser units, properties that are not inherent to any of the constituent parts. The very breath of life itself and the living world in all its richness demonstrate that the "whole" can be much more than the sum of its parts. We may be able to explain post facto why the structure of a given

Table 1 Criteria of behavioral homology used in several studies, in chronological order, beginning 1951.

Taxa	Traits used	Homology criteria[a]	Author
Diverse birds	Locomotion, ritual motions	Remane	Daanje (23)
4 genera of Balloon flies	Courtship	Connection by intermediates	Kessel (50)
General discussion		Morphology, motivation, Remane	Baerends (3)
Many cricket species	Songs	Special quality	Alexander (1)
Red jungle fowl	Entire repertoire	Ontogeny, special quality	Kruijt (53)1
22 families of caddisflies	Architecture, weaving	Special quality, ontogeny	Ross (85)
45 families of insects	Grooming	Morphology, special quality	Jander (49)
General discussion		Morphology, Remane	Atz (2)
General discussion		Morphology	Hodos (45)
12 subfamilies of Balloon flies	Courtship	Congruence with other data	Chvála (14)
4 genera of finches	Songs	Motivation, Remane	Mundinger (75)
7 subfamilies of spiders	Architecture, weaving	Special quality, morphology	Eberhard (28)
7 genera of sunfish	Feeding	Congruence with other data	Lauder (54)
5 genera of yellow-jackets	Architecture, social traits	Congruence with other data	Carpenter (11a)
7 species of fruit flies	Courtship	Congruence with other data	Grimaldi (39)
6 genera of social wasps	Architecture, social traits	Congruence with other data	Carpenter (12a)
7 genera of stickle-backs	Courtship	Congruence with other data	McLennan et al (67)
13 species of sandpipers	Songs	Special quality	Miller et al (72)
16 families of caddisflies	Architecture	Special quality	Wiggins & Wichard (111)
19 genera of spiders	Architecture, weaving	Congruence with other data	Coddington (18)
21 species of man-akins	Courtship	Congruence with other data	Prum (83)
24 families of caddisflies	Architecture	Congruence with other data	Weaver (102)
29 genera of paper-wasps	Architecture	Ontogeny, congruence with other data	Wenzel (105, 106)

[a] Remane's criteria (placement, special quality, connection by intermediates) are listed separately unless an author used more than one, in which case only "Remane" is listed. In many cases there was no explicit statement of criteria used.

system permits a certain behavior, but knowing the structure alone is usually not enough to predict how the system will behave. Understanding oral anatomy does not predict language, and the molecular code to build that anatomy is still further away from the spoken phrase. Although we must have a certain combination of genes to permit us to speak, there is likely no gene for speech.

Contributing to the emergent nature of behavior, and hence the need to be flexible in recognizing units, is the contextual significance of the act in question. Behaviors that originate in one context may gain additional significance in a different context, perhaps leading to an appropriation of the behavior for the new function. Tinbergen (93) and Daanje (23) found that for elements of displays co-opted from other parts of the behavioral repertoire, the same motion in a different context bears a different message. An old motor pattern in a new context can reflect an historical event that is informative about phylogeny. Here, it is not the motor pattern per se that provides the synapomorphy for a given group, but rather the new context in which the borrowed behavior was expressed and its new significance. Reducing the sequence of behaviors to its smallest components may discard the epiphenomenon that is the trait.

Russell et al (88) made a bold effort to classify behaviors according to the "fixed action pattern" (4), which they redefine as "Act," in an effort to outline precise concepts about the units of behavior. The scheme requires so much knowledge about neural aspects of messages and channels that it is impractical outside a physiological study of behavior. Furthermore, defining traits according to their functions is problematic and not necessarily related to phylogenetic history (see Special Quality, above), which is more the subject of the present review. Regardless of the enthusiasm Russell et al show for conceptual issues, the main problem with our understanding of behavior is not that we lack precise concepts, but rather that we lack adequate data from enough taxa.

Far more practical and useful as a guide for recording and documenting behavior is Miller's (73) clear, sophisticated discussion richly detailed with examples from studies of avian behavior. In agreement with the theme of the present paper, Miller states that "'natural' units are as elusive in ethology as in taxonomy" (73, p. 353). We cannot rely upon fixed action patterns alone because such rigid behaviors may exist mostly in the minds of observers. In reality, the defining characteristics of a fixed action pattern (that the behavior is innate, invariant, released by an external trigger stimulus, and that it cannot be reduced to successive responses that depend on different external stimuli for modulation) are rarely satisfied under thorough experimentation (56). Barlow (4, p. 222) concluded that "there is no single case where all the criteria have been properly tested and fulfilled." In any case, reliance upon fixed action patterns differs little from reliance upon morphological criteria of homology (above), so failure to demonstrate the same neuromuscular activity is inadequate for rejection of homology (above). I agree with Barlow (4, p. 217, 218) that "any unit is appropriate if it fits the needs of the study," and that "models based on the behavior of individual neurons can only be regarded as starting points."

The problem of units then relies on the needs of the study (95, p. 189 ff). Genetical studies can help to define components of a system (7), but dissolving everything to "ethons," the smallest units of behavior, will not likely serve all purposes. At the lower limit, four or fewer amino acid substitutions in a less conserved region of the *per* locus are known to control the difference between species-specific wing rhythms in *Drosophila melanogaster* and in *D. simulans* courtship display (109). However, variation in the genetic architecture between species does not address the question of whether or not rhythmic wing motion is homologous as such across these species, hence, a synapomorphy for these flies at a higher level in Drosophilidae. Rapid wing vibration is a primitively homologous state for the related genus *Zygothrica*. (39, Table 3, Character 3), and lack of vibration is derived separately in different species (39, fig. 115). These absences of wing vibration are not homologues, regardless of what the identity of nucleotides might be.

It seems likely that the appropriate units will be larger as a study encompasses more taxa. The exact motions of pollen collection would be relevant to a study of different genera or species of bees. For distinguishing families of bees from each other, the character will become larger such that details that converge or vary among many genera do not obscure basic, gross differences such as "pollen carried in crop" or "on legs" or "on ventral scopa." Still larger, "collects pollen" is a unit adequate to distinguish all families of bees from wasps, flies, and beetles. If bees appropriated ordinary insect grooming behaviors for the purpose of manipulating pollen, then pollen carrying and grooming are homologues and synapomorphic for bees, wasps, flies, and beetles to show they share a more recent common ancestry with each other than with spiders or crabs. The appropriate unit changes with the scale of the problem. As the units become increasingly coarse, the chance of lumping analogous traits increases. Furthermore, there is nothing to exclude the possibility that individual fine motor patterns may have broad phylogenetic significance.

Several levels of units may be required in a single study. Mowry et al (74) dissect in detail the behaviors surrounding oviposition in the onion fly, drawing a flow chart with probabilities assigned for stepping from one behavior to another. The study was not searching for homology, but the results are informative about the need for units in a variety of levels. The model includes a string of five separate boxes that are linked in sequence by probabilities of 100%. Such linkage makes the sequence a candidate for "fixed action pattern" in the classical sense (above), in which case it is a meaningful unit on its own even if the components are justified as individual elements by virtue of appearing separately elsewhere in the repertoire. Mowry et al also include variables to account for emergent properties only visible from afar and not inherent to any of the reduced components of the model, such as seven separate probabilities for repeating "inter-egg subsurface probing" based on whether the female is trying to lay her second, third, fourth, or up to eighth egg. This relationship between probability of repetition and the number of times the fly has cycled through the model becomes a trait by itself. Thus, Mowry et al have found at least three levels: the individual components, the combined fixed action pattern, and the dynamic nature of these behaviors under iteration.

The meaningful level of analysis from a phylogenetic viewpoint will be the level that shows characteristics shared among several but not all taxa. The appropriate units of comparison will have to be distributed more broadly than species-specific traits unless there is a credible transition series between the traits. It may eventually prove to be the case that many so-called species-specific behavioral traits are, in fact, shared by a number of species. This is true for architectural constructs (21), a class of behavioral characters (or at least surrogates of behavioral characters) for which variation is measured more easily, and perhaps interpreted more easily, than most. In speciose groups like sweat bees (31, 89), paper wasps (103, 105), or spiders (30), intraspecific architectural variation is broad and phyletic convergence of architecture may be rampant. Eberhard (30, p. 342) wrote of spider webs: "the impression of species-specificity may usually, however, be the product of lack of information . . . . Given the long-standing and repeated documentation of substantial *intra*specific variation . . . species-specificity will be uncommon" (emphasis in original). Marler (64) pointed out that bird calls with certain functions, such as aggressive and alarm calls, should be (and appear to be) convergent rather than distinctive because they are most useful when they are understood by many species.

The search for species-specific units is not only useless from the point of view of phylogenetic analysis (assuming the species are already identified), but it may also be ill-advised on the grounds that it deliberately or inadvertently obscures phylogenetically informative elements common to several species. Repeated claims of specificity of behavioral units in such well known systems as bird courtship rely in part on a philosophical argument that the ritual "requires each step to be performed in a specific manner and in a specific place in the order, failure of an individual to follow the script will deny parenthood" (5, p. 18). Note that this opinion dictates only conservation of established ritual, not necessarily species-specificity across many taxa. Indeed, Miller and collaborators (71, 72) have found that the breeding vocalizations of certain birds are a mix of species-specific and more broadly distributed traits. Moreover, Heinroth's (42) empirical work, arguably the origin of ethology, documents a heterospecific mated pair in which the female Egyptian goose (*Alopochen aegyptaicus)* and the male Muscovy duck (*Cairina moschata*) managed to combine very different body motions and to learn those of the mate in order to form their own ritual and a bond lasting for years (42, p. 667). Clearly, these animals were not restricted to expressing or recognizing unit behaviors unique to their own species or genera, even in the most "species-specific" context.

Learning

Learning presents both empirical and theoretical problems in behavioral biology. It is outside the scope of this paper to deal with the enormous literature on learning except to examine its relevance to the definition of a phylogenetic trait. The great "nature versus nurture" debate centers on the idea that if something is learned, then it cannot be genetically based. It follows that if a behavioral trait is not strictly heritable, then it is useless in phylogenetic reconstruction or interpretation of large-scale evolution because the determinations of homology will be weak or false; learned behaviors are poor characters, innate behaviors are good. This perspective is deficient for at least three reasons: (i) "learned" and "innate" have largely evaded definition; (ii) how something is learned rather than what is learned may be phylogenetically informative; (iii) criteria of homology rely more on simple persistence of the character across taxa than on details of genetic architecture and heritability that produce it. The first two of these objections will be examined immediately below, the third is discussed above (see UNITS OF BEHAVIOR).

The great debate several decades ago about the nature of learning seems to have been largely abandoned by behavioral biologists today. Originally, the opposite of learned behaviors were the innate behaviors defined by the "fixed action pattern." Lorenz (59, 61), and in the early days Tinbergen (94) argued that animals were capable of certain behaviors that were complete and distinct from the first time they were executed. These patterns of motion were "fixed" because they were performed in the presence of specific stimuli and were not subject to change by repeated performance. Lehrman (56) and others (41) attacked this idea because under certain experimental regimes the fixed action patterns failed to be expressed according to the rigid description required by Lorenz's model. Therefore, learning, or at least some undefined process of neural ontogeny related to experience, did play a role. Lorenz (61) defended his extreme position: "The naïve ethologist's assertion that [bird flight and orientation] are 'completely innate' is less inexact than the statement that a steam locomotive or the Eiffel Tower are built entirely of metal" (61, p. 27). Nonetheless, Lehrman's critique convinced Tinbergen (44) to retreat to a middle ground that included an interaction between inherent and environmental factors, eroding from one end the dichotomy between "innate" and "learned."

Recent work refers largely to learned behaviors such as those of Pavlov's dogs or Skinner's pecking pigeons, which are subject to classical conditioning. Skinner's school of psychology proposed that anything could be learned if an appropriate reward/punishment schedule was used, and the early success of this approach helped define modern psychology (92). However, dissent includes the work of Garcia et al (35), who showed that laboratory rats (relatively intelligent animals) can learn to associate flavor of pellets with illness, or size of pellets with shock, but not flavor with shock or size with illness (outwardly simple lessons). Whether or not the animal learns is in part due to an inherent propensity to learn given tasks (44). The innate/learned dichotomy is eroded from the other side. "Innate" cannot always be separated from "learned," and thorough studies of ontogeny of behavior show continual interaction (41, p. 127; 45).

At the same time that the definition of learning presents a problem for researchers who want to exclude learned behaviors from phylogenetic analysis, the details of the learning process provide additional fodder for researchers who want to include learned behaviors. The phenomenon of imprinting (57) has long been taken as a prime example of learning outside of classical conditioning (above), and Lorenz's success at getting geese to follow him as they would their mother is legendary. Here, one might say that imprinting is a "learned" phenomenon because the cues were learned by all the goslings. Such a declaration would be inaccurate, however, because only the cues are learned, not the process by which imprinting occurs. Imprinting per se is strictly controlled such that during a certain window of time anything presented appropriately will be then and forever identified as the relevant cue. In this case, discussions of homology should consider not only "learns cue," but also "anything observed in the imprinting window is learned to be the cue." The imprinting window is brought to the foreground for study, where details of the learning process may prove to be richer in information than is the observation that something is learned. Highly canalized learning via imprinting has been demonstrated in kin recognition in many groups (34), and behavioral biologists must be careful not to confuse the rigid (innate) process with the variable (learned) end point. In an unusually clear and explicit discussion of behavioral traits from a phylogenetic viewpoint, Mundinger (75) found that vocalizations in two large clades of perching birds—the oscines and sub-oscines—are largely learned or innate, respectively. He concluded that learned behaviors are valuable taxonomically: "call learning" is itself a character useful at the level of subfamily, while details of the calls are useful at lower taxonomic levels.

Perhaps the best demonstration that learning does not preclude postulates of homology comes from studies of the acquisition of intellect in humans. Piaget's (82) scheme, generally taken as the most thorough, coherent, and best supported by data (97), identified four stages through which children pass during the development of adult logical intelligence: (i) ten months to two years of age: sensory-motor exploration, Copernican revolution that "I am one object among many"; (ii) two years to seven or eight years: language acquisition, semiotic and symbolic development but without deductive logic; (iii) seven or eight years to 11 or 12: chains of causality but only for objects, hypothetical propositions still often unintelligible; (iv) 11 or 12 years and later: hypothetical reasoning, ability to see necessary consequences without having to decide upon validity or falsehood of component parts of a proposition. Only very few things are learned by direct experience alone, such as that the largest object is not always the heaviest. Piaget (82) observed the same process of learning in children that are somewhat mentally handicapped, or deaf-mute, or blind, or from cultures lacking written language, and recent research (68, 80) confirms the universality of this strict process for learning language. Thus, the ontogeny of intelligence is rather strictly controlled and similar in all children despite great differences in experience or absolute sensory and mental capacity. From the point of view of behavioral homology, the universality of this developmental pattern suggests that it has been inherited from a common ancestor, that learning language is a homologous process in all humans. There is no reason to discard any trait as useless to the illumination of homology and phylogeny because learning plays a role.

Ontogeny

Ontogeny continues to be a major and controversial resource for evolutionary biologists (37, 48). Behavioral ontogeny is generally regarded as a process of maturation by which an animal's age correlates with its performance. Learning or other interactions with the environment usually play some role (40). One approach is to document behaviors that appear at each age. Kruijt (53) showed how growing chicks acquire new behaviors, and that units of this inventory later interact to become composite behaviors in the adult repertoire. In some social insects, each adult specializes in a given task that differs with age; thus the adult repertoire itself paces through a schedule of separate behaviors necessary for colony maintenance (91, 113). An alternative approach, focussed more narrowly, is to describe the development of a single behavior by showing how experience or age effects a change in the animal's reactions toward various stimuli, change in the same motor patterns, or change in the perceptual system (41). Still more reductionistic is to show how a certain perceptual system changes with time at the cellular level (65).

Another perspective is to regard the sequential components of a complex behavior as if they were developmental steps toward the larger unit. The components can be compared across taxa as homologues according to their distinctive qualities and position in the sequence. Unlike processes of maturation (above), the sequence may be repeated many times, or the units reassembled to produce another complex behavior in a different context (5). Tinbergen (96, Figure 15) homologized the component steps of the greeting ceremony of various gulls. The greeting ceremony varies across taxa but can be shown to be composed of similar elements: "Thus all the Obliques are undoubtedly of common descent, and so are all forms of Jabbing, of Choking and of Facing Away." Ross (85) used spinning and construction sequences to discuss the evolution and elaboration of larval cases and cocoons in aquatic Lepidoptera and Trichoptera. Wenzel (105, 106) used classical ontogenetic changes of deletion, compression, and terminal addition in the nests of paper wasps to demonstrate hierarchical order loosely matching major clades inferred from morphological characters. None of these authors referred to the sequences as literal ontogenies, but they treated them just as morphologists have treated true ontogenies: steps are homologized across taxa, changes in speed or position constitute a departure from the ancestral pattern. Whether or not more explicit application of ontogenetic principles to large-scale behavioral problems (104–106) will be accepted or generally useful remains to be demonstrated.

Tinbergen (95) evidently felt that knowledge of the ontogeny of behavior would provide important understanding. Judging from the labors of morphologists, this may be optimistic. Similar ontogeny is often taken as evidence of morphological homology, but Wagner (98) argued that variability in developmental patterns is so great as to make irrelevant the information regarding cellular origins or inductive stimuli. Ontogenetic patterns are sometimes used as a substitute for the now widely used outgroup criterion (100) to infer the polarity of evolutionary change ("primitive" versus "derived") in morphological characters (76, 77, 108, 110). However, the ontogenetic polarization sometimes fails to match decisions derived by other means (51, 63), as one can easily imagine if the derived condition involves a deletion of terminal stages to produce a neotenic phenotype. Cases in which strict behavioral ontogeny, sensu Kruijt (53) or Hailman (41), is used to infer polarity among alternative states are rare (but see 29), although some can be found if "ontogeny" is broadened to include long sequences toward a composite whole (33, 85, 96, 105, 106). Nonetheless, such cases are unlikely to be any less controversial than those from morphology.

CONCLUSIONS

The tools that our predecessors left for us are adequate for determination of homology, although no single one is sufficient when used alone. Congruence with other data is probably the best support for a statement of homology when many traits in many taxa are already known. Strict morphological criteria for behavioral homology are inadequate when evolution acts separately on morphological and behavioral traits. Remane's criteria of placement, special quality, and connection by intermediates are useful in postulating behavioral homology. Functional or adaptive definitions of traits can be misleading. Arbitrary units of behavior are not a major problem. Analysis of relatively more learned behaviors is more complicated than for relatively more innate behaviors, but it is still possible from a phylogenetic viewpoint. Ontogeny of behavior has not yet contributed much to our understanding of larger evolutionary issues.

ACKNOWLEDGMENTS

Many colleagues contributed to this paper through correspondence and casual discussions over several years. Especially helpful were comments by B. A. Alexander, J. M. Carpenter, J. A. Coddington, W. G. Eberhard, R. Jander, G. V. Lauder, C. D. Michener and J. E. Miller. I thank B. A. Alexander, J. A. Coddington, C. D. Michener, and D. Willer for improving the manuscript. This paper was supported in part by NSF grant BSR-9006102 to J. M. Carpenter.

Literature Cited

1. Alexander, R. D. 1962. The role of behavioral study in cricket classification. *Syst. Zool.* 11:53–72
2. Atz, J. W. 1970. The application of the idea of homology to behavior. In *Development and Evolution of Behavior,* ed. L. R. Aronson, E. Tobach, D. S. Lehrman, J. S. Rosenblatt, pp. 53–74. San Francisco: Freeman
3. Baerends, G. P. 1958. Comparative methods and the concept of homology in the study of behavior. *Arch. Neerland. Zool. Suppl.* 13:401–17
4. Barlow, G. W. 1968. Ethological units of behavior. In *The Central Nervous System and Fish Behavior,* ed. D. Ingle, pp. 217–32. Chicago: Univ. Chicago Press
5. Beer, C. G. 1975. Multiple functions and gull display. In *Function and Evolution in Behaviour,* ed. G. Baerends, C. Beer, A. Manning, pp. 16–54. Oxford: Clarendon
6. Bentley, D. R., Hoy, R. R. 1970. Postembryonic development of adult motor patterns in crickets: a neural analysis. *Science* 170:1409–11
7. Bentley, D. R., Hoy, R. R. 1972. Genetic control of the neuronal network generating cricket *(Teleogryllus Gryllus)* song patterns. *Anim. Behav.* 20:478–92
8. Bock, W. J. 1981. Functional-adaptive analysis in evolutionary classification. *Am. Zool.* 21:5–20
9. Boyden, A. 1947. Homology and analogy. *Am. Midl. Nat.* 37:648–69
10. Brooks, D. R., McLennan, D. H. 1991. *Phylogeny, Ecology, and Behavior.* Chicago: Univ. Chicago Press
11. Bullock, T. H. 1983. Epilogue: neurobiological roots and neuroethological sprouts. See Ref. 47, pp. 401–12

11a. Carpenter, J. M. 1987. Phylogenetic relationships and classification of the Vespinae (Hymenoptera: Vespidae). *Syst. Entomol.* 12:413–31

12. Carpenter, J. M. 1988. Choosing among equally parsimonious cladograms. *Cladistics* 4:291–96

12a. Carpenter, J. M. 1988. The phylogenetic system of the Stenogastrinae (Hymenoptera: Vespidae). *J. New York Entomol. Soc.* 96:140–75

13. Carpenter, J. M. 1989. Testing scenarios: Wasp social behavior. *Cladistics* 5:131–44
14. Chvála, M. 1976. Swarming, mating and feeding habits in Empididae (Diptera), and their significance in evolution of the family. *Acta Entomol. Bohemoslov.* 73:353–66
15. Coddington, J. A. 1986. The monophyletic origin of the orb web. In

Spiders: Webs, Behavior, and Evolution, ed. W. A. Shear, pp. 319–63. Stanford, Calif: Stanford Univ. Press
16. Coddington, J. A. 1986. Orb webs in "non-orb weaving" ogre-faced spiders (Araneae: Dinopidae): a question of genealogy. *Cladistics* 2:53–67
17. Coddington, J. A. 1988. Cladistic tests of adaptational hypotheses. *Cladistics* 4:3–22
18. Coddington, J. A. 1990. Cladistics and spider classification: araneomorph phylogeny and the monophyly of orbweavers (Araneae: Araneomorphae; Orbiculariae). *Acta Zool. Fenn.* 190: 75–87
19. Coddington, J. A. 1990. Bridges between evolutionary pattern and process. *Cladistics* 6:379–86
20. Collias, N. E. 1964. The evolution of nests and nest-building in birds. *Am. Zool.* 4:175–90
21. Collias, N. E. 1964. Summary of the symposium on "The evolution of external construction by animals." *Am. Zool.* 4:241–43
22. Cracraft, J. 1981. The use of functional and adaptive criteria in phylogenetic systematics. *Am. Zool.* 21:21–36
23. Daanje, A. 1951. On locomotory movements in birds and the intention movements derived from them. *Behaviour* 3:48–98
24. de Pinna, M. C. C. 1991. Concepts and tests of homology in the cladistic paradigm. *Cladistics* 7:367–94
25. de Pinna, M. C. C., Salles, L. O. 1990. Cladistic tests of adaptational hypotheses: a reply to Coddington. *Cladistics* 6:373–77
26. Dobson, F. S. 1985. The use of phylogeny in behavior and ecology. *Evolution* 39:1384–88
27. Ducke, A. 1913. Uber Phylogenie und Klassifikation der Sozialen Vespiden. *Zool. Jahrb. Abt. Syst.* 36:303–30
28. Eberhard, W. G. 1982. Behavioral characters for the higher classification of orb-weaving spiders. *Evolution* 36: 1067–95
29. Eberhard, W. G. 1986. Ontogenetic changes in the web of *Epeirotypus* sp. (Araneae, Theridiosomatidae). *J. Arachnol.* 14:125–28
30. Eberhard, W. G. 1990. Function and phylogeny of spider webs. *Annu. Rev. Ecol. Syst.* 21:341–72
31. Eickwort, G. C., Sakagami, S. F. 1979. A classification of nest architecture of bees in the tribe Augochlorini (Hymenoptera: Halictidae; Halictinae), with description of a Brazilian nest of *Rhinocorynura inflaticeps*. *Biotropica* 11:28–37
32. Emerson, A. E. 1938. Termite nests—a study of the phylogeny of behavior. *Ecol. Monogr.* 8:247–84
33. Evans, H. E. 1958. The evolution of social life in wasps. *Proc. Tenth Int. Congr. Entomol., Montreal* 2:449–57
34. Fletcher, D. C., Michener, C. D. 1987. *Kin Recognition in Animals*. New York: Wiley
35. Garcia, J., McGowan, B. K., Ervin, F. R., Koelling, R. A. 1968. Cues: their relative effectiveness as a function of the reinforcer. *Science* 160:794–95
36. Ghiselin, M. T. 1976. The nomenclature of correspondence: A new look at "homology" and "analogy." See Ref. 65a, pp. 129–42
37. Gould, S. J. 1977. *Ontogeny and Phylogeny*. Cambridge, Mass: Harvard Univ. Press
38. Gould, S. J., Lewontin, R. C. 1979. The spandrels of San Marco and the adaptationist paradigm: A critique of the adaptationist programme. *Proc. R. Soc. London Ser. B* 205:547–65
39. Grimaldi, D. A. 1987. Phylogenetics and taxonomy of *Zygothrica* (Diptera: Drosophilidae). *Bull. Am. Mus. Nat. Hist.* 186:103–268
40. Groothuis, T. 1992. The influence of social experience on the development and fixation of the form of displays in the black-headed gull. *Anim. Behav.* 43:1–14
41. Hailman, J. P. 1967. The ontogeny of an instinct. *Behaviour Suppl.* 15:1–159
42. Heinroth, O. 1910. Beiträge zur Biologie, namentlich Ethologie und Psychologie der Anatiden. *Verh. Int. Ornithol. Kongr.* 5:589–701
43. Hennig, W. 1966. *Phylogenetic Systematics*. Chicago: Univ. Ill. Press
44. Hinde, R. A., Tinbergen, N. 1958. The comparative study of species-specific behavior. In *Behavior and Evolution,* ed. A. Roe, G. G. Simpson, pp. 251–68. New Haven, Conn: Yale Univ. Press
45. Hodos, W. 1976. The concept of homology and the evolution of behavior. See Ref. 65a, pp. 153–67
46. Hubbs, C. L. 1944. Concepts of homology and analogy. *Am. Nat.* 78:289–307
47. Huber, F., Markl, H., eds. 1983. *Neuroethology and Behavioral Physiology*. Berlin: Springer-Verlag
48. Humphries, C. J. 1988. *Ontogeny and Systematics*. London: Br. Mus. (Nat. Hist.)
49. Jander, U. 1966. Untersuchungen zur Stammesgeschichte von Putzbew-

egugen von Tracheaten. *Z. Tierpsychol.* 23:799–844

50. Kessel, E. L. 1955. The mating activities of balloon flies. *Syst. Zool.* 4:96–104
51. Kluge, A. G. 1985. Ontogeny and phylogenetic systematics. *Cladistics* 1: 13–27
52. Kravitz, E. A. 1988. Hormonal control of behavior: amines and biasing of behavioral output in lobsters. *Science* 241:1775–81
53. Kruijt, J. P. 1964. Ontogeny of social behavior in Burmese red junglefowl (*Gallus gallus spadiceus*) Bonnaterre. *Behaviour Suppl.* 12:1–201
54. Lauder, G. V. 1986. Homology, analogy, and the evolution of behavior. In *Evolution of Animal Behavior,* ed. M. H. Nitecki, J. A. Kitchell, pp. 9–40. New York: Oxford Univ. Press
55. Lauder, G. V. 1990. Functional morphology and systematics: Studying functional patterns in an historical context. *Annu. Rev. Ecol. Syst.* 21:317–40
56. Lehrman, D. S. 1953. A critique of Konrad Lorenz's theory of instinctive behavior. *Q. Rev. Biol.* 28:337–63
57. Lorenz, K. Z. 1935. Der Kumpan in der Umwelt des Vogels. *J. Ornithol.* 83:137–213, 289–413. See transl. Ref. 90, pp. 83–128
58. Lorenz, K. Z. 1941. Vergleichende Bewegungsstudien an Anatinen. *J. Ornithol.* 89:24–32
59. Lorenz, K. 1957. The nature of instinct. See Ref. 90, pp. 129–75
60. Lorenz, K. Z. 1958. The evolution of behavior. *Sci. Am.* 199:67–78
61. Lorenz, K. 1965. *Evolution and Modification of Behavior*. Chicago: Univ. Chicago Press
62. Lorenz, K. Z. 1972. The ritualization of display. In *Function and Evolution of Behavior,* ed. P. H. Klopfer, J. P. Hailman, pp. 231–59. London: Addison-Wesley
63. Mabee, P. M. 1989. An empirical rejection of the ontogenetic polarity criterion. *Cladistics* 5:409–16
64. Marler, P. 1957. Specific distinctiveness in the communication signals of birds. *Behaviour* 11:13–39
65. Masson, C., Arnold, G. 1984. Ontogeny, maturation and plasticity of the olfactory system in the workerbee. *J. Insect Physiol.* 30:7–14

65a. Masterton, R. B., Hodos, W., Jerison, H., eds. 1976. *Evolution, Brain, and Behavior*. Hillsdale, NJ: Lawrence Erlbaum

66. Matsuda, R. 1976. *Morphology and Evolution of the Insect Abdomen*. Toronto: Pergamon
67. McLennan, D. A., Brooks, D. R., McPhail, J. D. 1988. The benefits of communication between comparative ethology and phylogenetic systematics: a case study using gasterosteid fishes. *Can. J. Zool.* 66:2177–90
68. Meier, R. P. 1991. Language acquisition by deaf children. *Am. Sci.* 79:60–70
69. Michener, C. D. 1964. Evolution of the nests of bees. *Am. Zool.* 4:227–39
70. Michener, C. D. 1985. From solitary to eusocial: need there be a series of intervening species? In *Experimental Behavioral Ecology and Sociobiology,* ed. B. Hölldobler, M. Lindauer, pp. 293–305. New York: Sinauer
71. Miller, E. H. 1987. Breeding vocalizations of the surfbird. *Condor* 89:406–12
72. Miller, E. H. 1988. Breeding vocalizations of Baird's Sandpiper *Calidris bairdii* and related species, with remarks on phylogeny and adaptation. *Ornis Scand.* 19:257–67
73. Miller, E. H. 1988. Description of bird behavior for comparative purposes. *Curr. Ornithol.* 5:347–94
74. Mowry, T. M., Spencer, J. L., Keller, J. E., Miller, J. R. 1989. Onion fly (*Delia antiqua*) egg depositional behaviour: pinpointing host acceptance by an insect herbivore. *J. Insect. Physiol.* 35:331–40
75. Mundinger, P. C. 1979. Call learning in the Carduelinae: ethological and systematic considerations. *Syst. Zool.* 28:270–83
76. Nelson, G. 1978. Ontogeny, phylogeny, paleontology, and the biogenetic law. *Syst. Zool.* 27:324–45
77. Nelson, G. 1985. Outgroups and ontogeny. *Cladistics* 1:29–45
78. Owen, R. 1843. *Lectures on the comparative anatomy and physiology of the invertebrate animals, delivered at the Royal College of Surgeons, in 1843*. London: Longmans, Brown, Green & Longmans
79. Patterson, C. 1982. Morphological characters and homology. In *Problems in Phylogenetic Reconstruction,* ed. K. A. Joysey, A. E. Friday, pp. 21–74. London: Academic
80. Petitto, L. A., Marentette, P. F. 1991. Babbling in the manual mode: evidence for the ontogeny of language. *Science* 251:1493–96
81. Petrunkevitch, A. 1926. The value of instinct as a taxonomic character in spiders. *Biol. Bull. (Woods Hole, Mass.)* 50:427–32

82. Piaget, J. 1969. *Psychologie et Pédagogie*. Paris: Denoël
83. Prum, R. O. 1990. Phylogenetic analysis of the evolution of display behavior in the neotropical manakins (Aves: Pipridae). *Ethology* 84:202–31
84. Remane, A. 1952. Die Grundlagen des Natürlichen Systems der Vergleichenden Anatomie und der Phylogenetik. Leipzig: Geest und Portig K. G
85. Ross, H. H. 1964. Evolution of caddisworm cases and nets. *Am. Zool.* 4:209–20
86. Roth, V. L. 1988. The biological basis of homology. See Ref. 48, pp. 1–26
87. Rubenstein, D. I., Wrangham, R. W., eds. 1986. *Ecological Aspects of Social Evolution*. Princeton, NJ: Princeton Univ. Press
88. Russell, W. M. S., Mead, A. P., Hayes, J. S. 1954. A basis for the quantitative study of the structure of behaviour. *Behaviour* 6:153–205
89. Sakagami, S. F., Michener, C. D. 1962. *The Nest Architecture of Sweat Bees*. Lawrence, Kans: Univ. Kansas Press
90. Schiller, C. H. 1964. *Instinctive Behavior*. New York: Int. Univ. Press
91. Seeley, T. D. 1982. Adaptive significance of the age polyethism schedule in honeybee colonies. *Behav. Ecol. Sociobiol.* 11:287–93
92. Skinner, B. F. 1938. *The Behavior of Organisms*. New York: Appleton-Century
93. Tinbergen, N. 1940. Die Ubersprungbewegung. Z. Tierpsychol. 4:1–40
94. Tinbergen, N. 1942. An objectivistic study of the innate behaviour of animals. *Biblio. Biotheoret.* 1:39–98
95. Tinbergen, N. 1951. *The Study of Instinct*. Oxford: Clarendon
96. Tinbergen, N. 1959. Comparative studies of the behaviour of gulls (Laridae): a progress report. *Behaviour* 15:1–70
97. Tran-thong. 1970. *Stades et concept de stade de développment de l'enfant dans la psychologie contemporaine*. Paris: Librarie Philosophique J. Vrin
98. Wagner, G. P. 1989. The biological homology concept. *Annu. Rev. Ecol. Syst.* 20:51–69
99. Wanntorp, H. E., Brooks, D. R., Nilsson, T., Nylin, S., Ronquist, F., et al. 1990. Phylogenetic approaches in ecology. *Oikos* 57:119–32
100. Watrous, L. E., Wheeler, Q. D. 1981. The outgroup comparison method of character analysis. *Syst. Zool.* 30:1–11
101. Wcislo, W. T. 1989. Behavioral environments and evolutionary change. *Annu. Rev. Ecol. Syst.* 20:137–69
102. Weaver, J. S. 1992. Remarks on the evolution of Trichoptera: A critique of Wiggins and Wichard's classification. *Cladistics*. 8: In press
103. Wenzel, J. W. 1989. Endogenous factors, external cues and eccentric construction in *Polistes annularis* (Hymenoptera: Vespidae). *J. Insect Behav.* 2:679–99
104. Wenzel, J. W. 1990. Nest design and secondary functions of social insect architecture. In *Social Insects and the Environment,* ed. G. K. Veeresh, B. Mallik , C. A. Viraktamath, pp. 657–58. New Delhi: Oxford & IBH (Abstr.)
105. Wenzel, J. W. 1991. Evolution of nest architecture. In *Social Biology of Wasps,* ed. K. G. Ross, R. W. Matthews, pp. 480–519. Ithaca: Cornell Univ. Press
106. Wenzel, J. W. 1993. Application of the biogenetic rule to behavioral ontogeny: a test using nest architecture of paper wasps. *J. Evol. Biol.* In press
107. West-Eberhard, M. J. 1989. Phenotypic plasticity and the origins of diversity. *Annu. Rev. Ecol. Syst.* 20:249–78
108. Weston, P. H. 1988. Indirect and direct methods in systematics. See Ref. 48, pp. 27–56
109. Wheeler, D. A., Kyriacou, C. P., Greenacre, M. L., Yu, Q., Rutila, J. E., et al. 1991. Molecular transfer of a species-specific behavior from *Drosophila simulans* to *Drosophila melanogaster*. *Science* 251:1082–85
110. Wheeler, Q. D. 1990. Ontogeny and character phylogeny. *Cladistics* 6:225–68
111. Wiggins, G. B., Wichard, W. 1989. Phylogeny of pupation in Trichoptera, with proposals on the origin and higher classification of the order. *J. N. Am. Benthol. Soc.* 8:260–76
112. Wiley, E. O. 1981. *Phylogenetics: The Theory and Practice of Phylogenetic Systematics*. New York: Wiley-Intersci.
113. Wilson, E. O. 1976. Behavioral discretization and the number of castes in an ant species. *Behav. Ecol. Sociobiol.* 1:141–54

Annu. Rev. Ecol. Syst. 1992. 23:383–404

ON COMPARING COMPARATIVE METHODS

John L. Gittleman and Hang-Kwang Luh
Department of Zoology, University of Tennessee, Knoxville, Tennessee 37996-0810

KEY WORDS: comparative methods, evolution, statistics, adaptation, phylogeny

INTRODUCTION

Virtually every field in the biological sciences uses comparative, cross-taxonomic analysis. Unlike experimental study, comparative analyses have historically relied on simple correlation of traits across species. In the past ten years, especially since publication of a few landmark papers (e.g. 13, 24), this straightforward comparative methodology has become obsolete. Three simultaneous developments produced this change. First, accumulation of basic data on many phenotypic traits allows for more taxonomically complete and quantitative study; this is particularly evident in relatively modern fields of behavior and ecology (6, 7, 34, 48, 49, 96). Second, the essential framework of comparative study is more solid than ever: systematic biology has delivered a proliferation of classifications for most animal and plant taxa, including detail of a variety of systematic characters, topology of evolutionary radiations, and statistical boundaries on specific taxonomic schemes. Third, comparative studies now are unavoidably statistical; to paraphrase Felsenstein's (24) proclamation for bringing phylogeny into comparative study, "phylogenies [statistics] are fundamental to comparative biology; there is no doing it without taking them into account" (p. 14). Ignoring statistics in comparative study is tantamount to doing an experiment with no control; quite simply, comparative studies require careful selection of appropriate statistical techniques (66). Each of these developments has revitalized an old methodology which is entering a new phase of application in behavior, ecology, and evolution. Although quite exciting, this also brings considerable confusion to the practitioner who simply wants to search for general trends across species or to test a favorite evolutionary hypothesis against independently derived taxa, both reminiscent of classic comparative biology.

Therefore, the aim of this paper is to present some of the active and controversial issues that must now be confronted in any comparative study.

0066-4162/92/1120-0383$02.00

These include: (i) statistical issues related to detecting phylogenetic relations in cross-taxonomic traits prior to comparative tests; (ii) kinds of comparative data (forms of classification; trait variation) that influence selecting and executing comparative analyses; (iii) evolutionary assumptions inherent in various comparative techniques, using an illustration of two quite different comparative methods; and, (iv) an example of the type of simulation study useful for checking the effect of a particular comparative method on a comparative result. Coverage of these topics does not offer an exhaustive review. In particular, we do not describe methods for analyzing categorical or discrete traits (see 6, 16, 44, 54, 55, 60, 77), discuss philosophical and evolutionary debate about schools of systematics (27, 28, 51, 78, 87, 96), nor detail algorithms used in various statistical procedures. Recent volumes by Harvey & Pagel (44) and Brooks & McLennan (6) critically review much of this material. We do, however, point out relevant issues from these subjects. Throughout this article we stress the need to use comparative methods that provide diagnosis of the comparative data at hand and then verification of how a comparative method has transformed the data. We also encourage specification of the explicit assumptions of the models that underlie different comparative procedures (46, 70).

COMPARATIVE STUDY: GENERAL CONSIDERATIONS

Any comparative study can be stripped down to five elements:

1. The main hypotheses. Typically, these are causal explanations which involve ecological or evolutionary factors (e.g. mating system) influencing some phenotypic trait (e.g. body size dimorphism). Hypotheses may also be noncausal in the sense of covariation between two traits (e.g. body-brain size allometry).
2. The range of variation (standard deviation) of the trait(s) in question.
3. The presence, location, and form of any correlation relating to taxonomic or phylogenetic hierarchy.
4. The range of variation once the trait(s) have been transformed through some statistical procedure (logarithmic transformation; filtering through a comparative method for removing phylogenetic correlation).
5. Knowledge of evolutionary tempo and rate that will impinge on the divergence of traits. In most comparative studies such evolutionary information is based on an assumption (null model) for expected trait variation (e.g. Brownian motion).

Given these basic elements, deciphering problems in comparative study first involves asking about potential variability in different kinds of traits (e.g. morphological versus behavioral) and whether such variability might affect a comparative result. Most problems pertaining solely to the comparative traits are specific to each analysis. Issues involving spurious data points (e.g. error from data collection, crude trait measures, small sample sizes, standardizing), bias in taxonomic spread of the information, and random searches for causal relations have been discussed at length (14, 34, 41, 43, 52). An obvious problem, although it has received less attention, is the observed variability of a trait relative to an expected result. As Simpson et al (85) discussed, traits with small amounts of variance will reveal very different results than do those with large variance. Although variance may be instructive to ecological and evolutionary explanations, (e.g. changes of animal population density over time: 63, 72; taxonomic differences in slope of relative brain size: 69), it may also obscure problems in the data. Trait values of population, behavioral, or ecological variables often have coefficients of variation that lie between 25 and 250, whereas morphological characters (e.g. antler length; tooth row length) rarely exceed 20. Such inherent variability should be estimated prior to executing comparative analyses (see e.g. 10, 59).

The effect of classification systems (e.g. evolutionary taxonomy vs phylogenetic systematics) on a comparative study is beyond the scope of this paper. Nevertheless, at an immediate level, a comparative study rests on one of two types of classification: taxonomic levels or phylogenetic topology. In essence, "phylogeny is something that happened and classification [taxonomy] is an arrangement of its results" (84, p. 110). Taxonomy thus is a listing of hierarchical names. Depending on a particular taxonomy, one proceeds from lower specific listings to higher phyletic ones, with variability of monophyly and symmetry in between. The relevant point for comparative work is that a hierarchical structure, in effect, is a null model for comparative results: degrees of difference between trait variation and taxonomic listings determine a comparative result. Thus, if the distribution of points in a taxonomy is uneven and/or unreliable, then it may be useful either to run simulations of various taxonomies or to examine results against different taxonomic schemes (19). Obviously, this general precaution carries over to phylogenies as well. However, there are further potential problems with phylogenetic information. Phylogenetic trees provide additional information about branch lengths among taxa, and this conveys information about evolutionary time and genealogical relationships.

Finally, since most comparative studies do not have the luxury of phylogenetic information, we must ask how well a taxonomy reflects phylogeny. Harvey & Pagel (44) succinctly state the present dilemma: "Once the need to use phylogeny in comparative studies has been accepted, the natural procedure has been to refer to standard taxonomies. Unfortunately, it is usually the case that traditional taxonomies are not even meant to describe hypothetical phylogenetic relationships" (p. 51). Various techniques are available to estimate branch lengths from taxonomies (see 39, 44, 67). Grafen (39) has developed an algorithm that is particularly useful in comparative studies. Assuming that the expected variance of change along a branch is proportional to its length (a Brownian motion model), each node in a given phylogeny is assigned a height equal to one fewer than the number of species at or below the node. The length of a branch is then estimated as the difference in height between the top and the bottom of it. Thus, for example, a species will receive a zero, the next higher node is assigned one fewer than the number of species in it, and so on. Branch lengths are then calculated as the difference between successive nodes. The initial branch lengths are lengthened or shortened by a maximum likelihood procedure, depending on variance in the data. As Grafen emphasizes, while the branch lengths generated from this process are arbitrary, nevertheless they retain statistically tractable results. Grafen applies this algorithm only to species, but the entire tree may be filled in with a similar technique (75). Undoubtedly, phylogenetic reconstruction is the weakest link in most comparative studies to date. Considerable work remains before topological data can be estimated statistically from taxonomies. In the following sections we assume, as do others (44, 62), that our approximations of phylogenetic information are reasonably accurate.

COMPARATIVE PROBLEMS: DIAGNOSIS

Variation in quality of data or reliance on correlation to show causation are problems inherent in any comparative study. Although there are rules for avoiding or at least minimizing these problems (7, 14, 34, 43, 68–70), they are inherent to the approach in general because of disparate sources of data collection and nonmanipulation of variables. An individual researcher must therefore be sensitive to data usage and reasoning behind causal arguments. Similar concerns arise from statistical analysis. By contrast, however, there are some guidelines for special problems presented by the hierarchical nature of comparative data; namely, what association does phylogeny have with observed phenotypic differences across taxa?

In present context, the problem of phylogenetic association in comparative studies involves establishing *independence in the data* and examining *change in form* resulting from phylogenetic hierarchy (24, 38, 48, 79). As a hypothetical example, consider trait variation of two genera with three species in each (Figure 1). If we plotted the species points ignoring phylogenetic pattern we would lose potential relationships (Type I error: problems of form), whereas if generic clusters were plotted on a single line we would lack statistical independence and might erroneously claim correlation (Type II error: problems of independence). Thus, reflecting these potential problems, we define 'phylogenetic correlation' as simply the relationship between a trait and a given taxonomy or phylogeny. For empirical study, this issue can be diagnosed by three questions (38): (i) Is there phylogenetic correlation in the data (i.e. nonindependence between phylogenetic hierarchy and trait variation)? (ii) At what taxonomic level or phylogenetic distance does phylogenetic correlation occur? (iii) What procedures may be used to remove phylogenetic effect from the data? This section suggests ways to handle the first two questions; most of the remaining sections deal with phylogenetic removal using different comparative procedures.

The first two questions require descriptive statistics. It is necessary to find

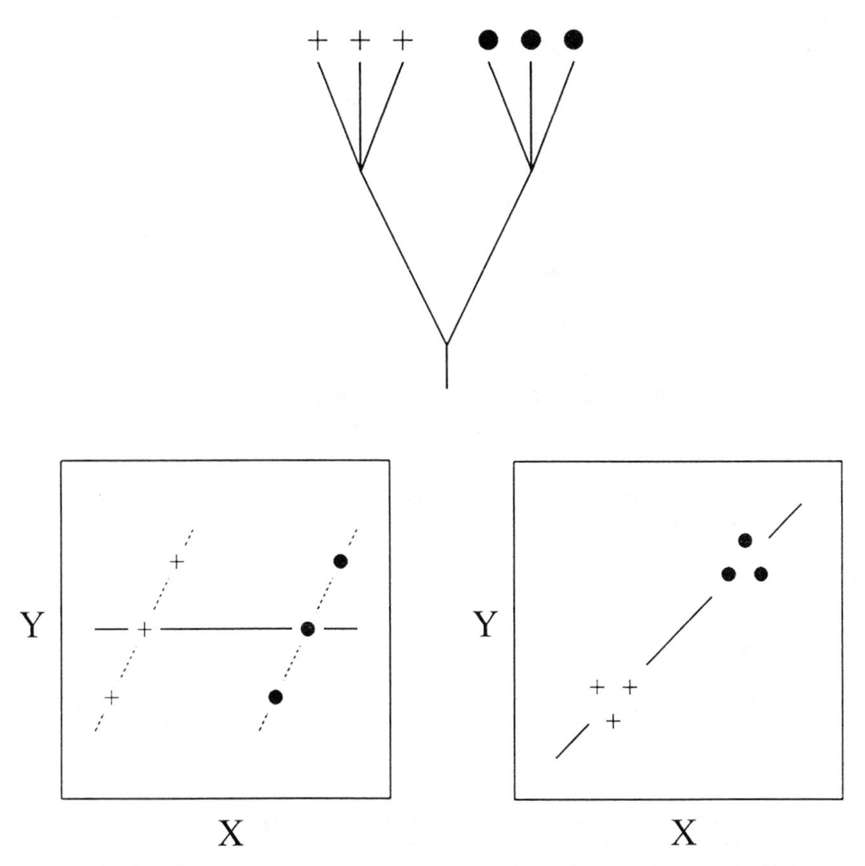

Figure 1 Hypothetical phylogeny for two genera containing three species in each. If phylogenetic relationships are ignored (bottom left), as reflected in the solid line, we risk the rejection of extant correlations (Type I error). On the other hand, if we ignore phylogeny (bottom right) by plotting monophyletic clusters, as in the solid diagonal line, we may claim correlation where none exists (Type II error). Illustration after Felsenstein (24), Gittleman & Kot (38), and Huey (48).

out if phylogenetic correlation is specifically involved with a given data set. That is, there is no quick recipe for showing phylogenetic association; separate analyses must be made with each different trait for each cross-taxonomic unit. Two statistics are useful. The Nested Analysis of Variance describes the percentage of variance in the data accounted for at successive taxonomic levels (91; for applications of this statistic, see 42, 43). For example, in a study of life history variation in primates (42), the distribution of variance in 11 variables was at least 72% at the subfamily level or above (see Table 1). Therefore, the use of subfamilial points (partially) justifies statistical independence and, more generally, reveals the likelihood of phylogenetic association at lower taxonomic levels (but see 30). With statistical independence of subfamilial points, these data are then analyzed for comparative tests. Similar usage of the Nested ANOVA statistic are in studies of brain size (32, 82), body size (31), life history patterns (33, 65, 80, 92, 98), metabolic rate (19), bird song (76), and sleep (20).

Another approach coopts the substantial literature on spatial autocorrelation for measuring effects that vary with distance (12, 94; see 89, 90 for applications of autocorrelation statistics in ecology and evolution). The Moran's *I* statistic estimates phylogenetic autocorrelation at different taxonomic levels or phylogenetic distances (38). In essence, the Moran's *I* is an autocorrelation coefficient which is a measure of the covariance of a given trait across taxa divided by a measure of the variance. Once *I*'s are calculated at different taxonomic levels (or indeed, at patristic distances), they may be plotted on a graph (a phylogenetic correlogram) to show where autocorrelation varies with taxonomy or phylogenetic distance. From the above example of primate life histories, estimating the degree and location of phylogenetic pattern requires several calculations of Moran's *I*: autocorrelation of species

Table 1 Comparison of descriptive statistics for example life history data in primates (from 42). Values represent relative variance (Nested ANOVA) adjacent to autocorrelation (Moran's *I*) calculated for four taxonomic levels.

Among: Within:	species genera	genera subfamilies	subfamilies families	families order
Life history		Nested ANOVA / Moran's *I*[a]		
Neonatal weight	3.7/.74	5.1/.82	16.9/.62	74.3/−.73
Age at sexual maturity	9.4/.47	3.9/.60	10.7/.47	76.0/−.58
Lifespan	6.7/.48	21.2/.50	25.8/.13	46.3/−.53

[a] Values of *I* are row normalized as in Gittleman & Kot (38).

within genera, genera within subfamilies, and subfamilies within families (see Table 1). Thus phylogenetic autocorrelation can be detected at different taxonomic levels. Interestingly, in some cases phylogenetic correlations do not drop-off with patristic distance: for example, with life history patterns the age at sexual maturity is not correlated among species within genera but is significantly correlated among genera within families (36). Also, extremely dissimilar traits are typically found among distantly related taxa (i. e. negative autocorrelation among families within orders, as in Table 1). Applying this autocorrelation statistic involves deciding how to measure distance and assumptions about normalization (38). Clearly, peculiarities in a given taxonomy or phylogeny (e.g. monotypic genera) will shift observed correlation to different points in a correlogram. When trait values indeed do not drop-off in an autoregressive manner, the autoregressive comparative method should be used with caution. The Moran's *I* statistic and associated correlograms have been used to study body size (38) and life history patterns (36) across carnivores.

In gauging the degree and location of phylogenetic correlation from a taxonomy, both the Nested Analysis of Variance and autocorrelation approach reveal similar information: high positive correlations of Moran's *I* show strong phylogenetic association that are reflected by low percentages of variance with the Nested ANOVA. Indeed, correlation of the example values in Table 1 indicates an *r*-value of -.95 between *I*'s and variances for associated taxonomic levels. Taken together, these are simply descriptive statistics that inform a comparative researcher whether and where phylogenetic correlation occurs with a given trait; they are not useful for inferring causation or process (see 44, 86). In some respects, however, diagnosing this problem is the most important analytical step in comparative study. If comparative data are statistically independent, showing no phylogenetic correlation by a Nested ANOVA procedure or autocorrelation statistic, then a comparative analysis often can be safely performed without more sophisticated comparative methods. Indeed, with many behavioral and ecological traits which show considerable plasticity, there is minimal phylogenetic correlation (e.g. home range size: 37; population density: 15). The fact that variation in cross-taxonomic traits often shows phylogenetic correlation should not be mistakenly assumed to indicate that this is always the case.

COMPARATIVE METHODS

At present, no fewer than seven comparative methods are available for analyzing continuous or quantitative data (44, 45): maximum likelihood procedures (58); nested analysis of covariance (2); phylogenetic regression

(39, 40; see also 74); minimum evolution method (50, 62); independent contrasts (8, 24, 25, 30, 44, 46, 75); autoregression (11, 38); nonphylogenetic analysis of simulated tips (62). Undoubtedly, given the fashion of comparative study, more will be forthcoming. All of these methods assume that comparative data are phylogenetically correlated, an assumption that should be empirically diagnosed (as above). However, given such correlation, comparative methods diverge by removing phylogenetic associations or in some fashion partitioning traits into phylogenetic and adaptive (nonphylogenetic) components (39, 44).

In the following we describe and simulate potential differences of two techniques, independent contrast and autoregressive methods, which reflect extreme forms of analysis. We have selected these methods in particular because previous simulation study (62) assessed minimum evolution and independent contrast techniques, which have similar built-in assumptions, without consideration of an autoregressive approach. Further, and more importantly, these methods: (i) reveal different assumptions about evolutionary models; (ii) both benefit from being flexible to various types of data yet represent different ways of analyzing traits that are correlated with phylogeny; and, pragmatically, (iii) are being used in the comparative literature.

INDEPENDENT CONTRAST COMPARISONS

The independent contrast method is based on a model first proposed by Felsenstein (24, 25). Felsenstein suggested a procedure that calculates comparisons between pairs of taxa at each bifurcation in a known phylogeny. This procedure results in N - 1 contrasts from N original tip species (see Figure 2 for an illustration of this method). If the ancestral nodes for the phylogeny are known, then each of the calculated differences between two taxa that share an immediate common ancestry is not confounded by phylogeny between them and therefore represents an independent evolutionary event. If two variables are related (e.g. brain size and body size), then a large calculated contrast with one variable at a specific node should be associated with a large contrast in the other. A series of contrasts, which are scaled by each expected standard deviation, are calculated across taxa for a group of variables which may then be used to test various comparative hypotheses.

Central to the independent contrasts method is the condition of a known phylogeny. Unfortunately, phylogenetic information (i.e. ancestral nodes and branch lengths) is usually unavailable and therefore it is necessary to estimate this unknown. Two general approaches may be adopted. First, when ancestral trait values are unknown but branch length information is available, the equation

$$X_k = \frac{\sum_{i=1}^{m} \left(\frac{1}{v_i}\right) X_i}{\sum_{i=1}^{m} \left(\frac{1}{v_i}\right)} \qquad 1.$$

may be used to estimate ancestral nodes, where X_i are traits of descendants, v_i are branch lengths, and m is the number of descendants (67). Second, when both branch lengths and ancestral values are unknown, evolutionary change may be assumed to follow a punctuational model in which changes only occur at a node. Since the variance of change is not directly related to the branch length, all branch lengths are equal to each other (44). In other words, the estimated ancestral node is the mean of its descendants.

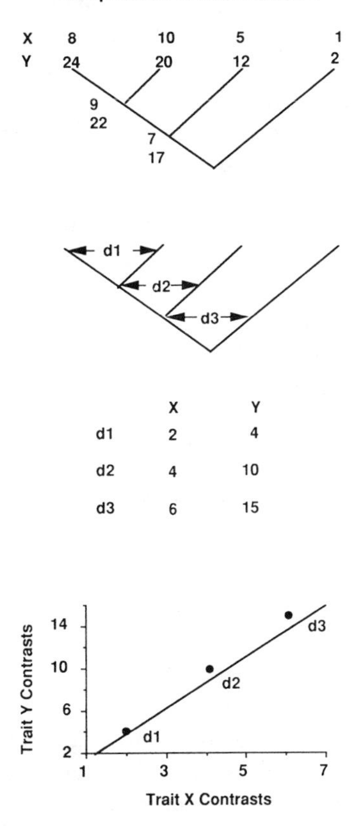

Figure 2 Simplified illustration of the procedure used in the "Independent Contrast Method" to produce and compare independent contrasts. After Harvey & Pagel (44).

The independent contrast method includes a number of important features:

(i) The method equally weights the number of evolutionary events at each node in a given phylogeny; for example, 20 species in one genus is calculated as one contrast value, the same as 5 species in another genus.

(ii) The procedure for making contrasts is explicitly model-dependent, typically that of a Brownian motion model of evolution (24, 25, 38a, 44, 46, 68, 70, 75; H-K. Luh and J. L. Gittleman, in preparation) which provides a process whereby trait variation at successive nodes are independent of one another (independent contrast methods may also accommodate punctuational effects or other evolutionary models, 62).

(iii) Each contrast must have a common variance via division by its standard deviation. The standard deviation is calculated by the square root of the sum of the branch lengths for a given contrast, where branch lengths are units of expected variance of change. Under a Brownian motion model expected variance of change is proportional to time, an assumption which is important to test empirically by evaluating standardization of contrasts (29).

(iv) The direction (sign) of each contrast is arbitrary, with regression through the origin being employed to examine linear relationships (24, 30, 44). To ensure that regression of two sets of variable contrasts is unique, it is suggested that the independent contrast plotted on the horizontal axis (i.e., the independent variable) be given a positive sign while doing the converse to the other independent contrast.

v. Generally, independent contrast methods use all of the variation in a trait to test for a comparative result, rather than partition traits into phylogenetic and nonphylogenetic components as in the autoregressive method (see below). Moreover, because contrasts are calculated at all hierarchical levels in a given phylogeny, comparative results may be detected at higher (e.g. familial and above) taxonomic levels.

Empirical examples using independent contrasts include: the evolution of limb differentiation rate (81); locomotion and morphology in lizards (30, 56, 57); metabolism and life histories among birds and mammals (45, 93); longevity and fecundity in parasitoid Hymenoptera (5); blood parameters among mammals (73); relative testis size and sperm competition in birds (64).

AUTOREGRESSIVE COMPARISONS

The autoregressive approach is based on a method developed by Cheverud et al (11) which partitions trait values into a phylogenetic (inherited) component and a specific component due to independent (adaptive) evolution. The rationale of separating trait values into phylogenetic heritage and specific adaptation is similar to the division of phenotypic values into additive genetic and environmental values; both use a variance decomposition model to analyze the influence of heritable and nonheritable factors (9). Cheverud et al's method (see Figure 3 for an illustration of this method) is derived from autoregressive techniques applied to problems of spatial pattern (12, 88). For comparative analyses, the model takes the form: $y = \rho Wy + \epsilon$ with W an $n \times n$ weighting matrix (see Figure 3 for an illustration of this procedure). The vector y of standardized trait values takes the linear combination ρWy as its phylogenetic component and the residual vector ϵ as its specific component. The autocor-

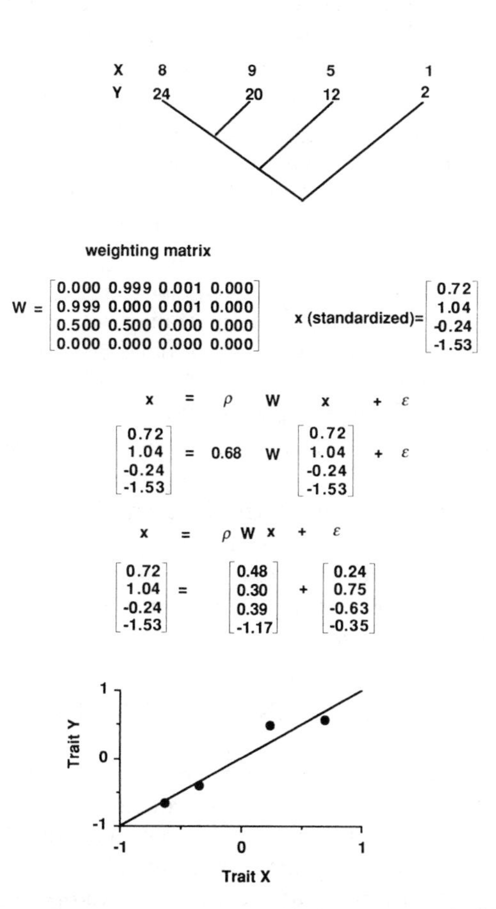

Figure 3 Simplified illustration of the procedure used in the "Autoregressive Method" to produce residual values that are used in comparative tests. The vector of $\psi W x$ is calculated from standardized values of the product Wx. Updated and modified after Cheverud et al (11), Gittleman & Kot (38), and Harvey & Pagel (44).

relation coefficient, ρ, measures the correlation between the phenotypic trait vector y and the purely phylogenetic value Wy. The residual ϵ depicts the independent evolution of each species.

As with the independent contrast method, there are a number of critical features in the autoregressive method.

(i) Residual values are calculated from the autoregressive procedure for each taxonomic unit (typically species) under study.

(ii) At the heart of autoregressive comparisons is the phylogenetic connectivity matrix **W**. The weighting matrix describes the relative phylogenetic distance of all pairs of species. More accurately, they reflect phenotypic similarities anticipated at these distances. In essence, the method views the qualities of an individual species as a weighted average of the qualities of all other species. Close relatives, because they share more recent ancestors, are expected to be most similar and are assigned the greatest weight; distant relatives are assigned low weights. In their study of sexual dimorphism in primates, Cheverud et al (11) chose to let similarity descend as the reciprocal of taxonomic distance. That is, all species within the same genus were assigned a value of 1.0, all species within the same family a value of 0.5, all species within the same superfamily 0.33, and so on.

(iii) Employing a variable weighting index (e.g. maximum likelihood procedure) provides greater flexibility in dealing with both conservative and plastic traits (38). Indeed, the Moran's I or other descriptive statistics reveal that phenotypic trait similarity often does not decay in a uniform manner, from trait to trait, with patristic distance (7, 36, 38). Using a variable weighting index also vastly improves the capacity to remove phylogenetic correlation in comparative data.

Empirical studies using autoregressive comparisons include: body size dimorphism in primates (10, 21); clutch size in nonpasserine birds (38); brain size and life history patterns across mammalian carnivores (35, 36); the evolution of cooperative breeding in perching birds (17); and, ornament dimorphism and mating system in birds (4).

DIVERGENT ASSUMPTIONS IN THE METHODS

Many differences exist between the independent contrast method and autoregressive method. Some of these (calculations; resulting sample size and degrees of freedom) are apparent from the analytical procedures themselves. Assumptions behind the models are more cryptic. It is useful to point out that many of the assumptions, procedures, and applications of comparative methods are similar to problems in the spatial time series literature (see 3, 12). Analogous methods for removing time trends in data are referred to as a filter (i.e. autoregressive method) or by the process of differencing (i.e. independent contrast method). As with these techniques, the following general assumptions differ with each comparative method.

MODEL The evolutionary process is the main underlying assumption separating the two methods. Built-in to independent contrasts is an expectation of trait variation through phylogeny, either by Brownian motion or some other process. Autoregression is a purely statistical method that removes phylogenetic correlation in the comparative data; expected variance is not a part of the method. Thus, assuming conditions in the data are constant (see below), when a specific evolutionary model is wedded to a particular comparative problem, adopting an independent contrasts approach would generally be more appropriate than using autoregression.

PROCEDURE The contrast method calculates a (presumed) independent evolutionary event at each node in a phylogenetic process. It assumes, therefore, that phylogenetic trends are pervasive, and this is likely for many comparative data sets. If, however, phylogenetic correlation is moderate or relatively minor and occurs only in certain segments of a tree and/or in a variety of directions (e.g. in a given family some genera have high correlation and others low), then the autoregressive method, via a weighting matrix, may be more flexible. It should be recognized that the weighting matrix, particularly when using a maximum likelihood procedure, may overestimate attributing phylogeny to trait variation.

SIMULATION STUDIES

In an important paper, Martins & Garland (62; see also 30) used computer simulation to compare statistical properties of various independent contrast and minimum evolution methods for analyzing quantitative traits. Simulations generated various models of evolutionary change (gradual; punctuational) along a known phylogeny, and random changes were added to this phylogeny until trait values were obtained for each species on the phylogeny (one thousand times). In brief, analyses using each comparative method revealed that: (i) simple correlation along tips of a phylogeny always yields high Type I error rates and relatively poor estimates of evolutionary relationships; (ii) given accurate information about phylogeny and evolutionary rate, Felsenstein's (24) method yields acceptable significance levels and relatively reliable estimates of evolutionary correlation; and, (iii) compared to Felsenstein's method, the minimum evolution technique generally shows high Type I error rates and lower statistical power. However, this approach may be useful for estimating certain types of evolutionary correlation (30). As described above, the autoregressive method represents fundamental differences from independent contrast approaches. Therefore, simulation analysis might show even more dramatic differences between these methods (H-K. Luh and J. L. Gittleman, in preparation; see also 38a).

The following simulations expand Martins & Garland's (62 see also 29, 30) study of phylogeny and character evolution (on known phylogenies) in order to further uncover potential systematic differences between autoregression and independent contrast approaches. We emphasize the importance of simulations because, given the number of uncertainties in present comparative studies, they are helpful in any empirical analysis to verify assumptions and procedures of a selected comparative method (see also 46). Simulations are also useful for indicating the statistical power of a comparative result with a given phylogeny (see also 71).

TAXONOMY/PHYLOGENY As mentioned previously, information on classifying organisms comes in two forms: taxonomic listings and phylogenetic trees. Although it is true that phylogenetic information is becoming more readily available (62), most groups of organisms are still only described by rudimentary taxonomies. Therefore, in contrast to Martins & Garland (62), we simulate conditions in which the comparative analysis uses either a pure taxonomy or a phylogeny that reflects true branch lengths and topology.

Both taxonomic and phylogenetic simulations are generated by random joining (61); simulation using a random joining process may be preferable to other models because it makes relatively few assumptions about speciation (83). For taxonomic simulation, taxa are randomly clustered in the lower taxonomic levels (e.g. species within genera) to form larger taxa in the higher levels (e.g. families within an order) until only one taxon is formed. To generate random phylogenetic multifurcating trees, a random number is simply assigned to each branch as its length. One restriction is necessary for generating phylogenies: branch lengths between higher nodes must be longer than the branch lengths between lower nodes. This assumption is, however, suitable for either comparative method. In autoregression, it is assumed that nodes nearer to one another are more closely related. In the independent contrasts, based on a Brownian motion model, the expected variance along a branch is related to its length; thus a contrast between higher nodes will have a higher variance than a contrast between lower nodes.

TRAITS Two continuous (quantitative) traits evolving along a phylogeny can be viewed as two random variables drawn from a bivariate normal distribution (62). The pattern of evolutionary change is determined by parameters of the probability distribution. The means of a bivariate normal distribution indicate the net change of evolution in a trait. The correlation of the bivariate distribution shows the general trend of two traits changing along a phylogeny.

If character evolution follows a gradual model of evolutionary change (as in the present simulations), then the total change between ancestral and descendant values will be normally distributed with a mean of zero and a unit variance which is proportional to the branch lengths. If the branch length is long, there is a high probability of a large expected variance with net change.

SIMULATED COMPARISONS Simulations involve three comparative problems which reflect various conditions and availability of information (see Table 2). All simulations assume that the two traits are not correlated with one another. For each of the following conditions, 50 trees are produced and, for each tree, 50 sets of two traits are generated. (i) There is no correlation between each trait and a hierarchical structure represented in the taxonomy; that is, both the taxonomy and the traits are randomized. (ii) As in the typical comparative problem which would require a statistical comparative method, there is correlation between the traits and a phylogeny (i.e. trait evolution follows a phylogeny), but branch lengths are unknown in the phylogeny (i.e. only taxonomic information is available). (iii) Traits are correlated with phylogeny, but unlike most comparative studies, a true phylogeny (i.e. with branch lengths) is known. For both simulations 2 and 3, the same phylogenetic information is used to simulate trait evolution but in simulation 2 only the taxonomy is employed.

RESULTS Prior to assessing the simulated results of each comparative method, it is necessary to diagnose whether the simulations actually represent

Table 2 Simulation study of linear contrasts and autoregressive comparative methods. Simulations show relative results under varying conditions of: phylogenetic correlation in the data, phylogenetic information, number of taxa, and degrees of freedom. Comparative results of each method are reported as relative rates of Type I error.

Simulation	Condition	Information	Method	Species Number (d.f.)	$\alpha = 0.05$ Type I error
1	No phylogenetic correlation	Taxonomy	Contrasts	10 (2–7)	.01
				50 (11–34)	.06
			Autoregression	10 (8)	.06
				50 (48)	.06
2	Phylogenetic correlation	Taxonomy	Contrasts	10 (2–7)	.18
				50 (13–30)	.12
			Autoregression	10 (8)	.08
				50 (48)	.13
3	Phylogenetic correlation	Phylogeny	Contrasts	10 (2–7)	.11
				50 (13–30)	.06
			Autoregression	10 (8)	.10
				50 (48)	.10

pre-established conditions. As prescribed for any empirical study, the Nested ANOVA or Moran's I can uncover phylogenetic correlation in the data. Thus, in simulation 1 the percentage of detecting correlation between each of two traits and phylogeny is very low (i.e. 10 species: 3.48% and 3.76%; 50 species: 4.08% and 3.12%) whereas in simulations 2 and 3 there is significant correlation, with larger samples carrying higher correlations (i.e. 10 species: 42.08% and 44.16%; 50 species: 76.60% and 77.04%). Thus, the diagnostic statistics verify the condition of phylogenetic correlation in the data.

To get a general perspective initially on how each method affects comparative data transformation, it is useful to consider overall distributions of correlation coefficients (between the two traits) in each of the three simulated conditions (see Figure 4). Three patterns are apparent, all of which appear to involve sample size effects. Distributions based on small sample sizes are more platykurtic than those including larger samples, regardless of different conditions or indeed the comparative method itself. The independent contrasts reveal more platykurtic distributions than the autoregressive method. Given phylogenetic correlation, conditions of small sample size and only taxonomic information produce the lowest capacity to detect the hypothetical relationship of traits (Figure 4c). On the surface, therefore, it appears that sample size may produce these results. However, this effect is more likely due to different calculations of the degrees of freedom. With independent contrasts, correlation is computed from contrasts at each node, so the degrees of freedom are set according to the number of nodes rather than total sample size. For example, a given tree may have many taxa but only three nodes (i.e. the smallest number of analyzable nodes) that would result in contrast values. When the degree of freedom is relatively low, then the correlation will be high (e.g. Figure 4a, c, e). Conversely, with the autoregressive model, degrees of freedom are determined by the number of species. Although sample sizes of 10 and 50 were simulated, degrees of freedom for the independent contrasts varied with the randomly generated tree topology (Table 2). The degrees of freedom in the autoregressive model, however, are always 8 and 48. Therefore, to assess the difference for rejecting the null hypothesis with each method, sample size must first be taken into account.

The more critical aspect of results from simulation are the actual differences in significance tests of correlation (rate of Type I error). For each comparative method, we count the number of correlation coefficients out of 2500 simulations that are significant at the 0.05 level (two-tailed t-test: 99). As with the general distributions described above from Figure 4, sample size appears to be an important factor influencing correlation coefficients after data transformation from comparative analyses (see Table 2). The significance test shows that the independent contrast method with higher degrees of freedom reduces the rate of rejecting the null hypothesis except in condition 1.

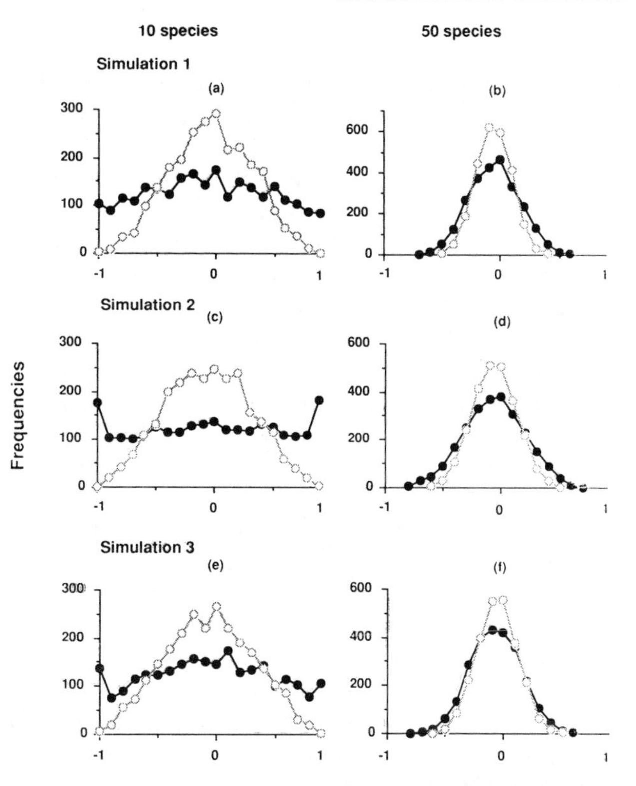

Figure 4 Distributions of bivariate correlations resulting from data transformation using the independent contrast (solid dots) and autoregressive (open dots) comparative methods. Simulations include 10 or 50 species under the following conditions: no phylogenetic correlation in the data and using only taxonomic information (Simulation 1); phylogenetic correlation in the data and using only taxonomic information (Simulation 2); phylogenetic correlation in the data and using true phylogenetic information with branch lengths (Simulation 3). Further details are given in the text.

On the other hand, the autoregressive method increases the rate of rejecting the null hypothesis when branch lengths are unavailable (simulation 2), but no difference is observed when branch lengths are available (simulation 3).

What factors influence these trends? To answer this question it is necessary to consider how the two methods use branch length information. For the independent contrasts branch length information is critical (simulation 3), as Felsenstein (24) originally stated. If this information is unavailable (simulation 2), then evolutionary change is assumed to be punctuational and only occurs at a node so that all branch lengths are set equal. This kind of data transformation loses information about variance of change between two nodes. Therefore, the independent contrast method will tend to overestimate the rate of rejection. As sample size increases, the independent contrast method will provide an acceptable rate (0.06) to reject the null hypothesis. When sample size is small, even when branch lengths are known, the relatively low degree of freedom may still cause a biased estimation. By contrast, the autoregressive method is designed to estimate the phylogenetic effect and remove it. Application of this method is not limited by the requirement of branch length information because the maximum likelihood technique (38) is able to estimate the proper weighting matrix from a taxonomy. Autoregression, however, suffers from the antipathic problem of independent contrasts: when true branch lengths are provided (simulation 3), the autoregressive model may overweight the phylogenetic correlation and increase the probability to invite Type I error. Further, with respect to sample size effects, the maximum likelihood technique elicits biased estimation with larger sample sizes because such a power law will likely make errors with a complex topology.

Aside from these results, simulations reveal a rather satisfying aspect of both methods. In the first simulation, where there is no phylogenetic correlation or correlation between traits, both comparative methods provide acceptable rates to reject the null hypothesis. Thus, forcing a comparative method on a situation where no phylogenetic correlation occurs does not artificially transform properties of the traits.

The purpose of these simulations is to extend Martins & Garland's (62) approach to diagnosing and verifying comparative data analysis. The primary modifications are the inclusion of the autoregressive method, the absence of phylogenetic correlation, and simulation of a more general situation of multifurcating trees. The simulations presented here suggest two general rules for analysis. First, it is necessary to use some descriptive statistic (Nested ANOVA or Moran's I) to evaluate phylogenetic correlation in the data set. If there is no correlation, data transformation using a comparative method is not necessary unless justified. Second, when there is phylogenetic correlation, sample size and branch length information are critical for deciding which comparative method to use. When sample size is large and branch lengths are available, the independent contrast method is probably the best candidate. When the sample size is small and branch lengths are unavailable, then the autoregressive method may be more appropriate. As a general rule though, for both scenarios, it is extremely useful to run simulations like Martins and Garland's or those illustrated here to verify the statistical power of rejecting a null hypothesis, especially while phylogenetic information is imperfect.

CONCLUDING REMARKS

Comparative methods test evolutionary pattern and process. Behind every comparative test, however, is embedded a null hypothesis about evolution. Thus future work should assess the appropriateness of specific evolutionary processes with comparative tests of specific traits (18). As an example, differences in heritability, variances, and evolutionary rates of behavioral versus morphological characters (e.g. 1, 22, 23) might require different comparative methods. Clearly, guidelines must be formulated for evaluating the accuracy of phylogenetic information used in comparative tests (26, 47). As Harvey & Pagel (44) end their volume, "Tell us how to reconstruct the past, and we shall perform the comparative analysis with precision" (p. 205).

With respect to comparative methods themselves, it is mandatory initially to diagnose the comparative data at hand. Is phylogenetic correlation a problem with a particular data set and, if so, what is the location? Nested ANOVA or Moran *I* type statistics will aid in these descriptive questions. If phylogenetic correlation is not observed, then comparative method procedures should not be adopted for needless data transformation. On the other hand, when data are phylogenetically correlated, comparative methods should be selected on the basis of whether an evolutionary model is assumed a priori, sample size of the comparative data, and availability of phylogenetic versus taxonomic information. Finally, if comparative data are filtered through some comparative method, it is informative to verify how much a comparative result is a function of the adopted comparative method. Until we indeed can reconstruct the past, we will have to rely on simulation studies to tell us how well our comparative methods are working.

ACKNOWLEDGMENTS

We are most grateful to Emília Martins and Ted Garland for providing computer simulation programs, Mark Pagel and Andy Purvis for copies of various independent contrast programs, Mark Kot, Paul Harvey, Joe Felsenstein, and Stuart Pimm for discussion of theoretical issues, and Ted Garland, Gordon Burghardt, Sandy Echternacht and Karen Holt for comments on the manuscript. Financial support was received from the Department of Zoology and Science Alliance at The University of Tennessee and the American Philosophical Society.

Literature Cited

1. Bateson, P. 1988. The active role of behaviour in evolution. In *Process and Metaphors in the New Evolutionary Paradigm,* ed. M.-W. Ho, S. Fox, pp. 191–207. Chichester: Wiley
2. Bell, G. 1989. A comparative method. *Am. Nat.* 133:553–71
3. Bennett, R. J. 1979. *Spatial Time Series.* London: Pion
4. Bjorklund, M. 1990. A phylogenetic interpretation of sexual dimorphism in body size and ornament in relation to mating system in birds. *J. Evol. Biol.* 3:171–83
5. Blackburn, T. M. 1991. A comparative examination of life-span and fecundity in parasitoid Hymenoptera. *J. Anim. Ecol.* 60:151–64
6. Brooks, D. R., McLennan, D. A. 1991. *Phylogeny, Ecology, and Behavior.* Chicago: Univ. Chicago Press
7. Burghardt, G. M., Gittleman, J. L. 1991. Comparative behavior and phylogenetic analyses: New wine, old bottles. In

Interpretation and Explanation in the Study of Animal Behavior: Comparative Perspectives, ed. M. Bekoff, D. Jamieson, pp. 195–225. Boulder: Westview
8. Burt, A. 1989. Comparative methods using phylogenetic independent contrasts. *Oxford Surv. Evol. Biol.* 6:33-53
9. Cavalli-Sforza, L., Piazza, L. 1975. Analysis of evolution: evolutionary rates, independence, and treeness. *Theor. Popul. Biol.* 8:127–65
10. Cherry, L. M., Case, S. M., Kunkel, J. G., Wyles, J. S., Wilson, A. C. 1982. Body shape metrics and organismal evolution. *Evolution* 36:914–33
11. Cheverud, J. M., Dow, M. M., Leutenegger, W. 1985. The quantitative assessment of phylogenetic constraints in comparative analyses: sexual dimorphism in body weight among primates. *Evolution* 39:1335–51
12. Cliff, A. D., Ord, J. K. 1981. *Spatial Processes: Models and Applications.* London: Pion
13. Clutton-Brock, T. H., Harvey, P. H. 1977. Primate ecology and social organisation. *J. Zool.* 183:1–39
14. Clutton-Brock, T. H., Harvey, P. H. 1984. Comparative approaches to investigating adaptation. In *Behavioural Ecology: An Evolutionary Approach,* ed. J. R. Krebs, N. B. Davies, pp. 7–29. Oxford: Blackwells. 2nd ed.
15. Damuth, J. 1981. Population density and body size in mammals. *Nature* 290:699–700
16. Donoghue, M. J. 1989. Phylogenies and the analysis of evolutionary sequences, with examples from seed plants. *Evolution* 43:1137–56
17. Edwards, S. V., Naeem, S. 1992. The phylogenetic component of cooperative breeding in perching birds. *Am. Nat.* In press
18. Eldridge, N., Cracraft, J. 1980. *Phylogenetic Patterns and the Evolutionary Process: Method and Theory in Comparative Biology.* New York: Columbia Univ. Press
19. Elgar, M. A., Harvey, P. H. 1987. Basal metabolic rates in mammals: allometry, phylogeny and ecology. *Funct. Ecol.* 1:25–36
20. Elgar, M. A., Pagel, M. D., Harvey, P. H. 1988. Sleep in mammals. *Anim. Behav.* 36:1407–19
21. Ely, J., Kurland, J. A. 1989. Spatial autocorrelation, phylogenetic constraints, and the causes of sexual dimorphism in primates. *Int. J. Primatol.* 10:151–71
22. Emerson, S. B., Arnold, S. J. 1989. Intra- and interspecific relationships between morphology, performance, and fitness. See Ref. 95, pp. 295–314
23. Endler, J. A. 1986. *Natural Selection in the Wild.* Princeton, NJ: Princeton Univ. Press
24. Felsenstein, J. 1985. Phylogenies and the comparative method. *Am. Nat.* 125: 1–15
25. Felsenstein, J. 1988. Phylogenies and quantitative characters. *Annu. Rev. Ecol. Syst.* 19:445–71
26. Fitch, W. M., Atchley, W. R. 1987. Divergence in inbred strains of mice: a comparison of three different types of data. In *Molecules and Morphology in Evolution: Conflict or Compromise?,* ed. C. Patterson, pp. 203–16. Cambridge: Cambridge Univ. Press
27. Friday, A. 1987. Models of evolutionary change and the estimation of evolutionary trees. *Oxford Surv. Evol. Biol.* 4:61-88
28. Funk, V. A., Brooks, D. R. 1990. Phylogenetic systematics as the basis of comparative biology. *Smithson. Contrib. Bot.* 73:1–45
29. Garland, T. Jr., Harvey, P. H., Ives, A. R. 1992. Procedures for the analysis of comparative data using phylogenetically independent contrasts. *Syst. Biol.* In press
30. Garland, T. Jr., Huey, R. B., Bennett, A. F. 1991. Phylogeny and coadaptation of thermal physiology in lizards: a reanalysis. *Evolution* 45:1969–75
31. Gittleman, J. L. 1985. Carnivore body size: ecological and taxonomic associations. *Oecologia* 67:540–54
32. Gittleman, J. L. 1986. Carnivore brain size, behavioral ecology, and phylogeny. *J. Mammal.* 67:23–36
33. Gittleman, J. L. 1986. Carnivore life history patterns: Allometric, phylogenetic, and ecological associations. *Am. Nat.* 217:744–71
34. Gittleman, J. L. 1989. The comparative approach in ethology: Aims and limitations. In *Perspectives in Ethology,* ed. P. P. G. Bateson, P. H. Klopfer, 8:55–83. New York: Plenum
35. Gittleman, J. L. 1991. Carnivore olfactory bulb size: allometry, phylogeny, and ecology. J. Zool. 225:253–72
36. Gittleman, J. L. 1992. Carnivore life histories: a reanalysis in light of new models. In *Mammals as Predators,* ed. N. Dunstone, M. Gorman. Oxford: Oxford Univ. Press. In press
37. Gittleman, J. L., Harvey, P. H. 1982. Carnivore home-range size, metabolic needs, and ecology. *Behav. Ecol. Sociobiol.* 10:57–63

38. Gittleman, J. L., Kot, M. 1990. Adaptation: statistics and a null model for estimating phylogenetic effects. *Syst. Zool.* 39:227–41
38a. Gittleman, J.L., Luh, H.-K. 1993. Phylogeny, evolutionary models, and comparative methods: a simulation study. In *Pattern and Process: Phylogenetic Approaches to Ecological Problems,* ed. P. Eggleton, D. Vane-Wright. London: Academic Press. In press
39. Grafen, A. 1989. The phylogenetic regression. *Philos. Trans. R. Soc. London* 326:119–57
40. Grafen, A. 1992. The uniqueness of the phylogenetic regression. *J. Theor. Biol.* 156:405–23
41. Harvey, P. H. 1982. On rethinking allometry. *J. Theor. Biol.* 95:37–41
42. Harvey, P. H., Clutton-Brock, T. H. 1985. Life history variation in primates. *Evolution* 39:559–81
43. Harvey, P. H., Mace, G. M. 1982. Comparisons between taxa and adaptive trends: problems of methodology. See Ref. 53, pp. 343–61
44. Harvey, P. H., Pagel, M. D. 1991. *The Comparative Method in Evolutionary Biology.* Oxford: Oxford Univ. Press
45. Harvey, P. H., Pagel, M. D., Rees, J. A. 1991. Mammalian metabolism and life histories. *Am. Nat.* 136:556–66
46. Harvey, P. H., Purvis, A. 1991. Comparative methods for explaining adaptations. *Nature* 351:619–24
47. Hillis, D. M., Bull, J. J., White, M. E., Badgett, M. R., Molineux, I. J. 1992. Experimental phylogenetics: Generation of a known phylogeny. *Science* 255:589–92
48. Huey, R. B. 1987. Phylogeny, history, and the comparative method. In *New Directions in Ecological Physiology,* ed. M. E. Feder, A. F. Bennett, W. Burggren, R. B. Huey, pp. 76–98. Cambridge: Cambridge Univ. Press
49. Huey, R. B., Bennett, A. F. 1986. A comparative approach to field and laboratory studies in evolutionary biology. In *Predator-Prey Relationships: Perspectives and Approaches for the Study of Lower Vertebrates,* ed. M. E. Feder, G. V. Lauder, pp. 82–96. Chicago: Univ. Chicago Press
50. Huey, R. B., Bennett, A. F. 1987. Phylogenetic studies of coadaptation: preferred temperatures versus optimal performance temperatures of lizards. *Evolution* 41:1098–115
51. Hull, D. L. 1988. *Science as a Process.* Chicago: Univ. Chicago Press.
52. Jarman, P. J. 1982. Prospects for interspecific comparison in sociobiology. See Ref. 53, pp. 323–42
53. King's College Research Group. 1982. *Current Problems in Sociobiology.* Cambridge: Cambridge Univ. Press
54. Lauder, G. V. 1986. Homology, analogy, and the evolution of behavior. In *The Evolution of Behavior,* ed. M. Nitecki, J. Kitchell, pp. 9–40. Chicago: Univ. Chicago Press
55. Lauder, G. V., Liem, K. F. 1989. The role of historical factors in the evolution of complex organismal functions. See Ref. 95, pp. 63–78
56. Losos, J. B. 1990. The evolution of form and function: morphology and locomotor performance in West Indian Anolis lizards. *Evolution* 44:1189–203
57. Losos, J. B. 1990b. Ecomorphology, performance capability, and scaling of West Indian Anolis lizards: an evolutionary analysis. *Ecol. Monogr.* 60: 369–88
58. Lynch, M. 1991. Methods for the analysis of comparative data in evolutionary biology. *Evolution* 45: 1065–80
59. Lynch, M. 1991. The rate of morphological evolution in mammals from the standpoint of the neutral expectation. *Am. Nat.* 136:727–41
60. Maddison, W. P. 1990. A method for testing the correlated evolution of two binary characters: are gains or losses concentrated on certain branches of a phylogenetic tree? *Evolution* 44:539–57
61. Maddison, W. P., Slatkin, M. 1991. Null models for the number of evolutionary steps in a character on a phylogenetic tree. *Evolution* 45:1184–97
62. Martins, E. P., Garland, T. Jr. 1991. Phylogenetic analysis of the correlated evolution of continuous characters: a simulation study. *Evolution* 45:534–57
63. McArdle, B. H., Gaston, K. J., Lawton, J. H. 1990. Variation in the size of animal populations: patterns, problems and artefacts. *J. Anim. Ecol.* 59:439–54
64. Møller, A. P. 1991. Sperm competition, sperm depletion, paternal care, and relative testis size in birds. *Am. Nat.* 137:882–906
65. Murphy, M. T. 1989. Life history variability in North American breeding tyrant flycatchers: phylogeny, size or ecology? *Oikos* 54:3-14
66. Pagel, M. D. 1991. Review of Brooks and McLennan's *Phylogeny, Ecology, and Behavior. Q. Rev. Biol.* 66:336–37
67. Pagel, M. D. 1992. A method for the analysis of comparative data. *J. Theor. Biol.* 156:431–42
68. Pagel, M. D., Harvey, P. H. 1988.

Recent developments in the analysis of comparative data. *Q. Rev. Biol.* 63:413–40
69. Pagel, M. D., Harvey, P. H. 1988. The taxon level problem in mammalian brain size evolution: facts and artifacts. *Am. Nat.* 132:344–59
70. Pagel, M. D., Harvey, P. H. 1989. Comparative methods for examining adaptation depend on evolutionary models. *Folia Primatol.* 53:203–20
71. Pagel, M. D. Harvey, P. H. 1992. On solving the correct problem: wishing does not make it so. *J. Theor. Biol.* 156:425–30
72. Pimm, S. L. 1991. *The Balance of Nature?* Chicago: Univ. Chicago Press
73. Promislow, D. E. L. 1991. The evolution of blood parameters: Patterns and their implication. *Physiol. Zool.* 64:393–431
74. Promislow, D. E. L. 1991. Senescence in natural populations of mammals: a comparative study. *Evolution* 45:1869–87
75. Purvis, A. 1991. Comparative analysis by independent contrasts (C.A.I.C.) A statistical package for the Apple Macintosh Version TEST. Private Distribution. Dep. Zool., Univ. Oxford, England
76. Read, A. F., Weary, D. M. 1990. Sexual selection and the evolution of bird song: a test of the Hamilton-Zuk hypothesis. *Behav. Ecol. Sociobiol.* 26: 47–56
77. Ridley, M. 1983. *The Explanation of Organic Diversity: The Comparative Method and Adaptations for Mating.* Oxford: Clarendon
78. Ridley, M. 1986. *Evolution and Classification: The Reformation of Cladism.* London: Longman
79. Ridley, M. 1989. Why not to use species in comparative tests. *J. Theor. Biol.* 136:361–64
80. Saether, B.-E. 1988. Pattern of covariation between life-history traits of European birds. *Nature* 331:616–17
81. Sessions, S. K., Larson, A. 1987. Developmental correlates of genome size in Plethodontid salamanders and their implications for genome evolution. *Evolution* 41:1239–51
82. Sherry, D. F., Vaccarino, A. L., Buckhenham, K., Herz, R. S. 1989. The hippocampal complex of food-storing birds. *Brain Behav. Evol.* 34:308–17
83. Simberloff, D. 1987. Calculating probabilities that cladograms match: a method of biogeographical inference. *Syst. Zool.* 36:175–95
84. Simpson, G. G. 1980. *Why and How: Some Problems and Methods in Historical Biology.* New York: Pergamon
85. Simpson, G. G., Roe, A., Lewontin, R. C. 1960. *Quantitative Zoology.* New York: Harcourt, Brace
86. Slatkin, M., Arter, A. E. 1991. Spatial autocorrelation methods in population genetics. *Am. Nat.* 138:499–17
87. Sober, E. 1988. *Reconstructing the Past: Parsimony, Evolution, and Inference.* Cambridge: MIT Press
88. Sokal, R. R., Menozzi, P. 1982. Spatial autocorrelations of HLA frequencies in Europe support demic diffusion of early farmers. *Am. Nat.* 119:1–17
89. Sokal, R. R., Oden, N. L. 1978. Spatial autocorrelation in biology. 1. Methodology. *Biol. J. Linn. Soc.* 10:199–228
90. Sokal, R. R., Oden, N. L. 1978b. Spatial autocorrelation in biology. 2. Some biological applications of evolutionary and ecological interest. *Biol. J. Linn. Soc.* 10:229–49
91. Sokal, R. R., Rohlf, F. J. 1969. *Biometry.* New York: Freeman
92. Stearns, S. C. 1983. The influence of size and phylogeny on patterns of covariation among life-history traits in the mammals. *Oikos* 41:173–87
93. Trevelyan, R., Harvey, P. H., Pagel, M. D. 1990. Metabolic rates and life histories in birds. *Funct. Ecol.* 4:135–41
94. Upton, G. J. G., Fingleton, G. 1985. *Spatial Data Analysis by Example.* Chichester: Wiley
95. Wake, D. B., Roth, G., eds. 1989. *Complex Organismal Functions: Integration and Evolution in Vertebrates.* New York: Wiley
96. Wanntorp, H.-E., Brooks, D. R., Nilsson, T., Ronquist, F., Stearns, S. C., Wedell, N. 1990. Phylogenetic approaches in ecology. *Oikos* 57:119–32
97. Wiley, E. O. 1981. *Phylogenetics: The Theory and Practice of Phylogenetic Systematics.* New York: Wiley
98. Wootton, J. T. 1987. The effects of body mass, phylogeny, habitat, and trophic level on mammalian age at first reproduction. *Evolution* 41:732–49
99. Zar, J. H. 1984. *Biostatistical Analysis.* Englewood Cliffs, NJ: Prentice Hall. 2nd ed.

Annu. Rev. Ecol. Syst. 1992. 23:405–47

RESAMPLING METHODS FOR COMPUTATION-INTENSIVE DATA ANALYSIS IN ECOLOGY AND EVOLUTION

Philip H. Crowley
Center for Evolutionary Ecology, T. H. Morgan School of Biological Sciences, University of Kentucky, Lexington, Kentucky 40506-0225, and NERC Centre for Population Biology, Imperial College at Silwood Park, Ascot, Berkshire SL5 7PY United Kingdom

KEYWORDS: bootstrap, jackknife, Monte Carlo methods, permutation tests, randomization tests

INTRODUCTION

The advent of fast, relatively inexpensive (thus, widely available) microcomputers is transforming the way we analyze data in ecological and evolutionary research. Even more profound, however, are the associated changes in questions asked, empirical methods used, studies conducted, and interpretations offered. Now that an array of computation-intensive statistical methods is newly available for general use, it seems particularly important to assess their advantages and limitations, to note how they are currently being used, and then to consider implications for the future.

I focus in this review on four related techniques known in the statistical and biological literature as *randomization* (or *permutation*) *tests, Monte Carlo methods, bootstrapping,* and the *jackknife*. I refer to them collectively as resampling methods, because each involves taking several-to-many samples from the original data set (randomization, bootstrap, jackknife) or from a stochastic process like the one believed to have generated the data set (Monte Carlo). Each of these methods is actually an extensive family of techniques and specific applications that cannot be thoroughly examined here; instead, I briefly characterize the focal methods and then survey the recent literature in ecology and evolution to identify the issues most frequently associated with these techniques. It emerges that resampling methods are well represented in

0066-4162/92/1120-0405$02.00

data analyses related to some of the most important issues and intense controversies currently in these fields of research.

The specific objectives of this paper are:

1. to acquaint a wider array of ecologists and evolutionary biologists with these useful techniques, which, at least until very recently, have been underemphasized or ignored in statistical training;
2. to document the association between certain research questions and one or more of these resampling methods;
3. to emphasize the role of the focal methods in expanding the range of feasible experimental designs and in shifting the conceptual basis of data analysis;
4. to compare and contrast resampling methods with more standard approaches, noting assumptions and other key features that bear on their appropriateness for particular applications; and
5. to highlight methods that need clarification and development, in the hope that these will soon be addressed by statisticians and biometricians.

The review is intentionally biased toward ecological studies, in accord with my own research experience and interests. Simpler, univariate analyses are emphasized in the interest of clarity and also because a review of computation-intensive multivariate methods in ecology is in progress (165). To respect page limits on contributions to this volume, I have restricted the number of examples cited and emphasized more recent papers most likely to contain additional citations of relevant work. I assume here that readers are familiar with rudimentary statistical concepts and basic methods.

The present review proceeds as follows: First, I describe briefly the four resampling techniques, including a relatively straightforward example of each from the literature of ecology and evolution. Next, I summarize results of a systematic literature search for applications, including a computer search of biological journals and edited volumes published during 1985–1990, and my own search by hand through all issues of two prominent ecological journals for the period 1985–1991. Resampling techniques are used to test for temporal trends in the use of these methods and for differences in frequency of use between ecological and evolutionary studies. Publications identified in the search are classified by topic and subtopic, from which are distilled seven major issues considered with example applications in more detail. Focus then shifts to the relation of resampling applications to classical and actively developing statistical methodology. Finally, I discuss advantages, disadvantages, and implications of these methods, highlight methodological questions that deserve attention, and close with some specific recommendations.

BRIEF DESCRIPTIONS OF THE FOCAL METHODS

Lucid descriptions and examples of these methods and computer programs suitable for implementing them are available in recent books by Edgington (64), Noreen (186), and Manly (163).

Randomization

Referring to the use of randomization tests in analyzing data, R. A. Fisher once claimed that statistical "conclusions have no justification beyond the fact that they agree with those which could have been arrived at by this elementary method" (75). In a randomization test, the chance of type 1 error under the null hypothesis (i.e. the *p*-value) is determined by repeated random assignment of the data to treatment levels. The *p*-value is simply the proportion of all data arrangements yielding test statistics at least as extreme in magnitude as the value resulting from the arrangement actually observed (see Figure 1).

When the null hypothesis is that the observed magnitude of the test statistic is (say) not larger than would be expected by chance (a one-tailed hypothesis), then the extreme values to be counted in calculating the *p*-value are those greater than or equal to the observed test-statistic value. When the null hypothesis is that the observed magnitude of the test statistic is not different from chance expectation (a two-tailed hypothesis), then separate counts are made of values greater than or equal to the observed and of values less than or equal to the observed; the lower of these counts is doubled and divided by the total number of data arrangements to obtain the two-tailed *p*-value (subject to the constraint that $p < 1$).

Data resampling requires pooling all data from the treatment levels (i.e. experimentally established or "observed" groups) to be compared and then reassigning data randomly and without replacement to the treatment levels, keeping the number of observations per treatment level the same as in the original data. In some cases, all possible redistributions of data among treatment levels can be readily obtained, resulting in an "exact" randomization test. In other cases, generally when the potential number of different redistributions approaches or exceeds 10^4–10^5, some of these (often about 10^3) are sampled with replacement for the test, which is then known as a "sampled" randomization test.

Randomization tests are often based on standard test statistics (e.g. *t, F*); it is thus the method of resampling the data and of calculating *p* that are definitive, rather than the statistic used. But the potential to use special-purpose or ad hoc statistics is a particularly important advantage of the randomization approach (and of resampling more generally), since this may increase the statistical power to accept the relevant alternative hypothesis (64, 163).

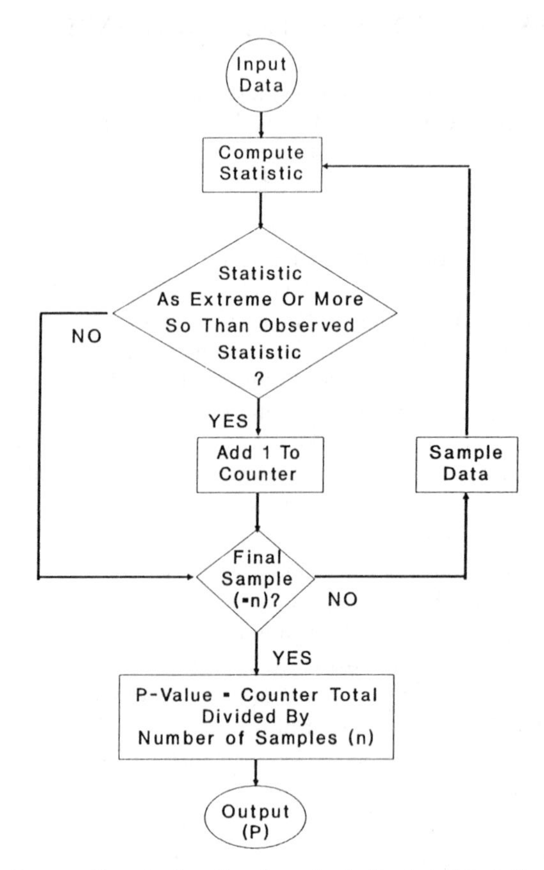

Figure 1 Flow diagram of hypothesis testing via randomization tests, Monte Carlo methods, and some types of bootstrapping techniques. Except for exact randomization when few samples are needed, this logic is usually implemented using a computer program that generates and processes a large number of samples n (typically n 1000); only rudimentary programming skills are needed, unless the statistic is particularly complex or the data structure requires a sophisticated sampling algorithm. See Edgington (64), Noreen (186), and Manly (163) for example programs. This diagram is a slight modification of one on the cover (and Figure 1, p. 51) of E.S. Edgington's book (64).

The basic rationale for randomization methods is that under the null hypothesis of, for example, no difference between treatment-level means, any of the possible distributions of data among treatment levels is equally probable. This equiprobability is assumed to follow from (i) random sampling of populations being compared (contemporary applications generally avoid this assumption, but those that invoke it are known as permutation tests), (ii) random assignment of experimental units to treatment level, or (iii) for nonexperimental studies, simply taking the data to be "exchangeable" among levels in the absence of treatment effects (see 163, 250). However, randomization tests of differences among means are sensitive to differences in variances and other moments (e.g. see 15, 219, 242; contrary to assertions in 64, p. vi, and others), implying that none of the enumerated assumptions is strictly sufficient. In lieu of defensible alternatives, a sufficient assumption is that observed distributions of data are identical except for the features actually compared in the test. See Table 1.

For example, consider the study by Loreau (156) on temporal niche differentiation in carabid beetles. Loreau was interested in whether species shifted their periods of activity seasonally such that niche overlap or another index termed "mean competitive load" is reduced, as might be expected from competition theory. The data consisted of biweekly activity levels by species and habitat (correlated here with successional stage) over a four-year period. Subject to some constraints on the timing of peak activity and on the boundaries of the active period that were intended to preserve biological realism, the observed temporal distribution of activity within a species was seasonally shifted at random, and the two indices were calculated for each species-habitat combination. For each species and habitat, this procedure was repeated systematically (i.e. by exact randomization) when the total number of distinct reorderings was less than 4000, or randomly with replacement of reorderings (i.e. by sampled randomization) 2000 times when the total number of distinct reorderings exceeded 4000. The p-values were determined as in Figure 1.

Results differed among habitats, among constraints imposed, and among species subsets considered. In the beechwood and pinewood habitats, p-values for both response indices tended to approach or achieve statistical significance as increasingly severe and realistic constraints were imposed on the randomization process. In these cases, niche overlap and mean competitive load calculated from the original data were lower than 95% or more of the corresponding values generated from the random seasonal shifts in activity pattern. It is difficult to imagine how these hypotheses could have been tested with these data by standard statistical methods. This example and the continuing debate over the interpretation of such carabid data (53, 266, and their references) indicate some of the challenges that can arise in attempting to operationalize the null hypothesis and specify the most suitable randomization algorithm. Nevertheless, the overall pattern of statistical significance in the present study does suggest increasingly distinct niche differentiation from successional to "climax" beechwood forest, as would be expected according to competition theory and some other possible interpretations.

Monte Carlo

In Monte Carlo methods, a particular random process (e.g. binomial coin flips or a complex stochastic simulation model) is assumed to underlie the observed data in determining confidence intervals or the expected response under the

Table 1 Assumptions and restrictions on parametric, standard nonparametric, and resampling methods.

Assumptions/restrictions[1]	Standard parametric methods	Standard nonparametric: Rank methods[2]	Standard nonparametric: Categorical methods[3]	Resampling methods: Randomization	Resampling methods: Monte Carlo	Resampling methods: Bootstrap	Resampling methods: Jackknife
GENERAL							
Statistically independent data	Yes[4]	Yes[4]	Yes	Yes[4]	No[5]	Yes	Yes[6]
Particular underlying distribution(s)	Yes[7,8]	No	Yes[9]	No	Yes[10]	No	Yes[7,8,11]
Empirical samples must be random	Yes	No	Yes	No	Yes	Yes	Yes
Relatively sensitive to outliers	Yes	No[12]	Yes	No[12]	Yes	No[12]	Yes[13]
Data ranked (or reduced to ranks)	No	Yes	No	No	No	No	No
Low values in data pose problems	No	No	Yes[14]	No	No	No	No
Ties in data pose problems	No	Yes[15]	Yes[15]	No	No	No	No
HYPOTHESIS TESTS FOR ≥ 2 SAMPLES							
Identical underlying distributions	Yes[7,16]	Yes	No	Yes[17]	Yes	No[18,19]	Yes[7,18]
Relatively sensitive to sample size differences	Yes	Yes	No	Yes[17]	No	Yes[18,19]	No[18]

[1] This list is not exhaustive, nor are the table entries definitive. The table is intended to provide general comparisons as guidelines for statistical practitioners.
[2] Mann-Whitney, Kruskal-Wallis, Friedman's ANOVA-by-ranks, Wilcoxon matched-pairs signed-ranks, etc.
[3] Chi-square, G-tests, etc.

[4] Using special techniques (e.g. time-series analysis), temporal or spatial autocorrelations may in some cases be taken into account without serious problems.

[5] The underlying stochastic process (e.g. a Markov chain) may adequately represent non-independence in the data.

[6] This assumption, in effect, is made and violated in all jackknife applications involving variance calculations, generally with unknown consequences (163).

[7] For the simplest and most common applications, much statistical research indicates robustness to moderate departures from this assumption when the underlying distribution in question is the normal.

[8] The Central Limit Theorem assures asymptotic agreement with this assumption for large sample sizes.

[9] Generally chi-square.

[10] Actually, the stochastic process that generates the data distribution is what must be known.

[11] For confidence intervals and hypothesis testing only. In these cases, the jackknifed *statistic*, not the measured variable, is assumed to be normally distributed. In some applications, the jacckknife is robust to underlying non-normality (e.g. 8), but not others (e.g. 179).

[12] Ranks imply limits on effects of extreme values. In randomization tests and bootstrapping, outliers alter both the observed data and the distribution used for comparison (see 64, 163).

[13] See Hinkley (112).

[14] The standard textbook guidelines for chi-square have long been that no expected frequency should be <1.0, and no more than 20% should be <5.0 (40). Expected frequencies in G-tests should be ≥ 5.0 (235). Generally, rows, columns, or both may need to be combined to satisfy these constraints. Results of recent statistical research on this issue are more equivocal (B. F. J. Manly, personal communication). Contingency-table tests based on small sample sizes can instead be conducted by Monte Carlo methods (e.g. see 153).

[15] Ties in rank tests require approximated p-values, and protecting the chance of type 1 error in these cases can reduce power somewhat (64). In categorical tests, "ties" can sometimes arise as intermediates between categories, which generally must be omitted from the test.

[16] Aspin-Welch t-tests are much more tolerant of unequal variances than are standard parametric tests (235, 242).

[17] (219, 242). Assuming identical underlying distributions implies that values from different samples are exchangeable under the null hypothesis. Random assignment to treatment is insufficient to satisfy the assumption; this depends in part on the *responses* to treatments.

[18] Hypothesis tests are not really standardized for these methods.

[19] Bootstrapping samples separately and comparing the observed statistic with the resulting bootstrapped distribution should at least partly relax this assumption. See Manly (162) for an example of this approach.

null hypothesis for hypothesis testing. This random process is then sampled repeatedly (e.g. many coin flips are simulated or the simulation model is run many times), with test statistics calculated in each case. For hypothesis testing (e.g. is a particular coin unequally likely to produce "heads" or "tails"? —or— do empirical observations differ significantly from the model's predictions?), the *p*-value is found from this frequency distribution of test statistics exactly as for randomization (see Figure 1). In fact, randomization is generally considered to be the special case of Monte Carlo tests in which the relevant random process simply samples the distribution of test-statistic values associated with equiprobable rearrangements of data among treatment levels (163). Nevertheless, some differences in assumptions and restrictions arise in comparing more typical Monte Carlo tests (i.e. those in which the observed data are not used to implement the random process) and randomization tests (see Table 1).

Monte Carlo methods are often used to generate confidence intervals (whereas this is possible but uncommon and usually cumbersome with randomization—e.g. see 163, p. 18–20). Though not particularly difficult, this is procedurally more complex than hypothesis testing because it requires accumulating and maintaining ordered arrays of extreme values of the statistic (corresponding to the tails of the distribution) as these are generated (see Figure 2, which illustrates the "percentile method"; more "adventurous" methods, with clear advantages in some cases, are described e.g. in 67 and 69; for recent work on Monte Carlo methods see 12, 82, 85, and their references).

Consider a demographic study of the colonial gorgonian *Leptogorgia virgulata* using projection matrices and Monte Carlo methods to analyze time-varying population growth (95). Field measurements of recruitment, colony growth, and survival for five size classes over 24 months, supplemented by other fecundity data, were used to construct 23 5 × 5 monthly projection matrices. Each entry represented the expected number of individuals in the row size class that arose by survival, growth, or reproduction from an individual in the column size class one month before. Multiplying this matrix by a column vector representing the numbers of individuals in each of the five size classes at the beginning of the month projected by the matrix yielded the numbers present in each size class a month later.

The less complex of two Monte Carlo applications in the paper concerns determinations of elasticity (i.e. proportional contributions by recruitment, growth, and survival rates to population growth rate) using the matrix techniques. The question of interest in the gorgonian study was whether the observed patterns in the data could be attributable simply to the general form of the matrices rather than their biological details. If so, then an arbitrary distribution of nonzero entries in the matrices should generate elasticity

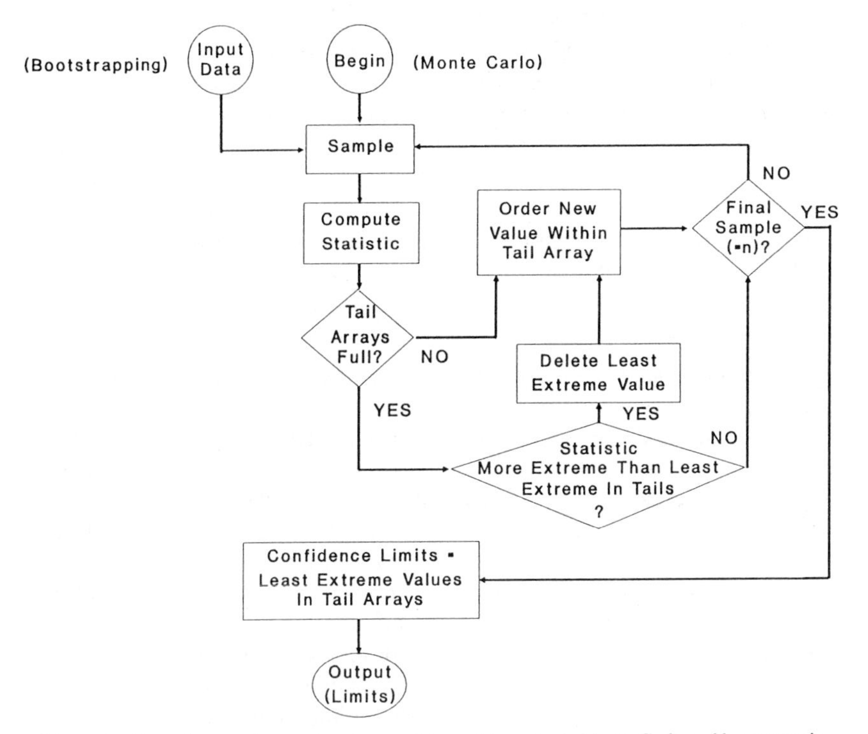

Figure 2 Flow diagram for constructing confidence intervals via Monte Carlo and bootstrapping according to the percentile method. This requires a computer program that uses "tail arrays" to collect and reorder the smallest and largest values of the statistics generated. If the confidence level of interest is $100(1\text{-}\alpha)\%$, where α is the corresponding significance level, and n values are computed to estimate the confidence interval, then each tail array for two-tail limits will contain $1 + \alpha n/2$ values (ignoring any fraction). Thus if $n = 10000$ and $\alpha = 0.05$, then the tail arrays ultimately hold the 251 largest and the 251 smallest values of the statistic. Initially, the two arrays are filled by the first 502 values, such that the larger values are ordered in one tail and the smaller values are ordered in the other. Then each subsequent value smaller than the largest in the lower tail or larger than the smallest in the upper tail is ordered within the appropriate array, and the least extreme value is eliminated; intermediate values, insufficiently extreme for either tail, are not stored. After all n values have been calculated and ordered appropriately, the interval defined by the least extreme value in each of the two tails is the confidence interval. One-tailed confidence intervals are handled similarly. Suitable programs are provided by Noreen (186). With bootstrapping, substantial bias can result from this straightforward approach in some cases (see 67 and 69 for some ways of dealing with this potential problem)

patterns and vital rates statistically indistinguishable from those observed. One thousand random projection matrices were constructed with the same zero elements as in the data matrices, but with all nonzero elements drawn from a uniform distribution ranging from zero to one. (Note that if the nonzero elements had repeatedly been randomly scrambled, rather than drawn from a particular statistical distribution, then this would have been classified as a randomization test.) Some of the observed pattern did indeed seem to be mimicked by the random matrices; but Monte Carlo tests showed that the gorgonian recruitment and population growth rates used as test statistics were significantly atypical of vital rates derived from the random matrices. So the tests helped separate general features of such matrices from the system-specific information contained in the data. As in the carabid randomization example above, it is difficult to see how any approach other than resampling could have been useful here.

The Jackknife

The jackknife, like its all-purpose namesake, was intended to offer crude but effective assistance when a more precise tool is unavailable (252). It provides systematic methods of resampling the actual data using relatively few calculations that can often be done efficiently on a calculator. The direct results of these calculations are an "improved" (i.e. less sampling-biased) estimate of some sample parameter (e.g. mean, kurtosis, intrinsic rate of increase) and often of the approximate variance and confidence interval associated with the estimate. The confidence interval is sometimes used in hypothesis tests (occasionally the jackknifed data themselves, known as pseudovalues, are used; see 114 and 180 for reviews on the jackknife).

Though higher-order versions may occasionally be useful (e.g. 194, 195), the first-order jackknife is by far the most commonly used and proceeds as follows: Suppose that the parameter of interest K (e.g. the true standard error of the mean for the underlying normal distribution of means) is estimated appropriately over the whole sample of m observations as k. Pseudovalues $\bar{\kappa}_i$ associated with each observation i are then obtained as $\kappa_i = k - (m - 1)(k_{-i} - k)$, where k_{-i} is just the standard parameter calculation with the ith observation deleted from the sample. The expression on the right-hand side of this equation is the sample parameter estimate minus a bias term, reflecting the deviation of the i-deleted estimate k_{-i} from the full sample estimate k. The mean of the pseudovalues $\bar{\kappa}$ is then the jackknife estimate of K. The difference $k - \bar{\kappa}$ measures the overall sampling bias of the original estimate k (bias can, for example, distort estimates of population density, particularly when the individuals are strongly clumped in space; see 67 for derivations of the above relationships).

Ignoring the correlations necessarily present among the pseudovalues, calculating their variance s^2 in the usual way, and dividing by the number of observations then generates the variance s^2/m of the jackknife estimate κ. Now the assumption that such jackknife estimates are based on normally distributed error yields the parametric confidence interval for the estimate: k

$\pm\ t^*s/m^{1/2}$, where t^* is the appropriate two-tailed critical value of the t-distribution with m-1 degrees of freedom.

Note that the variance estimate requires assuming that the correlations among pseudovalues are unimportant, but general conditions in which this might be valid have not been established (163). The normality assumption, though justifiable by the Central Limit Theorem for large sample sizes, is difficult to evaluate with the smaller-sample applications where it is more dubious. See Table 1.

In a study of interclutch intervals and reproductive success of feral pigeons nesting on a building at the University of Kansas, Johnson & Johnston (131) used the jackknife to test the relation between three selection parameters and four morphological features. Over 600 banded birds were included in the study; survival and reproductive activity were observed several times per week over a 17-month period and associated with measurements of body mass, tarsus length, bill length, and bill width. The three selection parameters of interest here were the standardized directional selection differential (i), the slope of relative fitness regressed on the morphological trait; the standardized stabilizing selection differential (C), for which a positive value indicates disruptive selection and a negative value implies stabilizing selection; and the standardized directional selection gradient (β), resulting from multiple regression of relative fitness on the combined morphological variables. Though this was not explicitly stated, the parameters were presumably considered significantly nonzero when the two-tailed confidence interval failed to include zero, following the procedure for constructing the interval that is outlined above.

Results indicated a highly significant directional selection differential (i) and gradient (β) for female body mass, and a significant directional selection differential for female bill length, interpreted here as correlated selection. Thus fecundity selection related to interclutch interval apparently "targets" female body mass. In this example, for parameters of unknown statistical distribution calculated over the whole dataset, only the jackknife and bootstrap and their close kin could readily estimate the sampling variation required for hypothesis tests.

The Bootstrap

Bootstrapping is a quite recent technique (66) that is still developing rapidly and attracting much attention in the statistical literature (e.g. 57, 68, 77). Like randomization and the jackknife, bootstrapping focuses on resampling the actual data to reveal some of the subtler patterns they imply (in fact, results obtained from the bootstrap are often closely approximated by those from the jackknife—66). Here, the basic notion is that the data themselves, viewed as a frequency distribution, represent the best available image of the frequency distribution from which they were drawn. Thus the bootstrap metaphor refers to the sense in which the data are used in their own statistical analysis. To bootstrap a confidence interval for a statistic (e.g. mean, skewness, or species diversity) calculated from a single data set of m observations, for example, simply requires m random draws with replacement per sample from the original data, calculating the statistic, and repeating the process many times according to the scheme illustrated in Figure 2. Here again, this simply specifies a particular random process that technically represents a special case of the Monte Carlo method.

Bootstrapping can also be used in hypothesis testing (e.g. 49, 76, 100); for example, with data from each treatment level (or data set to be compared) sampled with replacement separately, tests can be formulated according to the extent of overlap between confidence intervals or by combining the bootstrapped samples to calculate a test statistic (as in Figure 1; see 163, p. 28, and 186, p. 80). This can still be considered a special case of the Monte Carlo method, but since separate random processes are used to generate the separate samples from which the comparisons are made, the approach is quite different from the usual Monte Carlo approach. Bootstrapping is distinct from randomization, which redistributes the original data set over treatment levels, and it contrasts with the parametric and less computation-intense jackknife approach. By keeping the sampling process separate between the compared treatment levels, bootstrapping should be less dependent than most other statistical methods on similarity in underlying statistical distributions among treatment levels (B. F. J. Manly, personal communication; see Table 1).

In an extensive study of predator and parasitoid selection pressure on gall size of the goldenrod gall fly *Eurosta solidaginis*, Abrahamson et al (1) used bootstrapping to avoid problems with non-normality and correlated samples that arose in previous analyses. Selection intensities on gall size attributable to natural enemy attack were calculated as the difference between the mean gall diameter of the selected individuals and the population-mean gall diameter, divided by the population standard deviation. For each of 20 populations and two mortality sources (i.e. insects and birds), the observed number of linked observations (gall size in *mm*, survival from the relevant natural enemy as 0 or 1) were sampled with replacement from the original data, the selection intensity was calculated, and this process was repeated 1000 times to generate a two-tailed confidence interval, as in Figure 2. An observed selection intensity was considered significantly nonzero if its lower confidence limit was greater than zero. Two selection intensities were considered significantly different if their 95% confidence intervals did not overlap; this latter would be a very conservative approach to hypothesis testing, except that the p-values were not adjusted for the large number of comparisons implied.

The many significant selection intensities imposed by insects were all positive, and the few significant selection intensities imposed by birds were negative. Taken together, natural enemies therefore seem to generate stabilizing selection on gall size, though the parasitoid effect predominated, resulting in some overall directional selection for increasing gall size as well.

RECENT APPLICATIONS IN ECOLOGY AND EVOLUTION

Overview

To determine how randomization, Monte Carlo, jackknife, and bootstrapping methods are currently being used in the literature, I conducted a search of a large computer database (BIOSIS Previews on-line database, 2100 Arch Street, Philadelphia, Pennsylvania 19103-1399 USA) for the publication years 1985–1990 and directly examined all issues of the journals *Ecology* and *Oecologia* for 1985–1991.

During the period of interest, approximately 9000 biological journals were being abstracted by BIOSIS, apparently including all major journals in ecology and evolution. Searching titles, abstracts, and key words yielded 391 references from 154 journals and 11 chapters from books, once the few obvious mistakes were eliminated by a direct scan of the abstracts. I relied on the BIOSIS classification scheme to draw appropriate distinctions between references classified as "ecology," "evolution," or both. It is possible that the Monte Carlo category is somewhat inflated relative to the others, since the term is sometimes used for a wider range of simulation methods than just the statistical techniques of interest here; but the direct examination of journals indicated that any such effect would be minor.

Some results of the computer search are presented in Table 2. All four resampling methods are well represented in the recent literature, with Monte Carlo methods overall about twice as frequent as bootstrapping, which in turn was almost twice as common as either randomization tests or the jackknife. The hypothesis that resampling methods are becoming more common in the literature was corroborated statistically, though the evidence to support this for any particular method was more equivocal (randomization, Monte Carlo, bootstrap) or clearly contradictory (jackknife) (tested by randomization; see Table 2 and Appendix 1). Each of the four methods (and all taken together) was used disproportionately in evolutionary studies relative to ecological studies (Monte Carlo tests; see Table 2, Appendix 1), as suggested by the observed proportions v of evolutionary studies (v = 0.144–0.360) relative to the overall proportion in all papers published (0.098).

I scanned the methods and results sections and all figures and tables of the 1485 articles published in *Ecology* and the 2128 articles published in

Table 2 Number of ecology and evolution papers identified in a BIOSIS search by publication year that use one or more of the focal statistical techniques[1]

Method	Topic	Publication year						Statistical results[2]	
		1985	1986	1987	1988	1989	1990	Hypothesis 1	Hypothesis 2
Randomization	Ecology	5	2	6	3	7	14	b = 0.313	v = 0.2609
	Evolution	2	0	2	2	7	2	r = 0.617	p = 0.0003**
	Either	7	2	8	4	13	15	p = 0.1028	
Monte Carlo	Ecology	22	17	24	32	40	45	b = 0.787	v = 0.1443
	Evolution	5	3	12	6	5	6	r = 0.851	p = 0.0016*
	Either	26	20	33	36	44	50	p = 0.0194	
Bootstrap	Ecology	11	3	7	12	11	18	b = 0.595	v = 0.3596
	Evolution	2	5	2	5	6	17	r = 0.647	p = 0.0001***
	Either	13	8	8	16	15	34	p = 0.0750	
Jackknife	Ecology	11	8	11	6	4	4	b = −0.308	v = 0.2642
	Evolution	2	2	2	4	4	5	r = 0.830	p = 0.0001***
	Either	13	9	12	8	7	9	p = 0.9833	

Any of these	Ecology	44	28	44	51	62	77	b = 1.407	v = 0.2285
	Evolution	11	11	18	15	22	27	r = 0.736	p = 0.0001***
	Either	54	38	57	62	79	101	p = 0.472*	
Total sample searched	Ecology	36,525	39,175	44,880	46,862	45,552	48,511		
	Evolution	4,364	5,004	5,716	5,464	5,523	6,315		
	Either	39,631	42,801	48,897	50,654	49,679	52,752		

[1] Data were obtained using the on-line BIOSIS database on December 6 and 16, 1991, searching for the "words" randomi(s/z)ation and permutation (combined here as randomization), Monte Carlo, bootstrap-, and both jack-knif- and jackknif-. After removing a few inappropriate references (e.g. those on jackknife clams and on jellyfish found at Monte Carlo), the remaining 391 abstracts were used to classify the studies for the BIOSIS contributions to Table 3. A total of 154 biological journals and 11 edited volumes were represented in the 1985–1990 data. The eleven journals with eight or more papers in the sample are *Atmospheric Environment ($n = 9$), Biometrics (11), Canadian Journal of Fisheries and Aquatic Sciences (18), Ecology (17), Evolution (18), Genetics (10), Journal of Hydrology (8), Journal of Wildlife Management (8), Limnology and Oceanography (8), Oecologia (10), and Systematic Zoology (8).*

[2] Procedures are described in the Appendix. The hypotheses tested are that the focal methods considered in this review (i) are becoming more commonly used over the last several years and (ii) are used in differing frequency in ecology vs evolutionary studies. Results shown for Hypothesis 1 are based on frequencies normalized (i.e. divided) by the total sample searched in each year, indicating frequency per published paper. In each case, b is the slope of a least-squares regression line through the data, r is the product-moment correlation coefficient between normalized frequency and publication year, and p is the p-value of the randomization test. (Trends were in consistently better agreement with this hypothesis when frequencies were unnormalized, but this approach seemed less useful a priori.) For Hypothesis 2, v is the fraction of all exclusively ecology or evolution papers that are classified by BIOSIS as evolution. For both hypotheses, tests based on the "any of these" data were considered designed orthogonal comparisons, one-tailed for Hypothesis 1 and two-tailed for Hypothesis 2. Tests of the four individual methods were contingent on significance of the designed comparisons; in these cases, the Bonferroni adjustment (multiplying the observed p-value by the number of tests) was used. If p' is the p-value adjusted as appropriate for the number of tails and the number of tests, then significance levels are: ***, $p' < 0.001$; **, $0.01 > p' \geq 0.001$; *, $0.05 > p' \geq 0.01$; otherwise $p' \geq 0.05$.

Oecologia from 1985–1991. (For 1985–1990 only, there were 1270 articles in *Ecology* and 1841 in *Oecologia.*) Regardless of the authors' original designation, I classified methods as Monte Carlo when a mathematical random process was executed repeatedly to generate an estimate of biological variation (e.g. a confidence interval) or to test a hypothesis, following Figure 1. Analyses specifically classified as binomial tests were excluded (except as explicitly noted), though all were equivalent to exact randomization tests (see below).

During 1985–1990, 87 or 6.9% of the *Ecology* papers and 53 or 2.9% of the *Oecologia* papers included one or more of the focal resampling techniques. (In *Ecology,* allowing for the 8 papers each using 2 of the methods, there were 31 randomization, 33 Monte Carlo, 15 jackknife, and 16 bootstrap. In *Oecologia*, with 3 papers each using 2 of the methods, there were 21 randomization, 22 Monte Carlo, 5 bootstrap, and 8 jackknife.) The BIOSIS search was thus relatively inefficient overall (17/87 = 19.5% for *Ecology,* and 10/53 = 18.9% for *Oecologia,*) and the efficiencies probably differed among methods. (This is not particularly surprising, since statistical methods may not often warrant mention in the title, abstract, or key words, though some may be more likely to be mentioned than others.) The overall number of ecology/evolution articles during 1985–1990 that used these resampling methods can be very roughly estimated as the total number of articles identified by the BIOSIS search divided by the mean of these two efficiencies expressed as a decimal fraction, which to the nearest integer equals 2036. Clearly, a thorough and comprehensive review of this and more recent material would be overwhelming, both for reviewers and readers.

Combining the full 1985–1991 direct-examination data with the 1985–1990 BIOSIS results and classifying the papers by content generated Table 3. Notice that some particularly controversial issues in ecology and evolution (e.g. null models, size-ratio theory, detecting density dependence, phylogeny) are well represented here, perhaps mainly to exploit the considerable flexibility of resampling methods in applications involving nonstandard models and test statistics. This flexibility can be a mixed blessing, however, as I note below. The poor representation of behavior and behavioral ecology in the table is probably artifactual, reflecting the separation of behavior from ecology and evolution within BIOSIS.

Some Active Areas of Application

NULL MODELS, COMPETITION, AND COMMUNITY STRUCTURE Contemporary interest in competition as a mechanism underlying community structure led in the 1970s to the formulation of null (or neutral) models, with which statistical tests of predicted patterns could be conducted (35, 223, 229). Since the notion of a probabilistic model that can generate the statistical distribution consistent with a null hypothesis is the essence of Monte Carlo hypothesis testing, Monte Carlo tests (e.g. 201, 223) and closely related exact randomization tests (e.g. 50 and 247, which referred to them using the broader term "binomial tests") were soon prominent in these analyses. The controversy that erupted between those formulating null models for this purpose (42–44, 229, 247) and those considerably less enthusiastic about this approach (41, 56, 91–93) provides a cautionary tale: the potential for differing null models, misunderstandings of methods, procedural errors, and alternative interpretations of similar results can be high with resampling methods (cf Monte Carlo tests in 43, 44, and 93; see 262). Nevertheless, the null model approach seems to have taken hold in the recent literature, partly via Monte Carlo methods (Table 3; see 202, 265).

There is much relatively untapped potential to use resampling tests for detecting community organization (e.g. guild structure: 124, 129, 265) and community similarity (118, 263, 264), and for testing whether environmental factors can account for community structure (29, 83). In assessing species diversity and the variation associated with these estimates, the jackknife has been used most often (e.g. 110, 194, 195), but bootstrap applications (e.g. 23) may become more common.

Much of the null model controversy has addressed the distribution of species abundance and presence/absence on islands (214, 244, and analogous situations considered in 265). A good overview of this issue and problems associated with choosing appropriate constraints on randomization is provided by Manly (163, p. 233 ff).

Numerous recent attempts to test for niche differences (34, 109, 162) and to measure overlap (2, 107) have used resampling methods, particularly randomization (125, 152, 265; see 202). In other cases, temporal niche shifts have been tested by resampling (61, 206, and the binomial or exact randomization test in 50). Simberloff & Boecklen's forensic analysis of Santa Rosalia (230) stimulated several resampling tests (96, 271, and the equivalent of exact randomization in 19) of the constant-size-ratio hypothesis from Hutchinson's original paper (120).

Considerable recent work in plant ecology has focused on competition from immediate neighbors (27, 136, 251) and related distortion of the population size distribution (133, 143, 227). The geometry of access to resources and thus of potential response to competition has also been characterized (136, 215). These plant neighborhood-competition and size-distribution citations involve the gamut of resampling methods considered in this review, both for hypothesis tests (27, 251) and to estimate confidence intervals for the Gini coefficient (an indicator of size inequality; 133, 227) or to calculate skewness by jackknifing (143).

Table 3 Classification by research topic and statistical method of the relevant papers identified in the BIOSIS survey (1985–1990) and by direct examination of journals (*Ecology* and *Oecologia*, 1985–1991)[1]

TOPIC/Subtopic	Randomization	Monte Carlo	Bootstrap	Jackknife
Competition (total)	31	36	20	17
Null models	3	12	0	0
Niche differentiation, overlap & breadth	14	9	4	4
Size-ratio theory of niche displacement	1	3	0	1
Niche-shift dynamics & interaction intensities	3	1	1	0
Plant size hierarchies	1	0	7	3
Plant neighborhood competition	4	5	0	1
Community structure (total)	14	16	7	7
Detecting organization	2	5	2	0
Diversity	2	2	3	6
Community similarity	7	2	2	1
Temporal variability and stability	1	5	1	2
Detecting density dependence	7	4	0	0
Spatial patterns and processes (total)	15	33	3	9
Dispersion & spatial pattern	6	17	1	3
Dispersal & migration	5	7	1	4
Scale effects	4	3	0	1
Demography (total)	5	47	10	8
Population size of density	2	19	6	7
Vital rates	2	9	4	7
Growth, size & age relationships	1	7	0	0
Stock-recruitment relations	0	5	0	0
Agricultural/fisheries	0	8	0	2
Environmental factors (total)	4	58	10	3
Absorption & scattering of light	0	16	1	0
Air-quality models & indicators	0	10	4	0
Aquatic environmental quality/toxicology	2	8	3	2
Lake & stream acidification	1	4	1	0
Surface, soil & groundwater	1	13	0	1
Behavior/behavioral ecology (total)	10	16	2	2
Social organization	3	2	0	0
Foraging	6	7	2	1
Evolution/evolutionary ecology (total)	61	44	49	19
Selection intensity & response	0	10	3	4
Genetic differentiation & correlation	1	2	2	2
Mutation rates	0	7	0	0
Morphometric comparisons	3	2	2	3
Phylogeny	5	1	33	9

Statistical and modeling methods (total)	41	63	43	44
Analysis of variance	3	2	0	1
Regression & correlation	5	8	0	5
Mantel's test	7	0	0	0
Discriminant function analysis	2	2	0	5
Nearest-neighbor analysis	1	6	0	0
Sensitivity, error & uncertainty	0	14	1	3
Power analysis	0	6	0	1
Confidence intervals & variance	1	4	31	16
Bias estimation & reduction	0	10	7	8
Grand total	145	330	139	114

[1] Direct search of the two ecological journals located all of the references on detecting density dependence and almost all of the competition studies. The statistical and modeling methods references were derived both from BIOSIS and from the direct search. References for the remaining topics were obtained almost entirely or entirely from the BIOSIS search. Data shown here were collected using a lengthier list of topics and subtopics, but those categories accumulating fewer than five citations are not shown; topic totals include the additional citations from subtopics not shown, and the grand total includes citations from topics not shown. Many of the papers are tallied in more than one category.

An underutilized randomization method of very wide potential application, particularly in community analyses, is Mantel's test (119, 166; see the description and examples in 163 and 165). This flexible technique tests for correlation between two (or more) square distance matrices. Typically, entries in one matrix express Euclidian distances (or some alternative measure) between (say) species in quantitative multivariate features (e.g. in diets), and the other matrix may represent a postulated pattern among species (e.g. zeroes and ones indicating membership or not in the same guild). By randomly reassigning rows and columns of one matrix to species, recalculating the correlation between corresponding off-diagonal matrix elements (where the test statistic is the sum of the multiplicative products of these corresponding matrix elements), and then repeating this sequence many times, the statistical tendency in the original data for the postulated pattern to match the distance pattern can readily be assessed (e.g. see 198).

DETECTING DENSITY DEPENDENCE Another controversy of long standing in the ecological literature concerns the role of density dependence in population dynamics. (See e.g. 11, 55, 267, and their references for evidence that the controversy continues unabated.) Two important milestones were the initiation of experimental field tests of density dependence (70) and the formulation of statistical methods to detect density dependence in temporal sequences of density data (26). Resampling methods have proven useful in both of these approaches, particularly the latter (48, 51, 54, 203, 204, 210, 258, 259; an application to analysis of a field test is in progress—D. M. Johnson, T. H. Martin, L. B. Crowder, P. H. Crowley, in preparation).

Though concerns have been expressed about the potential for detecting density dependence in density sequences (84), recently developed methods, particularly the two tests based on randomization methods by Pollard et al (204) and Reddingius and den Boer (210), appear sufficiently powerful to be useful (48, 51, 259). Recent variations on this randomization theme have been used to evaluate bias in *k*-factor analysis (258), to extend the notion of detecting density dependence to the community level (48, 51), and to derive testable predictions about the direction of density changes (48, 51).

The Pollard et al (204) "randomization test" and the Reddingius and den Boer (210) "permutation test" both involve scrambling the order of the observed changes in log-transformed density (a measure of population growth over the time interval) for comparison with the observed sequence. Pollard et al (204) used the correlation coefficient between density at the start of each interval and the associated change in density during the interval as the test statistic (density dependence implies an inverse correlation; the randomization test avoids the problems inherent in the analogous parametric approach noted in 159 and 239). Reddingius and den Boer (210) used the log-range between the highest and lowest densities reached in the density sequence as the test statistic (density dependence implies a small log-range). Other test statistics may be more appropriate or powerful in particular cases (cf the "violation number" statistic in 48 and 51). It may often be helpful to use several different tests and test statistics on the same data set, since the test results are sometimes complementary (51, 259), though this may raise concerns about adequately protecting the chance of type 1 error over all tests.

SPATIAL PATTERNS AND PROCESSES Characterizing spatial patterns and processes is a major challenge in contemporary ecological research. A diverse array of resampling approaches has been used for this purpose. Descriptive methods include assessing the spatial distribution of sparsely sampled points and the spatial areas most closely associated with point locations (e.g. tree locations—236 and 136, respectively), and particularly spatial autocorrelation (149, 232). Hypothesis tests have been used to detect nonindependence of animal locations (237, 248); variations in territory size (249); differences in dispersion among size classes, species, and quadrats (105, an application of Mantel's test); differences in association of plant distribution and abundance with taxonomic composition vs vegetation structure (222, also via Mantel's test); and an association between spatial distribution and temporal dynamics (228). Monte Carlo methods have improved and extended the classic nearest neighbor analysis of Clark & Evans (39; see 33; 151; and 163, p. 21-23 and chapter 7).

In other cases, geographical limits of populations have been established (220), and the implications of spatial scale (4, 122, 215) and of environmental heterogeneity (4, 215, 222) have been addressed.

In considering insect dispersal processes, Monte Carlo simulation has been used to evaluate the need for a stochastic formulation to predict dispersal (253). Error associated with estimates of the diffusion coefficient (160, 188) or of the radius of patch detection (103) has been assessed primarily with the jackknife.

ESTIMATING POPULATION SIZE AND VITAL RATES Resampling methods, particularly Monte Carlo, are now in fairly common use to reduce bias and determine error associated with estimates of population density (111, 173, 181). The "smoothed bootstrap" (226) and randomization tests (171, 269) have been used to detect density changes, mainly in non-experimental studies.

One of those last randomization examples (i.e. 269) invoked an approach known as MRPP (multiresponse permutation procedures—176–178). MRPP, a special case of Mantel's test (163, p. 209), is conceptually consistent with graphical representations of the data and readily extends to multivariate problems. With this method, predefined groups (e.g. sites, treatments) can be tested for differences using standard statistical distance measures, from which a test statistic is derived and then assessed by ordinary randomization procedures (Figure 1). Interestingly, standard t and F tests and common nonparametric tests are special cases of MRPP, though practitioners argue that nonstandard formulations are generally more appropriate (269).

Following the comparison of jackknife and bootstrap methods by Meyer et al (175), there has been much recent interest in measuring and testing for differences in demographic costs of predator defense (13, 212, 260) and of other environmental factors (88, 139), as measured by the per-capita increase rate of zooplankton. (See 95 for an assessment of temporal changes in the per-capita increase rate derived from dominant eigenvalues of a matrix model.) Resampling methods have also been applied to tests and error estimates for other vital rates (birth rate: 58; mortality: 153; relative growth rate: 37; transmission rate of an insect virus: 63; many different vital rates: 254), reproductive effort (86), and extinction rate (199, 200).

ENVIRONMENTAL MODELING As the need for reliable environmental predictions and monitoring has steadily increased, a broad range of relatively realistic, quantitative models has appeared in the basic and especially in the applied ecological literature. A focal issue in many of these studies is evaluating the model's fit to data; for probabilistic models, Monte Carlo methods are often the best option and have commonly been used.

With regard to aquatic environments, resampling has been applied in toxicological models (20, 24, 221) and laboratory tests (205), time-series analysis of BOD data (197 via the "Bayesian bootstrap"), testing sensitivity of lakes to phosphorus loadings (28, 150), estimating an index of water quality

(97), and assessing the impact of acid deposition (62, 128, 132). Applications in soil and groundwater systems include those on soil hydraulic properties (117, 234), estimating runoff (10, 60, 99), and monitoring groundwater quality (155, 185, 240). In the atmosphere, resampling techniques have been used with models of carbon dioxide uptake and exchange (121, 138, 268), for impact assessment of radionuclide fallout (21, 211, 261), and especially in air-quality models and indicators (e.g. 25, 101, 108).

An important component of many climate and plant-growth models, both aquatic and terrestrial, is absorption and scattering of incident solar radiation. Monte Carlo applications are particularly common in these studies (e.g. 5, 38, 94).

EVOLUTIONARY PROCESSES AND RATES Resampling methods figure prominently in analyses of natural, sexual, and group selection. Examples include the introductory case studies of the jackknife and the bootstrap early in this review (1, 131), studies determining the magnitude of sexual selection (172, 187) and group selection (78, 102), and others concerned with various responses to selection (142, 182, 257).

Rates of evolution have been assessed and contrasted via resampling applications (78, 90, 140), as have mutation rates (79, 106, 189, 190) and evolutionary implications of genetic drift (189, 190, 217). In an analysis of taxon extinction rates, Raup & Sepkoski (209 and references therein) used randomization tests to identify significant periodicity of major extinction events in the geologic record (also 116; see 116 and 193 on speciation periodicity), but Quinn (207) argued that bootstrapping is more appropriate for this purpose (see 16 and the overview in 163, p. 192 ff).

All four resampling methods have been used to detect genetic differentiation between populations based on immunologic (225), electrophoretic (45, 59), and nucleotide-difference (216, 246) data. Discriminant function analysis, particularly with the help of Monte Carlo (225) or randomization (238) methods, can prove useful in such studies.

PHYLOGENY Phylogenetic analysis has evolved rapidly since the 1970s with widespread use both of molecular techniques and of computer simulation and data analysis. Pioneering simulation studies by Raup et al (208) demonstrated the possible importance of stochastic processes and potential biases in interpreting phylogenies. Some of the early work on null models emphasized biogeographic data (e.g. the binomial or exact randomization test of 247; also see 41 on the avoidable and unavoidable biases in such studies), and the usefulness of statistically contrasting proposed phylogenies against a null pattern is becoming more widely recognized (73; see the bootstrap approach of 89 and an exact Monte Carlo method in 233).

An important recent development was the formulation of techniques for establishing confidence intervals for monophyletic groups (jackknifing over taxa: 146; bootstrapping over characters: 72; see 196 for a comparative evaluation of these and related methods). Felsenstein's frequently used approach analyses character data contained in a species × character matrix. Bootstrap samples of characters (or, strictly, of the columns of species-specific values for particular characters) are used to construct alternative phylogenetic trees; the percentage of these containing a monophyletic group present in the tree based on the original data then estimates the confidence that the group is indeed monophyletic. Inherent assumptions that characters were sampled randomly and evolved independently raise some concerns about the method's validity (72, 73, 224) but apparently have not deterred applications of this and related bootstrapping approaches (e.g. 126, 127, 270). Bootstrapping clearly predominates in recent resampling studies of phylogeny (Table 3), including both cladistic analyses (e.g. 47, 145, 147) and the phenetic studies emphasized above.

Another issue of current interest concerns whether particular patterns derived from phylogenetic data can be considered nonrandom. Here, randomization tests have been used to scramble character values among species to determine whether the tree derived from the original data required significantly fewer evolutionary step-changes than the trees derived from scrambled character values (nonrandomness was detected in 6 but not in 7; also see an analogous cladistic analysis in 192).

Statistical Methodology

RELEVANT TYPES OF ANALYSIS There is much statistical and biometrical research in progress continuing the development of resampling methods (particularly the bootstrap). Moreover, to a greater extent than with other statistical approaches, each new application tends to extend the methodological possibilities because of the ad hoc nature of resampling analysis. Here, I note how resampling methods have been used to supplement or improve standard statistical methods and to stimulate or enhance new research initiatives as well.

Resampling methods avoid some of the more restrictive assumptions involved in standard regression and correlation analyses (e.g. see 64, p. 197), and there are now many published applications (e.g. regression: 30, 81, 170; correlation: 123, 249, 256). The useful generalized correlation methods known as Mantel's test and multiresponse permutation procedures (MRPP) have already been described and characterized in the literature summary above. Analysis of variance deserves special attention because of its central role in the design and analysis of experiments and because of restrictive assumptions that can proliferate with complexity of the design (191; 64, p. 58 ff; recent applications include 130, 162, and 168). Help needed from statistical researchers on problems associated with ANOVA is noted below.

Many multivariate methods are currently being revitalized and extended via resampling (e.g. cluster analysis: 148, 184, 255; discriminant function analysis: 14, 29, 80; principal components analysis: 241; indirect gradient analysis: 144). It is primarily the methods readily applicable to single samples (Monte Carlo, bootstrap, jackknife) that are of interest in this context, though randomization can be useful for discriminant function analysis (163, 238). See Manly (163 and especially 165) for thorough review of multivariate resampling applications.

Many other standard issues and approaches in experimental design and data analysis have been addressed with resampling methods. Some of the more important of these are assessing errors associated with sampling (2, 103, 256) or direct measurement (3), estimating the power of hypothesis tests (32, 87, 130), determining and reducing bias (2, 173, 183), and determining the appropriate sample size (22, 154, 164).

In empirical studies, the need for methods of analyzing the ecological response to large-scale perturbations (31, 169) has led to some resampling applications associated with intervention analysis (Monte Carlo: 158) or randomized intervention analysis (randomization tests: 32). The approach here is usually based on paired systems, one experimental and one control; each is monitored extensively before and after the experimental system is manipulated, so that some of these observations can be assumed essentially independent (though autocorrelation is directly assessed). This general approach or a successor may prove valuable, particularly where replicated experiments are infeasible, but additional care should be taken to ensure that the null hypothesis is tested against an appropriate alternative (e.g. by transforming to reduce heteroscedasticity or other distributional differences that may confound the test; see 71 and 242).

In modeling studies, there is much current interest in incorporating age or size structure (36, 174) or explicitly representing individuals (52) within population models. Moreover, optimization models now more commonly include stochastic elements (e.g. see 161) or parameter uncertainties that complicate interpretation. In these and similar cases, resampling methods can prove particularly useful in characterizing the model's behavior and evaluating its consistency with empirical observations (e.g. age structure: 137; individual-based model: 157; optimization: 213).

CLOSE RELATIONSHIPS WITH MORE STANDARD METHODS The resampling techniques of interest here are all closely related to the more standard and widely familiar statistical methods. By virtue of conceptual simplicity and the large number of nonparametric tests it has spawned, randomization can be considered fundamental to the standard methods (17, 46, 134). Transforming data to ranks is primarily a device to reduce data sets to a general form that permits construction of nonparametric significance tables, with entries at low sample sizes determined by randomization and at higher sample sizes by normal or chi-square approximations to the randomization results. The first three commonly used tests in Table 4 are examples of these; the sign test is also a kind of rank sum test. Fisher's exact test and the binomial (goodness-of-fit) test are directly calculated cases of exact randomization. The tests listed in the table are just a few of the more common nonparametric tests found in the *Ecology-Oecologia* sample.

Monte Carlo methods are generally used to derive statistical tables for tests based on data assumed to follow particular distributions, such as t, F, and χ^2 tests. In the Monte Carlo tests of interest here, the actual statistical distribution may be unknown, so long as the relevant stochastic process can be simulated according to the scheme in Figures 1 and 2. In some cases among the *Ecology* and *Oecologia* articles, a smaller number of simulations was used to draw conclusions without a formal test (e.g. 104, 245) or were compared with observations using standard categorical tests (e.g. 55, 141) or parametric tests (e.g. 74, 98). Such hybrid approaches may often prove useful where the underlying assumptions can be met, but in several of these cases, the standard Monte Carlo test might have been more defensible and straightforward.

Applications of the jackknife involving hypothesis tests or determination of confidence limits rely on parametric critical values and significance tables (see above). Though the bootstrap is not inherently tied to parametric methods, one area of active development is known as the parametric bootstrap, in which the standard error of the mean is bootstrapped and then used in parametric analyses as with the jackknife (e.g. see 67, 186, 231). Of course, the

Table 4 Percentages of papers published in *Oecologia* (1985–1991) featuring some common nonparametric tests, all of which are (or are equivalent to) randomization tests.

Test	Percentage
Mann-Whitney	11.3
Spearman rank correlation	7.1
Wilcoxon matched-pairs signed-ranks	4.7
Fisher's exact test	2.7
Sign test	1.3
Binomial test	1.2

asymptotic convergence of statistical sampling distributions on the normal distribution at sufficiently large sample sizes is implied by the Central Limit Theorem.

DISCUSSION

Advantages and Disadvantages of These Methods

An attempt to sort out the pros and cons of resampling methods relative to the more standard statistical techniques raises many issues of varying subtlely and complexity (Table 5). When their stringent assumptions are met, parametric procedures maximize power (i.e. the chance of rejecting a false null hypothesis in favor of a true alternative), for a specified type 1 error rate (i.e. chance of falsely rejecting a true null) (186). But rarely at small or moderate sample sizes can all of the assumptions be known or convincingly demonstrated to apply. The conservative approach is then to resort to standard nonparametric methods or resampling.

Nonparametric methods are generally slightly to considerably weaker than the stronger of parametric and resampling methods for several reasons. Essentially all nonparametric techniques in common use were necessarily designed for minimizing computation. In some cases, this resulted in inherently low power (e.g. the sign test—see 135). In others, a loss of power or inadequate protection of the type 1 error rate may be attributed to reducing data to ranks, approximations related to ties in rank tests, continuity corrections at low frequencies for categorical tests, or the possibility of inaccurate approximations in some tables at intermediate numbers of observations (64). Often, the

Table 5 Key features of three categories of statistical methods

Feature	Standard Parametric Methods	Standard Nonparametric Methods	Resampling Methods
Statistical power	High (when assumptions met)	Moderate	High
Known by researchers	Very widely	Widely	Sometimes & increasing
Acceptance	Widespread	Widespread	Common & increasing
Standardization	Very high	High	Moderate
Flexibility	Low	Moderate	High
Assumptions (see Table 1)	Moderate-strong (robust to some departures)	Moderate	Weak-moderate
Population or sample	Population	Sample	Population (except randomization)
Time & effort cost	Moderate	Somewhat lower	Higher & decreasing
Conceptual complexity	High	Moderate	Low-moderate

available tables seriously constrain the analyses by incompleteness (e.g. Friedman's ANOVA-by-ranks), by providing only one-tailed or only two-tailed significance values (Fisher's exact test and chi-square, respectively), or by providing only a rough indication of the *p*-value's magnitude (most nonparametric tests) (see 64).

In contrast, randomization tests yield about the same significance level as parametric methods when the parametric assumptions are met (115, 218) but may have more power than parametric methods when data are from non-normal distributions (64, p.94; 135). Less is known about circumstances in which Monte Carlo, bootstrap, and jackknife methods may be more powerful than standard parametric analysis (but see examples in 67).

An obvious current advantage of using standard parametric and nonparametric techniques is that they are widely known and accepted by editors and other researchers, though resampling methods are now clearly in common use as well. Yet bootstrapping may have been swept into the mainstream of ecological and particularly evolutionary research somewhat ahead of a full, balanced evaluation of its capabilities and shortcomings. Bootstrap confidence limits and hypothesis testing are not always reliable (e.g. see 67); some familiarity with current methodological advances and perhaps some ad-hoc checking could prove important, and use should be restricted to cases where randomization tests and parametric methods are inappropriate (163, 186).

Standardization and flexibility of statistical methods must trade off, to some extent. The psychological shift associated with the conceptually simple resampling approach, in which the data analyst necessarily controls and understands each step from hypothesis formation to designing a sufficiently powerful test statistic (and perhaps the test itself) to calculating an intuitively meaningful *p*-value, can be "liberating" (186). Otherwise infeasible experimental designs (e.g. those based on nonrandom sampling or requiring nonstandard response variables) become available with resampling methods. But this degree of versatility carries the cost that other defensible test statistics or procedures for conducting the test itself may lead to different conclusions (e.g. the survey of "null models" above)—or as a worst case even undermine the objectivity of the data analysis (see 9 and 113). It is thus important to consider a range of alternative test statistics and procedures and to justify the choices made, ideally before the data are analyzed. It should be clear that the appropriate alternative hypothesis would indeed be supported by rejecting the null; an instructive case is the randomization test of differences between means, in which the null hypothesis can be rejected for identical means but different variances (242).

The striking differences in assumptions underpinning the classical and resampling methods (Table 1; 186, p. 84-92) necessarily constrain the options to an extent that is often overlooked or ignored by editors, referees, and researchers. Making dubious assumptions should obviously be avoided where possible, but this should be balanced against the tacit assumptions involved in using some of the more speculative resampling methods (particularly the bootstrap and jackknife).

Randomization tests (and by implication their derivative nonparametric tests) apply only to the samples themselves, not to some underlying population that may have been sampled. This obviates the need for random samples and for certain assumptions about the population of origin, but it also lengthens the interpretive extrapolation from the observed results to the general situation or population of interest. Some such extrapolation is almost always necessary in any case, and this problem is of greater theoretical than applied relevance (64).

Finally, resampling methods currently require on average more expenditure of time and effort per analysis than do classical methods, largely because of the necessary computer programming. In fact, the required programming is often quite straightforward, and programs for many common applications are widely available in the literature (e.g. 64, 163, 186) or as shareware. Soon, commercial mainframe and microcomputer packages will be available (186).

Implications for Interpreting and Communicating Results

In hypothesis testing, the interpretation of results necessarily hinges on the *p*-value, or at least on its magnitude relative to the critical value. An attractive feature of resampling methods is that the direct calculation of the *p*-value obviates the discrete decision-theory distinction between significant and nonsignificant results, cleft sharply if arbitrarily at a knife-edge critical value. Instead, the *p*-value can simply be understood to measure the degree of consistency between the data and the null hypothesis, though the classical significance levels (0.05, 0.01, etc) retain their utility as benchmarks. Moreover, the directly calculated *p*-value may be much easier to communicate to nontechnical decision-makers; as noted in the introduction, in a randomization test of a difference between two means, the *p*-value is simply the proportion of random assignments of data to treatments that gives a difference between group averages at least as large as the difference obtained in the experiment (64, p. 10).

As with other statistical parameter estimates, an error estimate for the *p*-value is desirable, particularly where this error reflects only a moderate number of repetitions (e.g. 1000) for a resampling method (sampled randomization, Monte Carlo, or bootstrapping). In the latter case, the $100(1-\alpha)\%$ confidence interval is well approximated by $p \pm t_{1-\alpha}(p(1-p)\ /n)^{1/2}$ with an infinite number of degrees of freedom, where α is the significance level, $t_{1-\alpha}$ is the critical value of the t distribution for significance level α, and n is the number of repetitions (e.g. see 186, p. 34). Note that these error bounds on the p-value reflect variation derived from the intensity of resampling. More speculatively, it may be possible in some cases to obtain error bounds on the p-value associated with empirical sampling variation using the bootstrap or jackknife, but I am unaware that this has yet been attempted.

In research reports, error estimates via resampling generally require little fanfare—only the name of the method and the number of repetitions (if applicable). Hypothesis testing requires more information in the methods section, including specification and justification of the null and alternative hypotheses. Unless the approach is exotic or particularly central to the presentation, references to support resampling methods are unnecessary. The method, number of repetitions (if applicable), magnitude of the test statistic, and p-value (with confidence interval when appropriate) generally appear parenthetically with results of a hypothesis test.

Methodological Issues That Deserve Attention

Much statistical research remains to be done to develop, improve, and evaluate these resampling techniques. Under the general theme of robustness, both parametric and resampling methods need to be further examined and compared in their sensitivity to non-normality, non-equivalence of distributions (e.g. unequal variances), and sample size. Also, what are the implications for randomization tests of nonrandom assignment to treatment levels and for other resampling and parametric tests of nonrandom sampling? Ideally, such studies should focus on features typical of small-to-moderate-sized samples and the ways such data are actually gathered, rather than exclusively on the characteristics of large statistical populations from which such samples may be drawn (64).

Until resampling methods became generally feasible relatively recently, the available statistical methods were sufficiently constraining that the formulation of null and alternative hypotheses has been relatively straightforward and unambiguous. Now that the horizon for these hypotheses has widened considerably, perhaps new and useful guidelines can be devised that will help practitioners to match hypotheses more effectively to tests and test statistics.

Many of the more complex descriptive and hypothesis-testing techniques traditionally based on relatively assumption-bound parametric methods can be effectively refitted as resampling methods (see above and 163, 165). Of particular importance in ecology and evolution are methods that relate directly to common experimental designs, like analysis of variance. ANOVA has traditionally been considered robust to departures from the standard parametric assumptions, though not all agree (e.g. 18, 243), and more complex variations (e.g. factorial ANOVA, ANCOVA, MANOVA) can be more vulnerable to violations of parametric assumptions (e.g. 191). In parallel with further studies of robustness, the development of resampling ANOVA and its variations should continue (e.g. 8, jackknife ANOVA; 130, randomization ANOVA; 162, bootstrap ANOVA), and controversies like contrasting views on interaction terms in factorial ANOVA (see 64 vs 163) need to be resolved.

Further statistical research on Monte Carlo methods and especially on bootstrapping will continue to attract immediate interest and additional applications in ecology and evolution.

Some Specific Recommendations

1. *Resampling methods should be part of basic statistical training in ecology and evolution.* At least until these methods are incorporated into mainframe statistics packages, this will require some computer-programming skills as well. In exchange, perhaps less emphasis can be placed on standard nonparametric methods.
2. *Parameter estimates should be accompanied by estimates of the associated variation.* Resampling methods make it possible for this principle to be very broadly (if not universally) applied.
3. *With small-to-moderate sample sizes, maintain a healthy skepticism about the appropriateness of parametric analysis.* Even failures to reject normality and equal variances as null hypotheses are rarely conclusive, since the power of tests to evaluate them is low at the relevant sample sizes. When the random process that generated the data is statistically uncharacterized, the conservative approach is to use defensible methods making the fewest strong and unverifiable assumptions.
4. *Transformations should be used to improve the equivalence of distributions in randomization tests in essentially the same way that these are used in parametric analyses.* This should help neutralize a potential problem with standard randomization methods that has often been unrecognized.
5. *Where equivalence of distributions is unlikely to hold or to be achieved by transformation, multisample hypothesis tests can be conducted by bootstrapping.* With this approach (termed "bootstrapped randomization" in 186), data for different treatment levels are bootstrapped independently before the test statistic is calculated. In our present state of ignorance, bootstrapping should not ordinarily be used where parametric or randomization methods apply.
6. *For confidence intervals and hypothesis testing, other more theoretically defensible methods should generally be used instead of the jackknife.* The jackknife is particularly useful for eliminating bias in parameter estimates (67), as a check or extension of the bootstrap (68), or in cases where the other heavily computation-intensive methods are not feasible (67).
7. *Where possible, researchers should attempt to ensure that their exper-*

imental units constitute a random sample of some population of interest. Though it may often be unclear how to accomplish this, the object is to retain defensible options for the statistical analysis. When the case can be made, it may be useful to explicitly identify the population that has been sampled randomly.

8. *In the absence of random sampling, hypotheses comparing two or more samples with equivalent distributions should be tested by randomization.* Randomization can also be used to construct confidence intervals, though these tend to be relatively conservative (64).
9. *Use randomization, Monte Carlo, or bootstrapping methods instead of standard nonparametric methods, particularly when maximizing power is essential.* Any nonparametric test can be replaced by a potentially more flexible and powerful but otherwise equivalent resampling test.
10. *Whenever possible and appropriate, use a large number of repetitions in resampling tests (≥20000).* This is particularly important when it can influence the way the data are interpreted (e.g. when the p-value is near 0.05). For randomization, $n = 1000$ and $n = 5000$ are generally considered minimal for tests at the 5% and 1% significance levels, respectively (64, 65, 167).
11. *When using resampling methods, define null and alternative hypotheses with special care.* Justify these choices in the methods section of the research report.
12. *More attention should be paid to the applicability of assumptions underlying statistical analyses by researchers, editors, and referees.* With resampling methods becoming widely known and commonly used, standards for acceptably thorough and rigorous data analysis should continue to rise.

A summary of some common situations arising in data analysis and the most appropriate methods for dealing with them is presented in Table 6.

ACKNOWLEDGMENTS

It is a pleasure to thank some people who have aided the preparation of this review and development of the ideas herein: Peter Chesson, Dan Johnson, and Jerry Nagle for introducing me to randomization and bootstrapping; Eugene Edgington and Bryan Manly for encouragement and for breaking the ground so effectively; Ken Hudson for helping with the BIOSIS and journal searches of the literature; Bryan Manly, Brian McArdle, Cidambi Srinivasan, Clare Veltman, and participants in the 1992 course on experimentation in benthic ecology sponsored by the University of Umeå, Sweden, for stimulating discussions; Mick Crawley, Dan Johnson, John Lawton, Bryan Manly, Brian McArdle, and Clare Veltman for reading and commenting on the manuscript; Mick Crawley for letting me rummage at will through his journals; Brenda

Table 6 Recommended methods.

Purpose	Sampling	Empirical sample size[1]	Distribution(s) (or underlying model)	Preferred method
Confidence intervals and single-sample hypothesis tests	Random	Large	Any	Parametric
	Random	Small-moderate	Normal	Parametric
	Random	Small-moderate	Known, but non-normal[2]	Monte Carlo
	Random	Small-moderate	Unknown[2]	Bootstrapping
	Non-random	Any	Any	Randomization[3]
Multisample hypothesis tests	Random	Any	Equivalent	Parametric
	Random	Any	Non-equivalent[2]	Bootstrapping
	Non-random	Any	Equivalent	Randomization
	Non-random	Any	Non-equivalent[2]	- - -

[1] In practice, large and moderate sample sizes are generally distinguished subjectively.
[2] Transformations of the data can at least sometimes reshape the distribution(s) adequately for parametric analysis (with random sampling) or randomization tests (with non-random sampling), though this may prove difficult to demonstrate convincingly.
[3] Awkward to implement, and tends to yield conservative confidence limits.

and Chauncey Curtz for new perspectives on null hypotheses; Production Editor Nancy Donham for grace under pressure; Marcel Dekker, Inc., for permission to use a slightly modified version of E.S. Edgington's diagram as Figure 1; John Lawton for hosting my sabbatical year at the Centre for Population Biology; and Lillie, Sarah, and Martin for being patient and understanding when the crunch came. I acknowledge funding from the National Science Foundation grant INT 9014938 and a Visiting Research Fellowship from the Royal Society of London.

APPENDIX: HOW THE DATA OF TABLE 2 WERE ANALYZED

I used randomization and a Monte Carlo method to test two a-priori hypotheses concerning the data summarized in Table 2. The rationale and procedure used in each case are briefly described here as examples of how this general approach can be implemented.

Hypothesis 1: *The focal methods considered in this review are becoming more commonly used over the last several years in ecology and evolution.*

The BIOSIS literature search over the six publication years 1985–1990 on topics in ecology and evolution (Table 2) provides the basis for a test. To assure adequate sample sizes for each separate method, references fitting either or both of these categories for a particular publication year were pooled, yielding publication frequencies for each of the six years to be tested for trend.

Hypothesis 1 predicts a positive trend in publication frequency over years, generating a one-tailed test against the null hypothesis of no trend. I chose the linear regression coefficient b or least-squares slope of publication frequency as a function of publication year to be the response measure. (It can be demonstrated that using the correlation coefficient r for this purpose necessarily produces identical results; moreover, the much simpler metric $\Sigma x_i y_i$, where x_i and y_i are the coordinates of the ith data point, is equivalent to either of these and was the measure actually used in my tests—see 163, pp. 91–92.) Testing Hypothesis 1 then required calculating the observed $b = b_O$ (or the simpler equivalent) and demonstrating that a value as large or larger is very unlikely to have arisen by chance alone.

Now if a trend-free null process generated the observed sequence, then there should be nothing special about the order in which the publication frequencies were observed; any reordering should produce a statistically equivalent sequence (163, p. 92). There are $6! = 720$ different orderings of six numbers. Calculating a regression coefficient b (or equivalent) for each and comparing it with b_O indicated the proportion of these as large or larger than b_O. This proportion was taken to be the probability p that the observed sequence could have been generated by the same kind of trend-free process that produced the other 719 sequences. When p was smaller than the relevant significance level (generally 0.05), I rejected the null hypothesis in favor of the alternative (i.e. Hypothesis 1); otherwise I was unable to reject the null, and Hypothesis 1 was not supported.

This exemplifies systematic randomization, which determines exact p-values. My Pascal computer program (written in Turbo Pascal 6.0) to calculate b, r, and p was just over 100 lines long, the majority of which were needed to generate the 720 reorderings of the data. The solution could instead have been found by sampled randomization using a much simpler algorithm (see 163), half as many program lines, and some additional run time (roughly 30 sec rather than a fraction of a second on a typical 386/387 microcomputer).

Hypothesis 2: *The focal methods considered in this review are used in differing frequency in ecological versus evolutionary studies.*

The data of Table 2 permit this hypothesis to be tested for each method. Only the data in the "ecology" and "evolution" rows of the table were used. First, the few papers within years classified as both ecological and evolutionary were removed from the observed frequencies, eliminating a source of positive correlation. The residuals were then summed along rows (i.e. over years), and the overall proportion v_O of evolutionary studies out of all ecological and evolutionary studies for the given method was calculated as the evolution row sum divided by the sum of the evolution row sum and the ecology row sum. (This is of course equivalent to using the overall proportion of ecological studies, which is simply $1-v_O$.)

To determine whether v_O corroborated Hypothesis 2, I used a Monte Carlo method to generate a null distribution of v-values for comparison with v_O. If a given method were used just as frequently in both kinds of studies, then the observed frequencies in a particular year could have been generated as a sample from a binomial process based on the frequencies of all ecology and all evolution studies in that year. I thus randomly generated the number of studies observed in each year, with the chance of any particular study's being "evolutionary" equalling the "evolutionary" proportion in the given year, and I tallied the distribution between evolution and ecology. When these distributions had been simulated for all six years of the sequence, the overall proportion v of evolutionary studies was found as above from the evolution and ecology row totals. Each of 20,000 such v-values was determined, and the proportion as extreme or more extreme than v_O became the p-value estimate. Though this p-value was derived from a random sample of v-values and was therefore inexact, the large number of iterations assured adequate precision. (See 163, pp. 32–36. The precision of this p-value could be assessed using the confidence-interval calculation described in the Discussion subsection entitled Implications for Interpreting and Communicating Results.)

Special provision must be made for two-tailed tests of hypotheses like Hypothesis 2. I noted whether v_O was larger or smaller than the overall expected value or mean of v, setting the computer program to calculate p from v-values equally far or farther from the mean. As is typical for two-tailed tests, this required that the significance level be halved to 0.025. Thus if 499 or fewer of the v-values were as extreme or more extreme than v_O (as in all cases in Table 2), then the null hypothesis was rejected and hypothesis 2 corroborated. My Pascal program to calculate v_O and p was just over 50 lines and was straightforward to write; the substantial number of iterations required about 10–60 sec to run on a 386/387 microcomputer, depending on the observed publication frequencies for the method of interest.

Literature Cited

1. Abrahamson, W. G., Sattler, J. F., McCrea, K. D., Weis, A. E. 1989. Variation in selection pressures on the goldenrod gall fly and the competitive interactions of its natural enemies. *Oecologia* 79:15–22
2. Abrams, P., Nyblade, C., Sheldon, S. 1986. Resource partitioning and competition for shells in a subtidal hermit crab species assemblage. *Oecologia* 69: 429–45
3. Althawadi, A. M., Grace, J. 1986. Water use by the desert cucurbit *Citrullus colocynthis* (L.) Schrad. *Oecologia* 70:475–80
4. Antonovics, J., Clay, K., Schmitt, J. 1987. The measurement of small-scale environmental heterogeneity using clonal transplants of *Anthoxanthum odoratum* and *Danthonia spicata*. *Oecologia* 71:601–7
5. Antyufeev, V. S., Marshak, A. L. 1990. Inversion of Monte Carlo model for estimating vegetation canopy parameters. *Remote Sens. Environ.* 33:201–10
6. Archie, J. W. 1989. A randomization test for phylogenetic information in systematic data. *Syst. Zool.* 38:239–52
7. Archie, J. W. 1989. Phylogenies of plant families: A demonstration of phylogenetic randomness in DNA sequence data derived from proteins. *Evolution* 43:1796–800
8. Arveson, J. N., Schmitz, T. H. 1970. Robust procedures for variance component problems using the jackknife. *Biometrics* 26:677–86
9. Basu, D. 1980. Randomization analysis of experimental data: the Fisher randomization test. *J. Am. Stat. Assoc.* 75:575–82
10. Benjamini, Y., Harpaz, Y. 1986. Observational rainfall runoff analysis for estimating effects of cloud seeding on

water resources in northern Israel. *J. Hydrol.* 83:299–306
11. Berryman, A. A. 1991. Stabilization or regulation: what it all means. *Oecologia* 86:140–43
12. Besag, J., Clifford, P. 1991. Sequential Monte Carlo *p*-values. *Biometrika* 78: 301–4
13. Black, A. R., Dodson, S. I. 1990. Demographic costs of *Chaoborus*-induced phenotype plasticity in *Daphnia pulex*. *Oecologia* 83:117–22
14. Block, W. M., Brennan, L. A., Gutierrez, R. J. 1991. Ecomorphological relationships of a guild of ground-foraging birds in northern California, USA. *Oecologia* 87:449–58
15. Boik, R. J. 1987. The Fisher-Pitman permutation test: a non-robust alternative to the normal theory F-test when variances are heterogeneous. *Br. J. Math. Stat. Psychol.* 40:26–42
16. Boyajian, G. E. 1991. Taxon age and selectivity of extinction. *Paleobiology* 17:49–57
17. Bradley, J. V. 1968. *Distribution-Free Statistical Tests*. Englewood Cliffs, NJ: Prentice-Hall
18. Bradley, J. V. 1978. Robustness? *Br. J. Math. Stat. Psychol.* 31:144–52
19. Brandl, R., Topp, W. 1985. Size structure of *Pterostichus* spp. (Carabidae): aspects of competition. *Oikos* 44:234–38
20. Breck, J. E., DeAngelis, D. L., Van Winkle, W., Christensen, S. W. 1988. Potential importance of spatial and temporal heterogeneity in pH, aluminum, and calcium in allowing survival of a fish population: a model demonstration. *Ecol. Modell.* 41:1–16
21. Breshears, D. D., Kirchner, T. B., Otis, M. D., Whicker, F. W. 1989. Uncertainty in predictions of fallout radionuclides in foods and of subsequent ingestion. *Health Phys.* 57:943–53
22. Briggs, J. M., Knapp, A. K. 1991. Estimating aboveground biomass in tallgrass prairie with the harvest method: determining proper sample size using jackknifing and Monte Carlo simulations. *Southwest Nat.* 36:1–6
23. Brokaw, N. V. L., Scheiner, S. M. 1989. Species composition in gaps and structure of a tropical forest. *Ecology* 70:538–41
24. Brueggemann, R., Trapp, S., Matthies, M. 1991. Behavior assessment of a volatile chemical in the Rhine River. *Environ. Toxicol. Chem.* 10:1097–104
25. Brusasca, G., Tinarelli, G., Anfossi, D. 1989. Comparison between the results of a Monte Carlo atmospheric diffusion model and tracer experiments. *Atmos. Environ.* 23:1263–80
26. Bulmer, M. G. 1975. The statistical analysis of density dependence. *Biometrics* 31:901–11
27. Cain, M. L. 1990. Patterns of *Solidago altissima* ramet growth and mortality: the role of below-ground ramet connections. *Oecologia* 82:201–9
28. Canale, R. P., Effler, S. W. 1989. Stochastic phosphorus model for Onondaga Lake, New York, USA. *Water Res.* 23:1009–16
29. Capone, T. A., Kushlan, J. A. 1991. Fish community structure in dry-season stream pools. *Ecology* 72:983–92
30. Carnes, B. A., Slade, N. A. 1988. The use of regression for detecting competition with multicollinear data. *Ecology* 69:1266–74
31. Carpenter, S. R. 1990. Large-scale perturbations: opportunities for innovation. *Ecology* 71:2038–43
32. Carpenter, S. R., Frost, T. M., Heisey, D., Kratz, T. K. 1989. Randomized intervention analysis and the interpretation of whole-ecosystem experiments. *Ecology* 70:1142–52
33. Casas, J. 1990. Multidimensional host distribution and nonrandom parasitism: a case study and a stochastic model. *Ecology* 71:1893–903
34. Case, T. J. 1990. Patterns of coexistence in sexual and asexual species of *Cnemidophorus* lizards. *Oecologia* 83: 220–27
35. Caswell, H. L. 1976. Community structure: a neutral model analysis. *Ecol. Monogr.* 46:327–54
36. Caswell, H. 1989. *Matrix Population Models*. Sunderland, Mass: Sinauer
37. Chapin, F. S. III, Groves, R. H., Evans, L. T. 1989. Physiological determinants of growth rate in response to phosphorus supply in wild and cultivated *Hordeum* species. *Oecologia* 79:96–105
38. Chuh, H. T., Tan, H. S. 1990. A multiconstituent and multilayer microwave backscatter model for a vegetative medium. *Remote Sens. Environ.* 31: 137–54
39. Clark, P. J., Evans, F. C. 1954. Distance to nearest neighbor as a measure of spatial relationships in populations. *Ecology* 35:445–53
40. Cochran, W. G. 1954. Some methods for strengthening the common hi^2 tests. *Biometrics* 10:417–51
41. Colwell, R. K., Winkler, D. W. 1984. A null model for null models in biogeography. In *Ecological Communities: Conceptual Issues and the Evidence*,

ed. D. R. Strong, D. Simberloff, L. G. Abele, A. B. Thistle, pp. 344–59. Princeton, NJ: Princeton Univ. Press

42. Connor, E. F., Simberloff, D. 1979. The assembly of species communities: Chance or competition? *Ecology* 60: 1132–40
43. Connor, E. F., Simberloff, D. 1984. Neutral models of species' co-occurrence patterns. See Ref. 41, pp. 316–31
44. Connor, E. F., Simberloff, D. 1984. Rejoinder. See Ref. 41, pp. 41–43
45. Corbin, K. W., Livezey, B. C., Humphrey, P. S. 1988. Genetic differentiation among steamer-ducks (Anatidae: Tachyeres): an electrophoretic analysis. *Condor* 90:773–81
46. Cotton, J. W. 1973. Even better than before. *Contemp. Psychol.* 18:168–69
47. Cracraft, J. 1991. Patterns of diversification within continental biotas: hierarchical congruence among the areas of endemism of Australian vertebrates. *Aust. Syst. Bot.* 4:211–27
48. Crowley, P. H. 1992. Density dependence, boundedness, and attraction: Detecting stability in stochastic systems. *Oecologia.* 90:246–54
49. Crowley, P. H., Gillett, S., Lawton, J. H. 1988. Contests between larval damselflies: empirical steps toward a better ESS model. *Anim. Behav.* 36:1496–511
50. Crowley, P. H., Johnson, D. M. 1982. Habitat and seasonality as niche axes in an odonate community. *Ecology* 63: 1064–77
51. Crowley, P. H., Johnson, D. M. 1992. Variability and stability of a dragonfly assemblage. *Oecologia.* 90:260–69
52. DeAngelis, D. L., Gross, L. J., eds. 0000. *Individual-Based Approaches and Models in Ecology.* New York: Chapman-Hall
53. den Boer, P.J. 1989. Comment on the article "On testing temporal niche differentiation in carabid beetles," by M. Loreau. *Oecologia* 81:97–98
54. den Boer, P. J. 1990. On the stabilization of animal numbers. Problems of testing. 3. What do we conclude from significant tests? *Oecologia* 83:38–46
55. den Boer, P. J. 1991. Seeing the trees for the wood: random walks or bounded fluctuations of population size? *Oecologia* 86:484–91
56. Diamond, J. M., Gilpin, M. E. 1982. Examination of the "null" model of Connor and Simberloff for species co-occurrences on islands. *Oecologia* 52: 64–74
57. DiCiccio, T. J., Romano, J. P. 1989. A review of bootstrap confidence intervals. *J. R. Stat. Soc. B* 50:338–54
58. Dorazio, R. M. 1986. Estimating population birth rates of zooplankton when rates of egg deposition and hatching are periodic. *Oecologia* 69:532–41
59. Douwes, P., Stille, B. 1988. Selective versus stochastic processes in the genetic differentiation of populations of the butterfly *Erebia embla* Thnbg. (Lepidoptera: Satyridae). *Hereditas* 109:37–44
60. Dressing, S. A., Spooner, J., Kreglow, J. M., Beasley, E. O., Westerman, P. W. 1987. Water and sediment sampler for plot and field studies. *J. Environ. Qual.* 16:59–64
61. DuBowy, P. J. 1988. Waterfowl communities and seasonal environments: temporal variability in interspecific competition. *Ecology* 69:1439–53
62. Duff, K. E., Smol, J. P. 1991. Morphological descriptions and stratigraphic distributions of the chrysophycean stomatocysts from a recently acidified lake, Adirondack Park, New York, USA. *J. Paleolimnol.* 5:73–113
63. Dwyer, G. 1991. The roles of density, stage, and patchiness in the transmission of an insect virus. *Ecology* 72:559–74
64. Edgington, E. S. 1987. *Randomization Tests.* New York: Marcel Dekker. 2nd ed.
65. Edwards, D. 1985. Exact simulation-based inference: a survey, with additions. *J. Stat. Comput. Simul.* 22:307–26
66. Efron, B. 1979. Bootstrap methods: another look at the jackknife. *Ann. Stat.* 7:1–26
67. Efron, B. 1982. *The Jackknife, the Bootstrap and Other Resampling Plans.* Philadelphia: SIAM
68. Efron, B. 1992. Jackknife-after-bootstrap standard errors and influence functions. *J. Roy. Stat. Soc. B* 54:83–127
69. Efron, B., Tibshirani, R. 1986. Bootstrap methods for standard errors, confidence intervals, and other methods of statistical accuracy. *Stat. Sci.* 1:54–77
70. Eisenberg, R. M. 1966. The regulation of density in a natural population of the pond snail, *Lymnaea elodes. Ecology* 47:889–906
71. Faust, K., Romney, A. K. 1985. The effect of skewed distributions on matrix permutation tests. *Br. J. Math. Stat. Psychol.* 38:152–60
72. Felsenstein, J. 1985. Confidence limits on phylogenies: An approach using the bootstrap. *Evolution* 39:783–91
73. Felsenstein, J. 1988. Phylogenies from

molecular sequences: Inference and reliability. *Annu. Rev. Genet.* 22:521–65
74. Firbank, L. G., Watkinson, A. R. 1987. On the analysis of competition at the level of the individual plant. *Oecologia* 71:308–17
75. Fisher, N. I., Hall, P. 1990. On bootstrap hypothesis testing. *Aust. J. Stat.* 32: 177–90
76. Fisher, N. I., Hall, P. 1991. Bootstrap algorithms for small samples. *J. Stat. Plan. Infer.* 27:157–69
77. Fisher, R. A. 1936. The coefficient of racial likeness and the future of craniometry. *J. R. Anthropol. Inst.* 66:57–63
78. Fix, A. G. 1984. Kin groups and trait groups: population structure and epidemic disease selection. *Am. J. Phys. Anthropol.* 65:201–12
79. Foley, P. 1987. Molecular clock rates as loci under stabilizing selection. *Proc. Natl. Acad. Sci. USA* 84:7996–8000
80. Forina, M., Armanino, C., Leardi, R., Drava, G. 1991. A class-modelling technique based on potential functions. *J. Chemometrics* 5:435–54
81. Gallagher, E. D., Gardner, G. B., Jumars, P. A. 1990. Competition among the pioneers in a seasonal soft-bottom benthic succession: field experiments and analysis of the Gilpin-Ayala competition model. *Oecologia* 83:427–42
82. Garthwaite, P. H., Buckland, S. T. 1992. Generating Monte Carlo confidence intervals by the Robbins-Monro process. *Appl. Stat.* 41:159–71
83. Gascon, C. 1991. Population- and community-level analyses of species occurrences of central Amazonian rainforest tadpoles. *Ecology* 72:1731–46
84. Gaston, K. J., Lawton, J. H. 1987. A test of statistical techniques for detecting density dependence in sequential censuses of animal populations. *Oecologia* 74:404–10
85. Gates, J. 1991. Exact Monte Carlo tests using several statistics. *J. Stat. Comput. Simul.* 38:211–18
86. Gebhardt, M., Ribi, G. 1987. Reproductive effort and growth in the prosobranch snail, *Viviparus ater*. *Oecologia* 74:209–14
87. Gerrodette, T. 1991. Models for power of detecting trends—a reply to Link and Hatfield. *Ecology* 72:1889–92
88. Gilbert, J. J. 1990. Differential effects of *Anabaena affinis* on cladocerans and rotifers: mechanisms and implications. *Ecology* 71:1727–40
89. Gilinsky, N. L., Bamback, R. K. 1986. The evolutionary bootstrap: a new approach to the study of taxonomic diversity. *Paleobiology* 12:251–68
90. Gillespie, J. H. 1986. Variability of evolutionary rates of DNA. *Genetics* 113:1077–92
91. Gilpin, M. E., Diamond, J. M. 1982. Factors contributing to non-randomness in species co-occurrences on islands. *Oecologia* 52:75–84
92. Gilpin, M. E., Diamond, J. M. 1984. Are species co-occurrences on islands non-random, and are null hypotheses useful in community ecology? See Ref. 41, pp. 297–315
93. Gilpin, M. E., Diamond, J. M. 1984. Rejoinder. See Ref. 41, pp. 332–41
94. Gordon, H. R. 1991. Absorption and scattering estimates from irradiance measurements: Monte Carlo simulations. *Limnol. Oceanogr.* 36:769–77
95. Gotelli, N. J. 1991. Demographic models for *Leptogorgia virgulata,* a shallow-water gorgonian. *Ecology* 72: 457–67
96. Gotelli, N. J., Lewis, F. G. III, Young, C. M. 1987. Body-size differences in a colonizing amphipod-mollusc assemblage. *Oecologia* 72:104–8
97. G-Valdecasas, A., Baltanas, A. 1990. Jackknife and bootstrap estimation of biological index of water quality. *Water Res.* 24:1279–84
98. Haefner, J. W. 1988. Niche shifts in Greater Antillean *Anolis* communities: effects of niche metric and biological resolution on null model tests. *Oecologia* 77:107–17
99. Haith, D. A. 1986. Simulated regional variations in pesticide runoff. *J. Environ. Qual.* 15:5–8
100. Hall, P., Wilson, S. R. 1991. Two guidelines for bootstrap hypothesis testing. *Biometrics* 47:757–62
101. Hanna, S. R. 1988. Air quality model evaluation and uncertainty. *J. Air Pollut. Control Assoc.* 38:406–12
102. Harpending, H., Rogers, A. 1987. On Wright's mechanism for intergroup selection. *J. Theor. Biol.* 127:51–62
103. Harrison, S. 1989. Long-distance dispersal and colonization in the Bay Checkerspot butterfly, *Euphydryas editha bayensis*. *Ecology* 70:1236–43
104. Harvey, B. C., Stewart, A. J. 1991. Fish size and habitat depth relationships in headwater streams. *Oecologia* 87: 336–42
105. Harvey, L. E., Davis, F. W., Gale, N. 1988. The analysis of class dispersion patterns using matrix comparisons. *Ecology* 69:537–42
106. Hastings, A. 1987. Substitution rates

under stabilizing selection. *Genetics* 116:479–86
107. Hayward, G. D., Garton, E. O. 1988. Resource partitioning among forest owls in the River of No Return Wilderness, Idaho. *Oecologia* 75:253–65
108. Heidam, N. Z. 1987. Bootstrap estimates of factor model variability. *Atmos. Environ.* 21:1203–18
109. Heller, K.-G., von Helversen, O. 1989. Resource partitioning of sonar frequency bands in rhinolophoid bats. *Oecologia* 80:178–86
110. Heltshe, J. F., Forrester, N. E. 1985. Statistical evaluation of the jackknife estimate of diversity when using quadrat samples. *Ecology* 66:107–11
111. Henle, K. 1989. Population ecology and life history of the diurnal skink *Morethia boulengeri* in arid Australia. *Oecologia* 78:521–32
112. Hinkley, D. V. 1978. Improving the jackknife with special reference to correlation estimation. *Biometrika* 65:13–22
113. Hinkley, D. V. 1980. Discussion of Basu's paper. *J. Am. Stat. Assoc.* 75: 582–84
114. Hinkley, D. V. 1983. Jackknife methods. *Encyclopedia Stat. Sci.* 4:280–87
115. Hoeffding, W. 1952. The large-sample power of tests based on permutations of observations. *Ann. Math. Stat.* 23: 169–92
116. Holman, E. W. 1989. Some evolutionary correlates of higher taxa. *Paleobiology* 15:357–63
117. Hopmans, J. W., Guttierez-Rave, E. 1988. Calibration of a root water uptake model in spatially variable soils. *J. Hydrol.* 103:53–66
118. Horvitz, C. C., Schemske, D. W. 1990. Spatiotemporal variation in insect mutualists of a neotropical herb. *Ecology* 71:1085–97
119. Hubert, L. J., Schultz, J. 1976. Quadratic assignment as a general data analysis strategy. *Br. J. Math. Stat. Psychol.* 29:190–241
120. Hutchinson, G. E. 1959. Homage to Santa Rosalia, or why are there so many kinds of animals? *Am. Nat.* 93: 145–59
121. Ito, D., Oikawa, T. 1989. A simulation study on characteristics of light-interception and carbon dioxide assimilation in a densely planted mulberry field and their regional variation. *J. Sericul. Sci. Jpn.* 58:131–39
122. Jackson, D. A., Harvey, H. H. 1989. Biogeographic associations in fish assemblages: local vs. regional processes. *Ecology* 70:1472–84
123. Jackson, D. A., Somers, K. M. 1991. The spectre of "spurious" correlations. *Oecologia* 86:147–51
124. Jaksic, F. M., Medel, R. G. 1990. Objective recognition of guilds: testing for statistically significant species clusters. *Oecologia* 82:87–92
125. James, C. D. 1991. Temporal variation in diets and trophic partitioning by coexisting lizards (Ctenotus: Scincidae) in central Australia. *Oecologia* 85:553–61
126. Jansen, R. K., Holsinger, K. E., Michaels, H. J., Palmer, J. D. 1991. Phylogenetic analysis of chloroplast DNA restriction site data at higher taxonomic levels: an example from the Asteraceae. *Evolution* 44:2089–105
127. Jansen, R. K., Wallace, R. S., Kim, K.-J., Chambers, K. L. 1991. Systematic implications of chloroplast DNA variation in the subtribe Microseridinae (Asteraceae: Lactuceae). *Am. J. Bot.* 78:1015–27
128. Jenkins, A., Whitehead, P. G., Musgrove, T. J., Cosby, B. J. 1990. A regional model of acidification in Wales, UK. *J. Hydrol.* 116:403–16
129. Joern, A., Lawlor, L. R. 1981. Guild structure in grasshopper assemblages based on food and microhabitat resources. *Oikos* 37:93–104
130. Johnson, D. M., Pierce, C. L., Martin, T. H., Watson, C. N., Bohanan, R. E., Crowley, P. H. 1987. Prey depletion by odonate larvae: combining evidence from multiple field experiments. *Ecology* 68:1459–65
131. Johnson, S. G., Johnston, R. F. 1989. A multifactorial study of variation in interclutch interval and annual reproductive success in the feral pigeon, *Columba livia*. *Oecologia* 80:87–92
132. Jones, M. L., Minns, C. K., Marmorek, D. R., Heltcher, K. J. 1991. Assessing the potential extent of damage to inland lakes in eastern Canada due to acidic deposition. IV. Uncertainty analysis of a regional model. *Can. J. Fish. Aquat. Sci.* 48:599–606
133. Jurik, T. W. 1991. Population distributions of plant size and light environment of giant ragweed (*Ambrosia trifida* L.) at three densities. *Oecologia* 87:539–50
134. Kempthorne, O. 1955. The randomization theory of experimental inference. *J. Am. Stat. Assoc.* 50:946–67
135. Kempthorne, O., Doerfler, T. E. 1969. The behaviour of some significance tests under experimental randomization. *Biometrika* 56:231–48
136. Kenkel, N. C., Hoskins, J. A., Hoskins,

W. D. 1989. Edge effects in the use of area polygons to study competition. *Ecology* 70:272–74
137. Kimura, D. K. 1990. Approaches to age-structured separable sequential population analysis. *Can. J. Fish. Aquat. Sci.* 47:2364–74
138. King, A. W., O'Neill, R. V., DeAngelis, D. L. 1989. Using ecosystem models to predict regional carbon dioxide exchange between the atmosphere and the terrestrial biosphere. *Global Biogeochem. Cycles* 3:337–62
139. Kirk, K. L., Gilbert, J. J. 1990. Suspended clay and the population dynamics of planktonic rotifers and cladocerans. *Ecology* 71:1741–55
140. Kitchell, J. A., Estabrook, G., MacLeod, N. 1987. Testing for equality of rates of evolution. *Paleobiology* 13:272–85
141. Kjellsson, G. 1991. Seed fate in an ant-dispersed sedge, *Carex pilulifera* L.: recruitment and seedling survival in tests of models for spatial dispersion. *Oecologia* 88:435–43
142. Knapp, S. J., Bridges, W. C. Jr., Yang, M.-H. 1989. Nonparametric confidence interval estimators for heritability and expected selection response. *Genetics* 121:891–98
143. Knox, R. G., Peet, R. K., Christensen, N. L. 1989. Population dynamics in loblolly pine stands: changes in skewness and size inequality. *Ecology* 70: 1153–66
144. Knox, R. G., Peet, R. K. 1989. Bootstrapped ordination: a method for estimating sampling effects in indirect gradient analysis. *Vegetatio* 80:153–66
145. Krantz, T. S., Theriot, E. C., Zimmer, E. A., Chapman, R. L. 1990. The Pleurastrophyceae and Micromonadophyceae: a cladistic analysis of nuclear ribosomal RNA sequence data. *J. Phycol.* 26:711–21
146. Lanyon, S. M. 1985. Detecting internal inconsistencies in distance data. *Syst. Zool.* 34:397–403
147. Lavin, M., Doyle, J. J. 1991. Tribal relationships of *Sphinctospermum* (Leguminosae): integration of traditional and chloroplast DNA data. *Syst. Bot.* 16:162–72
148. Legendre, P., Dallot, S., Legendre, L. 1985. Succession of species within a community: chronological clustering with applications to marine and freshwater zooplankton. *Am. Nat.* 125:257–88
149. Legendre, P., Fortin, M. J. 1989. Spatial pattern and ecological analysis. *Vegetatio* 80:107–38
150. Lesht, B. M., Fontaine, T. D. III, Dolan, D. M. 1991. Great Lakes USA-Canada total phosphorus model post audit and regionalized sensitivity analysis. *J. Great Lakes Res.* 17:3–17
151. Lindberg, W. J., McElroy, D. A. 1991. Limitations in nearest neighbor tests of dispersion and association. *Fla. Sci.* 54:106–11
152. Linton, L. R., Edgington, E. S., Davies, R. W. 1989. A view of niche overlap amenable to statistical analysis. *Can. J. Zool.* 67:55–60
153. Loader, C., Damman, H. 1991. Nitrogen content of food plants and vulnerability of *Pieris rapae* to natural enemies. *Ecology* 72:1586–90
154. Lobel, P. S., Anderson, D. M., Durand-Clement, M. 1988. Assessment of Ciguatera dinoflagellate populations: sample variability and algal substrate preference. *Biol. Bull.* 175:94–101
155. Loganathan, G. V., Shrestha, S. P., Dillaha, T. A., Ross, B. B. 1989. Variable source area concept for identifying critical runoff-generating areas in a watershed. *Virginia Polytech. Inst. State Univ. Water Resour. Res. Center Bull.* I-XII:1–119
156. Loreau, M. 1989. On testing temporal niche differentiation in carabid beetles. *Oecologia* 81:89–96
157. Lumsden, C. J. 1988. Gene-culture coevolution: a test of the steady-state hypothesis for gene-culture translation. *J. Theor. Biol.* 130:391–406
158. Madenjian, C. P., Jude, D. J., Tesar, F. J. 1986. Intervention analysis of power plant impact on fish populations. *Can. J. Fish. Aquat. Sci.* 43:819–29
159. Maelzer, D. A. 1970. The regression of log N_{n+1} on log N_n as a test of density-dependence: an exercise with computer-constructed density-independent populations. *Ecology* 51:810–22
160. Mallet, J. 1986. Dispersal and gene flow in a butterfly with home range behavior: *Heliconius erato*(Lepidoptera: Nymphalidae). *Oecologia* 68:210–17
161. Mangel, M., Clark, C. W. 1988. *Dynamic Modeling in Behavioral Ecology*. Princeton, NJ: Princeton Univ. Press
162. Manly, B. F. J. 1990. On the statistical analysis of niche overlap data. *Can. J. Zool.* 68:1420–22
163. Manly, B. F. J. 1991. *Randomization and Monte Carlo Methods in Biology*. London: Chapman & Hall
164. Manly, B. F. J. 1992. Bootstrapping to determine sample sizes in biological studies. *J. Exp. Mar. Biol. Ecol.* In press
165. Manly, B. F. J. 1993. A review of

computer intensive multivariate methods in ecology. In *Multivariate Environmental Statistics*, ed. G. F. Patil, C. R. Rao, N. P. Ross. Amsterdam: North Holland/Elsevier. In press

166. Mantel, N. 1967. The detection of disease clustering and a generalized regression approach. *Cancer Res.* 27: 209–20
167. Marriott, F. H. C. 1979. Barnard's Monte Carlo tests: how many simulations? *Appl. Stat.* 28:75–77
168. Martin, T. H., Wright, R. A., Crowder, L. B. 1989. Non-additive impact of blue crabs and spot on their prey assemblages. *Ecology* 70:1935–42
169. Matson, P. A., Carpenter, S. R. 1990. Statistical analysis of ecological response to large-scale perturbations. *Ecology* 71:2037
170. McArdle, B. H. 1988. The structural relationship: regression in biology. *Can. J. Zool.* 66:2329–39
171. McKillup, S. C., Allen, P. G., Skewes, M. A. 1988. The natural decline of an introduced species following its initial increase in abundance: an explanation for *Ommatoiulus moreletii* in Australia. *Oecologia* 77:339–42
172. McLain, D. K. 1986. Null models and the intensity of sexual selection. *Evol. Theory* 8:49–52
173. Menkens, G. E. Jr., Anderson, S. H. 1988. Estimation of small-mammal population size. *Ecology* 69:1952–59
174. Metz, J. A. J., Diekmann, O., eds. 1986. *The Dynamics of Physiologically Structured Populations. Springer-Verlag Lecture Notes in Biomathematics* 68: 475–94
175. Meyer, J. S., Ingersoll, C. G., McDonald, L. L., Boyce, M. S. 1986. Estimating uncertainty in population growth rates: jackknife vs. bootstrap techniques. *Ecology* 67:1156–66
176. Mielke, P. W. 1984. Meteorological applications of permutation techniques based on distance functions. In *Handbook of Statistics,* Vol. 4, ed. P. R. Krishnaiah, P. K. Sen, pp. 813–30. Amsterdam: North-Holland
177. Mielke, P. W., Berry, K. J., Brier, G. W. 1981. Application of multiresponse permutation procedures for examining seasonal changes in monthly mean sea-level pressure patterns. *Monthly Weather Rev.* 109:120–26
178. Mielke, P. W., Berry, K. J., Brockwell, P. J., Williams, J. S. 1981. A class of nonparametric tests based on multiresponse permutation procedures. *Biometrika* 68:720–24
179. Miller, R. 1968. Jackknifing variances. *Ann. Math. Stat.* 39:567–82
180. Miller, R. G. 1974. The jackknife—a review. *Biometrika* 61:1–17
181. Minta, S., Mangel, M. 1989. A simple population estimate based on simulation for capture-recapture and capture-resight data. *Ecology* 70:1738–51
182. Mitchell-Olds, T., Shaw, R. G. 1987. Regression analysis of natural selection: statistical inference and biological interpretation. *Evolution* 41:1149–61
183. Mueller, L. D., Altenberg, L. 1985. Statistical inference on measures of niche overlap. *Ecology* 66:1204–10
184. Nemec, A. F. L., Brinkhurst, R. O. 1988. Using the bootstrap to assess statistical significance in the cluster analysis of species abundance data. *Can. J. Fish. Aquat. Sci.* 45:965–70
185. Newell, C. J., Hopkins, L. P., Bedient, P. B. 1990. A hydrogeologic database for ground-water modeling. *Ground Water* 28:703–14
186. Noreen, E. W. 1989. *Computer Intensive Methods for Testing Hypotheses: An Introduction*. New York: Wiley
187. Ocana, J., Ruiz de Villa, C., Ribo, G. 1991. Bias and efficiency in estimating sexual selection: Levene's Z index and some resampling alternatives. *Heredity* 67:95–102
188. Odendaal, F. J., Turchin, P., Stermitz, F. R. 1989. Influence of host-plant density and male harassment on the distribution of female *Euphydryas anicia* (Nymphalidae). *Oecologia* 78: 283–88
189. Ohta, T. 1988. Evolution by gene duplication and compensatory advantageous mutations. *Genetics* 120:841–48
190. Ohta, T. 1988b. Further simulation studies on evolution by gene duplication. *Evolution* 42:375–86
191. Olson, C. L. 1976. On choosing a test statistic in multivariate analysis of variance. *Psychol. Bull.* 83:579–86
192. Page, R. D. M. 1988. Quantitative cladistic biogeography: constructing and comparing area cladograms. *Syst. Zool.* 37:254–70
193. Page, R. D. M. 1991. Clocks, clades, and cospeciation: comparing rates of evolution and timing of cospeciation events in host-parasite assemblages. *Syst. Zool.* 40:188–98
194. Palmer, M. W. 1990. The estimation of species richness by extrapolation. *Ecology* 71:1195–98
195. Palmer, M. W. 1991. Estimating species richness: the second-order jackknife reconsidered. *Ecology* 72:1512–13
196. Pamilo, P. 1990. Statistical tests of

phenograms based on genetic distances. *Evolution* 44:689–97
197. Papadopoulos, A. S., Tiwari, R. C., Muha, M. J. 1991. Bootstrap procedures for time series analysis of BOD data. *Ecol. Model.* 55:57–66
198. Patterson, G. B. 1986. A statistical method for testing for dietary differences. *N. Zeal. J. Zool.* 13:113–15
199. Peltonen, A., Hanski, I. 1991. Patterns of island occupancy explained by colonization and extinction rates in shrews. *Ecology* 72:1698–708
200. Phillips, R. A., Neal, L., Johnson, D. M., Nagel, J. W., Laughlin, T. F., Lane, R. T. 1987. Simulating local extinction of trout populations. In *Tools for the Simulation Profession*, ed. R. Hawkins, K. Klukis, pp. 24–27. San Diego: Soc. Computer Simul.
201. Pimm, S. L. 1983. Appendix: Monte-Carlo analyses in ecology. In *Lizard Ecology: Studies on a Model Organism*, ed. R. B. Huey, E. R. Pianka, T. W. Schoener, pp. 290–96. Cambridge, Mass: Harvard Univ. Press
202. Pleasants, J. M. 1990. Null-model tests for competitive displacement: the fallacy of not focusing on the whole community. *Ecology* 71:1078–84
203. Poethke, H. J., Kirchberg, M. 1987. On the stabilizing effect of density-dependent mortality factors. *Oecologia* 74:156–58
204. Pollard, E., Lakhani, K. H., Rothery, P. 1987. The detection of density-dependence from a series of annual censuses. *Ecology* 68:2046–55
205. Pontasch, K. W., Smith, E.P., Cairns, J. Jr. 1989. Diversity indices, community comparison indices, and canonical discriminant analysis interpreting the results of multispecies toxicity tests. *Water Res.* 23:1229–38
206. Powell, G. L., Russell, A. P. 1985. Growth and sexual size dimorphism in Alberta populations of the eastern short-horned lizard, *Phrynosoma douglassi brevirostre*. *Can. J. Zool.* 63:139–54
207. Quinn, J. F. 1987. On the statistical detection of cycles in extinctions in the marine fossil record. *Paleobiology* 13:465–78
208. Raup, D. M., Gould, S. J., Schopf, T. J. M., Simberloff, D. S. 1973. Stochastic models of phylogeny and the evolution of diversity. *J. Geol.* 81:525–42
209. Raup, D. M., Sepkoski, J. J. 1988. Testing for periodicity of extinction. *Science* 241:94–96
210. Reddingius, J., den Boer, P. J. 1989. On the stabilization of animal numbers. Problems of testing. 1. Power estimates and estimation errors. *Oecologia* 78:1–8
211. Ribi, J., Tomczak, M. 1990. An impact assessment for the French nuclear weapon test sites in French Polynesia. *Mar. Pollut. Bull.* 21:536–42
212. Riessen, H. P., Sprules, W. G. 1990. Demographic costs of antipredator defenses in *Daphnia pulex*. *Ecology* 71: 1536–46
213. Ritchie, M. E. 1990. Optimal foraging and fitness in Columbian ground squirrels. *Oecologia* 82:56–67
214. Roberts, A., Stone, L. 1990. Island-sharing by archipelago species. *Oecologia* 83:560–67
215. Robertson, G. P., Huston, M. A., Evans, F. C., Tiedje, J. M. 1988. Spatial variability in a successional plant community: patterns of nitrogen availability. *Ecology* 69:1517–24
216. Rogers, D. S. 1990. Genic evolution, historical biogeography, and systematic relationships among spiny pocket mice, subfamily Heteromyinae. *J. Mammal.* 71:668–85
217. Rogers, A. R., Eriksson, A. W. 1988. Statistical analysis of the migration component of genetic drift. *Am. J. Phys. Anthropol.* 77:451–58
218. Romano, J. P. 1989. Bootstrap and randomization tests of some nonparametric hypotheses. *Ann. Stat.* 17:141–59
219. Romano, J. P. 1990. On the behavior of randomization tests without a group invariance assumption. *J. Am. Stat. Assoc.* 85:686–92
220. Root, T. 1989. Energy constraints on avian distributions: a reply to Castro. *Ecology* 70:1183–85
221. Rose, K. A., McLean, R. I., Summers, J. K. 1989. Development and Monte Carlo analysis of an oyster bioaccumulation model applied to biomonitoring data. *Ecol. Model.* 45:111–32
222. Rotenberry, J. T. 1985. The role of habitat in avian community composition: physiognomy or floristics. *Oecologia* 67:213–17
223. Sale, P. F. 1974. Mechanisms of coexistence in a guild of territorial fishes at Heron Island. In *Proc. 2nd Int. Coral Reef Symp.* 1:193–206. Brisbane: Great Barrier Reef Comm.
224. Sanderson, M. J. 1989. Confidence limits on phylogenies: the bootstrap revisited. *Cladistics* 5:113–29
225. Schill, W. B., Dorazio, R. M. 1990. Immunological discrimination of Atlantic striped bass stocks. *Trans. Am. Fish. Soc.* 119:77–85
226. Schluter, D., Repasky, R. R. 1991.

Worldwide limitation of finch densities by food and other factors. *Ecology* 72:1763–74
227. Schmitt, J., Eccleston, J., Ehrhardt, D. W. 1987. Density-dependent flowering phenology, outcrossing, and reproduction in Impatiens capensis. *Oecologia* 72:341–47
228. Setzer, R. W. 1985. Spatio-temporal patterns of mortality in *Pemphigus populicaulis* and *P. populitransversus* on cottonwoods. *Oecologia* 67:310–21
229. Simberloff, D. 1978. Using island biogeographic distributions to determine if colonization is stochastic. *Am. Nat.* 112:713–26
230. Simberloff, D., Boecklen, W. 1981. Santa Rosalia reconsidered: size ratios and competition. *Evolution* 35:1206–28
231. Simonoff, J. S. 1986. Jackknifing and bootstrapping goodness of fit statistics in sparse multinomials. *J. Am. Stat. Assoc.* 81:1005–11
232. Slatkin, M., Arter, H. E. 1991. Spatial autocorrelation methods in population genetics. *Am. Nat.* 138:499–517
233. Slowinski, J. B., Guyer, C. 1989. Testing the stochasticity of patterns of organismal diversity: an improved null model. *Am. Nat.* 134:907–21
234. Smettem, K. R. J., Collis-George, N. 1985. Prediction of steady-state ponded infiltration distributions in a soil with vertical macropores. *J. Hydrol.* 79: 115–22
235. Sokal, R. R., Rohlf, F. J. 1981. *Biometry: The Principles and Practice of Statistics in Biological Research*. New York: Freeman. 2nd ed.
236. Solow, A. R. 1989. Bootstrapping sparsely sampled spatial point patterns. *Ecology* 70:379–82
237. Solow, A. R. 1989. A randomization test for independence of animal locations. *Ecology* 70:1546–49
238. Solow, A. R. 1990. A randomization test for misclassification probability in discriminant analysis. *Ecology* 71: 2379–82
239. St. Amant, J. L. S. 1970. The detection of regulation in animal numbers. *Ecology* 51:823–28
240. Starks, T. H., Flatman, G. T. 1991. RCRA ground-water monitoring decision procedures viewed as quality control schemes. *Environ. Monit. Assess.* 16:19–38
241. Stauffer, D. F., Garton, E. O., Steinhorst, R. K. 1985. A comparison of principal components from real and random data. *Ecology* 66:1693–98
242. Stewart-Oaten, A., Bence, J. R., Osenberg, C. W. 1992. Assessing effects of unreplicated perturbations: no simple solutions. *Ecology*. In press
243. Still, A. W., White, A. P. 1981. The approximate randomization test as an alternative to the F test in analysis of variance. *Br. J. Math. Stat. Psychol.* 34:243–52
244. Stone, L., Roberts, L. 1990. The checkerboard score and species distributions. *Oecologia* 85:74–79
245. Stone, L., Roberts, L. 1991. Conditions for a species to gain advantage from the presence of competitors. *Ecology* 72:1964–72
246. Strobeck, C. 1987. Average number of nucleotide differences in a sample from a single subpopulation: a test for population subdivision. *Genetics* 117: 149–54
247. Strong, D. R., Szyska, L. A., Simberloff, D. 1979. Tests of community-wide character displacement against null hypotheses. *Evolution* 33:897–913
248. Swihart, R. K. and Slade, N. A. 1985. Testing for independence in animal movements. *Ecology* 66:1176–84
249. Temeles, E. J. 1987. The relative importance of prey availability and intruder pressure in feeding territory size regulation by harriers, *Circus cyaneus*. *Oecologia* 74:286–97
250. ter Braak, C. J. F. 1991. Permutation versus bootstrap significance tests in multiple regression and ANOVA. In *Bootstrapping and Related Techniques*, ed. K.-H. Jockel. Berlin: Springer-Verlag. In pressc
251. Thomas, S. C., Weiner, J. 1989. Including competitive asymmetry in measures of local interference in plant populations. *Oecologia* 80:349–55
252. Tukey, J. W. 1958. Bias and confidence in not quite large samples (Abstract). *Ann. Math. Stat.* 29:614
253. Turchin, P. 1987. The role of aggregation in the response of Mexican bean beetles to host-plant density. *Oecologia* 71:577–82
254. van Straalen, N. M. 1985. Comparative demography of forest floor Collembola populations. *Oikos* 45:253–65
255. Vassiliou, A., Ignatiades, L., Karydis, M. 1989. Clustering of transect phytoplankton collections with a quick randomization algorithm. *J. Exp. Mar. Biol. Ecol.* 130:135–46
256. Venrick, E. L. 1990. Phytoplankton in an oligotrophic ocean: species structure and interannual variability. *Ecology* 71: 1547–63
257. Verrier, E., Colleau, J. J., Foulley, J. L. 1990. Predicting cumulated response to directional selection in finite pan-

mictic populations. *Theor. Appl. Genet.* 79:833–40

258. Vickery, W. L. 1991. An evaluation of bias in *k*-factor analysis. *Oecologia* 85:413–18
259. Vickery, W. L., Nudds, T. D. 1991. Testing for density-dependent effects in sequential censuses. *Oecologia* 85: 419–23
260. Walls, M., Caswell, H., Ketola, M. 1991. Demographic costs of *Chaoborus*-induced defenses in *Daphnia pulex:* a sensitivity analysis. *Oecologia* 87:43–50
261. Whicker, F. W., Kirchner,T. B. 1987. PATHWAY: a dynamic food-chain model to predict radionucllide ingestion after fallout deposition. *Health Phys.* 52:717–38
262. Wilson, J. B. 1987. Methods for detecting non-randomness in species cooccurrences: a contribution. *Oecologia* 73:579–82
263. Wilson, J. B. 1989. A null model of guild proportionality, applied to stratification of a New Zealand temperate rain forest. *Oecologia* 80:263–67
264. Wilson, J. B., Hubbard, J. C. E., Rapson, G. L. 1988. A comparison of realized niche relations of species in New Zealand and Britain. *Oecologia* 76:106–10
265. Winemiller, K. O., Pianka, E. R. 1990. Organization in natural assemblages of desert lizards and tropical fishes. *Ecol. Monogr.* 60:27–55
266. Wolda, H. 1989. Comment on the article "On testing temporal niche differentiation in carabid beetles" by M. Loreau and the "Comment on the article of M. Loreau" by P. J. den Boer. *Oecologia* 81:99
267. Wolda, H. 1991. The usefulness of the equilibrium concept in population dynamics: A reply to Berryman. *Oecologia* 86:144–45
268. Yearsley, J. R., Lettenmaier, D. P. 1987. Model complexity and data worth an assessment of changes in the global carbon budget. *Ecol. Modell.* 39:201–26
269. Zimmerman, G. M., Goetz, H., Mielke, P. W. Jr. 1985. Use of an improved statistical method for group comparisons to study effects of prairie fire. *Ecology* 66:606–11
270. Zink, R. M., Dittmann, D. L., Rootes, W. L. 1991. Mitochondrial DNA variation and the phylogeny of *Zonotrichia*. *Auk* 108:578–84
271. Zwolfer, H., Brandl, R. 1989. Niches and size relationships in Coleoptera associated with Cardueae host plants: adaptations to resource gradients. *Oecologia* 78:60–68

Annu. Rev. Ecol. Syst. 1992. 23:449–80

PHYLOGENETIC TAXONOMY*

Kevin de Queiroz
Division of Amphibians and Reptiles, United States National Museum of Natural History, Smithsonian Institution, Washington, DC 20560

Jacques Gauthier
Department of Herpetology, California Academy of Sciences, Golden Gate Park, San Francisco, California 94118

KEYWORDS: cladistics, classification, nomenclature, systemics, taxon names

> Our classifications will come to be, as far as they can be so made, genealogies. . . . The rules for classifying will no doubt become simpler when we have a definite object in view. (Darwin 1859, p. 486)

> The doctrine of evolution is not something that can be grafted, so to speak, onto the Linnaean system of classification. (Woodger 1952, p. 19)

INTRODUCTION

During the century following the publication of Darwin's (24) *Origin of Species,* biological taxonomy waited for the revolution that should have followed upon acceptance of an evolutionary world view. Although the principle of common descent gained wide acceptance early in that era, it assumed a largely superficial role in taxonomy (131). Pre-existing taxonomies were explained as the result of evolution (16, 86, 124, 128), and evolutionary interpretations were given to long-standing taxonomic practices (111), but the principle of descent did not become a central tenet from which taxonomic principles and methods were derived (111, 115).

In the middle of the twentieth century, the late Willi Hennig outlined an approach (57–59) that represented a fundamental shift in outlook concerning the role of the concept of evolution in taxonomy, and which engendered significant changes in that discipline (see 3, 29, 30, 97, 133, and references therein). By deriving the principles and methods of his approach from the tenet of common descent, Hennig granted the concept of evolution a position

0066-4162/92/1120-0449$02.00

therein). By deriving the principles and methods of his approach from the tenet of common descent, Hennig granted the concept of evolution a position of central importance in taxonomy (114). The revolution initiated by Hennig is now well underway in systematic analysis, the field concerned with methods for estimating phylogenetic relationships (reviewed in 37, 129). But the same cannot be said about taxonomy proper, the discipline concerned with methods for communicating the results of systematic analysis. At the present time, biologists still have not developed a phylogenetic system of taxonomy. Hennig's distinction between monophyly and paraphyly was a crucial first step, but there is more to a phylogenetic system of taxonomy than the precept that all taxa must be monophyletic.

There is, after all, an important distinction between a phylogenetic taxonomy—an arrangement of names all of which refer to monophyletic taxa—and a phylogenetic *system* of taxonomy—a unified body of principles and conventions governing the use of those names. In order for a taxonomic system to be truly phylogenetic, its various principles and rules must be formulated in terms of the central tenet of evolution. It is this common evolutionary context that unifies the various principles and rules into a coherent system. But the concept of evolution still has not been granted such a central role in taxonomy. This is exemplified by the continued use of the Linnaean system, a taxonomic system based on a pre-Darwinian world view.

Even in taxonomies adopting Hennig's principle of monophyly, the Linnaean system generally has been taken for granted, and then modified or elaborated to accommodate the representation of phylogenetic relationships. Perhaps the most explicit example of this approach is Wiley's (132, 133) "annotated Linnaean system" (see also 20, 23, 30, 98). It is doubtful, however, that ad hoc modification of a body of conventions based on a pre-Darwinian world view is the most effective way to develop a phylogenetic system of taxonomy. Indeed, this preference for ad hoc modification, rather than reorganization starting from evolutionary first principles, demonstrates the resistance of biological taxonomy to the Darwinian Revolution. As long as this situation persists, a truly phylogenetic system is unlikely to be achieved.

Here we adopt an alternative approach intended to further the development of a phylogenetic system of taxonomy. Instead of taking the Linnaean system as a given and modifying it to facilitate the representation of phylogenetic relationships, we take the goal of representing phylogenetic relationships as primary. From this perspective, we outline the basic framework of a phylogenetic system of taxonomy by reformulating various taxonomic principles and rules in terms of the first principle of common descent. Some of the principles and rules of the phylogenetic system have already been formulated, and some current taxonomic practices are compatible with them. Nevertheless, there are important differences between the principles and rules of the phylogenetic system and those developed around the Linnaean system as formalized in the various codes of biological nomenclature (69–71). We emphasize some of these differences in order to illustrate the fundamental shift in both perspective and practice represented by a phylogenetic system of taxonomy.

BASIC TERMS, GENERAL PERSPECTIVE, AND SCOPE

In order to avoid ambiguity and misinterpretation, it is necessary to define some basic terms used throughout this paper. Our use of these terms differs in subtle ways from the same or similar terms as they are used by other authors. Some of these definitions have been stated above but are reiterated here for emphasis.

Phylogenetic taxonomy is the branch of *phylogenetic systematics* concerned with the representation—rather than the reconstruction or estimation—of phylogenetic relationships. The objective of phylogenetic taxonomy is to represent relationships of common descent using a system of names. A *phylogenetic taxonomy* is a particular system of names representing the entities that derive their existence from a particular set of phylogenetic relationships. Finally, a *phylogenetic system of taxonomy* (as contrasted with the *Linnaean system* or the *current taxonomic system)* is a body of principles and rules governing taxonomic practice, the components of which are unified by their relation to the central tenet of evolutionary descent.

The purpose of any taxonomy is communication, which implies that taxonomies contain some kind of information. The information contained in a phylogenetic taxonomy is information about phylogeny, that is, about common descent (20, 36, 59). That information is conveyed in two ways, through taxon names and their graphic arrangement, as in a branching diagram or an indented list. In order for a taxonomy to convey information most effectively, the meanings of taxon names should be stable, universal, and unambiguous (e.g. 23, 65, 69, 71, 76, 85, 87, 130, 133). Promoting these qualities is one of the primary purposes of a taxonomic system. In the case of a phylogenetic system of taxonomy, the rules should be designed to promote stable, universal, and unambiguous meanings of taxon names with regard to what they signify about common ancestry.

We limit our discussion to the principles and rules governing the taxonomy of monophyletic taxa (sensu Hennig). We do not reiterate arguments about monophyly and paraphyly here; they have received due consideration elsewhere (e.g. 3, 5, 9, 10, 12, 13, 20, 25, 30, 58, 59, 61, 95, 97, 99, 109, 111, 117, 133). Furthermore, populations of interbreeding organisms are not the subject of this paper; they are members of a fundamentally different category of biological entities than monophyletic taxa.

TAXA

Taxa are named entities, generally named groups of organisms. Under traditional interpretations, organisms belong to taxa because they possess certain characters (organismal traits). By contrast, taxa in phylogenetic taxonomy are historical entities (50, 133) resulting from the process of common evolutionary descent. This implies that taxa are composite wholes or systems (e.g. 46–50, 55, 59, 61, 66, 104, 111–114, 137) composed at one organizational level of organisms as their component parts. Therefore, organisms are parts of taxa not because they possess certain characters, but because of their particular phylogenetic relationships (59, 114, 116, 132, 134, 136).

Taxa as systems of common descent are unified by their common evolutionary history. For this reason, their organisms (parts) are not necessarily continuous with one another at any given time, although they exhibit historical continuity through lines of descent (47, 133). Neither is a phylogenetic taxon composed only of those organisms that are currently alive or are recognized as belonging to it. Because of its historical nature, a phylogenetic taxon is composed of *all* the organisms exhibiting the appropriate relationships of common descent, whether those organisms are living or dead, known or unknown (3, 59, 113, 114).

The nature of phylogenetic taxa was clarified by Hennig (e.g. 58, 59, 61), who restricted the concept of monophyly to complete systems of common ancestry—entities each consisting of an ancestor and all of its descendants—that is, clades. By deriving his concept of taxa from the principle of descent, Hennig took an important step in the development of a phylogenetic system of taxonomy. For our discussion, what is significant is that *the taxa of concern in a phylogenetic system are named clades* (46, 59). Consequently, *taxon names are the names of clades,* which convey information about the existence of monophyletic entities, and the nomenclatural aspect of phylogenetic taxonomy is concerned specifically with the naming of clades, or *clade nomenclature*. This perspective makes for a more straightforward terminology than that currently adopted, for example, by the *International Commission on Zoological Nomenclature* (ICZN) (Table 1). Despite the undeniable importance of Hennig's insight, the principle of monophyly is only one component of a phylogenetic system of taxonomy.

NESTED HIERARCHY

One component of the Linnaean system that is compatible with a phylogenetic system of taxonomy is the nested, hierarchical arrangement of taxa. Perhaps this feature accounts for the persistence of the Linnaean system well into the

Table 1 Comparison of terminology in the Linnaean and phylogenetic systems. Quotations are from the International Code of Zoological Nomenclature (1985).

Concept	Term: ICZN (1985)	Term: Phylogenetic System
"any taxonomic unit . . . whether named or not"	taxon	clade
a named taxonomic unit	taxon	taxon
"a nomenclatural concept denoted by an available name . . . but having no defined taxonomic boundaries"	nominal taxon	—
"A taxon . . . including whatever nominal taxa and individuals a zoologist at any time considers it to contain in his or her endeavour to define the boundaries of a zoological taxon"	taxonomic taxon	hypothesized content (approximately)
"The word or words by which . . . something is known"	name	name

era dominated by an evolutionary world view, for it can be deduced from the principle of common descent that phylogeny also forms a nested hierarchy of clades (24). According to Stevens (128), the nested, hierarchical structure of pre-Darwinian taxonomies may have hindered the development of phylogenetic taxonomy. Darwin used the nested, hierarchical structure of existing taxonomies as evidence supporting the theory of descent, which seemed to imply that these taxonomies were already phylogenetic.

The nested hierarchical structure of various pre-Darwinian taxonomies, however, does not have the same underlying basis as that of phylogenetic taxonomies. Humans commonly erect nested groups through a mental process of categorization in order to organize information (126), thus producing an arrangement similar in gross structure to the pattern of relationships resulting from common descent (59, 114). In most non-evolutionary taxonomies, however, taxa are treated either implicitly or explicitly as if they are abstract classes based on shared organismal traits (23, 46, 126). Consequently, the nested, hierarchical organization of such taxonomies reflects logical relationships among the abstract classes. In contrast, the nested hierarchical structure of phylogenetic taxonomies represents genealogical relationships among clades.

In any case, a hierarchical, nonoverlapping taxonomy is suited to the representation of phylogenetic relationships, although cases involving reticulate evolution require special conventions (98, 132, 133). Nested, hierarchical relationships among taxa are commonly represented by branching or Venn diagrams or by indented or sequenced lists. The arrangement of the names conveys information about the relationships of clades to one another, and in so doing it conveys information about the hypothesized content of taxa. Although nested, hierarchical arrangements of taxa are compatible with, and perhaps even integral to, a phylogenetic system of taxonomy, other conventions of the current taxonomic system are not.

LINNAEAN CATEGORIES

In the Linnaean system, a familiar set of categories (kingdom, phylum, class, etc) is used to convey information about the relative positions of taxa (i.e. rank) in the taxonomic hierarchy. Although Linnaean categories can be used for this purpose in phylogenetic taxonomies, they are unnecessary for conveying such information. If used only to represent relative position in a hierarchy, the Linnaean categories contain no information about common ancestry that is not present in a branching diagram or an indented list of names. Concomitantly, if used in conjunction with indentation or a branching diagram, the information provided by the categories is redundant. Moreover, the categorical assignments of taxa are, by themselves, insufficient to specify relationships. In order to determine the relationships among taxa from their categorical assignments, one must first know whether the taxa in question are internested or mutually exclusive. For example, simply knowing that one taxon is a family and another is an order does not indicate whether that family is nested within that order.

Traditional and Hennigian Approaches to Categorical Assignment

The traditional manner in which taxa were assigned to Linnaean categories was incompatible with phylogenetic taxonomy in that it often caused taxonomies to contradict phylogenetic relationships (133). Categories were assigned on the basis of the degree of distinctiveness or the importance attributed to certain characters (e.g. 85, 87). Because the taxonomic hierarchy is one of nested, mutually exclusive taxa, a conflict arose between the use of Linnaean categories for reflecting distinctiveness versus their use for reflecting common ancestry. For example, the mutually exclusive relationship implied by assigning Pongidae and Hominidae to the same Linnaean category (family) contradicts the nested relationship of the hominid clade within the clade stemming from the most recent common ancestor of the various pongids.

In an attempt to give the Linnaean categories meanings that were compatible with a phylogenetic system of taxonomy, Hennig (59) suggested that the categories be defined in terms of absolute time, and that taxa be assigned to categories on the basis of time of origin. Farris (35) advocated using time of differentiation instead of time of origin. Under either of these conventions the categories would convey information in addition to relative position in a hierarchy and thus would not be redundant if combined, for example, with indentation. Furthermore, in contrast with the traditional approach to categorical assignment, taxa assigned to the same category would be equivalent in at least one important respect—a respect that would facilitate comparisons of lineage diversity and other time-related phenomena (46, 59). Despite the advantages of basing the Linnaean categories on absolute time, this convention generally has not been embraced by taxonomists (e.g. 3, 23, 30, 56, 78, 123). Instead, most phylogenetic taxonomies have used the categories only to indicate relative divergence time, which requires nothing more than assigning sister taxa to the same category (but see below).

Mandatory Categories

Although Linnaean categories can be used in ways that are consistent with a phylogenetic approach to taxonomy, certain conventions associated with them are irreconcilable with that approach. One such convention is that certain categories—kingdom, phylum/division, class, order, family, and genus—are mandatory, although this is not stated in the zoological code; that is, every named species must be assigned to a taxon at each and every one of these levels (e.g. 85, 125). The problem with this convention is most evident in the case of organisms that are parts of ancestral populations, which are not parts of clades less inclusive than the one stemming from their own population (Figure 1, left). Thus, Hennig (59) noted that the stem species of birds is to be included in the taxon Aves, but not in any of the subgroups of Aves. If

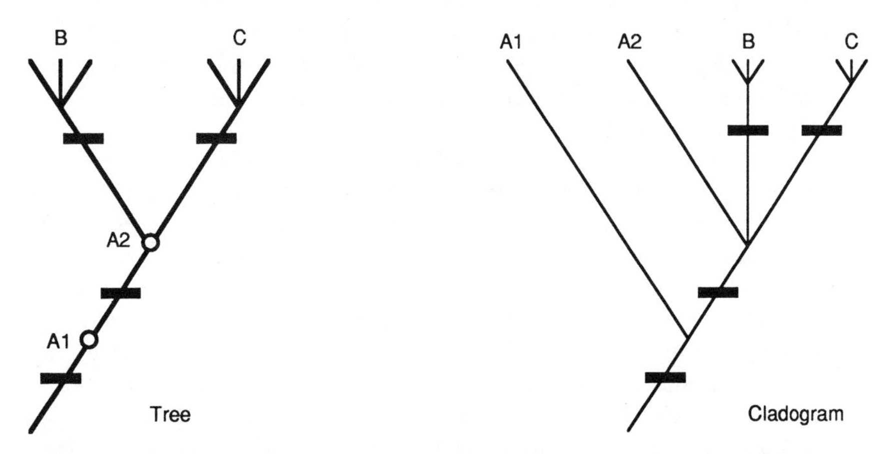

Figure 1 Ancestors cannot be assigned to monophyletic taxa less inclusive than those originating with them. *Left:* If the clade stemming from the most recent common ancestor of B and C (A2) is assigned the rank of class, then A1 and A2 are not parts of any monophyletic taxon assigned to a category of lower rank than class. *Right:* Even if no attempt is made to determine whether A1 and A2 are ancestors, they will lack the diagnostic apomorphies of subclades B and C and will not have any apomorphies unique to themselves. Consequently, they will not be referable to any taxon of lower rank than the category to which the clade stemming from the most recent common ancestor of B and C is assigned. Bars indicate apomorphies.

Aves is assigned to the class category, then the organisms in its stem species are not parts of monophyletic taxa below the class level.

The problem caused by mandatory categories is not dependent on identifying actual ancestors, the difficulty of which has been discussed by several authors (31, 35, 59, 97, 108). Ancestors, whether or not they can be identified as such, possess the derived characters of the clade originating with them but not those of any of its subgroups. Therefore, even if no attempt is made to identify ancestral organisms or populations *as ancestors,* they still will not be assignable to any of the monophyletic subgroups of the taxa originating with them (Figure 1, right). Consequently, recognizing taxa at all the mandatory categorical levels for such entities misrepresents their phylogenetic relationships by implying the existence of clades that do not, in fact, exist.

Wiley (132, 133) proposed to solve the problem posed by ancestors by assigning each ancestral species to a monotypic genus and placing the binomial in the taxonomic hierarchy in parentheses beside the name of the higher taxon stemming from it. This proposal has several problems. First, if the genus is interpreted as being equivalent to the higher taxon—suppose it is a class—then not only are the names of the genus and the class redundant, but the categories genus and class refer to the same level in the taxonomic hierarchy and thus the accepted relationship between the categories is violated. If the genus is not interpreted as being equivalent to the class, then the genus is nonmonophyletic. Moreover, Wiley's proposal does not solve the problem posed by ancestors within the context of mandatory categories, for it is, in spite of Wiley's avowed acceptance of that convention, based on an implicit rejection of mandatory categories. Continuing with the same example, the genus is not assigned to any taxon in the categories order and family.

The source of the problem, pointed out by Griffiths (54, 55, 56), is that phylogenies form truncated hierarchies, whereas the hierarchical structure of a taxonomic system with mandatory categories is not truncated. In other words, although various descendants are different distances (however measured) from the common ancestor at the base of the phylogeny, they are nevertheless forced into taxa representing all of the mandatory categories, which effectively imposes equal distances upon them. This problem applies both to ancestors and to extinct side branches relative to extant organisms, and it accounts for the supposed problem of including both fossil and Recent species in the same taxonomy (e.g. 12, 23, 59, 105). But that supposed problem is an artifact of taking the convention of obligatory Linnaean categories for granted. From the perspective of phylogenetic taxonomy, the problem is not with the systematization of fossils but with the convention of mandatory categories.

A comparable problem stems from the convention known as the principle of exhaustive subsidiary taxa (e.g. 30, 35, 125). According to that principle, if a nonmandatory categorical level is used within any taxon, then it is used for all members of that taxon. For example, if a subgroup of a family is assigned to the subfamily category, then new taxa are customarily recognized so that all other members of the family can be assigned to some taxon of the subfamily category. In effect, this convention makes a nonmandatory category mandatory within a given part of the taxonomic hierarchy. It therefore causes the same kinds of problems discussed above. In current practice, applying the principle of exhaustive subsidiary taxa, and that of mandatory categories in general, often has an additional undesirable consequence—the recognition of new taxa for which evidence of monophyly is lacking (see 18 for discussion of an example involving genera).

Because the convention of mandatory categories is incompatible with the representation of phylogenetic relationships, a phylogenetic system of taxonomy must abandon that convention. Farris (35) has already done this in his "indented-list classifications." Furthermore, leaving some specimens or species unassigned to taxa at particular levels in the taxonomic hierarchy is already practiced, albeit on a smaller scale, using the label "*incertae sedis*." Although commonly followed in practice, assignment to taxa representing Linnaean categories of higher rank than genus is not an absolute requirement of any of the current codes of nomenclature (30).

Proliferation of Categories

Within the context of the Linnaean system, the discovery of new clades resulting from increasing knowledge about phylogeny led to a proliferation of taxonomic categories (e.g. 1, 11, 40, 88). Although sometimes viewed as a problem peculiar to phylogenetic taxonomy (e.g. 8, 19, 23, 64, 87), the problem exists for any approach, phylogenetic or otherwise, that attempts to reflect fine levels of systematic resolution in a taxonomy (100). More importantly, if we consider the proliferation of taxonomic categories problematical, this implies that maintaining the traditional, limited number of categories supersedes the goal of representing phylogeny. Several ad hoc modifications of the Linnaean system have been proposed as solutions to this supposed problem, for example, using the sequence of names of equally ranked taxa to convey relationships (96, 98, 132, 133), using a special category, plesion, for extinct taxa (105, see also 135), and using various combinations of prefixes to generate new categories (35). The sequencing convention might be viewed as a means of avoiding the proliferation of taxon names as well as categories. However, given that the primary task is to represent phylogeny—and acknowledging that there are already more taxon names than anyone can remember—then naming clades seems preferable to leaving them unnamed (3, 107, 135, 136).

Taxonomy Without Linnaean Categories

In light of the many problems associated with the Linnaean categories and their associated conventions (see also Binomials, Synonymy and Priority, and Redundancy), and given that the categories are neither necessary nor sufficient for conveying phylogenetic relationships, *phylogenetic taxonomy may be best served by abandoning the Linnaean categories* (see also 3, 22, 43, 54–56, 101, 136). If so, then sequencing, the plesion category, means of generating additional categories, and other ad hoc modifications of the Linnaean system designed to cope with the proliferation of categories would all be unnecessary. Eliminating the Linnaean categories is not as radical as it might seem. Systematists commonly construct taxonomies without Linnaean categories in the form of branching diagrams, and several authors have explicitly avoided using Linnaean categories in taxonomies taking the standard form of indented lists (e.g. 3, 32, 42–44, 62, 63, 79, 82, 106, 110, 119; see Appendix).

Wiley (133) criticized the use of indentation without Linnaean categories to convey hierarchical relationships on the grounds that it is difficult to line up the names in taxonomies spanning more than one page. Hennig (60, 62, 63) solved this problem using a code of numeric or alphanumeric characters, which can be used with or without indentation, to indicate position in the hierarchy (see also 55, 56, 82). Criticisms of this proposal (e.g. 3, 30, 132, 133) are based on treating the numeric prefixes as formal substitutes for Linnaean categories rather than simply as a means of keeping track of hierarchical position within a given taxonomy.

Eliminating the Linnaean categories does not require changes in the names of taxa, that is, in the spelling of taxon names. This maintains continuity with previous work, thereby ensuring access to the literature. In contrast, traditional practice may obstruct access to the literature because changes in the categorical assignments of taxa cause changes in their names (56, 115; see Synonymy and Priority). In order to preserve the spellings of taxon names, the suffixes formerly associated with certain Linnaean categories (e.g. -iformes, -idae, -inae, etc.) would be retained, but these suffixes would no longer imply anything about Linnaean categories (56). But regardless of whether the Linnaean categories are retained or abandoned, *the development of a phylogenetic system of taxonomy amounts largely to replacing the Linnaean categories with the principle of common descent as the basis for taxonomic conventions*. This becomes evident upon our consideration of conventions related to the principles of synonymy and priority (see below).

BINOMIALS

Although the taxonomy of species is not our concern in this paper, the problem posed by mandatory categories has implications for the formation of species names. In the Linnaean system, the genus is effectively a mandatory category, because every species must be assigned to a genus in order to form its binomial name. Consequently, even if the Linnaean categories are otherwise abandoned, under the binomial convention used in the Linnaean system, the genus category would still be mandatory. A mandatory genus category faces the same problem as any other mandatory category, namely, that the ancestors of monophyletic taxa including more than one genus are not themselves parts of monophyletic genera.

For this reason, *a phylogenetic system of taxonomy cannot retain the Linnaean method of forming binomials; specifically, the names of genera cannot be parts of species names* (56). This does not mean that binomials themselves must be abandoned. If they are retained, however, the first name of a binomial species name would not be the name of a genus or a clade of any rank (56). Instead, the first name would simply be one part of a two part species name; Griffiths (56) suggested calling it a *forename* or *praenomen*. Consequently, a given species would not necessarily be more closely related to other species having the same praenomen than to those with a different praenomen, and this would be a potential source of confusion as long as such names continued to carry connotations about genera. Another alternative is uninomial species names (e.g. 17, 92, 93).

A taxonomic system in which the names of species are independent of the names of higher taxa, whether uninomials or non-Linnaean binomials, would also contribute to stabilizing the names of species. As pointed out by Cain (17) and Michener (92, 93), such a system would eliminate the alterations in species names caused by changes in generic assignment (both those involving the binomial combination and changes to the specific epithet necessitated by secondary homonymy). This kind of instability is exacerbated by attempts to achieve a phylogenetic taxonomy within the constraints of the Linnaean system, because eliminating paraphyletic taxa provides another reason for changing generic assignments. Species names that are independent of the names of higher taxa (genera) also permit use of Wiley's (132, 133) convention for including ancestral species in a taxonomy (see Mandatory Categories). Once species names are freed from any associations with the names of genera, the contradictions noted above vanish. (This convention assumes that the species category is a category of biological entities rather than one level or rank in the hierarchy of Linnaean categories—e.g. 113). Modification of the Linnaean approach to forming species names is not only necessary for phylogenetic taxonomy, it would also promote nomenclatural stability, one of the primary functions of the current codes.

TYPES

In the current taxonomic system, name-bearing or nomenclatural types provide objective standards of reference by which the application of names is determined. According to the principle of typification, every nominal taxon at or below a particular categorical level has such a type (69–71). Traditionally, types are of two basic kinds: organisms (type specimens) and taxa (type species, type genera). In the following discussion, we address only types of the second kind. The use of type specimens for species taxa does not appear to be problematical; in any case, it is outside of the scope of the present paper.

Under the current system, the names of type taxa of the genus category contain the word stems that serve as the bases for the names of taxa assigned to higher categories. Thus, the name of a zoological taxon assigned to the family category is formed by adding the suffix "-idae" to the stem of the name of its type genus (71). Many rules concerning types are tied to Linnaean categories, and the use of nomenclatural types in forming taxon names is itself a taxonomic convention rather than a logical or biological necessity (23). Therefore, basing a new taxonomic system on the principle of descent rather than on the Linnaean categories necessitates a reevaluation of the principle of typification.

Typification is not incompatible with the naming of clades. Although tied to Linnaean categories in the current taxonomic system, the use of nomenclatural types need not be so tied. In the absence of Linnaean categories, one might refer to type populations, or type (sub)clades, or simply type (nominotypical) taxa. Moreover, the name of a clade can be based on the name of one of its subclades or component populations regardless of whether any of the taxa involved are assigned to Linnaean categories. If such a convention is adopted, however, it would be useful to state the definition of each taxon name in terms of a specified relationship to the type (see Definitions of Taxon Names). For example, the name "Lepidosauromorpha" is defined as Lepidosauria and all saurians sharing a more recent common ancestor with Lepidosauria than with Archosauria (43). Such a convention would ensure that the nested relationship between a taxon and its nomenclatural type is preserved in the face of changing ideas about phylogenetic relationships.

DEFINITIONS OF TAXON NAMES

Reformulation of the manner in which taxon names are defined is central to developing a phylogenetic system of taxonomy because it provides a basis for the derivation of secondary principles and rules concerning the use of taxon names. Under the current system, the definitions of taxon names are stated in terms of characters, that is, organismal traits. For example, according to the zoological code, a definition "purports to give characters differentiating a taxon" (71, p. 253). Definitions of taxon names based on organismal traits are fundamentally non-evolutionary. Such definitions were in use long before the widespread acceptance of an evolutionary world view, and furthermore, they make no reference to common descent or any other evolutionary phenomenon (112, 115).

In the context of a phylogenetic approach to taxonomy, several authors have proposed that the definitions of taxon names are to be based on phylogenetic relationships (e.g. 43, 49, 111, 116, 118, 132, 134). Nevertheless, it was only recently that concrete methods for doing this were devised. De Queiroz & Gauthier (115; see also 112) identified three classes of phylogenetic definitions, that is, three ways of defining the names of taxa in terms of phylogenetic relationships (Figure 2). A *node-based definition* specifies the meaning of a taxon name by associating the name with a clade stemming from the immediate common ancestor of two designated descendants (Figure 2, left). A *stem-based definition* specifies the meaning of a name by associating the name with a clade of all organisms sharing a more recent common ancestor with one designated descendant than with another (Figure 2, middle). And an *apomorphy-based definition* specifies the meaning of a name by associating the name with a clade stemming from the ancestor in which a designated character arose (Figure 2, right). Examples of these three classes of phylogenetic definitions are given by de Queiroz & Gauthier (115) and references cited in that paper.

Phylogenetic definitions clarify other taxonomic issues in the context of a phylogenetic approach to taxonomy. For one, they clarify the distinction between definitions and diagnoses (115; see also 45, 46, 118). *Definitions* are statements specifying the meanings of taxon names (words); they are stated in terms of ancestry. *Diagnoses* are statements specifying how to determine whether a given species or organism is a representative of the taxon (clade) to which a particular name refers; they are most commonly stated in terms

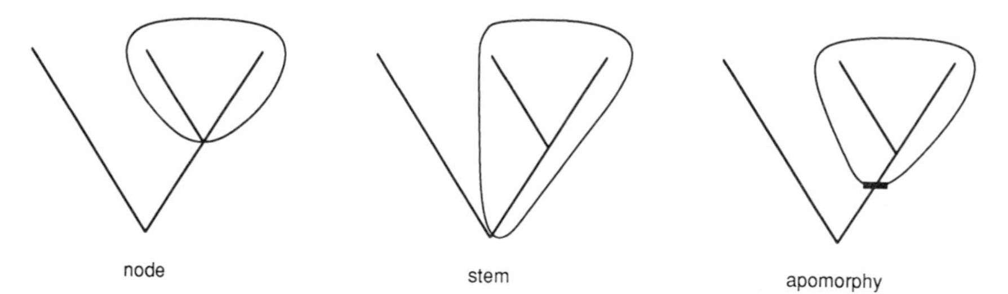

Figure 2 Three possible ways of defining taxon names phylogenetically. *Left:* The name is defined as referring to the most recent common ancestor of two designated taxa and its descendants (node-based definition). *Middle:* The name is defined as referring to all organisms sharing a more recent common ancestor with one designated taxon than with a second such taxon (stem-based definition). *Right:* The name is defined as referring to the first ancestor to evolve a designated character (bar) and its descendants (apomorphy-based definition). After de Queiroz & Gauthier (115); see that reference for examples.

of characters. Phylogenetic definitions also permit one to define any given name as referring to a monophyletic taxon and not to a paraphyletic or polyphyletic group or a metataxon (115). Although it is always possible to make mistakes about the contents (included species and organisms) and diagnostic characters of a taxon, a definition such as "the most recent common ancestor of archosaurs and lepidosaurs *and all of its descendants*" necessarily refers to a monophyletic taxon.

Phylogenetic definitions illustrate what it means for a taxonomic principle to be stated in terms of the central tenet of evolution. In contrast with character-based definitions, which make no reference to any evolutionary phenomenon, *phylogenetic definitions of taxon names are stated in terms of common ancestry relationships and the historical entities (clades) resulting from those relationships*. Phylogenetic definitions are thus thoroughly evolutionary in that the concept of common ancestry is fundamental to the meanings of the names they are used to define. Taxon names thus have explicit evolutionary meanings, and this bears on several other taxonomic issues.

SYNONYMY AND PRIORITY

Two taxonomic issues upon which phylogenetic definitions have direct bearing are synonymy and priority (115). Reevaluating these issues from an evolutionary perspective illustrates the difference between a taxonomic system based on the Linnaean categories and one based on the principle of descent. It also reveals fundamental incompatibilities between the current system and a phylogenetic approach to taxonomy.

In the current system, the concept of synonymy is tied directly to the Linnaean categories. Taxon names are considered synonymous if they are based either on the same nomenclatural type (homotypic, objective, or nomenclatural synonyms) or on different nomenclatural types (heterotypic, subjective, or taxonomic synonyms) considered to belong to a single taxon assigned to a particular Linnaean category (70, 71). Not only is this concept of synonymy based on a non-evolutionary criterion, it is also difficult to reconcile with phylogenetic interpretations of the meanings of taxon names.

As a result of being tied to the Linnaean categories, the criterion of synonymy adopted in the current system is responsible for considerable instability in the phylogenetic meanings of taxon names, which vary as the result of changes in, or differences in opinion about, the assignment of taxa to Linnaean categories. For example, if sister taxa originally considered to form two families are later judged to represent a single family, then their names are treated as synonyms. Each name thus shifts its association from a less inclusive to a more inclusive clade. In this particular example, the change in meaning results from an arbitrary decision that the taxa in question do not deserve to be ranked as separate families. Although such a change in categorical assignment may be judged unnecessary, a similar problem results from attempts to achieve a phylogenetic taxonomy while retaining a criterion of synonymy based on the Linnaean categories. That is, elimination of paraphyletic taxa provides a reason for the so-called lumping or splitting of taxa, which leads to changes in the associations of their names with particular clades (Figure 3).

In pre-Hennigian taxonomies, it is often the case that a particular taxon is paraphyletic with respect to another taxon assigned to the same Linnaean category. One means of eliminating the paraphyletic taxon is to "unite" that taxon with its derivative taxon into a single inclusive taxon at a given categorical rank, the action commonly known as "lumping." In such cases, the two names are treated as synonyms. Paraphyly of the family Agamidae (Figure 3, left), for example, can be rectified by recognizing a single family for the species formerly included in Agamidae and its derivative family Chamaeleonidae (Figure 3, right) (38). The names of the two previously recognized families are then judged to be synonymous. Associating the name of the paraphyletic taxon with the inclusive clade changes its meaning in terms of content (included species). That association might, however, be justified by the prior implicit association of the name with the ancestor of the inclusive clade (115). In any case, treating the name of the derivative taxon

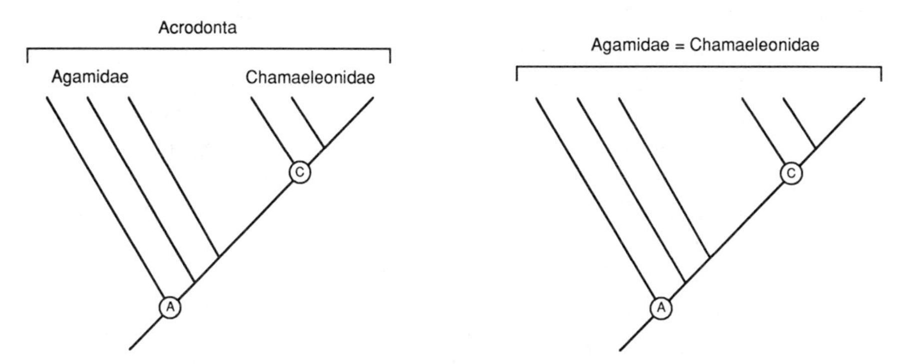

Figure 3 Change in the meaning of a taxon name resulting from an attempt to achieve a phylogenetic taxonomy within the constraints of a criterion of synonymy based on Linnaean categorical assignment. *Left:* Agamidae is paraphyletic if Chamaeleonidae is assigned to the same Linnaean category (family). *Right:* If all species in question are considered to belong to a single taxon assigned to the family category, "Agamidae" and "Chamaeleonidae" are treated as synonyms, and the latter name shifts its association from the clade stemming from ancestor C to that stemming from ancestor A. According to the principle of priority, "Chamaeleonidae" is the valid name of the clade stemming from ancestor A, even though "Agamidae" was originally associated with that ancestor and "Chamaeleonidae" was not. Furthermore, "Acrodonta," which was originally associated with the clade stemming from ancestor A, is rejected as the name of that clade because it is not a family name.

as a synonym changes the association of that name from a less inclusive to a more inclusive clade (Figure 3). Thus, the criterion of synonymy used in the current system causes instability in the phylogenetic meaning of "Chamaeleonidae" (115).

The instability noted above is highlighted by the manner in which the *valid* name of a taxon is most commonly determined in the current taxonomic system. According to the principle of priority, the valid name of a taxon is the oldest name applied to it (e.g. 70, 71). Because the various names applied to a taxon are synonyms, and because synonymy is currently tied to the Linnaean categories, the issue of priority is also ultimately tied to the Linnaean categories. For example, in zoological nomenclature, the valid name of a taxon assigned to the family category is the oldest family-group name based on the name of one of the genera included within that taxon.

The non-evolutionary criterion of synonymy adopted in the current taxonomic system can thus lead to drastic changes in the accepted names of clades. Continuing with the previous example (Figure 3), the valid name of the inclusive monophyletic taxon would be "Chamaeleonidae," because that is the oldest name of the family group based on one of the genera included within the taxon. Consequently, not only does "Chamaeleonidae" change its association from a less to a more inclusive clade (Figure 3), but that change in meaning occurs despite two other important facts. First, the taxon with which the name "Chamaeleonidae" was originally associated is still considered monophyletic and now either must be renamed or go unnamed. And second, the more inclusive clade already has a name, "Acrodonta," which is now ignored simply because it is not a family name. In general, the currently accepted criterion of synonymy combined with the practices of lumping and splitting is a major source of taxonomic instability. Lumping several taxa causes names to change their associations from less inclusive taxa to a more inclusive one, and splitting a single taxon results in restricting its name to a less inclusive taxon.

These problems stem from a fundamental incompatibility between the Linnaean and phylogenetic meanings of taxon names. Under the current system, categorical assignment partly determines the meanings of taxon names, because it determines the spellings of the names (i.e. their suffixes) and hence the taxa with which particular names are associated. This situation grants a non-evolutionary tradition primacy over the concept of evolution. That is to say, the association of a taxon name with a particular Linnaean category is, in effect, considered more important to the meaning of that name than its association with a particular clade or ancestor. This is unacceptable from the perspective of phylogenetic taxonomy. Names such as "Agamidae" and "Chamaeleonidae" have very different meanings when judged by their original reference to different sets of species, and thus implicitly to entities stemming from different common ancestors; in phylogenetic terms, they are not synonymous.

Furthermore, the very acts of lumping and splitting—which are intimately tied to changes in the meanings of taxon names under the current system—are difficult to interpret in phylogenetic terms. Taxonomists can neither lump nor split taxa as named clades, for clades are not things that taxonomists form, erect, unite, or divide, but rather things to which they give names. Outside the context of Linnaean categories, the notions of lumping and splitting make little sense, and they are irreconcilable with the phylogenetic meanings of taxon names.

In short, the criterion of synonymy used in the current taxonomic system is incompatible with the goals of phylogenetic taxonomy, that is, stable meanings of taxon names in terms of what they signify about common ancestry. Under the current system, different authors use the same name for different clades and different names for the same clade, and this can happen as the result of subjective differences concerning assignments to Linnaean categories even when the authors are in full agreement about what organisms and species make up the taxa (56). In the phylogenetic system, the Linnaean categories have nothing to do with the meanings of taxon names; *taxon names are synonymous if and only if they refer to the same clade* (115). Under phylogenetic definitions of taxon names, synonymy can be assessed unambiguously within the context of an accepted phylogeny by determining whether the names refer to clades stemming from the same ancestor (115).

Similarly, *priority in the phylogenetic system is not based on first use of a name in association with a particular Linnaean category but on first use of a name in association with a particular clade* (115). This is not to say that the valid name must always be established by priority. Indeed, recent movements advocating nomenclatural reform within the context of the Linnaean system seek to constrain the use of priority in establishing the validity of taxon names (122, 127). But regardless of whether or to what extent priority is used to establish the valid name of a taxon, in a phylogenetic system of taxonomy, the criterion of priority must be based on ancestry.

REDUNDANCY

Phylogenetic definitions of taxon names also bear on the problem of taxonomic redundancy. Linnaean taxonomies often contain monotypic taxa, which appear to be equivalent in content with the single included taxon at the next lower categorical level. In pre-Hennigian taxonomies, monotypic taxa were recognized in order to reflect distinctiveness (e.g. 14, 85, 121, 125). Phylogeneticists rejected distinctiveness as a justification for assigning a taxon to a Linnaean category of high rank, at least in cases where this practice resulted in the recognition of paraphyletic taxa (e.g. 59, 133). Nevertheless, monotypic taxa are also common in phylogenetic taxonomies adopting a Linnaean framework because of the constraints imposed by mandatory categories and exhaustive subsidiary taxa (35, 55). The apparent equivalency of, for example, a monotypic order and its single included family seems to contradict the hierarchical relationships between the Linnaean categories as well as the fact that taxonomists consider the two names to refer to different taxa. These seeming contradictions have been discussed by various authors under the name Gregg's Paradox (14, 15, 33–35, 52, 53, 67, 73, 120, 121). Furthermore, if the different names in fact refer to the same taxon, then they are redundant. One name is sufficient for a single clade.

Eliminating the convention of mandatory Linnaean categories (or use of the categories altogether) would solve these problems by removing the reason for recognizing monotypic taxa. But regardless of whether the Linnaean categories are retained, the problems are more apparent than real. On the one hand, apparent equivalency is often an artifact of restricting considerations to extant organisms (55). For example, Simpson (125) and Ruse (121) treat the taxa *Orycteropus*, Orycteropodidae, and Tubulidentata as if they are all monotypic because they are all made up of one and the same living species. There are, however, several known species of fossil aardvarks that have been referred to these taxa in such a way (102, 103) that only Tubulidentata can be considered monotypic in the sense of being made up of the same set of known organisms as Orycteropodidae.

Even taxa composed of the same known organisms are not necessarily identical. Using phylogenetic definitions of taxon names, different names can be defined so that they refer to different clades in a series of increasing inclusiveness, that is, clades stemming from successively more remote ancestors (Figure 4) (see also 59). For example, "Orycteropodidae" might be defined as the clade stemming from the immediate common ancestor of *Orycteropus* and the extinct *Plesiorycteropus*, whereas "Tubulidentata" might be defined as all those mammals sharing a more recent common ancestor with *Orycteropus* than with other extant mammals. Although all known tubulidentates are orycteropodids, the taxa are not necessarily equivalent. As noted above (see Taxa), a taxon as a named clade consists not only of those organisms that we recognize as its members; it consists of an ancestor and *all* of its descendants, extant and extinct, known and unknown. Therefore, although Orycteropodidae and Tubulidentata may appear equivalent when considerations are restricted to known or extant organisms, the definitions of the names refer to different clades and thus imply a difference in actual content. That difference can potentially be demonstrated by the discovery of currently unknown organisms.

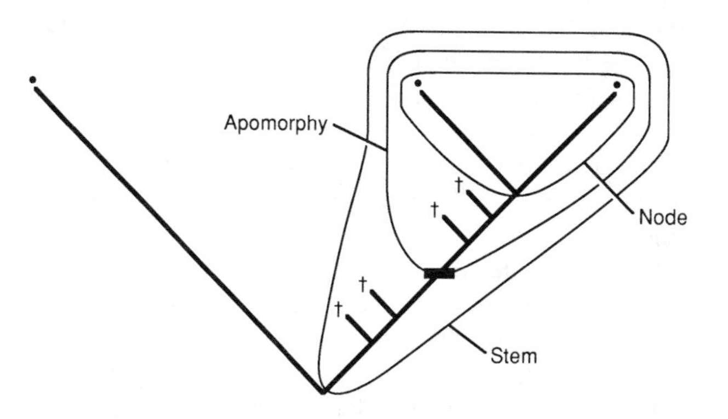

Figure 4 Although taxa may be identical in terms of known content, their names can be defined phylogenetically so that they refer to clades differing in actual content. The nested bubbles illustrate, in order of decreasing inclusiveness, stem-, apomorphy-, and node-based taxa in a nested series. Dots represent extant and/or known species; daggers represent extinct and/or unknown species. See text for example.

STANDARD NAMES

Although not necessary for developing a phylogenetic system of taxonomy, it is worthwhile to consider how the meanings of certain names might be standardized in a way that is most useful for comparative biology. Standardization would make the meanings of names universal, which is important to their function in communication (71, 76, 85, 87). Proposals for standardizing certain higher taxon names not covered by the zoological code have recently been made within the context of the Linnaean categories (e.g. 27, 28, 122); however, it is useful to consider the matter from a phylogenetic perspective.

Inconsistencies in the Current Use of Taxon Names

Of the vast numbers of taxon names that have been coined by taxonomists, some are more widely known and used than others. It is particularly important to standardize the meanings of widely known names to ensure that all biologists who use them are referring to the same entities. Currently, the supposed meanings of those names are inconsistent with the way they are most often used. The nature of that inconsistency is revealed by the following example.

Since its discovery (89, 90), the fossil taxon *Archaeopteryx* has been recognized as an outgroup to the clade stemming from the immediate common ancestor of extant birds (41, 68, 84). *Archaeopteryx* predates members of that clade by some 50 million years, and it retains several ancestral characters relative to extant birds. Nevertheless, that fossil has, with few exceptions, been referred to the taxon Aves because it has the "key" avian character (84). Indeed, to the extent that taxon names have phylogenetic connotations, they usually appear to be apomorphy-based. Thus, "Aves" is usually thought of as being associated with the clade stemming from first vertebrate possessing feathered wings.

Despite general agreement that Aves includes *Archaeopteryx,* biologists commonly use the name "Aves" ("birds") when making generalizations that apply to extant birds alone. Thus, supposed meaning and actual use are inconsistent. Furthermore, that inconsistency persists despite the existence of a less well known name coined for the specific purpose of making the relevant distinction. Most biologists continue to use the widely known name "Aves" when referring to the taxon explicitly associated (e.g. 21) with the name "Neornithes." A comparable situation holds for the fossil taxon *Ichthyostega,* the widely used name "Tetrapoda," and Gaffney's (39) less well known name "Neotetrapoda."

The reason for such inconsistencies seems to be that most well known names have gained wide use through their associations with distinctive groups of extant organisms. Those widely known names associated with clades have therefore most commonly been associated to one degree or another with crown clades—clades stemming from the immediate common ancestor of sister groups with extant representatives (74; see also 60, 62, 63). Distinctiveness is often a consequence of an incomplete fossil record, but even as the gaps are bridged by fossil discoveries, those gaps effectively persist because most biologists study systems that are not preserved in fossils. Nevertheless, discoveries of extinct outgroups to crown clades point to the existence of more inclusive clades, a fact that raises two alternatives concerning the associations of the original names. Either those names can be associated with the crown clades, or they can be associated with one of the larger clades including various extinct relatives. But even when the original names are explicitly defined so as to include at least some of the fossil outgroups, as they most often are, the majority of comparative biologists ignore the fossils. Consequently, the original names tend to be used as if they refer to crown clades, and new names coined specifically for the crown clades seldom gain wide use except among paleontologists.

In addition to the inconsistencies described above, the meanings of widely used taxon names also vary considerably among systematists explicitly addressing the taxonomy of fossils. These latter inconsistencies reflect differing preferences for what are, in effect, node-based, stem-based, and apomorphy-based definitions. For example, some authors use the name "Mammalia" for a crown clade (42–44, 119), whereas others use it for that crown clade and all extinct amniotes more closely related to it than to other extant amniotes (2–4, 81). Still others associate the name with an intermediate clade diagnosed by possession of a dentary-squamosal jaw joint, thus including the crown clade as well as a few of its extinct outgroups (e.g. 77, 91). And such differences in use often persist in spite of considerable agreement about phylogenetic relationships.

Standard Names for Crown Clades

No scientific enterprise, least of all one that considers the promotion of nomenclatural universality as one of its primary objectives, can accept the inconsistencies and ambiguities current in biological taxonomy. In some sense, progress in any scientific discipline can be measured in terms of further refinement, rather than escalating imprecision, in vocabulary. Therefore, it is imperative that biological taxonomy adopt rules for standardizing the meanings of important names. Because of the way in which those names are most often used, *the phylogenetic meanings of many widely known taxon names are most effectively standardized by tying them to clades within which both branches of the basal dichotomy are represented by extant descendants*. Names can be associated unambiguously with crown clades using node-based definitions (see Definitions of Taxon Names).

Restricting widely used names to crown clades standardizes their meanings in a way that is most useful to the largest number of comparative biologists. Although it will entail changes in the taxonomy of various fossils—*Archaeopteryx,* for example, will no longer be considered part of the taxon named "Aves"—this emphasis on extant organisms is not meant to imply that extant organisms are more important than fossils for establishing relationships. On the contrary, it is clear that phylogenetic relationships are best analyzed by considering both fossil and Recent organisms (26, 44). Nevertheless, most biologists study extant organisms, if for no other reason than that many aspects of the biology of extinct organisms are not only unknown but perhaps unknowable.

Because fossils are so often ignored, this proposal will bring the definitions of widely known names into agreement with the manner in which those names are most often used. It will also ensure that when neontologists and paleontologists use the same name, they will be referring to the same clade. Although paleontologists will have to restrict their use of various names, this is a more effective way to achieve universal meanings than forcing the vast majority of comparative biologists to learn and use more obscure names. Paleontologists will still have to use less well known names, but that is appropriate because the distinctions embodied in those names are of concern mainly to paleontologists. Furthermore, associating widely known names with crown clades is often consistent with a liberal interpretation of a phylogenetic criterion of priority (see Synonymy and Priority). Although most widely known names did not originally have explicit phylogenetic definitions, they tend to be old names that were used for groups of extant organisms prior to the discovery of their extinct outgroups.

Another advantage of this proposal is that it should discourage biologists from making unsupported generalizations about extinct outgroups while at the same time enabling them to make the greatest number of supportable inferences about the extinct members of taxa associated with widely used names. Biologists commonly make generalizations about the characters of entire taxa based on surveys of their extant representatives; however, it is not justifiable to extend those generalizations to fossil organisms sharing more remote common ancestors with the extant forms. For example, if the name "Tetrapoda" is defined to include *Ichthyostega,* one should not assume that the features common to the limb development of extant tetrapods (e.g. 94) characterize tetrapods as a whole. Therefore, restricting widely known names to crown clades will discourage biologists from making unsupported generalizations about the characters of extinct outgroups such as *Ichthyostega.* At the same time, it will also permit biologists to make the greatest number of inferences about the extinct representatives of taxa associated with widely used names based on properly conducted surveys of extant organisms. For example, if the name "Tetrapoda" is defined as the clade stemming from the most recent common ancestor of amphibians and amniotes, then features common to the limb development of diverse extant amphibians and amniotes can reasonably be inferred to have been present in the ancestral tetrapod.

Standard Names for More Inclusive Clades

Despite the advantages of restricting widely used names to crown clades, it is equally important to name the more inclusive clades, each consisting of a crown clade and all extinct taxa sharing with it common ancestors not shared with any other crown clade. These "total groups" (74, 75) or "panmonophyla" (80) are particularly important because they are stem-based taxa (not to be confused with paraphyletic stem-groups), and only stem-based taxa can be true sister groups, which are equivalent in age (Figure 5). Consequently, *for each standard name defined as the name of a crown clade, there should also be a standard name for the more inclusive clade consisting of the crown clade plus its extinct outgroups.* Such names can be associated unambiguously with the appropriate clade using stem-based definitions (see Definitions of Taxon Names).

Because of their equivalence in age, it is critical to use stem-based sister taxa in comparisons where such equivalence is important, as it is in investigations of taxonomic diversity. For example, it may be inappropriate to invoke an apomorphy of Aves to explain why that taxon is more speciose than is Crocodylia. On the one hand, the crown clades Aves and Crocodylia

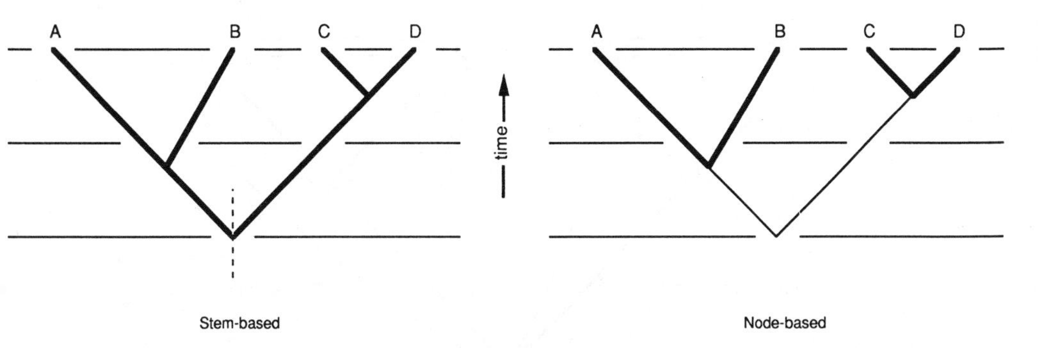

Figure 5 Comparison of stem-based and node-based taxa. *Left:* Stem-based taxa originating from a single cladogenetic event are true sister groups and are equivalent in terms of their age. *Right:* Node-based taxa, even those that are one anothers' closest relatives in terms of known or extant organisms (terminal taxa A-D), are not true sister groups and generally are not the same age. Thick lines are used to indicate the clades under consideration.

may not be of equal age. If Aves is older than Crocodylia, then the difference in species diversity may simply reflect that difference in age. On the other hand, the sister taxa of which Aves and Crocodylia are respective parts, that is, Ornithosuchia and Pseudosuchia, may have differed in species diversity from shortly after their initial divergence. If Ornithosuchia has always been more speciose than Pseudosuchia, then this difference cannot be explained by an apomorphy that arose in the most recent common ancestor of extant birds. If explicable by an apomorphy at all, then it must be one that arose early in the history of Ornithosuchia.

Since Hennig (58, 59), systematists have been aware of the distinction between the divergence of a lineage from its sister group and its subsequent diversification into lineages represented by extant species. Nevertheless, they have generally used the same name to refer to both entities, distinguishing between them using some notation, for instance, marking the name of the crown clade with an asterisk (e.g. 62, 63, 74, 75, 80). But use of the same name for different clades is likely to generate confusion, and such conventions have not gained wide use. For these reasons, different names should be applied to different clades (115). Thus, we (e.g. 42) use well-known names such as "Mammalia" and "Reptilia" for crown clades, while using less widely known names, in this instance "Synapsida" and "Sauropsida," for the larger clades including the extinct outgroups of Mammalia and Reptilia, respectively (Figure 6). The reasons detailed above underlie recent redefinitions of various taxon names by us and our colleagues (e.g. 6, 7, 32, 41–44, 51, 110, 119; see Appendix).

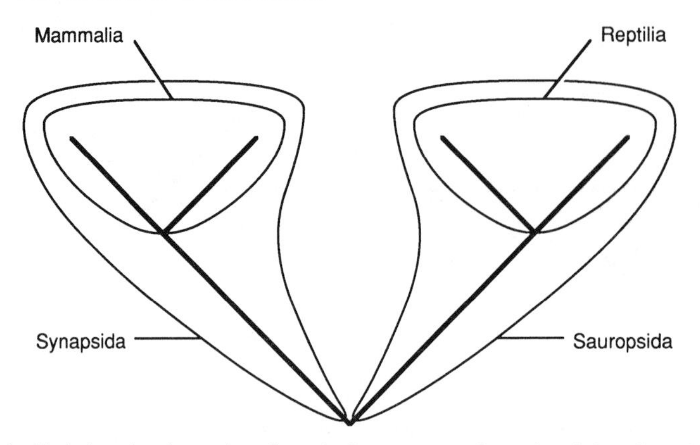

Figure 6 Node-based and stem-based standard taxon names for major clades of amniotes. The widely known names "Mammalia" and "Reptilia" are used for the crown clades in order to promote consistent use by paleontologists and neontologists. The less widely known names "Synapsida" and "Sauropsida" are used for more inclusive clades stemming from the initial divergence of the lineages leading to the crown clades; these taxa are important for making comparisons because of their equivalence in age (see *Figure 5*).

CONCLUSION

A taxonomic system is fundamental to comparative biology. Taxonomies are practical reference systems that permit communication and facilitate access to the literature. They also provide a theoretical context within which to make meaningful comparisons. In order to carry out these functions most effectively, biological taxonomy must be governed by a body of principles and rules designed to accomplish the practical goals within an appropriate theoretical context.

The taxonomic system developed by Linnaeus, and formalized in the various codes of biological nomenclature, has governed taxonomic practices admirably for over 200 years. Indeed, it is a tribute to a taxonomic system based on non-evolutionary principles that it has persisted for well over 100 years into the era dominated by an evolutionary world view—an era in which taxonomy is purported to be evolutionary. But biological taxonomy must eventually outgrow the Linnaean system, for that system derives from an inappropriate theoretical context. Modern comparative biology requires a taxonomic system based on evolutionary principles.

The late Willi Hennig and his followers initiated the development of such a system by granting the principle of descent a central role in establishing the nature of taxa. However fundamental, that advancement represents only the first step in the development of a phylogenetic system of taxonomy, because the system remains constrained by non-evolutionary Linnaean traditions. We have attempted to further the development of a phylogenetic system of taxonomy by reformulating specific taxonomic principles and rules in terms of its most general principle, the tenet of common descent. Some of our specific principles and rules may not be favorably received, but we hope that their shortcomings will not detract from the general perspective within which they were formulated. That general perspective is one in which the concept of evolution is granted a central role in taxonomy. As such, it embodies a change in the basis of the taxonomic system in which the Linnaean categories are replaced by the tenet of evolutionary descent.

Acknowledgments

We thank D. C. Cannatella, M. J. Donoghue, S. V. Fink, W. L. Fink, D. R. Frost, T. M. Gosliner, D. M. Hillis, D. L. Hull, R. Mooi, R. J. O'Hara, K. Padian, T. Rowe, M. Russell, and an anonymous reviewer for discussions of ideas and/or comments on earlier drafts of this paper. R. Etheridge suggested that our reformulation of certain taxonomic concepts (115) required a new set of rules, of which this paper represents the foundation. We dedicate this paper to the late Richard Estes, an inspiring teacher and colleague and a coauthor of the paper (43) in which many of these ideas originated. Our work was partially supported by a Tilton Postdoctoral Fellowship from the California Academy of Sciences to K. de Queiroz and National Science Foundation grant BSR 87–09455 to J. Gauthier.

Appendix: A Phylogenetic Taxonomy of Craniata

In order to illustrate the simplicity of a phylogenetic system of taxonomy, we present (Figure 7) a phylogenetic taxonomy of craniates based on the conventions proposed in this paper and the relationships proposed by Maisey (83) and Gauthier et al (42). This taxonomy is not intended to be complete, but it includes the taxa traditionally considered in introductory texts on vertebrate comparative anatomy. Consequently, it suffers from the problem of differential resolution (101), which is to say that the subordinate taxa within sister taxa do not always receive equal attention (e.g. Sauria vs. Chelonia). This reflects a longstanding bias in craniate taxonomy, in which the lineages leading to birds and mammals have received disproportionate emphasis.

The taxonomy is constructed according to the following conventions:

1. All names refer to clades.
2. The names of subordinate taxa within each clade are indented to indicate hierarchical relationships.
3. Linnaean and other categories (i.e. plesion) are not used.
4. Names formerly associated with the genus category are treated the same as those of all other clades; that is, they are capitalized but not italicized (e.g. "Sphenodon").

- [unnamed] - **Craniata**
  - Myxini - **Myxinoidea**
  - Myopterygii - **Vertebrata**
    - Petromyzontida - **Petromyzontidae**
    - [unnamed] - **Gnathostomata**
      - [unnamed] - **Chondrichthyes**
        - [unnamed] - **Holocephali**
        - [unnamed] - **Elasmobranchii**
      - Teleostomi - **Osteichthyes**
        - Crossopterygii - **Sarcopterygii**
          - Actinistia - **Latimeria**
          - Rhipidistia - **Choanata**
            - Porolepida - **Dipnoi**
            - Osteolepida - **Tetrapoda**
              - Temnospondyli - **Amphibia**
                - Apoda - **Gymnophiona**
                - Paratoidea - **Batrachia**
                  - Urodela - **Caudata**
                  - Salientia - **Anura**
              - Anthracosauria - **Amniota**
                - Synapsida - **Mammalia**
                  - Prototheria - **Monotremata**
                  - Theriiformes - **Theria**
                    - Metatheria - **Marsupialia**
                    - Eutheria - **Placentalia**
                - Sauropsida - **Reptilia**
                  - Anapsida - **Chelonia**
                  - Diapsida - **Sauria**
                    - Lepidosauromorpha - **Lepidosauria**
                      - Rhynchocephalia - **Sphenodon**
                      - Lacertilia - **Squamata**
                    - Archosauromorpha - **Archosauria**
                      - Pseudosuchia - **Crocodylia**
                      - Ornithosuchia - **Aves**
        - [unnamed] - **Actinopterygii**
          - Cladistia - **Polypterus**
          - [unnamed] - **Actinopteri**
            - [unnamed] - **Chondrostei**
            - Neopterygii - **Holostei**
              - Ginglymodi - **Lepisosteidae**
              - [unnamed] - **Halecostomi**
                - Halecomorphi - **Amia**
                - [unnamed] - **Teleostei**

Figure 7 A phylogenetic taxonomy of Craniata.

5. No redundant names are used.
6. For each pair of sister taxa, the first listed has fewer extant species than the second.
7. Two names are given on each line. The first is the stem-based name of a clade consisting of the crown clade named on the same line plus all extinct taxa more closely related to it than to any other crown clade. For example, the name "Temnospondyli" refers to the clade including Amphibia plus all known and unknown tetrapods sharing a more recent common ancestor with Amphibia than with Amniota. The second name is the node-based name of a crown clade defined as the clade stemming from the immediate common ancestor of its two immediately subordinate crown clades. For example, the name "Amphibia" refers to the clade stemming from the last common ancestor of Gymnophiona and Batrachia.
8. Widely known names are restricted to crown clades, while less well known names are used for stem clades.

Although some names in each group were formerly associated with paraphyletic taxa (e.g. "Reptilia"), we have nevertheless retained them because of their implicit associations with particular ancestors and, in the case of those associated with the ancestors of crown clades, because of their familiarity. In some cases, the hierarchical relationships between the names of crown and stem clades are reversed from traditional use. That reversal results from our applying names according to widespread uses, even when it contradicts supposed meanings that accord with less common uses. The name "Vertebrata," for example, is most commonly used for a crown clade and less commonly for a more inclusive clade encompassing several extinct forms which, in accordance with the supposed meaning of that name, also have vertebrae. Paleontologists (e.g. 72) thus use "Vertebrata" for the latter clade and apply the obscure name "Myopterygii" to what is otherwise known as "Vertebrata." In such instances, we elected to reverse the names in order to associate the well-known name with the crown clade.

Because of the nested relationship between corresponding crown and stem clades, there are advantages to basing the names of both clades on the same word stem. In keeping with the spirit of phylogenetic taxonomy, which emphasizes common ancestry over characters, the stem of the name of the crown clade could be combined with *gens* or *genea,* the respective Latin and Greek suffixes meaning clan, to form the name of the stem clade. Even if another suffix were used, that practice would simplify the taxonomy, and thus facilitate its memorization, because the names of corresponding crown and stem clades would differ only in their suffixes. Because of space limitations, we have not provided an alternative taxonomy using stem-based names of this kind.

Literature Cited

1. Anderson, S. 1975. On the number of categories in biological classifications. *Am. Mus. Novit.* 2584:1–9
2. Ax, P. 1985. Stem species and the stem lineage concept. *Cladistics* 1:279–87
3. Ax, P. 1987. *The Phylogenetic System: The Systematization of Organisms on the Basis of their Phylogenesis.* Chichester: Wiley. 340 pp.
4. Ax, P. 1989. The integration of fossils in the phylogenetic system of organisms. *Abh. naturwiss. Ver. Hamburg* 28:27–43
5. Ball, I. R. 1975. Nature and formulation of biogeographical hypotheses. *Syst. Zool.* 24:407–30
6. Benton, M. J., Clark, J. M. 1988. Archosaur phylogeny and the relationships of the Crocodylia. In *The Phylogeny and Classification of the Tetrapods,* Vol. 1. *Amphibians, Reptiles, and Birds,* ed. M. J. Benton, 8:295–338. Oxford: Clarendon. 377 pp.
7. Berta, A., Ray, C. E., Wyss, A. R. 1989. Skeleton of the oldest known pinniped, *Enaliarctos mealsi. Science* 244:60–62
8. Bock, W. J. 1977. Foundations and methods of evolutionary classification. In *Major Patterns in Vertebrate Evolution,* ed. M. K. Hecht, P. C. Goody, B. M. Hecht, pp. 851–95. New York: Plenum. 908 pp.
9. Bonde, N. 1975. Origin of "higher groups": Viewpoints of phylogenetic systematics. In *Problèmes Actuels de Paléontologie-EGRvolution des Vertébrés, Colloq. Int. CNRS (Paris)* 218: 293–324
10. Bonde, N. 1977. Cladistic classification as applied to vertebrates. See Ref. 8, pp. 741–804
11. Boudreaux, H. B. 1979. *Arthropod Phylogeny with Special Reference to Insects.* New York: Wiley. 320 pp.
12. Brundin, L. 1966. Transantarctic relationships and their significance, as evidenced by chironomid midges, with a monograph of the subfamilies Podonominae and Aphroteniinae and the austral Heptagyliae. *Kungl. Svenska Vetenskapsakademiens Handlinger* 11:1–471
13. Brundin, L. 1968. Application of phylogenetic principles in systematics and evolutionary theory. In *Current Problems of Lower Vertebrate Phylogeny,* ed. T. Ørvig, pp. 473–95. New York: Interscience. 539 pp.
14. Buck, R., Hull, D. L. 1966. The logical structure of the Linnaean hierarchy. *Syst. Zool.* 15:97–111
15. Buck, R., Hull, D. L. 1969. Reply to Gregg. *Syst. Zool.* 18:354–57
16. Cain, A. J. 1959. Deductive and inductive methods in post-Linnaean taxonomy. *Proc. Linn. Soc. London* 170:185–217
17. Cain, A. J. 1959. The post-Linnaean development of taxonomy. *Proc. Linn. Soc. London* 170:234–44
18. Cannatella, D. C., de Queiroz, K. 1989. Phylogenetic systematics of the anoles: Is a new taxonomy warranted? *Syst. Zool.* 38:57–69
19. Colless, D. H. 1977. A cornucopia of names. *Syst. Zool.* 26:349–52
20. Cracraft, J. 1974. Phylogenetic models and classification. *Syst. Zool.* 23:71–90
21. Cracraft, J. 1986. The origin and early diversification of birds. *Paleobiology* 12:383–99
22. Craske, A. J., Jefferies, R. P. S. 1989. A new mitrate from the Upper Ordovician of Norway, and a new approach to subdividing a plesion. *Palaeontology* 32:69–99
23. Crowson, R. A. 1970. *Classification and Biology.* New York: Atherton. 350 pp.
24. Darwin, C. 1859. *On the Origin of Species by Means of Natural Selection.* London: John Murray. 513 pp. Reprinted 1974. Cambridge, Mass: Harvard Univ. Press
25. Donoghue, M. J., Cantino, P. D. 1988. Paraphyly, ancestors, and the goals of taxonomy: A botanical defense of cladism. *Bot. Rev.* 54:107–28
26. Donoghue, M. J., Doyle, J. A., Gauthier, J., Kluge, A. G., Rowe, T. 1989. The importance of fossils in phylogeny reconstruction. *Annu. Rev. Ecol. Syst.* 20:431–60
27. Dubois, A. 1984. La nomenclature supragénérique des amphibiens anoures. *Mém. Mus. Natl. d'Hist. Nat. Paris, série A, zoologie* 131:1–64
28. Dundee, H. A. 1989. Higher category name usage for amphibians and reptiles. *Syst. Zool.* 38:398–406
29. Dupuis, C. 1984. Willi Hennig's impact on taxonomic thought. *Annu. Rev. Ecol. Syst.* 15:1–24
30. Eldredge, N., Cracraft, J. 1980. *Phy-*

logenetic Patterns and the Evolutionary Process. New York: Columbia Univ. Press. 349 pp.
31. Engelmann, G. F., Wiley, E. O. 1977. The place of ancestor-descendant relationships in phylogeny reconstruction. *Syst. Zool.* 26:1–11
32. Estes, R., de Queiroz, K., Gauthier, J. 1988. Phylogenetic relationships within Squamata. In *Phylogenetic Relationships of the Lizard Families: Essays Commemorating Charles L. Camp*, ed. R. Estes, G. K. Pregill, pp. 119–281. Stanford, Calif: Stanford Univ. Press. 631 pp.
33. Farris, J. S. 1967. Definitions of taxa. *Syst. Zool.* 16:174–75
34. Farris, J. S. 1968. Categorical ranks and evolutionary taxa in numerical taxonomy. *Syst. Zool.* 17:151–59
35. Farris, J. S. 1976. Phylogenetic classification of fossils with Recent species. Syst. Zool. 25:271–82
36. Farris, J. S. 1979. The information content of the phylogenetic system. *Syst. Zool.* 28:483–519
37. Felsenstein, J. 1982. Numerical methods for inferring evolutionary trees. *Q. Rev. Biol.* 57:379–404
38. Frost, D. R., Etheridge, R. 1989. A phylogenetic analysis and taxonomy of iguanian lizards (Reptilia: Squamata). *Misc. Publ. Univ. Kans. Mus. Nat. Hist.* 81:1–65
39. Gaffney, E. S. 1980. Tetrapod monophyly: A phylogenetic analysis. *Bull. Carnegie Mus. Nat. Hist.* 13:92–105
40. Gaffney, E. S., Meylan, P. A. 1988. A phylogeny of turtles. See Ref. 6, 5:157–219
41. Gauthier, J. 1986. Saurischian monophyly and the origin of birds. In *The Origin of Birds and the Evolution of Flight*, ed. K. Padian, 8:1–55. Calif. Acad. Sci. Mem.
42. Gauthier, J., Cannatella, D., de Queiroz, K., Kluge, A. G., Rowe, T. 1989. Tetrapod phylogeny. In *The Hierarchy of Life*, ed. B. Fernholm, K. Bremer, H. Jörnvall, 25:337–53. Amsterdam: Elsevier. 499 pp.
43. Gauthier, J., Estes, R., de Queiroz, K. 1988. A phylogenetic analysis of Lepidosauromorpha. See Ref. 32, pp. 15–98
44. Gauthier, J., Kluge, A. G., Rowe, T. 1988. Amniote phylogeny and the importance of fossils. *Cladistics* 4:105–209
45. Ghiselin, M. T. 1966. An application of the theory of definitions to systematic principles. *Syst. Zool.* 15:127–30
46. Ghiselin, M. T. 1969. *The Triumph of the Darwinian Method*. Berkeley: Univ. Calif. Press. Reprinted 1984. Chicago: Univ. Chicago Press. 287 pp.
47. Ghiselin, M. T. 1980. Natural kinds and literary accomplishments. *Mich. Q. Rev.* 19:73–88
48. Ghiselin, M. T. 1981. Categories, life, and thinking. *Behav. Brain Sci.* 4:269–313
49. Ghiselin, M. T. 1984. "Definition," "character," and other equivocal terms. *Syst. Zool.* 33:104–10
50. Ghiselin, M. T. 1985. Narrow approaches to phylogeny: A review of nine books of cladism. *Oxford Surv. Evol. Biol.* 1:209–22
51. Good, D. A. 1988. The phylogenetic position of fossils assigned to the Gerrhonotinae (Squamata: Anguidae). *J. Vert. Paleontol.* 8:188–95
52. Gregg, J. R. 1954. *The Language of Taxonomy: An Application of Symbolic Logic to the Study of Classificatory Systems*. New York: Columbia Univ. Press. 70 pp.
53. Gregg, J. R. 1968. Buck and Hull: A critical rejoinder. *Syst. Zool.* 17:342–44
54. Griffiths, G. C. D. 1973. Some fundamental problems in biological classification. *Syst. Zool.* 22:338–43
55. Griffiths, G. C. D. 1974. On the foundations of biological systematics. *Acta Biotheor.* 23:85–131
56. Griffiths, G. C. D. 1976. The future of Linnaean nomenclature. *Syst. Zool.* 25:168–73
57. Hennig, W. 1950. *Grundzüge einer Theorie der phylogenetischen Systematik*. Berlin: Deutscher Zentralverlag. 370 pp.
58. Hennig, W. 1965. Phylogenetic systematics. *Annu. Rev. Entomol.* 10:97–116
59. Hennig, W. 1966. *Phylogenetic Systematics*. Urbana: Univ. Ill. Press. 263 pp.
60. Hennig, W. 1969. *Die Stammesgeschichte der Insekten*. Frankfurt: Kramer. 436 pp.
61. Hennig, W. 1975. "Cladistic analysis or cladistic classification?": A reply to Ernst Mayr. *Syst. Zool.* 24:244–56
62. Hennig, W. 1981. *Insect Phylogeny*. Chichester: Wiley. 514 pp.
63. Hennig, W. 1983. Stammesgeschichte der Chordaten. *Fortschr. zool. Syst. Evolutionsforsch.* 2:1–208
64. Heywood, V. H. 1988. The structure of systematics. In *Prospects in Systematics*, ed. D. L. Hawksworth, 3:44–56. Oxford: Clarendon. 457 pp.

65. Hubbs, C. L. 1956. Ways of stabilizing zoological nomenclature. *Proc. XIV Int. Congr. Zool., Copenhagen* 1953: 548–53
66. Hull, D. L. 1978. A matter of individuality. *Philos. Sci.* 45:335–60
67. Hull, D. L., Buck, R. 1967. Definitions of taxa. *Syst. Zool.* 16:349
68. Huxley, T. H. 1868. On the animals which are most nearly intermediate between the birds and reptiles. *Geol. Mag.* 5:357–65
69. International Association of Microbiological Societies. 1975. *International Code of Nomenclature of Bacteria, and Statutes of the International Committee on Systematic Bacteriology, and Statutes of the Bacteriology Section of the International Association of Microbiological Societies.* Washington: Am. Soc. Microbiol. 180 pp. 1976 revision
70. International Botanical Congress. 1988. *International Code of Botanical Nomenclature.* Königstein: Koeltz Scientific Books. 328 pp. Edition adopted by the 14th Int. Botanical Congress, Berlin, July-August 1987
71. International Commission on Zoological Nomenclature. 1985. *International Code of Zoological Nomenclature.* London: Int. Trust for Zool. Nomenclature. 338 pp. 3rd ed.
72. Janvier, P. 1984. The phylogeny of the Craniata, with particular reference to the significance of fossil "agnathans." *J. Vert. Paleontol.* 1:121–59
73. Jardine, N. 1969. A logical basis for biological classification. *Syst. Zool.* 18:37–52
74. Jefferies, R. P. S. 1979. The origin of chordates—a methodological essay. In *The Origin of Major Invertebrate Groups,* ed. M. R. House, 17:443–77. London: Academic. 515 pp.
75. Jefferies, R. P. S. 1986. *The Ancestry of the Vertebrates.* London: Br. Mus. (Nat. Hist.). 376 pp.
76. Jeffrey, C. 1973. *Biological Nomenclature.* New York: Crane, Russak. 69 pp. 2nd ed. 1977
77. Kemp, T. S. 1988. Interrelationships of the Synapsida. In *The Phylogeny and Classification of the Tetrapods,* Vol. 2: *Mammals,* ed. M. J. Benton, 1:1–22. Oxford: Clarendon. 329 pp.
78. Kraus, O. 1976. Phylogenetische Systematik und evolutionäre Klassifikation. *Verh. Deutsch. Zool. Ges.* 69:84–99
79. Laurin, M. 1991. The osteology of a Lower Permian eosuchian from Texas and a review of diapsid phylogeny. *Zool. J. Linn. Soc.* 101:59–95
80. Lauterbach, K.-E. 1989. Das Pan-Monophylum—Ein Hilfsmittel für die Praxis der phylogenetischen Systematik. *Zool. Anz.* 223:139–56
81. Loconte, H. 1990. Cladistic classification of Amniota: A response to Gauthier et al. *Cladistics* 6:187–90
82. Løvtrup, S. 1977. *The Phylogeny of Vertebrata.* London: Wiley. 330 pp.
83. Maisey, J. G. 1986. Heads and tails: A chordate phylogeny. *Cladistics* 2: 201–56
84. Martin, L. D. 1985. The relationship of *Archaeopteryx* to other birds. In *The Beginnings of Birds: Proceedings of the International* Archaeopteryx *conference Eichstätt 1984,* ed. M. K. Hecht, J. H. Ostrom, G. Viohl, P. Wellnhofer, pp. 177–83. Eichstätt: Freunde des Jura-Museums Eichstätt. 382 pp.
85. Mayr, E. 1969. *Principles of Systematic Zoology.* New York: McGraw-Hill. 428 pp.
86. Mayr, E. 1982. *The Growth of Biological Thought. Diversity, Evolution, and Inheritance.* Cambridge, Mass: Harvard Univ. Press. 974 pp.
87. Mayr, E., Ashlock, P. D. 1991. *Principles of Systematic Zoology.* New York: McGraw-Hill. 475 pp. 2nd ed.
88. McKenna, M. C. 1975. Toward a phylogenetic classification of the Mammalia. In *Phylogeny of the Primates: A Multidisciplinary Approach,* ed. W. P. Luckett, F. S. Szalay, 2:21–46. New York: Plenum. 483 pp.
89. von Meyer, H. 1861. Vogel-Federn und *Palpipes priscus* von Solnhofen. *Neues Jahrb. Mineral. Geol. Palaeontol.* 1861:561
90. von Meyer, H. 1861. *Archaeopteryx lithographica* (Vogel-Feder) und *Pterodactylus* von Solnhofen. *Neues Jahrb. Mineral. Geol. Palaeontol.* 1861:678–79
91. Miao, D. 1991. On the origins of mammals. In *Origins of the Higher Groups of Tetrapods,* ed. H-P. Schultze, L. Trueb, 16:579–97. Ithaca: Cornell Univ. Press. 724 pp.
92. Michener, C. D. 1963. Some future developments in taxonomy. *Syst. Zool.* 12:151–72
93. Michener, C. D. 1964. The possible use of uninomial nomenclature to increase the stability of names in biology. *Syst. Zool.* 13:182–90
94. Müller, G. B., Alberch, P. 1990. Ontogeny of the limb skeleton in *Alligator mississippiensis:* Developmental invariance and change in the evolution of archosaur limbs. *J. Morphol.* 203:151–64
95. Nelson, G. 1971. "Cladism" as a phi-

losophy of classification. *Syst. Zool.* 20:373–76
96. Nelson, G. 1972. Phylogenetic relationship and classification. *Syst. Zool.* 21:227–31
97. Nelson, G. 1972. Comments on Hennig's "Phylogenetic Systematics" and its influence on ichthyology. *Syst. Zool.* 21:364–74
98. Nelson, G. 1973. Classification as an expression of phylogenetic relationships. *Syst. Zool.* 22:344–59
99. Nelson, G. 1974. Darwin-Hennig classification: A reply to Ernst Mayr. *Syst. Zool.* 23:452–58
100. Nelson, G. 1978. The perils of perfection: A reply to D. H. Colless. *Syst. Zool.* 27:124
101. O'Hara, R. J. 1992. Telling the tree: Narrative representation and the study of evolutionary history. *Biol. Philos.* 7:135–60
102. Patterson, B. 1975. The fossil aardvarks (Mammalia: Tubulidentata). *Bull. Mus. Comp. Zool., Harvard Univ.* 147:185–237
103. Patterson, B. 1978. Pholidota and Tubulidentata. In *Evolution of African Mammals,* ed. V. J. Maglio, H. B. S. Cooke, 12:268–78. Cambridge, Mass: Harvard Univ. Press. 641 pp.
104. Patterson, C. 1978. Verifiability in Systematics. *Syst. Zool.* 27:218–22
105. Patterson, C., Rosen, D. E. 1977. Review of ichthyodectiform and other Mesozoic teleost fishes and the theory and practice of classifying fossils. *Bull. Am. Mus. Nat. Hist.* 158:81–172
106. Patterson, D. J. 1988. The evolution of Protozoa. *Mem. Inst. Oswaldo Cruz, Rio de Janeiro* 83:580–600
107. Platnick, N. I. 1977. The hypochiloid spiders: A cladistic analysis, with notes on the Atypoidea (Arachnida, Araneae). *Am. Mus. Novit.* 2627:1–23
108. Platnick, N. I. 1977. Cladograms, phylogenetic trees, and hypothesis testing. *Syst. Zool.* 26:438–42
109. Platnick, N. I. 1978. Gaps and prediction in classification. *Syst. Zool.* 27: 472–74
110. de Queiroz, K. 1987. Phylogenetic systematics of iguanine lizards. A comparative osteological study. *Univ. Calif. Publ. Zool.* 118:1–203
111. de Queiroz, K. 1988. Systematics and the Darwinian revolution. *Philos. Sci.* 55:238–59
112. de Queiroz, K. 1992. Phylogenetic definitions and taxonomic philosophy. *Biol. Philos.* 7:295–313
113. de Queiroz, K., Donoghue, M. J. 1988. Phylogenetic systematics and the species problem. *Cladistics* 4:317–38
114. de Queiroz, K., Donoghue, M. J. 1990. Phylogenetic systematics or Nelson's version of cladistics. *Cladistics* 6:61–75
115. de Queiroz, K., Gauthier, J. 1990. Phylogeny as a central principle in taxonomy: Phylogenetic definitions of taxon names. *Syst. Zool.* 39:307–22
116. Ridley, M. 1986. *Evolution and Classification: The Reformation of Cladism.* London: Longman. 201 pp.
117. Rosen, D. E. 1974. Cladism or Gradism?: A reply to Ernst Mayr. *Syst. Zool.* 23:446–51
118. Rowe, T. 1987. Definition and diagnosis in the phylogenetic system. *Syst. Zool.* 36:208–11
119. Rowe, T. 1988. Definition, diagnosis, and origin of Mammalia. *J. Vert. Paleontol.* 8:241–64
120. Ruse, M. E. 1971. Gregg's Paradox: A proposed revision to Buck and Hull's solution. *Syst. Zool.* 20:239–45
121. Ruse, M. E. 1973. *The Philosophy of Biology.* London: Hutchinson Univ. Library. 231 pp.
122. Savage, J. M. 1990. Meetings of the Int. Commission on Zool. Nomenclature. *Syst. Zool.* 39:424–25
123. Scott-Ram, N. R. 1990. *Transformed Cladistics, Taxonomy and Evolution.* Cambridge: Cambridge Univ. Press. 238 pp.
124. Simpson, G. G. 1959. Anatomy and morphology: Classification and evolution: 1859 and 1959. *Proc. Am. Philos. Soc.* 103:286–306
125. Simpson, G. G. 1961. *Principles of Animal Taxonomy.* New York: Columbia Univ. Press. 247 pp.
126. Sneath, P. H. A., Sokal, R. R. 1973. *Numerical Taxonomy.* San Francisco: Freeman. 573 pp.
127. Stace, C. A. 1991. Naming names in botany. *Nature* 350:466
128. Stevens, P. 1984. Metaphors and typology in the development of botanical systematics 1690–1960, or the art of putting new wine in old bottles. *Taxon* 33:169–211
129. Swofford, D. L., Olsen, G. J. 1990. Phylogeny reconstruction. In *Molecular Systematics,* ed. D. M. Hillis, C. Moritz, 11:411–501. Sunderland, Mass: Sinauer. 588 pp.
130. Tubbs, P. K. 1991. The International Commission on Zoological Nomenclature: What it is and how it operates. *Bull. Zool. Nomen.* 48:295–99

131. Turrill, W. B. 1942. Taxonomy and phylogeny (Parts I–III). *Bot. Rev.* 8:247–70, 473–532, 655–707
132. Wiley, E. O. 1979. An annotated Linnaean hierarchy, with comments on natural taxa and competing systems. *Syst. Zool.* 28:308–37
133. Wiley, E. O. 1981. *Phylogenetics: The Theory and Practice of Phylogenetic Systematics*. New York: Wiley. 439 pp.
134. Wiley, E. O. 1989. Kinds, individuals, and theories. In *What the Philosophy of Biology Is*, ed. M. Ruse, pp. 289–300. Dordrecht: Kluwer. 337 pp.
135. Willmann, R. 1987. Phylogenetic systematics, classification and the plesion concept. *Verh. naturwiss. Ver. Hamburg* 29:221–33
136. Willmann, R. 1989. Palaeontology and the systematization of natural taxa. *Abh. naturwiss. Ver. Hamburg* 28:267–91
137. Woodger, J. H. 1952. From biology to mathematics. *Br. J. Philos. Sci.* 3:1–21

Annu. Rev. Ecol. Syst. 1992. 23:481–506

POPULATION VIABILITY ANALYSIS

*Mark S. Boyce**

Center for Ecological Sciences and Center for Theoretical Studies, Indian Institute of Science, Bangalore, 560 012 India

INTRODUCTION

Population viability analysis (PVA) is a process. It entails evaluation of data and models for a population to anticipate the likelihood that a population will persist for some arbitrarily chosen time into the future (125, 128). A closely related concept is minimum viable population (MVP) analysis. An MVP is an estimate of the minimum number of organisms of a particular species that constitutes a viable population. Reference is also made to population vulnerability analysis which is a negative appellation for PVA. PVA embraces MVP, but without seeking to estimate the absolute minimum population necessary to keep a species viable (136).

In the United States, the US Forest Service has a mandate to preserve viable populations on its lands under the National Forest Management Act (158). Likewise, the US Fish and Wildlife Service and the National Marine Fisheries Service have been evaluating PVAs for many species or populations proposed for listing under the Endangered Species Act (152). Establishing criteria for what constitutes a viable population is no longer strictly an academic pursuit.

PVAs have been attempted for at least 35 species; perhaps the most celebrated are those for the grizzly bear (*Ursus arctos horribilis*) (126, 129, 144), and the northern spotted owl (*Strix occidentalis caurina*) (18, 79, 95, 98a). Most PVAs are simulation studies that remain unpublished, or when published, they may only include outlines of model structure (95, 126, 131). Others invoke analytical methods or "rules of thumb," always burdened with severe assumptions (31, 152). PVAs vary according to the ecology of the species, the expertise of the modelers, and the extent of available data.

*Current address: Department of Zoology and Physiology, Laramie, Wyoming 82071-3166

0066-4162/92/1120-0481$2.00

There are no guidelines on what constitutes a valid PVA, and because each case is unique, I am loathe to devise any. Any attempt is qualified that involves a population simulation or analysis with the intent of projecting future populations or estimating some extinction or persistence parameter, e.g. time to extinction, probability of extinction, probability of persisting for 100 years, etc. Definitions and criteria for viability, persistence, and extinction are arbitrary, e.g. ensuring a 95% probability of surviving for at least 100 years. Discussion of such criteria can be found in Mace & Lande (92) and Thompson (152).

Collecting sufficient data to derive reliable estimates for all the parameters necessary to determine MVP is simply not practical in most cases. It is further questionable how well ecologists can predict the future (34), particularly over time horizons necessary to project extinctions. On the other hand, Soulé (136) suggests that managers have the right to expect population biologists to project the number of animals necessary to ensure the long-term viability of a population. But to do so has proven to be dangerous ground (83) which risks damaging the credibility of conservation biologists.

My purpose in this review is an attempt to place PVA and MVP on more comfortable ground by identifying a realistic domain. I maintain that PVA ought to be an integral part of any species management plan, but rather than being so presumptuous as to claim that we can actually use modeling to define a MVP, or to estimate the probability of extinction, I use it as a forum to champion the adaptive management approach of Holling (67) and Walters (162). For those all-too-frequent cases that cannot wait for a full-blown PVA, I review empirical evidence suggesting that use of rules-of-thumb for MVPs may not be unrealistic.

MODELING EXTINCTION

Fundamental to MVP is the fact that small populations are more likely to go extinct than larger ones due to inbreeding depression and genetic drift, or simply the threat of chance birth or death events (demographic stochasticity). Under high environmental variance or catastrophes, however, maintaining MVP may not be as effective a conservation target as would be managing for spatial configuration and location of habitats (38, 85, 106, 128). Irrespective of the target, the objective of PVA is to provide insight into how resource management can change parameters influencing the probability of extinction. This change may entail lengthening the expected time to extinction, $E(T)$, or reducing the probability of extinction within some time frame.

The most appropriate model structure for a PVA depends on the availability of data and the essential features of the ecology of the organism. In this section

I review structural features of PVA and extinction modeling. I begin with a discussion of stochastic variation, a necessary element in any consideration of extinction processes, and then review approaches focusing on genetics, demography, and ecology. I conclude with the argument that all of these elements ought to occur together in the same model. Because of the complexity of such a model, most PVAs will enter the realm of computer simulation modeling.

Stochasticity

Random events can be extremely important in extinction, especially for small populations, the target of most PVA (121). In a sense, the distinction between deterministic and stochastic processes in ecology is artificial because all ecological processes are stochastic. Maynard Smith (97) points out that "the use of deterministic rather than stochastic models can only be justified by mathematical convenience." For heuristic purposes, use of deterministic models is appropriate. For example, our understanding of a simple model of competition or predation would only be obfuscated by the complex mathematics of a stochastic version. Because they are more tractable, deterministic models may yield robust results; stochastic models are often too complex to be solved analytically and thus require use of simulation methods.

Usually, conclusions from stochastic models in ecology are strikingly different from deterministic ones (26). This is generally true because of (i) large variances, (ii) nonlinear functions, and (iii) highly skewed or otherwise non-normally distributed variables. The significance of high variance is easy to understand in the context of extinction (85). Less transparent is the fact that virtually all ecological processes are nonlinear (17, 122). When the system contains nonlinearities, its behavior may differ markedly due to Jensen's inequality which states that for any concave function, ϕ, of a random variable χ, $E[\chi)] \leq \phi[E(X)]$; the reverse inequality applies for a convex function (72). The greater the magnitude of nonlinearity, say as measured by the second derivative, the greater will be the effect of randomness.

Sampling from a skewed distribution can yield peculiar behaviors in stochastic models. This is illustrated by exponential population growth in a random environment, i.e. random growth rate (88). Under such a model, population sizes at some future time are lognormally distributed (31, 154). This distribution arises because a series of good years will lead to extraordinarily large population sizes due simply to the geometric nature of population growth. Sequential sampling from such skewed distributions of $N(t)$ results in the most likely population sizes (mode) being less than the mean. As a consequence we obtain the seemingly paradoxical observation that the growth rate for a typical sample path "will in general be less than the growth rate of average population" (154). Because population dynamics generally involve intrinsically multiplicative processes (83), we may expect complications due to distributional properties to be common.

Environmental stochasticity or "noise" is handled poorly in most PVA models. Environmental stochasticity in nature is not the unstructured "white" noise of a random number generator or "epsilon" term. Rather environmental stochasticity carries structure, such as autocorrelation and distributional properties, stemming from the manner in which errors are propagated through the system (101, 102, 108, 140). It really makes a big difference which variables in the system are fluctuating due to the environment, because this can affect the structure and dynamics of the system. Error propagation (including sampling error) can be examined by simulation methods (104), yet there has been little study of this problem in ecology.

This all calls for detailed understanding of the variance structure of populations (55). To understand the role of stochasticity in population extinction, we must understand how environmental variability affects the organism. Thus it will be a fundamental challenge in any PVA to decide how to model environmental stochasticity. Because data are limited, sampling variance may often overwhelm attempts to decompose variance into individual and environmental components. If sampling variance is included in a simulation model, projected variability will be much larger than in the true population.

Genetics

The ultimate objective behind PVA is to develop prescriptions for species survival for the purpose of preserving genetic diversity (136); thus it seems appropriate that models of genetic variation ought to contribute to the formulation of a PVA. We know that small population size can result in inbreeding depression in some populations, which may increase the risk of extinction for the population (81, 111, 112). We also know that small population size can reduce genetic variation through drift, thereby reducing the raw material for evolutionary change, and genetic variation can be essential to ensure preadaptation to disease, competition, or predation (45). But what we do not know is how much and what type of genetic variation is most important to preserve.

Templeton (149) makes a convincing argument for placing priority in conservation on unique evolutionary lineages such as species or subspecies. But even within a taxonomic group there are many forms of genetic variation, which may respond differentially to particular conservation strategies. Genetic variation is revealed by restriction site analysis of mitochondrial DNA, karyotypy, electrophoresis of allozymes, heritability of quantitative traits (40), and morphological variation (but see 69). And it is possible for many of these measures to vary almost independently of one another (80, 165). Genetic variation within populations often is measured by mean heterozygosity or the proportion of alleles that are heterozygous. Yet, if preadaptation to future insults from other species (disease, parasites, competitors, predators) is the reason to preserve genetic variation, it may be important to focus on preserving rare alleles (45). Perhaps the number of alleles per locus is a more important measure of genetic variation (2a).

Because quantitative traits are most frequently the target of natural selection, Lande & Barrowclough (81) argue that heritability should be monitored as a measure of genetic variation for conservation. Yet, from a pragmatic perspective, we know that estimates of heritability are often difficult to interpret because the response to selection can be greatly complicated by maternal effects (3). Relatively low levels of genetic variation may confer substantial heritability to some quantitative traits (90). There is also the difficulty of deciding which quantitative traits should be measured. Following Lande & Barrowclough's (81) rationale, the most important traits ought to be those that are most frequently the target of natural selection. Yet, these are exactly the traits expected to bear the lowest heritability as a consequence of selection (16, 40).

How genetic variation is structured within populations can also bear on conservation strategies (12). Many population geneticists believe that spatial heterogeneity is one of the most important mechanisms maintaining genetic variation in natural populations (63). Whether or not this pertains to the importance of inbreeding in natural populations has become the source of a fascinating debate (112, 130), but irrespective of this, there is no question that spatial variation in genetic composition of populations can be substantial. We are just beginning to understand the role of population subdivision on genetic structure and heritability (161). How significant is local adaptation? How important is coadaptation of gene complexes (149)? Although spatial structuring of genetic variation is complex and interesting, it is not clear that our understanding is sufficient to use it as a basis for manipulating populations for conservation. Attempts to manage the species by transplanting individuals between subpopulations is an effective tool to maintain or increase genetic variation within populations (57) but may destroy variance among populations.

The solution to this dilemma may ultimately entail foreseeing the sorts of threats a species is likely to encounter. If local subpopulations are likely to be threatened by habitat destruction or political unrest, it may be extremely important to maintain geographic variants to ensure that the species can continue to survive in other localities (see 142, 149). However, if future threats due to diseases and parasites are expected, there may be a premium on ensuring the maximum allelic diversity throughout the population (not necessarily all in one place—96).

Is there an optimum amount of mixing among subpopulations that maximizes total genetic variance in the population? Because different genes or gene complexes are found in within-vs-among subpopulation components of genetic variation, a general answer to this question probably does not exist. Furthermore, the genetic variance within populations is a convex function of dispersal (migration) with maximum variance at the highest possible dispersal. Likewise, variance among populations is a convex function of dispersal but with maximum variance at zero dispersal (25). The sum of these two functions is similarly convex, and no intermediate maxima exist. Thus, we need to minimize dispersal among sites to preserve geographic variation while ensuring large enough numbers in subpopulations to avoid inbreeding loss of genetic variance.

An approach commonly used in trying to determine a genetic basis for MVP is to examine effective population size, N_e (113). N_e gives insight into the potential consequences of genetic drift to loss of genetic diversity, but as is the case for measures of genetic variation, we have numerous measures of effective population size, depending upon the mechanisms affecting drift. For example, Ewens (37) reviews the calculation of N_{ei} relative to inbreeding, N_{ev} for the variance in gene frequencies among subpopulations, N_{ee} targeting the rate of loss of genetic variation, and N_{em} for mutation effective population size. Still more measures may be derived. For example, $N_e^{(meta)}$ defines the effective population size in a metapopulation experiencing repeated extinction-recolonization events (49). Each of these basic measures of N_e is then subject to adjustment for unequal sex ratio, age structure (65, 66), and variable population size (59). There is no sound basis for selecting one of these basic measures of N_e over another, yet as Ewens (37) shows, they can lead to much different conclusions about MVP.

I conclude by agreeing with Shaffer (128) and Lande (80) that modeling genetics is not likely to be as important as modeling demographic and ecological processes in the formulation of a PVA. This does not imply that genetic considerations are not important; rather, in many cases we do not yet understand the genetics well enough to use it as the basis for management. There is an urgent need for research on the link between genetics and demography (80, 94; cf 111). This conclusion also does not imply that models of genetic variability should not form the basis for PVAs. Indeed, I think this would be a novel approach for species in which erosion of genetic variability is likely to be an important consideration in the future management of a species, such as the African wild dog (*Lycaon pictus*) which has a highly subdivided population (51). But as Lande has emphasized (80), demography and associated ecology are likely to be of more practical significance than genetics in most PVAs.

Birth-Death and Demography

BIRTH AND DEATH PROCESSES Possibly the simplest approach to modeling extinction is a stochastic birth-death process (53, 91, 116), assuming independent, Poisson-distributed births and deaths. Demographic "accidents" are most likely in small populations due to sampling effects, i.e. simply because individuals do not survive for the same length of time, and individuals vary in the number of offspring they bear. This approach has been used to solve for *E(T)* as a function of population size, *N*, given density-dependent per capita birth, b_N, and death rates, d_N:

$$E(T) = \sum_{i=1}^{N} \sum_{j=i}^{N_m} (1/jd_j) \prod_{n=i}^{j-1} (b_n/d_n) \qquad 1.$$

up to a maximum possible population size, N_m.

As one might expect, such sampling effects are extremely sensitive to population size (53, 91), and these effects usually can be ignored if the population is larger than about 30 individuals (depending on age structure). The exception is where a population is divided into a large number of subpopulations, each so small that it faces a risk of chance demographic extinction. If recolonization is slow, there can be a significant risk of losing the entire population by demographic stochasticity alone (103).

Environmental stochasticity is much more significant than sampling or demographic stochasticity, except for very small populations (54, 55, 74, 85, 129). Recalling the assumption of Poisson distribution for births (b_N) and deaths (d_N), the variance in per capita growth rate at population size *N* ($r_N = b_N - d_N$) is simply

$$\mathrm{Var}(r_N) = (b_N + d_N)/N \qquad 2.$$

Recognizing this, Leigh (85) and Goodman (54, 55) rewrote the birth-death process model (Eq. 1) to make E(*T*) a function of the variance in r_N, Var(r_N):

$$E(T) = \sum_{i=1}^{N} \sum_{j=i}^{N_m} \{[2/j(j\mathrm{Var}[r_j]-E[r_j]) \prod_{n=i}^{j-1} [n\mathrm{Var}(r_n)+E(r_n)]/[n\mathrm{Var}(r_n)-E(r_n)]\} \qquad 3.$$

Here, Var(r_N) somehow becomes environmental variance (152), albeit still a function of the magnitude of b_N and d_N. The important outcome of the modified birth-death process model is that *E(T)* increases with population size more slowly when environmental variance is high (55). Goodman (55) validated predictions of the modified birth-death process model by simulating a more complex population that included density dependence and age structure, and he generally found good concordance between simulation results and analytical predictions from the birth-death model. However, ecological applications of birth-death process models have been criticized for several reasons.

Early interpretations that populations of more than 20–30 individuals were unlikely to risk extinction (91, 116, 132) were a concern given their basis solely in stochastic demography (129, 164). The use of a reflecting boundary, N_m, for maximum population size is unrealistic (30, 94), although this is resolved by Goel & Richter-Dyn (53). Additionally, the fact that the models are in continuous time renders it "highly questionable" (83) because of the importance of seasonal structure in the population. The assumption at Eq. 2 which is the basis for Eq. 3 (152) also merits consideration. For constant $E(r_N)$, increasing variance in r_N is accomplished by increasing birth and death rates. This is reasonable enough given explicit assumptions in the birth-death model but bears rather heavily on Belovsky's (6) attempt at empirical verification for Eq. 3 which draws on comparative analyses of the maximum demographic potential for r and b in mammals.

Another matter of concern in all models that predict the time to extinction, in general, is that $E(T)$ can be a misleading characterization of the likelihood of extinction (31, 41, 55). The distribution of time to extinction is positively skewed in each of these models, as well as in the age-structured model (82). The $E(T)$ is substantially greater than the median or mode of the distribution, because a few populations take extraordinarily long times to become extinct. The time to extinction most likely to occur (mode) or the middle of the distribution (median) may be more meaningful measures than the mean.

Despite these difficulties and restrictive assumptions, Leigh (85) and Goodman (54, 55) made an important point by clarifying that "demographic uncertainty" is most likely to be a concern only at low population sizes, whereas environmental uncertainty can pose significant risks for considerably larger populations (cf 101, 102 for similar results based on branching processes). Understanding the variation in population parameters attributable to environmental fluctuations is clearly fundamental to any PVA (55).

DEMOGRAPHIC PROJECTIONS Although "demographic uncertainty" is usually approached using birth-death process models, these models, in fact, do not contain age structure. This may be a serious shortcoming because age structure per se can have a significant effect on population trajectories and thereby on the probability of extinction (82, 154, 156).

Many PVAs employ projection matrices, such as the Leslie matrix, as age-structured models of population growth (95, 126). The Leslie matrix and similar stage-structured models have mathematical properties that give great insight into processes of population growth (24).

For example, sensitivity of population growth rate, r, to perturbations in vital rates (P_x, F_x) for a Leslie/Lefkovitch matrix can be solved analytically (i.e. dr/dP_x or dr/dF_x)(24, 79, 99). Understanding the response of growth rate to perturbations at various points in the life table may yield insight into how one should target management (79). For long-lived species, such as the spotted owl, adult survival is a very sensitive demographic parameter, whereas in species with shorter generation times, fecundity can be much more important (83, 99).

In nature, the elements of a projection matrix are random variables (14, 153) or functions of the environment (134, 160). Forecasted trajectories of population size depend not only on the schedule of vital rates, but also on the variance in these rates (156). It is important to note that variation in vital rates creates disequilibrium in age structure that further complicates the dynamics. The variance in population growth rate is thus attributable to both the variation in vital rates and the variance in population structure. For demography of humans in the United States in 1960, approximately two thirds of the variance in growth rate can be attributed to variance in vital rates, whereas about one third is due to fluctuations in age structure (154).

Projection matrices in their simplest form are models of exponential population growth. As such, there are essentially two possible outcomes of these models: they increase exponentially to infinity, or decrease to extinction. If the dominant eigenvalue for the average projection matrix is less than one, extinction is assured. But even when the average projection matrix might predict an increasing population, extinction may also occur when vital rates vary (14, 74, 156). Of course, exponential growth models are strictly unrealistic on time scales necessary to explore extinction probabilities.

FORECASTING METHODS Development of theory and applications of forecasting mostly have occurred in economics, but the opportunity exists to apply many of these procedures to population projections. Projecting a stochastic process into the future poses problems. First, one must assume that the mechanisms generating the historical data remain intact and unchanged in the future. Second, one must select the correct structural model that drives the population process or risk serious errors in prediction. Third, errors in predictions are magnified progressively into the future such that usually only a few time intervals can be predicted with any confidence or reliability (31a).

A time series model commonly used for forecasting is the ARIMA (auto-regressive integrated moving average) model (10, 15, 19, 31a, 154). Least-squares regression is used to calculate the dependence of $N(t)$ on lags of the entire time series, $N(t\text{-}1)$, $N(t\text{-}2)$, . . . $N(t\text{-}p)$. Differencing is employed to remove trends, and moving averages can be calculated to smooth out high-frequency noise. The resulting equation is then extrapolated into the future beginning from the last observed data point. The population's trajectory determines the forecast. Thus, if the trajectory does not show a population decrease, the forecast may continue to increase without bound. Confidence intervals around the forecasted value will, nevertheless, include 0 at some future time, but this will include sampling error and will be much shorter than $E(T)$. Many statistical software packages include programs that perform ARIMA forecasting.

A simplistic approach to forecasting can be derived from a diffusion model without age structure (31, 64). Itô calculus is used to solve a stochastic differential equation model of exponential population growth. The probability distribution function of extinction is the inverse Gaussian distribution (similar to the lognormal), and $E(T) = (x_0 - x_E)/|a|$, where x_0 is the $\log_e$ of the initial population size, x_E is the positive population size defined to constitute extinction (e.g. 1 in sexually reproducing species), and a is the average growth rate for the population. A maximum likelihood estimator (MLE) for $\hat{a} = ln(n_q/n_0)/(t_q - t_0)$, which only requires knowledge of the initial, n_0, and final censuses, n_q, at times t_0 and t_q respectively. Or alternatively one may use a linear regression approach (31). MLEs for σ^2 are also easily calculated (31, 64).

Although easy to use, one must imagine that the population trajectory observed thus far will also apply into the future. Also, any structural features of the population process, e.g. density dependence, which are reflected in the time series are overlooked in the estimator of a (154).

Building on the results of Tuljapurkar & Orzack (156) and Heyde & Cohen (64), Lande & Orzack (82) also modelled stochasticity as a diffusion process for exponential age-structured populations. Simulation trials were used to validate their estimators for $\hat{a}$ and σ^2. Although Lande & Orzack (82) emphasize that only three parameters are needed to use their model, one of these parameters is initial total reproductive value which requires complete life history and age-structure data!

A third approach to forecasting is to characterize the time series of vital rates with ARIMA, and then to insert these models into a projection matrix (84, 154). Such a "time-series matrix" retains more of the dynamic consequences of age structure, and therefore population fluctuations ought to be more realistic. I am unaware of any applications of this method in conservation biology.

For each of these forecasting models we assume a density-independent population. If density dependence acts in an age-specific manner (which it usually does: 43, 44), the complications to the age structure make it difficult to derive analytical results for the distribution of extinction times. Given that density dependence exists in natural populations (133), even in "density vague" populations (143), I am skeptical about using density-independent formulations except in two cases: (i) very small populations where density dependence may be inconsequential relative to demographic stochasticity (55), or (ii) for short-term forecasting. Further research is needed to develop forecasting models with ecologically realistic structures, e.g. with density dependence. But in the meantime, Monte Carlo simulation procedures (60, 118) can be used to generate estimates and distributions for extinction parameters for models with density dependence.

Sensitivity to initial conditions ensures that long-term forecasting will be unsuccessful for chaotic populations (122), but nonlinear forecasting methods may improve short-term forecasts for populations embedded in complex ecosystems (145). Populations experiencing fluctuations due to time-lagged processes may be particularly suited to one of several nonlinear methods reviewed by Casdagli (23). Software for generating forecasts and calculating confidence intervals is described by Schaffer & Tidd (123). Unfortunately, these methods work best for time series longer than are available for most endangered species.

Ecology

Although much of the literature on PVA has focussed on issues of genetics and stochastic demography, it is clear that ultimate causes and threats of extinction are primarily ecological. Loss or degradation of habitat is the most significant factor threatening species extinctions in the future (107, 164). For avian taxa currently endangered by extinction, 82% are associated with habitat loss, 44% with excessive take, 35% by introductions, and another 12% are threatened by chemical pollution or the consequences of natural events (148).

Most PVAs have ignored fundamentals of ecology such as habitat, focusing instead on genetics or stochastic demography. Although ecological factors influence demographic variables, seldom is our understanding sufficient to isolate these effects. A more appropriate approach for many species may be to model the habitat for the species and various strategies for managing this habitat. For example, Foin & Brenchley-Jackson (42) modelled the salinity, transpiration, and soil moisture of *Spartina* salt marshes in southern California, which is essential habitat for the endangered light-footed clapper rail. Reliable demographic details for the rail were unavailable, and the only connection between the bird and the habitat was a linear relationship between the biomass of Pacific cordgrass, *Spartina foliosa,* and the number of rails.

Most demographic PVAs performed thus far do not model ecological consequences of other species, e.g. predators, competitors, parasites, disease. In particular, exotic species can be a major threat in some systems (159). For example, invasions of exotics may be less likely in communities that possess a diversity of native taxa (114). In some species, dynamics of disease may be the most significant consideration in a PVA (35, 96). Understanding such relationships is necessary to predict population viability.

Regrettably PVAs often do not explicitly include management (136). Consider, for example, application of a forecasting method to a population trajectory for a species whose decline can be attributed to habitat loss. It makes no sense to extend such a trajectory if all remaining habitat for the species is now protected. Yet, this is precisely the inference that one would draw in applying a forecasting model (cf 31, 82). Leaving management out of a PVA is unfortunate because one of the greatest values of PVA modeling is the opportunity to evaluate the efficacy of various management options (67).

Indeed, it is the absence of ecology and management from most attempts at PVAs that is their biggest weakness. These processes ought to be the nuts and bolts of such modeling exercises! The power of ecological modeling rests in our ability essentially to play with nature to anticipate the consequences of various management scenarios (56, 139, 147). Some aspects of ecology such as density dependence, spatial heterogeneity, and the Allee effect are of particular significance to PVA because they have major consequences to the probability of extinction.

DENSITY DEPENDENCE The simplest possible model of population growth is an exponential population growth model. It has no ecology. The simplest possible ecological model is a density-dependent model such as the logistic. The existence of negative feedbacks in compensatory density dependence dampens population fluctuations and can greatly reduce the probability of extinction (52). In model selection, the principle of Occam's Razor is commonly invoked, whereby one requires statistical evidence for the existence of density dependence before adopting the more complex density-dependent model. I submit that, instead of requiring statistical demonstration of density dependence, one should test for deviations from a null model of logistic density dependence.

This is not to say that estimating parameters for a density dependent model is not without its difficulties (98, 141), in particular, lack of independence in a time series of census data (20, 110). One can avoid some of these problems by examining density dependence in survival or fecundity, while controlling for key environmental parameters (83). Elements of a projection matrix can be made functions of density, yielding dynamics qualitatively similar to the logistic (134, 160). Because sufficient data are seldom available to do justice to characterizing a density dependent function, one may need to resort to using a form consistent with that observed for similar taxa (43, 44). Because of the difficulties with parameterization of density dependence, it has been argued that it may be most conservative to use density-independent models because they were thought to bear higher probabilities of extinction (41, 52). But this is not necessarily true; for example, extinction under density dependence is imminent if all habitat for a species has disappeared. Furthermore, I do not accept this rationale on the grounds that reasonable behavior of the model should be a high priority in model validation (56).

THRESHOLDS At low densities, an Allee effect creates a positive relationship between per capita population growth rate and population size. This can be caused by difficulties in finding mates (30), difficulty in fending off predators or competitors (11, 28), social or physiological facilitation (80), or reduced predation efficiency (8). Similarly, low density consequent to habitat fragmentation may result in high juvenile mortality during dispersal (78, 79).

The consequences of Allee effects for PVA are exceedingly important because these mechanisms create threshold or critical population sizes below which extinction is much more probable or inevitable. As an example, Lande (78) presents a model where limitations to juvenile dispersal can create an extinction threshold in territorial species, which has been used in PVAs for the spotted owl (79, 151).

However, the mechanisms creating an Allee effect are not well understood except in a few species (11, 75). As a result we do not know the generality of the phenomenon. It has, however, been postulated to occur in a broad diversity of taxa (30, 78—80). Paucity of empirical evidence is in part due to the difficulty of studying populations at low densities. Experimental work such as that by Crowell (29) should help to identify the characteristics of species most likely to experience Allee effects and afford a more objective basis for incorporating relevant statistical functions into PVAs (cf 30).

Inbreeding depression can be modeled demographically as an Allee effect because its effect becomes more severe as population density becomes less. However, inbreeding is more complex because inbreeding depression is expected to erode with time as deleterious alleles are eliminated by a combination of drift and selection (80, 81). Dennis (30) reviews models that can be used to describe the Allee effect, and their statistical properties.

SPATIAL STRUCTURE "Habitat fragmentation is the most serious threat to biological diversity and is the primary cause of the present extinction crisis," Wilcox & Murphy observe (164). It would seem, therefore, that spatial structure should be incorporated into most PVAs. Yet, because partitioning a population into spatial subunits can be complex to model and parameterize, it is often ignored.

Spatial heterogeneity and dispersal can stabilize population fluctuations (46, 68) but can also have complex consequences depending on nonlinearities in the system (27). Asynchrony can average out fluctuations caused by demographic or environmental stochasticity, and if spatially removed, asynchrony may ensure species survival in the face of catastrophes (47). Of course, correlated fluctuations among subpopulations can drastically reduce $E(T)$ (48).

Incorporation of spatial structure into ecological models has involved a diversity of approaches including reaction-diffusion equations (86), discrete interacting subpopulations envisioned on a grid (157), and Markov transition matrices (1). Diffusion usually has a stabilizing effect on the dynamics of single-species models, tending to average population fluctuations in space. But when spatial structure is combined with ecological interactions, e.g. competition or predation, then instability and spatial patterns can emerge. Spatial models in discrete time also show the potential for very complex dynamics, using integrodifference equations (76), or predator-prey difference equations with dispersal simulated on a grid (61).

Spatial structure adds so much complexity to ecological models that generalities can be difficult to obtain (27, 39). For example, depending upon the species in question, corridors among habitat units may be either beneficial or detrimental. Corridors can reduce consequences of inbreeding or demographic stochasticity by facilitating dispersal among sites; but they can also serve as transmission routes for disease, exotic species, or predators (127, 132a).

Recently, considerable work has focussed on metapopulation models, where the occupancy of discrete habitat patches or islands is a consequence of a balance between dispersal and extinction processes (1, 49, 87). Fahrig & Paloheimo (39) show how inter-patch dynamics is fundamental in determining population size in spatially structured models. As the distance between habitat patches increases, say as a consequence of habitat fragmentation, inter-patch dispersal is expected to decrease. The ability of dispersers to detect new habitat patches can be an exceedingly important factor determining population size, patch occupancy, and probability of extinction (39, 78). This has been the guiding principle behind PVAs for the spotted owl by Lande (79) and Thomas et al (151).

The geometry of habitat can be critical (157) and virtually impossible to model with analytical techniques. In a simulation model for the spotted owl, the landscape of suitable habitat has been mapped on a geographic information system (GIS) and imported into a dispersal simulation model (98a). This technology has great potential for coping with the complex problem of identifying an appropriate spatial structure for population modeling.

MULTI-SPECIES SYSTEMS To construct a PVA, it is fundamental to develop an understanding of the mechanisms regulating population size (133). Single species models of populations are probably unrealistic characterizations of most populations, because population regulation actually entails dynamic interactions among species, e.g. plant-herbivore, predator-prey, parasite-host interactions. Herein lies a serious dilemma for PVA. We do not understand multispecies processes well enough for most species to incorporate such complexity into a PVA.

Modeling ecological processes ideally should include the interface with demography. For example, the dynamics of disease in a population can ultimately be determined by demographic processes. Demographic disequilibria sustained by stochastic perturbations in vital rates can result in sustained epidemiological fluctuations (155). Indeed, such interactions between time delays created by age structure and by ecological interactions may be a key to understanding dynamic behavior in general (62). But, of course, to model such processes requires detailed information on the age specificity of the ecological process.

Ecological processes are inherently nonlinear. This fact, along with the destabilizing effects of environmental seasonality and trophic-level interactions, means that complex dynamics, including chaos, are to be expected in many biological populations. It has been argued that chaos seldom occurs in ecological systems because species would be expected to go extinct when chaotic fluctuations reduce populations to low levels (9, 119). This is not necessarily true because a variety of mechanisms can ensure persistence. In particular, refugia and spatial heterogeneity (2, 71) can buffer local populations against extinction. And even if chaotic fluctuations were to cause local extinction, areas may be repopulated in the sense of a metapopulation (61, 119).

If anything, however, recent advances in nonlinear dynamics have made it clear that even simple ecological systems can possess remarkably complex dynamics. The implication is that such complex dynamics may frustrate our ability to predict long-term trajectories necessary to estimate extinction times. If we are to consider PVAs for chaotically fluctuating populations, the only hope may be to focus attention on the mechanisms that bound a systems dynamics, e.g. refugia, spatial heterogeneity, switching to alternative prey.

These remarks only give a glimpse into the true complexity of ecological systems. In performing PVAs we do not yet know how much complexity is necessary to capture the essence of the system. Deciding how much complexity is necessary should be based on advice from field biologists and managers, who have the best sense, and invariably on the availability of data.

Interactions Among Mechanisms

Because several mechanisms can contribute to extinction, and because each is complex in its own right, the usual approach has been to consider the mechanisms only piecewise, one or two at a time. In this approach, one might learn which mechanism appears to be most sensitive and which requires the largest MVP.

Unfortunately this approach is flawed because the interaction among components may yield critical insight into the probability of extinction. Indeed, if Gilpin & Soule's (50) idea of extinction vortices has any validity, the synergism among processes—such as habitat reduction, inbreeding depression, demographic stochasticity, and loss of genetic variability—is exactly what will be overlooked by viewing only the pieces.

It is feasible and straightforward to build a simulation model containing both demographic and environmental stochasticity, postulated consequences of inbreeding depression, Allee effects, habitat trajectories, and consequent ramifications to carrying capacity, etc, all in the same model. Then one can conduct a sensitivity analysis to learn which parameters have the most significant consequences, and one can simulate management alternatives to view their expected consequences. Furthermore, simulation offers the power to explore propagation of variances and the effects of various types of randomness for complex systems which cannot be understood except in the full-blown model (104, 118). All these things are possible, but in practice our ability to predict the behavior of complex ecological systems has been less than exemplary (34).

WHAT MAKES A GOOD PVA AND WHO DECIDES?

PVA may be a more ominous proposition than population biologists are accustomed to, given that PVAs have been challenged in court (5, 151). Some of the lessons learned from previous court challenges of population models may be instructive. For example, the statistical reliability of population projections is likely to be scrutinized, and it is therefore important that parameter estimation, robustness of models, and confidence limits for projections be carefully considered. How defensible is time to extinction, for example, if it carries confidence intervals spanning two orders of magnitude?

Presentation of results can be a delicate matter. Even though the model may be complex, it is essential that explanation be clear and understandable to nonscientists. Substantial testimony in one of the spotted owl hearings was ignored by the judge in her ruling because she did not understand the modeling. Yet, to oversimplify may risk misrepresentation.

There exists a delicate trade-off between building a model that realistically captures the essential ecology of the organisms and keeping the model simple enough that the number of parameters for estimation is reasonable (33). One of the beauties of some of the forecasting methods is that straightforward methods exist for extrapolating the confidence intervals and distributional properties of forecasts (31, 31a, 64). But use of such simple models requires bold assumptions about exponential population growth and perpetuation of population trajectories. For models that are more ecologically realistic, however, such variance estimators do not exist, requiring use of simulation methods, e.g. bootstrapping and jackknifing (100). There are no rules, but a strategic modeler will use the simplest possible model that still retains the essential features of the system's ecology.

To anticipate extinction probabilities, it is essential to understand the structure of variance, particularly environmental variance (cf 55, 99, 101, 102). Yet, obtaining good estimates of variances for environmental and demographic parameters requires vast amounts of data. Most PVAs conducted thus far have been unable to do justice to variance estimation. Shaffer (126) was unable to distinguish between sampling and environmental variance for demographic parameters of grizzly bears. Lande (79) estimated only the binomial component of sampling variance surrounding demographic parameters for the spotted owl.

For time series of population size, variances often increase as the sample size increases (108). At the very least, it would appear from data presented by Pimm & Redfearn (108) that 8 years of data are needed to stabilize the variance in insect census data, and 30–40 years for birds and mammals (150). To characterize the autocorrelation structure in a time series will require even more data, yet autocorrelation is known to be important in predicting extinction probabilities (101, 102).

A large literature exists on the philosophy and methods for simulation modeling in ecology (56, 58, 104, 118, 139, 147). Likewise, there are several book-length treatises on estimation of population and ecological parameters (77, 115, 117, 124, 138, 163). Some parameters such as survival or coefficients for multispecies interactions can present serious estimation problems. And in many cases there is little hope because data are unavailable or insufficient. In these instances, one may use data from similar species or areas, use a simpler model encumbered by unrealistic assumptions on the structure of the system, or explore the behavior of the system over a range of reasonable parameter values.

Deriving statistically reliable estimates for MVP is clearly a difficult if not impossible task. But it can be an even tougher task to extrapolate from the MVP into estimating the area of habitat necessary to support such a population, which requires a detailed understanding of a species' habitat requirements (13). Patches of habitat must not only be larger than some critical size (80), they must also be in a suitable geometric configuration to ensure dispersal among habitat units. Management for spotted owls (151) provides a complex case study.

Grant (56) suggests four important components for validating any PVA model. First, does the model address the problem? Because the "problem" is usually a management issue, it may be useful to interface the PVA with risk analysis (93). Second, does the model possess reasonable structure and behavior? The third step is to attempt a quantitative assessment of the accuracy and precision of the model's outputs and behavior. And fourth is to conduct a sensitivity analysis of the model by changing selected parameters in the model by an arbitrary amount and then studying system response and behavior.

Given careful consideration of the audience (i.e. who decides?), these validation approaches offer useful baseline criteria for evaluating a PVA model. But still, finding the appropriate balance between complexity and statistical reliability will be arbitrary and difficult to evaluate. Following these validation criteria, some approaches are fundamentally insufficient as PVAs, for example, simple calculations of $N_e > 50/500$, or projections of Leslie matrices until extinction. Examples of PVAs that have been particularly successful at stimulating enlightened management include those for grizzly bears (126,144) and spotted owls (98a, 151).

GETTING EMPIRICAL

We cannot expect that simulation PVAs will be conducted for most endangered species. Data are often insufficient, time is critical, PVAs can be costly, and there are simply too many species needing attention (136). For these reasons, there continues to be great interest in the formulation of "rules of thumb" for MVP, and the $N_e > 50/500$ guidelines for short-term versus long-term MVPs are commonly cited (136, 152).

The original formulation of these rules of thumb was genetically based, but not based upon defensible criteria (37, 81). Nevertheless, there is empirical evidence that such rules of thumb may be of appropriate magnitude. Studies of extinction of bighorn sheep (*Ovis canadensis*)(7), and birds on oceanic or habitat islands (70, 109, 137) consistently show that $N < 50$ is clearly insufficient and the probability of extinction was higher or even certain for such small populations. Populations of $50 < N < 200$ were marginally secure, and when $N > 200$, populations were apparently secure over the limited time frames of these studies (see 150). Clearly applications of such limited observations are restricted to particular taxa, and we would expect much larger population sizes to be necessary for insect populations, for example. And there are obvious advantages to maintaining three or more replicate populations (136).

There is opportunity to expand the empirical basis for PVA and rules of thumb (150). This should include extinction studies based upon empirical observations from islands, and experimental work with replicated small populations (29). We need to understand which factors contribute to extinction probabilities for various taxa. For example, social behavior has been shown to be an important contributor in primates (32). And it is of great interest to know whether herbivores undergo greater population fluctuations than do predators (6). Will simple single-species models suffice for herbivores whereas models incorporating trophic-level dynamics are essential for predators (94)?

Which demographic components are most critical in determining extinction probabilities? Karr (73) found that forest undergrowth bird species that have gone extinct on Barro Colorado Island (BCI) have, on average, lower survival rates, and that species with lower survival rates disappeared earlier. There was little evidence that N contributed significantly to these extinctions on BCI. Other life history traits may also be important, e.g. do smaller species undergo more violent population fluctuations, thereby predisposing them to a higher probability of extinction (108, 150)?

If conservation biology is truly going to be a scientific discipline (105), it must become more actively involved in experimental research. Experimental manipulation of habitats to determine the consequences to species richness, extinction, population turnover, and dispersal are on target (e.g. 89, 120). Likewise, much could be gained by performing PVAs for species in no danger of extinction, where populations could be manipulated experimentally to rigorously test the predictions of the model.

CONCLUSIONS: VIABILITY OF PVA?

Constructing models to include many of the complexities of the ecology of organisms presents no particular difficulties, but we simply do not have sufficient data to validate such models for most endangered species. It is seldom, if ever, that replications exist (34); thus conclusions cannot be robust (83). A great danger exists that resource managers may lend too much credence to a model, when they may not fully understand its limitations.

Nevertheless, there is too much to be gained by developing a stronger understanding of the system by modeling, to shirk modeling for fear of its being misinterpreted. PVA as a process is an indispensable tool in conservation, and it involves much more than feeble attempts to estimate MVP or probabilities of extinction. PVA entails the process of synthesizing information about a species or population, and developing the best possible model for the species given the information available. When done properly this involves working closely with natural resource managers to develop a long-term iterative process of modeling and research that can reveal a great deal about how best to manage a species. Done properly PVA can be a variation on Holling (67) and Walter's (162) notion of adaptive management, which has proven to be a powerful tool in many areas of resource management.

Adaptive management proposes application of different management tactics in time and space to develop a better understanding of the behavior of the system (162). For application to endangered species problems, when possible, implementation of various management strategies should be attempted in spatially separated subpopulations. By so doing, one can evaluate the efficacy of various conservation strategies. Active manipulation must be part of such a program, i.e. habitat manipulation, predator or disease control, manipulation of potential competitors, provisioning, transplanting individuals from other subpopulations to sustain genetic variation, supplementation of population with releases of captive stock, etc. Monitoring of the genetic and population consequences of such manipulations then provides data to validate and/or refine the PVA model.

PVA raises a large number of exciting research questions in population ecology and genetics. One promising theoretical area appears to be expanding theory and applications of extinction processes in age-structured populations, e.g. developing applications from the general theory outlined by Aytheya & Karlin (4)(cf 83, 101, 102, 154). Existence of true threshold populations, i.e. a definitive MVP, depends upon the existence of a density-dependent mechanism such as the Allee effect or inbreeding depression (111). We have distressingly little empirical data on these processes in natural populations to provide a basis for parameterization of models (30, 78). This must be one of the most urgent research needs for PVA.

Most important, I am confident that PVA will prove to be a valuable tool as we face the extinction crisis (114). Time is not available to perform PVAs for all of the species for which it is warranted (128, 136). Indeed, single-species approaches to conservation are too limited in scope for most applications in tropical conservation (22; contra 21, 132). We must choose species for PVAs wisely, because protecting diverse communities and keystone species may afford disproportionate benefits (137a).

ACKNOWLEDGMENTS

I thank M. Gadgil, J. M. Gaillard, E. Merrill, J. Meyer, U. Seal, M. Soule, M. Shaffer, and R. Sukumar for comments. J. Brown, K. McKelvey, R. Lamberson , U. Seal, and S. Temple kindly provided copies of unpublished reports or manuscripts prior to publication. I received support from American Institute of Indian Studies, US Educational Foundation in India Fulbright program, National Council for Air and Stream Improvement, and Wyoming Water Research Center.

Literature Cited

1. Akcakaya, H. R., Ginzburg, L. R. 1991. Ecological risk analysis for single and multiple populations. In *Species Conservation: A Population-Biological Approach*, ed. A. Seitz, V. Loeschcke, pp. 73–87. Basel: Birkhauser Verlag
2. Allen, L. J. S. 1983. Persistence and extinction in Lotka-Volterra reaction-diffusion equations. *Math. Biosci.* 65: 1–12
2a. Allendorf, F. W. 1986. Genetic Drift and the loss of alleles versus heterozygosity. *Zoo Biol.* 5:181–90
3. Atchley, W. R., Newman, S. 1989. A

quantitative genetic perspective on mammalian development. *Am. Nat.* 134:486–512
4. Athreya, K. B., Karlin, S. 1971. On branching processes with random environments. I. Extinction probabilities. *Ann. Math. Statist.* 42:1499–1520
5. Barthouse, L. W., Boreman, J., Christensen, S. W., Goodyear, C. P., VanWinkle, W., Vaughan, D. S. 1984. Population biology in the courtroom: The Hudson River controversy. *BioScience* 34:14–19
6. Belovsky, G. E. 1987. Extinction models and mammalian persistence. See Ref. 136, pp. 35–57
7. Berger, J. 1990. Persistence of different-sized populations: An empirical assessment of rapid extinctions in bighorn sheep. *Conserv. Biol.* 4:91–96
8. Berryman, A. A., Dennis, B., Raffa, K. F., Stenseth, N. C. 1985. Evolution of optimal group attack, with particular reference to bark beetles (Coleoptera: Scolytidae). *Ecology* 66:898–903
9. Berryman, A. A., Millstein, J. A. 1989. Are ecological systems chaotic—and if not, why not? *Trends Ecol. Evol.* 4:26–28
10. Binkley, C. S., Miller, R. S. 1988. Recovery of the whooping crane *Grus americana*. *Biol. Conserv.* 45:11–20
11. Birkhead, T. R. 1977. The effect of habitat and density on breeding success in the common guillemot *(Uria aalge)*. *J. Anim. Ecol.* 46:751–64
12. Boecklen, W. J. 1986. Optimal design of nature reserves: Consequences of genetic drift. *Biol. Conserv.* 38:323–28
13. Boecklen, W. J., Simberloff, D. 1986. Area-based extinction models in conservation. In *Dynamics of Extinction,* ed. D. K. Elliot, pp. 247–76. New York: Wiley. 294 pp.
14. Boyce, M. S. 1977. Population growth with stochastic fluctuations in the life table. *Theor. Popul. Biol.* 12:366–73
15. Boyce, M. S. 1987. Time-series analysis and forecasting of the Aransas/Wood Buffalo Whooping Crane population. *Proc. Int. Crane Workshop* 4:1–9
16. Boyce, M. S. 1988. Evolution of life histories: theory and patterns from mammals. In *Evolution of Life Histories of Mammals,* ed. M. S. Boyce, pp. 3–30. New Haven, Conn: Yale Univ. Press. 373 pp.
17. Boyce, M. S., Daley, D. J. 1980. Population tracking of fluctuating environments and natural selection for tracking ability. *Am. Nat.* 115:480–91
18. Boyce, M. S., Irwin, L. L. 1990. Viable populations of Spotted Owls for management of old growth forests in the Pacific Northwest. In *Ecosystem Management: Rare Species and Significant Habitats,* ed. R. S. Mitchell, C. J. Sheviak, D. J. Leopold, 471:133–35. Albany, NY: NY State Mus.
19. Boyce, M. S., Miller, R. S. 1985. Ten-year periodicity in Whooping Crane census. *Auk* 102:658–60
20. Bulmer, M. G. 1975. The statistical analysis of density-dependence. *Biometrics* 31:901–11
21. Burgman, M. A., Akcakaya, H. R., Loew, S. S. 1988. The use of extinction models for species conservation. *Biol. Conserv.* 43:9–25
22. Burks, K. A., Brown, J. H. 1992. Using montane mammals to model extinctions due to global change. *Conserv. Biol.* (In press)
23. Casdagli, M. 1989. Nonlinear prediction of chaotic time series. *Physica D* 35:335–66
24. Caswell, H. 1989. *Matrix Population Models: Construction, Analysis and Interpretation*. Sunderland, Mass: Sinauer
25. Chesser, R. K. 1991. Influences of gene flow and breeding tactics on gene diversity within populations. *Genetics* 129:573–83
26. Chesson, P. L. 1978. Predator-prey theory and variability. *Annu. Rev. Ecol. Syst.* 9:323–47
27. Chesson, P. L. 1981. Models for spatially distributed populations: The effect of within-patch variability. *Theor. Popul. Biol.* 19:288–325
28. Clark, C. W. 1974. Possible effects of schooling on the dynamics of exploited fish populations. *J. Conserv. Int. Explor. Mer.* 36:7–14
29. Crowell, K. L. 1973. Experimental zoogeography: Introductions of mice to small islands. *Am. Nat.* 107:535–58
30. Dennis, B. 1989. Allee effects: Population growth, critical density, and the chance of extinction. *Nat. Res. Model.* 3:481–538
31. Dennis, B., Munholland, P. L., Scott, J. M. 1991. Estimation of growth and extinction parameters for endangered species. *Ecol. Monogr.* 61:115–43
31a. Diggle, P. J. 1990. *Time Series: A Biostatistical Introduction*. Oxford: Clarendon. 257 pp.
32. Dobson, A. P., Lyles, A. M. 1990. The population dynamics and conservation of primate populations. *Conserv. Biol.* 3:362–80
33. Eberhardt, L. L. 1987. Population pro-

jections from simple models. *J. Appl. Ecol.* 24:103–18
34. Eberhardt, L. L., Thomas, J. M. 1991. Designing environmental field studies. *Ecol. Monogr.* 61:53–73
35. Edwards, M. A., McDonnell, U. 1982. *Animal Disease in Relation to Animal Conservation*. London: Academic. 336 pp.
36. Deleted in proof
37. Ewens, W. J. 1990. The minimum viable population size as a genetic and a demographic concept. In *Convergent Issues in Genetics and Demography,* ed. J. Adams, D. A. Lam, A. I. Hermalin, P. E. Smouse, pp. 307–16. Oxford: Oxford Univ. Press
38. Ewens, W. J., Brockwell, P. J., Gani, J. M., Resnick, S. I. 1987. Minimum viable population size in the presence of catastrophes. See Ref. 136, pp. 59–68
39. Fahrig, L., Paloheimo, J. 1988. Determinants of local population size in patchy habitats. *Theor. Popul. Biol.* 34:194–213
40. Falconer, D. S. 1981. *Quantitative Genetics*. London: Longman
41. Ferson, S., Ginzburg, L., Silvers, A. 1989. Extreme event risk analysis for age-structured populations. *Ecol. Model.* 47:175–87
42. Foin, T. C., Brenchley-Jackson, J. L. 1991. Simulation model evaluation of potential recovery of endangered light-footed clapper rail populations. *Biol. Conserv.* 58:123–48
43. Fowler, C. W. 1981. Density dependence as related to life history strategy. *Ecology* 62:602–10
44. Fowler, C. W. 1987. A review of density dependence in populations of large mammals. *Curr. Mammal.* 1:401–41
45. Futuyma, D. 1983. Interspecific interactions and the maintenance of genetic diversity. In *Genetics and Conservation,* ed. C. M. Schonewald-Cox, S. M. Chambers, B. MacBryde, W. L. Thomas, pp. 364–73. Menlo Park, Calif: Benjamin/Cummings. 722 pp.
46. Gadgil, M. 1971. Dispersal: Population consequences and evolution. *Ecology* 52:253–61
47. Gilpin, M. E. 1987. Spatial structure and population viability. See Ref. 136, pp. 125–39
48. Gilpin, M. E. 1990. Extinction of finite metapopulations in correlated environments. In *Living in a Patchy Environment,* ed. B. Shorrocks, I. R. Swingland, pp. 177–86. Oxford: Oxford Univ. Press. 246 pp.
49. Gilpin, M. E., Hanski, I. 1991. *Metapopulation Dynamics*. London: Academic. 336 pp.
50. Gilpin, M. E., Soulé, M. E. 1986. Minimum viable populations: Processes of species extinction. See Ref. 135, pp. 19–34
51. Ginsberg, J. R., Macdonald, D. W. 1990. *Foxes, Wolves, Jackals, and Dogs: An Action Plan for the Conservation of Canids*. Morges, Switzerland: IUCN/SSC Canid Specialist Group
52. Ginzburg, L. R., Ferson, S., Akçakaya, H. R. 1990. Reconstructibility of density dependence and the conservative assessment of extinction risks. *Conserv. Biol.* 4:63–70
53. Goel, N. S., Richter-Dyn, N. 1974. *Stochastic Models in Biology*. New York: Academic. 269 pp.
54. Goodman, D. 1987. Considerations of stochastic demography in the design and management of biological reserves. *Nat. Res. Model.* 1:205–34
55. Goodman, D. 1987. The demography of chance extinction. See Ref. 136, pp. 11–34
56. Grant, W. E. 1986. *Systems Analysis and Simulation in Wildlife and Fisheries Science*. New York: Wiley. 338 pp.
57. Griffith, B., Scott, J. M., Carpenter, J. W., Reed, C. 1989. Translocation as a species conservation tool: Status and strategy. *Science* 245:477–80
58. Hall, C. A. S. 1988. What constitutes a good model and by whose criteria. *Ecol. Model.* 43:125–27
59. Harris, R. B., Allendorf, F. W. 1989. Genetically effective population size of large mammals: An assessment of estimators. *Conserv. Biol.* 3:181–91
60. Harris, R. B., Maguire, L. A., Shaffer, M. L. 1987. Sample sizes for minimum viable population estimation. *Conserv. Biol.* 1:72–76
61. Hassell, M., Comins, H. N., May, R. M. 1991. Spatial structure and chaos in insect population dynamics. *Nature* 353:255–58
62. Hastings, A. 1986. Interacting age structured populations. In *Mathematical Ecology,* ed. T. G. Hallam, S. A. Levin, pp. 287–94. New York: Springer-Verlag. 457 pp.
63. Hedrick, P. W. 1987. Genetic polymorphism in heterogeneous environments: A decade later. *Annu. Rev. Ecol. Syst.* 17:535–66
64. Heyde, C. C., Cohen, J. E. 1985. Confidence intervals for demographic projections based on products of random

matrices. *Theor. Popul. Biol.* 27:120–53
65. Hill, W. G. 1972. Effective size of population with overlapping generations. *Theor. Popul. Biol.* 3:278–89
66. Hill, W. G. 1979. A note on effective population size with overlapping generations. *Genetics* 92:317–22
67. Holling, C. S. 1978. *Adaptive Environmental Assessment and Management.* New York: Wiley
68. Huffaker, C. B. 1958. Experimental studies on predation: Dispersion factors and predator-prey oscillations. *Hilgardia* 27:343–83
69. James, F. C. 1983. Environmental component of morphological differentiation in birds. *Science* 221:184–86
70. Jones, H. L., Diamond, J. M. 1976. Short-time-base studies of turnover in breeding bird populations on the California Channel Islands. *Condor* 78: 526–49
71. Joshi, N. V., Gadgil, M. 1991. On the role of refugia in promoting prudent use of biological resources. *Theor. Popul. Biol.* 40:211–29
72. Karlin, S., Taylor, H. M. 1975. *A First Course in Stochastic Processes.* New York: Academic
73. Karr, J. R. 1990. Avian survival rates and the extinction process on Barrow Colorado Island, Panama. *Conserv. Biol.* 4:391–97
74. Keiding, N. 1975. Extinction and exponential growth in random environments. *Theor. Popul. Biol.* 8:49–63
75. King, C. E., Dawson, P. S. 1972. Population biology and the Tribolium model. *Evol. Biol.* 5:133–227
76. Kot, M., Schaffer, W. M. 1986. Discrete-time growth-dispersal models. *Math. Biosci.* 80:109–36
77. Krebs, C. J. 1989. *Ecological Methodology.* New York: Harper & Row. 654 pp.
78. Lande, R. 1987. Extinction thresholds in demographic models of territorial populations. *Am. Nat.* 130:624–35
79. Lande, R. 1988. Demographic models of the northern spotted owl (*Strix occidentalis caurina*). *Oecologia* 75:601–7
80. Lande, R. 1988. Genetics and demography in biological conservation. *Science* 241:1455–60
81. Lande, R., Barrowclough, G. F. 1987. Effective population size, genetic variation, and their use in population management. See Ref. 136, pp. 87–123
82. Lande, R., Orzack, S. H. 1988. Extinction dynamics of age-structured populations in a fluctuating environment. *Proc. Natl. Acad. Sci. USA* 85:7418–21
83. Lebreton, J.-D., Clobert, J. 1991. Bird population dynamics, management, and conservation: the role of mathematical modelling. In *Bird Population Studies: Relevance to Conservation and Management,* ed. C. M. Perrins, J-D. Lebreton, G. J. M. Hirons, pp. 105–25. Oxford/New York: Oxford Univ. Press
84. Lee, R. D. 1974. Forecasting births in post-transition populations. *J. Am. Statist. Assoc.* 69:607–14
85. Leigh, E. G. Jr. 1981. The average lifetime of a population in a varying environment. *J. Theor. Biol.* 90:213–39
86. Levin, S. A. 1976. Population dynamic models in heterogeneous environments. *Annu. Rev. Ecol. Syst.* 7:287–310
87. Levins, R. 1970. Extinction. In *Some Mathematical Questions in Biology,* ed. M. Gerstenhaber, pp. 75–107. Providence, RI: Am. Math. Soc.
88. Lewontin, R. C., Cohen, D. 1969. On population growth in a randomly varying environment. *Proc. Natl. Acad. Sci. USA* 62:1056–60
89. Lovejoy, T. E., Bierregaard, R. O. Jr., Rylands, A. B., Malcolm, J. R., Quintela, C. E., et al. 1986. Edge and other effects of isolation on Amazon forest fragments. See Ref. 135, pp. 287–85
90. Lynch, M. 1985. Spontaneous mutations for life-history characters in an obligate parthenogen. *Evolution* 39: 804–18
91. MacArthur, R. H., Wilson, E. O. 1967. *The Theory of Island Biogeography.* Princeton, NJ: Princeton Univ. Press
92. Mace, G. M., Lande, R. 1991. Assessing extinction threats: Toward a reevaluation of IUCN threatened species categories. *Conserv. Biol.* 5:148–57
93. Maguire, L. A. 1991. Risk analysis for conservation biologists. *Conserv. Biol.* 5:123–25
94. Mangel, M. 1990. Book review: *Viable Populations for Conservation. Nat. Res. Model.* 4:255–71
95. Marcot, B. G., Holthausen, R. 1987. Analyzing population viability of the spotted owl in the Pacific Northwest. *Trans. N. Am. Wildl. Nat. Res. Conf.* 52:333–47
96. May, R. M. 1986. The cautionary tale of the black-footed ferret. *Nature* 320: 13–14
97. Maynard Smith, J. 1974. *Models in Ecology.* Cambridge, England: Cambridge Univ. Press
98. McCullough, D. A. 1990. Detecting

density dependence: Filtering the baby from the bathwater. *Trans. N. Am. Wildl. Nat. Res. Conf.* 55:534–43

98a. McKelvey, K., Noon, B. R., Lamberson, R. 1992. Conservation planning for species occupying fragmented landscapes: The case of the northern spotted owl. In *Biotic Interactions and Global Change,* ed. J. Kingsolver, P. Karieva, R. Huey. Sunderland, Mass:Sinauer

99. Meyer, J. S., Boyce, M. S. 1992. Life historical consequences of pesticides and other insults to vital rates. In *The Population Ecology and Wildlife Toxicology of Agricultural Pesticide Use: A Modeling Initiative for Avian Species*. In press

100. Meyer, J. S., Ingersoll, C. G., McDonald, L. L., Boyce, M. S. 1986. Estimating uncertainty in population growth rates: Jackknife vs. bootstrap techniques. *Ecology* 67:1156–66

101. Mode, C. J., Jacobson, M. E. 1987. A study of the impact of environmental stochasticity on extinction probabilities by Monte Carlo integration. *Math. Biosci.* 83:105–25

102. Mode, C. J., Jacobson, M. E. 1987. On estimating critical population size for an endangered species in the presence of environmental stochasticity. *Math. Biosci.* 85:185–209

103. Mode, C. J., Pickens, G. T. 1985. Demographic stochasticity and uncertainty in population projections—A study by computer simulation. *Math. Biosci.* 79:55–72

104. Morgan, B. J. T. 1984. *Elements of Simulation*. London: Chapman & Hall

105. Murphy, D. E. 1990. Conservation biology and scientific method. *Conserv. Biol.* 4:203–4

106. Murphy, D. E., Freas, K. E., Weiss, S. B. 1990. An "environment-metapopulation" approach to population viability analysis for a threatened invertebrate. *Conserv. Biol.* 4:41–51

107. Pimm, S. L., Gilpin, M. E. 1989. Theoretical issues in conservation biology. In *Perspectives in Ecological Theory,* ed. J. Roughgarden, R. M. May, S. A. Levin, pp. 287–305. Princeton, NJ: Princeton Univ. Press

108. Pimm, S. L., Redfearn, A. 1988. The variability of population densities. *Nature* 334:613–14

109. Pimm, S. L., Jones, H. L., Diamond, J. 1988. On the risk of extinction. *Am. Nat.* 132:757–85

110. Pollard, E., Lakhani, K. H., Rothery, P. 1987. The detection of density-dependence from a series of annual censuses. *Ecology* 68:2046–55

111. Ralls, K., Ballou, J. D., Templeton, A. 1988. Estimates of lethal equivalents and the cost of inbreeding in mammals. *Conserv. Biol.* 2:185–93

112. Ralls, K., Harvey, P. H., Lyles, A. M. 1986. Inbreeding in natural populations of birds and mammals. See Ref. 135, pp. 35–56

113. Reed, J. M., Doerr, P. D., Walters, J. R. 1988. Minimum viable population size of the red-cockaded woodpecker. *J. Wildl. Manage.* 52:385–91

114. Reid, W. V., Miller, K. R. 1989. *Keeping Options Alive: The Scientific Basis for Conserving Biodiversity*. Washington, DC: World Resourc. Inst. 128 pp.

115. Richter, O., Söndgerath, D. 1990. *Parameter Estimation in Ecology: The Link Between Data and Models*. Weinheim, Germany: VCH Verlagsgesellschaft mbH

116. Richter-Dyn, N., Goel, N. S. 1972. On the extinction of a colonizing species. *Theor. Popul. Biol.* 3:406–33

117. Ricker, W. E. 1975. Computation and interpretation of biological statistics of fish populations. *Fish. Res. Board Can. Bull. 191*.

118. Ripley, B. D. 1987. *Stochastic Simulation*. New York: Wiley

119. Ritchie, M. E. 1992. Chaotic dynamics in food-limited populations: Implications for wildlife management. In *Wildlife 2001,* ed. D. A. McCullough. London: Elsevier. In press

120. Robinson, G. R., Quinn, J. F. 1988. Extinction, turnover and species diversity in an experimentally fragmented California annual grassland. *Oecologia* 76:71–82

121. Samson, F. B., Perez-Trejo, R., Salwasser, H., Ruggiero, L. F., Shaffer, M. L. 1985. On determining and managing minimum population size. *Wildl. Soc. Bull.* 13:425–33

122. Schaffer, W. M. 1988. Perceiving order in the chaos of nature. See Ref. 16, pp. 313–50

123. Schaffer, W. M., Tidd, C. W. 1990. *NLF: Nonlinear Forecasting for Dynamical Systems*. Tucson, Ariz: Dynamical Systems. 154 pp.

124. Seber, G. A. F. 1982. *The Estimation of Animal Abundance*. London: Griffen. 2nd ed.

125. Shaffer, M. L. 1981. Minimum population size for species conservation. *BioScience* 31:131–34

126. Shaffer, M. L. 1983. Determining minimum viable population sizes for the grizzly bear. *Int. Conf. Bear Res. Manage.* 5:133–39

127. Shaffer, M. L. 1985. The metapopula-

tion and species conservation: The special case of the Northern Spotted Owl. In *Ecology and Management of the Spotted Owl in the Pacific Northwest,* ed. R. J. Gutierrez, A. B. Carey, pp. 86–99. *US For. Serv. Tech. Rep. PNW-185*

128. Shaffer, M. L. 1987. Minimum viable populations: coping with uncertainty. See Ref. 136, pp. 69–86
129. Shaffer, M. L., Samson, F. B. 1985. Population size and extinction: A note on determining critical population size. *Am. Nat.* 125:144–52
130. Shields, W. M. 1982. *Philopatry, Inbreeding, and the Evolution of Sex.* Albany, NY: State Univ. New York Press
131. Simberloff, D. 1988. The contribution of population and community biology to conservation science. *Annu. Rev. Ecol. Syst.* 19:473–511
132. Simberloff, D., Abele, L. G. 1982. Refuge design and island biogeographic theory: Effects of fragmentation. *Am. Nat.* 120:41–50

132a. Simberloff, D., Cox, J. 1987. Consequences and costs of conservation corridors. *Conserv. Biol.* 1:62–71

133. Sinclair, A. R. E. 1989. The regulation of animal populations. In *Ecological Concepts,* ed. J. M. Cherrett, pp. 197–241. Oxford: Blackwell
134. Smouse, P. E., Weiss, K. M. 1975. Discrete demographic models with density-dependent vital rates. *Oecologia* 21:205–18
135. Soulé, M. E. 1986. *Conservation Biology.* Sunderland, Mass: Sinauer
136. Soulé, M. E. 1987. *Viable Populations for Conservation.* Cambridge/New York: Cambridge Univ. Press. 189 pp.
137. Soulé, M. E., Bolger, D. T., Alberts, A. C., Wright, J., Sorice, M., Hill, S. 1988. Reconstructed dynamics of rapid extinctions of chaparral-requiring birds in urban habitat islands. *Conserv. Biol.* 2:75–92

137a. Soulé, M. E., Simberloff, D. 1986. What do genetics and ecology tell us about the design of nature reserves? *Biol. Conserv.* 35:19–40

138. Southwood, T. R. E. 1978. *Ecological Methods.* London: Methuen. 2nd ed.
139. Starfield, A. M., Bleloch, A. L. 1986. *Building Models for Conservation and Wildlife Management.* New York: Macmillan
140. Steele, J. H. 1985. A comparison of terrestrial and marine ecological systems. *Nature* 313:355–58
141. Stiling, P., Throckmorton, A., Silvanima, J., Strong, D. R. 1991. Does spatial scale affect the incidence of density dependence? A field test with insect parasites. *Ecology* 72:2143–54
142. Stromberg, M. R., Boyce, M. S. 1986. Systematics and conservation of the swift fox, *Vulpes velox,* in North America. *Biol. Conserv.* 35:97–110
143. Strong, D. R. 1986. Density-vague population change. *Trends Ecol. Evol.* 2:39–42
144. Suchy, W., McDonald, L. L., Strickland, M. D., Anderson, S. H. 1985. New estimates of minimum viable population size for grizzly bears of the Yellowstone ecosystem. *Wildl. Soc. Bull.* 13:223–28
145. Sugihara, G., May, R. M. 1990. Nonlinear forecasting: An operational way to distinguish chaos from measurement error. *Nature* 344:734–41
146. Deleted in proof
147. Swartzman, G. L., Kuluzny, S. T. 1987. *Ecological Simulation Primer.* New York: Macmillan. 370 pp.
148. Temple, S. A. 1986. The problem of avian extinctions. *Curr. Ornithol.* 3: 453–85
149. Templeton, A. R. 1986. Coadaptation and outbreeding depression. See Ref. 135, pp. 105–16
150. Thomas, C. D. 1990. What do real population dynamics tell us about minimum viable population sizes? *Conserv. Biol.* 4:324–27
151. Thomas, J. W., Forsman, E. D., Lint, J. B., Meslow, E. C., Noon, B. R., Verner, J. 1990. *A Conservation Strategy for the Northern Spotted Owl.* Portland, Ore: US Govt. Print. Off.
152. Thompson, G. G. 1991. *Determining Minimum Viable Populations under the Endangered Species Act. US Dep. Commerce, NOAA Tech. Memo NMFS F/NWC-198.* 78 pp.
153. Tuljapurkar, S. D. 1989. An uncertain life: Demography in random environments. *Theor. Popul. Biol.* 35:227–94
154. Tuljapurkar, S. D. 1990. Population dynamics in variable environments. *Lecture Notes in Biomathematics No. 85.* New York: Springer-Verlag. 154 pp.
155. Tuljapurkar, S. D., John, A. M. 1991. Disease in changing populations: growth and disequilibrium. *Theor. Popul. Biol.* 40:322–53
156. Tuljapurkar, S. D., Orzack, S. H. 1980. Population dynamics in variable environments. I. Long-run growth rates and extinction. *Theor. Popul. Biol.* 18:314–42
157. Turner, M. G. 1989. Landscape ecology: The effect of pattern on process. *Annu. Rev. Ecol. Syst.* 20:171–77

158. USDA Forest Service. 1986. *Draft Supplement to the Environmental Impact Statement for an Amendment to the Pacific Northwest Regional Guide,* Vols. 1, 2. Portland, Ore: USDA For. Serv.
159. Usher, M. B. 1988. Biological invasions of nature reserves: A search for generalizations. *Biol. Conserv.* 44:119–35
160. Van Winkle, W., DeAngelis, D. L., Blum, S. R. 1978. A density-dependent function for fishing mortality rate and a method for determining elements of a Leslie matrix with density-dependent parameters. *Trans. Am. Fish. Soc.* 107: 395–401
161. Wade, M. J. 1991. Genetic variance for rate of population increase in natural populations of flour beetles, Tribolium spp. *Evolution* 45:1574–84
162. Walters, C. J. 1986. *Adaptive Management of Renewable Resources.* New York: Macmillan
163. White, G. C., Anderson, D. R., Burnham, K. P., Otis, D. L. 1982. *Capture-Recapture and Removal Methods for Sampling Closed Populations.* Los Alamos, New Mex: Los Alamos Natl. Lab. LA-8787-NERP
164. Wilcox, B. A., Murphy, D. D. 1985. Conservation strategy: The effects of fragmentation on extinction. *Am. Nat.* 125:879–87
165. Zink, R. M. 1991. The geography of mitochondrial DNA variation in two sympatric sparrows. *Evolution* 45:329–39

Annu. Rev. Ecol. Syst. 1992. 23:507–36

DEFINITION AND EVALUATION OF THE FITNESS OF BEHAVIORAL AND DEVELOPMENTAL PROGRAMS

Marc Mangel

Department of Zoology and Center for Population Biology, University of California, Davis, California 95616

Donald Ludwig

Departments of Mathematics and Zoology, University of British Columbia, Vancouver, British Columbia V6T 1Y4, Canada

KEY WORDS: evolution, oviposition, sex change, ontogeny, life history theory

INTRODUCTION

> Whereas population genetics underrates the organism, life history theory underrates the gene. These fields are limited, in part because they ignore each other and in part because they ignore development. Thus the perspective of life history motivates a new look at development because developmental mechanisms could connect population genetics with life history theory to form a predictive theory of evolution more powerful than either of the first two attempted. (Bonner 1982, p. 238)

> . . . evolution can no longer be looked at solely as changes in gene frequencies within populations, or as fossil lineages: it is now essential to consider simultaneously the roles of genetics, development, ecology, and behavior (Bonner 1988, p. 24)

The definition and evaluation of fitness in terms of expected lifetime reproduction is central to our understanding of natural selection (118). In principle, fitness is defined in terms of a range of phenotypes that are the product of a single genotype; studies of such phenotypes are usually restricted to morphology. However, behavioral and developmental traits are subject to

0066-4162/92/1120–507$02.00

selection as well. Darwin used arguments concerning genetics (variation), development (embryology), ecology (geographical variation), and behavior (instinct) to support his theory. Here, we focus on dynamic programming methods to account for selection on such traits. Our intent is to explain the basic ideas of dynamic programming and to explore their applicability to the difficult problems of linking developmental, behavioral, and evolutionary phenomena. We shall show that, in principle, one can relate measurable physiological and ecological variables such as developmental times and behavioral traits to the fitness of an organism. We are far from being able to apply such methods to the calculation of the fitness of behavioral and developmental programs in most organisms, but sufficient progress has been made to warrant an exposition.

The linkage between development and evolution has consistently received attention (2, 9–11, 107). It has long been recognized (10, p. 179) that natural selection may modify developmental stages. Developmental biology typically focusses on the state of the organism and mechanisms that cause changes in this state, usually with little concern for evolutionary consequences. In contrast, in most of evolutionary biology, "fitness" is treated as an abstract concept, and the consequences of differences in fitness on gene frequencies are studied. A framework that joins these two is needed. It must perforce link states of organisms, ecological conditions, and an evolutionary measure of fitness within the constraints created by history and development.

Schmaulhausen (96) emphasized the role of development in molding the phenotype; Bonner (9) discussed (p. 165) the role of environmental cues in timing of developmental processes and considered animal behavior the pinnacle of biological complexity (10). For example, voltinism and diapause depend partially upon genetic control and partially upon environmentally controlled variables such as size and nutritional state (3, 7, 8, 14, 29, 36, 78, 100). Wilbur & Collins (117) proposed that size and growth rate together determine the initiation of amphibian metamorphosis, but they provided no means for predicting the thresholds for the onset of metamorphosis. Such a predictive theory would show how to link short-term behaviors such as foraging with life history (6, 103).

The uses of optimality go back to antiquity: when Dido founded the city of Carthage she had to enclose the greatest area she could in a bull's hide. Her solution was to cut the hide into a thin strip and form a semicircle, with part of the coast of the Mediterranean Sea as a boundary (64a, Book I, v. 519 ff). Optimality principles have been an outstanding success in physics. A prominent example is the principle of Fermat: Light rays always follow the path of shortest time connecting two points. This single principle explains curvature of light rays in a medium of varying velocity, Snell's law of refraction, the laws of reflection, and a qualitative theory of diffraction. Classical optimality theory (generalizing the work of Fermat) was extended to the analysis of a sequence of interdependent decisions, often under the designation of "dynamic programming." These methods have been applied with great success to problems arising in industry, economics, renewable resource management (20), and behavior (68, 69). Some critics of optimality theory claim that its assumption that genotypes or phenotypes are being optimized may not be correct. We show that many advantages of optimality theories can be realized without any assumption of optimality for the genotypes or phenotypes under consideration (also see 74).

Since the time of the neo-Darwinian synthesis, fitness defined as the expected lifetime reproductive output has been favored as an organizing concept in biology (31). The abstract principle of maximization of fitness is plausible and has many fruitful outcomes, but the calculation of its consequences is full of difficulties. A significant obstacle is the necessity to relate fitness (which depends upon complete life histories, perhaps over several generations) to a multitude of localized processes at various stages. To avoid these difficulties, many of the earliest applications of optimality ideas used a surrogate for the Darwinian fitness.

One attempt to link behavior and evolution used the paradigm of maximizing the rate of intake of energy. This led to the development of *optimal foraging theory* (OFT) (47, 56, 79, 82, 83, 98, 99, 104). The theory has branched widely from the original purposes of the prediction of diet choice and patch residence. Theories based on rate maximization have not met the challenge described above due to a number of limitations.

1. The theories focus almost exclusively on optimal behavior rather than fitness. A consequence of attempts to predict "optimal" behavior is that any variation "disproves" the theory (40, 45, 79). Since, as we shall see below, a static optimum may not always contribute to an organism's fitness in the same way as a dynamic optimum, focus on the static optimum can lead to a mistaken impression of the relative importance of behavioral traits.
2. The theories assume that fitness is an increasing function of energy intake (or some other "currency") but provide no way to determine how crucial it is to achieve the optimum. Changes in the fitness corresponding to deviations from the optimum cannot be assessed.
3. The theories are concerned only with flows (of energy, matings, etc) and usually ignore the internal states of the organism.
4. In stochastic situations, the theories assume that the time period of interest is long enough that a (possibly discounted) mean rate of energy intake

solely determines behavior but is short enough that behavior is fixed and not facultative over this interval.

5. There is no easy way to deal with multiple determinants of fitness, such as those that arise in the predation-starvation trade-off. Some recent work has attempted to remedy this problem (1, 19, 97).

In summary, the difficulties of OFT and related theories are a focus on flows rather than states of organisms, the prediction of no variability at selective equilibrium, an inability to treat disparate consequences of behavior in a consistent manner, and no method for assessing evolutionary consequences of short-term behaviors. More recent theories of behavior have addressed the difficulty of assessing evolutionary consequences of short-term behaviors by means of the concept of a "common currency," a focus on states of organisms, and use of stochastic dynamic programming (49, 61, 62, 70).

Ecological applications of dynamic programming were reviewed in 1978 (111). At that time, solution techniques required large mainframe computers. Partly for this reason, such methods have been regarded as esoteric by many biologists. Rapid development of hardware and software now allows solution of relatively complex problems on microcomputers, using simple languages such as BASIC. The difficulty remains that only problems involving small numbers of state and decision variables can be handled by computers; even supercomputers can be bogged down by plausible biological problems.

Our central concept is a fitness landscape, analogous to the fitness surface developed by Sewall Wright. Instead of considering such a surface whose height (the fitness) is associated with all possible allelic combinations, we consider fitness landscapes associated with different programs of development or behavior (examples are given in Figures 1 and 2 below). The definition of the fitness landscape first involves the specification of a variety of genetic programs that control development or behavior. One must then evaluate the expected reproductive success for an ensemble of organisms which all employ the same program, but which may encounter different environments due to random influences. Dynamic iteration (62) methods make it possible to define and compute the expected reproductive success, without any assumption of optimality.

We conclude this work with some examples where ideas of dynamic programming have been used to construct optimal strategies for oviposition of parasitoids (behavioral programs), ontogenetic niche shifts and sex change (developmental programs). These strategies are more complicated than those obtained by ignoring internal states of the organism, but some of the qualitative predictions have been verified by experiments.

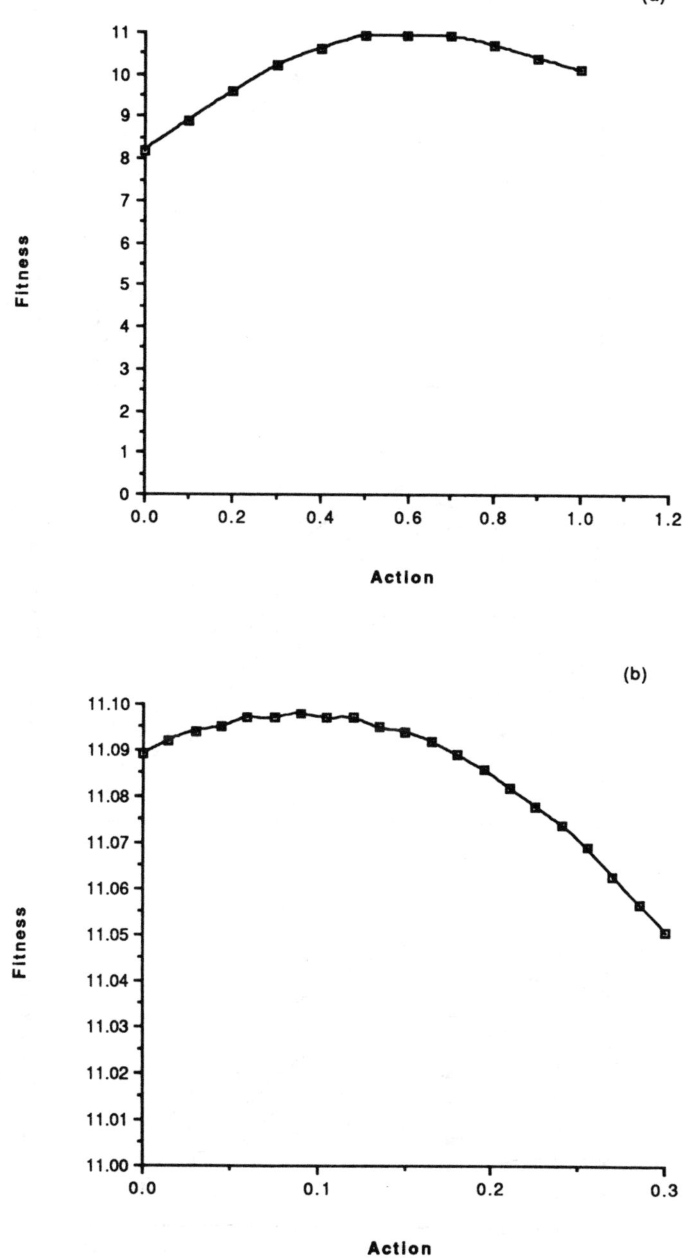

Figure 1 Fitness landscapes. (*a*) The dependence of the fourth column of Table 1 on a_1. (*b*) Fitness associated with the program Eq. 7 at $n' = 5$. Values of a_5 in the vicinity of .1 are favored, but the slight differences on the vertical axis indicate that only a slight improvement may be expected. (See discussion p. 519).

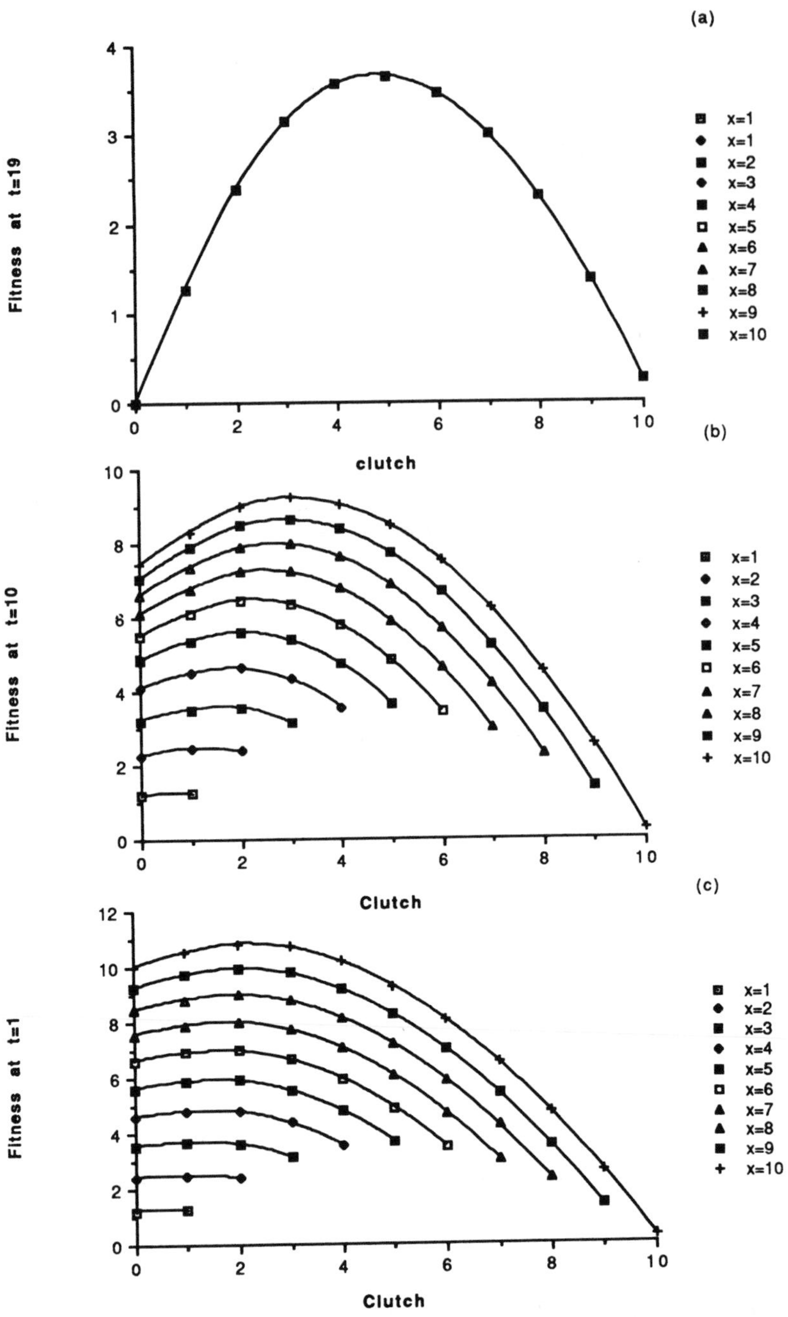
(a)
Fitness at t=19
clutch
x=1
x=1
x=2
x=3
x=4
x=5
x=6
x=7
x=8
x=9
x=10
(b)
Fitness at t=10
Clutch
x=1
x=2
x=3
x=4
x=5
x=6
x=7
x=8
x=9
x=10
(c)
Fitness at t=1
Clutch
x=1
x=2
x=3
x=4
x=5
x=6
x=7
x=8
x=9
x=10

THREE PRINCIPLES OF DYNAMIC PROGRAMMING

Our exposition is based upon three concepts:

1. The fitness function (expected lifetime reproduction) corresponding to a given program of development or behavior can be evaluated most efficiently by a "field" approach in which one constructs solutions to problems by comparing one solution with another, rather than by constructing solutions one at a time using simulation methods. Field methods were an important component of the intellectual revolution in physics (12, pg 679), and they can contribute new understanding to biological problems. Such methods require us to think in terms of families of programs or strategies instead of a single program in isolation.
2. Expected lifetime reproduction is insensitive to small deviations in programs in the vicinity of optimal programs. (In fact, this property is commonly used to determine optimal programs, as in setting the derivative of a function equal to 0 to find a maximum or minimum). The principle has the important biological implication that selection will be weak in the vicinity of optimal programs. The closer the program is to optimal, the weaker the selection (cf 47a).
3. Expected lifetime reproduction provides a "common currency" for the comparison of different programs.

None of these three concepts is new, but their generality and power are not always appreciated. We illustrate each of these concepts with a simple example, and each is developed at length in the sequel.

First Concept: The Field Approach and States of Organisms

We can illustrate the essentials of the field idea with a simple example where fitness is equated with survival, i.e. reproduction is independent of the condition of the organism (102).

EXAMPLE 1 Suppose that n represents the nutritional state of the organism, and that death results if $n = 0$, but survival is certain if n reaches a value N. Assume that during a given time interval, the probability of encountering food

←

Figure 2 The fitness landscape for insect oviposition. (*a*) Expected reproduction at $t = T-1$ corresponds to the fitness from oviposition in a single host, since there are no future opportunities. There is no dependence upon egg complement (x), except that the clutch $\leq x$. (*b*) The fitness landscape at t = 10 as a function of egg complement and clutch. Fitness is now composed of two components: reproduction in the current period plus expected future reproduction so that fitness depends upon both x and c. (*c*) The fitness landscape at $t = 1$. (See discussion p. 528) (and increasing n by 1) is α, and the probability of not encountering food (and decreasing n by 1) is $1 - \alpha$. In this way, we have linked a physiological state with a major component (survival) of expected reproduction. We want to find the probability of guaranteed survival (i.e. the chance that the nutritional state reaches N before it reaches 0) given that the current nutritional state of the organism is n.

To answer this question, let f_n be the probability of survival, for an individual with current nutritional state n. Since death results if $n = 0$, $f_0 = 0$. When $n = 1$ there are two possibilities: either (i) food is found, or (ii) food is not found. The probability of survival if food is found is f_2. The probability of survival if food is not found is 0, since $f_0 = 0$. Since the probability of finding food is α,

$$f_1 = \alpha f_2. \qquad 1.$$

At this stage both f_1 and f_2 are unknown. What is known is that they are related by Eq. 1. In general, for an individual starting at nutritional state n

$$f_n = \alpha f_{n+1} + (1 - \alpha) f_{n-1}, \; n = 1, \ldots, N-1, \qquad 2.$$

In order to express the assumptions that death is certain when n reaches 0 and survival is certain when it exceeds the value N, we add the conditions

$$f_0 = 0 \text{ and } f_N = 1. \qquad 3.$$

With the addition of conditions Eq. (3), the system Eq. (2) has the same number of equations as unknowns. Such a system can be solved without difficulty using packages available for microcomputers such as MINITAB or MATHEMATICA[1].

[1]An explicit solution can be found if we rewrite the equations in terms of the differences $g_n = f_n - f_{n-1}$. The resulting solution is:

$$f_n = (1 + R + R^2 + \ldots + R^{n-1}) \frac{1-R}{1-R^N}$$

where $R = \dfrac{1-\alpha}{\alpha}$

The approach we have taken is a standard technique in the theory of stochastic processes, whose roots are in the Hamilton-Jacobi theory of classical mechanics and optics (27). In physics, this approach has been identified with the idea of a "field," which describes the influence of distant objects upon a given object. In optics, a field describes the totality of light rays emerging from a given point. There is an analogous field concept in the more general mathematical theory of the calculus of variations. The field method eliminates action at a distance and sometimes makes it possible to solve many problems at the same time.

MacArthur & Wilson (58) used an approach similar to this one for the problem of extinction of island populations. The most important feature of such an approach is that we obtain the solutions of many problems simultaneously. In fact the method rests upon determining and solving relations between solutions rather than direct calculation of any but the simplest solutions. For example, if we were to attempt a direct solution by simulation methods, many hundreds or thousands of trials would be required to produce only limited accuracy in the answer to a single problem, with specific values of n, N, and α. The approach described here produces the solution $\{f_n\}$ of a family of problems at once; this solution is parameterized by n, N, and α.

Second Concept: Insensitivity of the Fitness Function to Changes in Programs near Optimal Programs

There is considerable debate concerning whether selection over a long enough period will lead to perfect optimization. This debate has centered on problems of matching constraints and shifting target optima. Ignoring the issue of constraints upon selection, the principle of insensitivity implies that response to selection may be expected to be slow for populations not too far from an optimal program. Thus we should not expect to observe populations following perfectly optimal programs, even in the absence of any genetic constraints.

EXAMPLE 2 In order to illustrate this phenomenon, suppose that fitness $f(a)$ depends only upon a certain behavior a, and not upon the state of the organism. Suppose further that the fitness is low at extreme choices of a, and higher for moderate choices of a. This is a behavioral analogue of stabilizing selection. In general, there might be many local maxima of f, just as there are possibly many local peaks of a Wrightian fitness surface. For the moment, let us consider a single such peak, and suppose that it is achieved at a behavior a^*, the "optimal behavior." The maximum possible fitness is $f(a^*)$, which will be termed the "optimal fitness." In the vicinity of a^*, the fitness curve is relatively horizontal. That is, the fitness of a behavior that is near a^* is close to the optimal. On the other hand, far from a^*, the fitness is substantially lower than optimal (as in Figure 1 below).

We conclude that selection will be strong on such behaviors far from a^*, but weak on nearby behaviors. For instance, if $f(a) = 1 - (a-a^*)^2$ and selection removes all but the top 1% of actions, we retain such actions as are within 10% of a^*. In order to achieve a maximum difference of 1% in actions, selection would have to remove all but the top .01% of actions. Therefore, we cannot expect that selection over a moderate time period will result in an optimal action, but only that the fitness of behaviors after selection will be close to optimal.

A related property of the fitness function near the optimal behavior is that small deviations of behavior from the optimal have still smaller effects. This property makes optimal actions robust to a great variety of variations. For instance, if the environment varies, then actions that are optimal with respect to a given environment will have effects upon fitness that are not far from optimal, provided that the environment doesn't change too much.

Although one may not expect to find optimal behaviors in nature, a compact qualitative description of the fitness landscape is obtained by finding the local maxima and describing the behavior in the vicinity of the maxima. We then have an idea of how many peaks are present in the fitness landscape and the shape ("steepness") of those peaks. This is the sort of procedure advocated when trying to sketch a complicated curve. It is usually worthwhile to compute the local optima for use as landmarks, and to provide a standard for comparison of other behaviors or programs.

Thus optimal programs are distinctive landmarks in the landscape of all programs. Even though genetic constraints may prevent the attainment of these special programs, study of them is instructive for a number of disparate reasons.

1. A by-product of the calculation of optimal solutions is a collection of dimensionless quantities that enter into the solution. That is, a key factor in determination of an optimum may be combinations of environmental or physiological parameters. Examples are the "minimize $\frac{\mu}{g}$ rule" (115; described below) and the quantities involving marginal returns.
2. Information is used by the organism to determine an optimal solution. In nature, information is necessarily inaccurate to a certain extent. The optimality principle assures that small inaccuracies in information have a very small effect.
3. Constraints may have large effects, but not if the optimal solution nearly satisfies the constraint.
4. The fitness of the programs that contain only small deviations from the optimal program can be determined rapidly and easily.

Third Concept: Expected Lifetime Reproduction as a "Common Currency"

Frequently a behavior has both beneficial consequences such as increased food intake, and harmful consequences such as increased risk of predation. To compare such qualitatively different consequences in a consistent manner, we combine the two models described above.

EXAMPLE 3 Suppose that the chance of finding food (denoted by α) is affected by a certain behavior a; we write $\alpha = \alpha(a)$. We also assume that there is a mortality risk associated with a, which is different in character from starvation, as in the case where foraging results in an increased hazard of predation. We denote this additional probability of mortality by $\mu(a)$. The analogue of Eq. 2 is

$$f_n = [1-\mu(a)]\ [\alpha(a)\, f_{n+1} + (1\text{-}\alpha(a))\, f_{n\text{-}1}], \qquad 4.$$

valid for $1 \leq n \leq N - 1$.

The new factor $1 - \mu(a)$ expresses the fact that survival is no longer certain. After adding the conditions of Eq. 3, we can obtain a numerical solution of Eq. 4. As yet we have not specified the behavior a. In principle a different behavior could be taken for each value of nutritional state n. As we shall see below, the most effective choices of behaviors have such a general form. They are called "feedback behaviors," since the physiological state affects behavior. The result of such calculations is a fitness vector $\{f_n, n = 1, \ldots N\}$ associated with a vector describing the program of behaviors $\{a_n, n = 1, \ldots N\}$. Although the fitness vector implicitly depends upon the behaviors and thus could be written as $f_n(\{a_n\})$, we shall not do so for notational convenience.

Envision a population that follows the behaviors $\{a_n\}$ and an invader that follows a deviant strategy at certain nutritional levels. In order to determine the effect of changes in the behaviors $\{a_n\}$ at a specified nutritional state n_1, we can examine the right-hand side of Eq. 4. One behavior may result in a greater fitness than the other, and such differences can be attributed to particular terms involving a, μ, and the fitness function f. Thus the fitness function offers a direct means of evaluating the trade-offs involved. In effect, the fitness function carries the information about the likely effects of given behavior at a given nutritional level. This is the greater benefit of the "field" approach.

PROGRAMS AND FITNESS

Programs Link Genes, Organisms and Selection

The genetic code can be interpreted as a "program" for constructing an organism from a linear sequence. This interpretation has been very successful in developmental and structural biology. We shall define a program as a collection of rules that determines the response of an organism to a set of internal and external states.

Successful programs generally utilize internally maintained states to mediate responses, instead of merely responding directly to external influences. We expect that successful organisms will have differing responses, depending upon physiological states such as hunger or fatigue. They may even respond to a lack of information about the environment by behaving in a manner that acquires information, with no other apparent benefit.

The Fitness Landscape of Programs

The fitness landscape of programs is constructed by calculating the fitness function corresponding to a certain collection of behavioral or developmental programs. We first discuss the concept of a program in more detail and then show how optimal or locally optimal programs may be constructed. Such optimal programs are to be considered as important landmarks in the fitness landscape; we do not assume that optimality is actually realized in nature.

There are two main methods for determining the fitness associated with programs. The first is the method of simulation. Given a program, simulations provide insight into individual trajectories or the distribution of certain types of trajectories. In simulations, time is run forward, and the entire program must be specified from the outset. The fitness value of a particular program is determined by averaging over the results of simulations. Simulations emphasize the trajectory space. Evaluation of the marginal returns from incremental changes in programs (which are important in evaluating take-offs and in formulating costs and benefits in a common currency) requires rerunning the entire simulation.

The second approach is the backward or field method described in this paper. This method enables one to calculate directly the expected returns for programs and the marginal returns from incremental changes in programs. The backward approach emphasizes the program space rather than the trajectory space: to each program is associated a single family of trajectories which fills a portion of trajectory space.

EXAMPLE 4 To illustrate these ideas, we adopt a somewhat more complicated model than Eqs. 2 or 4. Suppose that the probabilities of transition from one nutritional state to another depend upon both the current state n and the behavior a_n:

$$\begin{aligned}
&\Pr[n \text{ increases to } n+1, \text{ given behavior } a] = \alpha(n,a) = a(1 - \tfrac{n}{N}),\\
&\Pr[n \text{ decreases to } n-1, \text{ given behavior } a] = \beta(n,a) = 1 - a,\\
&\Pr[n \text{ does not change, given behavior } a] = \gamma(n,a) = 1 - \alpha(n,a) - \beta(n,a).
\end{aligned} \qquad 5.$$

These assumptions imply that it is impossibe for n to exceed N, regardless of the action taken. In the previous case, we considered survival as a measure of fitness. We also assume that there is a reproductive payoff $R(n)$ at each time step, which is proportional to the nutritional level.

When calculating the expected lifetime reproduction, it may be that because of population growth the relative genetic contribution of future offspring is less than that for present offspring (34, p. 27 ff). This effect can be included by multiplying the expected reproduction in the next period by e^{-r}, where r is the population growth rate, so that the analogue of Eq. 4 is

$$f_n = e^{-r} [1 - \mu(a_n)] [\alpha(n,a) f_{n+1} + \beta(n,a) f_{n-1} + \gamma(n,a) f_n] + R(n) \quad 6.$$

The program of behavior can be represented by a vector a in which the i-th component of the vector is the behavior when the nutritional state is i. We first consider behaviors that do not depend upon n, so that the program specifies $a_n = a_1$ for all values of n (Table 1). Each row corresponds to a particular choice of a_1, and each column corresponds to a single starting value for the nutritional state n. For example, the program of behavior a = (1, 1, 1, 1, 1, 1, 1) leads to fitness vector f = (4.37, 6.29, 8.22, 10.14, 12.06, 13.99, 15.91) while the program of behavior a = (.4, .4, .4, .4, .4, .4, .4) leads to fitness vector f = (2.28, 4.91, 7.71, 10.59, 13.52, 16.47, 19.43). From examination of the table, we see that no single choice of a_1 is best for all values of n. For example, $a_1 = 1$ is best for $n = 1$ but $a_1 = .2$ is best for $n = 7$ (Figure 1a, p. 511; this figure is a simple example of a fitness landscape). In harmony with Concept 2, there is very little difference in the fitness associated with nearly optimal values of a_1. For example, at $n = 4$, fitness ranges between 10.43 and 10.93 as a_1 ranges from 0.4 to 0.9. If selection were to act upon a population that is capable only of such simple programs, selection would proceed rather slowly.

Table 1 Fitness when behavior is independent of nutritional state[1]

Action	Nutritional State n						
	1	2	3	4	5	6	7
0.00	1.00	2.80	5.24	8.19	11.55	15.24	19.19
0.10	1.23	3.26	5.86	8.90	12.24	15.82	19.57
0.20	1.52	3.78	6.51	9.57	12.83	16.24	19.74
0.30	1.87	4.34	7.15	10.15	13.27	16.46	19.69
0.40	2.28	4.91	7.71	10.59	13.52	16.47	19.43
0.50	2.71	5.43	8.14	10.85	13.57	16.28	18.99
0.60	3.15	5.85	8.41	10.93	13.44	15.94	18.44
0.70	3.56	6.15	8.53	10.86	13.19	15.51	17.83
0.80	3.90	6.31	8.51	10.68	12.85	15.02	17.18
0.90	4.17	6.35	8.40	10.43	12.47	14.50	16.54
1.00	4.37	6.29	8.22	10.14	12.06	13.99	15.91

[a]Parameters are $\mu = .3a$, $e^{-r} = .8$, $R(n) = n$, and $N = 7$.

Suppose now that a program is introduced that reacts to low levels of nutrition by increasing the value of a_1. For example, if

$$a = (1, 1, 1, .4, .4, .4, .4), \qquad 7.$$

then the fitness resulting from the numerical solution of Eq. 6 is

$$f = (4.47, 6.48, 8.62, 11.09, 13.80, 16.63, 19.52). \qquad 8.$$

This simple change in program results in fitnesses all of which are larger than the corresponding fitnesses when nutritional state is ignored, except at the highest value of n. Such a program, if it is available for selection, would be favored over the simpler one which is independent of nutritional state. This elementary model illustrates how "feedback programs," i.e. programs which take differing actions depending upon the state of the organism, may be much more effective than simpler ones.

Three parameters are required to specify a program of the type of Eq. 7, in contrast to the single parameter a. Therefore visualization of the fitness landscape is more difficult. In the present case, by varying the value 0.4 in the later components Eq. 7, we can verify that 0.4 is close to the best choice, and that small changes in this value make little difference in the fitness. Similarly we find that it is best to make the break between the third and fourth components, but the results are not sensitive to this choice. Thus the fitness landscape looks like a dome centered over the behaviors described by Eq. 7.

Allowance having been made for the possibility of feedback programs, it is natural to ask whether substantial improvement may be made by allowing all possible programs. We suppose that some organisms in the population are capable of adjusting their responses to each level of nutrition. Each possible program consists of a specification of a_n and hence seven parameters. It is rather tedious to attempt to explore such a high dimensional space. We may instead use the fitness associated with a given program as a "common currency" and ask how the program may be improved. For instance we might take the program Eq. 7 with the fitness Eq. 8. At a given value, say $n' = 5$ we may examine the behavior of the right-hand side of Eq. 6 as a function of $a_{n'}$ (Figure 1b). Clearly values of a_5 in the vicinity of .1 are favored. But the slight differences on the vertical axis indicate that only a slight improvement may be expected. In a similar way, we may guess an improvement at each of the other points leading to

$$a = (1, 1, 0.75, .48, .09, 0, 0) \qquad 9.$$

with corresponding fitness

$$f = (4.50,\ 6.53,\ 8.72,\ 11.17,\ 13.95,\ 17.16,\ 20.73), \qquad 10.$$

which represents a modest improvement over Eq. 8, even though the behaviors in Eqs. 7 and 9 differ considerably for some values of nutritional state. If this process is carried out a few more times, the results quickly settle down to

$$a = (1,\ 1,\ 0.76,\ .47,\ .16,\ 0,\ 0). \qquad 11.$$

To the accuracy shown in Eq. 10, there is no improvement in the fitness. Although the program Eq. 11 actually is optimal, there is little improvement in fitness over the simpler program Eq. 7. We conclude that selection is unlikely to produce a population following the "optimal" program Eq. 11. On the contrary, there is a large amount of freedom to vary programs, with little loss of fitness. The implication for testing behavioral theories is that it is not sufficient to compare predicted and observed behaviors. If those behaviors differ one cannot reject the optimality model without consideration of the fitness of the observed behavior (60).

Seasonal Environments

We now consider the role of time, which has been neglected up to this point.

EXAMPLE 5 Reproduction may be postponed until a final time T and a certain nutritional state n_1 may be necessary in order to reproduce. In such a case, the fitness function $f_{t,n}$ will depend upon both the nutritional state n of the organism and the time t remaining until reproduction. We define the fitness at the final time T by

$$f_{T,n} = \begin{cases} 0 \text{ if } n < n_1 \\ r(n) \text{ if } n_1 \leq n \leq N \end{cases} \qquad 12.$$

We still assume that starvation results if $n = 0$:

$$\mathrm{f}_{\mathrm{t,0}} = 0 \text{ for } 1 \leq t \leq T. \qquad 13.$$

When Eq. 5 is used, no assumption need be made about the fitness at $n = N$; the values $f_{t,N}$ emerge as part of the solution. Now we require relations between the fitness at earlier and later times.

The Backward Equations

Eq. 4 generalizes to a dynamic relationship

$$f_{t,n} = [1 - \mu(a)]\,[\alpha(a)\,f_{t+1,n+1} + \beta(a)\,f_{t+1,n-1} + \gamma(a)\,f_{t+1,n}],$$

$$1 \leq t \leq T - 1,\ 1 \leq n \leq N - 1. \qquad 14.$$

These equations may be solved by the following scheme ("backwards induction"): $f_{T,n}$ is given by Eq. 12. Then $f_{T-1,n}$ is obtained from Eq. 14 with $t = T - 1$. The remaining $f_{t,n}$ are obtained in succession by continuing this process for decreasing t. Note that the equations for expected reproduction are dealt with most conveniently by taking time backwards. To see why this is so, note that the Eq. 12 describes reproduction at the final time T rather than any previous time. These conditions force a solution to be computed in the backwards direction, since otherwise we would have more equations than unknowns.

This example illustrates three principles which apply to the method of backwards evaluation ("Stochastic Dynamic Programming"—SDP) of the fitness of programs:

1. *Inclusion:* With probability equal to 1, something must happen during the next interval of time.
2. *Optimality:* If one knows the optimal program from time t to T, then the optimal program is known for all $s > t$ to T as well.
3. *Sequential Coupling* (62): The fitness of a program at the end of one stage of the life cycle equals the fitness at the start of the next stage of the life cycle times the probability of surviving to the next stage. For example, the system Eqs. 13 and 14, is sufficient to determine the variables $f_{t,n}$. Each of these variables is itself the fitness associated with a particular behavioral problem. Thus the "field concept" holds in this more general case.

This example also illustrates that in general fitness will depend upon nutritional status n and the time T remaining until reproduction. Thus Eqs. 13 and 14 permit a direct calculation of the fitness, without time-consuming simulations. The key is Eq. 14, which uses relationships between the fitnesses, rather than a separate calculation ab initio for each case.

A Simple Dynamical Example

Eq. 14 is incomplete without a specification of the action $a_{t,n}$ corresponding to each time t and nutritional state n. A great variety of such actions are possible. A simple description of behavior would be to forage as intensely as possible if below a value n_1, and not to forage at all if $n > n_1$, in analogy with Eq. 7. In such a case, the action is described by

$$a_{t,n} = \begin{cases} 1 & \text{if } n \leq n_1, \\ 0 & \text{if } n > n_1. \end{cases} \qquad 15.$$

Once the program of behaviors Eq. 15 and the reproductive output $R(n)$ are given, the solution of Eq. 14 is obtained by numerical methods (62, Chapter 2). The output of such a numerical procedure is a matrix showing fitness as a function of time and nutritional state. The result if the final reproductive output $R(n) = (0, 0, 0, 0, 1, 1, 1, 1, 1, 1)$ is shown in Table 2; each row corresponds to a single time t, and each column corresponds to a single nutritional state n. The triangular array of 1's in the upper right-hand appears since $n(T) > 3$ if, for example $n(5) = 10$, and so forth. Likewise, the final state cannot exceed 3 if $n(7) = 1$. Hence the triangle of 0's in the upper left-hand corner of Table 2. Fitness always increases from left to right across rows, since the probability of starvation or a low final state decreases in the same direction. At the nutritional states $n > 3$, fitness increases as t increases but for $n = 1$ or 2, fitness first increases and then decreases with time.

The Optimal Program

It is difficult to judge the adequacy of the program Eq. 15 without a standard for comparison. The best possible program for this case is obtained in analogy with Eq. 9: at each time t and for each nutritional state n, compute $a_{t,n}$ by maximizing the right-hand side of Eq. 14. The numerical output now consists of two matrices: the first is the optimal program as a function of time and nutritional state; the second is the fitness of such optimal programs (Table 3).

The first line of the program described in Table 3 has the same consequences as the program Eq. 15, since the fitness in the lines with $t = 9$ are

Table 2 Fitness associated with the program equation (15)

	Nutritional State									
Time	1.00	2.00	3.00	4.00	5.00	6.00	7.00	8.00	9.00	10.00
10.00	0.00	0.00	0.00	0.00	1.00	1.00	1.00	1.00	1.00	1.00
9.00	0.00	0.00	0.00	0.42	0.70	1.00	1.00	1.00	1.00	1.00
8.00	0.00	0.00	0.21	0.41	0.60	0.70	1.00	1.00	1.00	1.00
7.00	0.00	0.12	0.24	0.37	0.45	0.60	0.70	1.00	1.00	1.00
6.00	0.07	0.15	0.23	0.29	0.37	0.45	0.60	0.70	1.00	1.00
5.00	0.10	0.15	0.19	0.24	0.29	0.37	0.45	0.60	0.70	1.00
4.00	0.10	0.13	0.16	0.19	0.23	0.29	0.37	0.45	0.60	0.70
3.00	0.09	0.11	0.12	0.15	0.18	0.23	0.29	0.37	0.45	0.60
2.00	0.07	0.08	0.10	0.12	0.14	0.18	0.23	0.29	0.37	0.45
1.00	0.06	0.07	0.08	0.09	0.11	0.14	0.18	0.23	0.29	0.37

Table 3 The optimal program (a) and fitness (b) by which the simple program (15) can be evaluated

	Nutritional State									
Time	1.00	2.00	3.00	4.00	5.00	6.00	7.00	8.00	9.00	10.00
(a)										
9.00	0.00	0.00	0.00	1.00	1.00	0.00	0.00	0.00	0.00	0.00
8.00	0.00	0.00	1.00	1.00	1.00	0.50	0.00	0.00	0.00	0.00
7.00	0.00	1.00	1.00	1.00	0.83	0.42	0.36	0.00	0.00	0.00
6.00	1.00	1.00	1.00	1.00	0.59	0.48	0.19	0.29	0.00	0.00
5.00	1.00	1.00	1.00	0.69	0.55	0.36	0.29	0.00	0.24	0.00
4.00	1.00	1.00	0.74	0.60	0.45	0.34	0.18	0.13	0.00	0.21
3.00	1.00	0.74	0.62	0.50	0.39	0.26	0.17	0.00	0.00	0.00
2.00	1.00	0.63	0.52	0.42	0.32	0.22	0.10	0.00	0.00	0.00
1.00	1.00	0.56	0.44	0.35	0.26	0.16	0.06	0.00	0.00	0.00
(b)										
10.00	0.00	0.00	0.00	0.00	1.00	1.00	1.00	1.00	1.00	1.00
9.00	0.00	0.00	0.00	0.42	0.70	1.00	1.00	1.00	1.00	1.00
8.00	0.00	0.00	0.21	0.41	0.60	0.72	1.00	1.00	1.00	1.00
7.00	0.00	0.12	0.24	0.37	0.46	0.61	0.73	1.00	1.00	1.00
6.00	0.07	0.15	0.23	0.30	0.38	0.48	0.61	0.74	1.00	1.00
5.00	0.10	0.15	0.19	0.25	0.31	0.39	0.48	0.61	0.74	1.00
4.00	0.10	0.13	0.16	0.20	0.25	0.31	0.39	0.48	0.61	0.75
3.00	0.09	0.11	0.14	0.17	0.21	0.26	0.31	0.39	0.48	0.61
2.00	0.08	0.09	0.11	0.14	0.17	0.21	0.26	0.31	0.39	0.48
1.00	0.06	0.08	0.10	0.12	0.14	0.17	0.21	0.26	0.31	0.39

the same in Table 2 and the fitness array in Table 3b. For values of $t < 9$ the optimal actions differ more and more from the simple program, and the optimal fitness is accordingly higher. However, these differences need not be very large. For example at $t = 1$ we have:

Behavioral Program	Fitness for n =									
	1	2	3	4	5	6	7	8	9	10
Eq. 15	.06	.07	.08	.09	.11	.14	.18	.23	.29	.37
Optimal	.06	.08	.10	.12	.14	.17	.21	.26	.31	.39

We conclude that the simple program Eq. 15 can be improved by a program which depends upon t as well as n, but it might require considerable time for the simple program to be displaced by another. It is also clear that a great variety of programs will perform nearly as well as the optimal one. Therefore, one should not expect selection to achieve much more than a qualitative agreement between programs in nature and the optimal one.

Stationary and Nonstationary Programs

Programs like Eq. 15 that are independent of time are called "stationary," whereas those that depend upon time are called "nonstationary." Time did not appear explicitly in our example Eq. 4, and hence we considered only the possibility of stationary programs. What would happen to that example if it were made into a time-dependent problem? Suppose, for example, that we set $f_{T,n} = 0$ for all values of the nutritional variable. Including population growth in our computation of fitness, Eq. 14 becomes

$$f_{t,n} = e^{-r}[1 - \mu(a)]\,[\alpha(a)f_{t+1,n+1} + \beta(a)\,f_{t+1,n-1} + \gamma(a)\,f_{t+1,n}] + R(n),$$
$$1 \le t \le T - 1,\ 1 \le n \le N - 1. \qquad 16.$$

Once again, the numerical solution of Eq. 16 determines optimal actions and fitness associated with those actions (Table 4). We find that as t approaches 1, the actions approach the optimal stationary program Eq. 9, and the fitness also approaches the fitness that corresponds to the stationary program. The fitness corresponding to the stationary policy agrees with the fitness computed from Eq. 16, except in the second decimal digit for a few particular

Table 4 The optimal program (*a*) and fitness (*b*) when population growth is considered

	Optimal Actions Nutritional State						
Time	1.00	2.00	3.00	4.00	5.00	6.00	7.00
(*a*)							
9.00	0.00	0.00	0.00	0.00	0.00	0.00	0.00
8.00	1.00	1.00	1.00	0.62	0.11	0.00	0.00
7.00	1.00	1.00	0.92	0.55	0.17	0.00	0.00
6.00	1.00	1.00	0.85	0.53	0.18	0.00	0.00
5.00	1.00	1.00	0.81	0.51	0.17	0.00	0.00
4.00	1.00	1.00	0.79	0.50	0.17	0.00	0.00
3.00	1.00	1.00	0.78	0.49	0.17	0.00	0.00
2.00	1.00	1.00	0.77	0.48	0.16	0.00	0.00
1.00	1.00	1.00	0.77	0.48	0.16	0.00	0.00
(*b*)							
10.00	0.00	0.00	0.00	0.00	0.00	0.00	0.00
9.00	1.00	2.00	3.00	4.00	5.00	6.00	7.00
8.00	2.04	3.52	5.00	6.53	8.20	10.00	11.80
7.00	2.85	4.56	6.29	8.17	10.24	12.56	15.00
6.00	3.42	5.25	7.14	9.22	11.55	14.19	17.05
5.00	3.79	5.69	7.69	9.90	12.40	15.24	18.35
4.00	4.04	5.99	8.05	10.34	12.94	15.92	19.19
3.00	4.20	6.18	8.28	10.63	13.30	16.36	19.73
2.00	4.30	6.30	8.43	10.82	13.53	16.64	20.08
1.00	4.37	6.38	8.53	10.94	13.68	16.82	20.31

values of time and nutritional state. We thus see that, except for values of t close to the terminal time T, behavior is either weakly dependent on time or is independent of time and that increments in lifetime fitness decrease as the time to go $T-t$ increases. Some of the strongest predictions of this theory involve changes in behavior as a time constraint is approached (57, 59, 92, 95).

The Connection with Quantitative Genetics

Recent studies of the genetic basis of behavior (80) have revealed three main features. (*a*) There is a clear genetic basis of and influence on behavior. (*b*) Heritability estimates derived from the magnitude of response to selection are nearly always less than 50%. That is, in many cases most behavioral variability is not genetic in origin. (*c*) Many genes appear to affect behavior, which suggests that methods of quantitative genetics are appropriate.

Theoretical work in quantiative genetics (4, 5, 43, 54, 55, 108) has laid the foundation for the study of the evolution of behavior and development. These theories predict how the distribution of reaction norms will change over time in response to selection. The theories, however, are not self-contained since fitness is treated as an (usually ad hoc) input. The methods developed here provide a natural way to determine the fitness function for use in the theories of quantitative genetics. One need no longer assume that lifetime fitness is normally distributed. The assumption of normality simplifies the analysis of evolution because the treatment of heredity is simplified, but sacrifices biology for this simplification. Behavioral ecology allows us to compute fitnesses associated with behavioral or developmental programs. The interaction of behavioral ecology and quantitative genetics will strengthen each.

EXAMPLES OF BEHAVIORAL AND DEVELOPMENTAL PROGRAMS AND THEIR ANALYSIS

The methods described above aid in understanding the selective forces that act upon behavior and development. The analysis usually requires some level of numerical computation to determine a solution. Such specific numerical examples aid intuition and often provide general insights.

A Behavioral Program: Clutch Size in Insect Parasitoids (41, 51, 59)

Temperate insect parasitoids (109) are ideal organisms for testing many of the ideas developed here. Because the adult parasitoids usually die at the onset of winter, the terminal time T (as in Eq. 12) is clearly defined. In many species, adults emerge with essentially a full egg complement. Thus, the physiological state variable n can be interpreted as the current egg complement. Recent empirical work (e.g. 26, 35, 77) has shown that the number of mature eggs that a female carries clearly affects host selection.

Such parasitoids reproduce by oviposition in the larvae or pupae of other insects. The action a is the clutch of the ovipositing parasitoid. Because host volume is limited, each host can support a limited number of parasitoid larvae. In addition, the fecundity of an adult parasitoid is often directly related to its size, which in turn is directly related to resources available to it as a larvae. Thus, the per-egg fitness which an ovipositing female accumulates from oviposition in a host is usually a decreasing function of the number of eggs laid. Thus, a plot of total fitness per host (fitness per egg times the number of eggs laid) versus clutch size is "domed" (18, 59), and for a single host, there is an "optimal clutch," which maximizes the fitness obtained by the mother from oviposition in that host. The optimal clutch usually increases with host volume.

The dynamics of the state variable are

$$n_{t+1} = \begin{cases} n_t & \text{if no host is encountered, or a host is encountered but no clutch is laid, or} \\ n_t - a_t(n) & \text{if a host is encountered and a clutch of size } a_t(n) \text{ is laid.} \end{cases} \qquad 17.$$

These relations are used in Eq. 16 with appropriate modification (59, 62). The other parameters that enter into Eq. 16 are adult mortality $\mu(a)$ and the increment in fitness from oviposition in a single host. These can be determined empirically. Oviposition involves a "trade-off" between immediate reproduction (e.g. in a poorer host) and future expected reproduction (e.g. in a superior host).

The theory leads to a number of strong, qualitative predictions (59, 62). Optimal clutch sizes for a sequence of ovipositions are in general smaller than the optimal clutch size for a single host and clutches that are close to optimal produce nearly identical effects on lifetime fitness. Furthermore, increased adult survival or encounter rates with hosts will decrease optimal clutch sizes. As t approaches T, optimal clutch sizes increase. Most of these phenomena have been observed in recent experiments (92, 94).

What would we predict for the distribution of clutch sizes, in uncontrolled field observations, for which host volume varies and when parasitoids encounter hosts at different points in their lives, with different egg complements? We anticipate that any clutch smaller than the optimum clutch for a single host could be observed, with small clutches predominating (59, 62). This is exactly the case (18); the considerable variability observed in the field is consistent with the qualitative understanding provided by the optimality model.

Visualization of the behavioral landscape (Figure 2—see p. 512) requires solution of a particular problem, with particular values of parameters. We assumed that with the time horizon $T = 20$ periods, (*a*) the ovipositing insect has at most 10 eggs, (*b*) the probability of encountering a host in a single period is 0.3, (*c*) the probability of survival from one period to the next is .99, and (*d*) the expected reproduction associated with a clutch of size c is a domed curve, as shown in Figure 2a. Even if time horizon, maximum egg load, encounter rates, and mortality rates are held constant, the behavioral landscape is a function of time, current egg load (x), and clutch laid (c). When $t = T-1$ (Figure 3a), there is no expected future reproduction, so that the fitness landscape is identical to the expected reproduction from oviposition in a

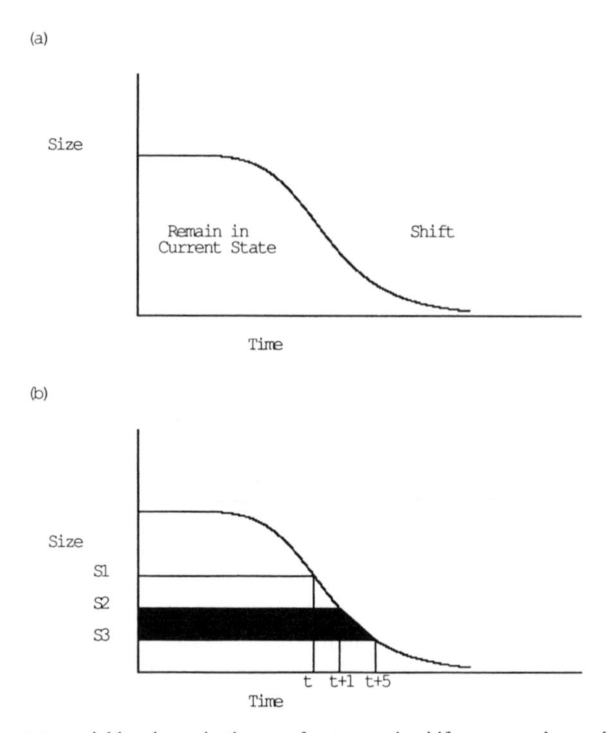

Figure 3 The state variable, dynamic theory of ontogenetic shifts or sex change leads to a "switching curve" $n_s(t)$ which describes the optimal pattern of development. For any time t before T, if the physiological variable is below the switching curve, the organism should remain in its current state whereas if the value of the physiological variable is above the switching curve, the organism should initiate a change. This theory can be tested (panel *b*) if we can find a cue which indicates that time has "jumped forward."

single host. As the time to go, $T-t$, increases, the balance between current and expected future reproduction affects the shape of the fitness landscape (Figure 2b,c). Note first the loss of fitness (expected reproduction) by laying a clutch that is smaller than optimal is generally less than the loss of fitness by laying a clutch that is larger than optimal (cf 28). This leads to the prediction that if clutches are not optimal, they are more likely to be smaller than optimal rather than larger than optimal. Second, for a fixed value of t the marginal increase in expected lifetime reproduction decreases as egg complement increases.

A Developmental Program: Ontogenetic Switching Curves (32, 33, 50, 57)

Many animals exhibit complex life cycles in which individuals undergo abrupt ontogenetic transformation (66, 113). For example, it is common that metamorphosis is associated with a shift in habitat usage. When habitats vary in risk of predation and growth rate, optimal programs often require habitat shift during ontogeny. In the first set of models of such ontogenetic habitat shifts (38, 113, 114, 115), workers concluded that the optimal behavioral program was to minimize the rate of mortality divided by the rate of growth. This is called the "minimize μ/g" strategy. This strategy can be derived from the assumption that reproduction is continuous and extends over an infinite time horizon. The predicted behavior is that all individuals will shift habitat at the same size.

Theories of sex change in sequential hermaphrodites (17) similarly assume populations with overlapping generations and a stable age distribution, corresponding to an infinite time horizon. These theories predict a single size at sex change.

Such general rules, for either habitat shift or sex change, may be broadly appropriate for interspecific comparisons. It is common, however, that reproduction is restricted to certain time periods and that the time horizon is not infinite. Considerable intraspecific variation in size dependent behaviors is common (32, 33, 105, 112, 116), and individuals often exhibit substantial variation in behavior during the course of their lives (16). In these cases, a dynamic, state variable theory of development is appropriate. Dynamic, state variable theories have been proposed for the ontogenetic niche shift (57, 95) and for sex change (33). In each case, the prediction of the theory (Figure 3a) is a "switching curve" $n_s(t)$ which indicates, at any given time, the size for an individual to initiate the developmental change.

The predictions of such an optimality theory can be tested experimentally if we can manipulate a cue which indicates that time has "jumped forward." That is, consider some intermediate time t (Figure 3b), with corresponding switching value S_1. In the following time period, the switching size will be S_2.

On the other hand, if the animals are given a cue indicating that the following period were $t+5$ instead of $t+1$, the switching value would be S_3. We thus predict that without the cue, individuals whose size was between S_3 and S_2 would not initiate developmental change, whereas those individuals who received the cue and whose size was in this interval (darkened region in Figure 3b) would initiate developmental change.

Fernandes (32) conducted such an experiment using the dusky slug *Arions subfuscus*. This slug is a protandrous hermaphrodite living in temperate climates. Essentially all individuals die at the onset of the first frost, so that T is well defined. The appropriate physiological state variable is body size, and the theory is an extension of the size-advantage hypothesis (37) to deal with time-limited life histories and intraspecific variation. Fernandes found that light cycle was a sufficient cue for developmental change, in the sense described above. As predicted by the theory (Figure 3b) when a population of slugs was subdivided randomly into a group receiving a normal light cycle (control) and a group receiving a light cycle indicating that time had shifted forward (experimental), slugs in the experimental group initiated sex change at a smaller size than slugs in the control group.

Additional Studies

In Table 5, we provide examples of a variety of other studies of behavioral and developmental programs that use dynamic, state variable approaches. This list is not intended to be comprehensive (also see Ref. 22 for additional studies).

Table 5 Other studies using dynamic state variable models[1,2]

Reference	Subject	Predictions	Exps[1]	Fitness
13	Effect of imperfect knowledge of predation rates on fitness	A wide range of estimates of predation rate may lead to the same fitness		+
15	Physiological integration in clonal plants	Abandonment of ramets as a function of size		
21	Foraging group size in lions	Group size as a function of physiological reserves	+	
23	Diel migration by juvenile salmon	Timing of diel migration	+	
24, 25	Risks and costs of parenthood Case study of Atlantic puffins dovekie	Parental defense, provision to nestlings as function of state	+	
39	Foraging group size in lions; sensitivity analysis of dynamic models	Foraging group size as a function of physiological reserves	+	+
42	Diet selection by guppies with/without predators present	Diet composition (3 prey types)	++	

(continued)

Table 5 (continued)

Reference	Subject	Predictions	Exps[1]	Fitness[2]
46	Timing of breeding of birds in temperature environments	Role of daylength temperature, food supply and predation in the timing of breeding	+	+
47b	Diet choice in fish	Comparison of static and dynamic models of diet choice	++	
48	Nest defense in plethodontid salamanders	When a parent should defend the nest, as a function of body size and eggs	+	+
52	Mate desertion in Cooper's hawks	Desertion as a function of physiological and ecological variables	++	+
63	Role of information in insect host choice	Superparasitism as a function of egg load and experience	+	
71	Partial preferences in foraging	Considerations associated with state variables can lead to predictions of partial preferences		
72	State-dependent contests for food	ESS behavior in the hawk-dove and war of attrition behavioral games may be state dependent and need not depend monotonically on state		
73	Daily routines of singing and foraging to attract a mate	Many, including that a peak in singing at dawn can result from variability in overnight energy expenditure even in the absence of circadian patterns	++	
76	Colony formation in social hymenoptera	Site selection for location of the colony	+	+
81	Territorial bequeathal by female red squirrels to their young	Prediction of females which will bequeath territories as a function of state and time of the season		
84	The way in which individuals acquire information about prospective mates	With search and sampling costs, sequential search for mates dominates the best-of-n strategy	+	+
90	Role of accurate information about host encounter rates in host choice	Behaviors with inaccurate information may lead to identical fitness as the case of perfect information	+	+
91	Density dependence of parasitism by fruit flies	Emergence of inverse density dependent parasitism in communities when individuals have direct density dependent parasitism	+	
93	Host (blood) and energy (nectar) seeking by mosquitoes	Seek blood under a wide variety of conditions but seek nectar only if crop volume, concentration, and energy are correct	+	
106	Behavior of small bird during the winter	Flocking as a function of energy reserves, predation risk, and mating status. Prediction of when to sing (to attract a mate) rather than flock	+	+
119	Juvenile alcid life histories	Mass of nestlings at fledging	+	

[1]A ++ indicates the experiments or field studies were conducted to test the state variable model; a + indicates comparison of the state variable model with previous empirical work.

[2]A + indicates that a fitness landscape, in addition to optimal behaviors, was computed.

CONCLUSIONS

A framework in which physiological (state) variables characterizing the organism are linked to reproductive success is required in order to account for selection on behavioral or developmental traits. Dynamic programming and related methods provide a framework for such accounting even if no assumptions about optimization are made. These methods are based on a field approach and work backwards in time, since terminal fitness can usually be assessed as a function of physiological state. The backward approach emphasizes behavioral and developmental programs rather than individual trajectories of state variables. Such methods explicitly consider the state of the organism and take development into account.

Developmental and behavioral programs are not static, but they may be stationary and thus depend only upon the physiological state of the organism. Simple rules such as rate maximization, which are common programs for the analysis of behavioral traits, may apply if there are no important internal states. Our methods allow the computation of a landscape of programs, which is a more operational version of the fitness landscape. Optimal programs, or more generally, programs whose value is insensitive to small changes, are distinctive landmarks in the landscape of all programs. Even though genetic constraints or slow selection near the optimum may prevent the attainment of these special programs, study of them is instructive.

Diversity underlies the uniqueness and fascination of biology. Dynamic, state variable methods help provide a framework in which this great diversity can be understood. Although dynamic state variable modeling as a tool in behavioral ecology was elucidated only five years ago (61, 70), these methods have been applied to a wide variety of biological situations (Table 5). We expect that empirical studies that are tests of models linking phenotypes and fitness will expand at every level (30, 89, 110) and that the description of the fitness landscape of behavior or development will become increasingly common.

ACKNOWLEDGMENTS

This work was begun while M. Mangel was Visiting Professor of Biology, Simon Fraser University. The work of M. Mangel is partially supported by NSF grants BSR 86-1073, 91-17603 and OCE 91-16895; of D. Ludwig by NSERC grant A9239. We thank J. Bonner, P. E. Bradshaw, R. Grosberg, P. J. B. Hart, R. Lalonde, M. Maxwell, and L. Rowe for comments.

Literature Cited

1. Abrahams, M. V., Dill, L. M. 1989. A determination of the energetic equivalence of the risk of predation. *Ecology* 70:999–1007
2. Arthur, W. 1988. *A Theory of the Evolution of Development*. New York: Wiley
3. Bargmann, C. I., Horvitz, H. R. 1991. Control of larval development by chemosensory neurons in *Caenorhabditis elegans*. *Science* 251:1243–46
4. Barton, N. H., Turelli, M. 1989. Evolutionary quantitative genetics: how little do we know. *Annu. Rev. Genet.* 23:337–70
5. Barton, N. H., Turelli, M. 1991. Natural and sexual selection on many loci. *Genetics* 127:229–55
6. Berry, R. J. 1989. Ecology: where genes and geography meet. *J. Anim. Ecol.* 58:733–59
7. Blakley, N. 1981. Life history significance of size-triggered metamorphosis in milkweed bugs (*Oncopeltus*). *Ecology* 62:57–64
8. Blakley, N., Goodner, S. R. 1978. Size-dependent timing of metamorphosis in milkweed bus (*Oncopeltus*) and its life history implications. *Biol. Bull.* 155:499–510
9. Bonner, J. T. 1974. *On Development*. Harvard Univ. Press
10. Bonner J. T. 1988. *Evolution of Complexity*. Princeton, NJ: Princeton Univ. Press
11. Bonner, J. T. 1982. *Evolution and Development*. New York: Springer-Verlag
12. Boorstin, D. J. 1985. *The Discoverers*. New York: Vintage
13. Bouskila, A., Blumstein, D. 1991. Rules of thumb for predations hazard: Predictions from a dynamic model. *Am. Nat.* 139:161–76
14. Brown, G. C., Berryman, A. A., Bogyo, T. P. 1978. Simulating codling moth population dynamics: model development, validation and sensitivity. *Environ. Entomol.* 7:219–27
15. Caraco, T., Kelly, C. K. 1991. On the adaptive value of physiological integration in clonal plants. *Ecology* 72:81–93
16. Caro, T. M., Bateson, P. 1986. Organization and ontogeny of alternative tactics. *Anim. Behav.* 34:1483–99
17. Charnov, E. L. 1982. *The Theory of Sex Allocation*. Princeton, NJ: Princeton Univ. Press
18. Charnov, E. L., Skinner, S. W. 1984. Evolution of host selection and clutch size in parasitoid wasps. *Fla. Entomol.* 67:5–21
19. Chesson, P., Rosenzweig, M. L. 1991. Behavior, heterogeneity, and the dynamics of interacting species. *Ecology* 72:1187–95
20. Clark, C. W. 1976. *Mathematical Bioeconomics*. New York: Wiley
21. Clark, C. W. 1987. The lazy, adaptable lions: a Markovian model of group foraging. *Anim. Behav.* 35:361–68
22. Clark, C. W. 1991. Modeling behavioral adaptations. *Behav. Brain Sci.*14:85–117
23. Clark, C. W., Levy, D. A. 1988. Diel vertical migrations by juvenile sockeye salmon and the antipredation window. *Am. Nat.* 131:271–90
24. Clark, C. W., Ydenberg, R. C. 1990. The risks of parenthood. I. General theory and applications. *Evol. Ecol.* 4:21–34
25. Clark, C. W., Ydenberg, R. C. 1990. The risks of parenthood. II. Parent-offspring conflict. *Evol. Ecol.* 4:312–25
26. Collins, M. D., Dixon, A. F. G. 1986. The effect of egg depletion on the foraging behaviour of an aphid parasitoid. *J. Appl. Entomol.* 102:342–52
27. Courant, R., Hilbert, D. 1962. *Methods of Mathematical Physics*. Vol. II. New York: Wiley Intersci.
28. Dhondt, A. A., Andriasensen, F., Matthysen, E., Kempenaers, B. 1990. Nonadaptive clutch sizes in tits. *Nature* 348:723–25
29. Dingle, H., Mousseau, T. A., Scott, S. M. 1990. Altitudinal variation in life cycle syndromes of California populations of the grasshopper, *Melanoplus sanguinipes* (F.). *Oecologia* 84:199–206
30. Dykhuizen, D. E., Dean, A. M. 1990. Enyzme activity and fitness: evolution in solution. *Trends Ecol. Evol.* 5:257–62
31. Endler, J. A. 1986. *Natural Selection in the Wild*. Princeton, NJ: Princeton Univ. Press
32. Fernandes, D. M. 1988. *The adaptive significance of protandry in the terrestrial slug,* Arion subfuscus. PhD thesis. Dep. Biol., Princeton Univ.
33. Fernandes, D. M. 1990. Sex change in terrestial slugs: social and ecological factors. In *Sex Change and Sex Allocation: Experiments and Models,* ed. M. Mangel. Providence, RI: Am. Math. Soc.
34. Fisher, R. A. 1958. *The Genetical Theory of Natural Selection*. New York: Dover
35. Fitt, G. P. 1986. The influence of a shortage of hosts on the specifity of

oviposition behavior in species of *Dacus* (Diptera, Tephritidae) *Physiol. Entomol.* 11:133–43
36. Forrest, T. G. 1983. Insect size tactics and developmental strategies. *Oecologia* 73:178–84
37. Ghiselin, M. T. 1969. The evolution of hermaphroditism among animals. *Q. Rev. Biol.* 44:189–208
38. Gilliam, J. F. 1982. *Habitat use and competitive bottlenecks in size structured fish populations*. PhD thesis. Mich. State Univ., East Lansing, Mich.
39. Gladstein, D. S., Carlin, N. G., Austad, S. N., Bossert, W. H. 1991. The need for sensitivity analyses of dynamic optimization models. *Oikos* 60:121–27
40. Glasser, J. W. 1984. Is conventional foraging theory optimal? *Am. Nat.* 124:900–5
41. Godfray, H. C. J., Ives, A. R. 1988. Stochasticity in invertebrate clutch-size models. *Theor. Popul. Biol.* 33:79–101
42. Godin, J-G. 1990. Diet selection under the risk of predation. In *Behavioural Mechanisms of Food Selection*, ed. R. N. Hughes, pp. 739–70. New York: Springer-Verlag
43. Gomulkiewicz, R., Kirkpatrick, M. 1992. Quantitative genetics and the evolution of reaction norms. *Evolution*. In press
44. Gould, S. J. 1989. A developmental constraint in Cerion, with comments on the definition and interpretation of constraint in evolution. *Evolution* 43:516–39
45. Gould, S. J., Lewontin, R. C. 1979. The spandrels of San Marco and the Panglossian paradigm: a critique of the adaptationist programme. *Proc. R. Soc. London Ser. B* 205:581–98
46. Goutis, C., Winkler, D. 1992. Hungry chicks and mortal parents: a state variable approach to the breeding season of birds. *Bull. Math. Biol.* 54:379–400
47. Gray, R. D. 1987. Faith and foraging: a critique of the paradigm argument from design. In *Foraging Behavior,* ed. A. C. Kamil, J. R. Krebs, H. R. Pulliam, pp. 69–140. New York: Plenum
47a. Hart, P. J. B. 1986. The influence of sex, patch quality, and travel time on foraging decisions by young adult *Homo sapiens* L. *Ethol. Sociobiol.* 7:71–89
47b. Hart, P. J. B., Gill, A. B. 1992. Choosing prey size: a comparison of static and dynamic foraging models for predicting prey choice by fish. In *The Behavioural Ecology of Fishes,* ed. F. A. Huntingford, P. Torricelli. London: Harwood. In press
48. Hom, C. L., Willits, N. H., Clark, C. W. 1990. Fitness consequences of nest defense in Plethodontid salamanders: predictions of a dynamic optimization model. *Herpetologica* 46:304–19
49. Houston, A. I., Clark, C. W., McNamara, J. M., Mangel, M. 1988. Dynamic models in behavioural and evolutionary ecology *Nature* 332:29–34
50. Iwasa, Y. 1991. Sex change evolution and cost of reproduction. *Behav. Ecol.* 2:56–59
51. Iwasa, Y., Suzuki, Y., Matsuda, H. 1984. Theory of oviposition strategy of parasitoids. I. Effect of mortality and limited egg number. *Theor. Popul. Biol.* 26:205–27
52. Kelly, E. J., Kennedy, P. L. 1992. A dynamic stochastic optimization model of mate desertion in Cooper's hawks. *Ecology*. In press
53. Kirkpatrick, M. 1988. The evolution of size in size-structured populations. In *Size Structured Populations, ed. B. Ebenman, L. Persson, pp. 13–28. New York: Springer-Verlag*
54. Kirkpatrick, M., Heckman, N. 1989. A quantitative genetic model for growth, shape, reaction norms, and other infinite-dimensional characters. *J. Math. Biol.* 27:429–50
55. Kirkpatrick, M., Lofsvold, D., Bumler, M. 1990. Analysis of the inheritance, selection, and evolution of growth trajectories. *Genetics* 124:979–93
56. Krebs, J. R., Stephens, D. W., Sutherland, W. J. 1983. Perspectives in optimal foraging. In *Perspectives in Ornithology: Essays Presented for the Centennial of the Am. Ornithilogists' Union,* ed. A. H. Bush, G. A. Clark Jr., pp. 165–221. Cambridge, UK: Cambridge Univ. Press. 560 pp.
57. Ludwig, D., Rowe, L. 1990. Life-history strategies for energy gain and predator under time constraints. *Am. Nat.* 135:686–707
58. MacArthur, R. A., Wilson, E. O. 1967. *Theory of Island Biogeography*. Princeton, NJ: Princeton Univ. Press
59. Mangel, M. 1987. Oviposition site selection and clutch size in insects. *J. Math. Biol.* 25:1–22
60. Mangel, M. 1991. Adaptive walks on behavioral landscapes and the evolution of optimal behavior by natural selection. *Evol. Ecol.* 5:30–39
61. Mangel, M., Clark, C. W. 1986. Towards a unified foraging theory. *Ecology* 67:1127–38
62. Mangel, M., Clark, C. W. 1988. *Dynamic Modeling in Behavioral Ecolo-*

gy. Princeton, NJ: Princeton Univ. Press
63. Mangel, M., Roitberg, B. D. 1989. Dynamic information and host acceptance by a tephritid fruit fly. *Ecol. Entomol.* 14:181–89
64. Manly, B. F. J. 1985. *The Statistics of Natural Selection*. New York: Chapman & Hall
64a. Maro, P. V. (Virgil) 19BC. *The Aeneid*. (Republished by Bantan Books, New York, 1985) 465 pp.
65. Matsuda, R. 1982. The evolutionary process in talitrid amphipods and salamanders in changing environments, with a discussion of "genetic assimilation" and some other evolutionary concepts. *Can. J. Zool.* 60:733–49
66. Matsuda, R. 1987. *Animal Evolution in Changing Environments with Special Reference to Abnormal Metamorphosis*. New York: Wiley Intersci.
67. Maynard Smith, J. 1982. *Evolution and the Theory of Games*. Cambridge, UK: Cambridge Univ. Press
68. McFarland, D. J. 1977. Decision making in animals. *Nature* 269:15–21
69. McFarland, D. J., Houston, A. I. 1981. *Quantitative Ethology—the State Space Approach*. London: Pitman
70. McNamara, J. M., Houston, A. I. 1986. A common currency for behavioral decisions. *Am. Nat.* 127:358–63
71. McNamara, J. M., Houston, A. I. 1987. Partial preferences and foraging. *Anim. Behav.* 35:1084–99
72. McNamara, J. M., Houston, A. I. 1989. State-dependent contests for food. *J. Theor. Biol.* 137:457–79
73. McNamara, J. M., Mace, R. H., Houston, A. I. 1987. Optimal daily routines of singing and foraging in a bird singing to attract a mate. *Behav. Ecol. Sociobiol.* 29:399–405
74. Mitchell, W. A., Valone, T. J. 1990. The optimization research program: studying adaptations by their function. *Q. Rev. Biol.* 65:43–65
75. Mitchell-Olds, T., Rutledge, J. J. 1986. Quantitative genetics in natural plant populations: a review of the theory. *Am. Nat.* 127:379–402
76. Nonacs, P. 1989. Competition and kin discrimination in colony founding by social Hymenoptera. *Evol. Ecol.* 3:221–35
77. Odendaal, F. J., Rausher, M. D. 1990. Egg load influences search intensity, host selectivity, and clutch size in *Battus philenor* butterflies. *J. Insect Behav.* 3:183–93
78. Palmer, J. O. 1984. Environmental determinants of seasonal body size variation in the milkweed leaf beetle, *Labidomera clivicollis* (Kirby) (Coleoptera: Chrysomelidae). *Ann. Entomol. Soc. Am.* 77:188–92
79. Pierce, G. J., Ollason, J. G. 1987. Eight reasons why optimal foraging theory is a complete waste of time. *Oikos* 49:111–18. Also see the reply directly following the article
80. Plomin, R. 1990. The role of inheritance in behavior. *Science* 248:183–88
81. Price, K. 1992. Territorial bequeathal by red squirrel mothers: a dynamic model. *Bull. Math. Biol.* 54:335–54
82. Pyke, G. H. 1984. Optimal foraging theory: a critical review. *Annu. Rev. Ecol. Syst.* 15:523–75
83. Pyke, G. H., Pulliam, H. R., Charnov, E. L. 1977. Optimal foraging: a selective review of theory and tests. *Q. Rev. Biol.* 52:137–54
84. Real, L. 1990. Search theory and mate choice: I. Models of single-sex discrimination. *Am. Nat.* 136:376–404
85. Real, L., Caraco, T. 1987. Risk and foraging in stochastic environments. *Annu. Rev. Ecol. Syst.* 17:371–90
86. Riechert, S. E., Hammerstein, P. 1983. Game theory in the ecological context. *Annu. Rev. Ecol. Syst.* 14:377–409
87. Riska, B. 1986. Some models for development, growth, and morphometric correlation. *Evolution* 40:1303–11
88. Roe, A., Simpson, G. G. 1958. *Behavior and Evolution*. New Haven, Conn: Yale Univ. Press
89. Roff, D. 1980. On defining sufficient causes for natural selection to occur. *Evol. Theory* 4:195–201
90. Roitberg, B. D. 1990. Optimistic and pessimistic fruit flies: evaluating fitness consequences of estimation errors. *Behaviour* 114:65–82
91. Roitberg, B. D., Friend, W. G. 1992. A general theory for host seeking decisions in mosquitoes. *Bull. Math. Biol.* 54:401–12
92. Roitberg, B. D., Mangel, M., Lalonde, R., Roitberg, C. A., van Alphen, J. J. M., Vet, L. 1991. Dynamic host shifts by a parasitoid. *Behav. Ecol.* 3:156–65
93. Roitberg, B. D., Mangel, M., Tourigny, G. 1990. The density dependence of parasitism by tephritid fruit flies. *Ecology* 71:1871–85
94. Rosenheim, J. A., Rosen, D., Rossler, Y. 1990. Foraging and oviposition decisions in the parasitoid *Aphytis lingnanesis:* distinguishing the influence of egg load and experience. *J. Anim. Ecol.* 60:873–94
95. Rowe, L., Ludwig, D. 1991. Size and timing of metamorphosis in complex life

cycles: time constraints and variation. *Ecology* 72:413–27

96. Schmaulhausen, I. I. (1949). 1986. *Factors of Evolution. The Theory of Stabilizing Selection.*
97. Schmitz, O. J., Ritchie, M. E. 1991. Optimal diet selection with variable nutrient intake: balancing reproduction with risk of starvation. *Theor. Popul. Biol.* 39:100–14
98. Schoener, T. W. 1971. Theory of feeding strategies. *Annu. Rev. Ecol. Syst.* 2:369–404
99. Schoener, T. W. 1987. A brief history of optimal foraging ecology. In *Foraging Behavior,* ed. A. C. Kamil, J. R. Krebs, H. R. Pulliam, pp. 5–68. New York: Plenum
100. Scott, S. M., Dingle, H. 1990. Developmental programmes and adaptive syndromes in insect life-cycles. In *Insect Life Cycles,* ed. F. Gilbert, pp. 69–85. New York: Springer-Verlag
101. Slatkin, M. 1987. Quantitative genetics of heterochrony. *Evolution* 41:799–811
102. Slobodkin, L. 1964. The strategy of evolution. *Am. Sci.* 52:342–57
103. Southwood, T. R. E. 1988. Tactics, strategies and templets. *Oikos* 52:3–18
104. Stephens, D. W., Krebs, J. R. 1986. *Foraging Theory.* Princeton, NJ: Princeton Univ. Press
105. Sweeney, B. W., Vannote, R. L. 1978. Size variation and the distribution of hemimetabolous insects: two thermal equilibrium hypotheses. *Science* 200:444–46
106. Szekely, T., Sozou, P. D., Houston, A. I. 1991. Flocking behaviour of passerines: a dynamic model for the non-reproductive season. *Behav. Ecol. Sociobiol.* 28:203–13
107. Thomson, K. S. 1988. *Morphogenesis and Evolution.* New York/Oxford: Oxford Univ. Press
108. Turelli, M., Burton, N. H. 1990. Dynamics of polygenic characters under selection. *Theor. Popul. Biol.* 38:1–57
109. Waage, J., Greathead, D. 1986. *Insect Parasitoids.* New York: Academic
110. Wade, M. J., Kalisz, S. 1990. The causes of natural selection. *Evolution* 44:1947–55
111. Walters, C. J., Hilborn, R. 1978. Ecological optimization and adaptive management. *Annu. Rev. Ecol. Syst.* 9:157–88
112. Warner, R. R. 1988. Sex change in fishes: hypotheses, evidence and objections. *Environ. Biol. Fishes* 22:81–90
113. Werner, E. E. 1986. Amphibian metamorphosis: growth rate, predation risk and the optimal size at transformation. *Am. Nat.* 128:319–41
114. Werner, E. E. 1988. Size, scaling and the evolution of complex life cycles. In *Size-Structured Popululations,* ed. B. Ebenman, L. Persson. New York: Springer-Verlag .
115. Werner, E. E., Gilliam, J. F. 1984. The ontogenetic niche and species interactions in size structured populations. *Annu. Rev. Ecol. Syst.* 15:393–425
116. Wilbur, H. M. 1980. Complex life cycles. *Annu. Rev. Ecol. Syst.* 11:67–93
117. Wilbur, H. M., Collins, J. P. 1973. Ecological aspects of amphibian metamorphosis. *Science* 182:1305–14
118. Williams, G. C. 1966. *Adaptation and Natural Selection.* Princeton, NJ: Princeton Univ. Press
119. Ydenberg, R. C. 1989. Growth-mortality trade-offs and the evolution of juvenile life histories in the alcidae. *Ecology* 70:1494–506

SUBJECT INDEX

D

E

H

I

CUMULATIVE INDEXES

CONTRIBUTING AUTHORS, VOLUMES 19–23

CHAPTER TITLES, VOLUMES 19–23

ANNUAL REVIEWS SERIES *Volumes not listed are no longer in print*		Prices, postpaid, per volume: Until 12-31-91 USA & Canada / elsewhere	Prices, postpaid, per volume: After 1-1-92 USA / other countries (incl. Canada)	Regular Order Please send Volume(s):	Standing Order Begin with Volume:
Annual Review of **BIOPHYSICS AND BIOMOLECULAR STRUCTURE**					
Vols. 1-11	(1972-1982)	$33.00/$38.00			
Vols. 12-18	(1983-1989)	$51.00/$56.00	$55.00/$60.00		
Vols. 19-20	(1990-1991)	$55.00/$60.00			
Vol. 21	(avail. June 1992)	$59.00/$64.00	$59.00/$64.00	Vol(s).______	Vol._____
Annual Review of **CELL BIOLOGY**					
Vols. 1-3	(1985-1987)	$33.00/$38.00			
Vols. 4-5	(1988-1989)	$37.00/$42.00	$41.00/$46.00		
Vols. 6-7	(1990-1991)	$41.00/$46.00			
Vol. 8	(avail. Nov. 1992)	$46.00/$51.00	$46.00/$51.00	Vol(s).______	Vol._____
Annual Review of **COMPUTER SCIENCE**					
Vols. 1-2	(1986-1987)	$41.00/$46.00	$41.00/$46.00		
Vols. 3-4	(1988, 1989-1990)	$47.00/$52.00	$47.00/$52.00	Vol(s).______	Vol._____

Series suspended until further notice. Volumes 1-4 are still available at the special promotional price of $100.00 USA /$115.00 other countries, when all 4 volumes are purchased at one time. Orders at the special price must be prepaid.

Volumes	Years	Until 12-31-91	After 1-1-92	Regular Order	Standing Order
Annual Review of **EARTH AND PLANETARY SCIENCES**					
Vols. 1-10	(1973-1982)	$33.00/$38.00			
Vols. 11-17	(1983-1989)	$51.00/$56.00	$55.00/$60.00		
Vols. 18-19	(1990-1991)	$55.00/$60.00			
Vol. 20	(avail. May 1992)	$59.00/$64.00	$59.00/$64.00	Vol(s).______	Vol._____
Annual Review of **ECOLOGY AND SYSTEMATICS**					
Vols. 2-18	(1971-1987)	$33.00/$38.00			
Vols. 19-20	(1988-1989)	$36.00/$41.00	$40.00/$45.00		
Vols. 21-22	(1990-1991)	$40.00/$45.00			
Vol. 23	(avail. Nov. 1992)	$44.00/$49.00	$44.00/$49.00	Vol(s).______	Vol._____
Annual Review of **ENERGY AND THE ENVIRONMENT**					
Vols. 1-7	(1976-1982)	$33.00/$38.00			
Vols. 8-14	(1983-1989)	$60.00/$65.00	$64.00/$69.00		
Vols. 15-16	(1990-1991)	$64.00/$69.00			
Vol. 17	(avail. Oct. 1992)	$68.00/$73.00	$68.00/$73.00	Vol(s).______	Vol._____
Annual Review of **ENTOMOLOGY**					
Vols. 10-16, 18	(1965-1971, 1973)				
20-32	(1975-1987)	$33.00/$38.00			
Vols. 33-34	(1988-1989)	$36.00/$41.00	$40.00/$45.00		
Vols. 35-36	(1990-1991)	$40.00/$45.00			
Vol. 37	(avail. Jan. 1992)	$44.00/$49.00	$44.00/$49.00	Vol(s).______	Vol._____
Annual Review of **FLUID MECHANICS**					
Vols. 2-4, 7	(1970-1972, 1975)				
9-19	(1977-1987)	$34.00/$39.00			
Vols. 20-21	(1988-1989)	$36.00/$41.00	$40.00/$45.00		
Vols. 22-23	(1990-1991)	$40.00/$45.00			
Vol. 24	(avail. Jan. 1992)	$44.00/$49.00	$44.00/$49.00	Vol(s).______	Vol._____
Annual Review of **GENETICS**					
Vols. 1-12, 14-21	(1967-1978, 1980-1987)	$33.00/$38.00			
Vols. 22-23	(1988-1989)	$36.00/$41.00	$40.00/$45.00		
Vols. 24-25	(1990-1991)	$40.00/$45.00			
Vol. 26	(avail. Dec. 1992)	$44.00/$49.00	$44.00/$49.00	Vol(s).______	Vol._____
Annual Review of **IMMUNOLOGY**					
Vols. 1-5	(1983-1987)	$33.00/$38.00			
Vols. 6-7	(1988-1989)	$36.00/$41.00	$41.00/$46.00		
Vol. 8	(1990)	$40.00/$45.00			
Vol. 9	(1991)	$41.00/$46.00	$41.00/$46.00		
Vol. 10	(avail. April 1992)	$45.00/$50.00	$45.00/$50.00	Vol(s).______	Vol._____

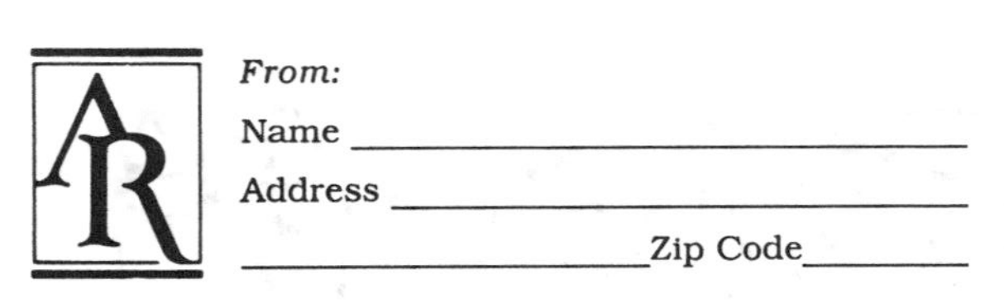

From:
Name ______________________
Address ______________________
______________________ Zip Code __________

ANNUAL REVIEWS INC.
4139 EL CAMINO WAY
P.O. BOX 10139
PALO ALTO, CA 94303-0897

ANNUAL REVIEWS SERIES *Volumes not listed are no longer in print*			Prices, postpaid, per volume Until 12-31-91 USA & Canada / elsewhere	After 1-1-92 USA / other countries (incl. Canada)	Regular Order Please send Volume(s):	Standing Order Begin with Volume:
Annual Review of PSYCHOLOGY						
Vols.	4, 5, 8, 10, 13-24	(1953, 1954, 1957, 1959, 1962-1973)				
	26-30, 32-38	(1975-1979, 1981-1987) ...	$33.00/$38.00	$40.00/$45.00		
Vols.	39-40	(1988-1989)	$36.00/$41.00			
Vols.	41-42	(1990-1991)	$40.00/$45.00			
Vol.	43	(avail. Feb. 1992)	$43.00/$48.00	$43.00/$48.00	Vol(s).______	Vol.____
Annual Review of PUBLIC HEALTH						
Vols.	1-8	(1980-1987)	$33.00/$38.00	$45.00/$50.00		
Vols.	9-10	(1988-1989)	$41.00/$46.00			
Vols.	11-12	(1990-1991)	$45.00/$50.00			
Vol.	13	(avail. May 1992)	$49.00/$54.00	$49.00/$54.00	Vol(s).______	Vol.____
Annual Review of SOCIOLOGY						
Vols.	1-13	(1975-1987)	$33.00/$38.00	$45.00/$50.00		
Vols.	14-15	(1988-1989)	$41.00/$46.00			
Vols.	16-17	(1990-1991)	$45.00/$50.00			
Vol.	18	(avail. Aug. 1992)	$49.00/$54.00	$49.00/$54.00	Vol(s).______	Vol.____

SPECIAL PUBLICATIONS		Prices, postpaid, per volume Until 12-31-91 USA & Canada / elsewhere	After 1-1-92 USA / other countries (incl. Canada)	Regular Order Please send:
The Excitement and Fascination of Science				
Volume 1	(1965) Softcover	$25.00/$29.00	$25.00/$29.00	________ Copy(ies).
Volume 2	(1978) Softcover	$15.00/$19.00	$25.00/$29.00	________ Copy(ies).
Volume 3	(1990) Hardcover	$90.00/$95.00	$90.00/$95.00	________ Copy(ies).
(Volume 3 is published in two parts with complete indexes for Volumes 1, 2, and both parts of Volume 3. Sold as a two-part set only.)				
Intelligence and Affectivity: Their Relationship During Child Development, by Jean Piaget				
	(1981) Hardcover	$8.00/$9.00	$8.00/$9.00	________ Copy(ies).

Send To: ANNUAL REVIEWS INC., a nonprofit scientific publisher
4139 El Camino Way • P. O. Box 10139
Palo Alto, CA 94303-0897 USA

Please enter my order for the publications indicated above. Prices are subject to change without notice.

Date of order ______________________________

Institutional purchase order No. ____________

Individuals: Prepayment required in U.S. funds or charge to bank card below.

❑ Amount of remittance enclosed $ __________

Or Charge my ❑ VISA

❑ MasterCard ❑ American Express

❑ Proof of student status enclosed

❑ California order, must add applicable sales tax.

❑ Canadian order must add 7% GST.

❑ Optional UPS shipping (domestic ground service except to AK or HI), add $2.00 per volume. UPS requires a street address. No P. O. Box, APO, or FPO.

Account Number ____ Exp. Date _____ / _____

Signature ____

Name ____
Please print

Address ____
Please print

____ Zip Code ______________

____ Send free copy of current ***Prospectus*** ❑

Area(s) of interest

ARI Federal I.D. #94-1156476